vademecum

Emergency Medicine

Sean O. Henderson, M.D.
Department of Emergency Medicine
Keck School of Medicine
University of Southern California
Los Angeles, California, U.S.A.

LANDES
BIOSCIENCE
Georgetown, Texas
U.S.A.

VADEMECUM
Emergency Medicine
LANDES BIOSCIENCE
Georgetown, Texas, U.S.A.

Printed in the U.S.A.

Please address all inquiries to the Publisher:
Landes Bioscience, 810 S. Church Street, Georgetown, Texas 78626, U.S.A.
Phone: 512/ 863 7762; Fax: 512/ 863 0081

ISBN: 1-57059-668-9

Cover artwork by Kristen Shumaker

Library of Congress Cataloging-in-Publication Data

Emergency medicine / [edited by] Sean O. Henderson.
p. ; cm. -- (Vademecum)
Includes bibliographical references and index.
ISBN 1-57059-668-9
1. Emergency medicine. I. Henderson, Sean O. II. Title. III. Series.
[DNLM: 1. Emergency Treatment--Handbooks. 2. Emergencies--Handbooks. WB 39 E5321 2006]
RC86.7.E5794 2006
616.02'5--dc22

2006005514

While the authors, editors, sponsor and publisher believe that drug selection and dosage and the specifications and usage of equipment and devices, as set forth in this book, are in accord with current recommendations and practice at the time of publication, they make no warranty, expressed or implied, with respect to material described in this book. In view of the ongoing research, equipment development, changes in governmental regulations and the rapid accumulation of information relating to the biomedical sciences, the reader is urged to carefully review and evaluate the information provided herein.

Dedication

Dedicated to the aspiring Emergency Physician.

Thanks to Drs. Greenspan, Calder, Swadron and Brown for their invaluable aid, and to all the authors, for their efforts on behalf of, and patience with, this project.

Sean O. Henderson, M.D.

Contents

Editor

Sean O. Henderson
Department of Emergency Medicine
Keck School of Medicine
University of Southern California
Los Angeles, California, U.S.A.
Email: sohender@usc.edu
Chapter 6

Associate Editor

Kirsten Calder
Department of Emergency Medicine
Keck School of Medicine
University of Southern California
Los Angeles, California, U.S.A.
Email: kcalder@usc.edu
Chapters 14, 15

Assistant Editors

Lance Brown
Division of Pediatric Emergency Medicine
Department of Emergency Medicine
Loma Linda University Medical Center and Children's Hospital
Loma Linda, California, U.S.A.
Email: lbrownmd@aol.com
Chapter 22

Jason Greenspan
Department of Emergency Medicine
Keck School of Medicine
University of Southern California
Los Angeles, California, U.S.A.
Email: greenspanj@adelphia.net
Chapter 2

Stuart P. Swadron
Department of Emergency Medicine
Keck School of Medicine
University of Southern California
Los Angeles, California, U.S.A.
Email: swadron@usc.edu
Chapters 1, 2

Contributors

Peter C. Benson
Department of Emergency Medicine
Keck School of Medicine
University of Southern California
Los Angeles, California, U.S.A.
Email: pbenson@sbcglobal.net
Chapter 1

Kathryn Challoner
Department of Emergency Medicine
Keck School of Medicine
University of Southern California
Los Angeles, California, U.S.A.
Email: challone@usc.edu
Chapter 12

Glenn W. Currier
Department of Psychiatry
University of Rochester
Rochester, New York, U.S.A.
Email: glenn_currier@urmc.rochester.edu
Chapter 17

Jeffrey Denham
LAC+USC Medical Center
University of Southern California
Los Angeles, California, U.S.A.
Email: jdwmd@aol.com
Chapter 9

Wendy Denham
Huntington Memorial Hospital
Pasadena, California, U.S.A.
Email: jdwmd@aol.com
Chapter 9

Deborah B. Diercks
Department of Emergency Medicine
University of California
Davis Medical Center
Sacramento, California, U.S.A.
Email: dbdiercks@ucdavis.edu
Chapter 3

Stephen D. Docherty
Department of Clinical
Emergency Medicine
Keck School of Medicine
University of Southern California
Los Angeles, California, U.S.A.
Email: magicdocs@aol.com
Chapters 7, 16

Olly Duckett
Department of Emergency Medicine
WakeMed
University of North Carolina - Chapel Hill
Raleigh, North Carolina, U.S.A.
Email: Duckett@weppa.org
Chapter 22

Sabrina Grassl
Department of Emergency Medicine
Keck School of Medicine
University of Southern California
Los Angeles, California, U.S.A.
Email: grasslss@hotmail.com
Chapter 6

Alicia Haglund
LAC+USC Medical Center
University of Southern California
Los Angeles, California, U.S.A.
Email: aliciahaglund@hotmail.com
Chapter 10

Matt Hendrickson
Cedars-Sinai Medical Center
and
Department of Emergency Medicine
Harbor-UCLA
Los Angeles, California, U.S.A.
Email: mhendrickson@sbcglobal.net
Chapter 18

Marilyn Hicks
Department of Emergency Medicine
WakeMed
University of North Carolina - Chapel Hill
Raleigh, North Carolina, U.S.A.
Email: lynhicks@earthlink.net
Chapter 22

Mark Hollinger
Department of Emergency Medicine
LAC+USC Medical Center
Los Angeles County College of Nursing
Los Angeles, California, U.S.A.
Email: mhollinger@dhs.co.la.ca.us
Chapter 21

Russell Karten
Department of Emergency Medicine
Lankenau Hospital
Wynnewood, Pennsylvania, U.S.A.
Email: Russell_Karten@hotmail.com
Chapter 22

Andrew S. Kassinove
Encino-Tarzana Medical Center
Tarzana, California, U.S.A.
Email: kassinove@yahoo.com
Chapter 5

Carrie S. Korn
Department of Emergency Medicine
Keck School of Medicine
and
LAC+USC Medical Center
University of Southern California
Los Angeles, California, U.S.A.
Email: ckorn@usc.edu
Chapter 17

David Leader
Department of Emergency Medicine
WakeMed
University of North Carolina - Chapel Hill
Raleigh, North Carolina, U.S.A.
Email: Leader@weppa.org
Chapter 22

Chi Lee
Department of Emergency Medicine
Keck School of Medicine
University of Southern California
Los Angeles, California, U.S.A.
Email: chiyonglee@hotmail.com
Chapter 6

Fred A. Lopez
Department of Medicine
Louisiana State University
and
Medical Center of Louisiana
New Orleans, Louisiana, U.S.A.
Email: alopez1@lsuhsc.edu
Chapter 11

William K. Mallon
Department of Emergency Medicine
Keck School of Medicine
University of Southern California
Los Angeles, California, U.S.A.
Email: wkmallon@yahoo.com
Chapter 1

Diku Mandavia
Department of Emergency Medicine
Cedars-Sinai Medical Center
and
Keck School of Medicine
University of Southern California
Los Angeles, California, U.S.A.
Email: mandavia@usc.edu
Chapters 19, 20

Sujal Mandavia
Department of Emergency Medicine
Cedars-Sinai Medical Center
and
Keck School of Medicine
University of Southern California
Los Angeles, California, U.S.A.
Email: smandavia@mac.com
Chapters 19, 20

Steven Offerman
Department of Emergency Medicine
University of California
Davis Medical Center
Sacramento, California, U.S.A.
Chapter 3

James Rhorer
Our Lady of the Lake Hospital
Baton Rouge, Louisiana, U.S.A.
Chapter 11

Mark G. Richmond
Department of Emergency Medicine
Loma Linda University Medical Center
and
Children's Hospital
Loma Linda, California, U.S.A.
Email: mrichmond@cox.net
Chapter 8

Jason Ruben
LAC+USC Medical Center
University of Southern California
Los Angeles, California, U.S.A.
Email: Ruben@usc.edu
Chapters 14, 15

Jacquelyn Hasler Salas
Lompoc District Hospital
Lompoc, California, U.S.A.
Email: salasjh@hotmail.com
Chapter 4

Paul Silka
Department of Emergency Medicine
Cedars Sinai Medical Center
and
Keck School of Medicine
University of Southern California
Los Angeles, California, U.S.A.
Email: paul.silka@cshs.org
Chapter 13

Ellen M. Slaven
Department of Emergency Medicine
Louisiana State University Health Sciences Center
Charity Hospital
New Orleans, Louisiana, U.S.A.
Email: eslaven@aol.com
Chapter 11

Peter E. Sokolove
Department of Emergency Medicine
University of California Davis Medical Center
Sacramento, California, U.S.A.
Email: pesokolove@ucdavis.edu
Chapter 3

Susan Stone
Department of Emergency Medicine
Keck School of Medicine
University of Southern California
Los Angeles, California, U.S.A.
Email: sstone_90064@yahoo.com
Chapter 5

Shahram Tabib
LAC+USC Medical Center
University of Southern California
Los Angeles, California, U.S.A.
Chapter 2

Mark Thoma
Department of Emergency Medicine
Davis Medical Center
University of California
Davis, California, U.S.A.
Email: togden@hotmail.com
Chapter 3

Adam J. Trenton
Department of Psychiatry
University of Rochester
Rochester, New York, U.S.A.
Chapter 17

Carla Valentine
Department of Emergency Medicine
Santa Barbara Cottage Hospital
Santa Barbara, California, U.S.A.
Email: carlavalentine@gmail.com
Chapter 16

Susan Zapalac
Department of Emergency Medicine
Keck School of Medicine
University of Southern California
Los Angeles, California, U.S.A.
Email: zapalac@cox.net
Chapter 8

Preface

The idea when we started was to collect the core Emergency Medicine information and present it in an abbreviated, succinct manner, useful to housestaff and medical students. As we progressed it became obvious that the very breadth of the specialty prevented any one person from accomplishing this task. It also became obvious that our specialty had advanced past the point where succinctness was possible. We peeled, boiled and pared, and came up with this. We hope you find it useful.

Acknowledgments

To Carrie S. Korn, R.N., for her help in keeping track of all that paper.

CHAPTER 1

Emergency Resuscitation

Stuart P. Swadron, Peter C. Benson and William K. Mallon

What Is Resuscitation?

Resuscitation, a word derived from the Latin word meaning "to set in motion", is the term most commonly used to describe the emergent treatment of the most severely ill and injured patients. To the emergency physician, the term encompasses not only attempts to reanimate those patients in cardiopulmonary arrest, but the treatment of virtually any diseases in the extremes of presentation. Resuscitation is an active process that is intervention-oriented and often invasive. The emergency physician (EP) confronted with a resuscitation must multitask and "set into motion" a team of health care workers which includes nurses, technologists and consultants.

Resuscitation and the Downward Spiral of Disease

Most disease processes move through stages of severity, beginning with an asymptomatic phase and progressing toward their end-stage. Generally speaking, disturbances in one physiologic function lead to disturbances in others and, through a sort of pathologic "multiplier effect", diseases gain momentum as they progress. Diseases that have reached their end-stage often have such momentum that they require intensive and rapid intervention if there is to be any hope of reversing the underlying pathology. Although patients may present to the emergency department at any stage in the continuum, it is those patients at the bottom of the spiral, those with decompensated and end-stage disease, that will require resuscitation.

In general, attempts are made to tailor the treatment of a particular patient to the tempo of their disease. The treatment of these processes should ideally occur at a similar pace, because abrupt changes may cause additional risk to the patient. Nonetheless, the momentum of end-stage disease will often force the emergency physician to use drastic and potent therapy, and such therapy is usually not without adverse consequences. The effect of the unwanted effects of therapy, together with the powerful and synergistic downward forces of multiply deranged physiologic functions, make resuscitation among the most challenging tasks of the emergency physician.

Shock: The Final Common Pathway

The final common pathway of most severe disease states is that of shock. Simply defined, shock is the failure of the circulation to provide adequate tissue perfusion. Although shock may not be present in all patients requiring emergent resuscitation, if untreated or treated inadequately, most will eventually deteriorate into a shock state. Once an illness progresses to a shock state, further deterioration involves a complex interaction between the underlying disease, host factors and the pathophysiology of the shock state itself.

Because of its central role in severe decompensated disease, a working knowledge of the classification and approach to shock is essential. When the diagnosis is known,

Table 1.1. Classification and causes of shock

Cardiogenic (inadequate pump function)
Cardiac rupture
Congestive heart failure
Dysrhythmia
Intracardiac shunt (e.g., septal defect)
Ischemia/infarction
Myocardial contusion
Myocarditis
Valvular dysfunction
Distributive (misdistrubution of the circulating volume)
Adrenal crisis
Anaphylaxis
Capillary leak syndromes
Neurogenic
Sepsis
Toxicologic
Obstructive (extracardiac obstruction to circulation)
Air embolism
Cardiac tamponade
Massive pulmonary embolus
Tension pneumothorax
Hypovolemic (Inadequate circulating volume)
Adrenal crisis
Hemorrhage
Severe dehydration

treatment is directed at both the underlying cause as well as the shock state itself. For those patients in whom the diagnosis is unknown, general resuscitative measures and treatment of shock proceeds alongside the diagnostic evaluation. Table 1.1 outlines the major classes of shock and gives examples of individual etiologies of each class. Many patients have compound presentations when more than one root cause is present.

The Recognition of Occult Shock

Many of the traditional clinical indicators of shock, such as blood pressure (BP) and heart rate (HR), lack the sensitivity to identify all patients in shock. In fact, more sophisticated indices, such as pooled venous oxygen saturation measured through a central catheter, can demonstrate a mismatch between the delivery of oxygen to the tissues and its consumption in some patients with normal or elevated BPs. Moreover, evidence suggests that using such indices to guide therapy in septic shock (not simply the BP) results in better outcomes. Thus, the early identification of shock before the traditional vital signs are grossly deranged (in its so-called "occult" form) is essential to management and disposition.

In the ED, shock is still most often recognized by the presence of persistent hypotension (e.g., systolic BP of <90 mm Hg in an adult) Nonetheless, there are many other clinical indicators that when considered together can alert the clinician to the presence of early shock, leading to appropriately vigorous resuscitation. Table 1.2 gives a list of clinical parameters that can assist in making the diagnosis of early or "occult" shock.

1

Table 1.2. Clinical parameters in the diagnosis of shock

Parameter	Comment
Heart rate	Tachycardia (HR >100 in non-pregnant adults) is present in most patients with shock; however, its presence may be masked by multiple factors including spinal cord injury, medications, intra-abdominal catastrophe, older age and cardiac conduction abnormalities.
Blood pressure	Hypotension (arbitrarily systolic BP <90) is a late finding in shock. In early shock, it may actually be transiently elevated. Measurements, in particular with standard BP cuff, become less accurate in shock states. A narrow pulse pressure may be present in hypovolemic shock. A wide pulse pressure may be seen in distributive shock.
Shock index	Heart rate/systolic blood pressure. An index of >0.9 is a more sensitive indicator of shock than either blood pressure or heart rate alone.
Pulsus paradoxus	A wide variation of blood pressure with respiration (>10 mm Hg) may indicate obstructive shock (e.g., cardiac tamponade)
Respirations	Either high (>24/min) or low (<12/min) rates may suggest a shock state, as may very shallow or deep breathing
Skin signs	Cool and clammy skin is often an indicator of a shock state although certain distributive shock states may have warm and dry skin (neurogenic and early septic shock). Delayed capillary refill (>2 seconds) is another sign of shock.
Urine output	Most often reduced (<30 ml/h) in shock states.

Diagnosis and Treatment in the Critically Ill Patient

In the classical medical model, the physician performs a history and physical examination before proceeding to diagnostic tests and then treatment. But the ED patient often requires treatment emergently and often in the absence of a diagnosis. This paradigm is taken to its extreme in the setting of resuscitation. There is clearly no time for history-taking or detailed physical examination in a patient who is pulseless and apneic. Treatment of this patient, regardless of the underlying diagnosis, must be immediate and maximal at the onset of the patient encounter (in this case by securing an airway, providing rescue breathing and performing chest compressions).

Because there are final common pathways for most disease processes, (e.g., the loss of spontaneous circulation and profound coma), the approach to any resuscitation always begins with general supportive measures that may not be specific to the underlying disease process. As more data is gathered, both by assessing the patient's response to therapy and obtaining incremental data from the ongoing history, examination and bedside laboratory testing, the resuscitation becomes more specific, focusing therapy to the most likely pathologies. Such upward reversal of disease momentum mirrors its downward spiral—powerful, broad therapies are used to reverse the intense downward momentum of end-stage disease, followed by more specific and focused therapy as the curve of disease momentum becomes less steep.

Overview of a Resuscitation

In resuscitation, multiple interventions, both diagnostic and therapeutic, occur simultaneously. These interventions are artificially separated out for the purposes of analysis and education, beginning with the primary survey, the so-called ABCDEs, a rapid evaluation and management of cardiopulmonary and cerebral function. The primary survey focuses the clinician on the critical early interventions. After the primary survey, a "resuscitation" phase is begun, which focuses on the acquisition of

Table 1.3. Overview of a resuscitation

Preparation	
Primary survey	
Airway	
Breathing	
Circulation	
Disability	
Exposure	
Resuscitation phase	
Continuation of interventions and monitoring begun in primary survey	
History	
Bedside diagnostic investigations	Blood tests
	Electrocardiography
	Diagnostic imaging
Advanced monitoring	Central venous pressure monitoring
	Pulmonary catheter monitoring
	Intraarterial catheter monitoring
	Cardiotocodynamometry (in pregnancy)
	Non-invasive monitoring (biothoracic impedence)
	Nasogastric and urinary catheter placement
Secondary survey	
Head-to-toe examination (inspection, percussion, palpation and auscultation)	
Definitive care phase	
Diagnostic specific therapy	
Consultations	
Preparations for patient disposition	Transfer
	Operating room
	Specialized intensive care unit
Family conference	
Resource utilization	Respiratory therapist
	Social services
	Religious support services
	Organ procurement agencies
	Law enforcement and forensic services
	Sexual assault and domestic violence

additional data and the continuation of emergent interventions initiated in the primary survey. The secondary survey which follows is a head to toe examination that may reveal abnormalities not yet noted and that may alter further therapy. Finally, the resuscitation enters the definitive care stage, which may continue for the course of several days, as the patient moves from the ED to the operating room, intensive care unit and/or other inpatient treatment areas. Table 1.3 gives the organizational framework of a resuscitation.

Preparation

The few minutes immediately prior to a major resuscitation may be precious. Any advanced notification that a critically ill patient is en route to hospital may be used to assemble the resuscitation team and to ready necessary equipment. This time is not to be taken for granted; if not utilized correctly, it may directly contribute to poor outcomes and inefficient resuscitation. Table 1.4 outlines critical steps that should occur before a patient(s) even arrives.

1

Table 1.4. Preparation for a resuscitation

Prehospital data analysis		
Number of patients		
Age and gender of patients (possibility of obstretrical, pediatric and/or neonatal resuscitation)		
Description of illness/injury		
C-spine precautions in place		
Level of consciousness		
Airway status		
Vital signs		
Estimated time of arrival		
Need for decontamination (hazardous materials), sterile field (burns)		
Need for law enforcement, security personnel (possible compromise of patient, staff safety)		
Team assembly		
Physicians	Team leader	
	Airway management	
	Procedural support	
	Consultant staff	-surgical -interventional cardiology -gastroenterology -neonatal/pediatric -obstetrical
Nursing	Charting and timing	
	Monitor placement	
	Venous access and emergency lab specimens	
	Catheter placement (Foley, NG tubes) and other procedures	
	Resuscitation medications and fluids ready and available	
	Circulating	
	Family and visitor management	
Respiratory therapist	Assistance with airway management	
	Ventilator and non-invasive ventilation techniques	
Radiation technologist	Perform and develop STAT portable X-rays	
	CT scan running and available	
Laboratory personnel	Blood bank readiness	
	STAT laboratory testing	
Clerical staff	Provide emergent identification and prepare hospital ID	
Equipment		
Airway management	Functioning suction with catheter attached	
	Flowing oxygen	
	Airway adjuncts including	-an array of sizes of oral airways -bag-valve-mask setup -ET tubes with stylets placed and balloons tested -laryngoscopes with bulbs tested -rescue airway adjuncts ready -medications for rapid sequence intubation ready
Other procedural	Dress and gloves for universal precaution maintenance	
	Blood warming and rapid transfusion equipment	
	Ready procedure trays for	-central venous access -thoracostomy and thoracotomy -surgical airway management (cricothyroidotomy)
	Warmed fluids/blankets, non-invasive warming equipment	
	Adequate tubing for blood/blood product administration	
	Emergency ultrasound at bedside	

1

The Primary Survey

During the primary survey, the critical therapeutic efforts of resuscitation are initiated. At the same time, the signs of the various shock states are unmasked and clues to the underlying diagnosis may be elicited. Although a definitive diagnosis is often not made initially, it is almost always possible to direct resuscitative efforts toward a particular class of shock.

When problems are encountered in the primary survey, they should be addressed immediately. Each element may be managed with either temporizing or definitive maneuvers. For example the airway may be *temporarily* managed with the chin-lift and bag-valve-mask ventilation, or *definitely* managed with endotracheal intubation.

A—Airway

When approaching the airway, the clinician ensures that cervical spine *precautions* are in place if trauma is a possibility and determines whether the airway is *patent*, *protected* and *positioned* adequately. The clinician:

Observes for level of consciousness, drooling and secretions, foreign bodies, facial burns, carbon in sputum

Palpates for any facial or neck deformities and checks for a gag reflex, and

Listens for hoarseness or stridor.

Findings	Diagnostic Implication
Drooling, stridor	Upper airway obstruction
Decreased level of consciousness	Unprotected airway
Diminished gag	
Facial burns	Unstable airway (potential obstruction)
Facial instability	

Airway management in the primary survey may be as simple as positioning of the airway using the chin lift or jaw thrust maneuvers (used when cervical spine instability is a concern). It may also involve the placement of nasopharyngeal or oral airway devices and the application of supplemental oxygen. In cases of obstruction, foreign bodies may need to be dislodged using basic life support maneuvers or manually with suctioning and Magill forceps. Definitive airway intervention, such as oral endotracheal intubation (with or without rapid sequence technique), nasotracheal intubation or a surgical airway (e.g., cricothyroidotomy) may be required.

B—Breathing

To assess the adequacy of the breathing apparatus, the clinician:

- *Observes* for signs of tracheal deviation, jugular venous distention (JVD), Kussmaul's sign (increased JVD with inspiration), respiratory distress (such as indrawing, splinting and use of accessory musculature) and trauma (contusions, flail segments, open wounds)
- *Palpates* for bony crepitus, subcutaneous air or tenderness
- *Auscultates* to assess air entry, symmetry, adventitial sounds (crackles, wheezes and rubs), and
- *Percusses*, if necessary, for hyperresonance or dullness on each side.

Findings	Diagnostic Implication
JVD, unilaterally absent breath sounds	Obstructive shock (tension pneumothorax)
JVD, clear lung fields	Obstructive shock (cardiac tamponade, massive pulmonary embolism) Cardiogenic shock (right ventricular myocardial infarction)

JVD, diffuse crackles	Cardiogenic shock (cardiogenic pulmonary edema)
JVD, diffuse or localized wheezes	Status asthmaticus, COPD exacerbation, aspiration syndrome
Kussmaul ("air hunger") breathing	Metabolic acidosis
Chaotic, irregular breathing	Central nervous system insult
Abdominal breathing, failure of chest expansion	High spinal cord injury

Possible interventions during the breathing segment of the primary survey include bag-valve mask ventilation, the administration of naloxone for narcotic induced apnea, placement of thoracostomy needles and tubes and the application of positive pressure ventilation, by either non-invasive or invasive means.

C—Circulation

To assess the circulation, the clinician:

- *Palpates* the pulse for rate, regularity, contour and strength. Pulses should be checked in all four extremities, and if absent, central pulses (femoral and carotid) are palpated. Also, palpates the skin for temperature, moisture and the briskness of capillary refill in the extremities.
- *Observes* for signs of obvious hemorrhage such as visible exsanguination, a distending abdomen, an unstable pelvis or long bone deformities.
- *Measures* the blood pressure, notes *pulse pressure*, and if necessary, compares BP among the extremities.
- *Auscultates* the precordium for the clarity of heart tones, listening for any extra sounds, murmurs, rubs or Hammon's crunch (pneumomediastinum)

Findings	Diagnostic Implication
Sinus tachycardia, hypotension, JVD cool, pale extremities	Obstructive shock (cardiac tamponade, tension pneumothorax, massive pulmonary embolism) Cardiogenic shock (right ventricular myocardial infarction)
Sinus tachycardia, hypotension cool, pale extremities	Hypovolemic shock
Hypotension, relative bradycardia warm, pink extremities	Distributive shock (neurogenic shock from spinal cord injury)
Tachycardia, hypotension, gallop rhythm (S3, S4)	Cardiogenic shock (left ventricular failure)
Tachycardia, hypotension, loud systolic murmur	Cardiogenic shock (acute mitral regurgitation or ventricular septal defect)
Central cyanosis	Hypoxia Methemoglobinemia

Interventions during the circulation segment of the primary survey include placing the patient on a cardiac and pulse oximetry monitor and the establishment of vascular access. They may also include the administration of fluids and blood products, electrical and pharmacological therapy for dysrhythmias, pericardiocentesis and, in some cases, such as penetrating trauma, emergency thoracotomy.

D—Disability

Disability represents the neurological assessment in the primary survey. If at all possible, it is desirable to obtain a cursory assessment prior to use of paralyzing agents. The clinician:

- *Assesses* the level of consciousness, using the Glasgow Coma Scale.

	Eye Opening	*Motor*	*Verbal*
1	None	No movement	No sounds
2	To pain	Decerebrate postutre	Moans
3	To command	Decorticate posture	Words
4	Spontaneous	Withdrawal from pain	Confused
5		Localize to pain	Oriented
6		To command	

Minimum Score = 3 (severe coma); Maximum Score = 15

- *Observes* the pupils for size, symmetry and reactivity to light, and *observes* all four extremities for their gross movement
- *Palpates* rectal tone by digital examination

Findings	**Diagnostic Implication**
Coma, unilateral dilated pupil, hemiparesis	Cerebral herniation
Pinpoint pupils	Opiate, cholinergic or clonidine overdose Pontine lesion
Dilated, reactive pupils	Sympathomimetic overdose
Dilated, unreactive pupils	Anoxia Anticholinergic overdose
Deviation of eyes to one side	Ipsilateral cortical lesion Contralateral brainstem lesion
Decreased rectal tone	Spinal cord injury Other neurological insults, seizures, toxins
Rigid extremities	Neuroleptic malignant syndrome Serotonin syndrome Tetanus, strychnine poisioning

Interventions in the disability segment of the primary survey are often limited to airway, breathing and circulation, as these all affect neurological function. Once these are addressed, attention can be directed toward interventions such as cranial CT, the administration of mannitol and hyperventilation for suspected acute brain herniation, and surgical decompression. Pharmacologic therapy is directed at causes of altered levels of consciousness, such as the administration of glucose for hypoglycemia, naloxone for suspected opiate overdose and thiamine for Wernicke-Korsakoff syndrome.

E—Exposure

Often described as "strip, flip, touch and smell", exposure means not only completely undressing the patient, but also looking for other important clues. The clinician should:

- *Expose* the entire surface area of the patient
- *Inspect and palpate the back* for abnormalities, using cervical spine precautions to roll the patient if there is a possibility of trauma. Also, *inspect* the skin for rashes, other obvious lesions and signs of trauma
- *Note* any particular odors about the patient, and
- *Measure* a rectal temperature

Findings	Possibile Diagnostic Significance
Hyperthermia/Hypothermia	Hypovolemic (severe dehydration) Distributive shock (e.g., septic) Cardiogenic shock
Unsuspected wounds (especially in axilla, back, neck, perineum)	Hypovolemic shock (hemorrhagic shock from occult trauma)
Odors:	
Fetid urine	Distributive shock (urosepsis)
Bitter almonds	Cyanide toxicity
Garlic	Organophosphate or arsenic toxicity
Fruity	Ketoacidosis, isopropyl alcohol toxicity
Alcohol	Complications of alcohol abuse (trauma, multiorgan toxicity)
Track marks of IV drug use	Distributive shock (sepsis) Cardiogenic shock (valvular disease) Opiate overdose Non-cardiogenic pulmonary edema
Dialysis shunt (AV fistula)	Cardiogenic shock (volume overload) Obstructive shock (pericardial tamponade) Hyperkalemia Uremic encephalopathy
Cullen's or Gray-Turner signs (periumbilical or flank ecchymosis)	Hypovolemic shock (retroperitoneal hemorrhage from ruptured aortic aneurysm, ectopic pregnancy, hemorrhagic pancreatitis and other abdominal catastrophes)
Diffuse purpuric rash	Distributive shock (meningococcal sepsis)
Diffuse maculopapular rash	Distributive shock (toxic shock syndrome)
Unilateral lower extremity edema	Obstructive shock (massive pulmonary embolism)

The most important intervention in the exposure segment of the primary survey is often the measurement of rectal temperature and the maintenance of euthermia. This may be as simple as placing a warm blanket on the patient or as involved as invasive rewarming procedures in the unstable hypothermic patient. In some resuscitations, hypothermia may be maintained or deliberately induced. Hyperthermic patients may simply receive acetaminophen, or, in the case of severely elevated temperatures (>105° F), aggressive mechanical cooling measures may be necessary. Sterile dressings should be applied to patients with burns.

Resuscitation Phase

History

Historical information should be elicited from either prehospital personnel, family members as well as patients themselves. Historical elements may point to a particular class of shock or underlying process. Some findings, however, such as altered mental status and chest pain, may be simply a result of inadequate tissue perfusion and not the key to determining the cause. Identification of a class of shock present will help guide the initial resuscitation. For example, a history of bleeding, vomiting, diarrhea or trauma will immediately alert the clinician to the possibility of hypovolemic shock and the need for rapid volume administration. A history of heart disease, especially with the symptoms of paroxysmal nocturnal dyspnea or orthopnea, are highly suggestive of a cardiogenic shock state. A history of infection, fever

Table 1.5. Important clues on history

Critical historical elements	
Bystander resuscitation	Rescue breathing
	Chest compressions
	Automated external defibrillation
Medical alert/identification bracelets	
Medications brought in by paramedics	
Old medical records/electrocardiograms	
Organ donor identification/drivers license	
Paramedic and bystander observations	
Patient's clothing/belongings for medications/devices/recreational drug paraphernalia	
Presence of possible toxins on scene	
Response to prehospital interventions	Oxygen
	Fluid challenge
	Electrical therapy
	Medications
	Positioning
Vital sign trends and neurological status changes in prehospital phase	
Historical clues in shock states:	
History	**Possible Class Of Shock**
Preceding chest pain, shortness of breath	Cardiogenic Obstructive
Orthopnea	Cardiogenic
Any new medication	Distributive (anaphylactic)
Vomiting and diarrhea	Hypovolemic
Hemorrhage	Hypovolemic
Rash	Distributive (anaphylactic, septic)
Intravenous drug use	Distributive (septic) Cardiogenic
Indwelling devices (catheters, lines)	Distributive (septic)
Chronic debility/neurologic disease	Distributive (septic) Hypovolemic
Trauma	Hypovolemic (hemorrhage)
Diagnostic investigations	

or the use of a new medication may suggest distributive shock. Table 1.5 reviews critical historical elements and clues that may lead to life-saving interventions in the resuscitation of the critically ill patient.

Bedside Diagnostic Investigations

The nature of emergency resuscitation precludes the type of methodical diagnostic workup that is possible in less critically ill patients. Each diagnostic tool must be evaluated for its ability to change the course of the resuscitation. Near immediate results are essential, and tests should not interfere with life-saving interventions. Table 1.6 outlines diagnostic investigations that are useful during the initial phases of resuscitation.

Secondary Survey

As the severity patient's condition on presentation increases, so does the relative importance of the physical examination. Thus, both primary and secondary surveys in resuscitation are primarily directed at physical findings. There is a significant

Table 1.6. Diagnostic investigations in resuscitation

Continuous monitoring	
Pulse oximetry	Pulse oximetry is considered "a fifth vital sign". It is tremendously helpful when it can be recorded accurately; however, in severe shock states diminished pulses and cool extremities may make it impossible to obtain. Pulse oximetry probes can be placed on the earlobes as well as the extremities. Falsely reassuring readings may occur with abnormal hemoglobins, such as with CO toxicity or methemoglobinemia.
Neurological status	Mental status has also been referred to as a vital sign. A progressive alteration in mental status has a broad differential diagnosis, but within the context of an individual resuscitation its significance is often clear. In shock states, it may represent worsening cerebral perfusion or hypoxia and the need for more aggressive resuscitative efforts. In patients with intracranial pathology, it may represent brain herniation and the need for lowering intracranial pressure, especially when combined with localizing signs. When toxic, metabolic and endocrinologic derangements are present, worsening electrolyte abnormalities or hypoglycemia may be present and a multitude of interventions, ranging from simple dextrose administration to hemodialysis may be necessary.
Pain scales	Signs of pain, both verbal and non-verbal, should not be ignored. These may indicate the need to search for an occult injury such as a fracture or penetrating trauma that may change the direction of the resuscitation. Pain can also be used as a guide to the success of resuscitation, as is the case when chest pain and dyspnea resolve with adequate treatment of myocardial ischemia or pulmonary edema.
Continuous cardiac monitor	Continuous telemetry is essential in any resuscitation to monitor for life-threatening dysrhythmias and responses to treatment.
Electrocardiography	
12-lead EKG	The 12-lead EKG is enormously helpful in resuscitation. It has utility in both cardiac and non-cardiac emergencies. EKG findings may be either the cause or result of the underlying condition requiring resuscitation. Attention is directed at signs of *myocardial infarction* and *ischemia, electrolyte derangements* and clues to other life threatening pathologies such as decreased voltage in *cardiac tamponade* or signs of acute right-sided heart strain in *pulmonary embolus*. Certain *drug toxicities* have characteristic EKG findings as well.
Additional EKG leads	Right-sided precordial leads (RV3 and RV4) may be critical in identifying the cause of cardiogenic shock as right ventricular MI. Posterior leads (V8 and V9) may unmask the presence of posterior MI.
Bedside laboratory tests	
Blood glucose	Critically low blood glucose results from many different life-threatening processes and must be addressed immediately. The finding of high blood glucose is similarly important and may help tailor early resuscitative efforts. Blood glucose should be measured in all patients with altered mental status and, when abnormal, frequent rechecks are indicated.

continued on next page

1

Table 1.6. Continued

Hemoglobin or hematocrit	Both of these tests express hemoglobin concentration and, as such, can appear misleadingly high in acute hemorrhage before volume resuscitation has occurred. These tests are subject to error, and repeat and serial values should be obtained when they are utilized to guide resuscitation.
Pregnancy test	A positive serum or urine pregnancy test may lead to a diagnosis of the underlying pathology in a critically ill female. In addition, this finding *may* affect decisions made during resuscitation with respect to monitoring, emergent procedures, the selection of medications and imaging studies and disposition.
Blood type and crossmatch	This is an essential test that must be performed to facilitate treatment with blood and blood products in a multitude of resuscitations, both traumatic and non-traumatic. The infusion of fresh frozen plasma and platelets also requires crossmatching.
Bedisde electrolytes	The availability of blood electrolyte analysis at the bedside is increasing and very helpful. Knowledge of the electrolytes in the first few minutes may enable critical interventions to be started early. In some cases, such therapies should be started even before electrolytes are available (e.g., giving emergent treatment for hyperkalemia in the presence of a typical EKG and history)
Arterial blood gases	Although an assessment for hypoxia and hypercarbia should be made clinically, arterial blood gases have a special role when pulse oximetry is not possible or unreliable, to assess for certain toxins such as carbon monoxide and methemoglobin, and to assist with mechanical ventilation management. The pH and base excess values obtained from blood gases (including venous gases) may also be used as an adjunct to gauge the severity of shock states and response to resuscitative efforts.
Pooled venous oxygen levels	Requires the placement of central venous line with a special probe. May be used to gauge the severity and response to resuscitation.
Other bedside assays	Although there are many potential pitfalls in their application and interpretation, bedside assays may be extremely helpful. In some cases, *elevated cardiac markers* may confirm suspicion of an MI. A variety of *toxicological tests* are now available, and, in the appropriate circumstances, bedside screening assays for various *bioterrorism agents.*
Diagnostic imaging	
Chest film	An early portable chest X-ray is of paramount importance. It may, by itself, identify the type of shock state present (e.g., the finding of cardiomegaly and pulmonary edema in cardiogenic shock, tension pneumothorax in obstructive shock, hemothorax or pleural effusion in hypovolemic shock). It may also be helpful in pulmonary embolism—less for the presence of rare signs such as Hampton's Hump and Westermark's sign than for the *absence* of significant findings pointing to alternative diagnoses such as pulmonary edema and pneumonia. A widened or abnormal mediastinum may represent aortic rupture or dissection.
Cervical spine films	The presence of cervical spine trauma may help explain the findings of shock, neurological deficits and ventilatory failure.

continued on next page

Table 1.6. Continued

Pelvis	This is an important film that may identify a source of hemorrhage and occult trauma.
Lateral soft tissue neck	This film may identify mechanical airway obstruction, a source of septic shock or foreign bodies.
Abdominal films	Although rarely helpful in resuscitation, a single abdominal film may show a pattern of calcification of the aorta in the case of a ruptured aortic aneurysm and the presence of radiopaque toxic ingestions such as iron, phenothiazines and enteric release tablets.
Ultrasonography	Bedside ultrasound is ideal for use in resuscitation because of its availability, repeatability and speed. Bedside echocardiography can be used to reveal the presence of various shock states by identifying cardiac tamponade, global hypokinesis or right ventricular outflow obstruction. In the future, it may be utilized by emergency physicians to evaluate valvular lesions and dyskinesis. It can also assist with the distinction between pulseless electrical activity and cardiac standstill (electromechanical dissociation). This may help to determine when resuscitation efforts should be terminated. Abdominal ultrasound may quickly identify free-fluid (most importantly, hemorrhage) in the peritoneal cavity. Hemothorax, as well as pleural effusions, may also be identified during the focused assessment with sonography for trauma (FAST) examination. The aorta may be quickly imaged to assess for abdominal aortic aneurysm. Pelvic ultrasonography in the female patient with intraperitoneal hemorrhage may further delineate the source of shock. The absence of an intrauterine gestation in a pregnant female may represent ectopic pregnancy, whereas its presence may indicate a bleeding cyst, heterotopic ectopic pregnancy or occult trauma. Ultrasonography also has a role in assisting with emergency procedures, such as line placement and pericardiocentesis.
Cranial CT	Of all CT studies, cranial CT, because of its speed and lack of need for contrast, may be performed even in the unstable patient. It may identify the need for emergent surgical decompression, measures to lower intracranial pressure or the search for other causes of altered mental status, all which may change the course of a resuscitation.

overlap in the examination during the primary and secondary surveys, but the secondary survey tends to reveal those features which would be missed unless specifically looked for. In the context of an individual resuscitation, some of these findings may be very important or even critical.

Simply stated, the secondary survey is a complete, compulsive physical examination. Once resuscitative measures are underway, every critically ill patient should have such an examination. Several examples of secondary survey findings that may alter acute management are given below:

1

Findings	Possible Diagnostic Significance
Impaired visual acuity	Occult trauma Arterial thromboembolism (cerebrovascular accident, aortic dissection)
Hemotympanum	Occult head trauma
Nuchal rigidity, meningeal signs	Meningitis Subarachnoid hemorrhage
Thyroidectomy scar	Myxedema coma
Right ventricular heave	Obstructive shock (acute right heart strain, massive pulmonary embolism)
Absent bowel sounds	Distributive or hypovolemic shock (peritonitis) Occult spinal cord injury
Retained vaginal foreign body	Distributive shock (toxic shock syndrome)
Dysmetria, ataxia limb movements	Cerebellar lesion
Unilateral upgoing plantar response	Cerebrovascular accident Non-convulsive status epilepticus

Definitive Care Phase

Definitive care phase of the resuscitated patient may begin in the ED and continue in various inpatient settings (the operating room, intensive care unit, cardiac catheterization or interventional radiology suite, etc.). Transfer to another facility for specialized care may be necessary.

The importance of the family of the critically ill patient should not be forgotten. Many patients requiring extensive resuscitation in the ED may have been previously well. In the case of patients with chronic, controlled disease, family members may be quite shaken by the sudden decompensation in their loved one's condition. It is the responsibility of the EP and other members of the primary resuscitation team (nursing staff, social services) to make themselves available to the family as soon as it is possible. Early communication with family and friends serves several purposes: to obtain additional relevant history, to explain the current condition and resuscitative efforts that are taking place, to clarify any advance directives or previously expressed wishes of the patient, and to express the concern and support of the resuscitative team. Although controversy exists as to whether family members should be permitted to view resuscitative efforts, there is little doubt that interacting with family members in these situations is a skill that requires training, practice and flexibility.

Other individuals that may become involved as indirect members of the resuscitation team include religious or spiritual counselors, organ procurement specialists, law enforcement, forensic specialists, sexual assault and domestic violence personnel. It is the responsibility of the EP to understand the reporting requirements for victims of violence, abuse, neglect and organ procurement in their respective practice jurisdictions.

Ethical and Legal Aspects of Resuscitation

Many ethical issues are magnified and intensified during a resuscitation. How aggressive should resuscitation efforts be when there is a low likelihood of survival? How should resources in the ED be distributed between critically ill patients with poor prognoses and less severely ill patients? Under what circumstances is a patient that is still communicating in a position to refuse resuscitative efforts when they are emergently needed? What process should be followed to obtain consent for organ donation?

There are, however, certain legal realities that the EP and other members of the resuscitation team need to be compliant with. Moreover, laws and guidelines that apply to medical emergencies differ from jurisdiction to jurisdiction. These include laws relating to Do Not Resuscitate (DNR) orders, advance directives, living wills, consent or refusal of treatment and mandatory reporting laws to police, coroner and various social agencies.

In general, resuscitative efforts should not be initiated when obvious signs of death are apparent, such as dependent lividity, rigor mortis or trauma inconsistent with life. Although statutes regarding DNR directives vary, the fundamental right of an individual to make decisions about their medical care, including end-of-life care, should be honored by medical personnel. This right was recognized in the United States by the Patient Self-Determination Act of 1991. Similar legislation exists in other countries.

The decision of when to cease resuscitative efforts once they have begun is often more difficult. Survival after prolonged loss of spontaneous circulation and, perhaps more importantly, survival with neurological function that would be acceptable to the patient, becomes less likely as time elapses, with the rare exception of miraculous survival such as sometimes occurs with victims of accidental hypothermia. Ultimately, a judgment must be made by the responsible physician, weighing the likelihood of benefit against the disadvantages of continuing aggressive resuscitative efforts.

CHAPTER 2

Cardiovascular Disorders

Jason Greenspan, Shahram Tabib and Stuart P. Swadron

Part A: Hypertension and Hypertensive Emergencies

Hypertension is one of the most common conditions affecting patients in developed countries. As the population ages and the emergency department continues to serve populations without access to appropriate primary care, issues regarding hypertension will become more important. Emergency Physicians must be comfortable in evaluating and treating patients with conditions associated with an acute rise in blood pressure, conditions secondary to long-standing hypertension, as well as with the complications of medications used to control hypertension.

Definitions

- Essential Hypertension is a persistently elevated blood pressure measured on two separate occasions. The Joint National Committee on Prevention, Detection, Evaluation, and Treatment of High Blood Pressure has classified hypertension based on the degree of elevation (Table 2A.1).
- Hypertensive Urgency is the presence of an elevated BP without signs or symptoms of end organ damage. Blood pressure should be reduced gradually over 24-48 h in hypertensive urgencies.
- Hypertensive Emergency is the presence of an elevated BP with evidence of end-organ damage. Table 2A.2 lists conditions regarded as true hypertensive emergencies. These conditions necessitate the careful reduction of blood pressure in minutes to hours.

Epidemiology/Pathophysiology

- The majority of hypertensive emergencies occur in previously hypertensive patients. In these patients, the ability of the body to autoregulate blood pressure is adjusted to accommodate for the chronic elevation of blood pressure. A hypertensive emergency occurs with an acute elevation in blood pressure over baseline.
- While the actual blood pressure is important in the evaluation and diagnosis of these conditions, it is the presence of end-organ damage and not the actual blood pressure measurement that indicates the need for emergent lowering of blood pressure.
- The rate of elevation of the blood pressure may be more important in the pathogenesis of end-organ damage than the actual blood pressure.

Diagnosis and Evaluation

History and Physical Examination

- The evaluation of the hypertensive patient involves a careful history focused on evaluating the presence of symptoms suggestive of end-organ damage, the risk of developing subsequent end-organ damage if untreated, and any past treatment for hypertension or associated conditions.

Table 2A.1. JNC-VI classification of blood pressure in adults

Category	Systolic BP	Diastolic BP
Optimal	<120	<80
Normal	<130	<85
High-Normal	130-139	85-89
Hypertension		
Stage 1	140-159	90-99
Stage 2	160-179	100-109
Stage 3	>180	>110

- Patients should also be questioned regarding foods, medicines, or the use of medications or illicit drugs which may contribute to blood pressure elevation.
- Physical exam begins with the proper measurement of blood pressure in multiple extremities with a proper sized blood pressure cuff.
- Examination should include a complete neurologic evaluation including fundoscopy to rule out retinal hemorrhages or papilledema, cardiovascular exam including evaluation of new murmurs, an S3 or S4, pulses in multiple extremities, pulmonary exam listening for findings indicative of pulmonary edema, and an abdominal exam to evaluate for bruits or aneurysms.

Laboratory and Studies

- The laboratory examination in severely hypertensive patients is geared towards the evaluation of any presenting emergent condition.
- Baseline studies should include a CBC to evaluate for the microangiopathic hemolytic anemia associated with malignant hypertension, evaluation of electrolytes and renal function, evaluation of cardiac specific enzymes, and urinalysis to evaluate for proteinuria and/or hematuria.
- An EKG and CXR should be obtained to evaluate cardiovascular emergencies.
- Other studies including CT scans, abdominal ultrasound, or aortography are done as needed.

Specific Hypertensive Emergencies

Certain disease processes are discussed in greater detail in other sections of this handbook. However, treatment with respect to blood pressure control is discussed here.

Table 2A.2. Hypertensive emergencies

Hypertensive encephalopathy
Stroke (ischemic or hemorrhagic)
Myocardial infarction or unstable angina
Pulmonary edema/congestive heart failure
Aortic dissection
Acute renal failure
Preeclampsia/eclampsia
Microangiopathic hemolytic anemia
Catecholamine excess

2

Hypertensive Encephalopathy

- When the acutely elevated blood pressure exceeds the CNS's ability for autoregulation, uncontrolled cerebral blood flow occurs. This results in vasospasm, vascular damage and leakage, and potential cerebral hemorrhage, ischemia, and/or edema.
- Patient complaints include headache, vomiting, drowsiness, confusion, visual changes, or focal neurologic changes. Papilledema is present and severe retinopathy is often present.
- Untreated, patients will progress to coma and death.
- Treatment involves the acute lowering of blood pressure with titratable IV medications. By definition, patients with hypertensive encephalopathy lack cerebral autoregulation. Precipitous lowering of the systemic blood pressure may cause a dangerous fall in cerebral perfusion pressure, leading to further cerebral ischemia. Therefore, careful monitoring of patients is necessary to evaluate for neurologic deterioration. Systemic blood pressure reduction should be done slowly with published guidelines of approximately 15-20% reduction in diastolic pressure within 1 h or a diastolic pressure of 110 mm Hg as therapeutic goals.

Hypertensive Stroke

- Severe, uncontrolled hypertension is frequently an etiologic factor in patients with strokes. Patients with this degree of hypertension have cerebral autoregulatory set points changed to accommodate the degree of chronic hypertension. Therefore, overaggressive lowering of blood pressure may cause a dangerous lowering of cerebral perfusion pressure and extend ischemic zones of the brain.
- Patients with hemorrhagic infarctions may present with severe hypertension as a response to the bleed, as well as an etiology of the infarction. This may resolve without treatment.
- With the above considerations, there is debate in the literature regarding the appropriate management of severe hypertension in stroke. Many authors recommend lowering for diastolic pressures >120, while others recommend that blood pressure never needs acute lowering in the emergency setting.

Hypertension with Ischemic Coronary Syndromes

- Severe hypertension is an etiologic factor in atherosclerotic heart disease. Many patients presenting with acute coronary syndromes (myocardial infarction or unstable angina) have chronic hypertension which may be severe or uncontrolled. Acute elevations in blood pressure may exacerbate coronary ischemia by increasing ventricular strain and myocardial oxygen demand.
- Decreasing afterload with the use of IV nitrates decreases myocardial work. IV β-blockers also decrease myocardial work by decreasing blood pressure and heart rate.
- The use of direct vasodilators, which may cause a reflex tachycardia, is contraindicated in the setting of ischemia.

Hypertension with Pulmonary Edema or Congestive Heart Failure

- Severe hypertension is an etiologic factor for congestive heart failure but may also be secondary to the catecholamine response to pulmonary edema.
- Treatment for these patients includes nitroglycerin and diuretics but may also require the addition of morphine sulfate and nitroprusside. The use of morphine sulfate is effective in decreasing the catecholamine response to heart failure. Both nitroglycerin and nitroprusside produce vasodilatation in the capacitance vessels thus improving cardiac hemodynamics. Nitroprusside has a more pronounced effect on arterioles, thus reducing afterload. However, a reflexive tachycardia and increased inotropy may counteract the decrease in afterload and even lead to an increase in cardiac workload. Nitroprusside may also cause coronary steal in patients with coronary artery disease. It is therefore not the first line drug in cardiac failure with severe hypertension.

Aortic Dissection

- The control of hypertension is essential in the emergency stabilization of a patient with an aortic dissection.
- The use of both β-blockers and nitroprusside is recommended to both decrease the systemic blood pressure and to decrease the shearing force of the systolic pulse on the weakened aortic wall. Standard medications used are nitroprusside plus labetolol or esmolol.
- Blood pressure should be decreased to the lowest possible level without exacerbating ischemic symptoms.

2

Hypertension and Acute Renal Failure

- Patients with long-standing uncontrolled hypertension often develop renal failure. Acute elevations in blood pressure may lead to intrarenal vascular injury, glomerular ischemia, and subsequent hematuria, proteinuria and loss of renal function.
- The management of acute renal failure secondary to acute hypertension is focused on maintaining normal volume status, renal perfusion, and minimizing secondary complications.
- IV β-blockers or calcium channel blockers are the drugs of choice. They must be used with caution, and euvolemia must be maintained in order to not decrease renal perfusion to a level which exacerbates instead of alleviating renal damage. The use of diuretics may be either beneficial, if used in hypervolemic or euvolemic patients to increase GFR, or problematic if used in hypovolemic patients. Nitroprusside, while effective for decreasing blood pressure, is problematic in patients with renal dysfunction because the thiocyanate metabolite of the drug may accumulate, leading to cyanide toxicity.
- Dialysis may be needed in severely symptomatic or hypertensive patients.

Preeclampsia/Eclampsia

- Preeclampsia and eclampsia represent diffuse end-organ damage secondary to pregnancy induced hypertension.
- Most patients with preeclampsia/eclampsia are vasoconstricted and hypovolemic.
- Hydralazine is the standard antihypertensive used in preeclampsia. However, IV β-blockers or calcium channel blockers may also be used.
- Careful management of volume status is important as these patients have renal and often cerebral vascular injury and are therefore prone to develop edema with overaggressive hydration. The treatment for preeclampsia/eclampsia is delivery of the fetus and placenta and close communication with an obstetric specialist is required.

Microangiopathic Hemolytic Anemia

- The endovascular damage associated with hypertensive crises results in fibrin deposition in arterioles and ultimately *fibrinoid necrosis*. This fibrin deposition may lead to a hemolytic anemia which is diagnosed by the presence of schistocytes on peripheral blood smear. This anemia is rarely seen in isolation in hypertensive emergencies and management is based on end-organ damage in other organ systems.

Catecholamine Excess

- Excess catecholamines may lead to hypertensive emergencies. Causes include pheochromocytomas; ingestions of stimulant medications or drugs, such as cocaine or amphetamines; withdrawal syndromes as seen in the rebound hypertension with clonidine withdrawal; or the ingestion of tyramine rich foods while taking MAOIs. These conditions all result in an increased α-adrenergic tone.
- Treatment requires α blockade with phentolamine. B-blockade alone is contraindicated as it leads to unopposed α stimulation. In the case of stimulant drug ingestions, anxiolytics such as lorazepam or valium may be effective in lowering blood pressure as well as treating associated hyperactivity.

2

Medications Used to Treat Hypertensive Emergencies and Urgencies

Pharmacologic treatment of hypertensive emergencies requires medications which are rapid acting, easily titratable, and which lack significant side effects. Intravenous medications are the most appropriate for true emergencies. Patients must be closely monitored during the use of these medications for adverse reactions including hypotension or worsening of the underlying condition.

In hypertensive urgencies, the blood pressure may be reduced slowly. Oral medications may be appropriate in these situations.

Sodium Nitroprusside

- Nitroprusside is the drug of choice for most hypertensive emergencies.
- Nitroprusside decreases both afterload and preload by direct arterial and venous dilatation. The blood pressure response is dose dependent.
- It is administrated as a light-sensitive IV solution at doses beginning at 0.25 mcg/kg/min. Onset of action is within 1-2 min. The half-life is 3-4 min which allows the pharmacologic effect to be quickly discontinued in patients with adverse reactions.
- As discussed above, nitroprusside may have multiple cardiovascular complications including an increase in cardiac work as well as coronary steal in atherosclerotic coronaries. However, in patients with congestive heart failure, nitroprusside has been shown to be effective in increasing cardiac output.
- Theoretically, nitroprusside may adversely effect cerebral perfusion. Nitroprusside's potent vasodilatation may cause dilation in the cerebral vasculature, thus increasing cerebral blood flow and intracerebral swelling. However, the decrease in systemic blood pressure counteracts this effect, making nitroprusside the drug of choice in patients with hypertensive encephalopathy.
- Nitroprusside is metabolized into thiocyanate and may therefore lead to cyanide toxicity if administered for a prolonged time or to patients with either liver or kidney disease.

Nitroglycerin

- Nitroglycerin is a direct vasodilator that acts predominantly on the venous circulation. At higher doses it has some effect on arterial tone.
- Doses begin at 5 mcg/min. Onset is within 1-5 min.
- Nitroglycerin dilates coronary arteries and thus, it is used primarily as an anti-anginal medication in patients with acute coronary syndromes. Nitroglycerin decreases preload which improves cardiac mechanics in failing hearts.
- Nitroglycerin may cause hypotension, especially in the face of hypovolemia, as well as headache, flushing and nausea.

Labetolol

- Labetolol produces both α and β blockade.
- It can be given both IV and PO. Therapeutic effect can be seen in approximately 2-5 min and peaks in approximately 10 min. Initial loading dose of 20 mg over 2 min can be repeated in 10 min intervals until a response is noted.
- IV labetolol produces orthostatic hypotension.
- It does not cause a decrease in cerebral or peripheral blood flow and is safe in patients with coronary artery disease.
- Contraindications are those for all β-blockers as a group and include bronchospasm, CHF, or heart block. Labetolol may also cause postural hypotension.

2

Table 2A.3. Medications for hypertensive emergencies

Emergency	Medications of Choice	Contraindicated Medications
Hypertensive Encephalopathy	Nitroprusside	
Stroke	Medication not indicated	
Ischemic coronary syndromes	Nitroglycerin	Hydralazine
CHF/Pulmonary edema	Nitroglycerin	
Aortic dissection	Nitroprusside plus β-blocker, Esmolol	Hydralazine
Acute renal failure	Nicardipine	β-blocker
Preeclampsia/Eclampsia	Hydralazine	
Catecholamine excess	Phentolamine, Lorazepam	β-blocker alone

Esmolol

- Esmolol is an ultra-short acting β-blocker with a rapid onset of action and cessation of action when discontinued.
- It is used IV and is most suitable for relatively unstable patients with supraventricular dysrhythmias or aortic dissection.
- Contraindications include those previously mentioned for β-blockers.

Nicardipine

- Nicardipine is a titratable IV calcium channel blocker which may also be given orally.
- The onset is 5-15 min with duration of action approximately 30 min. The starting dose is 5-15 mg/h.
- Nicardipine decreases afterload without reducing heart rate or cardiac output. It has a favorable effect on failing hearts by improving ejection fraction and acts as an anti-anginal by dilating coronary arteries.
- Adverse effects include hypotension, headache, and nausea.

Hydralazine

- Hydralazine produces direct vasodilatation and is the drug of choice in hypertensive emergencies associated with pregnancy.
- As with all direct arterial dilators, hydralazine causes a reflex tachycardia causing an increase in cardiac work. Hydralazine is therefore not used in patients with cardiac ischemia and severe hypertension

Phentolamine

- Phentolamine is an IV short acting α-adrenergic blocker with rapid onset.
- It is given as an initial bolus of 5-10 mg and repeated as necessary.
- Adverse effects include the development or exacerbation of coronary ischemia.
- Its use is limited to settings with catecholamine excess.

Nifedipine

- Oral nifedipine was used commonly in the treatment of hypertensive emergencies.
- It causes a potent arteriole dilatation 5-10 min after administration. Unfortunately, this reduction in blood pressure is often uncontrolled, leading to adverse effects on cerebral blood flow as well as adverse cardiac effects secondary to reflexive tachycardia. It is contraindicated in hypertensive emergencies because of these dangers.

2

Clonidine

- Clonidine is a central acting α antagonist.
- Reductions in blood pressure can be seen 30 min after oral administration with maximal effect in 2-4 h.
- While it appears to be safe, it has a slower onset than the IV drugs listed above. It is also not easily titratable and thus may be dangerous if adverse effects occur.
- Clonidine also causes drowsiness and thus difficulties in monitoring mental status.

Common Effects of Anti-Hypertensive Medications

Anti-hypertensive medications are prescribed commonly. A functional knowledge of the use and side-effect profile of these drugs is important when managing patients taking these medications.

Diuretics

- Diuretics are an excellent choice for initial therapy in hypertension. If they are not the initial medication used, they are indicated as a secondary medication as they have an additive effect on blood pressure when used in combination. They are especially effective in African Americans.
- The most common side effects of the diuretics are due to their effect on fluid and electrolyte balance. Dehydration is the most obvious side effect. Hypokalemia and hyperkalemia may both occur depending on the agent used.
- Diuretics may also cause increases in uric acid and thus precipitation of gouty arthritis.

Beta Blockers

- Beta blockers are recommended as monotherapy for hypertension. They are especially useful in patients with ischemic heart disease, tachydysrhythmias, essential tremor, or migraines.
- Beta blockade may cause exacerbation of asthma or other bronchospastic diseases. It may worsen CHF or cause heart block especially when combined with calcium channel blockers. Beta blockade may mask the symptoms of hypoglycemia in diabetics. Depression and sexual dysfunction may also occur.

Calcium Channel Blockers

- Calcium channel blockers are especially effective in African Americans.
- Vasodilatation by calcium channel blockade may precipitate orthostatic hypotension or peripheral edema, especially in the elderly. CHF may be worsened by the negative inotropic effect of these medications. Heart block may also occur.

ACE Inhibitors

- ACE inhibitors are recommended as initial therapy in diabetics, patients with heart failure and patients with renal insufficiency (creatinine <3 mg/dl).
- ACE inhibitors may cause angioedema or cough. Hyperkalemia may also occur as well as worsening of renal failure especially in patients with renal artery stenosis.

Suggested Reading

1. Abdelwahab W, Frishman W, Landau A. Management of hypertensive urgencies and emergencies. Journal of Clinical Pharmacology August 1995; 35(8):747-62.
2. Grossman E, Ironi AN, Messerli FH. Comparative tolerability profile of hypertensive crisis treatments. Drug Safety August 1998; 19(2):99-122.
3. Jackson R. Hypertensive emergencies. In: Tintinalli JE, Kelen GT, Stapczynski JS, eds. Emergency Medicine. A Comprehensive Medicine Study Guide. McGraw-Hill, 2000.
4. Kitiyakar C, Guzman NJ. Malignant hypertension and hypertensive emergencies. Journal of the American Society of Nephrology January 1998; 9(1):133-42.

5. Mathews J. Hypertension. In: Marx JA, Hockberger RS, Walls RM, eds. Emergency Medicine. Concepts and Clinical Practice. Mosby, 2002:2.
6. Murphy C. Hypertensive emergencies. Emergency Medicine Clinics of North America November 1995; 13(4):973-1007.
7. Thach AM, Schultz PJ. Nonemergent hypertension. New perspectives for the emergency medicine physician. Emergency Medicine Clinics of North America November 1995; 13(4):1009-35.
8. Tietjen CS, Hurn PD, Vlatowski JA et al. Treatment modalities for hypertensive patients with intracranial pathology: Options and risks. Critical Care Medicine February 1996; 24(2):311-22.
9. The sixth report of the joint national committee on prevention, detection, evaluation, and treatment of high blood pressure. Archives of Internal Medicine November 1997; 157(21):2413-46.
10. Varon J, Marik P. The diagnosis and management of hypertensive crises. Chest July 2000; 118:214-227.
11. Vaughn CJ, Delanty H. Hypertensive emergencies. Lancet July 2000; 356(9227):411-7.

Part B: Acute Coronary Syndromes

Coronary artery disease is the most common cause of death in the United States, accounting for approximately 600,000 deaths annually. Of 6.0 million ED visits per year for chest pain, about 1.2 million people are diagnosed with myocardial infarction and another million with unstable angina. It has been estimated that the overall cost of coronary artery disease exceeds 100 billion dollars annually in the U.S. There is also a significant cost in terms of malpractice claims, with missed myocardial infarction and acute coronary syndromes continuing to constitute a large percentage of both claims and costs. Mortality and morbidity continue to decrease with advances in therapy. There was a 54% reduction in age-adjusted mortality from myocardial infarction in the U.S. from 222/100,000 in 1963 to 101/100,000 in 1990.

Definitions

- Coronary artery disease (CAD) is a spectrum of disease that ranges clinically from asymptomatic or "silent" to one of the following clinical syndromes: stable angina, variant angina or acute coronary syndrome.
- Stable angina is episodic, exertional chest pain lasting approximately 5-15 min. EKG changes occur <50% of the time. Cardiac enzymes are not elevated.
- Variant or Prinzmetal angina is uncommon and occurs primarily at rest without provocation, typically at the same time of the day. ST elevation can be seen on EKG.
- Acute coronary syndromes comprise the entire spectrum of disease from unstable angina through to myocardial infarction.
  - Unstable angina is defined as new onset angina or angina that is increasing in frequency, intensity, duration, sensitivity to exercise or nitroglycerin (NTG) requirement. EKG changes occur in 50% or more of patients and cardiac troponin may be mildly elevated.
  - Acute myocardial infarction (AMI) occurs when there is frank necrosis of myocytes. Chest pain is usually sustained >20 min. EKG changes occur in 50% or more of cases and cardiac serum markers are elevated. AMI is classically divided into transmural (Q-wave) or subendocardial (non Q-wave) MI, although there is significant overlap between these entities. A more important classification of AMI in the ED is to distinguish ST elevation MI (STEMI) from non-ST elevation MI (non-STEMI).

Pathophysiology

- In most cases, coronary artery disease is caused by atheromatous plaques in the lumina of the epicardial vessels. However, other sources of occlusion include thrombus formation associated with arterial dissections as well as thrombi from heart chambers and prosthetic valves. Inflammatory processes, such as those associated with Kawasaki disease and systemic lupus erythematosis are uncommon causes of coronary artery disease.
- Stable angina is caused by chronic, fixed nonulcerating plaques. Angiography usually reveals 50-75% obstruction.
- Variant angina is primarily due to vasospasm. Angiography reveals normal coronaries in one-third of the cases and CAD in conjunction with vasospasm in the remaining two-thirds.
- Acute coronary syndromes share the common pathophysiology of a fissuring or "unstable" plaque (often with <50% occlusion) that becomes a nidus for the aggregation of platelets and fibrin. Vasospasm at the site of the fissured plaque also occurs.
- Mechanisms of cardiac ischemia and infarction independent of coronary artery obstruction include those from metabolic, hematologic and toxicologic conditions. Hypoxia, anemia, hypotension and carbon monoxide toxicity all tend to result in injury in a global cardiac distribution rather than in the distribution of a particular epicardial artery, but may manifest first in territories of vessels with preexistent CAD.

Risk Factors

- Major risk factors for CAD include age, male gender, diabetes mellitus, chronic hypertension, family history of premature CAD, cigarette smoking, hyperlipidemia and lack of hormone replacement after menopause.
- Because these well known risk factors are based on lifelong risks in populations, they are less important in the ED than they are in the primary care setting. In the ED, the individual patient's presenting symptoms (e.g., "crushing retrosternal chest pain versus sharp, intermittent pain") and the appearance of the 12-lead EKG (e.g., ST segment/T wave changes versus normal) overshadow any predictive value of the classic risk factors.

Diagnosis and Evaluation

History

- The classic presentation of symptomatic CAD is that of left-sided or retrosternal chest pain of a pressure-like nature. However, many variations exist including burning pain, pain akin to indigestion (approximately 20% of patients) and sharp, stabbing pain (5-20% of patients). Pain may radiate to the jaw, neck, back or down either upper extremity, corresponding to the C8-T5 dermatomes. It is important to note that lack of classic characteristics of pain cannot be used to rule out CAD as a cause for chest pain.
- The temporal pattern of pain and its relation to activity will help to classify the presentation into one of the clinical categories outlined above.
- Common associated symptoms include nausea and vomiting (especially with inferior wall ischemia), diaphoresis, shortness-of-breath and lightheadedness.
- In diabetic and elderly patients, chest pain itself may be absent. In some of these cases an "anginal equivalent" such as shortness-of-breath, lightheadedness or nausea may be present. In the oldest subset of patients, CAD may present very nonspecifically with weakness, malaise or simply with the complications of an acute coronary event such as congestive heart failure (CHF), dysrhythmia or cerebrovascular accident.
- 5-15% of AMIs are completely asymptomatic or "silent". These occur mostly in elderly and diabetic patients.

Table 2B.1. Relationship of EKG leads to location of ischemia related epicardial vessel

EKG Lead	Location of Ischemia/Infarction	Epicardial Vessel
II, III, and AVF	Inferior Wall	Right Coronary Artery
V1, V2	Septal Wall	Left Anterior Descending Artery
	Posterior Wall*	Right Coronary Artery (in most cases)
I,AVL, V5 and V6	Lateral Wall	Left Circumflex Artery
V1, V2,V3,V4,V5	Anterior Wall	Left Anterior Descending Artery
Special Leads		
V8, V9	Posterior Wall	Right Coronary Artery
RV4	Right Ventricle	Right Coronary Artery

* Note: In posterior wall ischemia/infarction, changes in leads V1 and V2 will be opposite to those expected in other leads. Because ST segment changes in leads V1 and V2 can represent both STEMI in the posterior wall as well as non-STEMI in the anterior and septal locations, additional leads (e.g., V8 or V9) or an echocardiogram may be obtained to help clarify the situation. The presence of a tall R wave in lead V1 is the equivalent of a Q wave in the other infarct locations.

2

Physical Examination

- Physical examination is rarely diagnostic of symptomatic CAD. Although signs such as the new onset of an S4 gallop and a paradoxical spilt of S2 have been described with AMI, these are unreliable.
- The physical examination serves two other important roles:
  - The exclusion of other life-threatening causes of chest pain such as pneumothorax (asymmetrical resonance to percussion), aortic dissection (asymmetry of pulses, neurological signs) and pulmonary embolus (signs of right-sided heart strain)
  - The identification of complications of myocardial ischemia and infarction such as CHF (jugular venous distension, S3 gallop, rales, edema, hypotension), acute papillary muscle rupture and myocardial rupture (new onset murmur with signs of severe CHF) and pericarditis (friction rub).

EKG Findings

- The EKG remains the most important diagnostic tool in the evaluation of acute chest pain and guides the early management of patients with CAD. A 12-lead EKG should be obtained as soon as possible.
- There is a great degree of variability in the patterns of change on the 12-lead EKG during symptomatic CAD. In unstable angina and subendocardial MI, when a partial obstruction to flow in a coronary artery exists, T wave inversions and ST segment depressions are most commonly seen. ST elevations in the vascular territory of a particular epicardial artery (see above) is characteristic of transmural MI, when there is complete obstruction.
- Because of the time-dependent nature of revascularization treatments for STEMI, the most important distinction that must be made in the initial evaluation of a patient with symptomatic CAD is that between those patients with significant ST elevation and those without.
- Since ST elevation may not be present on the initial EKG, it is important to obtain frequent serial EKGs, especially when there is a dynamic nature to the patients presenting symptoms. Conversely, ST elevation on the initial EKG may completely resolve with medical therapy, again with major implications on therapy. For similar

2

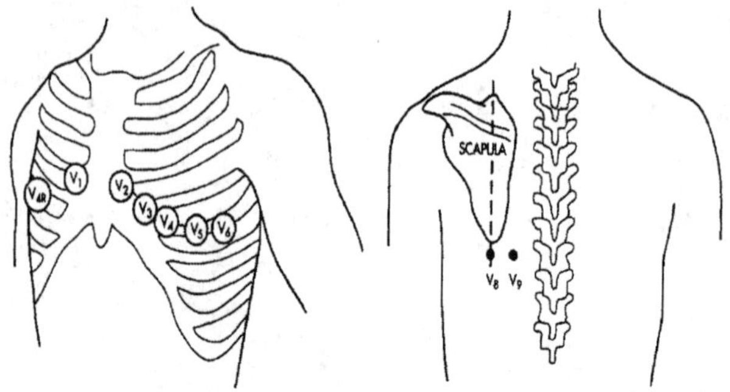

Figure 2B.1. The placement of additional leads V8, V9 and RV4 (the 15-lead EKG)

reasons, a previous EKG tracing of the patient prior to the acute episode may be extremely valuable in making emergency management decisions.

- Although variable, the typical progression of EKG changes in STEMI begin with the brief (minutes) appearance of giant or "hyperacute" T waves, followed by ST elevations in the distribution of the occluded or "infarct-related" vessel. With time (typically several hours), there is the development of Q waves. T wave changes are variable.
- The presence of a left bundle branch block (LBBB) is extremely important in the evaluation of ACS. Its presence indicates a poor prognosis and the need for more aggressive management. If the LBBB is documented or presumed to be new in onset, the treatment of ACS follows similar lines as that of STEMI.
- The progression of nonSTEMI is even more variable than with ST elevation MI. Q waves do not usually occur.
- The distribution of ischemia/infarction within the heart is important to predict likely complications as well as to predict prognosis.
- The additional leads V8 and V9 may be placed when there is suspicion for acute posterior wall MI, in which case ST segment elevation in these leads will be seen. Suspicion for posterior wall MI exists when there are changes suggestive of ischemia in leads V1 and V2.
- The additional lead RV4 may be placed when there is suspicion for acute right ventricular MI. Acute RV infarction is confirmed by the presence of ST segment elevation. Right ventricular MI may be suspected either on clinical grounds (e.g., because of an exaggerated hypotensive response to preload reduction therapy such as nitrates) or when ischemic changes are present in the inferior leads (II,III and avL).
- Changes of a dynamic nature (e.g., changes from previous EKG recordings on the same patient or changes over serial EKGs in the ED during the initial hours of the presentation) are more suggestive of ACS and predict a higher mortality and need for aggressive therapy.
- A normal EKG cannot be used to rule out the presence of ACS or AMI. However, in those patients with normal or minimally abnormal EKGs, the risk of mortality and complications is much lower.

Serum Cardiac Markers

- Cardiac markers have a role in the diagnosis as well as the risk stratification of patients with ACS.
- At the present time, no perfect marker of ACS exists, although there ongoing advances in technology and research continue to improve the utility of cardiac markers.

Table 2B.2. Serum markers in acute coronary syndromes*

Marker	Time to Elevation (hours)	Time to Peak (hours)	Time to Normalization (days)
Myoglobin	2	8	1
CK (creatine phosphokinase)	6-8	24-30	3-4
CK-MB isoenzyme	3-4	18-24	2
Cardiac troponin I and T	3-6	18-24	7-14

*The above figures are estimates. Characteristics of individual assays, as well as normal and diagnostic values, vary with technique and manufacturer. Laboratories will be able to provide specific information on marker assays used in individual institutions.

- Most institutions have a protocol for the serial measurement of cardiac markers in the evaluation of possible ACS. Regardless of the protocol used, a knowledge of the benefits and limitations of each marker is important.
- Myoglobin, CK and cardiac troponin are all released into the blood after cardiac cell injury and death. Table 2B.2 compares the temporal characteristics of these markers in AMI.
- Myoglobin is elevated quickly and is very sensitive. However, its total lack of specificity limits its diagnostic value.
- CK-MB assays are generally less specific and less sensitive than cardiac troponin, although differences in technique account for significant variations in reported accuracy.
- To date, cardiac troponin I and T are the most cardiac specific markers and have been adopted by many institutions in the U.S.
- Before the advent of cardiac troponin, the commonly accepted definition of AMI included an elevation in CK-MB. A new group of patients has been identified in whom there is a mild elevation of cardiac troponin without an increase in CK-MB, making the precise definition of AMI less clear. However, identification of this group of patients with ACS and "microinfarction" has proved important clinically, because they have poorer outcomes than patients with no marker elevations and warrant more aggressive management strategies.
- Although there is some degree of elevation of cardiac troponin in renal failure, it still appears to have prognostic value.
- In certain cases it may be helpful to combine markers because of their different temporal profiles. For example, the addition of CK to cardiac troponin may be helpful to detect early reinfarction in the week following an AMI when troponin values are still elevated from the initial event.

Other Laboratory Values

- When suspicion for ACS is high, additional laboratory investigations are appropriate and may include CBC, basic chemistry, urine toxicology and blood typing (when fibrinolytic or invasive procedures are contemplated).

Imaging Studies

- Emergent portable chest X-ray helps to rule out other important causes of chest pain such as aortic dissection and pneumothorax. Many institutions have protocols to ensure that a chest X-ray is reviewed prior to the administration of fibrinolytic therapy so that signs suggestive of aortic dissection are not overlooked. Chest X-ray is also important to detect cardiomegaly and other signs of congestive heart failure that may complicate or coexist with ACS.

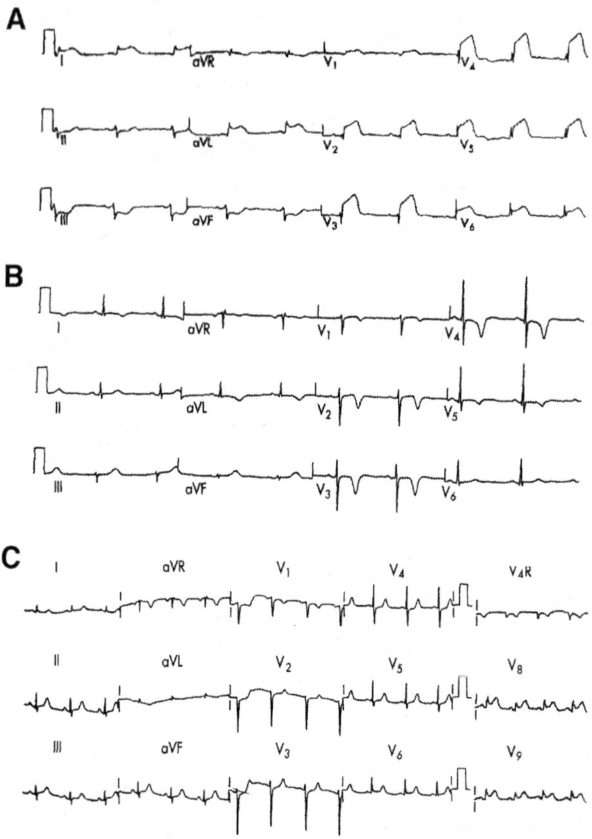

Figure 2B.2. Important EKG findings in acute coronary syndromes. A) Acute anterior MI showing ST elevations in the precordial leads, with reciprocal ST depression in the inferior leads (reciprocal changes increase diagnostic certainty of AMI and also indicate a graver prognosis). B) Deep, symmetrically inverted T waves in the anterior leads in setting of cardiac symptoms are strongly suggestive of proximal left anterior descending obstruction and carry a poor prognosis without prompt therapeutic intervention. Although emergent cardiology referral and admission to the CCU are appropriate, this patient does not meet criteria for the administration of fibrinolytics. C) A 15-lead EKG performed in a patient with ST segment depressions in leads V1 and V2. In this case, the finding of ST elevation in leads V8 and V9 confirms that the changes in V1 and V2 actually represent an acute posterior infarction rather than only septal ischemia—this patient now meets criteria for revascularization therapy with fibrinolytics or percutaneous coronary intervention (PCI)

- Emergent echocardiography has an important role in cases where the diagnosis of ACS or MI is still in question after the initial assessment and EKG. Echocardiography gives a great deal of information that may guide management, including an assessment of the global functioning if the heart (ejection fraction), assessment of regional wall motion and the function of the cardiac valves.

- In the setting of ACS with cardiogenic shock, echocardiography is also useful to rule out mechanical complications such as papillary muscle, septal and free-wall rupture that mandate emergent surgical consultation. At present, bedside ultrasound by EPs is limited to the detection of pericardial effusion, which may by itself be helpful in some cases.

ED Management

- As with every patient in the ED, management begins with the primary survey and emergent resuscitation as required.
- All patients being evaluated for ACS must be placed on a cardiac monitor with a pulse oximeter. IV access should be obtained in all patients. Intubation and resuscitation equipment, including an external pacer and defibrillator, must be checked and ready for use at the bedside.
- Treatment of symptomatic CAD is tailored in each case to the severity of the presenting syndrome. All patients with presumed ACS receive the most basic treatments, with the most aggressive therapies reserved for massive STEMIs. In general:
  - The treatment of STEMI or ACS with presumed new LBBB focuses on immediate revascularization therapy with fibrinolytics, percutaneous coronary intervention (PCI) or surgery.
  - The treatment of non-STEMI ACS focuses on reduction in myocardial necrosis and the prevention and treatment of complications.
  - The treatment of patients with stable angina or chest pain of uncertain etiology focuses on the elimination of ACS from the differential diagnosis and risk stratification for future coronary events.

Oxygen

Oxygen should be given to all patients. It may be provided as 2-4 L/min by nasal cannula or by face mask in the presence of hypoxia.

Aspirin (ASA)

Aspirin (ASA) decreases platelet aggregation by inhibition of cyclooxygenase, which results in decreased production of thromboxane A2.

- ASA is the single most effective medication in treatment of CAD and reduces mortality by over 20% in ACS. All patients should receive ASA except those with true allergies or active hemorrhage. Concurrent antacid use can decrease GI upset.
- The dose in ACS is 160-325 mg PO (lesser doses may be acceptable for chronic therapy, but not for ACS)
- Ticlopidine (Ticlid) and clopidogrel (plavix) are alternative agents for patients with severe (anaphylactoid) reactions to ASA. These agents have a slower onset of action, more adverse reactions and are more expensive than ASA. Their role in the ED is very limited at this point, although an additive benefit of clopidogrel to ASA therapy was identified in one recent trial.

Nitrates

Nitrates cause vasodilatation of coronary arteries, relieve vasospasm, and decrease preload and afterload, which in turn decreases myocardial oxygen demand.

- Evidence for their benefit in ACS is indirect at best. Nonetheless, nitrates have an important role in relieving symptoms and improving hemodynamics in ACS.
- Patients with systolic BP>90 Hg mm and ongoing chest pain should be treated with nitroglycerin (NTG).
- Caution should be exercised when using the sublingual dose (0.4 mcg SL Q3-5 min) because it can cause profound hypotension, especially in right ventricular infarction.
- Nitrates are contraindicated when sildenafil (Viagra®) has been used in the preceding 48 h.

- If pain continues after the initial 3 doses of sublingual NTG, it may be administered by IV infusion, starting at 10 mcg/min and quickly titrated upward to resolution of pain while maintaining systolic BP >90.
- If symptoms are mild, NTG may be given by paste applied to the chest wall. Doses range from 1/2-2 inches of topical paste.

2

Morphine

Morphine is a potent analgesic and sedative. Evidence for its effectiveness in AMI is indirect, but it is very effective at decreasing patient anxiety and fear. Similar to nitrates, caution is advised when systolic BP is low. Even small doses (e.g., 2 mg) may be beneficial.

Beta-Blockers

Beta-blockers exert their effects by decreasing afterload, contractility, overall myocardial oxygen demand and myocardial irritability.

- Studies have shown that early use of β-blockers in AMI reduces infarct size and reduce mortality significantly.
- Beta-blockers should be administered to all patients without contraindications, which include cardiogenic shock, hypotension (systolic BP<90), reactive airway disease (asthma and COPD), allergy, advanced (2nd or 3rd degree) AV blocks and bradycardia
- Atenolol and metoprolol are B1 selective IV agents and are preferred. Metoprolol can be given in 5mg increments IV q 5min, to a usual total dose of 15 mg. Esmolol, which is an IV titratable agent with a half-life of under 10 min, can be given in cases where reversibility is desired (e.g., in the presence of relative or uncertain contraindications)

Unfractionated Heparin (UFH) and Low Molecular Weight Heparins (LMWH)

UFH and LMWH are antithrombin agents. Their effectiveness in ACS and AMI is supported by some studies but is controversial. In addition, the benefits of antithrombin agents appear to disappear once they are discontinued. UFH and LMWH are also not without risk, with a very significant incidence of life-threatening bleeding complications. Because of the potential for hemorrhage, risk/benefit ratios for the use of these agents must be considered in each individual case.

- Although some experts recommend their use in all cases of ACS in the absence of a specific contraindication, they may be more appropriately limited to a subset of patients at highest risk for mortality and complications (e.g., those with significant EKG changes) as a "bridge" to more definitive revascularization therapy.
- Unfractionated heparin (UFH) is administered according to ideal body weight with the usual dose of 60 μ/kg bolus followed by an infusion of 12 μ/kg/h (maximum 4,000 u bolus and 1,000 μ/h infusion). PTT must be monitored regularly starting at 4-6 h after the initiating of the infusion) with a goal of maintaining PTT at 2x the upper limit of normal.
- Low molecular weight heparins (LMWH) have been used in many trials. They are as effective as UFH and have several advantages, including ease of administration (subcutaneous injections versus IV infusion), more consistent therapeutic effects, and a lack of need for routine therapeutic monitoring. It is also possible to use LMWH in patients for whom PCI is anticipated, although heparin is still preferred by some in this setting because of its easier titratability. A commonly used LMWH is enoxaparin, with a dose of 1mg/kg SQ q12h.

Glycoprotein IIb/IIIa Receptor Inhibitors

Glycoprotein IIb/IIIa receptor inhibitors (abciximab, eptifibatide, tirofiban) are very potent anti-platelet agents that block the final common pathway of platelet aggregation (the formation of fibrinogen bridges between platelets). They are administered as IV infusions in ACS.

- Currently, there is good evidence that these agents are beneficial in high-risk ACS patients who are undergoing PCI, including those who receive a stent. Their use in patients not receiving interventional therapy is controversial at this time and they are not recommended for general use in patients with unstable angina or nonSTEMI.

Fibrinolytic (Thrombolytic) Agents

Fibrinolytic (thrombolytic) agents promote clot lysis through activation of plasminogen. Available agents include streptokinase (SK), tissue plasminogen activator (t-PA), anistreplase (APSAC), reteplase and tenecteplase (TNK-t-PA)

- Fibrinolytic therapy is indicated in STEMI and LBBB of new onset in the clinical setting of ACS, when the duration of symptoms is <12 h. In some cases, when the onset of symptoms is unclear or symptoms are stuttering in nature, it may be possible to extend this window.
- Specific EKG criteria for ST segment elevation include the presence of at least 1 mm of elevation in two leads of the same vascular territory (2 mm in the precordial leads).
- Benefit is greatest with early administration (<30 min), with progressively less efficacy as each h passes after the onset of symptoms.
- The similarities among these agents are greater than the differences—the efficacies of the various available agents are all comparable. The choice of agent is more likely dependent on the institutional and departmental availability and practice, and the ease and safety of the protocol is an important consideration in this choice.
- SK and APSAC may cause immune reactions and hypotension—they should be avoided in patients with allergy or recent exposure to either agent.
- Absolute contraindications to fibrinolytic therapy include active internal bleeding, known intracranial malignancy, previous hemorrhagic CVA, ischemic CVA in the last 12 mo and suspected or known aortic dissection.
- Relative contraindications include uncontrolled hypertension (>200/110 mm Hg), pregnancy, active peptic ulcer, internal bleeding within the last 4 wk, recent major trauma or surgery, recent puncture of a major noncompressible vessel, current use of anticoagulants, proliferative diabetic retinopathy, history of prior CVA and the presence of occult blood on rectal examination.
- Because of the potent nature and serious consequences of adverse reactions, protocols for fibrinolytic administration should be in place in every ED, including drug preparation instructions, dosing schedules and a checklist to ensure that no contraindications are overlooked prior to the initiation of the protocol. Informed consent should be obtained after a discussion of the risks and potential benefits of therapy with the patient and family members.
- Patients undergoing fibrinolytic therapy require extremely close monitoring for resolution of symptoms and 12-lead EKG changes, dysrhythmias, bleeding complications (especially intracranial hemorrhage) and other adverse reactions.
  - During the reperfusion phase, an accelerated idioventricular rhythm (usual rate 100-120) is often seen which usually does not require therapy. All other arrhythmias should be treated according to ACLS protocols.
  - In the event of life-threatening hemorrhage or any suspicion of intracranial hemorrhage (sudden severe headache, focal neurologic signs or decreasing level of consciousness), fibrinolytic and heparin infusions must be discontinued immediately. Emergent brain CT and neurosurgery consultation should be obtained. Fibrinogen

may be replaced by cryoprecipitate (10-20 μ IV) and coagulopathy may be reversed with fresh frozen plasma (2-4 units FFP IV). Protamine can be used to reverse heparin (dosing varies with amount of heparin infused). Platelets, beginning with 1 single-donor unit can also be infused. In refractory cases, aminocaproic acid, a fibrinolysis inhibitor, may be used.

Primary Percutaneous Coronary Intervention (PCI)

PCI is becoming increasingly popular in U.S. centers and in experienced hands provides outcomes superior to fibrinolytic therapy for STEMI, particularly when coronary stenting is employed.

- Primary PCI has an advantage over fibrinolytic therapy when cardiogenic shock complicates AMI. It may be the only option for revascularization when there are absolute contraindications to fibrinolytic therapy. It also may be preferred in patients >75 yr of age who have higher rates of complications with fibrinolytics.
- The most important factor in outcome remains time to revascularization. Therefore, any benefit of PCI may be lost if there is significant delay in bringing the patient to the catheterization laboratory. A delay of more than 90 min is not likely warranted, and if such a delay is anticipated, administration of fibrinolytics should proceed. A reasonable goal is to perform revascularization by interventional or pharmacological means within 60 min or less from the time that the patient arrives in the ED.
- In some cases, PCI may be considered after fibrinolytics, especially when there is a poor response to therapy.

Complications

The two complications of ACS responsible for the majority of deaths are dysrhythmias and congestive heart failure/cardiogenic shock.

Arrhythmias

- Arrhythmias are very common after an ischemic event. The incidence of premature ventricular contractions (PVCs) is virtually 100% and sinus tachycardia has an incidence of 40-60%. Ominous rhythms such as ventricular tachycardia and/or ventricular fibrillation have an incidence of 5-10%.
- The risk of a life-threatening dysrhythmia complicating ACS generally increases with infarct size and is greatest in infarcts of the left anterior descending artery (anterior) distribution.
- Anterior wall MIs are more likely to be complicated by severe bradydysrhythmias (e.g., Mobitz II and third degree AV block) due to injury of the conducting system. These rhythms may not respond to atropine and preparations for external and/or invasive pacing should be made immediately.
- Inferior wall MIs are often complicated by less severe bradydysrhythmias (e.g., first degree AV block or Wenkebach patterns) due to an increase in vagal tone. These are usually transient and responsive to IV atropine.
- Tachydysrhythmias increase myocardial oxygen demand and should be treated according to ACLS protocols.
- It should be noted that the presence of low-grade ectopy in ACS, such as intermittent premature ventricular contractions (PVCs), is not routinely treated with antidysrhythmic agents as was once common practice.

Left Ventricular Failure/ Cardiogenic Shock

- Left ventricular failure and cardiogenic shock are more common after anterior wall AMI because of the typically larger infarct size.

- Vasoactive catecholamines such as dopamine, dobutamine, epinephrine and norepinephrine all have a role in management, but all increase cardiac irritability and oxygen consumption, making them less than optimal treatment modalities. An arterial line and Swan-Ganz catheter should be used to guide therapy when these agents are used.
- Intra-aortic balloon pump (IABP) is another important tool that may be used in cases of refractory shock, especially as a bridge to revascularization or transplantation surgery.
- Ventricular septal defect and papillary muscle rupture should be sought as causes of shock and sudden decompensation; 50% of all cases occur in the first 5 days of an MI and 90% in the first 14 days. If these are suspected, emergent cardiovascular surgical consultation is indicated.

Disposition

- All patients with suspicion for ACS should be admitted to telemetry or a chest pain unit for observation.
- For those patients in whom AMI is ruled out by serial cardiac markers and EKGs (usually over a time period of 8-12 h), the diagnosis of unstable angina is still a possibility. In the case of a patient with significant risk factors for CAD or typical symptoms, further risk stratification can occur with provocative testing, such as stress echocardiography or nuclear medicine studies, on an inpatient basis. For those patients with atypical presentations of chest pain with few or no risk factors for CAD that are felt to have a very low likelihood of occlusive disease, timely outpatient follow-up (2-3 days) with a primary care provider is appropriate.
- Patients with a confirmed diagnosis of AMI should be admitted to a cardiac or intensive care unit (CCU or ICU).

Suggested Reading

1. Aufderheide TP, Brady WJ, Gibler WB. Acute ischemic coronary syndromes. In: Marx JA, Hockberger RS, Walls RM, eds. Emergency Medicine. Concepts and Clinical Practice. Mosby, 2002:2.
2. Brady WJ. ST segment elevation: Causes and diagnostic accuracy. J Emerg Med 1998; 16:797.
3. Hamm CW et al. Emergency room triage of patients with acute chest pain by means of rapid testing for cardiac troponin T or troponin I. N Engl J Med 1997; 337:1648.
4. Hollander JE. Acute coronary syndromes: Unstable angina, myocardial ischemia, and infarction. In: Tintinalli JE, Kelen GT, Stapczynski JS, eds. Emergency Medicine. A Comprehensive Medicine Study Guide. McGraw-Hill, 2000.
5. Hunink MGM et al. The recent decline in mortality from coronary heart disease, 1980-1990. The effect of secular trends in risk factors and treatment. JAMA 1997; 227:535.
6. Kirk JD et al. Evaluation of chest pain in low-risk patients presenting to the emergency department: The role of immediate exercise testing, Ann Emerg Med 1998; 32:1.
7. Lee T et al. Acute chest pain in the emergency room: Identification and examination of low risk patients. Arch Intern Med 1985; 1456:65.
8. McCarthy BD et al. Missed diagnoses of acute myocardial infarction in the emergency department: Results from a multicenter study. Ann Emerg Med 1993; 22:579.
9. Marin MM, Teichman SL. Use of rapid serial sampling of creatine kinase MB for early detection of myocardial infarction in patients with acute chest pain. Am Heart J 1992; 123:354.
10. The CAPTURE Investigators: Randomized placebo-controlled trial of abciximab before and during coronary intervention in refractory unstable angina: The CAPTURE study. Lancet 1997; 349:1429.
11. The GUSTO Angiographic Investigators: The effects of tissue plasminogen activator, streptokinase, or both on coronary-artery patency, ventricular function, and survival after acute myocardial infarction. N Engl J Med 1993; 329:1615.

Part C: Congestive Heart Failure

Congestive heart failure (CHF) is one of the most commonly encountered entities in the Emergency Department. Because the prevalence of CHF increases with age, the enormous burden of this disease is also on the increase. Currently, CHF costs the health care system about $20 billion per year in the U.S. and it accounts for more hospitalizations than any other disease in patients older than 65.

Definitions

- Congestive heart failure (CHF) exists when the heart is unable to pump sufficient blood to meet the metabolic requirements of the body's tissues. Because of natural compensatory mechanisms in response to heart failure, it most commonly is associated with abnormal retention of fluid.
- CHF is classified in many ways. Although in some respects, these classifications are artificial because they do not exist as independent entities, they are nonetheless very helpful distinctions to make in the evaluation and treatment of CHF patients:
    - High output failure versus low output failure: In low output failure there is an inherent problem with the contractility of the heart. In high output failure, an intact myocardium is unable to keep up with excess functional demands secondary to hypermetabolic states such as thyrotoxicosis, anemia or AV shunts. Low output failure is much more common.
    - Left-sided versus right-sided failure: Left-sided failure is usually due to mechanical overload or ischemia. The most common cause of right-sided failure is pulmonary hypertension secondary to left-sided failure.
    - Systolic versus diastolic failure: Systolic failure is more common and is due to impaired contractility during systole. Diastolic failure occurs when impaired relaxation prevents adequate filling of the ventricles during diastole. Diastolic failure is less well understood and appears to be due to hypertension, as well as other less common causes such as restrictive cardiomyopathy or aortic stenosis.
- Backward versus forward failure: Backward failure refers to the accumulation of fluid behind the ventricles (e.g., edema and hepatic congestion in right-sided failure, pulmonary edema in left-sided failure). Forward failure refers to the failure of the heart to provide adequate perfusion of the tissues, which is usually manifested by some degree of hypotension, whether relative or absolute.
    - Acute and chronic heart failure essentially involve the same process, although when heart failure develops over a short period of time, it tends to involve more forward failure and hypotension and less accumulation of fluid. In chronic failure, there is more time for the evolution of compensatory processes.
- Cardiogenic pulmonary edema refers to the accumulation of fluid in the interstitial and alveolar spaces as a result of CHF. It is a severe form of left-sided CHF.

Epidemiology

At the present time, nearly 5 million patients are diagnosed with CHF in the U.S. with 500,000 new cases identified each year. Almost 300,000 patients die from CHF or its complications every year.

Pathophysiology

- The most common underlying pathology in CHF is ischemic heart disease. Other causes of low output failure include valvular disease, myocarditis, chronic hypertension and cardiomyopathies (such as those caused by ethanol and cocaine abuse). Anemia and thyrotoxicosis are causes of high output failure. Table 2C.1 lists common processes that

Table 2C.1. Common precipitants of congestive heart failure

Myocardial infarction/ischemia	Dietary and fluid excesses
Dysrhythmia	Valvular heart disease
Infection	Non-compliance with medications
Thyrotoxicosis	Uncontrolled hypertension
Pulmonary embolism	

2

can precipitate CHF. The causes in the first column are especially important to identify because they have specific therapy that can be initiated in the ED.

- In CHF, elevated left ventricular filling volumes exceeds the threshold over which increased preload increases efficiency and instead, the heart begins to pump more inefficiently, with decreasing stroke volume. This relationship is described by the Frank-Starling curve. Decreasing stroke volume leads to the clinical effects of both backward and forward failure.
- Neurohormonal compensatory responses play an important role in the pathophysiology of CHF. Alterations in adrenergic tone redistribute blood flow to the brain and heart and reduce flow to other organs. The reduction in flow to the kidneys increases stimulation of renin-angiotensin-aldosterone axis which leads to an increased secretion of ADH, renin and angiotensin II. This leads to increased sodium and water retention and a rise in both preload and afterload. Increased adrenergic tone leads to arteriolar vasoconstriction, which also contributes to increased afterload.

Diagnosis and Evaluation

Clinical Features

- Left-sided failure symptoms include dyspnea (particularly with exertion), orthopnea, paroxysmal nocturnal dyspnea, fatigue and nocturia.
- Physical examination features in left-sided failure include tachypnea, tachycardia, pulmonary rales and/or wheezing ("cardiac asthma"), dullness to percussion, diaphoresis, poor peripheral perfusion causing pale and cool extremities, S3 and S4 gallop rhythms and in severe pulmonary edema secondary to CHF, abnormal breathing patterns, such as Cheyne-Stokes.
- Right-sided failure symptoms include lower extremity edema and right upper quadrant pain and anorexia due to liver congestion.
- Examination features in right-sided failure include jugular venous distention, hepatojugular reflux, hepatomegaly, RUQ abdominal tenderness, and peripheral edema.

Diagnostic Studies

- In addition to establishing the diagnosis of CHF, the EP must focus on the identifying and correction of its underlying causes. Important precipitating causes are listed in Table 2C.1.
- Chest X-ray (CXR) findings are mostly a function of left-sided failure, which results in increased pulmonary venous pressures.
- Because of the higher resistance to flow in the lower regions, blood flow is initially shunted to the upper pulmonary vasculature. This appearance of engorged vessels in the upper lung fields is known as "cephalization" and is the earliest finding of CHF on the chest X-ray.
- As the vascular pressure increases, fluid starts to move to the gravity-dependent lower zones of the lungs, which is known as "caudalization"
  - Interlobular edema creates short horizontal linear markings in the lung periphery at the bases which are known as Kerley B lines.

- With increasing severity, a bat-wing pattern may be seen, representing interstitial edema in both upper and lower regions of the lungs
- In its most severe form, edema fluid spills from the interstitial spaces into the air-spaces, causing opacification of the lung fields on CXR
- Cardiomegaly is usually evident on the CXR, especially in patients with chronic CHF. In those cases where CHF occurs acutely in a previously healthy individuals, the EP must distinguish cardiogenic from noncardiogenic causes of pulmonary edema, such as the adult respiratory distress syndrome (ARDS)
- Pleural effusion may also be seen in CHF. It is often bilateral.
- A 12-lead EKG should be obtained immediately in patients presenting with signs and symptoms suggestive of CHF because an important precipitant of CHF is ischemia and infarction.
- In chronic CHF, changes such as ventricular hypertrophy and conduction abnormalities can be seen. These changes invariably make the identification of acute coronary syndromes (ACS) more difficult, underscoring the importance of performing serial EKGs in the ED and obtaining previous EKGs for comparison. The presence of dynamic or new changes in the 12-lead EKG should raise the suspicion for ACS.
- Bedside echocardiography performed by EPs may be very valuable in ruling out the presence of a pericardial effusion in the setting of a patient with signs of failure and an enlarged heart on CXR. Formal echocardiography yields a great deal of valuable information to help guide the management of patients with CHF but is not necessarily required during the course of ED management.
- Basic laboratory investigations such as CBC, electrolytes, urea and creatinine are indicated in CHF. Anemia is an important contributing factor in CHF. Electrolyte disturbances are common and potentially life-threatening. These are often the result of chronic diuretic therapy and renal insufficiency associated with CHF.
- In most cases, serum troponin or other cardiac markers should be measured during the ED evaluation because of the strong association between CHF and ischemic heart disease.
- A digitalis level should be obtained in those patients receiving digoxin therapy.
- B-type natriuretic peptide (BNP) is an endogenous cardiac peptide produced in the ventricles that is released from the heart in response to fluid overload. Recently, an assay of BNP has been introduced to assist in the diagnosis of CHF. It is both sensitive and specific for the presence of CHF; however, its clinical role has yet to be determined. At present, it appears that this test may have a role in a limited subset of patients in whom the diagnosis of CHF may be clouded by coexisting pulmonary disease such as chronic obstructive pulmonary disease (COPD) or sleep apnea.

ED Management

Emergency Resuscitation

- In severe cases, emergent management may involve aggressive intervention to maintain airway, breathing and circulation.
- All patients require vascular access, cardiac monitoring and pulse oximetry.
- All patients should be placed on oxygen therapy. The specific technique of delivery (nasal cannula, face mask, non-rebreather mask) should be guided by the severity of presentation and pulse oximetry values.
- Aggressive airway management should be initiated in the setting of severe respiratory distress, with or without hypoxia. Endotracheal intubation will be necessary in patients who become so somnolent that they are unable to protect their airway or in those who have tired of breathing and may have begun to have periods of apnea.
- Noninvasive ventilation techniques such as continuous positive airway pressure (CPAP) or biphasic positive airway pressure (BiPAP) have been used in many instances where

intubation and mechanical ventilation were once considered inevitable. They are used as a bridging therapy in the patient with severe cardiogenic pulmonary edema until pharmacologic therapy takes effect. These devices provide continuous or biphasic positive airway pressure through a tight-fitting face mask. This positive pressure improves cardiac output by decreasing venous return and excessive preload. It also improves oxygenation by increasing alveolar recruitment and decreases the work and metabolic requirement of breathing.

- Heart rate is extremely important in CHF and both inappropriately slow and fast rates will decrease the ability of the heart to work efficiency. Brady- and tachydysrhythmias should be treated according to ACLS guidelines in the setting of CHF.
- When frank shock is present, vasopressor therapy may need to begin immediately. An inotropic agent such as dopamine is preferred. These agents all increase myocardial oxygen consumption and ischemia, creating an additional risk for dysrhythmias.
- Although pressors may raise arterial blood pressure, this alone does not guarantee sufficient perfusion, if cardiac output is too low. Adequate perfusion must always be assessed by using clinical indicators such as urine output, skin appearance and mental status.
- Interventional therapies may be possible in the event that acute exacerbation is due to ACS (e.g., percutaneous coronary intervention) or a related structural problem (e.g., valvular or myocardial rupture repair).
- Intra-aortic balloon counterpulsation (IABP) can be used as a bridge to other interventional therapies such as cardiac transplantation.

Specific Pharmacotherapy

- The use of pharmacotherapeutic agents is aimed at improving the unfavorable hemodynamics of acute CHF. In general, these agents share the ability to precipitate or worsen hypotension. As each successive agent is introduced (at a pace commensurate with the severity of symptoms), close attention must be paid to vital signs, fluid balance and symptomatic response to therapy.
- Nitrates cause both arterial and venous vasodilatation, reducing both preload and afterload. Nitroglycerin (NTG) is the initial agent of choice in acute CHF.
  - Nitrates can be administered through many routes including sublingual, transdermal, and intravenous.
  - Caution should be exercised with sublingual NTG because of its ability to cause precipitous drops in blood pressure.
  - Transdermal paste (1/2-2 inches applied to chest wall) gives a slower, lower dose release of NTG but may not be effective in severe CHF when cutaneous perfusion is compromised.
  - IV dosing is required in severe CHF, and IV infusions may be started at 10-20 mcg/min and titrated to relief of symptoms while maintaining systolic BP <90 mm.
  - Nitrates predictably cause headache, which will often require treatment with analgesics.
- Diuretics reduce volume overload. Loop diuretics are the preferred class, and initial treatment should begin with furosemide (Lasix) 20-80 mg IVP, depending on the patient's previous exposure and severity of symptoms. Even with IV administration, full effect may take up to 30 min, underscoring the importance of using diuretics together with nitrates. Adverse effects of diuretics include electrolyte imbalances, prerenal azotemia, contraction alkalosis and hypotension.
- ACE inhibitors decrease afterload. Because of several trials that demonstrate reduction in morbidity and mortality in CHF, ACE inhibitors should be used in the treatment of chronic CHF unless a contraindication exists. Several agents can be given in the ED, however, captopril (Capoten) 6.25-12.5 mg PO q8h is preferred because it is the

2

fastest acting of the various agents available. Angiotensin II receptor blockers (ARBs) inhibit the effects of angiotensin II. They decrease afterload and improve diuresis. Currently, they should only be used in place of ACE inhibitors in patients who are intolerant or allergic to ACE inhibitors.

- Morphine decreases catecholamine levels by decreasing anxiety, theoretically causing a decrease in preload and afterload. However, little evidence for the effectiveness of morphine exists and its use is controversial. Respiratory depression with morphine may result in unnecessary intubation—a small dose (e.g., 2 mg IV) may be beneficial without impairing respiratory effort.
- Beta-blockers have been shown in multiple well-designed, randomized studies to improve the symptoms and altered hemodynamics in chronic CHF. They inhibit the neurohormonal cascade and improve symptoms, especially in the setting of cardiac ischemia. They have also been shown to decrease mortality. However, they are not generally indicated in the treatment of acute heart failure and may result in acute decompensation. When CHF occurs within the setting of AMI, the decision to add β-blockers to therapy should be made with the guidance of a cardiologist.
- Digoxin enhances contractility and reduces afterload by blunting the cardiac sympathetic response. It also causes AV nodal blockade which may be beneficial in patients with atrial dysrhythmias. Currently, there are no conclusive data to show that digoxin reduces mortality in CHF although it may reduce hospitalizations and improve quality of life. Because it is slow to act, its use in the ED for CHF is on the decline. IV loading may be initiated in the ED with 0.25 mg followed by further IV or PO doses over the next several hours.
- Calcium channel blockers are not currently recommended in the ED treatment of CHF, unless they are required for rate control in accordance with ACLS guidelines.
- Dobutamine is a synthetic catecholamine that unlike other pressor agents causes vasodilatation in addition to inotropy. For this reason, it is the one vasoactive catecholamine sometimes used in treatment of CHF in the absence of cardiogenic shock. Dobutamine must be given with close hemodynamic monitoring and may initially result in hypotension.
- Nesiritide (Beta natriuretic peptide) is an endogenous substance that acts both a vasodilator and a diuretic. It has recently been demonstrated to be effective in the treatment of acute CHF. However, its effects are more prolonged and it is less titratable than NTG. Its role as a first-line agent in the treatment of CHF and superiority over NTG has yet to be demonstrated.

Prognosis

Prognosis in CHF decreases proportionately with severity. Two commonly used classification systems for severity of heart failure are given in (Table 2C.2). Both classifications demonstrate the typical progression of chronic CHF from backward to forward failure.

Disposition

Patients diagnosed with acute heart failure or acute exacerbation of chronic heart failure most commonly require admission to hospital. Moderate or severe presentations require admission to the cardiac care unit (CCU) or monitored unit. Mild exacerbations may be admitted to unmonitored settings in the absence of suspected ACS. It may also be appropriate to admit patients with mild exacerbations to short-stay or observational units or even to discharge them home with close follow-up when no acute serious underlying pathology is suspected and symptoms have resolved.

2

Table 2C.2. Classification of heart failure severity and prognosis

Killip Classification (Clinical)	Definition	Incidence (%)	1 yr Mortality (%)
Class I	No signs of pulmonary edema	30	5
Class II	Mild failure (rales, S3)	40	15-20
Class III	Frank pulmonary edema	10	40
Class IV	Cardiogenic shock	20	80

Diamond-Forrester Classification (Based on invasive monitoring)	Cardiac Index (L/min/m^2)	Pulm. Artery Wedge Pressure (mm Hg)	1 yr Mortality (%)
Class I	>2	<18	3
Class II	>2	>18	9
Class III	<2	<18	23
Class IV	<2	>18	51

Suggested Reading

1. Cairns CB. Heart failure and pulmonary edema. In: Tintinalli JE, Kelen GT, Stapczynski SJ, eds. Emergency Medicine. A Comprehensive Medicine Study Guide. McGraw-Hill, 2000.
2. Falk JL, O'Brien JF, Shesser R. Heart Failure. In: Marx JA, Hockberger RS, Walls RM, eds. Emergency Medicine. Concepts and Clinical Practice. Mosby, 2002:2.
3. Hunt SA, Baker DW, Chin MH et al. ACC/AHA guidelines for the evaluation and management of chronic heart failure in the adult: A report of the American College of Cardiology/ American Heart Association Task Force on Practice Guidelines. 2001 American College of Cardiology Web site: Available at http://www.acc.org/clinical/guidelines/failure/hf_index.htm. (Accessed 3/28/2002.).
4. Packer M, Cohn JN. Consensus recommendations for the management of chronic heart failure. AM J Cardiol 1999; 83:1A-38A.
5. The ELITE II Investigators. Effect of losartan compared with captopril on mortality in patients with symptomatic heart failure. Randomized trial—The Losartan Heart Failure Study ELITE II. Lancet 2000; 355:1582-1587.
6. The SOLVD Investigators. Effect of enalapril on mortality and the development of heart failure in asymptomatic patients with reduced left ventricular ejection fractions. N Engl J Med 1992; 327:685-691.
7. Tsutamoto T, Wada A, Maeda K et al. Attenuation of compensation of endogenous cardiac natriuretic peptide system in chronic heart failure: Prognostic role of plasma brain natriuretic peptide concentration in patients with chronic symptomatic left ventricular dysfunction. Circulation 1997; 96:509-516.

Part D: Endocarditis

Epidemiology/Pathophysiology

- *Endocarditis* is an infection of the heart valves which can present either acutely or as a chronic disease. It is a life threatening infectious disease that is difficult to diagnose with certainty in the Emergency Department.
- Intravenous drug abusers, immunocompromised patients, patients with a history of rheumatic heart disease, and patients who have undergone valve replacement are at heightened risk for developing endocarditis. Other patients at risk include those with intracardiac devices (pacemakers, defibrillators), those with a history of endocarditis,

2

those with mitral valve prolapse and regurgitation, and patients with certain congenital heart defects.
- *Streptococci* are the most common cause of native valve endocarditis.
- *Staph* species are responsible for the majority of IVDU-related endocarditis and coagulase negative *Staph* is responsible for the majority of prosthetic valve endocarditis.
- Other etiologic agents include *Enterococcus*, Gram-negative bacteria, the *HACEK* organisms (*Haemophilus, Acetinobacillus, Cardiobacterium, Eikenella,* and *Kingella species*), *Candida* and *Aspergillus*.

Diagnosis and Evaluation

- Diagnosis of endocarditis has traditionally been based on clinical findings and bacteriologic criteria from blood cultures. The development and increased utilization of echocardiography has provided increased ability to diagnose endocarditis. The Duke criteria describes clinical, bacteriological and echocardiographic diagnostic criteria for endocarditis.

History and Physical Exam

- Endocarditis presents with a variety of clinical complaints.
- The triad of fever, anemia, and a heart murmur is highly suggestive of endocarditis but is not routinely present.
- Fever, malaise and altered mental status are common historical features.
- Patients may also present with embolic sequelae, which occur in approximately 25-50% of patients with endocarditis. 65% of emboli involve the central nervous system, usually in a MCA distribution.
- Physical findings include heart murmurs, most commonly of the mitral and aortic valves. However, less than one-third of IVDU patients with endocarditis will have murmurs.
- Osler's nodes, painful erythematous nodules found on the palmar fingertips, are rarely found. Janeway lesions, flat nontender macular areas on the palms are also rarely seen.

Laboratory and Studies

- Laboratory findings include positive blood cultures as discussed above and a mild anemia. Most patients have microscopic hematuria as well.
- Echocardiography is a powerful tool to aid in the diagnosis of endocarditis.
  - Transthoracic echo has a poor sensitivity (approximately 60%) but excellent specificity if vegetations are seen.
  - Transesophageal echocardiography is more sensitive (>85% sensitivity) for visualization of vegetations.

ED Management

- Blood culture results are often unavailable to the Emergency Physician and thus empiric antibiotic coverage is required. Blood cultures should be drawn before beginning antibiotics as antibiotics reduce the bacterial recovery rate of cultures by approximately one-third.
- Patients with an acute presentation of endocarditis, a history of IVDU or a prosthetic heart valve should receive vancomycin, an aminoglycoside (gentamycin), and rifampin. Patients with a subacute course and native heart valves should receive either penicillin and an aminoglycoside or a penicillinase-resistant penicillin (nafcillin) and an aminoglycoside.
- Occasionally, patients may require surgical therapy. Indications for surgery include severe congestive heart failure, recurrent emboli, fungal endocarditis, and failure of IV antibiotic therapy.

Table 2D.1. The Duke criteria for the diagnosis of endocarditis

Major Criteria	Positive blood cultures with typical microorganisms consistent with endocarditis
	Evidence of endocardial involvement with a positive echocardiogram or new valvular regurgitant murmur
Minor Criteria	Predisposing heart condition or IVDU
	Fever
	Vascular phenomena including emboli, Janeway lesions, pulmonary infarctions, mycotic aneurysms, or, splinter hemorrhages
	Immunologic phenomena including Osler's nodes, Roth spots, glomerulonephritis
	Blood cultures suggestive of endocarditis but not meeting the major criteria
	Echocardiographic evidence suggestive of endocarditis but not meeting major criteria

The presence of 2 major criteria, 1 major criterion and 3 minor criteria, or 5 minor criteria establishes a diagnosis of endocarditis.

- Patients with conditions predisposing them to endocarditis should receive antibiotic prophylaxis prior to dental, GI, or GU procedures. Amoxicillin or erythromycin should be given 1 h before the procedure for proper prophylaxis.

Suggested Reading

1. Bayer A et al. Diagnosis and management of infective endocarditis and its complications. Circulation Dec 1998; 98(25).
2. ACC/AHA guidelines for the management of patients with valvular heart disease. A Report of the American College of Cardiology/American Heart Association Task Force on practice guidelines. J Am Coll Cardiol 1998; 32(5):1486-588.
3. Cline D. Valvular Emergencies and Endocarditis. In: Tintinalli JE, Kelen GT, Stapczynski JS, eds. Emergency Medicine. A Comprehensive Medicine Study Guide. McGraw-Hill, 2000.
4. Dunmire S. Infective Endocarditis and Acquired Valvular Heart Disease. In: Marx JA, Hockberger RS, Walls RM, eds. Emergency Medicine. Concepts and Clinical Practice. Mosby, 2002:2.

Part E: Pericardial Diseases

Pericardial disease is an important consideration in patients presenting with cardiopulmonary symptoms. The understanding of pericardial disease is important as these diseases not only cause significant morbidity, but they may also mimic other diseases which may require alternative treatment.

Definitions

- Pericarditis is an inflammation of the pericardial layer surrounding the heart.
- A pericardial effusion is an abnormal accumulation of fluid in the pericardial space.
- Cardiac tamponade refers to an impairment of cardiac output caused by a pericardial effusion.
- Scarring and thickening of the pericardium may cause constrictive pericarditis, a rare complication of pericarditis.

Epidemiology/Pathophysiology

- Pericarditis has multiple etiologies including infection, malignancy, uremia, connective tissue disorders, trauma, myocardial infarction, or medications. Pericarditis may also be idiopathic, but it is unclear if these represent undiagnosed viral pericarditis.
  - Pericarditis that occurs post myocardial infarction is called Dressler's syndrome.
  - Dressler's syndrome is associated with large, anterior infarctions and may indicate a poor long-term prognosis.
- Pericardial inflammation causes an increase in pericardial fluid beyond the 15-50 ml normally present. As pericardial fluid accumulates, intrapericardial pressure increases exponentially.
- Cardiac tamponade occurs when intrapericardial pressures rise to a level such that diastolic filling of the heart is impaired. This leads to an increase in central venous pressure and a decreased cardiac output.
- Constrictive pericarditis may lead to similar hemodynamic abnormalities as cardiac tamponade.

Diagnosis and Evaluation

History and Physical Examination

- Pericarditis may be asymptomatic. Presenting symptoms correlate with the etiology of pericarditis.
  - Acute viral pericarditis typically presents with fevers, myalgias, and fatigue.
- Chest pain is common in acute pericarditis. The pain is classically sharp and pleuritic. It is often exacerbated by leaning forward and relieved by supine positioning. Radiation of the pain to the trapezius ridge is a specific finding.
- As a pericardial effusion enlarges and tamponade begins, symptoms of increased venous pressure and decreased cardiac output (dyspnea, orthopnea, syncope) present.
- A pericardial friction rub is the classic physical finding of pericarditis. Friction rubs have been described in many ways such as "creaky" or "velcro-like." They are best heard with the bell of the stethoscope and can be heard anywhere over the pericardium. Friction rubs are transient and change in quality over time. Thus, the absence of a friction rub does not rule out pericarditis.
- As cardiac tamponade develops, signs of increased venous pressure (JVD, hepatomegaly) and decreased cardiac output (hypotension) develop.
  - Beck's triad of hypotension, JVD, and muffled heart sounds is specific for cardiac tamponade, but rarely present.
  - Pulsus paradoxus, a decrease of systolic blood pressure of at least 10 mm Hg during inspiration, is another sign of cardiac tamponade, but its presence is not sensitive enough to rule out the diagnosis.

Laboratory and Studies

- Laboratory tests do not play a major role in the evaluation of pericardial disease.
  - An increased ESR is a nonspecific test that is usually elevated in pericarditis.
  - Cardiac enzymes may be elevated in pericarditis indicating a concurrent inflammation of the underlying myocardium.
- The electrocardiogram is an important tool in the evaluation of patients with pericardial disease. As these patients often present with chest pain that may not be indistinguishable from ischemic chest pain, subtle differences in the EKG may dramatically change treatment.
- The EKG typically evolves through four stages in acute pericarditis:
  - Stage 1, hours to days after symptom onset, demonstrates a diffuse concave upward elevation of the ST segment in all leads but AVR and V1. The PR segment is depressed in 80% of patients. There are generally no T-wave abnormalities.

    - In stage 2 there is a normalization of the above ST and PR segments
    - Stage 3 demonstrates diffuse deep, symmetric T-wave depressions.
    - Stage 4 is a normalization of the T-wave depressions.
- Electrical alternans, the finding of changing QRS polarity with every other beat, is a specific finding for large, chronic pericardial effusions most commonly caused by malignancy.
- Chest X-ray may demonstrate cardiomegaly or a "water bottle" heart with large pericardial effusions.
- Echocardiography is a fast, reliable method to diagnose pericardial effusions. Echocardiography can detect as little as 15 ml of fluid and can be done at bedside in unstable patients. Echocardiography is a sensitive test for signs of cardiac tamponade including right ventricular diastolic collapse and IVC dilation with lack of inspiratory collapse.

ED Management

- Treatment of acute pericarditis includes pain control and control of inflammation with NSAIDS. Treatment may continue on an outpatient basis in stable patients, but may require inpatient management in patients with severe pain, significant pericardial effusions, or any signs of hemodynamic instability.
- Treatment of pericardial effusions is dependent on the etiology. Uremic pericarditis with effusion, for example, is an indication for dialysis.
- Cardiac tamponade is treated with drainage of pericardial fluid via pericardiocentesis or operative drainage.
- Constrictive pericarditis is treated with operative removal of the pericardium.

Suggested Reading

1. Fowler N. Cardiac Tamponade, a clinical or an echocardiographic diagnosis? Circulation 1993; 87:5.
2. Pawsat D, Lee J. Inflammatory disorders of the heart. Pericarditis, Myocarditis, and Endocarditis. Emerg Med Clin North Am 1998; 16:3.
3. Reddy P, Curtiss E. Cardiac tamponade. Cardiology Clinics November 1990; 8:4.
4. Shabeti R. Acute pericarditis. Cardiol Clin 1990; 8:4.
5. Spodick D. Pathophysiology of cardiac tamponade. Chest 1998; 113:5.

Part F: Structural Heart Disease

The Emergency Physician must be comfortable managing patients with structural heart disease. Although the incidence of rheumatic fever has decreased, thus decreasing the incidence of rheumatic heart disease, many patients with structural heart disease of other etiologies are being increasingly recognized and living longer. These patients have special needs that the Emergency Physician must include in the criteria for urgent or emergent operative repair and the need for prophylactic antibiotics for procedures. This section discusses the presentation and treatment of structural heart disease as well as the potential complications of prosthetic valves.

Aortic Stenosis

Epidemiology/Pathophysiology

- Aortic stenosis is caused by rheumatic heart disease or valvular calcification. Calcific aortic stenosis presents in the fifth to sixth decade of life in patients with a congenital bicuspid valve and in the seventh to eighth decade of life in patients with normal tricuspid valves.
- The pathophysiology of aortic stenosis is due to a narrowing of the cardiac outflow with compensatory increase in left ventricular size and subsequent diastolic dys-

2

function. The progressive narrowing of cardiac outflow leads to a decrease in cardiac output with progressive systolic dysfunction causing the classic symptom triad of angina, syncope, and heart failure.

- Angina—Left ventricular hypertrophy leads to increased myocardial oxygen demand, causing angina despite normal coronary arteries; 50% of patients with aortic stenosis and angina have coexisting significant coronary artery disease.
- Syncope—The inability to increase cardiac output past a stenotic valve leads to exertional dizziness and ultimately syncope. Other explanations include a dysfunction of ventricular baroreceptors leading to inappropriate peripheral dilatation despite decreased cardiac output.
- Heart failure—Symptoms of heart failure are secondary to diastolic dysfunction from a hypertrophied left ventricule. Only late in the disease does systolic function decompensate to a pathologic level. The inability to increase cardiac output during times of exertion will also cause dyspnea on exertion.

Diagnosis and Evaluation

- Classic presentation of critical aortic stenosis includes the above trilogy: angina, syncope, and heart failure.
- Other symptoms of aortic stenosis include symptoms of atrial fibrillation or sudden cardiac dysrhythmias.
- The murmur of aortic stenosis is classically described as a crescendo-decrescendo systolic best heard at the right upper sternal border and radiating to the carotids. A single second heart sound is present and there is a delayed carotid upstroke.
- Echocardiography is the test of choice for patients with symptoms or signs suggestive of aortic stenosis. It can delineate the severity of outflow obstruction, discover concurrent valvular abnormalities (80% of patients with aortic stenosis have concurrent aortic regurgitation), and evaluate left ventricular response to the stenotic valve.

ED Management

- Treatment of aortic stenosis is surgical.
- Medical treatment is reserved for treating concurrent diseases (i.e., coronary artery disease) and the routine use of antibiotics for endocarditis prophylaxis.
- Valve replacement is indicated for patients with symptomatic aortic stenosis. Without surgical correction, the 2-yr survival rate is <50%.

Aortic Insufficiency

Epidemiology/Pathophysiology

- Aortic insufficiency may be acute or chronic.
- Acute aortic insufficiency is due to aortic root disease with dissection or valvular destruction secondary to endocarditis. Acute aortic insufficiency causes a rapid increase in cardiac afterload secondary to the regurgitant blood volume as well as an acute increase in pulmonary vascular pressure. Symptoms include severe dyspnea with signs of pulmonary edema. Other symptoms are those of the underlying disease such as tearing chest pain in patients with aortic dissection.
- Chronic aortic insufficiency is most likely secondary to rheumatic heart disease. The damaged valve leads to a backflow of blood during diastole, thus increasing the stroke volume. This increases afterload as the ventricle attempts to push the increased blood volume against the regurgitant flow. Preload is also increased as the regurgitant flow significantly increases the volume load of the left ventricle. Early in the disease, the left ventricle is able to compensate with hypertrophy and dilatation, with minimal symptoms. However, as the disease progresses, the ventricle is unable to compensate, ejection fraction decreases, and symptoms of heart failure develop. Other symptoms include

anginal chest pain as the hypertrophied heart outgrows its blood supply or palpitations from the increased systolic outflow.

Diagnosis and Evaluation

- The murmur of aortic insufficiency is a "blowing" diastolic murmur heard best at the left sternal border of the heart. There may be an accompanying systolic ejection murmur heard as the increased volume is expelled during systole. Other findings include a "water-hammer" pulse with a fast upstroke and an abrupt collapse.
- The electrocardiogram is variable in acute disease, often reflecting the underlying cause. Chronic aortic insufficiency will cause signs of LVH to be present on the EKG.
- The diagnostic test of choice is the echocardiogram, which can assess the severity of the valvular dysfunction as well the left ventricular function. In acute disease, the echocardiogram is also important for evaluation of concurrent life threatening complications of the underlying disease, especially pericardial tamponade associated with aortic dissections.

ED Management

- Treatment of aortic insufficiency is surgical.
- Acute disease is a cardiothoracic surgical emergency requiring immediate operative repair of both the valve and often the aortic root.
- Chronic disease develops over years with surgical correction suggested for most symptomatic patients and most patients with severe disease and/or concurrent left ventricular dysfunction despite clinical symptoms.
- Medical management includes the use of afterload reducing agents in order to decrease the regurgitant volume, hopefully retarding the onset of left ventricular dysfunction. However, medical management is not a replacement for surgical valve replacement or repair. Other medical management issues include the treatment of concurrent diseases including coronary artery disease, atrial fibrillation, and the prevention of endocarditis.

Mitral Stenosis

Epidemiology/Pathophysiology

- The most common cause of mitral stenosis is rheumatic heart disease.
- The mitral valve becomes thickened, calcified, and fused. The increased pressure required to force blood across the stenotic valve leads to an elevation of left atrial pressures and subsequent left atrial dilatation. With progression of disease, the pressure column backs into the pulmonary circulation, leading to pulmonary hypertension, tricuspid valve dysfunction, and right heart failure. The disease progresses slowly over years but may be accelerated by conditions increasing the demand for flow across the damaged valve, such as atrial fibrillation, pregnancy, infection, or other stressors.

Diagnosis and Evaluation

- Mitral stenosis presents with symptoms of congestive heart failure and pulmonary hypertension usually in the fifth to sixth decade of life.
- Patients may also present with complications of previously unrecognized disease. This includes atrial fibrillation in 30-40% of patients, hemoptysis secondary to pulmonary hypertension and erosion of bronchial veins, chest pain (despite the lack of concurrent coronary artery disease), or embolic disease in 10-20% of patients.
- The murmur of mitral stenosis is a low-pitched diastolic rumble best heard at the apex of the heart. Other findings include an opening snap. Auscultatory findings are best heard with the patient in the left lateral position with the bell of the stethoscope.
- Electrocardiographic findings include "P-mitrale," a widened p-wave in the limb leads.

- Echocardiography is the test of choice to visualize the abnormal mitral valve, assess the severity of valvular obstruction and evaluate its effect on the pulmonary circulation.

ED Management

- Treatment for mitral stenosis is surgical via mitral valve replacement or balloon valvotomy.
- Medical treatment is reserved for the treatment of concurrent diseases including atrial fibrillation with rate control medications and anticoagulation, as well as antibiotic prophylaxis against endocarditis.

Mitral Regurgitation

Epidemiology/Pathophysiology

- Mitral regurgitation may be acute or chronic.
- Acute mitral regurgitation is secondary to infective endocarditis with erosion of the valve or acute ischemia of the chordae tendinae/papillary muscle support system. Acute mitral regurgitation causes an abrupt increase in the pulmonary vascular pressure leading to acute pulmonary edema. This often progresses to cardiogenic shock and cardiac arrest.
- Chronic mitral regurgitation has multiple etiologies, most commonly rheumatic heart disease. Chronic mitral regurgitation leads to a compensatory enlargement of both the left atrium and ventricle in order to handle the regurgitant blood volume. Early in the course of the disease, the contractile force of the left ventricle is preserved and stroke volumes are supranormal with both the normal stroke volume and the regurgitant volume expelled during systole. With disease progression, the left ventricle enlarges to a point that compromises the contractile function, lowering the ejection fraction. The decrease in forward flow leads to an increase of pulmonary pressures and symptoms of heart failure.

Diagnosis and Evaluation

- The murmur of mitral regurgitation is holo-systolic at the apex of the heart that radiates to the axillae. In acute mitral regurgitation, the murmur is typically harsh with signs of pulmonary edema present. In chronic mitral regurgitation, there may also be a diastolic murmur heard at the apex indicative of the increased regurgitant flow across the valve.
- Physical signs of cardiac enlargement are usually present.
- The electrocardiogram typically demonstrates LVH and LAE in mitral regurgitation. However, these may be absent in cases of acute valvular dysfunction.
- The diagnostic test of choice is the echocardiogram, which can assess the severity of the valvular dysfunction as well as the left ventricular function.

ED Management

- Treatment of acute mitral regurgitation is surgical.
- Medical support is geared toward treatment of acute pulmonary edema and the underlying etiology of the dysfunction (i.e., ischemia).
- Surgical repair is indicated in most patients with decompensated chronic mitral regurgitation.
- Medical treatment options for patients with chronic mitral regurgitation are complex. Afterload reduction with ACE inhibitors and other vasodilators may reduce the regurgitant flow and thus decrease symptoms and prolong the time between diagnosis and the need for operative repair. Medical treatment is also geared toward the prophylaxis and treatment of potential complications including atrial fibrillation and endocarditis.

Mitral Valve Prolapse

Epidemiology/Pathophysiology

- Mitral valve prolapse is the most common form of valvular disease.
- Etiologies include congenital valve degeneration and connective tissue disorders.
- The overall prognosis for most patients with mitral valve prolapse is excellent, with most symptoms being benign.
- Sudden death is a rare complication, occurring in 1-2% of patients.

Diagnosis and Evaluation

- The presentation of mitral valve prolapse is diverse. Patients are usually asymptomatic, with the diagnosis made on auscultatory findings alone.
- Common symptoms include chest pain and palpitations. The chest pain is often times atypical, nonanginal type chest pain but can occasionally have angina-like characteristics. The palpitations of mitral valve prolapse are usually due to occasional premature ventricular complexes but can occasionally be due to more troubling dysrhythmias such as SVT or rarely ventricular tachycardias.
- Mitral valve prolapse is also associated with autonomic hyperactivity symptoms. Patients may present with anxiety, panic attacks, or other symptoms of concurrent psychiatric disease.
- Neurologic complications such as TIAs and migraine headaches are also associated with mitral valve prolapse.
- The auscultatory finding of mitral valve prolapse is a midsystolic click. Mitral valve prolapse can progress to significant mitral regurgitation at which time the systolic murmur of MR is heard.
- Diagnosis is confirmed by echocardiography.

ED Management

- Treatment is focused on the relief of symptoms and the prevention of complications. Beta-blockers are used to control palpitations and anxiety symptoms.
- Antibiotics are used for the prevention of endocarditis.
- Aggressive rhythm control is required for patients with more severe dysrhythmias.

Tricuspid Valve Disease

Epidemiology/Pathophysiology

- Tricuspid valve disease is usually found with concurrent left-sided valvular diseases and pulmonary hypertension.
- Congenital abnormalities and endocarditis, seen most commonly in abusers of intravenous drugs, cause isolated right-sided valvular disease.
- Weight loss medications may also result in tricuspid valve disease.

Diagnosis and Evaluation

- In patients with tricuspid disease and associated left-sided valve dysfunction or pulmonary hypertension, symptomatology is dominated by the concurrent diseases.
- Isolated tricuspid disease presents with symptoms of right-sided heart failure including hepatomegaly, ascites, and peripheral edema.
- The murmur of tricuspid regurgitation is pansystolic at the left lower sternal border. Tricuspid stenosis inconsistently produces a high-pitched diastolic murmur. Prominent jugular venous pulsations are present in both abnormalities.
- Diagnosis is confirmed by echocardiography.

ED Management

- Treatment is geared toward concurrent disease processes and prevention of complications such as endocarditis.
- Surgical correction may be needed.

2

Prosthetic Valve Dysfunction and Complications

- Valve replacement surgery is very common with more than 40,000 replacements done per year.
- Prosthetic heart valves are either bioprosthetic (usually bovine, porcine, or human cadaveric) or mechanical.
- The evaluation of a patient with a prosthetic valve begins with an understanding of the type of valve placed and the reason for placement. Patients are instructed to carry a card describing the prosthetic valve.
- Mechanical valves have a mechanical click and systolic murmurs. Bioprosthetic valves normally have only slight murmurs.
- Any change in clinical status or auscultatory findings in patients with prosthetic valves requires an evaluation of the valve including echocardiography.
- Complications of prosthetic valves include valve thrombosis, infection, and degeneration of the valve or surgical site.
    - Thrombosis of a prosthetic valve can be acute or chronic and is much more common in mechanical valves. Acute valvular thrombosis is a cardiothoracic surgical emergency, presenting with acute heart failure and cardiogenic shock. Thrombi may also occur chronically and present with either progressive valvular dysfunction and symptoms of worsening valvular disease, or with embolic phenomena. Treatment includes valve replacement but may also include fibrinolytic therapy.
    - Endocarditis is common in prosthetic valves. Within the first 2 mo after placement, *Staph* species are common. After the first 2 mo, the etiology is similar to native valve endocarditis. Treatment with antibiotics as in patients with native valve endocarditis is required. Surgical valve replacement may be required.
    - Valve dysfunction occurs more commonly with bioprosthetic valves with approximately 30% requiring replacement within 10 yr. All prosthetic valves have some regurgitation and stenosis inherently. Degeneration of the valve itself, the perivalvular surgical site, or the diseased myocardium can lead to worsening regurgitation or stenosis. Valve failure presents with symptoms of either the stenosis or regurgitation of the diseased valve. It is often difficult to distinguish valve dysfunction from progression of underlying cardiac disease in these patients, thus requiring liberal use of echocardiography in the evaluation of these patients.
- Mechanical prosthetics may also cause chronic hemolysis and subsequent anemia. Hemolysis from mechanical valves or perivalvular degeneration is usually compensated and asymptomatic. Severe dysfunction may lead to a more severe anemia and subsequent symptoms.

Suggested Reading

1. Otto C. Aortic Stenosis: Clinical evaluation and optimal timing of surgery. Cardiol Clinics 1998; 16:3.
2. Bruce C, Nishimura R. Clinical assessment and management of mitral stenosis. Cardiol Clinics 1998; 16:3.
3. Quinones M. Management of mitral regurgitation: Optimal timing for surgery. Cardiol Clinics 1998; 16:3.
4. Bonow R. Chronic aortic regurgitation: Role of medical therapy and optimal timing for surgery. Cardiol Clinics 1998; 16:3.
5. Safi H, Vinnerkvist A, Subramaniam M et al. Management of the patient with aortic root disease and aortic insufficiency. Cardiol Clinics 1998; 16:3.

6. Barbetseas J, Zoghbi W. Evaluation of prosthetic valve function and associated complications. Cardiol Clinics 1998; 16:3.
7. Blaustein A, Ramanathan A. Tricuspid valve disease: Clinical evaluation, physiopathology, and management. Cardiol Clinics 1998; 16:3.
8. Carabello B, Crawford F. Medical progress: Valvular heart disease. New Engl J Med 1997; 337.
9. ACC/AHA Guidelines for the management of patients with valvular heart disease. A report of the American College of Cardiology/American Heart Association Task Force on Practice Guidelines. J Am Coll Cardiol 1998; 32(5):1486-588.
10. Cline D. Valvular Emergencies and Endocarditis. In: Tintinalli JE, Kelen GT, Stapczynski JS, eds. Emergency Medicine. A Comprehensive Medicine Study Guide. McGraw-Hill, 2000.
11. Dunmire S. Infective Endocarditis and Acquired Valvular Heart Disease. In: Marx JA, Hockberger RS, Walls RM, eds. Emergency Medicine. Concepts and Clinical Practice. Mosby, 2002:2.

2

Part G: Aortic Emergencies

Diseases of the aorta are occurring more frequently with the aging of the population. Imaging with CT and ultrasound has led to increased recognition of aortic pathology. These diseases present in both dramatic and subtle ways, but left untreated they are almost universally fatal. Therefore, the Emergency Physician must be aware of the various presentations of aortic emergencies and have a complete understanding of their management.

Aortic Dissection

Epidemiology/Pathophysiology

- The most important cause of aortic dissection is long-standing systemic hypertension. The forceful ejection of the cardiac output results in repeated sheer stress on the intimal wall, ultimately leading to the wall disruption that causes dissection.
- Patients with Marfan's syndrome or Ehlers-Danlos syndrome have a congenitally weakened aortic wall, thus predisposing these patients to aortic dissection.
- Dissection of the thoracic aorta is caused by a disruption of the intimal wall of the aorta. Blood is transmitted through the tear creating a false lumen in the aortic wall. Once the medial wall is weakened by the false lumen, the dissection can rupture through the remainder of the outer wall, rupture through the side branches of the aorta, or rarely rupture back into the true lumen of the aorta.
- Occasionally, dissections of the ascending aorta may also damage the coronary arteries or the aortic valve.

Diagnosis and Evaluation

History and Physical Exam

- The presentation of aortic dissection usually involves the acute onset of severe chest pain. Classically, the pain is described as "tearing" and is most severe at its onset.
- Occasionally, dissections will present with pain in other locations due to the migration of the dissection and damage to side branch vessels.
- Neurologic symptoms may also be present if the carotid or spinal arteries are involved.
- Syncope may be a presenting symptom and is usually associated with ascending aorta dissections.
- On clinical exam, hypertension is usually present, unless there are pulse deficits in the arms being measured.

2

- Hypotension in the face of an aortic dissection is indicative of a dissection into the pericardium causing tamponade or severe hypovolemia secondary to hemorrhage.
- Pulse deficits may be present on exam and should be sought initially and on reexamination to check for propagation of the dissection.
- If the dissection spreads proximally, dissection through the aortic valve may occur, resulting in the findings of acute aortic insufficiency and congestive heart failure.
- As mentioned above, neurologic deficits may be present.

Laboratory and Studies

- Laboratory results are variable in aortic dissections; the only lab test of true importance is the type and cross.
- The electrocardiogram usually indicates evidence of long-standing hypertension. Proximal dissections may disrupt coronary blood flow, thus causing myocardial infarctions with related EKG changes. Most commonly, the right coronary artery is involved, leading to inferior myocardial infarctions.
- The chest X-ray is almost always abnormal. Most commonly, mediastinal widening is present. Other X-ray signs include obliteration of the aortic knob, right-sided deviation of a nasogastric tube, depression of the left mainstem bronchus, or a small left-sided pleural effusion, or a left apical cap.
- Definitive diagnosis of an aortic dissection involves direct imaging of the aorta. The gold standard remains aortography, which allows complete aortic visualization but is being rapidly replaced by other modalities.
- Contrast CT scanning allows visualization of the extent of the dissection, as well as pericardial and pleural effusions. CT scanning does not precisely localize intimal tears, unreliably demonstrates side branch involvement, and is not able to define aortic regurgitation. Furthermore, the contrast load needed for the CT scan is substantial and can have adverse consequences, especially for patients with renal insufficiency.
- Transesophageal echocardiography (TEE) is highly accurate for proximal aortic dissections. Benefits of TEE include speed and safety. TEE is also able to quickly evaluate for aortic insufficiency and pericardial effusions and evaluate myocardial function. Drawbacks to TEE include the need for an experienced operator and the inability to evaluate the descending aorta in its entirety or the side branch arteries.

ED Management

- The anatomic location of a dissection is the major determinant for therapy. Two classification schemes are widely used to describe dissections, the Stanford and DeBakey classifications.
  - The Stanford classification labels dissection as type A if the ascending aorta is involved and type B if there is no ascending aorta involvement.
  - The DeBakey classification describes dissections as Type I involving both the ascending and descending aorta, Type II if the dissection involving the ascending aorta only, and Type III involving the descending aorta only.
- The treatment of an aortic dissection begins with control of hypertension. These patients can be very sensitive to blood pressure manipulation. Short acting, titratable medications are therefore appropriate in this setting. β blockade (esmolol, labetolol, or propranolol are appropriate) is used to decrease the sheer stress placed on the aorta by the systolic pulse. Nitroprusside is used in conjunction with β blockade to control hypertension. The goal of blood pressure control is to lower the blood pressure to the lowest level which still allows organ perfusion.
- Definitive treatment is based on anatomy.
  - Dissections involving the ascending aorta are treated surgically with replacement of the involved segment. Concurrent aortic insufficiency or coronary insufficiency can be corrected surgically during the procedure.

- The treatment of isolated dissections of the descending aorta is intensive blood pressure control alone. Indications for surgical management include uncontrollable hypertension, rupture, or involvement of a major aortic branch with subsequent end-organ ischemia.

Aortic Aneurysm

Epidemiology/Pathophysiology

- Aneurysms are defined as dilatation of an aortic segment >3 cm.
- They are true aneurysms, indicating a dilatation of all three layers of the aortic wall.
- Men older than 70, smoking, hypertension, and a family history of aneurysms are predisposing factors.
- Most aortic aneurysms are diagnosed in the seventh decade of life.
- Abdominal aortic aneurysms (AAA) are most commonly located in an infrarenal location, but may occur at any level of the aorta.
- Aneurysms enlarge at a rate of 0.5 cm per year on average, with larger aneurysms enlarging at a faster rate.
- The risk of aneurysm rupture is largely based on size. While small aneurysms may rupture, the risk of rupture increases dramatically as aneurysms enlarge to >5cm.
- The mortality rate of ruptured aortic aneurysms is approximately 90% with over half of patients dying before reaching the hospital.

Diagnosis and Evaluation

History and Physical Examination

- The most important factor in diagnosing AAA is entertaining the diagnosis. Middle-aged patients presenting with abdominal or flank pain should always have the diagnosis entertained.
- Unruptured aneurysms are usually asymptomatic.
- Chronic abdominal pain, back pain, and ureteral colic-like symptoms are common presentations of aneurysms.
- Physical exam findings may include palpation of a pulsatile abdominal mass, but this finding is neither sensitive nor specific enough to rule in or out aneurysmal disease.
- A ruptured aneurysm classically presents with pain, hypotension and a pulsatile mass. However, many patients present without these findings. The location and quality of pain is variable, most commonly presenting as acute, severe abdominal, back, or flank pain. Hypotension is a late and grave finding. Other presentations include syncope and altered mental status.

Laboratory and Studies

- Lab work is not helpful except for a type and cross which is imperative in cases of ruptured AAA.
- Imaging studies are indicated and the choice of study is largely dependent on the clinical stability of the patient.
  - Plain films of the abdomen can be helpful if positive but are not sensitive enough to rule out aneurysms. Calcification of the aortic wall can be seen with obvious enlargement of the aorta. These studies are almost universally available and can be done rapidly in the case of an unstable patient.
  - Ultrasound is the test of choice to detect aneurysmal disease. It is very sensitive and can evaluate the size of an aneurysm as well as identify intraperitoneal free fluid indicative of rupture. Bedside ultrasound can be done rapidly, making it especially useful for unstable patients. Drawbacks of ultrasound include operator-dependent accuracy and difficulty in visualization of the aorta in patients with excess bowel gas or obesity.

- CT scanning is almost 100% sensitive for the detection of AAA. It has better sensitivity than ultrasound for the detection and location of rupture and is better in defining the surrounding anatomy. IV contrast is helpful, but not imperative for the examination. The main disadvantage of CT scanning is the inability to monitor critically ill patients during the exam.

2

ED Management

- Treatment of AAA depends on the stability of the patient. Asymptomatic aneurysms discovered during physical exam or during evaluation for other problems may be referred for further evaluation and treatment. Generally, aneurysms <5 cm. are observed with repeat ultrasound evaluation and elective surgical management based on patients' concurrent illnesses and the size of the aneurysm.
- Newer intravascular stent techniques are beginning to be used and are appropriate for a select group of patients.
- Treatment of ruptured AAA includes rapid medical resuscitation with blood products as needed.
- Definitive treatment is surgical and arrangements for surgery should be made as quickly as possible.
- Complications of AAA repair include infection, thrombosis, erosion, and dilation of the graft. These are often life-threatening.
  - Infections may occur immediately postoperatively or from hematogenous spread of other infections later in life. *Staph* species are generally responsible. Treatment includes IV antibiotics and removal and repair of the infected graft.
  - Thrombotic complications present in multiple ways with embolic phenomena or ischemic symptoms. Evaluation includes visualization of the aorta and anticoagulation and repair as necessary.
  - Erosion of the graft may lead to rupture and/or fistula formation. Aorto-enteric fistulae are caused by erosion into the GI tract. The duodenum is the most common site of fistulae. Fistulae present with GI bleeding and can range from a slow, chronic process, to an acute life-threatening hemorrhage. Many patients with aorto-enteric fistulae also have septic complications necessitating antibiotics. Treatment for fistulae is surgical.

Suggested Reading

1. Rebel R. Diseases of the thoracic aorta. Heart Aug 2001; 86(2).
2. Pretre R, Segesser L. Aortic dissection. Lancet 1997; 349:1461-64.
3. Dmowski A, Carey M. Aortic dissection. Am J Emerg Med 1999; 17(4).
4. Hals G, Pallaci M. The clinical challenges of abdominal aortic aneurysm: Rapid, systematic detection and outcome-effective management. Emerg Med Reports 2000; 21(12).
5. Thompson MM, Bell PRF. Arterial aneurysms. Brit Medl J 2000; 320(7243).
6. Hallett J. Management of abdominal aortic aneurysms. Mayo Clin Proc 2000; 75(4).
7. Glover J. Thoracic and abdominal aortic aneurysms. In: Tintinalli JE, Kelen GT, Stapczynski JS, eds. Emergency Medicine. A Comprehensive Medicine Study Guide. McGraw-Hill, 2000.
8. Bourland M. Aortic Dissection. In: Marx JA, Hockberger RS, Walls RM, eds. Emergency Medicine. Concepts and Clinical Practice. Mosby, 2002:2.
9. Bessen H. Abdominal Aortic Aneurysm. In: Marx JA, Hockberger RS, Walls RM, eds. Emergency Medicine. Concepts and Clinical Practice. Mosby, 2002:2.

CHAPTER 3

Pulmonary Emergencies

Deborah B. Diercks, Steven Offerman, Mark Thoma and Peter E. Sokolove

Basic Anatomy and Physiology

- The trachea, bronchi, and bronchioles are the conducting airways and consist of a series of branching tubes that become narrower and shorter as they penetrate into the lungs. These airway structures have no diffusion capacity and represent about 150 ml of lung volume. Eventually the terminal bronchioles lead to the alveoli that form the actual gas-exchange interface. The alveolar surface consists of approximately 3000 ml of lung volume.
- The walls of the conducting airways contain smooth muscles which, when contracted, cause airway narrowing and increased resistance to airflow. These smooth muscles respond to both sympathetic and vagal input. $Beta_2$-receptor stimulation causes muscle relaxation, while α-receptor and vagal stimulation result in bronchoconstriction. Constriction is also reflexive and may be initiated by irritants, temperature, and psychogenic causes.
- Contraction of the diaphragm and intercostal muscles increases the volume of the thoracic cavity creating a negative intrathoracic pressure. This bellows action draws air into the airways and alveoli by bulk flow. Expiration is a passive process that occurs as the elastic lung tissue returns to its preinspiratory volume.

Part A: Acute Respiratory Failure (ARF)

- Definition: ARF is an impairment of oxygen (O_2) or carbon dioxide (CO_2) gas exchange that results in immediate or impending breakdown of cell metabolism. There are two fundamental mechanisms (see Table 3A.1):
  - Failure to oxygenate: Room air PaO_2 <60 mm Hg.
  - Failure to ventilate: $PaCO_2$ >50 mm Hg.
- An increased work of breathing may signal impending ventilatory failure prior to the development of hypercarbia or hypoxia.
- Patients who chronically retain CO_2 have baseline elevation of their $PaCO_2$ but usually have a normal pH via metabolic compensation. In these patients, ventilatory failure is identified by an increase in $PaCO_2$ above baseline and a corresponding decrease in pH.

Diagnosis

- Dyspnea is the most common symptom and is almost universal in awake patients.
- Crucial aspects of the physical evaluation include general appearance, vital signs, and pulmonary examination.
  - General appearance
    - Signs of increased work of breathing include diaphoresis, tripod positioning, intercostal muscle retractions, nasal flaring, and audible grunting. Patients may be unable to speak full sentences. Significant increases in work of breathing indicate acute or impending respiratory failure. Agonal respirations are slow, shallow breaths that identify impending respiratory arrest.

Emergency Medicine, edited by Sean Henderson. ©2006 Landes Bioscience.

3

Table 3A.1. Common causes of acute respiratory failure

Failure to Oxygenate	
	Ventilation-perfusion mismatch (pneumonia, aspiration, ARDS*, pulmonary embolism)
	Decrease in FiO_2
	Intra/extrapulmonary shunting
	Diffusion defects (emphysema, interstitial lung disease)
	Restrictive lung disease
	Ventilatory failure
Failure to Ventilate	
	Depressed mental status (drugs, stroke, sepsis, seizures)
	Upper airway obstruction (croup, epiglottitis, burns, cancer, trauma)
	Lower airway obstruction (asthma, COPD$^+$, cancer)
	Chest wall disorders (flail chest, kyphosis, muscular dysfunction)

*ARDS: acute respiratory distress syndrome
$^+$COPD: chronic obstructive pulmonary disease

- Altered mental status (AMS) is an important indicator of ARF. Confusion, somnolence and agitation may occur secondary to hypoxia and/or hypercarbia. The presence of decreased mentation in patients with respiratory distress indicates the need for immediate intervention.
- Vital signs
  - Respiratory rate (RR) is typically abnormal and may be elevated or depressed. Tachypnea occurs secondary to stimulation of central respiratory centers in patients with hypoxia or hypercarbia. Hypopnea results from drug ingestion, stroke, seizures, hypothyroidism, and other causes of impaired brainstem function.
  - Patients with ARF usually have tachycardia as a result of underlying hypoxia and/or an adrenergic response. However, severe hypoxia may also cause bradycardia.
  - Pulse oximetry: all patients with oxygen saturation <90% should be considered severely hypoxic (see following section for full discussion).
- Pulmonary examination
  - Stridor is associated with upper airway obstruction (larynx or trachea) and is audible without a stethoscope. Inspiratory stridor is classically seen with supraglottic obstruction and expiratory stridor with subglottic pathology.
  - Wheezes are associated with lower airway obstruction. Bronchspasm is the most common cause but other etiologies include foreign body and pulmonary edema. Some patients with bronchospasm or airway obstruction may have little or no wheezing if airflow is severely reduced.
  - Rhonchi occur with airflow through areas narrowed by inflammation, smooth muscle contraction, or mucous.
  - Rales are suggestive of alveolar inflammation or fluid.
  - Decreased or absent breath sounds may signify nonventilated lung segments, pleural effusion or pneumothorax.

- Pulse Oximetry (Pox) provides rapid, noninvasive measurement of oxygenation and correlates well with measured PaO_2. A Pox reading of 90% corresponds to a PaO_2 of approximately 60 mm Hg. Carbon monoxide (CO) poisoning may result in falsely elevated readings. Dark nail polish, peripheral vascular disease, hypoperfusion, and anemia may cause falsely depressed readings. Note that a normal Pox does not rule out the presence of hypercapnia.

- Arterial blood gas (ABG) may aid in the diagnosis in patients with suspected CO poisoning. It also allows the physician to assess the degree of hypoxia and hypercapnia but is not a necessary study in patients with a clinical picture consistent with ARF.
- Portable chest radiograph (CXR) is indicated in all patients with acute or impending respiratory failure. Findings are often useful for identification of the underlying cause and may have treatment implications. However, the decision to intubate or administer other airway interventions is nearly always based on clinical, rather than radiographic criteria. CXR should also be obtained after endotracheal intubation to assess tube placement.
- Laboratory results rarely affect management. However these patients are often critically ill with comorbid illness. Basic studies including complete blood count (CBC), electrolytes, blood urea nitrogen (BUN), creatinine and glucose as well as an electrocardiogram (EKG) should be obtained in most patients with ARF. Other studies may be indicated depending on the presentation.

Treatment

- Supplemental oxygen increases the delivered FiO_2 with each liter of oxygen increasing FiO_2 by approximately 4%. Many delivery devices are available but nasal cannulae and masks are the most commonly used.
  - Nasal cannula delivers up to 44% FiO_2. Oxygen administered at 1 to 6 L/min. Nasal cannula may be used for patients with mild hypoxia but is not appropriate in the setting of severe respiratory distress.
  - Nonrebreather mask (NRB) delivers up to 98% FiO_2 (almost 100%). Oxygen is generally administered at 15 L/min. NRB may be used in patients with moderate to severe hypoxia or as a bridge to more definitive therapy.
- Noninvasive Positive Pressure Ventilation (NPPV)
  - NPPV provides positive pressure to airways using either a nasal or face mask. Both inspiratory pressure (IPAP) and expiratory pressure (EPAP) can be controlled. NPPV is probably most effective in disorders where treatment may be expected to result in rapid improvement of respiratory status, such as asthma, COPD, or pulmonary edema. Use of NPPV may avoid endotracheal intubation. The vast majority of patients who will fail treatment do so within the first 12 h.
  - Most of the studies regarding NPPV have focused on COPD patients. The bulk of evidence is positive. Several controlled trials have shown improved gas exchange and lower intubation rates among patients treated with NPPV. Asthma and acute pulmonary edema have also been treated successfully with NPPV.
  - NPPV does not provide airway protection. In order to be a candidate, a patient must have a clear sensorium, be able to initiate breaths, and be able to tolerate the mask.
  - NPPV should be used in conjunction with a respiratory therapist, nurse, or physician who is skilled in its use. Once instituted, IPAP and EPAP are set independently. IPAP is adjusted to decrease the work of respiratory muscles and is titrated to the desired $PaCO_2$. Avoid peak pressures >20 cm H_2O. Oxygenation is controlled by adjusting the FiO_2 and EPAP. Common initial settings are an EPAP of 3-5 cm H_2O and IPAP of 10 cm H_2O.
- Endotracheal intubation is the gold standard for respiratory support and airway management. Placement of an endotracheal tube (ETT) provides the maximum control of ventilation, oxygenation, and airway patency.
  - Indications for intubation include:
    - Severe or progressive hypoxemia
    - Severe or progressive acute hypercapnia
    - Severe or progressively increased work of breathing
    - Airway protection

        - Acute or impending airway occlusion
        - Pulmonary support in the critically ill or injured patient
        - Need for life-saving diagnostic studies or therapies in uncooperative patients
    - Ventilator management varies depending upon the underlying mechanism. A detailed discussion of ventilator management is beyond the scope of this text.
        - Defect in oxygenation: adjust FiO_2 and/or positive end expiratory pressure (PEEP) to achieve desired pO_2.
        - Defect in ventilation: adjust RR and/or tidal volume to achieve desired pCO_2.

3 - Specific treatment: once the patient's respiratory status is stabilized, directed therapy can begin. This might include medical therapy, surgical intervention, and/or specific ventilator strategies.

Disposition

- All patients with respiratory failure should be admitted to an intensive care unit (ICU). Patients with impending respiratory failure should be admitted to either an ICU or another closely monitored bed (e.g., step-down unit).
- Patients with a stable respiratory status who are at very low risk for deterioration can be admitted to a ward bed.

Part B: Asthma

Asthma is a chronic disease characterized by increased airway responsiveness to various stimuli. This causes widespread narrowing of the lower airways that reverses either spontaneously or with treatment. Although the exact pathophysiology of asthma is complex and poorly understood, inflammation is thought to play a central role. Pathologic changes that occur in asthma include smooth muscle hypertrophy, mucosal edema, and mucous plugging. Asthma affects 4-5% of adults and 10% of children. Onset usually occurs in children and young adults.

Etiology and Risk Factors

- Asthma is commonly classified as allergic (extrinsic) or nonallergic (intrinsic).
- Allergic asthma is more common and is responsible for the majority of childhood asthma and a significant portion of adult disease. These patients are sensitive to specific inhaled allergens. Patients with allergic asthma frequently have a personal and family history of allergic diseases, including allergic rhinitis and atopic dermatitis. In contrast to patients with intrinsic asthma, those with allergic asthma have increased levels of immunoglobulin E (IgE). Inhalation of an allergen induces a response in two phases.
    - The early response usually begins within minutes of exposure and lasts up to period of several hours. Caused by mast cell degranulation. Mediator release subsequently induces bronchoconstriction and an inflammatory reaction.
    - The late response is characterized by airway inflammation that results in further bronchoconstriction and mucous production. Symptoms may persist for days to weeks after the initial exposure.
- Nonallergic asthma is associated with numerous stimuli including exercise, emotion, air pollution, cigarette smoke, medications, and occupational exposures.

Diagnosis

- A definitive diagnosis is made via pulmonary function tests (PFTs) that demonstrate reversible airway obstruction. PFTs are not practical for use in the emergency department (ED) where the diagnosis is made clinically.

- In stable patients, historical information can be obtained that may help guide therapy and disposition. There are several factors associated with poor outcome.
  - History of prior intubations for ARF secondary to asthma
  - Multiple or recent hospitalization(s) for asthma exacerbation
  - Recent use of corticosteroids
- Other important information includes the time of onset, inciting factors, and use of medications prior to arrival.
- Patients generally complain of dyspnea and cough. Severity ranges from mild to life threatening. Note that dyspnea is likely unrelated to hypoxia and may not resolve with supplemental oxygen. The cough can be either dry or productive. Patients with "cough-variant asthma" present with a nonproductive cough that tends to be nocturnal. They may not have audible wheezing. These patients have ventilatory impairments demonstrable with PFTs and usually experience relief with bronchodilator therapy.
- General appearance, vital signs, and pulmonary evaluation should be assessed as previously discussed (see ARF). Patients often have tachypnea and tachycardia that should improve with appropriate treatment. Pulsus paradoxus is associated with acute asthma but is not a practical aspect of the ED evaluation. Common auscultatory findings include wheezing, decreased breath sounds, and prolongation of the expiratory phase. Absence of wheezing may be indicative of severe airway obstruction. Reexamination after bronchodilator therapy in such patients is often notable for increased wheezing. AMS, increased work of breathing, hypoxia, and hypercarbia indicate ARF and mandate immediate intervention.
- While asthma is mainly a clinical diagnosis, various diagnostic modalities can contribute to management and disposition.
  - Pox—Saturation should be continuously monitored on all asthma patients (see *Part A, Diagnosis*, for discussion of Pox). Any patient with a saturation <90% should be considered severely hypoxic and treated accordingly.
  - ABG—Blood gas assessment is not routinely indicated but can help guide ventilator management and determine the degree of hypercarbia/hypoxia in patients with severe exacerbation. During an acute exacerbation, the ABG usually shows a respiratory alkalosis. Normal or increasing pCO_2 reflects deterioration in ventilation although this should also be clinically evident.
  - Pulmonary function testing/Peak flow—As previously discussed, PFTs are not a routine aspect of the ED evaluation. Peak expiratory flow rate (PEFR) provides a means of assessing pulmonary function at the bedside although patient cooperation is required. PEFR values do not correlate well with prognosis, and there is no absolute value that mandates admission. It is best used serially to monitor the effects of therapy on patients with mild to moderate disease. Ideally, PEFR in the ED is compared against the patient's known baseline. Measurements can also be compared to predicted levels using nomograms that consider age, sex, and height. In general for adults patients, PEFR <300 indicates a mild exacerbation, PEFR <200 a moderate exacerbation, and PEFR <100 a severe exacerbation.
  - CXR—In mild to moderate asthma exacerbations, routine CXR is not necessary. It is helpful for identification of complications such as pneumothorax (PTX). CXR is also indicated if the patient does not improve with therapy or has fever, focal findings on pulmonary exam, pleuritic chest pain, or hypoxia. Patients presenting with a first episode of wheezing and those with an unclear diagnosis should have CXR to evaluate for underlying pathology.
  - Laboratories rarely influence management and are not routinely indicated. The decision to obtain laboratory screening should be based upon the patient's age, medication use, and other comorbid conditions. An increased leukocyte count

(WBC) is consistent with both infection and steroid use. However, a normal WBC does not exclude infection. Also consider a theophylline level in patients taking that medication.

- Differential Diagnosis
  - The emergency physician (EP) must always search for a cause of acute exacerbation especially in those patients with severe symptoms. Allergen exposure is the most likely but respiratory infection, PTX, and pulmonary embolism (PE) are important and potentially fatal problems that must be identified.
  - The EP should remember that "all that wheezes is not asthma." Other conditions to consider include COPD, congestive heart failure, allergic reaction, airway obstruction, and pulmonary embolism. A directed history and physical examination, along with proper use of diagnostic testing, will help to differentiate these entities.

Treatment

Respiratory Support

- Oxygen can be given liberally since asthmatic patients do not chronically retain CO_2. The amount and route primarily depend upon the patient's symptoms and degree of hypoxia.
- NPPV has been shown to be effective in improving gas exchange and avoiding intubation in some asthmatic patients (see "Acute Respiratory Failure"). NPPV is not appropriate for patients with AMS and an obviously ineffective respiratory effort.
- Endotracheal Intubation
  - The decision to intubate is purely clinical. There are no ABG parameters or CXR findings that mandate this intervention. Patients with severe refractory hypoxia, altered mental status, severely increased work of breathing, and/or ineffective respirations are candidates for immediate intubation.
  - In the conscious patient, rapid sequence induction (RSI) is the safest method for endotracheal tube placement. Agents commonly used for RSI include a benzodiazepine in conjunction with a paralytic agent. Ketamine, in addition to its analgesic and anesthetic properties, is a bronchodilator and therefore should be considered the induction agent of choice in young asthmatics. In older patients with coronary artery disease, the cardiovascular risks of ketamine may outweigh the benefits. There are no specific contraindications to paralytic agents during RSI of the acute asthmatic patient.
  - Preoxygenation should be attempted prior to intubation. However, this is sometimes not possible for patients in extremis. Furthermore, oxygen saturation may decline very rapidly during intubation.
  - Oral intubation is preferred to nasotracheal intubation because a larger tube can be inserted. This allows for adequate suction and for bronchoscopy if needed.
- Ventilator Management
  - The intubated asthmatic patient is at risk for barotraumatic complications such as PTX, pneumomediastinum, or subcutaneous emphysema. The goal of mechanical ventilation is to supply the lowest minute ventilation that yields adequate gas exchange keeping peak airway pressures (PAP) below 35 cm H_2O. Suggested initial adult ventilator settings include FiO_2 of 100%, tidal volume of 6-8 ml/kg, ventilatory rate of 10, and inspiratory time/expiratory time (I/E) ratio of 1:3 or 1:4. FiO_2 can be titrated based on Pox and/or ABG.
  - Permissive hypercapnia is a strategy sometimes used to help control PAP. Patients are intentionally hypoventilated and airway pressures minimized via low tidal volume and RR. $PaCO_2$ is permitted to rise and pH to fall, generally to a level of around

7.25. In severe cases, pH can be further decreased and a sodium bicarbonate infusion initiated. Oxygenation is maintained via high FiO_2.

- After intubation, continued sedation and paralysis allow for maximal relief of the respiratory muscles and for permissive hypercapnia. This can be achieved with longer acting benzodiazepines and paralytic agents. Subsequent dosing of ketamine is also appropriate.
- Patients may also develop hypotension secondary to increased intrathoracic pressure and impaired venous blood return to the right ventricle resulting in decreased cardiac output. This must be differentiated from tension PTX. Both will also cause an elevation of PAP. If the former is suspected, the patient should be disconnected from the ventilator and manually ventilated at a slower rate (6 to 8 breaths per minute). This will allow for exhalation of trapped air. In addition, a CXR should be ordered and the patient suctioned. If a tension PTX is suspected, immediate decompression of the affected side by needle thoracostomy is indicated.

Medications

- Beta$_2$ Agonists
  - Inhaled β_2 agonists are the mainstay of therapy for acute asthma exacerbation. These agents relax bronchial smooth muscles and reverse bronchospasm. Albuterol is the most commonly used agent and is generally delivered by nebulizer. Onset of action is <5 min, and repetitive administration produces incremental effect. Nebulized albuterol is usually given in 2.5-5 mg increments every 15-20 min as needed. It can be given continuously for patients with severe symptoms. There is no defined maximum dose. Administration is usually limited only by symptoms (tremor, tachycardia, nausea).
  - Studies have shown that metered dose inhalers (MDIs) are as effective as nebulizers. However proper MDI use is essential and a severe exacerbation may preclude proper use.
  - Remember that intubation does not cure asthma. Intubated patients should continue to receive aggressive in-line β_2 agonists.
  - Levalbuterol is the single (R) isomer preparation of albuterol, as opposed to traditional racemic albuterol, which is a 50/50 mixture of the (R) and (S) isomers, the (S) component being inactive. It is thought to have similar efficacy to racemic albuterol but fewer nonrespiratory side effects. The cost of levalbuterol is about five times that of racemic albuterol. The benefit of this new preparation over standard albuterol is debatable.
  - Although multiple studies have demonstrated the advantages of inhaled β_2 agonists over systemic medications, it has been suggested that inhaled medications may be unable to reach critical areas in patients with severe asthma and profoundly impaired airflow. In these severe circumstances, terbutaline or epinephrine may be administered subcutaneously. The adult dosing for epinephrine is 0.3-0.5 mg of 1:1000 solution every 15-20 min as needed, up to 3 doses. Terbutaline dosage is 0.25-0.5 mg every 15-20 min as needed. Epinephrine should be used with caution in elderly patients and those with cardiovascular disease.
- Inhaled Anticholinergics
  - These agents block muscarinic receptors preventing smooth muscle contraction and diminishing mucous gland secretions.
  - Ipratropium bromide is the most commonly used agent. It is not to be used as single agent therapy but has been shown to be effective for the treatment of severe asthma when added to albuterol. Inhaled ipratropium does not appear to result in significant systemic side effects unlike other anticholinergic agents such as atropine.

3

- Ipratropium is delivered by nebulization and can be mixed with albuterol. The dose is 0.25-0.50 mg every 15-20 min, up to three doses.
- Most of the studies that support the beneficial effect of ipratropium used with small to moderate doses of albuterol. Thus, it is possible that the same benefit may be obtained by simply using higher doses of albuterol without ipratropium.

- Corticosteroids
  - Corticosteroids suppress inflammation and have been shown to improve patient outcomes, prevent relapses, and prevent hospital admission. There is little immediate benefit because of the delayed onset of these agents (about 6 h). Corticosteroids are probably unnecessary in mild asthma, but should be given in moderate to severe cases and in any patient who does not respond promptly to inhaled β_2 agonists.
  - Oral prednisone and intravenous (IV) methylprednisolone are the most commonly used agents. There is no difference between these two medications in terms of efficacy or onset of action. Prednisone is less expensive and more easily administered and should be given to the majority of patients. Methylprednisolone is preferred in patients who are unable to take oral medications due to vomiting or respiratory distress. The adult dose of prednisone is 60 mg and methylprednisolone 125 mg.
  - Patients discharged from the ED after being treated with corticosteroids should be continued on outpatient therapy for 5-7 days. This dosing regimen does not require tapering.
  - Inhaled corticosteroids have few systemic side effects and are beneficial in long-term management but currently have little use in the ED.
- Methylxanthines
  - The methylxanthines have multiple effects including bronchodilation and enhancement of diaphragmatic contraction. The mechanisms of action are not well defined. Use of methylxanthines in the ED is discouraged due to side effect profiles, complicated dosing, and lack of established efficacy.
  - If the decision is made to use intravenous aminophylline for a patient with refractory disease, frequent serum levels are necessary in order to avoid toxicity. Note that elimination rates are highly variable. For patients already taking theophylline, a baseline level is mandatory before beginning acute therapy.
- Magnesium sulfate (Mg) is a weak bronchodilator and a second-line agent for asthma exacerbations. The most recent data suggests that Mg benefits only the most severe asthmatic patients and should not be given routinely. Mg is inexpensive and safe in patients with normal renal function. The adult dose is 2 g IV over 10-15 min.
- Heli-ox is a combination of helium and oxygen that is usually administered in a 70%/30% mixture. Heli-ox is thought to improve laminar gas flow through airways, resulting in improved gas exchange and decreased work of breathing. Although early case reports were positive, subsequent clinical studies have shown little benefit and use of heli-ox remains controversial. Heli-ox is safe and inexpensive, and many physicians use this as adjunctive therapy in severe cases or in intubated patients with elevated PAP.

Disposition

- Disposition is dependent upon the patient's response to therapy. In general, patients with complete or near-complete resolution of symptoms and a PFR of at least 300 (or near the patient's baseline) can be discharged. Patients who don't meet discharge criteria who have mild to moderate symptoms can be admitted to a ward bed. Patients with more severe symptoms should be admitted to a monitored bed where timely respiratory assessment and therapy is available. Intubated patients and those with the potential for respiratory failure require ICU admission.
- All patients discharged from the ED should receive bronchodilator therapy ± corticosteroids.

Part C: Chronic Obstructive Pulmonary Disease

- Chronic obstructive pulmonary disease (COPD) is defined as progressive, chronic airflow obstruction due to chronic bronchitis, emphysema, or both. The majority of patients have components of both, although one of these entities will frequently dominate the clinical picture.
  - Emphysema—airspace enlargement distal to the terminal bronchioles due to destruction of alveolar septa.
  - Chronic bronchitis—chronic airway inflammation and bronchospasm. Clinically defined as productive cough lasting for at least 3 mo over 2 consecutive years.
- Although COPD is irreversible, patients with acute exacerbations do have reversible bronchospastic and inflammatory components.

Etiology and Risk Factors

- Cigarette smoking, including passive exposure to cigarette smoke, is by far the leading cause.
- Occupational exposures and hereditary α-1 antitrypsin deficiency are less common.

Diagnosis

- Clinical diagnosis is based on the presence of dyspnea, wheezing, and/or cough in a patient with a history of causative exposure and chronic, progressive symptoms. Patients usually present in the fifth or sixth decades of life. Alpha-1 antitrypsin deficiency should be suspected in any patient younger than 40 yr old with signs and symptoms of COPD.
- Presentation may separated into two syndromes, depending on the predominate pathologic process.
  - "Pink puffer" (emphysema dominant)
    - Patient is barrel-chested with thin build.
    - Cough is nonproductive or has scant sputum only. Exam remarkable for decreased breath sounds.
    - Hypoxemia and hypercarbia occur only in end-stage disease.
    - CXR shows hyperinflation, flattened diaphragms, and a small heart.
  - "Blue bloater" (bronchitis dominant)
    - Patient is overweight with stocky build.
    - Patients have prominent, productive cough with wheezes and rhonchi on examination.
    - Patients usually retain CO_2. Hypoxemia and hypercarbia occur early in disease.
    - CXR shows increased vascular markings and a large heart.
- Patients may present with varying degrees of respiratory difficulty. Dyspnea, worsening cough, and chest tightness are common complaints.
- Physical examination is similar to asthma with varying degrees of audible wheezing, decreased breath sounds, and prolonged expiratory phase. Patients may also have other signs such as a barrel chest and stigmata of chronic pulmonary disease such as clubbing.
- Diagnostic Studies
  - Pox and ABG—All COPD patients should have continuous Pox monitoring. Unlike patients with asthma, many with COPD have baseline oxygen saturations well below 95%. ABG is helpful in critically ill patients and those requiring mechanical ventilation. Note that COPD patients often have an elevated pCO_2 at baseline. In these patients, ventilatory insufficiency is indicated by a decreased pH in conjunction with a high pCO_2. ABG may be helpful in assessing the severity of an exacerbation if a baseline pCO_2 is available in the patient's chart.

3

- PEFR is sometimes used to monitor the effects of maintenance therapy in patients with mild to moderate disease but is of little use in the ED.
- CXR—The EP should consider CXR in most patients with COPD exacerbation, the exception being those with mild exacerbation and prompt response to therapy. CXR is helpful in diagnosing an underlying source of for acute symptoms including PTX and pulmonary infiltrates.
- Laboratory testing—As with asthma, routine laboratory evaluation contributes little to management (see "Asthma"). If a CBC is obtained, the EP may note polycythemia secondary to chronic hypoxia. The EP may consider electrolytes, renal function studies, and/or cardiac enzymes as indicated by presentation and comorbidities.
- EKG—Patients in moderate to severe distress require continuous EKG monitoring. The 12-lead EKG often has findings consistent with right heart strain. An EKG should be obtained in those patients with chest pain, severe hypoxia, suspected dysrrhythmia or acute coronary syndrome.

- Differential Diagnosis
  - The diagnosis of COPD is usually not difficult. However, the EP should determine the cause of the acute exacerbation. Respiratory infections, allergen exposure, continued cigarette smoking, air pollution, and patient noncompliance are common causes.
  - Acute PTX, lobar atelectasis, and PE are the most potentially deadly causes of exacerbation. Unfortunately, PTX and PE can be difficult to diagnose in the COPD patient but should be suspected in all patients with exacerbation especially those with acute onset of symptoms.
  - Pneumonia occurs frequently in patients with COPD. This diagnosis should be considered based on clinical findings since CXR may or may not reveal an infiltrate.

Treatment

- To a large degree, this mirrors therapy for asthma (see "Asthma") with some variations as discussed below. The most important aspect of therapy is to initiate rapid intervention for those patients with acute or impending respiratory failure.
  - Respiratory support
    - Concern exists that aggressive oxygen therapy may thus worsen hypercarbia by suppression of hypoxic respiratory drive. This concern is somewhat theoretical and less important in the ED where ventilatory support is immediately available. A safe approach in the nonintubated patient is to titrate oxygen to achieve saturation between 90-92%.
    - Application of NPPV, endotracheal intubation, and ventilator management in COPD patients is similar to use as described in "Asthma" section.
  - Medications
    - $Beta_2$ agonists—The EP should follow the same dosing recommendations as previously described but should keep in mind that many patients with COPD are elderly and have cardiovascular comorbid disease. As a result, administration of β_2 agonists is more likely to be limited by adverse side effects.
    - Inhaled Anticholinergics—these agents are very effective in COPD both alone and in conjunction with β_2 agonists. Ipratropium should be used in all patients with COPD exacerbation. Dosing is the same as for asthma.
    - Corticosteroids, methylxanthines, and magnesium—Indications and dosing are discussed in the asthma section.
    - Antibiotics
      - Although the role of bacterial infection in acute bronchitis is controversial, antibiotic therapy has been shown to improve outcomes for patients with purulent sputum and severe COPD exacerbation. Trimethoprim-sulfmethoxazole,

doxycycline, amoxicillin-clavulanate, azithromycin, or clarithromycin are appropriate choices for both acute bronchitis and outpatient pneumonia therapy.
- Empiric inpatient pneumonia treatment is with second or third generation cephalosporin and possibly a macrolide to cover atypical organisms. If possible, sputum cultures should be obtained for all admitted patients to guide future antibiotic therapy.

Disposition

- Patients who respond rapidly to therapy and return to baseline in the ED can be discharged with close outpatient follow-up. However, many patients with COPD exacerbation require admission. This is due to the relatively smaller reversible component of airway disease that exists in COPD. The EP should also maintain a low threshold for admission for those with pneumonia. Intubated patients and those at risk for decompensation require ICU admission.
- All discharged patients should receive appropriate therapy including bronchodilators ± anticholinergics, corticosteroids, and antibiotics.

Suggested Reading

1. Madison MJ, Irwin RS. Chronic obstructive pulmonary disease. Lancet 1998; 352:467-473.
2. Advanced Cardiac Life Support Textbook. Dallas, TX: American Heart Association, 1994.
3. Gammon RB, Strickland JH, Kennedy JI et al.: Mechanical ventilation: A review for the internist. Amer J Med 1995; 99:553-562.
4. McFadden ER. Asthma. In: Isselbacher KJ, ed. Harrison's Textbook of Medicine 13th Ed. New York: McGraw-Hill, 1994.
5. Honig EG, Ingram RH. Chronic bronchitis, emphysema, and airways obstruction. In: Isselbacher KJ, ed. Harrison's Textbook of Medicine. 13th Ed. New York: McGraw-Hill, 1994.
6. Mandavia DP, Dailey RH. Chronic obstructive pulmonary disease. In: Rosen Frakes MA, Richardson LE, eds. Magnesium therapy in certain emergency conditions. Am J Emerg Med 1997; 15:182-187.
7. West JB. Respiratory physiology-the essentials, 4th ed. Baltimore: Williams & Wilkins, 1990.
8. Emond SD, Camargo CA, Nowak RM. 1997 National Asthma Education and Prevention Program guidelines: A practical summary for emergency physicians. Ann Emerg Med 1998; 31(5):579-594.
9. Brenner B, Kohn MS. The acute asthmatic patient in the ED: To admit or discharge. Am J Emerg Med 1998; 16(1):69-75.
10. Panacek EA, Pollack CV. Medical management of severe acute asthma. In: Brenner BE, ed. Emergency Asthma. New York: Marcel Dekker Inc., 1999.
11. Stedman's Medical Dictionary. In: William R. Henyl, ed. Baltimore: William & Wilkens, 1990.

Part D: Pneumonia

Pneumonia is an infection of the gas exchange segments of the lung parenchyma. It can cause a profound inflammatory response leading to airspace accumulation of purulent debris. Pneumonia costs are $8 billion annually, accounts for nearly one-tenth of all hospital admissions, and remains a leading cause of mortality in the United States.

Etiology and Risk Factors

- There are numerous risk factors as discussed in (Table 3D.1).
- The pathogens involved vary depending upon the host (see Table 3D.2).

3

Table 3D.1. Risk factors for pneumonia

Risk Factor	Comments
Aspiration/absent gag reflex	Stroke, intubation, seizure, altered mental status, sedative use
Mucociliary clearance disorders	Smoking, alcohol, COPD, cystic fibrosis, chronic bronchitis, viral infections
Alteration of normal oral flora	Acute illness and antibiotic use
Immunocompromise	AIDS*, diabetes, transplant, steroid use, asplenia, sickle cell disease, uremia, neoplasia, chemotherapy, extremes of age, complement deficiency
Hematogeonous	Indwelling catheters, intrathoracic devices
Geography/environment	American southwest (Valley Fever), Ohio/ MississippiValleys (histoplasmosis, blastomycosis), Southeast Asia (tuberculosis), pigeon droppings (psittacosis), bovinesources (Q fever), buildings with contaminated water supply
Community dwelling	Dormitory, prison, barracks, nursing home

* AIDS: acquired immune deficiency syndrome

Diagnosis

- Pneumonia is sometimes divided into two categories depending upon the causative agent and presentation (see Table 3D.3). Note that considerable overlap exists between the two categories and differentiation in the ED may be difficult.
- Patients typically complain of dyspnea, cough, and fever. Depending upon the etiology, they may also have night sweats, weight loss, myalgias, and localized extrapulmonary symptoms. History should focus on acuity symptom onset, presence of associated symptoms, recent travel history, immunization history, and comorbidities. In certain populations such as the elderly, pneumonia can present with nonspecific symptoms such as weakness and fatigue.
- Physical exam findings depend upon the etiology and the extent of lung involvement. Pulmonary exam often reveals rales and decreased or bronchial breath sounds. Although sometimes difficult to assess in the ED, patients can also have dullness to percussion, tactile fremitus, and egophony. Associated findings include tachypnea, tachycardia, diaphoresis, AMS, and increased work of breathing. Note that the pulmonary examination sometimes does not correlate with CXR findings.
- Laboratory Studies
  - Sputum—Because of the low sensitivity of the sputum Gram stains, the clinical utility in the ED is controversial. This test is most helpful if a single predominant organism is identified and requires an adequate specimen (>25 WBCs and <10 epithelial cells

Table 3D.2. Common pathogens in pneumonia

Population	Causative Pathogen
Community acquired	*Streptococcus pneumoniae, Mycoplasma pneumoniae, viruses, Chlamydia pneumoniae, Haemophilus influenzae,* Legionella, *Staphylococcus aureus*
Nosocomial (>likely to be resistant to antibacterial therapy)	Gram-negative bacilli, *Staphylococcus aureus*, anaerobes, and *Streptococcus pneumoniae* (less frequent)

Table 3D.3. Typical and atypical pneumonias

Category	Pathogens	Presentation
Typical (usually bacterial)	*Streptococcus pneumoniae* *Haemophilus influenzae* *Staphylococcus aureus* *Klebsiella pneumoniae* Anaerobes *Psuedomonas aeruginosa*	Acute onset Shaking chills and high fever Cough with purulent sputum Dyspnea Pleuritic chest pain
Atypical	Mycoplasma pneumoniae Viruses Legionella *Chlamydia pneumoniae* *Mycobacterium tuberculosis* *Pneumocystis carinii*	Gradual onset Low grade fever Scant sputum Mild respiratory complaints Extrapulmonary complaints Mycoplasma: myalgias, headache, sore throat, rash Viral: upper respiratory symptoms Legionella: AMS, gastrointestinal symptoms

per high power field) as well as experienced laboratory personnel. Sputum cultures are helpful for critically ill or immunocompromised patients but are rarely of use to the EP and should not be routinely ordered. An acid fast (AFB) stain is indicated patients with risk factors or presentation consistent with tuberculosis (TB).

- Serum
  - There are no specific laboratories for pneumonia although CBC, electrolytes, and renal function studies are often ordered. These tests should be obtained routinely in patients who are critically ill or if significant comorbid disease is present. Note that presence of an elevated WBC does not identify a bacterial source. Nor does a normal WBC rule it out.
  - Serum antibody titers are available for Legionella, *Mycoplasma pneumoniae*, and viruses among others but are of little use in the ED.
- CXR
  - Ordered in nearly all patients with suspected pneumonia although studies debate the utility of this study in otherwise healthy people being treated empirically as an outpatient.
  - Certain radiographic patterns have been described depending upon the etiology (see Table 3D.4). These patterns sometimes vary and do not provide an accurate means of diagnosis.
  - Note that radiographic findings often lag behind clinical symptoms. Patients with early disease and immunosuppression may not have classic findings.

- Differential diagnosis includes COPD, bronchitis, asthma, allergic reaction, and PE among others.

Treatment

- Stabilization of cardiopulmonary status is the first priority. Depending on the disease severity, patients may have respiratory compromise and/or circulatory collapse that mandate immediate intervention.
- Early antibiotic treatment decreases morbidity and mortality. Empiric therapy should be started as soon as possible after appropriate resuscitative measures. Many patients are treated as outpatients, although certain groups are at risk for poor outcome and should be considered for hospital admission (see Table 3D.5). Admitted patients should

3

Table 3D.4. Radiographic presentation of pneumonia

Radiographic Pattern	Pathogens
Lobar	*Streptococcus pneumoniae*
	Klebsiella pneumoniae (classically RUL, bulging fissure)
Patchy	Atypical agents
	Haemophilus influenzae
	Staphylococcus aureus
	Fungi
	Viruses
Interstitial	*Mycoplasma pneumoniae*
	Viruses
	Pneumocystis carinii
Abscess	Tuberculosis and other fungi
	Staphylococcus aureus
Effusion	*Streptococcus pneumoniae*
	Staphylococcus aureus
	Mycoplasma pneumoniae
	Viruses
	Tuberculosis
Apical	Tuberculosis
	Klebsiella pneumoniae

receive IV antibiotics and outpatients appropriate oral therapy as indicated for their age, comorbid conditions, and suspected pathogen (see Table 3D.6).
- All discharged patients should follow-up with their primary care physician.

Suggested Reading

1. Feldman CF. Pneumonia in the elderly. Clin Chest Med 1999; 20(3):563.
2. Dean NC. Use of prognostic scoring and outcome assessment tools in the admission decision for community-acquired pneumonia. Clin Chest Med 1999; 20(3):521.
3. American Thoracic Society: Guidelines for the initial management of adults with community acquired pneumonia: Diagnosis, assessment of severity, and initial microbial therapy. Am Rev Respir Dis 1999; 148:1418.

Table 3D.5. High risk patients

Risk Factor	Comment
Abnormal vital signs	Tachypnea (>30/min)
	Hypotension (<70 mm Hg systolic)
	O_2 saturation <95% on room air
Extremes of age	<6 mos or >60 yr
Comorbid conditions or disease	Pregnancy
	Congestive heart failure
	Renal or hepatic insufficiency
	Immunosuppression: HIV, asplenia, diabetes, alcohol/drug abuse
Recent hospital admission	
Patients who fail initial therapy	
Risk of aspiration	Stroke, AMS, alcohol abuse
Pathogen	Suspected tuberculosis
	Gram-negative bacilli on sputum examination
Inability to care for self as outpatient	

Table 3D.6. Antimicrobial guidelines for pneumonia

Group	Treatment*	Alternatives
Outpatient therapy	Erythromycin	Levofloxacin
Adults 18-65 yr	Clarithromycin	Second generation cephalosporin
No comorbid disease	Azithromycin (5 days)	Doxycycline Amoxicillin/clavulanate
Outpatient therapy	Bactrim	
Adult >65	Doxycycline	
Alcohol/tobacco use	Azithromycin (5 days)	
	Levofloxacin	
Inpatient therapy[a]	Ceftriaxone or cefotaxime + macrolide	
General ward	Cefuroxime + macrolide	
	Levofloxacin	
Inpatient therapy	Azithromycin + ampicillin/sulbactam	
Suspected aspiration	Levofloxacin + clindamycin	
	Second or third generation cephalosporin + clindamycin	
Inpatient therapy	Ticarcillin/clavulanate + aminoglycoside	
Ventilated/ICU	Piperacillin/tazobactam + aminoglycoside	
	Ceftazidime + aminoglycoside	
	Imipenem	

* All regimens are for 7-14 days unless otherwise noted
[a] All medications for inpatient therapy via IV route

4. Emergency Medicine Reports: Community-acquired pneumonia (CAP) in the geriatric patient: Evaluation, risk-stratification, and antimicrobial treatment guidelines for inpatient and outpatient management 2000; 21(20).
5. The Sanford Guide to Antimicrobial Therapy, 31st edition. 2001.

Part E: Hemoptysis

- Definition—Expectoration of blood from the respiratory tract below the level of the larynx.
- The amount can vary from blood-tinged sputum to mild (<5 ml in 24 h) to moderate (5-600 ml in 24 h) to severe (>600 ml in 24 h).
- Mortality is often a result of hypoxemia secondary to impaired gas exchange and also depends upon the underlying disease process (see Table 3E.1).

Etiology

- Neoplasm and TB are responsible for a significant number of cases but there are many causes of hemoptysis (see Table 3E.1).

Diagnosis

- History should include symptom acuity, and quality/quantity of expectorate, presence of associated symptoms (i.e., weight loss, fever, etc), past medical history, risk factors for pulmonary disease (i.e., cigarette smoking), and recent travel history.
- Patients present with varying degrees of respiratory and/or circulatory compromise depending upon the severity of bleeding and the underlying cause. In cases of massive hemorrhage, the patient may present with the affected side recumbent to prevent blood from filling the uninjured lung.

3

Table 3E.1. Etiology of hemoptysis

Infectious	Chronic bronchitis
	Tuberculosis
	Fungal and parasitic infections
	Necrotizing pneumonia
	Pulmonary abscess
Neoplasia	Bronchogenic carcinoma
	Pulmonary metastasis
	Bronchial adenoma
Cardiopulmonary	Mitral valve stenosis
Vascular	Pulmonary embolus
	Alveolar arteriovenous malformation
Other	Trauma
	Foreign body
	Bronchiectasis
	Wegener's granulomatosis
	Goodpasture's syndrome
	Systemic lupus erythematosus
	Coagulopathy and use of anticoagulant medications
	Idiopathic hemosiderosis

- Both the pulmonary and extrapulmonary exams help identify the cause of the bleeding. Pulmonary findings may include rhonchi, rales, decreased breath sounds, egophony, or a pleural rub. Extrapulmonary findings may include a diastolic murmur of mitral valve stenosis, supraclavicular adenopathy suggestive of cancer, or digital clubbing in patients with chronic lung disease. Also look for mucosal or cutaneous changes in patients with vasculitic pathology.
- Diagnostic Studies
  - CXR is indicated in all cases and often aids identification of the etiology.
  - Sputum examination—True hemoptysis is identifiable by its characteristic bright red appearance and alkaline pH. Hematemesis is usually darker, has an acidic pH, and may contain food particles. However, aspiration of gastric hemorrhage may create confusion. An AFB stain and culture is mandatory in all patients for whom TB is suspected.
  - Laboratory studies—CBC with differential is the most important and commonly ordered test. Others including PT, electrolytes, glucose, BUN, creatinine, and blood type and screen may be performed depending upon the patient's history and presentation.
  - An EKG should be obtained in patients with suspected valvular or congestive heart disease.
  - Specialized radiography such as computerized tomography (CT) and ventilation/perfusion (V/Q) scans are ordered as needed for suspected neoplasm, bronchiectasis or PE.
  - Bronchoscopy is the gold standard for diagnosis and allows for clot removal and retrieval of material for biopsy and culture. This is often not possible with severe, uncontrolled bleeding.

Treatment

- Management of the patient's airway, breathing and circulatory status are paramount. Supplemental oxygen as well as crystalloid and/or blood product administration should be administered as needed. Patients with respiratory failure or difficulty maintaining a patent airway mandate intubation. Orotracheal intubation with a large (≥8.0) endotracheal tube is preferred. This facilitates suctioning and allows for subsequent bronchoscopy.

- Temporizing Measures for Hemorrhage Control in those with Severe Bleeding
  - Bronchoscopic balloon tamponade by a pulmonologist
  - Selective bronchus intubation
    - If the bleeding source is the left lung, selective intubation of the right mainstem bronchus is accomplished by advancing the tube 4-5 cm beyond the usual position.
    - Intubation of the left mainstem bronchus is more difficult. Rotating the endotracheal tube 90 degrees counter-clockwise so the tube concavity faces the left during intubation is sometimes successful. If available, a double-lumen endotracheal tube can be used although there are often complications and most physicians have little to no experience with the product.
- Definitive Hemorrhage Control
  - Treatment should address any underlying condition such as infection, vasculitis, or coagulopathy.
  - Patients with moderate to severe bleeding warrant emergent evaluation by a pulmonary specialist for bronchoscopy. Arterial embolization by interventional radiology is an option for those with uncontrolled hemorrhage or when bronchoscopy is not possible or not successful.
  - Some disease processes are amenable to surgical therapy and a thoracic surgery consult is indicated if other modalities fail to control bleeding.

Disposition

- All patients with respiratory compromise or unstable hemodynamics should be admitted to an intensive care unit. There is a high incidence of recurrence in patients with self-limiting massive hemoptysis and these patients also require intensive care admission.
- Patients with suspected TB should be admitted and kept in respiratory isolation until appropriate testing is completed.
- Patients with minor, self-limiting hemoptysis can be considered for discharge. Outpatient treatment should address the underlying etiology. All discharged patients should follow-up with their primary care provider or a pulmonologist.

Massive Hemoptysis

Expectoration of blood from lower respiratory tract (systemic bronchial vessels and low pressure pulmonary vessels) >50 ml per episode or 600 ml/24 h. It may be differentiated from hematemesis and bleeding from a ENT source (such as epistaxis) during the course of resuscitation, which must proceed emergently in severe cases.

Primary Survey

Airway: Endotracheal intubation with RSI technique is indicated. A large diameter ET tube should be used (8.0 or larger if possible) to provide pulmonary toilet and facilitate bronchoscopy

The ET tube should be advanced to the mainstem bronchus of nonbleeding lung, if there is persistent bleeding. The right mainstem is easily entered, the left requires specialized technique and/or equipment.

Until the airway is secured with endotracheal intubation, personnel should take precautions against respiratory spread of tuberculosis.

Breathing: Both before and after intubation, the patient should be positioned with bleeding lung dependent to maximize gas exchange and minimize the filling of the unaffected side with blood.

3

Sedation and paralysis should be considered to prevent coughing and retching that may dislodge clot and worsen hemorrhage.

Circulation: IV fluid resusciation may be initiated with normal saline through large bore IV access, followed by emergent blood transfusion as needed. Blood type and crossmatch is critical.

Fresh frozen plasma and platelets should both be considered when there is suspected coagulopathy or severe thrombocytopenia.

Massive, uncontrolled hemoptysis may require a spectrum of emergent specialty consultation, including cardiothoracic surgery, interventional radiology and pulmonary medicine.

Disability: A cursory neurological examination should be sought prior to paralysis and endotracheal intubation so the need to image the head for intracranial pathology can be assessed.

Resuscitation Phase

Critical Questions: Other coexistent conditions that may require other critical actions in the setting of massive hemoptysis:

Conditions	**Actions**
Advanced malignancy	Consider level of intervention Seek advance directives, family conference
Pneumonia	Sputum cultures and IV antibiotics
Valvular heart lesion	Emergent cardiac surgery consultation

Critical investigations: These may also include:

Emergent bronchoscopy	To localize and treat source of bleeding
Emergency bronchial arteriography	
CT Chest	

Suggested Reading

1. Cahill BC, Ingbar DH. Massive hemoptysis: Assessment and management. Clin Chest Med 1994; 15(1):147.
2. Dweik RA, Stoller JK. Role of bronchoscopy in massive hemoptysis. Clin Chest Med 1999; 20(1):89.
3. Goldman JM. Hemoptysis: Emergency assessment and management. Emerg Med Clin North Am 1989; 7(2):325.
4. Jean-Baptiste E. Clinical assessment and management of massive hemoptysis. Crit Care Med 2000; 28(5):1642.
5. Marshall TJ, Flower CDR, Jackson JE. The role of radiology in the investigation and managment of patients with haemoptysis. Clin Radiol 1996; 51:391.

Part F: Pleural Effusions

- The pleural space normally contains a minimal amount of fluid. A pleural effusion is an excessive collection of fluid in the pleural space resulting from an underlying disease process (see Table 3F.1).
- Effusions may be transudative (resulting from changes in hydrostatic or oncotic pressure) or exudative (secondary to alterations in capillary permeability or lymphatic/vascular obstruction).

Diagnosis

- Patients with small effusions are often asymptomatic. Common complaints with symptomatic effusions are dyspnea, pleuritic chest pain, or cough. Patients may also have complaints related to their underlying disease or give a history of cancer, heart failure, or other comorbidity.

Table 3F.1. Causes of pleural effusions

Transudative	Congestive heart failure
	Nephrotic syndrome
	Renal failure
	Cirrhosis
	Pulmonary embolism
Exudative	Pulmonary infections
	Pulmonary embolism
	Malignancy (primary or metastatic)
	Drug induced effusion
	Connective tissue disease
	Trauma
	Subdiaphragmatic abscess
	Esophageal perforation
	Pancreatitis

3

- The physician should note any increased work of breathing or obvious respiratory distress. Pulmonary exam may also reveal decreased breath sounds, dullness to percussion, and decreased tactile fremitus. A friction rub is sometimes noted with medium-sized effusions. Extrapulmonary findings are present depending upon the etiology and include peripheral edema, jugular venous distension, ascites, abdominal tenderness, and lymphadenopathy among others.
- Diagnostic Studies
  - CXR—As little as 175 ml is visualized as a blunting of the costophrenic angle on a routine film. A lateral decubitus view can identify even smaller amounts of fluid. Subpulmonic effusions appear as an elevated hemidiaphragm.
  - Laboratory—Selected studies often include a CBC, electrolytes, BUN, creatinine, and glucose depending upon the suspected etiology. If a thoracentesis will be preformed, additional tests should include a serum protein and lactate dehydrogenase (LDH).
  - Thoracentesis
    - Classification of pleural effusions as a transudate includes a ratio of pleural fluid protein to serum protein <0.5, pleural fluid LDH <200 IU/ml, a ratio of pleural fluid LDH to serum LDH <0.6, fluid protein <3 g/100 ml, and fluid pH <1.016. Effusions that exceed these values are classified as exudates.
    - Other tests to consider for exudative pleural fluid are cell count and differential, pH, glucose, Gram stain, bacterial culture, and cytology. Consider amylase if pancreatitis or esophageal rupture is suspected.

Treatment

- Initial treatment includes oxygenation, and ventilatory and circulatory support if needed. Large effusions causing respiratory compromise require emergent drainage.
- Patients with effusions should have a diagnostic thoracentesis unless the etiology is apparent (heart failure, pneumonia, etc). It has been recommended that no more than 1,000-1,500 ml is drained at one time in order to prevent reexpansion pulmonary edema. This complication is rare and is minimized by the avoidance of excessive negative pressure.
- Specific treatments are based on the underlying cause of the effusion as determined by clinical presentation and diagnostic thoracentesis.
- Chest tube placement is required for empyema and hemothorax.

Disposition

- The need for hospital admission is based on the degree of respiratory or circulatory impairment, as well as the cause of the effusion. Most patients are admitted to the hospital following a thoracentesis for observation and treatment of the underlying condition.
- In a minority of cases, well-appearing patients can be discharged home after thoracentesis following 4-6 h of observation. All patients who have had a thoracentesis must have a post-procedure CXR to rule out complications such as pneumothorax or hemothorax.

Suggested Reading

1. Berkman N, Kramer MR. Diagnostic tests in pleural effusion—an update. Postgrad Med J 1993; 69:12-8.
2. Heffner JE. Evaluating diagnostic tests in the pleural space. Differentiating transudates from exudates as a model. Clin Chest Med 1998; 19:277-93.
3. Light RW. Useful tests on the pleural fluid in the management of patients with pleural effusions. Curr Opin Pulm Med 1999; 5:245-9.
4. Strange C. Pleural complications in the intensive care unit. Clin Chest Med. 1999; 20:317-27.
5. Ross DS. Thoracentesis. In: Roberts JR, Hedges JR, eds. Clinical Procedures in Emergency Medicine, 2nd ed. WB Saunders Company, 1991.

Part G: Pneumothorax

- A simple pneumothorax (PTX) is an accumulation of air in the pleural space and no communication with the atmosphere. It can occur spontaneously or as a result of trauma. Tension PTX occurs when air continues to enter the pleural space via a one-way communication with the atmosphere. This causes a collapse of the lung, shifting of the mediastinum away from the PTX, and compression of the mediastinal vessels. If untreated, the result is decreased venous return, hypotension, and death.
- Spontaneous PTX is seen in patients with (secondary) and without (primary) underlying pulmonary disease. Primary spontaneous PTX occurs more commonly in tall men 20-40 yr of age and has a high rate of recurrence. Secondary spontaneous PTX is usually associated with chronic obstructive lung disease as well as other pulmonary disease states such as infection, asthma, neoplasm, and occupational disease.

Diagnosis

- Presentation
  - The most common symptoms are dyspnea and acute onset of ipsilateral chest pain.
  - Depending on the size of the PTX and the patient's pulmonary reserve, varying degrees of respiratory and circulatory distress are noted. Pulmonary exam can reveal decreased breath sounds on the affected side, crepitus, and hyperresonance to percussion.
  - Signs of a tension PTX include tachycardia, hypotension, hypoxia, agitation, decreased breath sounds, and jugular venous distension. Tracheal deviation is a late physical finding.
- Evaluation
  - Diagnosis of a tension PTX is based on history and clinical presentation. Relying on CXR for the diagnosis can result in a fatal delay in treatment.
  - CXR can confirm the diagnosis when a simple PTX is suspected. CXR will reveal hyperlucency with a lack of lung markings at the periphery of the lung on the affected side. An expiratory or lateral decubitus film is helpful in identifying a small PTX when inspiratory CXR is not diagnostic.

Treatment

- Oxygen administration will increase the rate of resorption of the air from the pleural space as well as improve oxygen saturation.
- Definitive therapy is release of air from the pleural space.
  - Tension PTX is treated by immediate needle thoracostomy in the second or third intercostal space at the midclavicular line. This is followed by tube thoracostomy.
  - Traumatic PTX is treated by placing a large (36-40 French) thoracostomy tube in the fourth of fifth intercostal space at the anterior axillary line.
  - Primary, spontaneous PTX can be treated by simple aspiration with a 16-gauge needle through the second intercostal space at the mid-clavicular line, via a small catheter with a one-way valve (Heimlich), or with a small thoracostomy tube. Aspiration of the PTX is more likely to succeed when <20% of the involved lung is collapsed. For a very small PTX, with only a rim of air visible on CXR, observation and close follow-up is appropriate.

Disposition

- Patients with traumatic PTX require close monitoring as indicated by other injuries.
- Patients with a small PTX who have successful needle aspiration or one-way valve placement may be discharged home after a period of observation and post-procedure CXR. 24 h follow-up is recommended.

Suggested Reading

1. Light RW. Management of spontaneous pneumothorax. Ann Rev Respir Dis 1993; 148:245.
2. Miller AC, Harvey JE. Guidelines for the management of spontaneous pneumothorax. Br Med J 1993; 307:114.
3. Tanaka F, Masatosi I, Esaki J et al. Secondary spontaneous pneumothorax. Ann Thoracic Surg 1993; 44:372.

Part H: Pulmonary Embolism

Pulmonary embolism (PE) is the third most common acute cardiovascular disease after ischemic heart disease and stroke. It is a potentially fatal disorder that is often difficult to recognize and diagnose.

Risk Factors

The strongest risk factor for PE is prior thromboembolic disease including past PE and deep-venous thrombosis (DVT). PE is detected on perfusion imaging in a majority of patients with documented DVT even in the absence of clinical findings. The classic triad of stasis, hypercoaguability, and endothelial injury forms the basis for the many other causes (see Table 3H.1).

Diagnosis

- Clinical presentation: depends on the size of the clot and the degree of subsequent hemodynamic compromise. Signs and symptoms can be extremely subtle and non-specific (even non-existent) and a high degree of suspicion is often necessary to make the diagnosis.
  - The classic presentation is acute onset of sharp, pleuritic chest pain with associated dyspnea. Other symptoms include cough, non-pleuritic chest pain, reproducible chest pain, anxiety, syncope, and hemoptysis.
  - Physical findings may include cyanosis, tachypnea, tachycardia, hypotension, diaphoresis, fever, S3 or S4, or clinical signs of a lower extremity DVT.

3

Table 3H.1. Risk factors for pulmonary embolism

Causative Factor	Comment
Stasis	Immobility (bed rest, casting, air/car travel), paralysis, obesity, heart failure, varicose veins, myocardial infarction
Hypercoagulability	Prior thromboembolic disease, malignancy, inflammatory disease, nephrotic syndrome, sepsis
Hematologic Disorder	Protein C & S deficiencies, antiphospholipid antibodies, antithrombin III deficiency, polycythemia
Increased Estrogen	Pregnancy and <3 mo postpartum, oral contraceptive use
Endothelial Injury	Trauma, intravenous drug use, surgery, central venous catheters

- Evaluation
  - CXR is normal in only 30% of patients with PE. While the diagnosis is rarely made by CXR alone, this study can help exclude other diseases with a similar presentation such as PTX, pneumonia, or pulmonary edema. A wedge-shaped, pleural-based density that points to the hilum (Hampton's hump) and a prominent central pulmonary artery with decreased distal pulmonary vessels (Westermark's sign) are fairly specific radiographic findings for PE but are not commonly seen. Other nonspecific findings include an elevated hemidiaphragm, pleural effusion, or atelectasis.
  - EKG should be performed on all patients with suspected PE. The most common abnormalities are sinus tachycardia and nonspecific ST-segment and T-wave changes. Evidence of right heart strain, such as right bundle branch block, right axis deviation, or a right ventricular strain pattern are seen in a minority of patients with PE. The most specific, although insensitive, EKG finding is a prominent S-wave in lead I, with an inverted T-wave and prominent Q-wave in lead III (S1Q3T3 pattern).
  - ABG can lend support to the diagnosis. Specific findings include hypoxia, hypocapnia or an elevated alveolar-arterial (A-a) gradient. In the PIOPED study, 85% of the patients had a PO_2 <90 mm Hg, while 80% had an A-a gradient >20. This test cannot be used to exclude the diagnosis since 5-15% of patients with PE have a normal ABG.
  - D-dimer assays have been suggested to have diagnostic utility. Unfortunately, these assays lack specificity, and there is a large range of sensitivity depending on the assay used (69-100%). The most sensitive is the ELISA assay. A normal D-dimer by ELISA assay decreases the likelihood that a patient has a PE, but by itself cannot exclude the diagnosis.
  - V/Q scan has classically been the diagnostic study of choice although helical CT is now being used in many centers. The results of V/Q scanning are reported as normal, low probability, intermediate probability, and high probability. Clinical interpretation of results depends on the degree of clinical suspicion for PE. A normal or scan in combination with a low clinical suspicion makes the diagnosis of PE unlikely. A high probability scan in combination with a high clinical suspicion makes the diagnosis of PE very likely although not definite. All other combinations of V/Q scan results and clinical suspicion are not definite enough to either rule in or out PE and additional diagnostic testing is indicated.
  - Helical CT scan can also be used to identify a PE. This scan may be useful in patients with significantly abnormal CXR when there is a high likelihood that the V/Q scan will be nondiagnostic. Although in most studies this test is very sensitive for diagnosing PE in the larger pulmonary vessels, it has been shown to lack sensi-

Table H2. Fibrinolytic dosing for pulmonary embolism

Urokinase	4400 IU/kg over 10 min then 4400 IU/kg/h for 12 h
Streptokinase	250,000 IU over 30 min then 100,000 IU/h for 24 h
Tissue Plasminogen Activator	100 mg over 2 h

tivity for emboli in the smaller, subsegmental branches. Therefore a negative CT scan does not rule out the diagnosis.

- Lower extremity venous studies may aid in the diagnosis for patients with intermediate probability V/Q scans. A lower extremity DVT is present 50-70% of all patients with a proven PE. Evaluation for DVT can be done by impedance plethysmography, doppler ultrasound, or venogram. The gold standard for diagnosis of a DVT is the venogram; however this is an invasive procedure and is technically difficult. The duplex ultrasound is the most commonly used tool for the diagnosis of a DVT. It has 93% sensitivity and 98% specificity for diagnosis of proximal DVT. The sensitivity is much lower for calf DVTs (60%). Note that a negative venous study does not rule out PE—the tests are only helpful if positive.
- The gold standard for diagnosis of PE is the pulmonary angiogram although even angiography can miss small, distal emboli. It is an invasive procedure with a low risk of mortality. However, in most institutions this study is not available 24 h a day. It is generally used only to confirm the diagnosis in patients with nondiagnostic V/Q scans or when results of V/Q scanning does not correlate with clinical suspicion.
- Echocardiography is useful when evaluating a hemodynamically unstable patient with suspected PE. This modality can help diagnose other potential etiologies and can identify changes consistent with PE such as right ventricular enlargement, pulmonary artery dilatation, and tricuspid regurgitation.

- Treatment
  - Support of airway, breathing, and circulation is the initial goal of therapy.
  - Treatment of the stable patient consists of anticoagulation. This should be started prior to final diagnosis when there is a high degree of clinical suspicion for PE. There are two options for anticoagulation. Low molecular weight heparin (LMWH) has been shown to effective in patients with PE. The most commonly used agent is enoxaparin 1 mg/kg SQ every 12 h. The alternative is standard heparin administered as an initial bolus of 80 units/kg followed by an intravenous drip of 18 units/kg/h. Relative contraindications to anticoagulation include recent stroke or major surgery, advanced liver or kidney failure, or bleeding diathesis.
  - Fibrinolytics should be considered in hemodynamically unstable patients. The dosing differs from the protocols used for myocardial infarction (see Table 3H.2).
  - Surgical embolectomy is the final option for hemodynamically unstable patients who have failed medical therapy or who have a contraindication to fibrinolytics.

Disposition

All patients with suspected PE should be admitted to a telemetry bed for monitoring and anticoagulation. Intensive care unit admission is necessary for patients with hemodynamic compromise.

3

Suggested Reading

1. Value of the ventilation/perfusion scan in acute pulmonary embolism. Results of the prospective investigation of pulmonary embolism diagnosis (PIOPED). The PIOPED Investigators. JAMA 1990; 263:2753-9.
2. Curtin JJ, Mewissen MW, Crain MR, Lipchik RJ. Postcontrast CT in the diagnosis and assessment of response to thrombolysis in massive pulmonary embolism. J Comput Assist Tomogr 1994; 18:133-5.
3. Fisman DN, Malcolm ID, Ward ME. Echocardiographic detection of pulmonary embolism in transit: Implications for institution of thrombolytic therapy. Can J Cardiol 1997; 13:685-7.
4. Goldhaber SZ. Treatment of pulmonary thromboembolism. Intern Med 1999; 38:620-5.
5. Kline JA, Johns KL, Colucciello SA, Israel EG. New diagnostic tests for pulmonary embolism. Ann Emerg Med 2000; 35:168-80.
6. Perrier A, Desmarais S, Miron MJ et al. Non-invasive diagnosis of venous thromboembolism in outpatients. Lancet 1999; 353:190-5.
7. Rathbun SW, Raskob GE, Whitsett TL. Sensitivity and specificity of helical computed tomography in the diagnosis of pulmonary embolism: A systematic review. Ann Intern Med 2000; 132:227-32.

CHAPTER 4

Neurologic Emergencies

Jacquelyn Hasler Salas

Part A: Headache

Anatomy

- Headache, or cephalgia, is defined as pain in various parts of the head, not confined to the area of distribution of any nerve.
- Headache is caused by distention, traction, displacement, inflammation, vascular spasm, dilation, or compression of the pain-sensitive structures in the head and neck. The pain-sensitive structures of the supratentorial space refer pain via the trigeminal nerve, which innervates the anterior scalp and face. Pain-sensitive structures in the infratentorial space refer pain via cranial nerves IX, and X, and the second and third cervical nerves; thus, pain originating in the posterior fossa may be referred to the ear or throat, or the posterior area of the head and neck.
- Intracranial sources of head pain include the cranial sinuses and afferent veins; the anterior and middle meningeal arteries; the trigeminal (V), glossopharyngeal (IX), and vagus (X) nerves; falx cerebri; the dura at the base of the skull; the major arteries at the base of the brain; the brain stem periaqueductal gray matter; and the sensory nuclei of the thalamus.
- Muscles frequently involved in extracranial causes of cephalgia include the masseter, frontalis, temporal, occipital, trapezius, sternocleidomastoid, and deep cervical muscles. In addition, the skin, the periosteum of the skull, the subcutaneous tissues and arteries, the eyes, ears, teeth, sinuses, oropharynx, and the mucous membranes of the nasal cavity may be sources of pain.

Etiology/Risk Factors

Primary headaches are benign, usually recurrent, and have no underlying cause. These include migraine, cluster, and tension-type headaches. Secondary headaches, on the other hand, are a symptom of underlying organic disease. Risk factors for secondary headache disorders include:

- "First or worst" headache
- Very sudden onset
- Nausea and vomiting
- Systemic illness
- Ocular findings
- History of head trauma
- History of immunodeficiency or cancer
- Increased frequency or severity
- Focal neurologic deficits, or altered mental status
- Onset after age 50, or before 3
- Fever, neck stiffness, or meningeal signs
- Headache that begins with exertion or is positional

Emergency Medicine, edited by Sean Henderson. ©2006 Landes Bioscience.

4

Diagnosis

History

- The most important task in evaluating a patient with headache is to identify or exclude underlying pathology based on the history and physical examination. A detailed history should be taken to include the following elements:
- Was the onset sudden, gradual, or subacute? Was the patient awakened from sleep? Are there any precipitating or aggravating factors (e.g., activity, stress, menses/hormonal therapy, medications/medication withdrawal, foods, cough/Valsalva maneuver, environmental exposures, position changes, trauma)? Episodes of migraine are often concentrated around the menstrual period. Exposure to tyramine- or amine-containing foods, nitrates, MSG, or ethanol may precipitate a migraine; stress, weather changes, changes in sleep patterns, and caffeine withdrawal are also potential triggers. Tension-type headaches are often stress-related. Alcohol is reported to precipitate cluster headaches. The use of cocaine is a risk factor for subarachnoid hemorrhage (SAH). Are there any alleviating factors?
- Is there a prodrome (e.g., visual, auditory, or olfactory aura or hallucinations; numbness, paresthesias or motor weakness; speech impairment)? An aura may precede a migraine headache by up to an hour; patients without aura may have other symptoms suggesting the onset of a migraine, including lethargy, depression, hyperactivity, or food craving.
- Where is the pain located (e.g., unilateral or bilateral; ocular/retro-ocular; paranasal; frontal or occipital; at the vertex; in the pharynx or external auditory meatus)? What is the character of the pain (e.g., sharp, stabbing; dull, steady ache; burning; lancinating; "worst headache ever")? Tension-type headaches are described as a diffuse pressure or tightness. Migraines are typically unilateral, with a pulsating quality. Cluster headaches are excruciating, sharp, unilateral headaches. Does the pain radiate?
- Are there any associated symptoms (e.g., fever/chills; nausea/vomiting; neck pain or stiffness; seizures; focal neurologic deficits; ataxia; speech deficit; dizziness/vertigo; visual changes or eye pain; altered mental status or transient loss of consciousness; jaw claudication; myalgias; weight loss)? Migraine headaches are frequently associated with photophobia or phonophobia. Symptoms accompanying a cluster headache include conjunctival injection, lacrimation, rhinorrhea, ptosis, myosis, and ipsilateral forehead sweating. Suspicion should be raised for SAH when nausea and vomiting, photophobia, neck stiffness, or loss of consciousness accompanies the headache.
- What is the frequency of the pain (e.g., intermittent, chronic, seasonal)? Are the headaches increasing in frequency or severity? How long does the pain last? Cluster headaches occur once or twice daily—generally at the same time each day—and last about 30 to 90 min. The symptoms may recur for several weeks, and then the patient may remain pain-free for months or years. Headaches that persist for more than 10 wk, without associated symptoms, are unlikely to be caused by a neoplasm.
- Is there a family history of headaches? Are there any ongoing medical problems or recent illnesses? The presence of polycystic kidney disease, Ehlers-Danlos or Marfan's syndrome, Grave's disease, fibromuscular dysplasia, coarctation of the aorta or abdominal aortic aneurysm, sickle cell disease, atherosclerosis, or hypertension places the patient at increased risk for SAH or unruptured aneurysm. Is there an HIV history or risk? Are there any close contacts with similar symptoms? Patients with mild carbon monoxide poisoning may complain of nonspecific headache and flu-like symptoms; frequently, members of the same household will have the same toxic exposure and thus present with similar symptoms.

Vital Signs

- Vital signs may reveal an elevated temperature or blood pressure; tachycardia; or tachypnea.

Physical Exam

- A detailed examination of the head may reveal tenderness of the scalp, temporomandibular joint (TMJ), temporal artery, or sinuses; evidence of head trauma; or disorders of the eyes, ears, nose, or teeth.
- Examine the neck for bruits, range of motion, and tenderness.
- Examination of the skin may reveal a focal cellulitis, a generalized rash, neurofibromas or café au lait spots, or cutaneous angiomas.
- A complete neurologic exam is essential, including level of consciousness, mental status, pupillary responses, cranial nerves, deep tendon reflexes, motor and sensory function, gait, cerebellar function, and pathologic reflexes.

Differential Diagnosis

Primary Headache

- Migraine
- Tension-type
- Cluster headache

Secondary Headache

- Associated with head trauma
  - Post-traumatic headache syndromes
- Associated with vascular disorders
  - Infarction, transient ischemic attack (TIA), SAH, unruptured vascular malformation, arteritis, carotid or vertebral artery pain (arterial dissection, carotidynia), intracranial hematoma, venous thrombosis, acute arterial hypertension
- Associated with nonvascular intracranial disorders
  - High and low CSF pressures, infection, intracranial mass, sarcoidosis, other granulomatous/inflammatory diseases
- Associated with exposure or use of substances or their withdrawal
- Associated with noncephalic infection
- Associated with metabolic disorders
  - Hypoxia, hypercapnia, dialysis, hypoglycemia, hypo- or hyperthyroidism, hypoadrenalism
- Associated with facial or cranial structures
  - Skull, neck, eyes, ears, nose, sinuses, teeth and jaws, mouth, TMJ joint, or other facial or cranial structures
    - Paget's disease, skull metastases, cervical arthritis, acute angle closure glaucoma, retro-orbital infection, sinusitis, dental caries
- Cranial neuralgias, nerve trunk pain, and deafferentation pain
  - Optic neuritis, zoster (post-herpetic neuralgia), trigeminal neuralgia, glossopharyngeal neuralgia, occipital neuralgia, pain of cranial nerve origin
- Other causes of headache pain
  - Idiopathic stabbing headache, external compression headache, cold stimulus headache, benign exertional headache, benign cough headache, associated with sexual activity
  - High-altitude, anemia, hypotension

Evaluation

Laboratory

- A CBC may be useful in cases of suspected infection, hematologic disorders, or vasculitis.

- A CD4 count of <500 or 200 in HIV-infected individuals increases the risk for:
  - Meningitis (cryptococcal, tuberculous, syphilitic, and lymphomatous);
  - Focal brain lesions (toxoplasmosis, CNS lymphoma, PML, abscess, crypotococcoma)
  - Diffuse brain lesions (CMV, HSV, toxoplasmosis).
- An erythrocyte sedimentation rate (ESR) of at least 55 mm/h—and usually over 100—is seen in 90% of patients with temporal arteritis.
- A carboxyhemoglobin level is obtained if carbon monoxide poisoning is suspected.
- A lumbar puncture with analysis of the cerebrospinal fluid (CSF) is essential in the evaluation of several headache disorders:
  - Meningitis—CSF may be cloudy, with elevated WBC and protein, and low glucose. Specimens should also be sent for stat Gram stain and culture, and—in specific cases—VDRL, India ink, bacterial antigens, and CSF antibody detection (for viral encephalitis).
  - Subarachnoid hemorrhage—opening pressure may be increased; fluid may be bloody or xanthochromic (discolored supernatant of centrifuged CSF, as a result of hemolysis), with increased RBC and protein, and normal glucose. (In the setting of a negative head CT, LP must still be performed to exclude small SAH not identified by CT.)
  - Benign intracranial hypertension—opening pressure is elevated (above 200 mm H_2O); other CSF components are normal; relief from headache with removal of fluid may be diagnostic.

Imaging

- A panorex radiograph may reveal a dental etiology in selected patients.
- A cervical spine series may be warranted in the setting of trauma or suspicion of cervical arthritis.
- Emergent neuroimaging is performed in order to identify treatable lesions (e.g., tumors, AVMs, SAH, cerebral sinus thrombosis, subdural and epidural hematomas, and hydrocephalus). Computed tomography (CT) scanning of the head without contrast is indicated when certain historical or physical "red flags" are identified in the evaluation of a new-onset headache:
  - Acute onset, severe headache
  - Any neurologic deficit
  - History of seizures
  - Ocular abnormalities including papilledema, visual impairment, or diplopia
  - Persistent or frequent vomiting preceded by recurrent headaches
  - Changing character of the patient's headache
  - Extremes of age (patients <3 or over 50)
  - Children with neurofibromatosis
  - Diabetes insipidus, HIV
  - Antecedent head trauma
- Patients with signs of increased intracranial pressure (e.g., papilledema or absent venous pulsations on funduscopic exam, altered mental status, or focal neurologic deficits) should be scanned prior to having a lumbar puncture.
- Contrast-enhanced head CT should be considered in patients with a history of immunodeficiency or malignancy (after a negative noncontrast CT).
- Magnetic resonance imaging (MRI) of the brain may be preferable to CT in cases in which the posterior fossa must be specifically evaluated but is generally not as readily available as CT.

Treatment

Primary Headache

- Migraine

Table 4A.1. Primary headache treatment options

Medication	Dose	
Acetaminophen	325 mg tablets, up to 6 PO	
Ibuprofen	200 mg tablets, 1-4 PO	
Naproxen sodium	275 mg tablets, 2-3 PO at onset; may repeat 1-2 tabs in 2 h	
Ketorolac	15-30 mg IV/IM	
Sumatriptan	6 mg SQ; may repeat in 1 h (maximum 2 doses/day)	
Prochlorperazine	5-10 mg slow IV push; may be repeated in 30-60 min	
Metoclopramide	10 mg PO/IM/IV	
Promethazine (Peds)	0.25 mg/kg	6-9 yr, max 25 mg
		>9 yr, max 50 mg
DHE	0.5-1 mg IV/IM; may repeat in 1 h, up to 3 doses in 24 h	
Pediatric dosing:	0.1-0.15 mg IM (6-9 yr); 0.2 mg IM (9-12 yr); 0.25-0.5 mg IM (12-16 yr)	
Prednisone	80 mg PO, rapid taper over 1 wk	
Dexamethasone	10 mg IV followed by 4 mg every 6 h; or 20 mg PO, rapid taper	

- Patients with new onset migraine may respond to simple analgesics such as acetaminophen or oral nonsteroidal anti-inflammatory drugs (NSAIDs). A maximal dose should be given for prompt relief. Parenteral NSAIDs (ketorolac) may be useful in patients who present later in the course of their symptoms or have nausea or vomiting.
- Serotonin (5-HT) receptors have been shown to modulate neurogenic peptide release and vasoconstrict dilated dural vessels. As a result they are the main focus of pain management. Many medications traditionally employed in the management of migraine headaches—dihydroergotamine (DHE), prochlorperazine, and metaclopramide—act at a variety of serotonin and other aminergic receptors. Metoclopramide and prochlorperazine also act as dopamine antagonists and are often successful in relieving both the pain and the nausea associated with migraine. Side effects include orthostatic hypotension, akathesia, and dystonic reactions.
- Ergotamine derivatives are potent vasoconstrictors and are contraindicated in pregnancy, renal or hepatic disease, uncontrolled hypertension, and in patients with known or suspected coronary artery disease or Prinzmetal's angina. In addition, they tend to cause nausea and are generally given in combination with an antiemetic.
- The triptans—selective 5-HT_1 receptor agonists—can be effective in reducing or eliminating migraine pain without significant sedation. Side effects include chest, neck, and/or throat tightness, heaviness, pressure, or pain; paresthesias, and flushing. They have the same contraindications as DHE; in addition, they should not be used in patients suffering from basilar or hemiplegic migraine.
- Corticosteroids may be useful, especially in aborting status migrainosus.
- Narcotic agents should be used only when other medications are contraindicated or ineffective. Rather than terminating the headache, they merely dull the pain. They are associated with nausea, sedation, and orthostatic hypotension.

- Tension-Type Headache
  - Tension-type headaches generally respond to NSAIDs. Other medications used to treat migraine headaches are generally effective.
- Cluster Headaches
  - The same medications used to treat migraine headaches are effective for cluster headaches. In addition, acute cluster attacks frequently respond to inhalation of 100% oxygen by nonrebreathing face mask over 10 to 15 min. For refractory pain, intranasal 4% lidocaine or dexamethasone (8 mg/day for 3-4 days) should be

tried. Patients who suffer from chronic cluster without remissions or episodic bouts lasting more than a few weeks may benefit from a 7- to 10-day course of prednisone (60 to 80 mg/day), with a tapering dose the following week.

Secondary Headache

- Patients treated with any of the previously mentioned medications may experience relief from their headache, despite having significant intracranial pathology as a cause of their pain. Thus, response to therapy should not be used to distinguish primary from secondary causes of headache.
- Treatment of secondary headache varies by etiology. Treatment of primarily neurologic diagnoses will be described here. Management of neurosurgical, infectious, and other causes is explained elsewhere.

- Giant-Cell Arteritis
  - Giant-cell arteritis (temporal arteritis) is typically seen in patients over age 50 (with a peak incidence in the 70s). The headache is usually in the temporal region (but may occur anywhere) and variously described as continuous or intermittent, throbbing or steady, boring, or aching. Associated symptoms include jaw claudication (virtually pathognomonic, when present), amaurosis, malaise, anorexia, weight loss, myalgias, arthralgias, fever, neuropathies, TIAs, and stroke. Patients often complain of scalp tenderness, and examination may reveal a tender, indurated, warm, temporal artery with reduced or absent pulse. If the diagnosis of temporal arteritis is suspected, treatment with prednisone (1 mg/kg) should be initiated; improvement in headache should be observed within 48 h. Intravenous methylprednisolone, 250 mg q6h, is recommended for patients with associated visual loss. Temporal artery biopsy should be accomplished within 72 h of initiating treatment; however, immediate treatment with steroids may prevent complications and should not be delayed pending confirmation of the diagnosis.
- Benign Intracranial Hypertension
  - Benign intracranial hypertension (pseudotumor cerebri) refers to a diffuse increase in intracranial pressure, producing a diffuse headache, papilledema, and visual changes. MRI or CT scanning reveals slit-like ventricles. Acetazolamide, with or without a diuretic, may be sufficient for mild cases. Otherwise, prednisone may be indicated. Refractory cases may require intermittent lumbar puncture or lumboperitoneal shunting. Maintaining intracranial pressure at relatively normal levels helps to protect against irreversible vision loss. The disease process is generally self-limited and resolves within several months.
- Trigeminal Neuralgia
  - Trigeminal neuralgia is described as unilateral jabs of lightning-like pain confined to the areas of the face supplied by the second and third divisions of the trigeminal nerve. Attacks may be precipitated by tactile or mechanical stimulation (e.g., brushing the teeth), and last seconds to minutes. Oral carbamazepine is highly effective for remission of symptoms (often within 24 h) but cannot be loaded; the initial dose is 100 mg PO BID, increased every 2 days by 100-mg increments (max. dose, 1.2 to 2 g/day). An acute attack may be aborted with intravenous phenytoin, 250 mg, and the patient discharged on 300 mg PO at bedtime. Alternatively, parenteral narcotics may be required.
- Post-Traumatic Headache
  - Patients may experience headache within hours to days following trauma—in the setting of a normal neurologic exam and CT scan—which may last for several weeks. When post-traumatic headache is associated with other nonspecific symptoms (e.g., dizziness, vertigo, nausea and vomiting, difficulty with concentration, and labile mood swings), the symptom complex is referred to as post-traumatic headache

syndrome. NSAIDs are generally used as first-line therapy, with the addition of sedatives, physiotherapy, psychotherapy, and—as a last resort—narcotics, if necessary. Symptoms generally resolve within months.

- Post-Lumbar Puncture Headache
  - Post-lumbar puncture headache is caused by leakage of CSF through the defect in the dura caused by the spinal needle. The pain is aggravated by sitting or standing and relieved with recumbency. Associated symptoms include orthostatic lightheadedness, tinnitus, photophobia, anorexia, nausea, and vomiting. The incidence of post-lumbar puncture headache is reduced by use of a small-gauge spinal needle, as well as minimizing the amount of CSF removed; however, bed rest and increased fluid intake immediately following the procedure have not been shown to influence the course or duration of symptoms. Treatment includes rest and analgesics, and symptoms are generally self-limited. For more severe cases, caffeine (500 mg in 1 L of normal saline administered over 1 h) may be effective. The dose may be repeated once, if necessary. Finally, in persistent cases, consultation with anesthesiology for an autologous epidural "blood patch" may be warranted.

Disposition

- Most patients who present to the ED with headache can be discharged home after adequate pain relief has been achieved.
- Patients with status migrainosus—or any patient whose pain is not adequately controlled in the ED—should be admitted.
- Patients with documented visual loss and suspected temporal arteritis should be admitted for intravenous steroid therapy.
- Any patient with new-onset headache should have close follow-up with his/her primary care provider (PCP) or the appropriate specialist. Patients with migraine or cluster headaches should be encouraged to follow-up with their PCPs for consideration of prophylaxis.
- All patients should receive instructions on appropriate use of discharge medications and explicit return precautions.

Part B: Motor Deficits

Anatomy

- Motor deficits can be caused by CNS or peripheral nervous system dysfunction.
- The motor unit is composed of an anterior horn cell, its motor axon, and muscle fibers. The motor nerve fibers and the muscle fibers make up the presyntaptic and postsynaptic components of the neuromuscular junction, respectively.
- Muscle contraction involves an action potential at the motor axon, causing an influx of calcium that releases acetylcholine into the synaptic cleft; this results in an action potential at the motor end plate, and subsequent depolarization of the postsynaptic membrane and contraction of the muscle cell.

Scope of the Problem

- Based on the anatomy of the neuromuscular system, peripheral nervous system causes of motor weakness correlate with involvement at 1 of 4 levels: the anterior horn cells, peripheral nerve, neuromuscular junction, or the muscle fiber itself.
- Progressive motor weakness may be accompanied by sensory and autonomic dysfunction, requiring rapid respiratory or hemodynamic stabilization.

Etiology

The following list is not exhaustive, but includes most etiologies encountered on an emergency basis. The patient may have a known neurologic disease and present

4

with an exacerbation or deterioration, or may have new-onset symptoms without a prior diagnosis.

- CNS
  - Multiple sclerosis (MS)
- Motor Neuron
  - Amyotrophic lateral sclerosis (ALS)
  - Poliomyelitis
- Peripheral Nerve
  - Guillain-Barre syndrome (GBS)
  - Tick paralysis
  - Porphyric polyneuropathy
  - Arsenic poisoning
  - Paralytic shellfish poisoning
  - Hypophosphatemia
- Neuromuscular Junction
  - Myasthenia gravis (MG)
  - Botulism
  - Lambert-Eaton (LEMS)
  - Hypermagnesemia
- Muscle
  - Myoglobinuric myopathy
  - Hypo- or hyperkalemia
  - Toxic myopathy
  - Dermato- and polymyositis
- Guillain-Barre syndrome is the most common form of acute inflammatory demyelinating polyradiculoneuropathy (AIDP). The most frequent impairment results from immunologic reaction against nerve roots, peripheral nerves, and cranial nerves. The typical clinical manifestation is motor weakness beginning in the legs and ascending to the arms, with symptoms evolving over a few days. Approximately 14% of patients will present with symptoms beginning in the cranial nerves or arms, descending to the legs.
- Myasthenia gravis (MG) is characterized by the formation of antibodies to the acetycholine receptor of the post-synaptic component of the neuromuscular junction. Weakness and fatiguability are the predominant symptoms, with a predilection for the ocular muscles. The proximal muscles of the limbs are affected more than distal muscles. Oropharyngeal muscle involvement may impair speech, swallowing, or chewing. Respiratory muscles (i.e., intercostals and diaphragm) are affected in one-third of patients, leading to respiratory failure.
- Multiple sclerosis (MS) is a demyelinating disease of the axons in the CNS. Clinical symptoms wax and wane and are attributed to discrete lesions in the CNS which are "scattered in time and space." In the presence of demyelinated axons, action potentials are not conducted normally: edema and inflammation may impair action potential propagation leading to "negative" symptoms (e.g., diplopia from CN VI paresis), while hyperexcitable demyelinated axons may produce "positive" symptoms (e.g., Lhermitte's sign).
- Idiopathic dysfunction of both upper and lower motor neurons (in the anterior horn cells) contributes to the motor weakness seen in patients with amyotrophic lateral sclerosis (ALS).

Risk Factors

- Two-thirds of patients with GBS recall an antecedent event, approximately 1 to 3 weeks prior to symptom onset. The most common is infectious (e.g., URI, flulike symptoms, or diarrhea), but immunizations or surgery may predate the disorder.

- Myasthenic crisis (i.e., exacerbation requiring ventilatory support) is often precipitated by infection, changes in medications, pregnancy (or just before menses), or surgery.
- Lambert-Eaton Myasthenic syndrome (LEMS) is an autoimmune disorder affecting presynaptic nerve terminals. Like myasthenia gravis, the characteristic weakness, fatiguability, and pain primarily affects proximal muscles; unlike myasthenia gravis, cranial nerves are only mildly impaired. An underlying malignancy (usually squamous cell carcinoma of the lung) is present in 75% of male and 25% of female patients.
- First-degree relatives of patients with MS have a 15- to 20-fold greater risk of developing the disease. Hyperthermia (resulting from underlying infection) may exacerbate previous neurologic deficits, or precipitate new symptoms.

Diagnosis

History

- Determine the time course and the distribution of symptoms.
  - Toxic or metabolic disturbances may have an abrupt onset; neoplastic, infective, or inflammatory disorders may progress over days to weeks; hereditary, endocrinologic, degenerative, or other neoplastic processes may cause symptoms which develop over a period of months to years.
  - Proximal muscle weakness causes difficulty getting up from a squatting position, climbing or descending stairs, or washing or brushing the hair. Distal weakness of the upper limbs may manifest as clumsiness or loss of fine motor skills (e.g., tying shoelaces or buttoning).
  - Patients with cranial nerve involvement may present with diplopia, dysarthria, or impaired chewing or swallowing, with nasal regurgitation.
- Does the patient have a preexisting neuromuscular or systemic disorder? Have there been any recent infections or immunizations?
  - Patients with dystrophies may present with clinical deterioration; likewise, ALS symptoms tend to worsen with concurrent illness (especially pneumonia). In addition, patients with neuromuscular junction disorders (e.g., myasthenia gravis) suffer exacerbations with illness or addition of pharmacologic agents. A history of similar symptoms suggests familial periodic paralysis or myoglobinuria. Finally, patients with connective tissue disorders are at risk for vasculitic neuropathy.
- The patient's medications, social history, and recent diet should be ascertained.
  - Ingestion of shellfish containing saxitoxin—specifically, mussels and clams from both U.S. coasts during the summer months—may produce sensory deficits and ascending paralysis within 30 min. Home-canned goods may contain botulinum toxin. Symptom onset temporally related to new medications (e.g., oral contraceptives, anti-epileptic drugs) may suggest porphyria. Patients on diuretics may become hypokalemic. In the setting of renal insufficiency, a patient with a neuromuscular junction disorder may experience increasing weakness after ingesting magnesium-containing antacids. Myotoxicity has been attributed to several medications, including cholesterol-lowering agents, colchicine, chloroquine, cyclosporine, and L-tryptophan.
- Determine whether the patient has had any known exposure to toxins, chemicals, solvents, or tick bites (e.g., in wooded areas).
- Are there any associated symptoms?
  - Patients with GBS commonly report dysesthesias in the hands and feet prior to the onset of weakness. They may also experience transient bladder paralysis (resulting in urinary retention) and paralytic ileus. Abdominal pain and mental status changes in the presence of motor weakness suggests porphyric polyneuropathy. Arsenic poisoning causes an encephalopathy, in addition to systemic signs and symptoms. The

pure motor weakness attributed to polymyelitis is accompanied by fever and meningeal signs.
- Is there a family history of neuromuscular disease?

Vital Signs

- Patients with sensory GBS are at increased risk for autonomic nervous system instability, manifesting as arrhythmias, orthostatic hypotension, and hypertension.

Physical Exam

- Note the patient's general appearance. Confusion, in a tachypneic patient using accessory muscles of respiration, suggests impending neuromuscular respiratory failure.
- Neck: The strength of the patient's neck flexors (and the trapezius, responsible for shoulder elevation) parallels that of the diaphragm.
- Chest exam may reveal clear breath sounds in a patient in respiratory distress from motor weakness. A patient's vital capacity can be grossly measured by asking him to count as high as possible with one breath (normal >50).
- Paradoxical abdominal movements indicate diaphragm weakness.
- Neurologic Exam
  - Upper motor neuron lesions are associated with weakness (or paralysis), spasticity, increased deep tendon reflexes (DTRs), and a Babinski response. Lower motor neuron lesions are characterized by weakness (or paralysis), hypotonia, loss of DTRs, and normal plantar reflexes. Disorders of neuromuscular transmission exhibit normal or reduced muscle tone; normal or diminished DTRs; and waxing and waning weakness in a patchy distribution (i.e., not attributed to a single, discrete lesion) with cranial nerve involvement; there are no sensory deficits. Myopathic processes typically cause proximal >distal weakness; DTRs are normal until late in the course; there is no impairment in sensation or sphincter function.
  - Patients with GBS generally have symmetric limb weakness, and decreased or absent deep tendon reflexes. Despite subjective paresthesias, there is minimal objective sensory loss.
  - Patients with myasthenia gravis who have ocular involvement have ptosis and extraocular muscle weakness. Fatiguability can be measured by asking the patient to look upwards for 2 min or by repeatedly testing the proximal muscles (e.g., deltoids and iliopsoas).
  - In contrast to myasthenia gravis, patients with LEMS show improvement with repeated testing. The extraocular muscles are spared.

Evaluation

- Forced vital capacity is recommended to determine a patient's respiratory status.
- Laboratory
  - Arterial blood gas is helpful in determining ventilatory status.
  - Urinalysis, BUN, and creatinine may reveal myglobin and renal insufficiency in the clinical setting of rhabdomyolysis. Likewise, the creatine kinase (CK) may be elevated in myopathies.
  - Electrolytes, including calcium, phosphorus, and magnesium should be obtained.
  - Patients with exacerbation of myasthenia gravis or MS should have a urinalysis and urine culture, as well as a CBC.
  - Lumbar puncture (generally not required in the ED) may be helpful in the diagnosis of GBS. The CSF may be normal in the first 48 h after symptom onset; however, within 1 wk, elevated CSF protein is noted (without pleocytosis).
  - EKG
    - EKG may reveal evidence of hypo- or hyperkalemia or an arrhythmia resulting from the autonomic dysfunction associated with GBS.

- Radiography
  - Chest radiography may show evidence of pneumonia, atelectasis, an elevated hemidiaphragm (resulting from weakness), or a malignancy. Occult respiratory infection should be sought in any patient with MG or MS exacerbation.
  - Consider head CT in patients with presumed myasthenia gravis (especially patients with ocular symptoms) to exclude an intracranial mass producing similar symptoms.
- Tensilon test
  - Edrophonium (Tensilon) is a short-acting acetylcholinesterase inhibitor; patients with myasthenia gravis generally have transient improvement in their motor strength within minutes.
  - Cholinergic side effects of edrophonium include excessive salivation, bradycardia, and nausea. Cautious use (with atropine at the bedside) is warranted in patients with heart or lung disease.
  - An initial test dose of 2 mg is given intravenously. If the patient tolerates the medication but shows no improvement over 1 min, an 8 mg dose is given and the patient is observed for the next 3 to 5 min for a response. Patients in myasthenic crisis will show improvement, while patients in cholinergic crisis will be worse.

Treatment

A rapid assessment of the patient's airway, breathing, and circulation is critical in patients presenting with motor weakness.

- Specific Treatment
  - Guillain-Barre syndrome
    - Hypotension generally responds to isotonic fluids. Hypertension should only be treated if severe and persistent. Short-acting, α-adrenergic blockers are recommended.
    - Plasmapheresis and intravenous immune globulin (IVIG) have been shown to be equally effective, and both are superior to supportive care alone.
  - Myasthenia gravis
    - Patients with myasthenia gravis may present in cholinergic crisis; in response to increasing weakness, the patient increases his anticholinesterase medications, which worsens the weakness. The presence of cholinergic signs—pallor, miosis, sweating, nausea/vomiting and diarrhea, salivation, and bradycardia—helps to distinguish cholinergic crisis from myasthenic crisis.
    - Patients with mild exacerbation of MG (i.e., no respiratory or oropharyngeal compromise) may be treated with prednisone, pyridostigmine, and an eye patch (for diplopia).
    - Moderate or severe MG is an indication for plasmapheresis (or IVIG), in addition to predisone. If the vital capacity is <12 to 15 mL/kg, the patient should be intubated. A source of infection should be sought and treated.
  - Multiple sclerosis
    - Any identifiable infection should be aggressively treated; in the presence of infection, corticosteroids are not warranted. Symptoms will generally resolve when the fever or infection is treated.
    - Clinical deterioration without an identifiable cause (e.g., infection) is an indication for high-dose corticosteroids (PO or IV). Optic neuritis warrants IV methylprednisolone.

Disposition

- Patients at risk for respiratory compromise should be admitted to the ICU. This includes patients with vital capacity <35 mL/kg, or other evidence of diaphragmatic weakness (e.g., weak neck flexors or trapezius muscles).

- Patients with any other evidence of autonomic nervous system instability should be admitted to the ICU.
- Patients with worsening symptoms of myasthenia gravis should be observed in the ICU for respiratory and bulbar difficulties, including nasal regurgitation and choking on food.
- Patients with MS exacerbations who are unable to tolerate oral medications or liquids, require parenteral medications (i.e., antibiotics or steroids), or lack home support should be admitted.
- Patients with mild MS exacerbations who are candidates for oral steroids or do not require treatment (e.g., viral infections), and those with mild infections amenable to outpatient management, may be discharged home. This decision should be made in consult with the patient's neurologist.

Part C: Altered Level of Consciousness

Scope of the Problem

The term "coma" is broadly used to refer to any alteration in consciousness. Normal consciousness requires the integration of both wakefulness (or arousal) and awareness (or cognition).

- The ascending reticular activating system, located in the pons, is primarily responsible for arousal of the cerebral cortex.
- Impaired cognition is caused by total or near-total dysfunction of both hemispheres of the cerebral cortex.

Etiology

Traumatic	Subdural, epidural hematoma
	Axonal shear injury
	Hypovolemic shock
Vascular	Hypertensive encephalopathy
	Subarachnoid hemorrhage (SAH), intracerebral hemorrhage
	Infarction
	Venous occlusion
	Basilar aneurysm, basilar migraine
Infectious	Abscess, meningitis, encephalitis
	Septic shock
Metabolic	Hypo-, hyperglycemia
	Wernicke's encephalopathy
	Hypoxia, CO_2 narcosis
	Hypo-, hypernatremia
	Hypo-, hypercalcemia
	Acidosis, alkalosis
	Uremia
	Hyperammonemia
Environmental	Hypo-, hyperthermia
Toxic	CO, cyanide, hydrogen sulfite poisoning
	Alcohols, narcotics, sedatives
	Anticonvulsants
	Psychotropics
	INH, heavy metals
Endocrine	Hyper-, hypothyroidism
	Pheochromocytoma

Hematologic	Anemia
	Porphyria
Neurologic	Neurogenic shock
	Seizures, postictal state
Neoplastic	CNS tumor
	Carcinoid meningitis
Cardiac	Cardiogenic shock
Autoimmune	Cerebral vasculitis
Allergic	Anaphylactic shock

Diagnosis

- History
  - The patient is usually unable to give a reliable history. Alternative sources of information include a purse or wallet, Medic-Alert bracelets or necklaces, prescription bottles, prehospital personnel, police, family, friends, and neighbors.
  - When, and under what circumstances, did the patient lose consciousness? When was the last time the patient was observed in his usual state of health? Vascular events frequently evolve over seconds or minutes. Some metabolic processes develop over minutes to hours, while infectious and other metabolic disorders progress over hours to days. A CNS tumor, abscess, or chronic subdural hematoma may result in symptoms developing over days to weeks.
  - Were there any preceding symptoms or events? A preceding state of confusion, without focal neurologic symptoms, usually suggests a metabolic etiology.
  - What are the patient's medications and allergies? Is there a history of drug or alcohol abuse?
  - What is the patient's medical history?
  - Have there been any previous similar episodes?
- Vital Signs
  - Rectal temperature is critical in patients with altered mental status. Hypothermia may be environmental, or accompany alcohol or sedative intoxication, hypoglycemia, sepsis, Wernicke's or hepatic encephalopathy, or myxedema. Hyperthermia may be due to heat stroke, seizures, malignant hyperthermia, anticholinergic intoxication, pontine hemorrhage, sepsis or thyroid storm.
  - The respiratory rate and character may vary but are only occasionally helpful in determining the pathology responsible for a patient's alteration in consciousness. Some of the commonly described respiratory patterns are: (1) hyperventilation—resulting in reduced pCO_2—associated with hypoxia, metabolic acidosis, brainstem herniation, salicylate overdose, and heptic encephalopathy; (2) hypoventilation—with CNS depressants, uremia, diabetic coma, and elevated ICP; (3) ataxic breathing—irregular rate and depth of inspirations and expirations—frequently precedes respiratory arrest; (4) Cheyne-Stokes—crescendo-decrescendo respiratory rate and depth, alternating with periods of apnea; and (5) apneustic breathing—bradypnea with a pause at end-inspiration.
  - Blood pressure will vary depending upon etiology and may be very low, normal or elevated. Likewise, the pulse may also vary. Bradycardia and elevated blood pressure—referred to as Cushing's reflex—is seen in patients with elevated ICP.
- Examination
  - C-spine immobilization should be maintained in any patient with suspected trauma, and patients at risk for cervical injury with minor manipulation (e.g., rheumatoid arthritis).
  - The initial exam should focus on the patient's stability. Is the patient maintaining his airway? Is the patient's breathing regular and effective? Does he have palpable pulses?

4

Is there any external evidence of life-threatening pathology that needs to be addressed immediately (e.g., dilated pupil suggesting herniation)?

- General appearance: Note the patient's posture. Patients with hemispheric dysfunction generally lie in normal positions. Brainstem lesions tend to be associated with abnormal postures. The patient's breath may smell of acetone (in diabetic ketoacidosis) or alcohol. Completely undress, carefully inspect, and palpate the patient.
- HEENT: Check the scalp for signs of trauma, surgical scars, or a ventricular shunt. Look at the conjunctivae for icterus or pallor. The ears should be examined for hemotympanum. Is there a tongue laceration or dental trauma indicating seizure activity?
- Neck: Listen for bruits. Check range of motion (only if the cervical spine has been cleared). The absence of rigidity does not exclude the possibility of SAH or meningitis. Does the patient have a surgical scar, suggesting a thyroidectomy (and possible incidental removal of parathyroid glands)?
- Chest: Palpate for evidence of trauma. Listen for equal breath sounds throughout the chest. Is there any evidence of acute or chronic pulmonary disease?
- Cardiac: Is there evidence of cardiogenic shock (e.g., dysrhythmias, murmurs, extra heart sounds)?
- Abdomen: Palpate for intraperitoneal fluid, organomegaly, and pulsatile masses.
- Extremities: Does the patient have equal pulses bilaterally? Are there "track marks," indicating intravenous drug abuse? Range all joints to exclude injuries (e.g., posterior shoulder dislocation incurred during a seizure).
- Skin: Examination of the skin may reveal evidence of renal failure (uremic frost); hyperthyroidism (warm, moist, smooth skin); or hypothyroidism (cool, dry skin with a yellow tint).
- Neurologic: The neurologic exam is limited to the patient's level of consciousness, eyes, spontaneous movements, and reflexes.
  - The Glasgow Coma Scale was originally designed for use in accurately describing the best response of a patient who sustained head trauma. However, despite its limitations, it is widely used in nontrauma settings as well. Scores range from 3 (worst) to 15, with coma defined as a score <8 (unless the patient has spontaneous eye opening).

Eyes	**Verbal**	**Motor**	**Score**
No opening	No sounds	No movement	1
Open to noxious stimulus	Unintelligible sounds	Extensor response	2
Open to verbal stimulus	Nonsensical speech	Flexor response	3
Open spontaneously	Confused	Flexion withdrawal	4
	Oriented	Localizes noxious stimulus	5
		Follows commands	6

  - Examine the eyes at rest. Disconjugate gaze may localize the lesion to the pons or cerebellum (vertical, "skew" deviation), CN VI (persistent adduction of one eye), CN III (persistent abduction of one eye), or diffuse injury (conjugate upward gaze is seen in anoxic encephalopathy). Note pupil size, symmetry, and reactivity. Small pupils suggest an interruption of the sympathetic pathway, organophosphate poisoning, opiate overdose, or a pontine lesion. Dilated pupils indicate compression of CN III, anticholinergic overdose, or sympathomimetic intoxication. Pupils that are normal in size but unreactive are seen with brainstem (midbrain) lesions. Examination of the eyelids may reveal bilateral or unilateral ptosis. Bilateral ptosis occurs in midbrain lesions (e.g., basilar artery embolism); unilateral ptosis is seen with Horner's syndrome and CN III palsy. Extraocular movements can be assessed by eliciting the oculocephalic or oculovestibular reflex. The oculocephalic (or doll's eyes) maneuver should not be performed if the patient is at risk for cervical injury; instead, the more sensitive oculovestibular (or cold-water calorics) reflex should be tested. A normal response (i.e., conjugate, horizontal eye movements away from the direction of head movement, or toward the side of the cold-water irrigation) indicates an intact brainstem. Abnormal responses (e.g., no eye movement or downward deviation of the eyes) suggest brainstem involvement or sedative drug intoxication.

- Cranial nerves: The corneal reflex involves both the afferent limb of CN V and the efferent limb of CN VII. The presence or absence of a gag reflex (CN IX and X), although important for airway protection, has no localizing value.
- Motor: Observe the patient for spontaneous movements. Unilateral movement may suggest a focal neurologic deficit. The only evidence of nonconvulsive status epilepticus may be subtle movement. Observe the patient's response to a noxious stimulus.
- Reflexes: Hyperreflexia and clonus suggest hemispheric dysfunction. Asymmetry of reflexes may localize the deficit.

Differential Diagnosis

- Locked-in syndrome: syndrome of intact consciousness, with voluntary movement restricted to opening and closing the eyes and moving the eyes in the vertical plane.
- Acute psychosis
- Psychogenic unresponsiveness
  - Conversion reaction
  - Malingering

Evaluation

- Pulse Oximetry
  - Rapid assessment of the patient's oxygenation status can be obtained with pulse oximetry.
- Labs
  - Bedside glucose testing should be done to exclude hypoglycemia, an easily treated cause of altered mental status.
  - An arterial blood gas will provide more information regarding the patient's ventilatory (pCO_2) and acid-base status (pH, HCO_3, and base deficit). In addition, a low pO_2 in the setting of a normal pulse oximetry value, as well as an elevated carboxyhemoglobin level, are indicative of carbon monoxide poisoning.
  - Electrolytes (including calcium), BUN and creatinine are routinely ordered. Serum osmolarity, serum drug levels, ammonia, and TSH may be indicated based on the history and physical exam.
  - Selective ordering of serum levels of ethanol and other toxic alcohols and toxicology screening may confirm or exclude intoxication. True coma is rarely caused by ethanol levels under 250 mg/dL; patients suspected of acute intoxication but with lower levels require further testing (e.g., head CT, LP) to search for another explanation for their symptoms.
  - Urinalysis and pregnancy testing should routinely be performed.
  - Consider blood, urine, and throat cultures in any case where sepsis is a concern.
  - Lumbar puncture for analysis of cerebrospinal fluid (CSF) is indicated in any patient suspected of having a CNS infection or subarachnoid hemorrhage (after negative head CT). This procedure is deferred when there is clinical evidence of increased ICP.
- Electrocardiogram
  - EKG may reveal a cardiac etiology (especially in the elderly) or evidence of other pathology: "cerebral" T waves in SAH, Osborne or "J" waves in hypothermia, or prolonged Q-T interval in hypocalcemia.

4

- EEG
  - Although not readily available, bedside EEG may be useful in suspected nonconvulsive status epilepticus.
- Imaging
  - Noncontrast head CT is the primary neuroimaging study and will identify intracranial hemorrhage and some abscesses, tumors, and noninfectious inflammatory disorders. Contrast-enhanced CT scans are more sensitive for abscess, tumor, and inflammatory processes; it will also identify infectious or neoplastic meningeal disease.
  - Chest radiography is useful in suspected pulmonary infections and malignancy and is performed after endotracheal intubation for confirmation of adequate tube placement.
  - MRI and CT are both insensitive in the first few hours after ischemic stroke. Diffusion-weighted MRI, however, will detect acute infarction.

Treatment

- Maintain an adequate oxygen saturation. Consider intubation for airway protection or maintenance of adequate ventilation.
- All patients with altered mental status of unknown etiology should receive thiamine 100 mg IV. Patients with hypoglycemia (also consider in patients with low-normal bedside results) should receive 25 g of 50% dextrose IV (pediatric dosing: 25% dextrose, 0.5 g/kg). Although some references recommend empiric administration of naloxone, selective use guided by the history, vital signs, and physical exam is acceptable.
- Flumazenil is indicated only for acute benzodiazepine overdose. Indiscriminate use may cause seizures in patients with cocaine or tricyclic toxicity or cause withdrawal seizures in chronic benzodiazepine users.
- Hypotension should be treated with isotonic fluids and vasopressor agents, as needed. An appropriate target mean arterial pressure is 90 to 100 mm Hg.
- Determining the etiology of hypertension, malignant hypertension, ischemic stroke, or response to elevated ICP is crucial to guiding treatment (if any).
- Elevated intracranial pressure should be recognized early and managed with head-of-bed elevation to 30 degrees (when practical), hyperventilation (target CO_2 30-35), mannitol 0.5 g/kg (20% solution) infused over 5-15 min, and furosemide 1 mg/kg IV. Use of medications will likely be in concert with neurosurgical consultation.

Disposition

- Patients in true coma require admission to an ICU with neurologic or neurosurgical consultation, as appropriate. Patients with no clear etiology for their altered mental status should be admitted, even if all symptoms have resolved.
- Patients with hypoglycemia, unrelated to insulin use, should be admitted for evaluation to determine the cause of the episode (e.g., stroke, MI, sepsis in the elderly).
- Although admission is sometimes recommended for narcotic overdose (because of the long duration of action of most narcotics compared to that of naloxone), this is usually not practical. A period of observation in the ED (i.e., 4-6 h) is sufficient to prevent recurrence of symptoms
- Acute ethanol intoxication requires observation until the patient is able to "walk and talk." This allows reassessment for concomitant illness or injury and ensures that the patient is safe to leave.
- Patients with psychogenic "coma" may be discharged after resolution of symptoms. Psychiatric consultation is indicated prior to discharge.
- Patients who are discharged home should have follow-up with a primary physician within 24-48 h.

Suggested Reading

1. Coma. In: Simon RP, Aminoff MJ, Greenberg DA, eds. Clinical Neurology 4th ed. Stamford: Appleton and Lange, 1999.
2. Feske SK. Coma and confusional states: Emergency diagnosis and management. Neurol Clin North Am 1998; 16:2.
3. Hamilton GC: Altered mental status—depressed level of consciousness. In: Hamilton GC, ed. Presenting Signs and Symptoms in the Emergency Department. Baltimore: Williams and Wilkins, 1993.
4. Wolfe R, Brown D. Coma. In: Rosen P, Barkin RM, eds. Emergency Medicine: Concepts and Clinical Practice. 4th ed. St. Louis: Mosby-Year Book Inc., 1996.

Part D: Infections of the Central Nervous System

Basic Anatomy

- The central nervous system (CNS) is encased within three membranous layers, called meninges. These meninges, from the outermost layer inward, are the dura mater, the arachnoid, and the pia mater. The dura adheres to the inner surface of the cranium; the arachnoid attaches to the inner surface of the dura; and the pia is attached to the brain, following all of its contours. The space between the arachnoid and pia—the subarachnoid space—is filled with cerebrospinal fluid (CSF).
- The cranial dura extends through the foramen magnum to become the spinal dura mater. The spinal epidural space is located between the periosteum of the vertebrae and the dura and is filled with fatty connective tissue and a vertebral venous plexus.
- The spinal arachnoid closely attaches to the inner surface of the dura, creating the subarachnoid space between itself and the spinal cord that, like the cranial subarachnoid space, is filled with CSF. The spinal cord ends (i.e., becomes the cauda equina) at about the level of the disk between the first and second lumbar vertebrae; however, the spinal dural sheath (and its arachnoid lining) ends at about the second sacral vertebra. The large subarchnoid cistern, between these two points, is the site at which sampling of CSF occurs (i.e., lumbar puncture) with relatively little risk of damage to the spinal cord.

Scope of the Problem

- Meningitis
    - Meningitis is inflammation of the membranes of the brain or spinal cord, which may accompany an infectious, neoplastic, toxic, or autoimmune process. Because the precise etiology may not be evident in the emergency department, empiric treatment for bacterial meningitis is of utmost importance.
    - Despite early and aggressive use of antibiotics, the overall mortality rate remains at 25% for bacterial and fungal meningitis.
    - The causative organism varies with the age, immune status, living conditions, travel history, and overall health of the individual. However, with the decline in frequency of *Haemophilus influenzae* meningitis as a result of the *H. influenzae* type b vaccine, *S. pneumoniae* is now the most common cause in adults and children over one month old. *N. meningitidis* is the second most common organism isolated in both age groups.
    - Antibiotic resistance is a frequently observed trend, with increasing resistance of *S. pneumoniae* to penicillin and third-generation cephalosporins.
    - Long-term sequelae of bacterial meningitis include cognitive deficits, seizure disorders, hearing loss, blindness, gait disturbances, focal motor deficits, and hydrocephalus.
    - "Aseptic" meningitis refers to conditions in which there is CSF pleocytosis and a clinical suspicion of meningitis, but with negative bacterial cultures. Typical etiologies include viral meningitis, fungal infections, and drugs (e.g., NSAIDs, TMP-SMX, and INH).

- Encephalitis
    - Encephalitis is inflammation of the brain parenchyma. It may coexist with viral meningitis or it may present as a distinct entity, caused most commonly by arboviruses, herpes viruses, and rabies. *Listeria* and cat-scratch disease are rare etiologies.
    - Encephalitides caused by certain arboviruses (Japanese, Eastern equine, and St. Louis encephalitides) are associated with high mortality rates and severe neurologic sequelae. The death rate from HSV encephalitis has been reduced by acyclovir; however neurologic deficits—including epilepsy, focal motor deficits, and altered mentation—are common.
- West Nile Virus
    - West Nile virus, an arthropod-borne virus (arbovirus), may cause encephalitis, meningitis, or meningoencephalitis. Patients at highest risk for symptomatic infection include persons over age 50 and the immunosuppressed. Associated symptoms may include fever, headache, nausea, vomiting, weakness, altered mental status, stiff neck, and an erythematous rash. West Nile virus is not cultured from the CSF or brain tissue, but IgM antibodies may be present in the CSF or serum. Alternatively, PCR testing of the CSF for West Nile virus RNA may be positive.
- CNS Abscess
    - CNS abscess denotes a circumscribed collection of purulent material, or a localized infection, which may exist within the brain parenchyma (brain abscess); within the meninges (epidural or subdural empyema); or within or surrounding the spinal cord (intramedullary or epidural spinal abscess). Complications of intracranial abscess include epilepsy, focal motor or sensory deficits, and intellectual deficits. Patients with spinal abscesses may have residual motor or sensory deficits, or bowel or bladder dysfunction.

Risk Factors

- Meningitis
    - As mentioned above, the most common pathogens in patients over one month of age are *S. pneumoniae* and *N. meningitidis*; risk factors for other organisms are shown in Table 4D.1.
- Encephalitis
    - The means of access to the CNS varies according to the virus (Table 4D.2).
- CNS Abscess
    - CNS abscesses develop as an extension of a contiguous infection (e.g., otitis media, sinusitis, dental infection), or by hematogenous seeding from a remote site (e.g., pulmonary, endocarditis, osteomyelitis). Other risk factors include intravenous drug abuse, neurosurgical procedures, and penetrating head injury. The causative organisms vary according to the primary source of the infection and the immune status of the patient (Table 4D.3).

Diagnosis

History

- The classic triad of fever, nuchal rigidity, and altered mental status is seen in approximately two-thirds of patients with community-acquired bacterial meningitis. All patients, however, will likely have at least one of these findings. Other signs and symptoms which should cause one to suspect meningitis include headache, chills, vomiting, myalgias/arthralgias, lethargy, malaise, focal neurologic deficits, photophobia, and seizures. Elderly patients may present with subtle findings, frequently limited to an altered sensorium. Fungal meningitides present with an atypical constellation of symptoms, including headache, low-grade fever, weight loss, and fatigue; similarly, tuberculous meningitis may be associated with fever, weight loss, night sweats, and malaise, with or without headache and meningismus.

Table 4D.1. Organisms causing meningitis

Population	Additional Potential Pathogens
Neonate (<1 mo)	Group B streptococci, *E. coli, Listeria monocytogenes*
1 mo to 50 yr	*H. influenzae* (rarely), *L. monocytogenes* (unlikely)
Adults (over 50 yr), alcoholics, other debilitating diseases	*L. monocytogenes, Enterobacteriaceae*
Closed head injury with CSF leak	*S. aureus, Enterobacteriaceae, P. aeruginosa*
Recent neurosurgical procedure or penetrating head injury	*S. aureus, S. epidermidis,* other *Streptococcus* species, *Bacteroides fragilis, Enterobacteriaceae*
CSF shunt infection	*S. epidermidis, S. aureus, Enterobacteriaceae,* diphtheroids, *P. acnes*
Splenectomy	*H. influenzae*
Chronic otitis media	*Streptococcus* species, *Enterobacteriaceae Bacteroides fragilis*
Malignant otitis externa (diabetes)	*Pseudomonas* species
Sickle-cell disease, diabetics	*Enterobacteriaceae*
Immunosuppressed host	*L. monocytogenes, P. aeruginosa, Enterobac teriaceae, S. aureus, H. influenzae,* Streptococci, anaerobes, *Mycobacterium tuberculosis, Actinobacter* spp., syphilis, Cryptococcus neoformans, toxoplasmosis, Herpes simplex, Cytomegalovirus

Table 4D.2. Encephalitis, causative organisms

Virus	Route of Entry
Arbovirus (California, W. Equine, E. Equine, St. Louis, West Nile)	Mosquito bite; hematogenous spread
Herpes virus	
Herpes simplex type 1	Skin lesions; retrograde neuronal spread
Varicella zoster	Skin lesions; retrograde neuronal spread
E-B virus	Mononucleosis
Rabies	Animal bite; retrograde neuronal spread
Measles, mumps	Post-infectious

Table 4D.3. Etiology of CNS abscess

Source of Infection	Likely Pathogen
Local or remote infection	
Sinuses, teeth	Streptococci
Otitis media, pulmonary infection	*Bacteroides*
Endocarditis	*S. aureus*
Other sources	*Enterobacteriaceae,* Nocardia (rarely)
Neurosurgical procedure, penetrating head injury	*S. aureus, Enterobacteriaceae*

- Other important historical factors include:
  - Duration of symptoms: a fulminant course indicates a bacterial meningitis or aggressive viral encephalitis, while a subacute presentation suggests a viral, fungal, or parasitic infection.

4

- Antecedent infection: recent otitis media, sinusitis, respiratory tract infection, pharyngitis, or intracranial abscess may suggest recent colonization with, or contiguous spread of, a particular organism.
- Recent course of antibiotics: may alter CSF analysis and clinical presentation.
- History of a penetrating or closed head injury, neurosurgical procedure (including VP shunt placement), or congenital dural defect.
- Living conditions or epidemic exposure: college dormitories, military barracks, and jails/prisons are typical areas for epidemics of *N. meningitidis*; exposure in day-care centers, or to family members with a specific infectious disease (e.g., *M. tuberculosis*) may suggest an otherwise atypical causative organism.
- Immune suppression: HIV, malignancy, splenectomy, or other immunologic deficits.
- Social history: alcohol or IV drug abuse, low socioeconomic status.
- Underlying medical conditions: sickle cell disease or thalassemia major, bacterial endocarditis, cirrhosis, diabetes.
- Barrier disruption: VP shunt, central IV lines, loss of cutaneous integrity (including prior varicella-zoster infection).
- History of mosquito or tick bite; exposure to animals at risk for rabies infection.

Examination

- Meningitis
  - Evaluate the patient's overall appearance and mental status.
  - HEENT exam should include a search for evidence of trauma, surgery, infections (otitis, mastoiditis, sinusitis, pharyngitis), or pupillary abnormalities. Note that papilledema takes time to develop, and this finding can be absent in the majority of patients with bacterial meningitis. In infants <12 mo of age, when meningeal signs are unreliable, the anterior fontanelle should be evaluated for bulging.
  - Test the neck for rigidity: Brudzinski's sign (if the neck is passively flexed, flexion of the hips occurs; or, on passive flexion of one hip, flexion of the other hip occurs); and Kernig's sign (resistance to passive extension of the knee). Neck stiffness is often absent at the extremes of age, or in patients with altered levels of consciousness, immunosuppressed, or partially treated disease.
  - Examination of the chest may reveal a concurrent pneumonia.
  - A new heart murmur may indicate endocarditis.
  - Examination of the abdomen may suggest an infectious process, and thus a source for bacteremia and meningitis or abscess.
  - A complete neurologic exam must be documented, revealing a number of potential abnormalities: isolated cranial nerve deficits (including ophthalmoplegia); focal motor or sensory deficits; cerebellar dysfunction; and increased deep tendon reflexes. Localizing signs are generally absent in bacterial meningitis; their presence suggests the possibility of a focal infection, such as an abscess. The level of consciousness may range from confusion or delirium to stupor or coma.
  - The skin should be examined for the petechial or hemorrhagic lesions suggestive of meningococcemia, or a rash characteristic of HSV, herpes zoster, or leptospirosis (purpura and petechiae on the oral, vaginal, or conjunctival mucosa).
  - Arthritis may be associated with *N. meningitidis* or, less commonly, other bacterial meningitides.
- Encephalitis
  - Clinical suspicion of encephalitis should be raised in the setting of new "psychiatric" symptoms, cognitive deficits (especially memory disturbances and aphasia), acute confusion, and movement disorders (e.g., choreoathetosis and parkinsonism).
- Abscess

- Patients with CNS abscess often experience a delay in diagnosis as a result of nonspecific presenting complaints.
- While headache is almost universally present, fever is seen approximately half of the time.
- One-third of patients may have focal neurologic signs, including hemiparesis and seizures.
- Increased intracranial pressure may cause vomiting, confusion, or altered levels of consciousness in 50% of patients.
- Meningismus is noted on <50% of exams, while papilledema is present in one-third of patients.

Evaluation

- Delay in the diagnosis of bacterial meningitis in the elderly, especially with nonspecific symptoms, is responsible for the high mortality in this population. While urinalysis or chest X-ray may indicate an infectious process outside the CNS, it is important to remember that the diagnosis of meningitis still needs to be suspected—and aggressively pursued—because of the risk of hematogenous spread of the involved organisms.

Laboratory

- Cerebrospinal Fluid
  - Lumbar puncture (LP) should be performed whenever meningitis is suspected. If there will be a delay in performing the LP, blood cultures should be drawn and antibiotics administered empirically. Immediate LP may be contraindicated in the following situations:
    - Suspected HIV disease
    - Focal neurologic exam
    - Evidence of increased intracranial pressure
    - Hemodynamic instability
    - Overlying infection at the LP site
    - Suspected coagulopathy
    - Suspect subarachnoid hemorrhage
  - An opening pressure, if measured, should be performed with the patient fully extended. Normal adult pressures are 5-19 cm H_2O, when the patient is in the lateral recumbent position. Opening pressure may be elevated in bacterial and fungal meningitis.
  - The cerebrospinal fluid is normally clear and colorless. Infection, inflammation, or bleeding may cause the fluid to be turbid. Note that fluid can be clear even when several hundred cells are present.
  - True CNS bleeding (e.g., a subarachnoid hemorrhage) may be distinguished from a traumatic tap by the presence of xanthochromia. In addition, RBC count will generally decrease in sequential tubes with a traumatic tap.
  - CSF analysis should include cell count and differential, glucose and protein, stat gram's stain and culture, and a fourth tube for special tests as indicated by the clinical scenario.
    - Normal adult CSF contains <6 WBCs/mm^3, with no more than one PMN. Early in the course of bacterial meningitis, lymphocytes may predominate.
    - CSF analysis in viral meningitis and encephalitis typically reveals <500 WBCs/mm^3, with almost 100% mononuclear cells (but early presentations may have PMN pleocytosis).
    - Brain abscess and parameningeal infections (e.g., subdural empyema, epidural abscess) have cell counts and differentials similar to those of viral meningitis.
  - The normal ratio of CSF:serum glucose is 0.6 (0-0.4 in the setting of severe hyperglycemia). CSF glucose may be decreased in bacterial, fungal and tuberculous

4

meningitis; carcinomatous meningitis, and (normal or decreased) in subarachnoid hemorrhage. Mild decreases in CSF glucose may be seen in viral and parameningeal processes.
- CSF protein levels are normally <45 mg/dl in adults. An elevated protein, usually >150 mg/dl, is suggestive of bacterial meningitis. Other causes of elevated CSF protein include any infectious meningitis, viral or parasitic encephalitis, carcinomatous meningitis, subarachnoid hemorrhage, CNS vasculitis, neurosyphilis, hepatic encephalopathy, and demyelination syndromes.
- Other tests to consider include viral cultures, acid-fast stain and culture for *M. tuberculosis*, India ink and cryptococcal antigen, VDRL, cytology, bacterial antigens for *S. pneumoniae, H. influenzae,* and *N. meningitidis* (especially in patients recently treated with antibiotics), and *Borrelia* antibodies in cases of suspected Lyme disease. Other more specialized tests are rarely ordered in the ED.

- Blood Tests
  - Although a CBC with differential may add to the clinical picture in a patient with suspected meningitis, a normal result does not exclude the diagnosis.
  - PT and PTT may be useful to exclude a suspected coagulopathy, or DIC.
  - Serum electrolytes, BUN/Cr, and glucose are routinely ordered.
  - Blood cultures (two specimens, collected 15 min apart) may identify the organism in up to 80% of cases depending upon the etiology.
- Imaging
  - A CT scan of the head should be performed prior to LP in the following situations:
    - Altered mental status
    - Focal neurologic exam (excluding ophthalmoplegia)
    - Evidence of increased intracranial pressure
    - Minimal or absent fever
    - Recent-onset seizure
    - Suspected subarachnoid hemorrhage or intracranial mass lesions
  - Contrast-enhanced CT is the study of choice for evaluation of possible CNS abscess. Although not readily available, MRI is equally sensitive.
  - MRI is the imaging study of choice when cranial epidural abscess or subdural empyema is suspected.
  - A chest X-ray may reveal a concomitant pneumonia, especially in cases of pneumococcal meningitis.
- EEG
  - In the setting of suspected herpes encephalitis, focal or lateralized EEG abnormalities may help pinpoint the diagnosis.

Treatment

- Meningitis
  - Administration of IV antibiotics should begin as soon as the suspicion of bacterial meningitis is entertained. Empiric therapy should be based on the suspected pathogen, taking into consideration the patient's age and risk factors for specific organisms. An infectious disease consultant may be helpful for information regarding local drug resistance patterns (Table 4D.4).
- Encephalitis
  - Herpes simplex and varicella encephalitis are the only treatable forms of encephalitis. Acyclovir is dosed at 10mg/kg IV q8 h.
- CNS Abscess
  - The mainstay of treatment of CNS abscess is antibiotics. Neurosurgical consultation is recommended for possible aspiration or excision (Table 4D.5).

Table 4D.4. Empiric antimicrobial therapy for meningitis

Population	Empiric Therapy
Neonate (0-7 days)	Ampicillin + Cefotaxime *or* Ampicillin + Gentamicin *or* Ampicillin + Amikacin
Neonate (8-28 days)	Ampicillin + Cefotaxime
1 mo to 50 yr	Vancomycin + Ceftriaxone *or* Vancomycin + Cefotaxime
Adults (over 50 yr)	Ampicillin + Ceftriaxone + Vancomycin *or* Ampicillin + Cefotaxime + Vancomycin
Recent neurosurgical procedure, penetrating head injury, CSF leak	Vancomycin + Ceftazidime
CSF shunt infection	Peds: Vancomycin + Cefotaxime *or* Vancomycin + Ceftriaxone Adults: Vancomycin + Rifampin
Immunosuppressed host	Ampicillin + Ceftazidime +Vancomycin
Suspected HSV 2 meningitis	Acyclovir

- Specific Treatment
  - Meningitis
    - Glucocorticoids, based on studies in children with predominantly *H. influenzae* type b meningitis, have been shown to reduce the morbidity associated with an inflammatory reaction to bacteriolysis. Dexamethasone, 0.15 mg/kg IV, may be considered in children as well as adults, especially in cases with positive Gram's stain (indicating a high bacterial load), altered level of consciousness, or increased intracranial pressure. If dexamethasone is given, benefit is greatest when started prior to or concurrent with initial antibiotic therapy.
    - In areas with high prevalence of drug-resistant *S. pneumoniae,* vancomycin is added empirically.
    - In penicillin-allergic patients, trimethoprim-sulfamethoxazole and vancomycin are recommended.
    - Amphotericin B and flucytosine are the drugs of choice for cryptococcal meningitis.
    - Initial therapy for tuberculous meningitis should include isoniazid (INH), rifampin, ethambutol, and pyrazinamide (PZA).

Table 4D.5. Antibiotic therapy of CNS abscess

Source	Empiric Therapy
Otogenic	Ceftriaxone + metronidazole *or* Cefotaxime + metronidazole
Sinogenic, odontogenic	Penicillin G + metronidazole
Remote infection	Penicillin G + metronidazole
Post-surgical, head trauma	(Cefotaxime *or* Ceftriaxone) + nafcillin *or* (Cefotaxime *or* Cefotaxime) + oxacillin *or* (Cefotaxime *or* Ceftriaxone) + vancomycin (MRSA)
Unknown source	Ceftriaxone + metronidazole *or* Cefotaxime + metronidazole

4

Table 4D.6. Meningitis prophylaxis

Organism	Treatment	
H. influenzae type b:	Rifampin, 20 mg/kg PO (up to 600 mg) q 12 h for 4 doses	
N. meningitidis:	Peds:	Rifampin 10 mg/kg PO q 12 h for 4 doses *or*
		Ceftriaxone 125 mg IM, single dose
	Adults:	Rifampin 600 mg PO q 12 h for 4 doses *or*
		Azithromycin 500 mg PO, single dose
		Ciprofloxacin 500 mg PO, single dose *or*
		Ceftriaxone 250 mg IM, single dose

- If the CSF pressure is >40 cm H_2O, consider giving mannitol. Only the CSF in the manometer should be collected, to reduce the risk of tentorial herniation. Patients with clinical and CT evidence of increased intracranial pressure also require appropriate therapy.
- Complications of meningitis should be anticipated and managed including dehydration, hypotensive shock, hyponatremia, coagulopathies, seizures, cerebral edema, loss of protective airway reflexes, respiratory failure, and stroke.
- Close contacts of patients diagnosed with *H. influenzae* type b or *N. meningitidis* should receive prophylactic treatment. Health care personnel coming into contact with respiratory droplets are also candidates for prophylaxis (Table 4D.6).
- Encephalitis
  - Post-infectious encephalomyelitis—an acute inflammatory, demyelinating process preceded by infection or immunization with influenza, measles, and varicella—is treated with high-dose intravenous methylprednisolone.
  - Bacterial causes of encephalitis include *Listeria*, Lyme disease, Rocky Mountain Spotted Fever (RMSF), Leptospirosis, and the Ehrlichioses.
  - Immunocompromised patients are at increased risk for amebic encephalitis, toxoplasmosis, and CMV, in addition to the herpes viruses. Empiric therapy of encephalitis—*in addition to acyclovir*—is shown in Table 4D.7.
- CNS abscess
  - Although not a true abscess, focal lesions in the brain parenchyma attributable to toxoplasmosis in the immunocompromised patient are treated with pyrimethamine plus sulfadiazine. Patients with sulfa allergies are treated with pyrimethamine plus clindamycin. If a fungal abscess is suspected, amphotericin B should be added to the empiric regimen.
  - Treatment of epidural abscess and subdural empyema includes surgical drainage and intravenous antibiotics, as described above for empiric therapy of CNS abscess.

Table 4D.7. Encephalitis, empiric therapy

Organism		Treatment
Lyme disease		Ceftriaxone
RMSF	Peds <8 yo:	Chloramphenicol
	Adults:	Doxycycline
Leptospirosis		Penicillin G
Ehrlichiosis		Doxycycline
Amebae		Amphotericin
CMV		Gancylcovir or foscarnet (more toxicity)

Disposition

- All patients with suspected bacterial meningitis, encephalitis, or CNS abscess are admitted.
- Patients definitively diagnosed with viral meningitis, if the social situation permits, may be discharged if associated symptoms (e.g., pain, vomiting) are controlled.

Suggested Reading

1. Disorders of consciousness approach to diagnosis and acute confusional states. In: Simon RP, Aminoff MJ, Greenberg DA, eds. Clinical Neurology 4th ed. Stamford, Appleton and Lange, 1999.
2. Gilbert DN, Moellering RC, Sande MA, eds. The Sanford Guide to Antimicrobial Therapy. 32nd ed. Hyde Park: Antimicrobial Therapy Inc., 2002.
3. Lavoie FW. Meningitis, Encephalitis, and Central Nervous System Abscess. In: Rosen P, Barkin RM, eds. Emergency Medicine: Concepts and Clinical Practice. 4th ed. St. Louis: Mosby Year Book Inc., 1996.
4. Pruitt AA. Infections of the Nervous System. Neurol Clin North Am 1998; 16:2

Part E: Cerebrovascular Emergencies

Basic Anatomy

- The anterior circulation, consisting of the paired internal carotid arteries and their branches (ophthalmic, anterior cerebral, and middle cerebral arteries), supplies most of the cerebral hemispheres and the deep cortical gray matter.
  - The anterior cerebral artery (ACA) supplies the parasagittal cerebral cortex, which includes portions of the motor and sensory cortex related to the contralateral lower limb and the so-called bladder inhibitory or micturition center.
  - The middle cerebral artery (MCA) supplies most of the remainder of the cerebral hemisphere, and deep cortical structures.
    - The cortical branches of the MCA include the superior division, which supplies the entire motor and sensory cortical representation of the face, and upper limb; and the expressive language (Broca's) area of the dominant hemisphere.
    - The inferior division supplies the visual radiations, the region of visual cortex related to macular vision, and the receptive language (Wernicke's) area of the dominant hemisphere.
    - Lenticulostriate branches supply the basal ganglia, as well as motor fibers related to the face, hand, arm, and leg.
- The posterior circulation, consisting of the paired vertebral arteries, the basilar artery, and their branches—the posterior inferior cerebellar (PICA), anterior inferior cerebellar (AICA), superior cerebellar, and posterior cerebral arteries—supplies the brainstem, cerebellum, thalamus, and the medial aspects of the occipital and temporal lobes.
- The anterior and posterior circulations join via the posterior communicating arteries to form the circle of Willis at the base of the skull.

Clinicoanatomic Correlation

- Anterior Circulation
  - Anterior circulation strokes rarely have associated symptoms; neurologic deficits accompanied by headache, nausea, and vomiting are more suggestive of intracerebral hemorrhage or posterior circulation stroke.
  - Clinical syndrome of ACA occlusion
    - Contralateral weakness and sensory loss affecting the lower limb.
    - Confusion and impaired cognition.
    - Urinary incontinence.

4

- Clinical syndrome of MCA occlusion
  - Superior division stroke results in contralateral weakness and sensory deficit in the face and upper limb.
    - If the dominant hemisphere is involved, Broca's aphasia occurs (impaired language expression with normal comprehension).
  - Inferior division stroke results in marked impairment of contralateral cortical sensory functions, contralateral homonymous hemianopsia, and contralateral neglect.
    - If the dominant hemisphere is involved, Wernicke's aphasia occurs (impaired comprehension and recall, with fluent—but often nonsensical—speech).
    - An acute confusional state may indicate involvement of the nondominant hemisphere.
  - Occlusion at the bifurcation or trifurcation of the MCA combines the features of superior and inferior division stroke.
- Posterior Circulation
  - Posterior circulation strokes frequently present more of a diagnostic challenge. In addition, complications of cerebellar infarcts, such as edema compressing brainstem structures, may cause rapid deterioration (i.e., within hours).
  - Symptoms associated with vertebrobasilar infarcts include syncope, vertigo or lightheadedness, nystagmus, dysphagia, dysarthria, facial sensory disturbance, ataxia, depressed consciousness or confusion, and incontinence.
  - Clinical syndrome of posterior cerebral artery (PCA) occlusion
    - Contralateral homonymous hemianopsia.
    - Ocular abnormalities (including oculomotor (III) nerve palsy and internuclear ophthalmoplegia).
    - With involvement of the dominant hemisphere, patients may be unable to name objects or to read but retain the ability to write.
  - Clinical syndrome of basilar artery occlusion
    - Thrombosis is often incompatible with survival.
    - Basilar artery thrombosis affects the pons and may be confused with pontine hemorrhage.
      - Coma is common.
      - Hemiplegia or tetraplegia is usually present.
      - The pupils are constricted but reactive.
      - Involvement of the dorsal portion of the pons (i.e., the tegmentum) produces unilateral or bilateral abducens (VI) nerve palsy; vertical nystagmus may be present.
      - Infarction of the ventral portion of the pons with sparing of the tegmentum causes tetraplegia in a patient who remains conscious (referred to as the locked-in syndrome). Voluntary eye opening, vertical eye movements, and ocular convergence are preserved.
    - Emboli usually stop at the top of the basilar artery, at the bifurcation into the PCAs.
      - Immediate loss of consciousness (or depressed LOC) is caused by impaired blood flow to the reticular activating system.
      - Hemiplegia or tetraplegia with decerebrate or decorticate posturing.
      - Ipsilateral or bilateral oculomotor (III) nerve palsies.
  - Clinical syndrome of PICA occlusion
    - Results in lateral medullary (Wallenburg's) syndrome
    - Symptoms may include vertigo, nausea, vomiting, dysphagia, dysarthria, hoarseness (due to vocal cord paralysis), and hiccup.

- Neurologic exam may reveal nystagmus; ipsilateral Horner's syndrome, paralysis of the soft palate and posterior pharynx, and limb ataxia; and impaired pain/temperature sensation in the ipsilateral face and contralateral limbs.
- The motor system is spared.
- Clinical syndrome of AICA occlusion
  - Produces ipsilateral facial weakness, gaze palsy, deafness, and tinnitus.
- Clinical syndrome of superior cerebellar artery occlusion
  - Hypertropia (deviation of the eyes in opposite directions equally).
  - Contralateral sensory disturbance.
- Lacunar Infarction
  - These infarcts occur in the deep gray and white matter structures (e.g., basal ganglia and internal capsule); the onset may be subacute with symptoms developing over hours or days.
  - Headache is absent or minor; there is no impairment in level of consciousness.
  - Pure motor hemiparesis: uniformly affects the contralateral face and limbs.
  - Pure sensory stroke: contralateral hemisensory loss (may be associated with paresthesia).
  - Dysarthria-clumsy hand syndrome: includes dysarthria, dysphagia, and contralateral facial and hand weakness and clumsiness.
  - Ataxic hemiparesis: consists of cerebellar ataxia and leg > arm > face weakness.

4

Scope of the Problem

- Disruption in the flow of blood to the brain results in ischemia and cell death. The central area of infarction is surrounded by a region of salvageable tissue, referred to as the penumbra.
- Mechanisms of ischemia include embolism, thrombosis, and hemorrhage. Identifying the etiology of the patient's symptoms is critical for determining therapy.
- Massive cerebral infarcts are typically associated with cerebral edema, which peaks 3 to 5 days after onset. Patients with such swelling are at risk for herniation.

Risk Factors

- Vascular Disorders
  - Atherosclerosis
    - Diastolic or isolated systolic hypertension
    - Hyperlipidemia (hypercholesterolemia)
    - Cigarette smoking
    - Oral contraceptive use
    - Diabetes mellitus
    - Hereditary predisposition (i.e., family history of ischemic vascular disease at age <60)
    - Excessive alcohol use
    - Physical inactivity
    - Age
    - Male gender
    - Ethnicity
  - Carotid or vertebral artery dissection
    - Signs and symptoms of carotid dissection may include jaw or neck pain, visual changes similar to those that accompany migraine headaches, and Horner's syndrome.
    - Vertebral or basilar artery dissection is associated with headache, posterior neck pain, and acute brainstem dysfunction.
  - Venous or sinus thrombosis

        - Patients with this disease process usually have a predisposing condition, such as contiguous infection (e.g., otitis, sinusitis), hypercoagulable state, or dehydration.
        - Clinical manifestations include headache, papilledema, depressed level of consciousness, seizures, and focal neurologic deficits.
    - Inflammatory disorders
        - Giant cell arteritis, systemic lupus erythematosis, polyarteritis nodosa, granulomatous angiitis (primary angiitis of the CNS), syphilitic arteritis, AIDS
    - Drug abuse
        - Cocaine, amphetamines, and heroin
        - Infective endocarditis caused by IV drug abuse may lead to embolic stroke.
        - Other forms of drug abuse are postulated to cause vasospasm, vasculitis, or rupture of preexisting aneurysms or vascular malformations.
    - Migraine
    - Fibromuscular dysplasia
- Cardiac Disorders
    - Mural thrombus
        - Generally associated with myocardial infarction (MI) or cardiomyopathy.
        - Rheumatic heart disease
    - Arrhythmias
        - Atrial fibrillation and bradycardia-tachycardia (sick sinus) syndrome
    - Endocarditis
    - Mitral valve prolapse: small increase in risk of stroke; massive strokes are rare
    - Paradoxic embolus
        - Embolic material from the systemic circulation may gain access to the brain through a pathologic communication between right and left sides of heart (e.g., ASD, patent foramen ovale).
    - Atrial myxoma
    - Prosthetic heart valves
- Hematologic Disorders
    - Thrombocytosis, polycythemia, sickle cell disease, leukocytosis (i.e., leukemia), hypercoagulable states

Diagnosis

History

- Attempt to identify, as accurately as possible, the onset, course, and type of symptoms, as well as the patient's activity at onset.
- Determine the patient's stroke risk factors, other potential causes for the patient's symptoms, and any contraindications to thrombolytics or other agents.
- Neurologic deficits that occur abruptly, and are maximal at onset, suggest an embolic stroke. Stepwise, incremental deficits are more indicative of thrombosis.
- Hemorrhagic strokes often have a rapid onset. Maximum deficit may be present immediately but worsening may occur. Consciousness may be impaired.
- Subarachnoid hemorrhage (SAH) is characterized by symptoms of variable onset and progression; severe headache and neck stiffness; and often impaired consciousness.

Vital Signs

- Hypotension may be the underlying cause of a stroke; markedly elevated blood pressure is suggestive but not diagnostic of a hemorrhagic stroke.
- An irregular cardiac rhythm may indicate chronic or new-onset atrial fibrillation and an embolic source.
- Increased ICP may be accompanied by bradycardia.

Physical Examination

- Focus on searching for an underlying systemic cause, especially a treatable one.
  - HEENT
    - Note any signs of trauma. Palpate the temporal arteries.
    - Fundoscopic exam may reveal evidence of embolization. Papilledema indicates increased ICP. Retinal hemorrhages may be noted in SAH.
    - Determine patency of the airway and ensure the airway is protected.
  - Neck: evaluate the carotid pulses and check for carotid bruits, Neck stiffness may or may not be present in SAH.
  - Cardiac exam: note any arrhythmias or evidence of valvular disease (i.e., murmurs).
  - Neurologic exam
    - The initial neurologic exam should be a brief search for signs of increased ICP or impending transtentorial herniation (e.g., dilated pupil, depressed consciousness, or posturing).
    - Neurologic deficits that do not conform to the distribution of any single cerebral artery suggest a hemorrhagic or mass lesion.
    - Assess the level of consciousness and mental status. Determine whether the patient has aphasia (expressive or receptive?) or dysarthria.
    - Evaluate cranial nerve function, with special attention paid to pupils, extraocular movements, visual fields, facial symmetry, gag reflex and corneal reflex.
    - The presence of a gaze palsy may help localize the lesion. A patient with a cerebral hemispheric stroke will typically gaze toward the side of the insult; a brainstem infarct will cause the patient to gaze away from the side of the lesion.
    - Motor exam: the most sensitive indicator of upper extremity weakness is the presence of a pronator drift. Whenever possible, lower extremity strength should be assessed by observing the patient's gait.
    - Sensory exam: peripheral sensation (light touch, pinprick, and vibration/position sense). Double simultaneous stimulation: assess sensation on both sides of the body simultaneously; patients with cortical infarcts will only notice the unaffected side.
    - Cerebellar exam: look for nystagmus, ataxia, or poor coordination.
    - Reflexes: recent stroke is accompanied by hyporeflexia (and flaccidity). Search for pathologic reflexes (e.g., presence of a Babinski reflex indicates an upper motor neuron disorder).

Evaluation

- Pulse Oximetry
  - Rapid determination of oxygen saturation may reveal impending respiratory failure and the need for mechanical ventilation. Supplemental oxygen may suffice.
- Laboratory
  - Bedside glucose testing is crucial to exclude hypoglycemia as a cause of focal neurologic deficits.
  - A complete blood count may reveal an underlying hematologic disorder presenting as a stroke.
  - Prothrombin time (PT) and partial thromboplastin time (PTT) are helpful in excluding coagulopathy.
  - Baseline serum electrolytes are recommended.
  - An elevated ESR may suggest giant cell arteritis or other vasculitis.
  - Blood for type and cross match (patients with SAH)
  - Consider a toxicology screen in patients suspected of drug abuse (although the results seldom change the patient's acute management).
  - Pregnancy test in females of child-bearing age.

- Lumbar puncture (LP) is indicated when CNS infection or SAH is suspected. In the case of SAH, LP is performed after a negative CT scan. LP may also suggest a venous or sinus thrombosis (CSF pressure is typically increased and pleocytosis may occur), or granulomatous angiitis (CSF pleocytosis occurs).
- EKG
  - May detect cardiac ischemia, or cardiac arrhythmias which predispose to stroke (e.g., atrial fibrillation). In addition, multiple abnormalities are described in patients with subarachnoid hemorrhage, including peaked or symmetric, deeply inverted T waves; shortened PR intervals; and tall U waves.
  - Chest radiograph
    - In selected patients, chest X-ray (CXR) may reveal an infectious process, malignancy, or evidence of heart failure. In intubated patients, CXR is used to confirm placement of the endotracheal tube.
  - Imaging
    - Noncontrast CT scan is the emergency radiologic study of choice when evaluating patients with suspected stroke, TIA, or SAH. CT is able to distinguish between ischemic and hemorrhagic strokes, and will detect blood in more than 90% of cases of aneurysmal rupture. In addition, early signs of cerebral edema are identified with noncontrast CT scanning.
    - When available, MRI is useful for early ischemic infarcts. MRI may also be more sensitive for ischemic strokes in the brainstem or cerebellum, as well as in detecting thrombotic occlusion of venous sinuses.
  - Electroencephalogram (EEG) is rarely useful in the acute evaluation of stroke. It may, however, help to differentiate between seizure disorder and TIA, or between lacunar and cortical infarcts.

Differential Diagnosis

- Vascular Disorder
  - Intracerebral hemorrhage, SAH
  - Subdural or epidural hematoma
  - Hypertensive encephalopathy
  - Complicated migraine
  - Arterial embolism to an extremity
  - Air embolism
- Structural Lesion
  - Abscess, neoplasm
  - Multiple sclerosis
- Metabolic Process
  - Hypoglycemia, hyperosmolar nonketotic hyperglycemia
- Infectious Process
  - Encephalitis, meningitis
- Peripheral Nerve Disorder
  - Bell's palsy, other peripheral nerve palsies
  - Peripheral vertigo
- Other
  - Acute angle closure glaucoma
  - Seizure with postictal (Todd's) paralysis

Treatment

- General Management Issues
  - As with all emergent conditions, evaluation of the patient with a neurologic deficit begins with airway, breathing, and circulation. Patients with severely depressed mental

status and patients with an unprotected airway may require intubation and mechanical ventilation.
- All patients should be placed on continuous pulse oximetry and cardiac monitoring, and have peripheral intravenous access established.
- Patients noted to have signs of increased ICP should be aggressively managed.

- Oxygen
  - Supplemental oxygen may be required to maintain oxygen saturation >95%. However, in the absence of hypoxia, supplemental oxygen has not been shown to affect outcome.
- Blood Pressure Control
  - In the first few hours following acute stroke, mild to moderate hypertension is commonly observed. Over the next few hours to days, the blood pressure generally declines spontaneously. The ischemic penumbra may be dependent upon a moderately increased blood pressure for adequate perfusion; thus, use of antihypertensive agents may exacerbate the patient's condition.
  - Cautious blood pressure control—with easily titratable, short-acting parenteral medications—is recommended in the following situations: (1) hypertensive encephalopathy; (2) cardiac ischemia or aortic dissection; (3) intracerebral hemorrhage; (4) when thrombolytic therapy is considered (see exclusion criteria below); and (5) the mean arterial pressure (MAP) is >130, or the SBP is >220.
- Anticoagulants
  - Heparin has been shown to benefit patients at risk for cardioembolic stroke (especially patients with ischemic or rheumatic heart disease). In addition, patients with venous sinus occlusion benefit from heparin use.
- Antiplatelet Agents
  - Aspirin has been shown to reduce the rate of nonfatal recurrent stroke and death after acute stroke (versus placebo). Aspirin is recommended in patients who are not candidates for thrombolytics or other anticoagulants.
  - Ticlopidine inhibits the ADP pathway of platelet aggregation; it also reduces blood fibrinogen levels. In patients with TIA or minor stroke, ticlopidine results in significant risk reduction in stroke recurrence or death versus aspirin. However, the hematologic side effects (TTP, thrombocytopenia, and neutropenia) have limited its recommended use to patients who have a contraindication to aspirin and who can be closely monitored in this setting; clopidogrel is preferred by many practitioners.
  - Clopidogrel is similar in action and effect to ticlopidine. It is recommended for aspirin-intolerant patients and has a significantly lower rate of TTP than ticlopidine.

Specific Treatment

Acute Ischemic Stroke

- Intravenous Thrombolytics (IV rt-PA)
  - In the NINDS rt-PA stroke study, patients treated with rt-PA within 3 h of symptom onset were at least 30% more likely (than patients given placebo) to have minimal or no disability at 3 mo on various clinical scales. However, 6.4% of patients in the treatment arm had a symptomatic ICH within 36 h of stroke onset, compared with 0.6% of patients given placebo. Mortality was similar in both groups. Based on these results, the FDA approved the use of IV rt-PA in acute ischemic stroke within 3 h of symptom onset. In addition, the AHA and the American Academy of Neurology have issued practice guidelines recommending the use of intravenous thrombolysis.
  - Other studies have failed to show improvement in outcome over placebo. For this reason, the use of thrombolytics in stroke remains controversial. However, because this modality is now being described by certain groups as the standard of care, the

4

inclusion and exclusion criteria are included here. It is important to note that both ACEP and ABEM have developed position statements which indicate that further studies are necessary before thrombolytics become a community standard of care.
- Dosing for IV rt-PA is as follows: 0.9 mg/kg total dose, with 10% given over 1 min as a bolus; the remainder is infused over 1 h.
- Consultation with a neurologist is strongly recommended prior to the initiation of IV rt-PA. Patients with very severe deficits, as well as those over age 75, are probably at increased risk for ICH. Accurate informed consent is critical.
- Inclusion criteria
  - Age >18 yr
  - Ischemic stroke with a clearly defined time of onset within 3 h of initiation of treatment
  - Neurologic deficit measurable on the NIHSS
  - Baseline CT scan of the brain with no evidence of intracranial hemorrhage
- Exclusion criteria
  - Patient history
    - Stroke or serious head trauma within the preceding 3 mo
    - Major surgery (or biopsy of a parenchymal organ) within the past 14 days
    - Prior intracranial hemorrhage; history of CNS neoplasm, aneurysm, or AVM
    - Arterial puncture at a noncompressible site within the past 7 days
    - Gastrointestinal or urinary tract hemorrhage within the previous 21 days
    - Lumbar puncture within the past 7 days
    - Recent myocardial infarction
    - History of known hereditary or acquired abnormal hemostasis
    - Pregnant or lactating female
  - Clinical features
    - Systolic BP >185 mm Hg or diastolic BP >110 mm Hg, or aggressive treatment is required to reduce blood pressure to the specified limits
    - Rapidly improving or minor symptoms
    - Symptoms suggestive of subarachnoid hemorrhage (SAH)
    - Seizure at the onset of stroke
    - Clinical presentation consistent with acute MI or post-MI pericarditis
  - Diagnostic studies
    - Major hypodensity or effacement of cerebral sulci in more than one-third of the territory of the MCA
    - Received heparin within 48 h of stroke onset, with elevated PTT
    - Prothrombin time (PT) >15 seconds
    - Platelet count <100,000/mm^3
    - 50 > glucose >400 mg/dL

TIA

- In the past, patients with neurologic symptoms that resolved within 24 h were said to have had a transient ischemic attack (TIA). However, recent evidence indicates that evidence of infarction is seen on CT of patients whose symptoms last more than 1 h. For this reason, many neurologists are describing symptoms that persist for more than 1 h as an acute stroke. In addition, the majority of TIAs resolve within minutes.
- Patients with TIAs are at increased risk for subsequent stroke and death.
- Aspirin, ticlopidine, and clopidogrel have all been shown to reduce the risk of future TIAs, strokes, and death.
- Anticoagulation is recommended for patients with cardiac embolus as a cause of their TIA symptoms.

Table 4E.1. Grading of patients with subarachnoid hemorrhage

Grade	Mental Status	Neurologic Deficits	Surgical Candidate
I	Normal	None	Yes
II	Mildly altered	Minimal, focal	Yes
III	Confused	Mild, focal	Yes
IV	Stuporous	Moderate to severe	No
V	Coma	Decerebrate posturing	No

Hemorrhagic Stroke

- Cautious blood pressure control is recommended. Various guidelines exist, ranging from keeping the diastolic blood pressure at approximately 100 mm Hg to basing the target systolic and diastolic levels on the patient's premorbid blood pressure.
- Look for and aggressively treat early signs of increased ICP with head-of-bed elevation, diuretics (furosemide, mannitol), and hyperventilation.
- Although most cases of ICH are not amenable to neurosurgical intervention, some patients benefit from surgical drainage. Potential surgical candidates are those with neurologic deterioration, superficial cerebral hemorrhages causing mass effect, and cerebellar hematomas. Neurosurgical consultation is recommended in all patients with intracerebral hemorrhage.

Subarachnoid Hemorrhage

- SAH frequently results from rupture of an aneurysm or AVM. Preventing rerupture, by maintaining adequate blood pressure control, is the mainstay of treatment. Elevation of the head of the bed, mild sedation, and analgesics (for headache) may suffice. The blood pressure should be reduced to approximately 160/100 mm Hg, using rapidly titratable, parenteral medications if necessary.
- Nimodipine, a calcium channel blocker, is indicated for the prevention of cerebral vasospasm in order to prevent subsequent ischemia; the dose is 60 mg enterally every 4 h.
- A prophylactic anticonvulsant is recommended; consider corticosteroids.
- Prompt neurosurgical consultation is recommended. The timing and outcome of surgical intervention are determined by the patient's clinical grade and medical stability, among other factors.
- Grading of patients with SAH (Table 4E.1).

Disposition

- Acute Stroke Syndromes
  - Patients with acute intracerebral hemorrhage require admission to an ICU or step-down unit for close monitoring.
  - All patients with acute ischemic strokes should be hospitalized for medical stabilization and evaluation of their functional independence as well as their rehabilitation potential. Patients who receive thrombolytics should be admitted to the ICU within 3 h.
  - Patients with multiple previous strokes may not require admission if they are medically stable and social support is adequate.
- TIA
  - Current recommendations are that patients with new TIAs be admitted for workup and initiation of treatment.

Suggested Reading

1. Barsan WG, Kothari R. Stroke. In: Rosen P, Barkin RM, eds. Emergency Medicine: Concepts and Clinical Practice. 4th ed. St. Louis: Mosby-Year Book, Inc., 1996.

2. Baumlin KM, Richardson LD. Stroke syndromes. Emerg Med Clin North Am 1997; 15:3.
3. Brott T et al. Measurements of acute cerebral infarction: A clinical examination scale. Stroke 1989; 20:864-870.
4. Headache and Facial Pain. In: Simon RP, Aminoff MJ, Greenberg DA, eds. Clinical Neurology. 4th ed. Stamford: Appleton and Lange, 1999.
5. Henneman PL, Lewis RJ. Is admission medically justified for all patients with acute stroke or transient ischemic attack? Ann Em Med 1995; 25:458-463.
6. Hickenbottom SL, Barsan WG. Acute ischemic stroke therapy. Neurol Clin 2000; 19:2.
7. Kasner SE, Grotta JC. Ischemic stroke. Neurol Clin North Am 1998; 16:2.
8. Llinas R, Caplan LR. Evidence-based treatment of patients with ischemic cerebrovascular disease. Neurol Clin 2001; 19:1.
9. Stieg PE, Kase CS. Intracranial hemorrhage: Diagnosis and emergency management. Neurol Clin North Am 1998; 16:2.
10. Stroke. In: Simon RP, Aminoff MJ, Greenberg DA, eds. Clinical Neurology. 4th ed. Stamford, Appleton and Lange, 1999.
11. The National Institute of Neurological Disorders and Stroke. rt-PA Stroke Study Group. Tissue plaminongen activator for acute ischemic stroke. N Engl J Med 1995; 333:1581-1587.
12. Younger JG, Barsan WG. Cerebrovascular emergencies. In: Howell JM, ed. Emergency Medicine. Philadelphia: WB Saunders Company, 1998.

Part F: Dizziness and Vertigo

Basic Anatomy

- The vestibular system provides input to the brain regarding movement of the head. The vestibular portion of the 8th cranial nerve is composed of the utricle, the saccule, and three semicircular canals that lie at right angles to each other.
- Maintenance of equilibrium is dependent upon input from the vestibular system and the eyes, as well as proprioception; in addition, the information must be properly integrated in the brain. Abnormal information, or improper integration, results in a complaint of "dizziness" or vertigo.

Scope of the Problem

- The complaint of "dizziness" can be attributed to numerous disorders, both benign and life-threatening. The first task is to define the patient's symptoms. Is it dizziness, vertigo or weakness? For patients with true vertigo, the next task is to determine the source of the symptoms; is it a peripheral or central vestibular process?
- Peripheral vestibular disorders involve structures at the level of the inner ear and the 8th cranial nerve. Central lesions involve the vestibular nuclei (located in the brainstem) and their connections (e.g., CN III and VI nuclei, MLF).

Etiology

- Peripheral Vestibular Disorders
  - Cerebellopontine angle tumor (e.g., meningioma, schwannoma)
  - Acute Ramsay Hunt syndrome
  - Benign paroxysmal positional vertigo (BPPV)
  - Acute vestibular neuronitis/neuritis or labyrinthitis
  - Perilymphatic fistula
  - Post-traumatic vertigo
  - Labyrinthine concussion
  - Meniere's disease

- Central Vestibular Disorders
  - Posterior fossa tumor
  - Vertebrobasilar ischemia/infarction
  - Multiple sclerosis
  - Cerebellar hemorrhage
  - Infections and parainfectious encephalomyelitis
  - Wernicke's encephalopathy

Diagnosis

History

- Ask the patient to describe the symptoms as accurately as possible: is there a spinning sensation (i.e., true vertigo); a sense of being off balance, without spinning (dysequilibrium); a faint feeling (near syncope); or an otherwise vague sensation of "lightheadedness?" Patients with vestibular lesions use decriptors such as spinning, feeling drunk or motion sick, or a sensation of imbalance (i.e., falling or tilting to one side). Patients with nonvestibular lesions explain the feeling as one of spinning inside the head, near-faint, floating, swimming in the head, or giddiness.
- Determine the duration of the attack—seconds, minutes, or hours to days? Peripheral vestibular disorders are associated with intermittent episodes of vertigo that are sudden, brief, and severe; symptoms from central lesions have a more gradual onset, last longer, and are less intense.
- Are the episodes provoked by specific movements, position changes, or maneuvers? Peripheral lesions are commonly initiated by turning the head to the side or tilting the head up, whereas central causes are not positional. Symptoms of vertebrobasilar insufficiency may be reproduced with neck movement. Valsalva's maneuver worsens vertigo associated with a perilymphatic fistula. Symptoms that are precipitated by stress or certain situations (e.g., driving a car) generally indicate a nonvestibular cause. Cervical vertigo refers to the symptom complex of neck pain, vertigo, and nystagmus that is worse with movement of the neck.
- Are there associated symptoms? Peripheral vestibular lesions are commonly associated with nausea or vomiting; hearing loss, tinnitus, or ear fullness; other neurologic deficits (e.g., diplopia, dysphagia, dysarthria, extremity weakness, or sensory impairment) almost always accompany a central lesion. Are there symptoms in the absence of vertigo? Loss of coordination between attacks indicates 8th cranial nerve or brainstem disease.
- Has there been a recent illness? Vestibular neuritis and acute labyrinthitis are commonly preceded by a viral illness. Is there any history of trauma? Post-traumatic positional vertigo, labyrinthine concussion, and perilymphatic fistula are all caused by head trauma. A rapid, twisting injury of the neck can cause vertebral artery dissection or occlusion.
- Does the patient's medical history suggest an etiology? Uremia, Parkinson's disease, diabetes, and chronic alcohol abuse are causes of peripheral neuropathy and orthostasis. Is the patient taking any medications that may cause orthostatic hypotension (e.g., diuretics, anti-hypertensives) or dysequilibrium (e.g., anti-convulsants, sedatives)? Is the patient on any new medications? Frequently, elderly patients with poor vision and sensation—especially after starting sedating medications—describe a feeling of being off balance and stumbling (dysequilibrium), without true vertigo or presyncopal symptoms.
- Is there a history of similar episodes? What type of work-up was done, and what were the results? Patients with ill-defined light-headedness often have extensive work-ups without an explanation for their symptoms.
- Is there a family history (e.g., spinocerebellar degeneration, Friedrich's ataxia, ataxia-teloangiectasia, Wilson's disease)?

4

Physical Exam

- HEENT: Note any signs of trauma. Inspect the ears for evidence of infection, trauma, or cerumen impaction. Are there vesicles (suggestive of VZV); is there fluid or a TM perforation? With the patient looking straight forward, ensure that the eyes are properly aligned. Funduscopic exam may reveal papilledema, suggesting increased intracranial pressure from a mass lesion (usually in the posterior fossa).
- Neck: evaluate range of motion; carotid bruits suggest atherosclerosis.
- Cardiac exam: evaluate the rhythm and listen for murmurs (may suggest an outflow obstruction).
- Neurologic Exam
  - Determine the patient's mental status.
  - Abnormal cranial nerve findings suggest a central process.
    - Evaluate the extraocular muscles in all six cardinal positions of gaze and look for nystagmus. Nystagmus is described by the position of gaze in which it is provoked and the direction of the fast component. Nystagmus associated with peripheral disorders is generally horizontal or rotatory and does not change directions, but is more pronounced when the patient looks in the direction of the fast component (away from the involved side).
    - Central disorders produce gaze pareses and nystagmus in any direction (including vertical) that may change direction if the patient looks away from the direction of the fast phase.
    - Check the patient's auditory acuity; perform Weber and Rinne tests.
    - If indicated by the history, test facial nerve function. The patient with a cerebellopontine angle tumor may have a depressed corneal reflex or facial nerve palsy ipsilateral to the lesion.
    - Lower brainstem disease may be accompanied by weakness of the tongue (CN XII), hoarseness, dysphagia, and weakness of the palate (CN IX and X).
  - Motor exam: Note any motor weakness or involuntary movements (e.g., asterixis, myoclonus, chorea).
  - Sensory exam: Proprioception and vibratory sensation are impaired in patients with sensory ataxia produced by neurosyphilis, vitamin B_{12} deficiency, and polyneuropathies.
  - Cerebellar exam: Tests of motor coordination evaluate cerebellar hemispheric function. These include finger-to-nose, finger-tapping and toe tapping, rapid alternating movements, and heel-knee-shin.
  - Stance: Romberg's sign is positive if the patient is unable to stand with the feet together when the eyes are closed. This indicates a sensory (proprioceptive) or vestibular disorder. Patients with vestibular lesions will fall toward the side of the lesion. In contrast, patients with cerebellar pathology are unable to compensate with visual cues and are unsteady with eyes open or closed.
  - Gait: A wide-based, staggering gait is noted in cerebellar ataxia; in addition, tandem gait is always impaired.
  - Reflexes: hypoactive reflexes accompany cerebellar disorders and polyneuropathies causing sensory ataxia. Multiple sclerosis, vitamin B_{12} deficiency, and focal brainstem lesions are associated with hyperactive reflexes and the Babinski sign.
  - Dix-Hallpike maneuver: The seated patient is rapidly lowered onto the exam table with the head hanging off the end at a 20-degree angle. Ask the patient if he has vertigo and examine the patient's eyes. Slowly return the patient to a sitting position and repeat the procedure with the head turned 45 degrees horizontally. Slowly return the patient to sitting with the head still at a 45-degree angle; repeat the procedure with the head turned to the other side.

- Nystagmus associated with a peripheral vestibular disorder has a latency period (i.e., begins several seconds after the position change); fatigues (i.e., terminates spontaneously if the position is maintained); habituates (i.e., repeated maneuvers will result in less pronounced symptoms); and is suppressed when the gaze is fixed. In contrast, central vertigo is typically not positional, has no latency period, does not fatigue or habituate, and is not suppressed by visual fixation.

Ancillary Evaluation

- Consider pulse oximetry when a systemic process is likely; in addition, patients with a history of trauma, possible stroke, or intracranial hemorrhage (ICH) may be hypoxic.
- A bedside glucose test should be performed on all patients with neurologic symptoms.
- Serum drug levels may be useful in patients on vertigogenic medications. Consider a toxicology screen in the appropriate clinical setting.
- Patients with dizziness or ill-defined symptoms may require further evaluation with a complete blood count, electrolytes and other laboratories as indicated.
- Emergent noncontrast CT scan of the head is indicated in patients with a suspected life-threatening central process (e.g., cerebellar hemorrhage, tumor, abscess, infarction). CT may also reveal bony erosion in cases of bacterial otomastoiditis with labyrinthine involvement.
- MRI is indicated in patients suspected of having multiple sclerosis, posterior fossa tumor, or brainstem infarction (after negative head CT).

Differential Diagnosis

- Syncope and near-syncope
- Hypovolemia from any cause
- Acute coronary syndrome
- Intoxication
- Hyperventilation syndrome
- Anxiety and affective disorder
- Dysequalibrium
- Metabolic disorders
- Sepsis
- Intracranial pressure

Treatment

- Peripheral Vestibular Disorders
  - Vestibular suppressants are useful in the acute period. Prolonged use may impede central compensation. Therefore, only a three-day supply should be prescribed (Table 4F.1).
  - Vestibular exercises facilitate the central compensation process and may be useful for patients with chronic vertigo or recurring BPPV. The patient should assume a position with his head that causes nystagmus, and then attempt to focus the eyes and move them in a position that maximizes his symptoms. As the nystagmus diminishes, the patient should begin to move the head up and down or from side to side while visually fixating on a target. He should attempt to stand and walk while the nystagmus is still present, and (as symptoms improve) should move the head from side to side or up and down while walking (first slowly, then quickly in all directions). Compensation may take several months.
  - Specific treatment
    - The Canalith repositioning maneuver is effective in cases of BPPV resulting from otoconia. Details of this procedure are described elsewhere. Note that patients may have an increase of symptoms as a result of repositioning maneuvers.

Table 4F.1. Treatment options in peripheral vertigo

Antihistamines	Meclizine	25 mg PO q6h
	Diphenhydramine	25-50 mg PO q6h
Phenothiazines	Prochlorperazine	5-10 mg PO q6-8h; 25 mg PR q12h
	Promethazine	25-50 mg PO/IM q4-6h;
Anticholinergics	Scopolamine	0.5 mg transdermally q 3 days
		0.4-0.8 mg PO q6-8h
Benzodiazepines	Diazepam	5-10 mg PO q6-8h
Serotonin antagonists	Ondansetron	4 mg PO q 8 h

4

Furthermore, feasibility of patient education in a busy ED may be an issue. Otolaryngology follow-up is advised.

- Acute peripheral vestibulopathy encompasses vestibular neuritis and acute labyrinthitis of unknown etiology (but commonly preceded by a viral infection). Both disorders are associated with acute onset vertigo and nystagmus, nausea, and vomiting that may last for 2 wk. The distinction between the two is based on the presence (labyrinthitis) or absence (neuritis) of concomitant hearing loss or tinntus. A 3-day course of a vestibulosuppressant with bed rest is recommended. In addition, a 10-day course of prednisone may shorten the course of the illness.
- Ramsay Hunt syndrome is a reactivation of varicella-zoster that can involve multiple cranial nerves (including V, VII, VIII, IX and X). Prednisone and acyclovir have been found to facilitate recovery, if treatment is initiated within 3 days (compared to more than 7 days) after symptom onset.
  - Prednisone 60 mg PO q day, tapered over 10 days
  - Acyclovir 400 mg PO 5x/day for 10 days
- Meniere's disease is thought to be caused by excessive fluid within the inner ear. It is characterized by acute attacks of vertigo and ear pressure lasting hours, associated with tinnitus and sensorineural hearing loss. In addition to vestibulosuppressants, patients may benefit from restricted sodium, caffeine, and nicotine intake. All patients with Meniere's disease should be referred to ENT.
- The patient with dysequilibrium caused by multiple sensory deficits functions better with improved lighting, elimination of sedating medications, and (in the short term) using a walker. He may benefit from a referral for vision refraction or rehabilitation, as indicated by his deficits.

Disposition

- Patients with neurologic deficits or suspected central disorders should be admitted. Patients with intractable vomiting or severe dehydration may require inpatient treatment.
- Patients with suspected peripheral vestibular disorders who are discharged home with medication should be instructed to follow-up with their PCP if symptoms persist more than a few days.

Suggested Reading

1. Baloh RW. Dizziness: Neurological emergencies. Neurol Clin North Am 1998; 16:2.
2. Disorders of equilibrium. In: Simon RP, Aminoff MJ, Greenberg DA, eds. Clinical Neurology 4th ed. Stamford: Appleton and Lange, 1999.
3. Herr RD. Dizziness and vertigo. In: Howell JM, ed. Emergency Medicine. Philadelphia: WB Saunders Company, 1998.
4. Tusa RJ. Vertigo. Neurol Clin 2001; 19:1.

Part G: Seizures

Etiology/Risk Factors

- Head Trauma
  - Depressed skull fracture
  - Subdural hematoma, intracerebral hemorrhage
- Infectious
  - Meningitis, encephalitis, abscess
- Vascular
  - Stroke or vascular malformations
  - Hypertensive encephalopathy
  - Eclampsia
- Environmental
  - Hyperthermia
- Toxic
  - Drug overdose or withdrawal
- Neoplasm
- Metabolic
  - Hypoglycemia
  - Hypo- or hypernatremia
  - Hypo- or hypercalcemia
  - Uremia; hepatic encephalopathy
  - Hyperosmolar states
- Congenital
  - Idiopathic epilepsy
- Hematologic
  - Porphyria
- Neurologic
  - Febrile seizures
  - HIV encephalopathy
  - Global cerebral ischemia

Scope of the Problem

- Nearly 1% of all emergency department visits are due to new-onset generalized seizures in adults. Seizure may be the sole presenting symptom of a life-threatening illness requiring immediate treatment.
- Seizures are defined as disordered discharges of cerebral neurons. The outward expression of a seizure may take many forms:
- Generalized seizures involve a loss of consciousness.
  - Tonic-clonic seizures are characterized by a phase of tonic muscle contractions causing extension of the limbs (and falling) and cessation of ventilatory effort, followed by a clonic phase of rhythmic muscle contraction and relaxation resulting in symmetric jerking of the limbs with return of spontaneous respirations. Urinary incontinence may occur. A postictal phase of unconsciousness or confusion is not uncommon. It usually clears within 30 min, but may last for hours.
  - Absence seizures are characterized by a brief (5-10 second) loss of consciousness, during which postural tone is maintained. Blinking or head turning may be the only motor manifestation of the seizure. There is no postictal period of confusion.
- Partial Seizures
  - Simple partial seizures begin within a specific region of the cortex, which determines the symptoms (i.e., sensory, motor, or autonomic). The symptoms may

sponteously resolve, recur, spread to contiguous cortical regions (jacksonian march), or become secondarily generalized. In the absence of generalization, there is no loss of consciousness.

- Complex partial seizures cause impaired consciousness. Patients experience the same symptoms during each successive ictal event. The episode classically begins with a blank stare, and (occasionally) loss of muscle tone, resulting in a fall. Epigastric sensations are most common, but affective, cognitive, or sensory symptoms also occur. Automatisms are common and can be simple (e.g., chewing, blinking, laughing) or complex (e.g., vocalizations or repetitive movments). Secondary generalization may occur so rapidly that the preceding partial component is not recognized, and only the altered mental status is observed. A postictal period is common but usually short (i.e., minutes).

4

Diagnosis

History

- If the seizure activity has terminated prior to the patient's arrival in the emergency department, a description of the event from a reliable witness is invaluable. Any history of trauma (recent or remote) should be elicited. A description of events immediately preceding the seizure activity should also be sought, including any complaints of pain or focal neurologic deficits. Attempt to determine whether the patient was injured during the episode (e.g., fall).
- Obtain the patient's medical history, if possible, including a prior history of seizures or other medical conditions, medications, or recent symptoms (e.g., infections). Ask about the use of drugs or alcohol, or exposure to other toxins.

Physical Examination

- Include a rectal temperature with the vital signs.
- Look for evidence of trauma, either as a cause or a result of the seizure. A detailed neurologic examination should be performed. If the patient has an altered level of consciousness, is he in a postictal state? Or is there another cause? Are there any focal neurologic deficits? Look for signs of increased intracranial pressure (ICP) (e.g., papilledema). Is there evidence of a CNS or systemic infection? Are there other signs of systemic illness? Is there evidence of a toxic exposure? Serial neurologic exams are critical.

Differential Diagnosis

Toxic and metabolic encephalopathies with fluctuating consciousness	Hypoglycemia; renal or hepatic dysfunction Recreational drug use; alcoholic blackouts Delerium tremens
Syncope	Neurocardiogenic; vasovagal; orthostatic; cardiac
Cerebrovascular	TIA (incl. vertebrobasilar insufficiency)
Movement disorders	Dystonias Tonic spasms with tetanus, strychnine, and camphor Tic disorders; tremor Benign nocturnal myoclonus Asterixis with hepatic and renal failure Rabies
Transient global amnesia	
Migraine (including acephalgic migraine)	
Paroxysmal endocrine disturbances	Pheochromocytoma; carcinoid syndrome
Sensory disturbances	Visual hallucinations with visual field loss Paroxysmal vertigo

Sleep disorders	Apnea Night terrors; sleep walking Narcolepsy; hypersomnia
Psychogenic	Hyperventilation; breath-holding spells in children Panic attacks; episodic dyscontrol; dissociative states Conversion disorder; hysteria; malingering; psychosis Obsessive-compulsive disorder Pseudoseizure

Evaluation

As with the work-up of any presenting sign(s) or symptom(s), the use of diagnostic tests should be guided by the history and physical examination of each patient who presents with seizure activity. When a differential diagnosis is formulated for a particular patient, the following studies may be helpful in ruling in or excluding specific etiologies:

- Laboratory
  - Glucose should be checked on all first-time seizure patients. Although commonly ordered, routine electrolytes, calcium and magnesium have low diagnostic yield in otherwise healthy patients with a first seizure. Consider these studies when clinically indicated.
  - A pregnancy test is indicated in all females of reproductive age.
  - Antiepileptic drug (AED) levels
  - A more extensive work-up is appropriate in patients with a history of alcohol abuse, to include CBC, PT, electrolytes, BUN, and creatinine. A blood alcohol level and toxicology screen should also be considered.
  - Magnesium levels should also be checked in patients with diabetic ketoacidosis.
  - Coagulation studies are recommended in patients on anticoagulants, with a known coagulopathy, or with a history of platelet disorders.
  - Lumbar puncture is indicated in the following situations:
    - Persistent alteration in mental status or status epilepticus (after patient is stabilized).
    - Signs of CNS infection (nuchal rigidity, petechiae)
    - Severe headache (i.e., when unruptured aneurysm or SAH is suspected)
    - In a patient with a history of cancer and negative CT scans (leptomeningeal metastases?)
    - History of immunosuppression, without an identifiable cause for the seizure (i.e., lab or radiographic abnormality)
    - Children with recent antibiotic use
    - Adults with fever, without an infectious source (neutropenic patients excluded)
- Imaging
  - Emergent noncontrast CT scan of the head is indicated in all first-time seizure patients without an identifiable, nonstructural cause (e.g., hypoglycemia, febrile seizure). The following factors increase the likelihood of an abnormal CT:
    - A focally abnormal exam or signs of increased ICP
    - Multiple or focal seizures
    - Higher likelihood of structural abnormalities (i.e., increased age, history of head trauma, HIV/other immunocompromised states, cancer, alcohol abuse, anticoagulation, vascular disease, demographic risk of cysticercosis)
    - Previous CNS disorders
  - Subtherapeutic antiepileptic drug levels are the most common cause of recurrent seizures. Indications for CT scan in patients with a previously diagnosed seizure disorder include:

        - Change in seizure pattern without a known cause
        - Persistently altered mental status or prolonged postictal confusion
        - New focal neurologic deficits
    - Contrast-enhanced head CT should be performed in immunocompromised patients and those with a history of malignancy (after a negative noncontrast CT).
    - MRI is recommended, on an elective basis, if the screening CT is negative.
- EEG
    - EEG is generally not readily available in the ED and usually is not required in the ED work-up of seizures. However, in the following cases, emergent EEG is indicated:
        - The seizure appears to have terminated, but the patient remains altered.
        - Status epilepticus is suspected (convulsive or nonconvulsive).
- EKG
    - Seizure may be the presenting symptom of hypoxia in a patient with a dysrhythmia resulting from myocardial ischemia. EKG should be considered in patients with known or suspected coronary artery disease.
    - Patients with long QT syndrome frequently present after a syncopal or ictal event. Congenital long QT syndrome is seen in children or young adults who may have a family history of syncope or early cardiac death, or a personal history of congenital deafness. Long QT syndrome may be acquired and is a side effect of tricyclic antidepressants, phenothiazines, and amiodarone. Associated EKG abnormalities include prolonged or abnormal T waves and bradycardia.

Treatment

- As with any patient in the ED, attention to the patient's airway, breathing, and circulation is paramount. The patient should be positioned in such a way as to protect the airway in case of vomiting, and suction should be readily available. Supplemental oxygen should be administered by nasal cannula or face mask, and the patient placed on continuous pulse oximetry.
- If trauma is a concern, the cervical spine should be immobilized.
- As soon as possible, intravenous or intraosseous access should be obtained so that fluids and medications can be given. The patient should be placed on a cardiac monitor.
- Underlying, correctable etiologies should be rapidly identified and treated.
    - Benzodiazepines are the agents of choice to acutely terminate seizure activity, although the necessity of this practice as routine is debatable. Most seizures are brief (<2 min) and there is no evidence that a single, brief seizure has deleterious central nervous system effects. Several agents are available (see Table 4G.1).
    - Diazepam is highly lipophilic and thus crosses the blood-brain barrier rapidly. It is usually administered via the intravenous route but is equally effective when given rectally. Other routes include oral and endotracheal. The median time to terminate seizure activity after injection is 2 min. However, its antiepileptic activity lasts only 20 to 30 min. Depending upon the clinical setting, a longer acting AED may be necessary after administration of diazepam.
    - Lorazepam is less lipid soluble than diazepam but has a similar time to seizure control (3 min). Its antiepileptic activity lasts 12-24 h, negating the need to administer an additional AED if seizure activity is terminated. Acceptable routes of administration include intravenous, rectal, sublingual, and oral.
    - Midazolam is used less frequently but is another option. Unlike diazepam and lorazepam, this agent is well absorbed when given via the intramuscular route because of its water-solubility. After administration, it becomes lipid soluble and, like diazepam, has rapid penetration of the blood-brain barrier as well as a short duration of antiepileptic activity. Acceptable routes of administration include intranasal, intramuscular, intravenous, rectal, and buccal.

Table 4G.1. Antiepileptic agents for status epilepticus

Medication	Dose (IV)	Side Effects
Midazolam	0.2 mg/kg then 0.05-1.0 mg/kg/h	Respiratory depression, hypotension, sedation Tachyphylaxis with prolonged infusions
Diazepam	0.2 mg/kg (up to 20 mg in adults)	Respiratory depression, hypotension, sedation
Lorazepam	0.1 mg/kg (usual maximum dose 8 mg)	Respiratory depression, hypotension, sedation
Phenytoin	20 mg/kg @ 50 mg/min May give additional 5-10 mg/kg	Hypotension, atrioventricular block, dysrhythmias Soft tissue necrosis, sterile abscess with IM use or extravasation
Fosphenytoin	20 mg/kg PE* @ 150 mg/min May use same dose IM	Hypotension and pruritis
Phenobarbital	20 mg/kg not to exceed 100 mg/min	Respiratory depression, apnea, hypotension
Propofol	1-3 mg/kg then 2-10 mg/kg/h	Respiratory depression, hypotension
Pentobarbital	3-5 mg/kg then 1-5 mg/kg/h	Respiratory depression, apnea, hypotension, myocardial depression, weakness during recovery period

* Phenytoin equivalent

- Initiation of a long-term AED for a first seizure should be considered in certain settings. Close neurologic follow-up is indicated for these patients. All AEDs have adverse side effects and "label" the patient. If there is any question as to the necessity of initiating chronic therapy, neurologic consultation is advised. Multiple AEDs are available—a comprehensive discussion is beyond the scope of this writing. Consider initiation of an AED in the following cases:
  - An underlying cause that can't be promptly treated (e.g., tumor)
  - A patient at risk for recurrent seizures (e.g., cysticercosis, penetrating head injury)
  - Presence of an identifiable, remote cause (debatable)
  - Patients with a history of recent (<2 yr) seizure that went untreated

Status Epilepticus (SE)

- SE has classically been defined as persistent seizure activity for 30 min or recurrent seizures without full recovery between events. Since most isolated ictal events last <2 min, a more practical definition for SE is seizure activity that persists for 5 min or more.
- SE is not a disease in itself, but rather a manifestation of another illness. One of the goals when treating SE is to identify and address acute precipitants. In adults, the most common cause of SE is noncompliance with AEDs. In children, congenital abnormality and infection are the most common.
- The earlier that treatment for SE is initiated, the easier it is to control. In addition, the following complications may be avoided.
  - Autonomic dysfunction including hypertension, tachycardia, and hyperthermia
  - Vertebral and other fractures; shoulder dislocations
  - Rhabdomyolysis
  - Aspiration pneumonia

4

  - Metabolic derangements
  - Cerebral edema
- Treatment (see Table 4G.1)
- All patients require appropriate supportive care. If intubation is required, short-acting paralytics are preferred in order to allow the practitioner to identify ongoing seizure activity.
- Benzodiazepines are first-line therapy for SE. Because of its duration of action, lorazepam is preferred. Adequate dosing is imperative. If SE is not controlled with an appropriate dose of benzodiazepine, it is unlikely that subsequent doses or use of another benzodiazepine will be effective.
- Phenytoin is a long-acting AED that is effective for both SE and chronic maintenance therapy. Phenytoin is generally the preferred second-line agent for SE if benzodiazepines have failed. Side effects of the intravenous preparation are attributed to the propylene glycol diluent. These are minimized by infusing at a rate not to exceed 1 mg/kg/min in children and 50 mg/min in adults. Fosphenytoin is a phosphorylated ester of phenytoin. It is highly water-soluble, and is rapidly converted to phenytoin after administration. It is rapidly and completely absorbed when given intramuscularly and can also be given intravenously at three times the rate of phenytoin. Because it has no intrinsic action before conversion, it is believed to have the same onset of action as phenytoin. Its primary disadvantage is cost, averaging twenty-fold more than phenytoin.
- Phenobarbital, a long-acting barbiturate, may lead to hypotension as well as profound respiratory depression and apnea. Rate of infusion should not exceed 100 mg/min. It is generally reserved for cases in which benzodiazepines and phenytoin have failed.
- Agents for refractory SE—All patients with refractory ictus require EEG monitoring as well as ventilatory support. Pressors may also be necessary in the setting of hypotension.
  - Midazolam—Discussed above
  - Propofol—This is a nonbarbiturate anesthetic agent that also has antiepileptic effects. It has a rapid onset of action and a quick recovery time after the drug is discontinued.
  - Pentobarbital—This barbiturate has more pronounced side effects than both midazolam and propofol. Patients will often require pressors because of significant hypotension and myocardial suppression.
- Adjunctive therapy—Includes both pyridoxine and magnesium.

Disposition

- All patients with SE require ICU admission.
- Patients with neurologic disorders, systemic disease or electrolyte abnormalities (e.g., neurosurgical lesions, CNS infection, hepatic or renal dysfunction, hyponatremia) require admission and management of the underlying disease process.

Suggested Reading

1. Bradford JC, Kyriakedes CG. Evaluation of the patient with seizures: An evidence based approach. Emerg Med Clin North Am 1999; 17:1.
2. Haafiz A, Kissoon N. Status epilepticus: Current concepts. Ped Emerg Care 1999; 15:2.
3. Hanhan UA, Fiallos MR, Orlowski JP. Status epilepticus. Ped Clin North Am. 2001; 48:3.
4. Krohmer JR, Sheets CA. Seizures. In: Hamilton GC, ed. Presenting Signs and Symptoms in the Emergency Department. Evaluation and Treatment. Baltimore: Williams and Wilkins, 1993.
5. Roth HL, Drislane FW. Seizures. Neurol Clin North Am 1998; 16:2.
6. Seizures and syncope. In: Simon RP, Aminoff MJ, Greenberg DA, eds. Clinical Neurology 4th ed. Stamford: Appleton and Lange, 1999.
7. Yoon Y, Jagoda A. New antiepileptic drugs and preparations. Emerg Med Clin North Am 2000; 18:4.

CHAPTER 5

GI Emergencies

Susan Stone and Andrew S. Kassinove

Introduction

Abdominal pain accounts for over five million visits annually to Emergency Departments. Between 20-40% of these patients will require inpatient admission. However, it is often difficult to arrive at the etiology of abdominal pain in the course of the ED visit. In particular, women who are of child-bearing age or pregnant, children and elderly patients create a diagnostic challenge (due to atypical presentations). Unfortunately, many of the disease processes share similar clinical presentations and may be difficult to sort out by history alone. Physical examination and laboratory evaluation can both lack sensitivity, making the job of the emergency physician difficult. Therefore, the diagnosis of abdomen pain of unclear etiology is a common diagnosis made in cases where the underlying pathology is not clear. This obligates the physician to provide patients with adequate reexamination to monitor the progression of the process.

While the discussion of abdominal disorders in this chapter is not exhaustive, the most common etiologies are reveiwed.

As a general rule, elderly patients presenting with abdominal pain form a unique group. Despite lack of identification of a focal disease process on initial presentation, at least half will have a disorder requiring surgical intervention. Physical examination will often lack sensitivity, as will laboratory evaluation. Women of child-bearing age and young children will have atypical presentations of common disorders, such as appendicitis and may require more observation time.

Despite technology in imaging and laboratory diagnostics, a large portion of patients will still have undifferentiated abdominal pain, requiring close follow-up and referral.

Mesenteric Ischemia

Risk Factors/Etiology

- Age >50 yr old
- Occlusive disease (80% of mesenteric ischemia) occurs 50% of the time from the sudden occlusion of the superior mesenteric artery originating from a proximal source and 25% of the time as local thrombosis. Mesenteric venous thrombosis is the cause only 5% of the time. Prolonged occlusion can result in both proximal and distal reactive vasospasm, further aggravating the insult to the intestines.
- Occlusive disease often results from the propagation of left atrial or ventricular thrombi that fragment during or after a dysrhythmia or from atherosclerotic disease at the origin of the SMA itself. Thus, the major risk factors are recent MI, peripheral vascular disease, and cardiac dysrhythmias. Mesenteric venous thrombosis occurs during the classic hypercoagulable states as well as during malignancy, abdominal trauma, and estrogen therapy.

Emergency Medicine, edited by Sean Henderson. ©2006 Landes Bioscience.

- Nonocclusive disease (20% of mesenteric ischemia) occurs during low flow states, such as cardiogenic shock, hypovolemia, or sepsis resulting in reactive vasoconstriction of the splanchnic circulation. Additionally, regional vasospasm can result from use of vasoactive medications, such as digoxin, diuretics, cocaine, or vasopressin.

Clinical Presentation and Diagnosis

- The historical factors may be nonspecific, but the diagnosis should be pursued in any person >50 yr old with sudden onset of acute abdominal pain and with an associated low flow, atherosclerotic, or hypercoaguable disease state.
- Historical factors include abdominal angina, recent weight loss, or recent change in bowel habits.
- Signs and symptoms include nausea and vomiting, colicky, severe, diffuse abdominal pain associated with repeated bowel movements, and a diffusely tender abdomen. As the disease progresses peritoneal signs occur and are an ominous finding. Abdominal distention and rectal bleeding may be the only initial complaint in up to 25% of the cases. The only initial abnormality on physical exam may be the presence of fecal occult blood, occurring in over half of the cases.

- Laboratory tests are nonspecific. An elevated WBC count is common, but a normal count does not exclude the diagnosis. Additionally, metabolic acidosis with a base deficit, an elevated amylase, and evidence of hemoconcentration are sensitive (present in more than half the cases) but nonspecific findings.
- Plain radiographs are often normal early on but may be used to rule out other pathology. They may show pneumatosis intestinalis, portal vein gas, or thumb printing in late disease. CT scan and ultrasound may show edema of the bowel wall and mesentery, ascites, abnormal gas patterns, and evidence of mesenteric venous thrombosis. CT is 82% sensitive and 93% specific and is better than angiography for venous obstruction. Ultrasound is only 28% sensitive.
- Angiography remains the diagnostic gold standard with a sensitivity of 88% (92% arterial and 50% venous). Angiography is contraindicated in shock states or with patients on vasopressor therapy because they confound the diagnosis of nonocclusive mesenteric ischemia. False negative studies are common with mesenteric venous thrombosis.
- The differential diagnosis is large and includes all sources of acute abdominal pain, especially ruptured abdominal aortic aneurysm, perforated dudodenal ulcer, strangulated bowel obstruction, and urolithiasis. Common initial misdiagnoses include constipation, gastroenteritis, ileus, and small bowel obstruction.

Treatment

- Stabilization of the patient is of the utmost priority, with initial focus on the ABCs. Hypovolemia is common and should be corrected. Vasoactive drugs should be discontinued. CHF and dysrhythmias should be managed. A nasogastric tube should be placed for early decompression. Broad spectrum antibiotics should be begun early.
- Surgical consultation should be obtained immediately.

Abdominal Aortic Aneurysm (AAA)

Risk Factors/Etiology

- The diagnosis of aortic aneurysms has increased, reflecting an aging population and more liberal use of CT scan and ultrasound. However this diagnosis is frequently overlooked.
- It is necessary to classify the aneurysm as intact, ruptured or contained rupture.
- This process occurs most frequently in elderly patients over 65 yr of age.

- Risk stratification shows that males, tobacco use, atherosclerosis and hypertension are important predisposing factors. By age 80, 6% of men will have AAA. Younger patients with collagen vascular disease are also at risk of aortic dissection.
- Blunt abdominal trauma less commonly results in an aortic aneurysm.
- Patients with preexisting aortic grafts and those with immunosuppression may develop mycotic infections, where the aorta becomes infected and may rupture.
- Aneurysms result from dilatation of the aortic wall with an increase in diameter by over 50%. This dilatation is a mechanism of the artery to compensate for a proximal stenosis. Once the aneurysm exceeds 5 cm, the risk of rupture is up to 40%. The 5 yr survival after repair is almost 70%.

Clinical Presentation

- Classical presentation of pulsatile mass in the patient with abdominal pain and pulse deficits is not always present.
- Syncopal episodes with hypotension or transient hypotension alone in the elderly patient with abdominal pain should raise suspicion.
- Hematuria may result from dissection of the renal arteries. Over 95% of AAA will be infrarenal.
- Pulse deficits
- Patients presenting with an abdominal aneurysm with abdominal pain are ruptured until proven otherwise and surgical consult is mandatory. Use caution with results of CT scanning, as this may miss a rupture or leak of the aneurysm.
  If an aneurysm is identified on ultrasound or CT scan, the patient needs to be followed by a vascular surgeon.

5

Treatment

Management is broken down into three categories:

1. Asymptomatic: serial ultrasounds and close follow-up
2. Rupture stable: CT scan and surgical repair
3. Rupture unstable: surgical repair

The differentiation between a stable and unstable rupture is trivial as the process is dynamic. Therefore early surgical consultation is critical.

The perioperative mortality is over 25% secondary acute myocardial infarction in emergent surgery compared to fewer than 5% for elective. Therefore it is preferred, but not always possible, to prime the patient for the operating room.

Bowel Obstruction

Risk Factors/Etiology

- Small bowel obstruction is typically caused by postoperative adhesions, hernias, or tumors.
- Large bowel obstruction is caused by carcinoma, diverticulitis, volvulus, inflammatory bowel disease, radiation colitis, or foreign bodies. It is primarily a disease of the elderly.
- Sigmoid volvulus in the United States occurs in debilitated elderly people secondary to chronic severe constipation.
- Cecal volvulus is most common in 25-35 yr olds but may occur at any age. It is likely due to a hereditary hypofixation of the cecum to the posterior abdominal wall. Risk factors include marathon running, pregnancy, and prior abdominal surgery.

Clinical Presentation and Diagnoses

- Acute onset of severe intermittent abdominal pain followed by nausea and vomiting is the common clinical manifestation. Vomiting may be absent in distal obstructions.

The abdomen is diffusely tender and becomes progressively distended. Obstipation may be absent early on or in a partial obstruction, and its absence does not exclude the diagnosis. Peritoneal signs or fever suggest strangulation or perforation.

- Signs include a tympanitic distended abdomen, high pitched "tinkling" bowel sounds, or a tender mass in closed loop obstructions.
- Laboratory values are nonspecific. An elevated WBC count may be present in both simple and strangulated obstructions. Electrolyte abnormalities are late findings. Hemoconcentration may reflect third spacing of fluid.
- Plain radiographs are often diagnostic, demonstrating small bowel obstruction (SBO) in 50-60% of cases and suggesting it in 20-30% more. A supine abdominal film along with either a lateral decubitus or upright abdominal films are minimally needed for diagnosis. An upright chest film may be added to search for free air under the diaphragm indicating a perforated viscous. CT scan is 94% sensitive and 83% specific in diagnosing SBO. Ultrasound is 88% sensitive and 96% specific in diagnosing SBO. Plain abdominal films are the test of first choice.

- In small bowel obstruction, distension of the small bowel is seen, often with distal collapse. The small bowel is differentiated from the large bowel by the presence of "valvulae conniventes" which are numerous, narrowly spaced and cross the entire lumen. A "string of pearls" sign is highly suggestive of small bowel obstruction and is described as a line of air pockets in a fluid filled small bowel. Air fluid levels in a stepladder pattern are also suggestive of a small bowel obstruction.
- A large bowel obstruction is suggested by dilation of bowel with "haustra", which are widely spaced, do not cross the entire lumen, and are less numerous than the "valvulae conniventes." A sigmoid volvulus is diagnosed by a single dilated loop of large bowel in the mid abdomen in the classic "bent inner tube" configuration, 80% of the time. If not, sigmoid volvulus can be diagnosed by the classic "birds beak" sign on barium enema. Distended large bowel in the left lower quadrant with absence of right-sided gas may indicate a cecal volvulus.
- Differential diagnoses include gastroenteritis, mesenteric ischemia, adynamic ileus, and incarcerated hernia. The intermittent nature of the pain is suggestive of bowel obstruction but is also present in mesenteric ischemia.

Treatment

- Early nasogastric decompression, aggressive fluid resuscitation, broad spectrum antibiotics including coverage of Gram negatives and anaerobes, and early surgical consultation are the mainstays of treatment of small and large bowel obstructions. Up to 75% of partial small bowel obstructions and up to one-third of complete small bowel obstructions will resolve with decompression and fluid resuscitation alone. Strangulated obstructions indicated by fever, tachycardia, and/or localized tenderness are operative cases. Uncomplicated obstructions are usually initially treated conservatively, with surgery reserved for treatment failures.
- Octreotide may be useful in nonoperative cases of bowel obstruction by decreasing GI secretions and motility.
- Sigmoid volvulus is treated with rectal tube decompression or surgery and usually does not require aggressive fluid resuscitation as there is little third spacing. Cecal volvulus often requires surgery.

Disposition

- These patients are all admitted to the hospital, almost always under the care of a surgeon.
- "Never let the sun rise or set on a bowel obstruction."

Acute Appendicitis

Risk Factors/Etiology

- Appendicitis has a yearly incidence of 1/1000 persons with a lifetime incidence of 7%. There is a slightly increased incidence in males. The highest incidence occurs in 10-30 yr olds, with atypical presentations more common in the very young or very old and women of child-bearing age.
- Appendicitis is the most common surgical emergency in children.

Clinical Presentation and Diagnoses

- The classic description is of periumbilical, epigastric, or diffuse dull pain migrating over several hours to McBurney's point in the right lower quadrant, with the pain changing in character from dull to sharp as the overlying peritoneum becomes inflamed. Peritoneal signs, including involuntary guarding, rigidity and diffuse percussion tenderness may indicate perforation.
- The pain is less likely to be appendicitis if it has been ongoing for more than 72 h.
- Associated symptoms which increase the likelihood of appendicitis are anorexia or nausea and vomiting following the onset of abdominal pain. Less specific and less frequently associated symptoms include fever, chills, diarrhea, dysuria and frequency, and constipation. Constipation is a more common symptom in the elderly.
- The location of the pain is highly variable. 20% of surgically proven appendicitis presents without right lower quadrant pain. Retrocecal appendices or those displaced in pregnancy may cause flank pain. A pelvic appendix may irritate the bladder, resulting in suprapubic pain or dysuria, while a retroileal appendix may irritate the ureter, causing testicular pain. More than two-thirds of appendices lie within 5 cm of McBurney's point, with more inferior and medial.
- Frequently associated signs include low-grade temperature, abdominal, rebound, rectal, or cervical motion tenderness. Less commonly present are the psoas and obturator signs or a palpable mass.
- Associated laboratory values include a WBC count >10,000 and less frequently pyuria (>5 WBC's/hpf). However, between 10-60% of patients will initially have a normal white count.
- C-reactive protein (CRP) has been shown to be elevated in several studies when used serially to be helpful in excluding appendicitis, with the diagnoses being rare with two normal values drawn 12 h apart. However, one isolated value is neither sensitive nor specific.
- Perforation rates are commonly quoted at 20%, with children and elderly incurring perforation >70% of the time. Perforation is the most common malpractice claim for abdominal emergencies and the fifth most expensive claim overall in emergency medicine.
- Diagnostic studies are merely ancillary and should not replace the clinical impression. Abdominal plain films have little or no utility and should not be routinely ordered, as even the finding of an appendicolith are neither sensitive nor specific for appendicitis. An upright CXR finding of free air is neither sensitive nor specific for perforated appendicitis.
- The two radiologic studies to consider in cases of diagnostic uncertainty are abdominal ultrasound and helical CT. Ultrasound has reported sensitivity up to 93% and specificity up to 95% and is the preferred test in children and pregnant women. Triple contrast oral, rectal, and IV CT of the abdomen has a sensitivity approaching 100% and specificity of 95-98% when used with the latest CT scanners and experienced readers. Noncontrast helical CT has a sensitivity of 90% and a specificity of 91%.

- Additionally, some institutions are using the laparoscope to quickly look inside the abdomen. However, this technique often leads to open laparotomy. Less widely available, Technetium-99m labeled WBC scans have a reported sensitivity of 98% and specificity of 95% but are time consuming.
- Serial abdominal exams over 6-12 h will often aid in the diagnoses as the pain of appendicitis is progressive, leading to the diffusely tender, often rigid abdomen of a perforation. Several studies have shown observation is safe and effective.
- A urine pregnancy test should always be obtained in all women of child-bearing age.
- Uncommon mimics of acute appendicitis include DKA and Streptococcal pharyngitis in young children. Other diagnoses to consider include testicular torsion, ruptured ectopic pregnancy, peptic ulcer disease, billiary tract disease, diverticulitis, abscesses, renal colic, pyelonephritis, bowel obstruction, and abdominal aortic aneurysm.

Treatment and Disposition

- High clinical suspicion: IV fluids if dehydrated, NPO, antibiotics (Cefotetan or Cefoxitan), early surgical consultation with early appendectomy.
- Moderate clinical suspicion: IV fluids if dehydrated, NPO, surgical consultation, adjunctive test (ultrasound or CT scan). If CT scan or ultrasound positive then early appendectomy and antibiotics. If negative, then observation with serial abdominal exams.
- Low clinical suspicion: surgical consultation, follow-up in 12 h or earlier if symptoms persist or worsen.
- Acceptable negative appendectomy rates are approximately 20%.

Colonic Diverticulitis

Risk Factors/Etiology

- 96% of patients are older than 40 yr of age.
- Most commonly confined to the sigmoid colon.
- Almost exclusively found in western civilizations, due to a relatively low fiber diet.
- 50% of people older than 65 will have diverticula, with 10-20% of those eventually developing diverticulitis.
- Inflammation occurs when fecal matter is trapped in a diverticular sac. Microperforations in the colon then occur producing a pericolic abscess or even peritonitis.

Clinical Presentation and Diagnoses

- Persistent abdominal pain, initially vague and diffuse, later localizing to the left lower quadrant is the most common presentation of sigmoid diverticulitis.
- Constipation, low grade fever, and malaise are common. Dysuria and frequency are also common due to irritation of the nearby bladder and ureter. Less common are nausea, vomiting, anorexia, or diarrhea.
- Exam will reveal left lower quadrant tenderness and may reveal peritoneal signs if significant spillage of bowel contents has occurred. A mass may occasionally be palpated. The rectal exam may reveal local tenderness and will often be fecal occult blood positive.
- An elevated WBC count with a left shift is common. Iron deficiency anemia is uncommon and should prompt a look for other causes, such as carcinoma.
- The diagnoses of sigmoid diverticulitis are often made on clinical grounds alone. Other diagnostic modalities include plain abdominal films to rule out obstruction, followed by CT with oral contrast, IV, and rectal contrast, to confirm the diagnoses (sensitivity and specificity approaching 100%), look for an abscess, and rule out other pathology such as appendicitis or carcinoma. An upright chest X-ray may also be obtained to look for free air under the diaphragm, signaling a perforated viscus.

Graded compression ultrasound has a sensitivity of 84-98% and a specificity of 93-97% for diverticulitis. Barium enema is not recommended as it may increase the chance of perforation.

- Cecal or right-sided diverticulitis is difficult to distinguish from appendicitis. Other diagnoses high in the differential include colon carcinoma with localized perforation, ischemic colitis, ulcerative colitis, and bacterial colitis.
- Complications include abscess or fistula formation, perforation, or colonic obstruction. Colovesicular fistulas present with pneumaturia, fecaluria, dysuria, frequency, or incontinence. Colonic obstruction is rarely complete.

Treatment

- Most episodes can be managed medically with admission to the hospital for bowel rest, IV antibiotics and fluids and analgesics.
- Antibiotic regimens must cover both anaerobic and Gram-negative bacteria. Regimens may include Cefoxitin 2-3 g IV q 8 h or triple therapy with Ampicillin, Gentamycin, and Metronidazole.
- Mild disease in reliable patients may be treated as an outpatient with cephalosporins or quinilones, pain medicine, a clear liquid diet and early follow-up.
- Emergent surgical consultation should be obtained if evidence of generalized peritonitis.

Acute Pancreatitis

Risk Factors/Etiology

- The underlying etiology of pancreatitis is most commonly due to gallstones or alcoholism. However the differential diagnosis is extensive.
- The presence of obstructing gallstones and cholangitis must be excluded as these entitities represent surgical emergencies.

Clinical Presentation and Diagnoses

- The typical presentation of pancreatitis is epigastric pain radiating to the back. Patients may also present with refractory vomiting and diffuse abdominal pain.
- Pancreatitis is a multisystem disorder which can lead to overwhelming organ failure and death.
  - Pulmonary: Hypoxemia, pleural effusions and pulmonary edema
  - Cardiac: Tachycardia, hypotension and shock
  - Neurologic: Confusion and coma
  - Renal: Azotemia and oliguria
  - Metabolic: Hypocalcemia, hyperglycemia, ketosis and hypertriglyceridemia
- Ranson's criteria indicators of morbidity and mortality:
  - 0-2- 2% mortality, 3-4- 15%, 5-6- 40%, 7-8- 100%
  - Criteria on admission:
    - age >55 yr
    - white blood cell count >16000 IU/L
    - blood glucose >200 mg/d
    - serum LDH >350 IU/L
    - SGOT >250 IU/dL
- At 48 h
  - Hematocrit fall >10%
  - BUN increase >8 mg/dl
  - Arterial PO_2 <60 mm hg
  - Base deficit >4 meq/L
  - Fluid sequestration >6 L

- Lipase has higher diagnostic accuracy and is more cost effective compared to amylase. Amylase is both of salivary and pancreatic origin, but most labs do not differentiate between the two. Therefore the results of an elevated amylase are misleading. Lipase has a higher sensitivity.
- Imaging studies are indicated in patients with intrabdominal sepsis and to rule out complications. CT scan and ultrasound can be used. If suspect pancreatitis but no lab diagnosis may be indicated.

Treatment

- Patients with pancreatitis may have baseline hypoxemia which progresses to ARDS if not managed early. Once ARDS develops, pancreatic injury is exacerbated. Therefore application of oxygen early on is preferred.
- Caution should be used with administration of insulin as there is exaggerated response with pancreatitis and profound hypoglycemia may result.

5

- These patients tend to be hypocalcemic and may benefit from supplementation.
- Nasogastric suction is no longer recommended as it prolongs hyperamylasemia and increases pain medication requirement. However, it may be necessary in the case of duodenal obstruction. Cimetidine, glucagon and atropine fail to show any benefit in alleviating symptoms or complications.
- Patients will require admission and monitoring for complications such as pseudocyst, gastrointestinal bleeding, cholangitis, ARDS, metabolic disturbances, encephalopathy and DIC.
- CT scanning in patients with overwhelming sepsis without a source may identify cases of pancreatitis. Autopsy studies have shown many missed cases of fatal pancreatitis in this subgroup of patients.
- Mild cases without evidence of biliary tract involvement or systemic involvement may be candidates for outpatient management if close follow-up is available.

Peptic Ulcer Disease and Gastritis

Risk Factors/Etiology

- *H. pylori* (most common cause of ulceration)
- NSAIDs, aspirin common cause of bleeding ulcers and gastritis
- Alcohol
- Bile reflux
- Pancreatic enzyme reflux
- Gastrinoma (Zollinger-Ellison syndrome)
- Severe stress (e.g., trauma, burns)

Clinical Presentation and Diagnosis

- Gastritis and PUD usually are indistinguishable in the ED without endoscopy.
- Typically epigastric/left upper quadrant burning pain that may radiate to the back after meals.
- May be relieved by food, antacids (duodenal), or vomiting (gastric).
- A type and cross match should be sent if the patient is actively bleeding.
- Helicobacter testing is useful in the primary care setting to guide treatment.

Treatment

- Iced saline lavage should never be performed. This is a *dangerous action*. It had been thought that cooling would cause vasoconstriction and accelerate cessation of bleeding, but this is now known to cause arrhythmias without decreasing bleeding.
- Nasogastric tube. While evacuation of blood from the gut may be one benefit from nasogastric lavage, this procedure can cause great discomfort. It may also be

nondiagnostic in a large number of cases, and ultimately the patient will need endoscopy. Therefore early consultation with the gastroenterologist is key.

- Etiology of gastrointestinal bleeds on endoscopy in those with history of varices found the majority of patients with bleeding were bleeding from acute gastric duodenitis followed by gastric ulcer then duodenal ulcer, varices, Mallory Weiss, esophagitis.

H_2 Blockers and Proton Pump Inhibitors

- Proton pump inhibitors and H_2 blockers are not effective in the acute phase of bleeding. However proton pump inhibitors may help decrease rebleeding after endoscopy.

Vasoconstrictors

- Vasopressin not too beneficial and risk of hypertension, stroke, coronary ischemia
- Dose-related decrease in coronary flow and cardiac output
- Octreotide has a similar mechanism as somatostatin but is more potent and longer acting decreasing splanchnic blood flow and inhibiting gastric acid secretion
- Patients required less blood transfusion, fewer required surgical and endoscopic intervention after a 100 mcg bolus followed by 25 mcg/h when compared to H_2 blockers
- Given to patients with hematemisis and/or tarry stool and evidence of bleeding peptic ulcer on early endoscopy—subset to which it is applicable

Outpatient Management of UGI

Low risk criteria that have been identified that may warrant outpatient management:

- No high risk endoscopic feature/varices/portal hypertenisve gastropathy
- No debilitation
- No orthostatic vital sign change
- No liver disease or concomitant disease
- No anticoagulation or coagulopathy
- No fresh hematemesis
- No severe anemia
- Adequate home support

High Risk Criteria

- **BLEED** criteria good screening tool to decide which groups are likely to develop related in-hospital complication. **B**leeding, **L**ow systolic blood pressure, **E**levated prothrombin time, **E**rratic mental status, **D**isease comorbid.
  - 33% of patients with BLEED criteria had complications.
  - Patients with an upper gastrointestinal bleed with signs of hypoperfusion such as syncope, confusion, dizziness or hypotension should have myocardial ischemia excluded.
- Upper GI bleed
  - Peptic ulcer disease:
    - Risk groups include alcohol use, cigarette use, medications such as NSAID use.
  - Variceal bleeds:
    - Risk groups include those with liver failure and portal hypertension.
    - Variceal bleeds are frightening but remember that most UGI bleeds in patients with varices are from sources other than the varices.
- If do have variceal bleed, drug therapy to decrease splanchnic flow is first line therapy (vasopressin and octreotide):
  - Vasopressin decreases cardiac output, increases systolic hypertension, arrhythmias and bradycardia. inhibits fibrinolysis and may therefore interfere with hemostasis.
    - Requires a continuous infusion due to short half life (20 min)

- Somatostatin is more effective in many regards—localized to splanchnic vasculature and produces fewer effects but no difference in all cause hospital mortality
  - When compared, long acting vasopressin and somatostatin (octreotide) equivalent
  - Given the longer half life of octreotide and fewer adverse risks then may be better
  - In last abstract initial control of bleeding achieved in 90% of patients with sclerotherapy and 84% octreotide group

Massive GI Hemorrhage Treatment

Massive upper GI bleeding is a far more common emergency than lower GI bleeding. It may present as hematemesis, melena or simply shock with a positive stool test for blood or NG aspirate. Sources include peptic ulcer and gastroesophageal varices. Massive lower GI bleeding is caused by angiodysplasia and diverticular disease.

Primary Survey Will Require Vigorous Suctioning of the Airway

- Wide open normal saline with multiple large bore IV's (16 gauge or greater) are indicated to treat shock. Type O blood should follow the first 2 L via rapid transfuser, until type-specific and then cross-matched blood is available.
- Fresh frozen plasma is indicated for suspected coagulopathy (e.g., with stigmata of chronic liver disease). NG tube for lavage and gastric emptying is indicated and may be used as one gauge of active bleeding. Emergency endoscopy for hemostasis can be performed by qualified personnel.
- In addition to emergent consultation for endoscopy, emergent surgical consultation should be sought.
- Octreotide (50-100 mcg IV bolus followed by an infusion of 50 mcg/h) is indicated as an adjunct to other therapies or when other therapies are unavailable for massive upper GI bleeding.
- Balloon tamponade via Sengstaken/Blakemore or similar tube should be attempted in the event of severe uncontrolled upper GI hemorrhage.
- Apparent massive lower GI hemorrhage may be from an upper GI source (which may be revealed by NG tube or upper GI endoscopy).
- If these are negative, either interventional radiological methods or laparotomy will be required to stop bleeding.
- Although not entirely reliable in acute hemorrhage, serial bedside hemoglobin determinations are helpful in guiding resuscitations.
- Hypothermia should be avoided by covering the patient with warm blankets and using warmed IV fluids and blood products.
- Stigmata of chronic liver disease as well as purpura and petechiae should be sought on initial survey. This may assist in predicting the site of bleeding (e.g., the possibility of esophageal and gastric varices) and need for factor replacement during the resuscitation.
- A history of aortic repair surgery may indicate a aortoenteric fistula, which is managed operatively.

Cholecystitis

Risk Factors/Etiology

- Common illness with over 15 million Americans affected.
- Present frequently for pain control.
- Typically in adult females. Tends to be associated with fatty meal. Incidence rises with age. While uncommon in children, presence may suggest underlying disorder (sickle cell, hemolytic anemia).

Clinical Presentation and Diagnoses

- May present with abdominal pain in right upper quadrant or epigastrium which tends to be constant and severe.
- Pain can radiate to shoulder or back.
- Nausea and vomiting common.
- Murphy's sign: inspiratory arrest during palpation of RUQ.
- If fever or refractory pain is present or consider cholecystitis.
- Gallstones are either cholesterol (most common), pigment (associated with hemolytic anemia).
- Elevated liver function studies suggest common duct obstruction, cholangitis, cholecystitis or hepatic involvement.
- Ultrasound should be initial imaging study since it is over 90% sensitive and allows visualization of common bile duct:
  - Duct size over 10 mm suggests obstruction
  - Gallbladder wall thickening over 5 mm or pericolic fluid suggestive of cholecystitis.
- HIDA scanning may detect obstruction or cholecystitis if stone is at neck of gallbladder and nonvisualized on ultrasound.
- Routine abdominal films are of no utility.

Treatment

- The first priority is analgesia. NSAIDs have been used with great success. Narcotic analgesics with antiemetic are also part of the initial therapy. Unsuccessful pain control or cholecystitis needs hospital admission.
- Surgical consultation is required in patients requiring admission or those diagnosed with cholangitis, common bile duct dilatation and/or cholecystitis.
- Antibiotics are indicated for acute cholecystitis, cholangitis, or common bile duct dilatation. First line antibiotics include amp/sulbactam, flouroquinolone and flagyl or pipercillin/tz.

Hernias

Risk Factors/Etiology

- A hernia occurs when a viscus internally or externally protrudes from its normal cavity.
- Risk factors include prematurity, family history, genitourinary abnormalities, ascites, peritoneal dialysis, ventriculoperitoneal shunt, cystic fibrosis, lung disease, pregnancy, or wound sites.
- Most hernias are inguinal and most occur in men.
- An indirect inguinal hernia occurs lateral to the inferior epigastric vessels. These are the most common hernias in both men and women.
- A direct inguinal hernia is an acquired defect medial to the inferior epigastric vessels and occurs mostly in older adults.
- A femoral hernia occurs below the inguinal ligament in the femoral canal, occurs mostly in women, is very rare, and often becomes strangulated.
- An umbilical hernia occurs in infants. Most close spontaneously by one year of age.
- Spigelian hernias occur just lateral to the rectus sheath in the abdominal wall and are very rare.
- A Richter hernia involves incarceration of the wall of a hollow viscus.
- An incisional hernia occurs at the site of a previous wound.

Clinical Presentation and Diagnosis

- An asymptomatic hernia presents without pain or tenderness, is reducible, and will enlarge with standing or increased intra-abdominal pressure.

- An incarcerated hernia is not reducible and is painful. It may or may not present with symptoms of bowel obstruction, including nausea and vomiting.
- A strangulated hernia presents as a toxic appearing incarcerated hernia. Systemic toxicity secondary to ischemic bowel may be present. If necrotic bowel is suspected do not attempt to reduce the hernia and return the dead bowel into the abdomen.
- A femoral hernia will present with medial thigh and or groin pain.
- Diagnosis is based on clinical exam. Abdominal radiographs are useful to exclude a bowel obstruction.
- A hernia must be distinguished from a scrotal hydrocoele, which will transilluminate with a light source, while a hernia will not.

Treatment

- An incarcerated hernia without signs of bowel obstruction may be reduced with pain meds and gravity and referred for surgical follow-up.
- Infants with inguinal hernias are at high risk for incarceration and should be urgently referred to a pediatric surgeon for repair. In contrast, umbilical hernias often have a benign course and do not require urgent referral.
- Strangulated hernias or any toxic-appearing patient should be started on broad spectrum antibiotics, IV fluids, nasogastric decompression, and receive emergent surgical consultation.

Esophageal Emergencies

Risk/Factors Etiology

- 75% of esophageal perforations are due to iatrogenic perforations.
- Boerhaave syndrome is the cause of 10% of esophageal perforations with 10% of these due to trauma.
- Boerhaave syndrome occurs because the esophagus lacks a serosal layer, Thus, the wall is weaker and may rupture at a lower intraluminal pressure, usually with severe vomiting. Most perforations occurs in the lower third of the esophagus in the left posterolateral region.
- All other cases of esophageal perforation are caused by foreign bodies or traumatic perforations.
- Boerhaave syndrome is more common in middle-aged men after a drinking and eating binge.
- Esophageal foreign bodies are most common in children. Most objects pass spontaneously. The most common site of obstruction is the cricopharyngeal narrowing (C6).

Clinical Presentation and Diagnosis

- Delay in diagnoses of Boerhaave syndrome is common and leads to a high mortality rate. Classic presentation for spontaneous rupture is vomiting followed by severe chest pain. However, many patients have less dramatic presentations. Patients may present with abdominal pain, dyspnea, and hematemesis.
- Subcutaneous emphysema may be palpable in the neck or chest. Hamman's crunch may be heard in 20% of cases over the heart. Patients usually appear acutely ill, hypotensive, and septic depending on the delay in presentation.
- Esophageal foreign bodies usually present in adults with retching, vomiting, coughing, dysphagia or choking. In kids, common symptoms also include refusal to eat and stridor.
- Diagnosis of foreign body ingestion includes plain radiographs. Patients with foreign body sensation and negative radiographs should undergo direct laryngoscopy. Specialty consultation should be obtained prior to contrast radiographs.
- Coins are visible on AP radiographs with the flat side visible if in the esophagus and on end if in the trachea.

Treatment

- Treatment of Boerhaave syndrome or a delayed iatrogenic perforation in the emergency department includes rescusitation of shock, broad spectrum antibiotics, and emergent surgical consultation.
- Esophageal foreign bodies require emergent surgical consultation if the patient appears toxic, has an acute abdomen, bloody stools, or persistent vomiting.
- Button battery ingestion requires emergent GI consultation as esophageal burns can occur within 4 h and perforation within 6 h.
- Food impaction can be treated conservatively if the patient can tolerate their own secretions. Intravenous glucagon, 1 mg, can be used to relax esophageal smooth muscle. Sublingual nitroglycerin can be used to relax lower esophageal sphincter pressure.

Anal/Rectal Disorders

Risk Factors/Etiology

- Hemorrhoids are more common in pregnancy and are associated with constipation and straining bowel movements, obesity, and chronic liver disease. Internal hemorrhoids are located above the dentate line and are painless. External hemorrhoids are located below the dentate line and are painful.
- Anal fissures are the most common cause of painful rectal bleeding.
- Pilonidal cysts usually occur in young people and are a chronic recurring reaction to an ingrown hair.
- Perirectal and perianal abscesses are common in diabetics and drug abusers.

Clinical Presentation and Diagnoses

- Internal hemorrhoids cause painless bright red blood with defecation. They are only visible through an anoscope.
- External hemorrhoids cause pain with defecation and are usually visible on exam.
- Rectal bleeding must be referred for further evaluation for malignancy.
- Other causes of rectal pain include foreign body, venereal proctitis, trauma, abscesses, and anal fissures.
- Anal fistulas present with malodorous bloody discharge through the fistula.
- Anal fissures present with painful bowel movements, with the pain resolving between bowel movements.
- Perianal and perirectal abscesses present as a tender red mass and may have concurrent fever and leukocytosis.

Treatment

- Most hemorrhoid patients may be managed conservatively with sitz baths, good hygiene, bulk laxatives, and stool softeners. Thrombosed external hemorrhoids should be referred to a surgeon, or may be excised and the clots removed in the ED if conservative measures have failed.
- Surgical referral is needed for anal fistulas.
- Anal fissures may be treated conservatively with sitz baths and local analgesics.
- All but the most simple perianal abscesses should be drained and followed by a surgeon. Perianal abscesses should be drained and packed, with antibiotics only necessary in diabetics and other immunocompromised hosts.
- Most rectal foreign bodies can be removed in the ED after adequate analgesia. Antibiotics and surgical or gastroenterological consultation should only be obtained in cases of high risk or perforation or with peritoneal or other toxic signs.
- Pilonidal cysts may be drained and packed in the ED, with surgical referral appropriate for definitive removal of the cyst.

5

Special Considerations

Elderly Patients with Abdominal Pain

Elderly patients presenting with abdominal complaints warrant a complete evaluation.

- Up to 50% of those over 65 yr of age will ultimately have a process evolving that requires medical or surgical intervention. Therefore, it is judicious to check laboratory and imaging studies.
- While physical examination is important, elderly patients frequently present atypically. They will often lack fever and leukocytosis. For instance, it is well documented that elderly patients will have perforated appendicitis at an early time and with a less obvious exam, than their younger counterparts. Higher morbidity and mortality from higher rates appendiceal perforation is the rule. After 50 yr of age perforation rates are from 32-70%!

Delays in the diagnosis are typically due to:

- Atypical presentations
- Anatomy
  - Less blood supply to appendix most likely
  - One-third of perforations are in those with symptoms <24 h.

Additionally, the risk of solid masses increases as does the presence of intra-abdominal infections. Elderly patients often will lack vital sign abnormalities. They may have hemoperitoneum with normal vital signs.

HIV and Abdominal Pain

The immunosuppression of HIV creates diagnostic challenges. All the usual etiologies should be considered such as appendicitis, cholelithiasis or nephrolithiasis. However, there are a few unique considerations.

Medication related etiologies of abdominal pain in this group include pancreatitis, acalculous cholecystitis and nephrolithiasis.

Immunosuppressive complications of their illness may include infection of the biliary tract with M.A.I. or abdominal lymphomas or Kaposi's sarcoma. They may also present with parasitic infections such as amebiasis and liver cysts.

The syndrome of HIV diarrhea is common and a chronic condition. They may also develop CMV infection through the gastrointestinal tract and have perforation.

Suggested Reading

1. American College of Emergency Physicians. Clinical Policy: Critical issues for the initial evaluation and management of patients presenting with a chief complaint of nontraumatic acute abdominal pain. Ann Emerg Med October 2000; 36:406-415.
2. Rao PM, Rhea JT, Novelline RA et al. Helical CT techniques for the diagnoses of appendicitis: Prospective evaluation of a focused appendix CT examination. Radiology 1997; 202:139-144.
3. Albu E, Miller BM, Choi Y et al. Diagnostic value of a C-reactive protein in acute appendicitis. Dis Col Rect 1994; 37(1):49-51.
4. Ramsden WH, Mannion RA, Simpkins KC et al. Is the appendix where you think it is—and if not, does it matter? Clin Radio 1993; 47(2):100-103.
5. Lyons D, Waldron R, Ryan T et al. An evaluation of the clinical value of the leukocyte count and sequential counts in suspected acute appendicitis. Br J Clin Pract 1987; 41:794-796.
6. Nase HW, Kovalcik PH, Cross GH. The diagnoses of appendicitis. Am Surg 1980; 46:504-507.

7. Wagner JM, McKinney WP, Carpenter JL. Does this patient have appendicitis? JAMA 1996; 276:1589-1594.
8. Colucciello SA, Lukens TW, Morgan DL. Assessing abdominal pain in adults: A rational cost-effective, and evidence-based strategy. Emergency Medicine Practice 1999; 1:1-16.
9. Trott AT, Lucas RH. Acute abdominal pain. In: Rosen, ed. Emergency Medicine Concepts and Clinical Practice. 4th ed. 1998:1888-1901.
10. Greenfield RH, Henneman PL. Disorders of the small intestine. In: Rosen, ed. Emergency Medicine Concepts and Clinical Practice. 4th ed. 1998:2005-2019.
11. Bitterman RA. Disorders of the large intestine. In: Rosen, ed. Emergency Medicine Concepts and Clinical Practice. 4th ed. 1998:2027-2030.

Chapter 6

Genitourinary Emergencies

Chi Lee, Sean O. Henderson and Sabrina Grassl

Urinary Tract Infections

- Urinary tract infections (UTIs) are one of the most frequently encountered infectious processes diagnosed by emergency physicians (EPs). The epidemiology of UTIs varies with sex and age.
  - It is estimated that women have about a 50% lifetime occurrence rate for a UTI.
  - The prevalence of UTIs in febrile infants is about 5%, while in the geriatric population the prevalence approaches 20%.
  - While the diagnosis of UTI is relatively straightforward, the treatment and ultimate disposition of patients depends on special circumstances such as extremes of age, pregnancy, treatment failures, and underlying medical conditions.
- UTIs are most often caused by Gram-negative aerobic bacilli, the most common of these being *E. coli* in 80% of cases.
  - The second most common pathogen (10-20%) is *Staphylococcus saprophyticus*, a coagulase-negative Gram-positive bacteria.
  - *Proteus mirabilis*, Klebsiella, and Enterococcus account for <5% of the remaining infections.

Clinical Presentation

- Classically the signs and symptoms of lower UTIs are dysuria, frequency, urgency, hesitancy, hematuria, and suprapubic pain.
- Upper UTIs (pyelonephritis) typically present with fever, chills, flank pain, nausea, vomiting, anorexia, and associated costovertebral angle (CVA) tenderness.
  - Some studies have shown that up to 50% of women with classical lower UTI symptoms have silent kidney involvement.
- In women, a history of vaginal discharge should always be elicited, and a pelvic exam, if indicated, will allow one to rule out PID, cervicitis, or vaginitis as the cause of dysuria.
- Males with dysuria and discharge should undergo a urethral swab, which should be sent for gonorrhea and chlamydia cultures.

Diagnosis

- The diagnostic mainstay of a UTI is the urinalysis (UA).
  - A UA from a properly obtained midstream, clean-catch specimen is as accurate as that of a catheterized specimen, except in debilitated patients, patients of extreme ages, or the morbidly obese. For such patients, a catheterized specimen may be necessary.
- An initial screening test is the urine dipstick. Leukocyte esterase (LE) and nitrites may be present in UTIs.
  - The urine dipstick for LE has a reported sensitivity of 75-96% with a specificity of 94-98% in detecting >10 leukocytes per high-powered field.
  - The nitrite test detects the presence of bacteria that produce nitrite reductase and is highly specific (92-100%) but not nearly as sensitive (35-85%).

- Empiric treatment is appropriate in symptomatic patients with a positive LE test.
- If the urine dipstick is negative, urine microscopy is not indicated.
- Although direct microscopy techniques lack standardization, it is presently accepted that the presence of 8 leukocytes or more per mL of uncentrifuged urine constitutes pyuria.
  - In a patient with pyuria without bacteriuria, the diagnosis of sexually transmitted urethritis should be considered.
  - Microscopic hematuria is more commonly caused by a UTI than a sexually transmitted disease (STD).
  - In patients with pyelonephritis the UA will often show white blood cell casts.
- Most cases of uncomplicated UTIs do not necessitate a urine culture. But there are several important risk factors for complicated UTI in which urine cultures should be obtained. These include:
  - All children, adult males, and debilitated elderly
  - Immunosuppressed patients (HIV, steroid use, solid organ transplant patients)
  - Pregnant women
  - Treatment failures, recurrent UTIs, or previous antimicrobial therapy within 2 wk
  - Hospitalized (or recently) patients
  - Patients with chronic indwelling catheters or recent instrumentation
  - Acute pyelonephritis
  - Patients with preexisting anatomic urologic abnormalities or urinary tract obstruction
  - Patients with serious medical diseases (DM, sickle cell anemia, cancer)
- Additional laboratory tests such as CBC, electrolytes, BUN and creatinine are optional, and should be tailored to each individual.
- Blood cultures are of little value.

Differential Diagnosis

- For lower UTI:
  - Urethritis
  - Cervicitis
  - PID/STDs
  - Vulvovaginitis
  - Prostatitis
  - Epididymitis
- For upper UTI:
  - All of the above
  - Nephrolithiasis
  - Renal abscess
  - Appendicitis
  - Cholecystitis
  - Lower lobe pneumonia
  - Diverticulitis

Special Circumstances

- Pyelonephritis
  - Clinical presentation is classic lower UTI symptoms (dysuria, frequency, etc) with associated CVA/flank pain, nausea, vomiting, dehydration or toxic appearance.
  - Obtain urine culture in all cases. CBC is optional, and blood cultures are not indicated.
  - Resuscitate early with 1-2 L of NS.
  - Early parenteral antibiotics
  - Antiemetics for vomiting, and analgesia for pain

- Admission criteria
  - Inability to tolerate oral intake (perisistent nausea/vomiting)
  - Pregnancy
  - Unstable vital signs and toxic appearance
  - Immunocompromised state (diabetes, cancer, transplant, patient AIDS, sickle cell disease)
  - Any underlying anatomical urinary tract abnormality or obstruction
  - Extremes of age
  - Poor social situation or unreliable follow-up
- Pregnancy
  - Asymptomatic bacteriuria (ASB) is defined as persistent colonization of the urinary tract without UTI symptomatology.
  - Untreated ASB is associated with increased incidence of preterm delivery and low birth weight infants. The progression of ASB to pyelonephritis is associated with significant maternal and fetal morbidity and mortality.
  - About 5-10% of pregnant women will have ASB.
  - ASB should always be treated with a 3-7 day course of oral antibiotics followed by culture at the end of treatment to ensure sterilization of the urine.
  - There is a paucity of literature on cystitis in pregnancy and its relationship to the risk of preterm birth, low birth weight, or pyelonephritis.
  - For cystitis, diagnosis can be obtained from urine culture. Treatment is the same as that for ASB, but should be extended to 7-10 days. Patients should be have a repeat urine culture done after treatment to ensure sterilization.
  - Acute pyelonephritis occurs in 1-2% of all pregnancies. Clinical signs and symptoms do not vary much from those of the nonpregnant population, but because of various anatomic and physiologic changes during pregnancy, a broader differential diagnosis must be considered. This includes normal back pain of pregnancy, gallbladder disease, renal abscess, nephrolithiasis, pulmonary embolism, and appendicitis.
  - Any evidence of renal involvement requires admission for IV antibiotics.
- Elderly
  - 20-50% of women over 65 yr of age have ASB. The incidence increases with age and is thought to be due to a combination of factors including changes in bladder emptying, increased incontinence (both fecal and urinary), and decrease in estrogen levels.
  - The elderly often lack the usual presenting signs and symptoms. They may present with fever, but hypothermia and euthermia are also not uncommon. The chief complaint may be altered mental status, nausea and vomiting, weakness, dizziness, abdominal pain, or respiratory distress.
  - In general, ASB in the elderly is not treated. However, elderly patients presenting with symptoms consistent with UTI, foul-smelling urine, or new symptoms of urge incontinence should be treated.
  - Acute pyelonephritis usually presents as a septic syndrome with fever, tachycardia and altered mental status.
  - Misdiagnosis of UTI in the geriatric patient is about 20-40% due to the wide range of presenting symptoms.
- Men
  - The incidence of bacteruria in the adult male is uncommon, but rises at the age of 50 with increasing incidence of prostatic hypertrophy. By age 65, the incidence of UTIs among males and females becomes equal.
  - UTIs in the male population are always considered complicated because the etiology is usually due to a structural or functional defect, which leads to incomplete voiding or obstruction.

- The diagnosis of UTI in the male is made by urine culture. Physical exam can assist in determining the cause as cystitis, prostatitis, or epididymitis.
- In males, a lower bacterial colony count (10^3-10^4 cfu/ml) is considered positive.
- Treatment is the same as that for women, but for the duration of 7-10 days. Nitrofurantoin, however, is not recommended because it does not achieve reliable tissue concentrations in the prostate.
- The presence of a UTI in males of any age warrants a genitourinary follow-up.

Treatment

- Treatment choice and duration is guided by several factors: anatomic localization of infection (lower vs. upper UTI), severity of symptoms, and special considerations mentioned above.
- Antibiotic selection should be guided by current and local resistance patterns to the most common urinary pathogens. For instance, resistance to Trimethoprim/Sulfamethoxazole (TMP/SMX) has been shown to be as high as 35% in parts of the West Coast, while resistances to ciprofloxacin and nitrofurantoin are 1% and 13% respectively.
- Adults
  - Acute uncomplicated UTI
    - TMP/SMX DS PO bid x 3 days
      - Only in areas where resistance is <10%
      - No history of recent hospitalization
      - No recurrent UTI in the past year
      - No recent use of TMP/SMX in the previous 6 mo
    - Ciprofloxacin 250 mg PO bid x 3 days
    - Levofloxacin 250 mg PO qd x 3 days
    - Amoxicillin clavulanate 875 mg PO bid x 3 days
    - Any oral cephalosporin x 3 days
    - Other fluoroquinolones x 3 days
    - Nitrofurantoin 100 mg PO bid x 7 days (caution in patients with G6PD deficiency)
    - Fosfomycin 3 g single oral dose
    - Treat dysuria with phenazopyridine 100-200 mg PO tid x 2 days, only. Warn patients of orange color change in urine and contact lenses.
  - Acute complicated UTI
    - Ciprofloxacin (or other fluoroquinolone) x 7 days
  - Acute uncomplicated pyelonephritis (outpatient)
    - Ciprofloxacin 500 mg PO bid x 7 days (preferred)
    - Levofloxacin 500 mg PO bid x 7 days
    - Other fluoroquinolones for 7 days
  - Acute uncomplicated pyelonephritis (inpatient)
    - All treatment x 14 days. Treat with IV antibiotics until afebrile for 24-48 h, then may be switched to oral meds for the remaining duration of therapy.
      - Ciprofloxacin 400 mg IV bid
      - Levofloxacin 250 mg IV qd
      - Ceftriaxone 1-2 g IV qd
      - Cefotaxime 1-2 g IV q 4-12 h
      - Ampicillin 150-200 mg/kg/day IV divided q 4 h + gentamicin 5-7 mg/kg qd
      - IV Pipercillin/tazobactam 3.375 g IV q 6 h
  - Complicated pyelonephritis, urosepsis, and indwelling catheter
    - Treatment duration 14 days
      - Ciprofloxacin 400 mg IV q 8 h + gentamicin 5-7 mg/kg qd

6

        - Ampicillin 150-200 mg/kg/day IV divided q 4 h + gentamicin 5-7 mg/kg qd
        - Pipercillin/tazobactam 3.375 g IV q 6 h
        - Ticarcillin/clavulinic acid 3.1 g IV q 6 h
        - Imipenim 0.5 g IV q 6 h
- Pregnancy
    - Cystitis/Lower UTI: (3-7 days)
        - Amoxicillin
        - Cephalexin
        - Nitrofurantoin (increased risk of hemolytic anemia with G6PD deficiency)
        - TMP/SMX (in areas with low resistance only, contraindicated in first and third trimesters)
    - Pyelonephritis: (10-14 days)
        - Ampicillin + gentamicin
        - Cephalosporins (cetazolin)
        - Aztreonam
        - TMP/SMX (contraindicated in first and third trimesters)

Emphysematous Pyelonephritis

- Emphysematous pyelonephritis (EPN) is a rare acute necrotizing parenchymal and perirenal infection caused by gas-forming uropathogens, namely *E. coli, K. pneumonia,* and *Proteus spp.*
- The condition predominantly affects diabetic patients, with high tissue glucose levels providing the substrate for carbon dioxide-producing microorganisms.
- EPN preponderantly affects females over males (6:1), which may be due to the increased susceptibility to UTI in females, and all the documented cases of emphysematous pyelonephritis have been in adults.
- The left kidney is more commonly affected than the right, reflecting the preponderance of left-sided urinary tract obstruction.

Clinical Presentation

- The most common presentation of EPN is fever, flank pain, and pyuria, a clinical picture not significantly different from a classic upper UTI.
- Thrombocytopenia, acute renal function impairment, disturbance in consciousness, and shock can be initial presentations.
- The overall mortality is described to be around 40%, with delay in diagnosis and treatment contributing to both morbidity and mortality.

Predictors of Outcome

- The most reliable predictor of outcome in EPN has been determined to be serum creatinine.
- Patients with serum creatinine levels >1.4 mg/dl had an increase in post-test probability of death from 69-92% in one study.
- Platelet counts 60,000/mm or less also indicated higher risk of mortality.
- Additionally, disturbance of consciousness and shock are associated with mortality and poor outcome, explained by expected poor prognosis of CNS and cardiovascular dysfunction.

Diagnosis

- The diagnosis of EPN is classically made by demonstrating the presence of gas in renal or perinephric tissue by plain abdominal X-ray film or by renal ultrasound.
- When present, a crescent shaped collection of gas over the upper pole of the kidney is more distinctive than mottled gas shadows, which are often mistaken for bowel gas.

- As the infection progresses, gas extends into the perinephric space and retroperitoneum. However, gas could be demonstrated only on one-third of plain abdominal radiographs in some studies.
- Ultrasonography usually demonstrates strong focal echoes, suggesting the presence of intraparenchymal gas; however, it may again be difficult to distinguish the necrotic gas-filled area from gas in the bowel. IVP is rarely of value as the affected kidney usually is nonfunctioning or poorly functioning.
- Obstruction has been demonstrated in approximately 25% of EPN cases.
- CT scan is the best means to localize gas and extent of infection.
  - The presence of streaky or mottled gas with or without bubbly and loculated gas appears to be associated with rapid destruction of renal parenchyma and a 50-60% mortality rate.
  - A gas pattern characterized by the presence of bubbly or loculated gas and the absence of streaky or mottled gas is associated with a more favorable prognosis.
- A renal CT scan should be performed to assess the degree of renal function impairment of the involved kidney and the status of the contralateral kidney.

Management

- Patients should be started on appropriate antimicrobial agents, and treatment of diabetes must be initiated.
- Obstruction of the affected kidney, if present, must be eliminated, and function of the contralateral kidney must be established, because of reported bilateral cases.
- At the same time, surgical intervention poses a substantial risk for patients with hemodynamic instability caused by fulminant infection and is not an appropriate option for bilateral kidney involvement.
- Previous studies have emphasized that surgical treatment must be complete extirpation. In more recent studies, CT-guided percutaneous drainage has proven successful in as high as 92% of patients in treating multiloculated, ill-defined, and extensive dissecting air and fluid collections in EPN, with 80% nephron-sparing.

Perinephric Abscess

- Perinephric abscess is a life-threatening but treatable process, consisting of suppurative material occupying the space between the renal capsule and the surrounding fascia.
- Most of the perinephric abscesses result from the rupture of an intrarenal abscess into the perirenal space, and are caused most commonly by *E. coli*, *Proteus species*, and *S. aureus*.
- Other sources include dissemination from other sites of infection including liver, gallbladder, pancreas, pleura, prostate, and the female reproductive tract.
- Much of the associated mortality is the result of failure to diagnose this entity in a timely fashion. This failure may be due to the nonspecific clinical picture on presentation.

Clinical Features

- The symptoms of perinephric abscess, including fever, flank pain, chills, nausea, vomiting, and dysuria, may develop insidiously, making early recognition difficult.
  - Fever is the most common symptom
  - Abdominal tenderness
  - Referred pain is also common to areas of the hip, thigh, and knee.
- The peripheral white blood cell count is usually elevated with a left shift.
- Urinalysis may be normal up to one-third of the time, and blood cultures as well as urine cultures may fail to identify correctly the bacterial pathogens responsible for the abscess.
- Distant extension of a perinephric abscess may result in a multitude of processes including empyema and colon perforation. While these extensions are rare, direct extensions into the flank or psoas muscles are more common, which may even extend to drain as a flank abscess.

Diagnosis

- A perinephric abscess should be in the differential of patients presenting with fever of unknown origin and with unexplained peritonitis, pelvic abscess, or empyema.
- Additionally, perinephric abscess should be considered in the differential diagnosis of any patient presenting with a urinary tract infection that fails to respond promptly to antibiotic therapy, particularly in those known to have anatomical abnormalities of the urinary tract or diabetes mellitus.
- Chest X-ray and abdominal films may show a range of findings, including subtle abnormalities, nonspecific findings, or nothing at all.
- Ultrasonography, however, will show a mass, often with thickened, uneven walls, with heterogeneous internal echoes. However, the ultrasound was falsely negative in as high as 36% of cases when compared to CT in one study. CT scan, therefore, is the diagnostic test of choice as it identifies the abscess and defines involvement of surrounding and distant structures.

Treatment

- Perinephric abscesses have been associated with mortality rates as high as 50%, although with early recognition by CT scan, prompt percutaneous drainage, and effective antimicrobial therapy, mortality has decreased.
- Unlike intrarenal abscesses, antibiotic therapy alone is not sufficient in treating perinephric abscesses.
- Percutaneous drainage under CT or ultrasound guidance with adjunctive antibiotics is recommended as the treatment of choice.
- If percutaneous drainage fails or is contraindicated, surgical drainage is performed.
- Empiric antimicrobial therapy should be directed mainly against common Gram-negative uropathogens and *S. aureus*.
  - An aminoglycoside (gentamicin or tobramycin) and an anti-staphylococcal β-lactam (oxacillin, nafcillin, cefazolin) are appropriate initial antibiotics.
  - An extended spectrum β-lactam may be used in place of an aminoglycoside for Gram-negative coverage in case of abnormal renal function.
  - Once cultures are done, antibiotic therapy should be modified accordingly.

Renal Abscesses

- Intrarenal abscesses are classified into renal cortical abscess and renal corticomedullary abscess.
- The treatment of renal abscesses is still under debate, perhaps because of the failure of many recent studies to distinguish between renal cortical abscesses and renal corticomedullary abscesses, which are different in their pathogenesis, prognosis, and therapies.
- Similar to perirenal abscesses, patients with intrarenal abscesses can present to the Emergency Department acutely with fever, flank pain, nausea, and vomiting, masking as a classic pyelonephritis. However, mortality rate for intrarenal abscess has been positively correlated with the timeliness of diagnosis.

Renal Cortical Abscess (Renal Carbuncle) vs Renal Corticomedullary Abscess

Pathogenesis

- Renal carbuncles result from hematogenous spread of bacteria from primary focus of infection elsewhere in the body, usually skin lesions such as cutaneous carbuncles, furuncles, paronychia, cellulitis, osteomyelitis, and endovascular infections.
- The most common cause is *S. aureas* infection, and conditions associated with an increased risk for staphylococcal bacteremia such as diabetes mellitus and intravenous drug use are predisposing factors.

- Most carbuncles are unilateral, single lesions occurring in the right kidney. Only 10% of these eventually rupture through the renal capsule to form a perinephric abscess.
- In contrast, renal corticomedullary infections occur most commonly as a complication of ascending urinary tract infections with or without accompanying urinary tract abnormalities, which include most commonly obstructive problems such as scarring from previous infections or renal stones, or genitourinary abnormalities associated with diabetes mellitus.
- Enteric aerobic Gram-negative bacilli, including *Escherichia coli, Klebsiella species*, and *Proteus* species are commonly responsible for renal corticomedullary infections.
- Another contrast between renal cortical abscesses and renal corticomedullary abscesses is that the Gram-negative corticomedullary infection frequently causes a severe parenchymal infection that may extend to and perforate the renal capsule, thus more commonly forming a perinephric abscess.

Clinical Features

- Unlike other intrarenal abscesses, renal carbuncles are approximately three times more common in men than in women.
- They occur most commonly between the second and fourth decades of life in patients presenting with chills, fever, back or abdominal pain, and few localizing signs.
- Although 95% of patients present with elevated white blood cell counts, most patients do not have bacteruria or dysuria as the infectious process is circumscribed in the cortex and generally does not communicate with the excretory passages.
- Consequently, urinalysis is usually normal. Likewise, blood cultures are usually negative.
- Although dysuria may not necessarily be present in renal corticomedullary patients, they may have a previous history of recurrent urinary tract infections, renal calculi, or a history of prior genitourinary instrumentation.
- Again, leukocytosis is generally present, but urinalysis is often abnormal in renal corticomedullary abscesses (70% of the time) with bacteriuria, pyuria, proteinuria, or hematuria because of drainage into the collecting system.

Diagnosis

- The nonspecific clinical presentation of fever, chills, and back pain may be seen with a variety of renal processes.
- Renal cortical abscesses can mimic renal tumors, cysts, renal corticomedullary abscesses, and perirenal abscesses. Furthermore, renal cortical abscesses are difficult to distinguish from renal medullary abscesses.
- Ultrasonography is useful in the diagnosis of cortical abscesses because it provides information about renal morphology and characterizes an intrarenal lesion as cystic, tumorous, or suppurative.
  - Furthermore, the ultrasound can provide information about the presence of an obstructive uropathy, retroperitoneal or intra-abdominal processes, and suppurative renal complications .
  - Although the ED ultrasound is used often to diagnose patients with intrarenal abscess, there are no current studies that describe the sensitivity and specificity of its use in the ED.
- To date, the CT scan provides the most anatomic information and is able to detect abscesses <2 cm in size.
  - Particularly if the ultrasound is equivocal or negative, CT scan may be of benefit in definitive diagnosis.
  - On CT, most abscesses appear as low-density masses with vascular enhancement of the wall. Gas within a low-density mass is pathognomonic for an abscess.

Prognostic Factors

- Bamberger et al demonstrated that poor prognostic factors were abscesses of diameter >5 cm, involvement by more than one organism, presence of Gram-negative bacilli, duration of therapy <4 wk, and use of aminoglycoside as the only antibiotic.
- Factors that bear resistance to antibiotic therapy alone include large abscesses, renal obstruction, advanced age, and urosepsis.

Treatment

- Unlike emphysematous pyelonephritis, renal abscesses are managed medically as firstline treatment.
- There is mounting evidence that the success of renal abscesses treated with antibiotics alone can be as high as 86% in large studies. Because *S. aureus* is usually the cause of the renal carbuncle, it responds to antistaphylococcal antibiotics, and surgical intervention is not required.
- If urinalysis shows no bacteria or Gram-positive cocci, oxacillin or nafcillin, 1-2 g every 4-6 h, is the therapy of choice.

  - For penicillin allergic patients, first generation cephalosporins provide adequate Gram-positive coverage.
  - Parenteral antibiotics should be continued for 10 days to 2 wk, and subsequent oral antistaphylococcal therapy for 2-4 wk.
  - The course of resolution includes defervescence after 5-6 days of IV antibiotics, and improvement of flank pain in <24 h.
- For renal corticomedullary abscesses, medical therapy is successful in most cases; however, smaller renal abscesses are more successfully treated with antibiotic treatment alone than larger abscesses, >5 cm in diameter.
  - In most cases, an intensive trial of appropriate antibiotic therapy should be attempted before considering surgical drainage for lesions localized to the renal parenchyma.
  - Antimicrobial therapy should target the most common bacterial organisms, including *E.coli, Klebsiella,* and *Proteus* species.
  - Monotherapy can be given with an extended spectrum penicillin, and extended spectrum cephalosporin or ciprofloxacin. Combination therapy has not been proven to be any more effective.
- Considerations for the ED physician in determining whether further surgical intervention is needed include failure of antibiotic therapy, large abscess >5 cm diameter, multifocal abscesses, obstructive uropathy, advanced age, deteriorating patient, and immunocompromised patient.
- Percutaneous drainage of the abscess combined with full course of parenteral antibiotics have been shown to be successful in those requiring drainage, offering the advantages of minimal invasiveness, favorable nephron-sparing, and minimal morbidity.
- If open drainage is required, incision and drainage, not total nephrectomy, are recommended when possible.
- Nephrectomy is reserved for elderly, septic patients with diffuse renal parenchymal injury requiring urgent intervention for survival.

Nephrolithiasis

- Urologic stone disease is one of the most common disorders of the urinary tract, and one of the most common diseases seen in the ED.
- About 2-5% of the population will form a urinary stone during their lifetime.
- Several factors correlate with an increased incidence of stone formation: men greater than women (3:1 ratio), age between 20 and 50 yr, a sedentary lifestyle, warm weather (peak incidence during the hottest 3 mo) and residence in the Southeastern United States.

Table 6.1. Types of renal stones

Stone Type	Composition	Incidence	Risk Factors
Calcium	Calcium oxalate or Calcium phosphate	75%	Hypercalciuria (secondary to increase ingestion of antacids, milk and vitamins C, A, D, bone disease, immobilization, PUD) Hyperparathyroidism Hyperoxaluria (secondary to small bowel disease, jejunoileal bypass surgery) Familial history Dehydration Diet—ingestion of high oxalate foods such as coffee, cola drinks, beer, citrus fruit, spinach, high dose vitamin C
Struvite	Magnesium-Ammonium-phosphate	15%	Chronic kidney infection UTI with urea-splitting bacteria (*Proteus, Klebsiella, Pseudomonas species)* Persistent alkaline urine (pH 7.6) Usual composition of "staghorn" calculi
Uric acid	Uric acid	10%	Increase urinary uric acid excretion (acidic urinary pH) secondary to diet high in purines such as organ meats, dried legumes,fish
Cystine	Cystine	1%	Associated with rare hereditary disorder-cystinuria

- In addition, there is increased genetic predisposition seen within families.
- Nephrolithiasis is a recurrent disease for most people with 37% developing another stone within 1 yr, 50% within 5 yr, and 70% within 9 yr.
- There are four basic types of renal stones (Table 6.1):
    - calcium
    - struvite
    - uric acid
    - cystine
- Retrieval and subsequent analysis of the stone is important to determine possible causes of stone formation and guide future therapy.

Clinical Presentation

- The classic presentation of renal colic is a history of abrupt onset of severe, crescendo flank pain that eventually radiates into the lower abdomen and ipsilateral testes or labia as the stone progresses down the ureter.
    - The patient is often found writhing in bed, unable to find a comfortable position.
    - This is in marked contrast to the patient with peritonitis who will be lying completely still avoiding any movements.
    - The pain is colicky in nature, waxing and waning, but rarely absent.
    - Nausea and vomiting are almost always present, and abdominal distension with an ileus is not uncommon.
    - A history of fevers or chills is suggestive of infection and should be aggressively pursued.
    - About one-third of patients will give a history of gross hematuria.

Diagnosis

- The most important laboratory test to obtain in this clinical setting is the urinalysis.

- Hematuria is almost always present, although there is no correlation between the degree of hematuria and the extent of ureteral obstruction. In fact, about 20% of patients with documented ureterolithiasis on IVP have no microscopic hematuria.
- Pyuria can be seen in the absence of infection and is probably the result of ureteral inflammation. However, the finding of bacteruria implies urinary tract infection and always requires further investigation, especially if fever and chills are present.
- A urine culture should always be sent when infection is suspected. A serum WBC >15,000 suggests an infectious etiology, while mild leukocytosis without concominant fever usually represents demargination.
- Some authors also recommend obtaining a serum uric acid level, as it will be elevated in 50% of all patients with uric acid stones, and a serum calcium to screen for hyperparathroidism and other disorders of calcium metabolism.
- Other important laboratory tests include CBC, serum electrolytes, BUN and creatinine.

Radiographic Studies

- The IVP has both high sensitivity and specificity, establishing the diagnosis of calculous disease 96% of the time.
- While a flat plate of abdomen (kidney, ureter, and bladder, or KUB) is the standard scout film done prior to an IVP, alone the KUB is not a reliable study to diagnose renal stone disease.
- Contraindications to radiocontrast medium are known allergy to contrast dye and renal insufficiency.
- Other studies that can be used in patients who cannot tolerate IV are helical CT, ultrasound and renal scan. (See *Renal Colic* on page 507 of this volume.)

Differential Diagnosis

- The key is not to miss a life-threatening condition, such as rupturing abdominal aortic aneurysm (Table 6.2).
- A careful history and physical exam can frequently elucidate the underlying pathology.

Management

- The mainstay of therapy for ureterolithiasis is IV hydration, analgesia, and antiemetics, if needed.
- Hydration can be initiated with IV crystalloid infusion of 1 L NS over 30-60 min, and then 200-500 ml/h.
- Patients presenting with renal colic are in severe pain and often require significant amounts of narcotic analgesics (morphine sulfate or meperidine in age and weight determined doses).
- NSAIDs, particularly ketorolac, are frequently used in conjunction with narcotics as they are thought to decrease pain by diminishing ureterospasm and renal capsular pressure.
- Pain medication should be administered promptly while awaiting the results of further tests.
- Most patients with uncomplicated renal stone disease, whose pain is adequately controlled and can tolerate oral fluids can be discharged home without patient urologic follow-up and careful instructions.
- All patients should be given a urinary strainer and instructed to strain all urine for up to 72 h following cessation of pain.
- Patients should be instructed to return to the ED immediately for any fever and chills, persistent nausea and vomiting, or for intractable pain, not relieved by prescription oral narcotics.
- Urologic follow-up should be arranged within 1-2 wk.

Table 6.2. Differential diagnosis of renal colic

Abdominal Aortic Aneurysm (dissecting or rupturing)	May have similar clinical presentation, with gross or microscopic hematuria More likely in older males May present with hypotension Palpate for pulsatile abdominal mass with focal tenderness Listen for bruit Palpate distant extremity pulses	Contrast CT scan or angiogram, if stable
Acute Pyelonephritis	Mild to severe flank pain, although typically not as acute as renal colic More prolonged prodrome, with fever Urinalysis shows pyuria and bacteruria CAUTION: renal obstruction with pyelonephritis is a urologic emergency requiring prompt consultation	IVP (or other radiographic imaging) if obstruction suspended
Papillary Necrosis	Secondary to passage of sloughed papillae down ureter Seen in patients with sickle cell disease, diabetes, NSAID abuse, or history of acute or chronic UTI UA can show hematuria and pyuria Requires urologic consultation with possible admission	IVP—may show sloughed renal papillae as a lucency within renal pelvis, but may also mimic a stone
Renal Infarction		
1. Renal artery embolism	Presents with acute flank pain and hematuria Most common acute renal vascular event Most often of cardiac origin—atrial fibrillation, acute bacterial endocarditis mural thrombus	IVP shows absent renal function
2. Renal artery aneurysm	Also presents with acute flank pain and hematuria Usually small and clinically not significant Dissection or rupture is rare—but will cause shock	Emergent angiography indicated
3. Renal vein thrombosis	Acute flank pain with hematuria and proteinuria Predisposing factors are nephrotic syndrome, malignancy, and pregnancy Emergent urologic consultation is required for all cases of renal infarction	IVP shows decreased renal function and increased renal size
Ectopic Pregnancy	Unilateral abdominal pain History of amenorrhea, abnormal vaginal bleeding May present with shock	Pregnancy test required in all women of child-bearing years
Appendicitis	Unilateral presentation Subacute prodrome Abdominal tenderness with guarding or rebound	Laboratory tests and physical exam; if diagnosis still in question—CT scan with oral contrast

- There are several situations in which admission is indicated:
  1. Acute obstruction with concurrent infection. The finding of fever, pyuria, or bacteruria in a patient with renal colic requires further work-up and admission. Urine and blood cultures should be obtained and intravenous antibiotics, covering the usual urinary pathogens, should be promptly started while in the ED. Urologic consultation is required.
  2. Solitary kidney with complete obstruction. Patients with only one kidney become essentially anephric with complete obstruction and may require surgical drainage. Emergent urologic consultation and admission are required.
  3. Uncontrolled pain. Patients whose pain can only be controlled by intravenous analgesia require admission.
  4. Intractable emesis. Patients who are unable to tolerate oral fluids must be admitted for IV hydration.
- Other scenarios, which may or may not require hospitalization, but should be discussed in consultation with a urologist are: patient with underlying renal insufficiency, dye extravasasation demonstrated on IVP, large stone size, and high grade obstruction.

Testicular Torsion

- The acute scrotum is a common urologic complaint, and the differential diagnosis rests between testicular torsion and other causes of pain.
- While epididymitis, orchitis, torsion of the appendix testis, hydrocoele and hernias all represent entities seen by emergency physicians, and are real emergencies to the patients that present with them, testicular torsion represents the true urologic emergency.

Epidemiology/Pathophysiology

- Testicular torsion occurs at a baseline frequency with two additional significant peaks occurring at infancy and puberty.
- The baseline frequency is due to the presence of a so-called "bellclapper's" deformity in a subset of the population.
- This lack of physical tethering system for the testicle places these individuals at a unique risk for the disease and is responsible for the reports of torsion in all age groups.
- In one series as many as 40% of individuals with torsion were found to have such an abnormality.
- The two peaks are seen at times during maturation when the testicle grows relatively faster than its tethering gubernaculum.
- The result is a testes that can rotate about its axis, pinching off the blood supply.
- When torsion occurs, the venous blood supply is obstructed resulting in edema and hemorrhage. These in turn lead to occlusion of the arterial blood supply to the gonad.
- Although reported in all age groups from neonates to the elderly, the peak incidence is in individuals between the ages of 12 and 18 yr, with an incidence of 1 in 4000 in those below the age of 25 yr of age.

Diagnosis

- The initial evaluation of a patient with acute scrotal pain or swelling should focus on ruling out the presence of testicular torsion.
- A history of an acute onset of pain, and the absence of dysuria, suggest torsion over such entities as epididymitis or orchitis.
- The pain of torsion is usually described as thunderclap in onset and is not associated with a discrete mass such as might be seen in an inguinal hernia.
- The testicle is tender over its entirety unlike torsion of the appendix testis, and the pain is continuous and unremitting.

- The absence of a high-riding testicle or the presence of a cremasteric reflex should not be used as evidence that torsion does not exist.
- Of interest to the emergency physician is the phenomenon of torsion/detorsion of the testicle.
- The classic presentation is of a young man who presents with a history of significant scrotal pain that has resolved by the time he arrives in the emergency department. He denies any dysuria or urethral discharge; however he states that he has two such episodes in the past two days.
- The emergency department workup is normal.
- The concept of an intermittently torsing testicle should be entertained in this setting and appropriate referral to Urology should be given as well as discharge instructions for immediate return in the face of any returning pain.
- From a laboratory standpoint, the only mandatory test would be a dip urinalysis looking for hematuria/pyuria. The presence of either of these might suggest an infectious etiology of the patient's pain. Urologic consultation should be obtained early as surgical exploration is definitive therapy.
- Imaging of the testicle and its blood supply may be accomplished using color flow doppler or radionuclide imaging.
- Although the former is considered the standard of care by many, studies have shown it has important limitations especially in the pediatric population.
- In one study, up to 40% of normal testes in a pediatric population scanned showed no blood flow. Overall the specificity is reported to be between 83-100% and the sensitivity 89-100% for decreased or absent testicular blood flow when compared to the contralateral testicle. Radionuclide imaging has been reported to have a sensitivity of 87-100% and a specificity of 93-100%. Despite these impressive numbers, radionuclide imaging is considered an alternative in those cases where doppler is not practical or the results are inconclusive.
- In one study, the color flow doppler had a sensitivity of only 57% for torsion. Therefore, it should be stressed once again that the diagnosis of this disease should not rely on any single test.
- The long term ramifications are of enough significance that clinical judgment and surgical intervention may be all that is necessary to make and confirm the diagnosis.

6

Treatment

- Emergency department therapy for patients with a suspected torsion focuses on analgesia and preparation for surgical exploration.
- In the mid-1990s there was a group of authors that suggested conservative management protocol of the patient with the acute scrotum. Overwhelmingly the follow-up literature was not supportive for such a treatment strategy.
- Orchipexy by the urologist should be done emergently and should involve both testicles. In some rare cases there have been reports of torsion after orchipexy, a phenomenon which appears to be related to the use of absorbable sutures.
- As an aside, in the event that no urologist is available to perform the surgery, and given the time sensitive nature of the disease (4-6 h of ischemia time), a general surgeon may be consulted to perform the procedure.

Prognosis

- Overall, testicular torsion carries a relatively poor prognosis with regards to the involved side.
- The literature reports a salvage rate of 55% to as low as 18% with time to surgery being the single most important prognostic factor. Within 10 h, >80% of testes are

lost and by 24 h the number reaches almost 100%. Reasons for delay include hesitation in seeking medical advice as well as misdiagnosis.

- Ipsilateral torsion not only places the affected testicle at risk but also the contralateral testicles. It has been demonstrated that torsion of one testicle leads to decreased blood flow to the contralateral side, with relative hypoxia and apoptosis. The subsequent decrease in germ cells has been implicated in decreased fertility of these patients.
- One suggested solution is to increase blood flow via capsaicin which has been shown to decrease the apoptosis which occurs after testicular torsion in rats.
- Glucocorticoids and nitric oxide have also been suggested as anti-apoptotic agents and have also been shown to be effective in rats, but no studies of these agents have yet been carried out in human subjects.

Periuethral Abscess

Background

- Periurethral abscess is a rare but life-threatening infection of the male urethra and periurethral areas.
- The spectrum of disease varies from a small isolated abscess to an extending necrotizing fasciitis.
- Periurethral abscess is classically known to be a complication of stricture disease and gonococcal urethritis.
- The most common predisposing factors include:
    - urethral stricture disease
    - urinary obstruction
    - previous periurethral abscess
    - gonorrhea
    - recent urinary tract infection
    - diabetes mellitus
    - urethral trauma or surgery
    - chronic indwelling Foley catheter
- The danger in this seemingly benign process is that if the abscess perforates Buck's fascia, extensive necrotizing fasciitis may ensue.

Diagnosis

- Clinical presentation depends largely on the area affected.
- If the abscess is confined to Buck's fascia, scrotal and penile swelling is the main symptomology.
- If Buck's fascia is perforated, necrotizing fasciitis may extend throughout the inguinal area.
- Fever is a common presenting symptom and may be accompanied by sepsis and shock, depending on the extent of the disease.
- Extravasation of infected urine has been linked to urethral strictures, which can result in subsequent urethral disruption.
- Therefore, it is no surprise that the organisms responsible for urinary tract infections are found on abscess evaluation, the most common being *E. coli, Proteus mirabilis, Enterococcus species, Morganella morganii, Pseudomonas aeruginosa, Staphylococcus epidermis* and *Bacteriodes species*
- Diagnosis of periurethral abscess is suggested largely by clinical examination.
- However, urethral sonography and doppler imaging have been shown to offer the advantages of avoiding radiation to the testes, providing real-time evaluation of the distensibility of the urethra, and having the capacity of assessing spongiosum and periurethral tissue involvement and urethral artery location in urethral strictures.

- Retrograde urethrography can show the presence of urine extravasation but provides overall much less information and diagnostic data than ultrasonography of anterior urethral strictures.

Management

- The mainstay of treatment for periurethral abscesses consists of surgical drainage and antibiotics.
- Wide debridement and immediate suprapubic urinary diversion must not be delayed.
- Empiric antibiotics with broad-spectrum coverage should be started immediately and should be modified appropriately once organisms are isolated with wound, urine, and blood cultures.
- Walther and colleagues showed that IV cephalosporin and aminoglycoside given for 8 days followed by oral antibiotics with prompt surgical intervention decreased mortality to 1.6% in 63 cases, whereas mortality was as high as 50% in the preantibiotic era, 6% with penicillin and sulfonamides, and 4% with oxytetracycline.

Fournier's Gangrene

Epidemiology

- First described in 1764 by Jean-Alfred Fournier as a gangrene of the penis and scrotum, necrotizing fasciitis of the perineum, genital or perianal area affects both sexes and is a true emergency.
- Failure to recognize the entity, delay in treatment and lack of aggressive medical and surgical therapy are all contributors to the high morbidity and mortality seen with this disease.
- The mortality rate varies by report from 3-45%, and causes of death have included sepsis, coagulopathy, renal failure, diabetic ketoacidosis and multiple organ failure.
- Older age, renal or hepatic dysfunction, and anorectal infectious source, are associated with higher mortality.
- Although glycosuria is present in over two-thirds of these patients, the presence of diabetes mellitus appears to have no affect on the outcome of the disease and is only associated with approximately 20% of the cases.
- In addition, the chronic use of alcohol has been associated with 25-50% of the cases.

Pathophysiology

- Although seen in children and women, Fournier's gangrene is more common in males (10:1) and was originally described as a severe gangrenous infection of the scrotum.
- The most common sources of infectious agents are the local skin (24%), anorectal (21%), and urologic (19%).
- The disease appears to affect the affluent as well as the socioeconomically depressed individuals, and the average age of presentation is older than originally described, in one recent series the mean age of patients was 50 yr.
- The bacteriology of confirmed cases reveals as a rule mixed aerobic and anaerobic organisms including *Clostridia, Klebsiella, Streptococcus,* coliforms and *Staphylcocci* species. These organisms work synergistically, with the aerobic bacteria keeping the oxygen tension low enough to allow anaerobic growth.
- Initially a cellulitis develops, superficial vessels are thrombosed and gangrene of the superficial skin and subcutaneous fat results.
- While extension of the infection into the muscle layers may lead to myonecrosis, this is not a characteristic of classic Fournier's gangrene.
- In addition, while the scrotum may often be affected, the underlying testicles, which receive blood from an independent source, are usually not affected.
- If such involvement does occur, it should prompt a search for a retroperitoneal or intra-abdominal source.

Diagnosis

- Clinical presentation of perineal necrotizing fasciitis, while often described as sudden, is more likely insidious over the span of several days.
- Given its location, some patients may present later than usual due to embarrassment.
- In its early stages, the disease presents with pain, erythema, and scrotal swelling.
- Advanced cases are described as rapidly advancing (up to 1 in/h) woody indurations extending up the anterior abdominal wall, associated with crepitus and purulent, malodorous discharge.
- Diabetes mellitus is a comorbidity in more than two-thirds of these patients.
- Laboratory findings include leukocytosis, anemia, thrombocytopenia and hyperglycemia as mentioned above. Hypocalcemia, caused by chelation of calcium by the bacterial lipases, has been reported as an important diagnostic clue, and hyponatremia may also be present.
- Imaging may reveal free air in the scrotum or dissecting upward through the fascial planes.
- Ultrasound may demonstrate gas in the scrotum, and CT scan may allow definition of the spread of the disease.
- In reality however, none of these imaging studies should delay the institution of therapy once there is suspicion of disease.

Treatment

- Treatment of Fournier's gangrene is aimed at stabilizing the thermodynamics of the patients and beginning antimicrobial infection as rapidly as possible.
- Appropriate antibiotics include any broad-based regimen.
    - Classically, penicillins were given to combat the streptococcal sp., metronidazole for anaerobic organisms and gentamicin or a third generation cephalosporin for coliforms.
    - Current recommendations replace this cocktail with medications such as imipenem or meropenem as single agents in the patient with confirmed polymicrobial infection.
- Urologic or surgical consultation is mandatory and should precede any laboratory or imaging results.
- Replacement of fluids, blood transfusion and antibiotics in isolation cannot replace surgical debridment of the infectious nidus.
- Hyperbaric oxygen, recommended by some as an adjunct after the initial debridement, has not been shown to improve outcomes when used in this setting.

Penile Emergencies

Phimosis

- Condition in which the foreskin cannot be retracted behind the glans penis
- By 3 yr of age, 90% of foreskins can be retracted
- Fewer than 1% of males have phimosis by age 17
- Usually not painful, but may produce urinary obstruction with ballooning of foreskin
- May occur as a result of recurrent balanitis
- May lead to chronic inflammation and carcinoma
- Treatment in boys older than 4 or 5 yr of age and in those who develop balanitis or balanoposthitis is topical corticosteroids (0.1% dexamethasone) to the foreskin three to four times daily for 6 wk. This loosens the phimotic ring in two-thirds of cases and usually allows the foreskin to be retracted manually.
- In uncircumcised boys older than 7 or 8 yr old with corticosteroid-resistant phimosis or in boys with ballooning of the foreskin or recurrent balanitis, circumcision or dorsal slit is recommended.

Paraphimosis

- Condition in which the foreskin has been retracted and left behind the glans penis, constricting the glans and causing painful vascular engorgement and edema of the foreskin distal to the phimotic ring.
- Can occur iatrogenically and frequently occurs after penis has been examined or urethral catheter has been inserted
- Can result in marked swelling of the glans penis such that the foreskin can no longer be drawn forward, which may lead to arterial compromise and gangrene
- Reduction of paraphimosis can be initiated with gentle, steady pressure to the foreskin to decrease the swelling. Elastic bandage wrap (2 x 2 in) used for 5 min may be helpful in some cases. Short-term ice-packing may help as an analgesic or a local anesthetic block of the penis may be indicated in marked discomfort.
- Marked or irreducible cases may necessitate an emergency dorsal slit or circumcision by a urologist. Reduced paraphimosis should be scheduled for a dorsal slit or circumcision at a later date, as paraphimosis tends to recur.

Balanitis

- Inflammation of the glans, which occurs usually as a result of poor hygiene, from failure to retract and clean under the foreskin.
- Usually responds to local care and antibiotic ointment. Occasionally oral antibiotic therapy may be necessary.
- Recurrent balanitis may result in phimosis.
- Balanitis in older patients may be a presenting sign of diabetes, in which cases, circumcision may be necessary.

Balanoposthitis

- Severe balanitis, in which the phimotic band is tight enough to retain inflammatory secretions, creating a preputial cavity abscess.
- Treatment includes cleansing and application of antifungal creams (clotrimazole bid)
- Urologic follow-up and possible circumcision may be indicated.
- In the presence of secondary bacterial infection, an oral cephalosporin should be prescribed.
- On occasion, an emergent dorsal slit is required.

Penile Fracture

- Acute tear of the tunica albuginea, presenting with acute swelling, discoloration, and tenderness.
- Usually caused by trauma during intercourse accompanied by a snapping sound.
- Urologic consultation is indicated.

Peyronie's Disease

- Condition that results in fibrosis of the tunica albuginea, the elastic membrane that surrounds each corpus cavernosum, producing curvature of the penis during erection.
- Difficult to diagnose in flaccid state; however patient's prior history of buckling trauma may establish the diagnosis.
- Physical exam reveals fibrous plaques or ridges along the dorsal shaft of the penis.
- Benign condition that may resolve or stabilize spontaneously without treatment.
- Complication may include erectile dysfunction.
- Reassurance and urologic follow-up are indicated.

Priapism

- Prolonged painful and tender erection that persists beyond or is not related to sexual activity

- Occurs most commonly in patients with sickle cell disease but can also occur in those with advanced malignancy or coagulation disorders, those on total parenteral nutrition, certain drug therapy, and after trauma or idiopathically.
- Classified as primary/idiopathic and secondary, or ischemic/veno-occlusive and nonischemic/arterial.
  - Ischemic priapism beyond 4 h is a compartment syndrome requiring emergent medical intervention
  - Nonischemic priapism is less common and is caused by unregulated cavernous inflow, which usually presents with an erection that is not fully rigid and is painless
- Drugs reported to cause priapism include but are not restricted to:
  - Antidepressants—bupropion, trazadone, fluoxetine, setraline, lithium
  - Antipsychotics—clozapine
  - Tranquilizers—mesoridazine, perphenazine
  - Anxiolytics—hydroxyzine
  - Psychotropics—chlorpromazine
  - Alpha-adrenergic blockers—prazosin
  - Hormones—GnRH
  - Anticoagulants—heparin, warfarin
  - Recreational drugs—cocaine, alcohol
- Complications include urinary retention, infection, and impotence
- Initial therapy includes terbutaline 0.25-0.5 mg subcutaneously in the deltoid area
- Corporal aspiration (corpus cavernosum) and irrigation with normal saline or α-adrenergic blocker is the next step by a urologist.

6

Suggested Reading

1. Ahlering, TE, Boyd SD, Hamilton CL et al. Emphysematous pyelonephritis: A 5-yr experience with 13 patients. J Urol 1985; 134:1086.
2. Alexander AA, Eschelman DJ, Nazarian LN et al. Transrectal sonographically guided drainage of deep pelvic abscesses. AJR 1994; 162:1227-1230.
3. Anderson KA, McAninch JW. Renal abscesses: Classification and review of 40 cases. Urology 1980; 16:333.
4. Andriole VT. Renal and perirenal abscess. In: Schrier, Gottschalk, eds. Diseases of the Kidney. 4th ed. Boston: Little Brown, 1987:1049-1063.
5. Baker WJ, Wilkey JL, Barson LJ. An evaluation of the management of periurethral phlegmon in 272 consecutive cases at the Cook County hospital. J Urol 1949; 61:943-8.
6. Bamberger DM. Outcome of medical treatment of bacterial abscesses without therapy drainage: Review of case reported in the literature. Clin Infect Dis 1996; 231:592-603.
7. Best CD, Terris MK, Tacker JR et al. Clinical and radiological findings in patients with gas forming renal abscess treated conservatively. J Urol 1999; 162(4):1273-6.
8. Brook I. The role of anaerobic bacteria in perinephric and renal abscesses in children. Pediatrics 1994; 93(2):261-4.
9. Caldamone AA, Frank IN. Percutaneous aspiration in the treatment of renal abscess. J Urol 1980; 123(1):92-3.
10. Campell MF. Complications of gonorrhea: Periurethral abscess, stricture, and arthritis. Am J Surg 1931; 12:277.
11. Campell MF. Periurethral phlegmon, 135 cases. Surg Gynecol Obstet 1929; 48:382-9.
12. Campell MF. Perinephric abscess. Surg Obstet Gynecol 1930; 51:674.
13. Capelli-Schellpfeffer M, Gerber GS. The use of hyperbaric oxygen in urology. J Urol 1999;162:647-654.
14. Cardinal AS, De Blay V, Gilbeau JP. Emphysematous pyelonephritis: Successful treatment with percutaneous drainage. AJR 1995; 164:1554.
15. Chen MT, Huang CN, Chou YH et al. Percutaneous drainage in the treatment of emphysematous pyelonephritis: 10-yr experience. J Urol 1997; 157(5):1569-73.

16. Chiou RK, Anderson JC, Tran T et al. Evaluation of urethral strictures and associated abnormalities using high-resolution and color doppler ultrasound. Urology 1996; 47(1):102-7.
17. Corr J, Gleeson M, Wilson G et al. Percutaneous management of emphysematous pyelonephritis. Brit J Urol 1993; 71:487.
18. Corriere Jr JN, Sandler CM. The diagnosis and immediate therapy of acute renal and perirenal infections. Urol Clin North Am 1982; 9:219.
19. Dalla Palma L, Pozzi-Mucelli F, Ene V. Medical treatment of renal and perirenal abscesses: CT evaluation. Clin Radiol Dec 1999; 54(12):792-7.
20. Edelstein H, McCabe RE. Perinephric abscess: Modern diagnosis and treatment in 47 cases. Medicine 1988; 67:118.
21. Eloubeidi MA. Fowler Jr VG. Images in clinical medicine. Emphysematous pyelonephritis. N Engl J Med 1999; 341(10):737.
22. Evans JA, Meyers MA, Mosniak MA. Acute renal and perirenal infections. Semin Roentgenol 1971; 6:274.
23. Finestone EO. Peri-urethral phlegmon: Clinical study of 32 cases. Ann Surg 1942; 116:109-21.
24. Finestone EO. Periurethral phlegmon: Pathogenesis and experimental study. Surg Gynecol Obstet 1941; 73:218-27.
25. Gelman ML, Stone LB. Renal carbuncle: Early diagnosis by retroperitoneal ultrasound. Urology 1976; 7:103.
26. Goichot B, Andres E. Emphysematous pyelonephritis. N Engl J Med 2000; 342(1):60-1.
27. Goldman SM, Minkin SD, Naraval DC et al. Renal carbuncle: The use of ultrasound in its diagnosis and treatment. J Urol 1977; 118:525.
28. Hoddick W, Jeffrey RB, Goldberg HI et al. CT and sonography of severe renal and perirenal infections. AJR 1983; 140:517.
29. Huang JJ, Sung JM, Chen KW et al. Acute bacterial nephritis: A clinicoradiologic correlation based on computed tomography. Am J Med 1992; 93(3):289-98.
30. Huang JJ, Tseng CC. Emphysematous pyelonephritis: Clinicoradiological classification, management, prognosis, and pathogenesis. Arch Intl Med 2000; 160(6):797-805.
31. Hudson MA, Weyman PJ, van der Vliet et al. Emphysematous pyelonephritis: Successful management by percutaneous drainage. J Urol 1986; 136:884.
32. Klein FA, Smith MJV, Vick III CW et al. Emphysematous pyelonephritis diagnosis and treatment. South Med J 1986; 79:41.
33. Landers DV, Sweet RL. Tuboovarian abscess: Contemporary approach to management. Rev Infect Dis 1983; 5:876-84.
34. Lyons RW, Long JM, Lytton B et al. Arteriography and antibiotic therapy of renal carbuncle. J Urol 1972; 107:524.
35. McDonough WD, Sandler CM, Benson GS. Acute focal bacterial nephritis: Focal pyelonephritis that may simulate renal abscess. J Urol 1981; 126(5):670-3.
36. McHugh TP, Albanna SE, Stewart NJ. Bilateral emphysematous pyelonephritis. Am J Emerg Med 1998; 16(2):166-9.
37. McNeeley SG, Hendrix SL, Mazzoni MM et al. Medically sound, cost-effective treatment for pelvic inflammatory disease and tuboovarian abscess. Am J Obstet Gynecol 1998; 178(6):1272-8.
38. Merenich WM, Popky GI. Radiology of renal infection. Med Clin North Am 1991; 75:425.
39. Michaeli J, Mogle P, Perlberg S et al. Emphysematous pyelonephritis. J Urol 1984; 131:203.
40. Mirhashemi R, Schoell WM, Estape R et al. Trends in management of pelvic abscesses. J Am Coll Surg 1999; 188:567-572.
41. Molino D, Anastasio P, Casoli E et al. Renal abscess: Recovery without hospitalization and drainage. Clin Nephrol Aug 2001; 56(2):169-71.
42. Murray HW, Molavi A. Perinephric abscess: An unusual presentation of perforation of the colon. Johns Hopkins Med J 1977; 140:15.

43. Mylonakis E, Nizam R, Freeman N. Periurethral abscess: Complication of UTI. Geriatrics Aug 1997; 52(8):86-88.
44. Najjar M, Gouda HE, Rodriguez P et al. Successful medical management of emphysematous pyelonephritis. Am J Med 2002; 113(3):262-3.
45. Niedrach WL, Lerner RM, Link CA. Penile abscess involving the corpus cavernosum: a case report. J Urol 1989; 141:374-5.
46. Patel NP, Lavengood RW, Fernandes M et al. Gas-forming infections in genitourinary tract. Urology 1992; 39(4):341-5.
47. Patterson JE, Andriole VT. Bacterial urinary tract infections in diabetes. Infect Dis Clin North Am 1997; 11(3):735-50.
48. Pedersen JF, Hanck S, Kristensen JK. Renal carbuncle: Antibiotic therapy governed by ultrasonically guided aspiration. J Urol 1973; 109:777.
49. Rauschkolb EN, Sandler CM, Patel S et al. Computed tomography of renal inflammatory disease. J Comp Assist Tomog 1982; 6:502.
50. Roberts JA. Management of pyelonephritis and upper urinary tract infections. Urolog ClinNorth Am 1999; 26(4):753-63.
51. Saiki J, Vaziri ND, Barton C. Perinephric and intranephric abscesses: A review of the literature. West J Med 1982; 136:95.
52. Salvatierra Jr O, Bucklew WB, Morrow JW. Perinephric abscess: A report of 71 cases. J Urol 1967; 98:296.
53. Sanders CJ, Mulder MM. Periurethral gland abscess: Aetiology and treatment. Sexually Transmitted Infections 1998; 74(4):276-8.
54. Scheinfeld J, Erturk E, Spataro RF et al. Perinephric abscess: Current concepts. J Urol 1987; 137:191.
55. Shokeir AA, El-Azab M, Mohsen T et al. Emphysematous pyelonephritis: A 15-yr experience with 20 cases. Urology 1997; 49(3):343-6.
56. Siegel JF, Smith A, Moldwin R. Minimally invasive treatment of renal abscess. J Urol 1996; 155(1):52-5.
57. Stein JP, Spitz A, Elmajian DA et al. Bilateral emphysematous pyelonephritis: A case report and review of the literature. Urology 1996; 47(1):129-34.
58. Tersala K, Heinonen PK, Punnonen R. Transvaginal ultrasound in the diagnosis and treatment of tubo-ovarian abscess. Br J Obstret Gynecol 1990; 97:178-180.
59. Throley JD, Jones SR, Sanford JP. Perinephric abscess. Medicine 1974; 53:41.
60. Timmons JW, Perlmutter AD. Renal abscess: A changing concept. J Urol 1976; 115:229.
61. Truesdale BH, Rous SN, Nelson RP. Perinephric abscess: A review of 26 cases. J Urol 1977; 118:910.
62. vanSonnenberg E, D'Agostino VS, Giovanna C et al. US-guided transvaginal drainage of pelvic abscess and fluid collections. Radiology 1991; 181:53-56.
63. Varhese JC, O'Neill MJ, Gervais DA et al. Transvaginal catheter drainage of tuboovarian abscess using the trocar method: technique and literature review. AJR 2001; 177 (1):139-44.
64. Vas W, Carlin B, Salimi Z et al. CT diagnosis of emphysematous pyelonephritis. Comput Rad 1985; 9:37.
65. Walther MM, Mann BB, Finnerty DP. Periurethral abscess. J Urol 1987; 138:1167-70.
66. Wan YL, Lo SK, Bullard MJ et al. Predictors of outcome in emphysematous pyelonephritis. JUrol 1998; 159(2):369-73.
67. Wan YL, Lee TY, Bullard MJ et al. Acute gas-producing bacterial renal infection: Correlation between imaging findings and clinical outcome. Radiology 1996; 198:433.
68. Yen DH, Hu SC, Tsai J et al. Renal abscess: Early diagnosis and treatment. American J Emerg Med 1999; 17(2):192-7.
69. Dokucu AI, Ozturk H, Ozdemir E et al. The protective effects of nitric oxide on the contralateral testis in prepubertal rats with unilateral testicular torsion. BJU International 2000; 85:767-771.
70. Kadish HA, Bolte RG. A retrospective review of pediatric patients with epididymitis, testicular torsion, and torsion of testicular appendages. Pediatrics 1998; 102:73-76.

71. Church JM, Yagahan RJ, Al-Jaberi TM et al. Fournier's Gangrene: Changing face of the disease. Dis Colon Rectum 2000; 43:1300-1308.
72. Eke N. Fournier's gangrene: A review of 1726 cases. Brit J Surg 2000; 87:718-728.
73. Dunne PJ, O'Loughlin BS. Testicular torsion: Time is the enemy. Aust N Z J Surg 2000; 70:441-442.
74. Rajfer J. Congenital anomalies of the testis. In: Walsh PC, Retik AB, Stamey TA, Vaughan ED, eds. Campbell's Urology 6th ed. Philadelphia: W.B. Saunders, 1992.
75. Stone SS, Avelar J, Henderson SO. Necrotising Fasciitis. (Unpublished data).
76. Watkin NA, Reiger NA, Moisey CU. Is the conservative management of the acute scrotum justified on clinical grounds? Brit J Urol 1996; 78:623-627.
77. Pearce I, Islam S, McIntyre IG et al. Suspected testicular torsion: A survey of clinical practice in North West England. JRSM 2002; 95:247-249.
78. Sarioglu-Buke A, Erdem S, Gedikoglu G et al. Capsaicin effectively prevents apoptosis in the contralateral testis after ipsilateral testicular torsion. BJU International 2001; 88:787-789.
79. Yazawa H, Sasagawa I, Suzuki Y et al. Glucocorticoid hormone can suppress apoptosis of rat testicular germ cells induced by testicular ischemia. Fertility and Sterility 2001; 75:980-985.
80. Harwood-Nuss AL, Etheredge W, McKenna I. Urologic Emergencies. In: Rosen P, Barkin R, eds. Emergency Medicine. 4th ed. St Louis: Mosby, 1998:2249-2258
81. Stewart C. Nephrolithiasis. Emerg Med Clin North Am 1988;6(3);617-629
82. Peacock WF, Urologic Stone Disease. In: Tintinalli JE, Ruiz E, Krome RL, eds. Emergency Medicine. 4th ed. New York: McGraw-hill, 1996:549-553.

CHAPTER 7

OB/GYN Emergencies

Stephen D. Docherty

Part A: Selected Obstetric Emergencies

Ectopic Pregnancy

Definition

- Any pregnancy occurring outside the uterine cavity

Location

- 95% of all ectopic pregnancies occur in the fallopian tubes with 5% being ovarian or abdominal pregnancies.
- Heterotopic pregnancy, described as simultaneous intrauterine and ectopic pregnancy may also occur.

Incidence

- Annually 70,000 cases of ectopic pregnancy occur in the United States with a current incidence of 20 ectopics per 1000 pregnancies.
- The incidence of heterotropic pregnancy is 1 in 4000 pregnancies.

Clinical Presentation

- Classic triad—Seen in <50% of patients
  - Abdominal or pelvic pain
  - Missed menstrual period with associated abnormal vaginal bleeding
  - Pelvic examination demonstrates a tender adnexal mass
- The pelvic pain when it is present is usually unilateral, severe and sudden, although there may be significant variability in quality, intensity, duration and location. Up to 10% of patients seen with ectopic pregnancy present with no pain.
- Adnexal tenderness is present in 96% of cases, and there may also be associated cervical motion tenderness.
- The uterus is perceived to be of normal size in approximately 71% of cases.

Table 7A.1. Risk factors

- Previous history of ectopic pregnancy
- Pelvic inflammatory disease (PID)
- Tubal surgery /pelvic surgery
- Assisted reproduction
- DES exposure
- Intrauterine contraceptive devices (IUD)

Table 7A.2. Differential diagnosis of ectopic pregnancy

- Appendicitis
- Salpingitis
- Ovarian torsion
- Threatened abortion
- Gastroenteritis
- Urinary tract infection in early pregnancy
- Urolithiasis in early pregnancy
- Dysfunctional uterine bleeding
- Normal intrauterine pregnancy
- Corpus luteum cyst

- The presence of a palpable adnexal mass or fullness with associated tenderness is present in up to two-thirds of patients however its absence does not rule out the possibility of an ectopic pregnancy.
- Vaginal bleeding is generally less than normal menses. Uterine decidual tissue casts may be passed in 5-10% of patients and can be mistaken for tissue from a spontaneous abortion.
- The presence of hypotension and/or tachycardia is the presenting sign in <5% of ectopic pregnancies and is usually associated with ectopic rupture. In the case of ectopic pregnancy rupture, peritoneal signs may be present on abdominal examination secondary to hemoperitoneum. In the unruptured ectopic pregnancy, the vital signs are more likely to be normal.

7

Rupture of an Ectopic Pregnancy

- Rupture of an ectopic pregnancy is associated with:
  - syncope
  - sudden onset of severe pelvic/abdominal pain
  - hypotension
- When an ectopic pregnancy ruptures, there occurs hemorrhage into the peritoneal cavity leading to peritoneal signs.
- Hemorrhage from ectopic pregnancy is the major cause of pregnancy-related death during the first trimester.
- Maternal death in these patients is often related to a delay in initial diagnosis.

Diagnostic Evaluation

- B-HCG—Beta Human Chorionic Gonadotropin
  - All female patients of child-bearing years who present with abdominal and/or pelvic pain should have an immediate bedside urine HCG performed. A HCG level >25 mIU/ml is considered positive.
  - In the case of a known pregnancy a quantitative serum B-HCG should be performed to assist with evaluation and serve as a baseline marker.
  - B-HCG levels can be useful in determining the optimal timing of ultrasound visualization of a gestational sac. A gestational sac can usually be visualized by transvaginal ultrasound when the quantitative B-HCG is approximately 2000 mIU/ml.
- Progesterone
  - Progesterone is produced by the corpus luteum during pregnancy. A progesterone level >25 ng/ml is consistent with a viable intrauterine pregnancy with a 97.5% sensitivity. Lower levels however do not reliably correlate with the location of the patient's pregnancy.
  - There is not sufficient data available at this time to recommend its routine use in the evaluation of ectopic pregnancy.

Table 7A.3. Ultrasound findings in ectopic pregnancy (one or more present)

- Empty uterus
- Decidual reaction (pseudogestational sac)
- Free fluid in the culdosac
- Cystic or complex adnexal mass
- Live embryo visualized in the adnexa

7

- Complete Blood Count (CBC)
  - Hemoglobin/hematocrit—Initial Hg/Hct serves as a baseline for later comparison. Initial values may be normal, however a low Hg/Hct initially or an acute drop over the first several hours is concerning when considering the possibility of ectopic pregnancy in your differential diagnosis.
  - White blood count (WBC)—Not useful in the diagnosis of ectopic pregnancy,
  - 50% of ectopic pregnancies have a normal WBC. May be helpful for identifying other potential entities in your differential diagnosis once ectopic pregnancy has been ruled out.
- Rh /Type and Screen
  - Should the patient have an ectopic pregnancy, especially a ruptured ectopic pregnancy and be hemodynamically compromised, they may need blood to be transfused along with appropriate resuscitation fluids.
  - If the patient is Rh negative, they will need to receive Rh immune globulin (RhoGAM) to prevent isoimmunization.
- Ultrasound
  - The primary purpose of ultrasound in the diagnosis of ectopic pregnancy is to demonstrate the location of the pregnancy. Ultrasound helps us to determine whether or not the pregnancy is an intrauterine pregnancy (IUP) or an ectopic pregnancy. The ultrasound results can be improved with the use of color doppler.
  - It is suggestive of ectopic pregnancy if the uterus is empty in appropriately advanced pregnancy by date of LMP or the quantitative HCG result.
  - Discrimination level—the HCG level necessary to visualize an intrauterine pregnancy by ultrasound.
    - Transvaginal = 1800-2000 mIU/ml
    - Transabdominal = 6000-6500 mIU/ml.

Emergency Department Management

- The stable patient with **low** clinical suspicion and inconclusive testing may be followed as an outpatient by OB/GYN consult with serial quantitative HCG and ultrasound evaluation.
- The stable patient with **high** clinical suspicion should have immediate OB/GYN consult. The possible options are laparoscopy with appropriate surgical intervention if an ectopic pregnancy is identified. Another possible option at the discretion of the OB/GYN consult is medical treatment with systemic methotrexate. Methotrexate inhibits DNA synthesis and cell multiplication in the developing embryo. Methotrexate can be used in the stable patient with an ectopic pregnancy <3.5 cm.
- The **unstable** patient should immediately have:
  - two large bore IVs with normal saline
  - be place on a cardiac monitor and pulse oximetry
  - receive supplemental oxygen
  - have a CBC, PT/PTT, type and cross for 4 units of packed red blood cells (PRBC) and a STAT bedside ultrasound
  - an immediate OB/GYN consultation

Table 7A.4. Indications for methotrexate usage in ectopic pregnancy

- Ectopic pregnancy unruptured and <3.5 cm in greatest diameter
- Desire for future fertility
- No active bleeding /hemoperitoneum
- Quantitative HCG level doesn't exceed 15,000 mIU/ml
- The patient has no contraindication to methotrexate

  - the operating room staff should also be made aware as the patient most likely will be admitted directly to the OR for laparoscopy/laparotomy.
- Should in a **rare circumstance** ultrasound not be available for patient evaluation then a culdocentesis may be performed. The procedure is done by aspiration of the contents from the pouch of Douglas entered by way of the posterior fornix. The aspiration of nonclotting blood is considered a positive test that is suspicious for ectopic pregnancy. If no blood is aspirated it is considered a nondiagnostic test. The false positive rate is approximately 5%.

Vaginal Bleeding in the First Half of Pregnancy

Forty percent of pregnant patients present with some degree of vaginal bleeding during early pregnancy. Approximately one-half of these will progress on to spontaneously abort. The vast majority of these spontaneous abortions occur prior to 8 wk of gestation. At least half of all spontaneous abortions are due to genetic abnormalities; the rest being due to a combination of factors such as uterine abnormalities, incompetent cervix, progesterone deficiency, tobacco or alcohol use.

Once again, one must always consider **ectopic pregnancy** in the differential when evaluating the pregnant patient with vaginal bleeding. The patient with unilateral pelvic pain and vaginal bleeding needs thorough evaluation to differentiate early abortion from ectopic pregnancy.

Definitions

- Threatened Abortion—Uterine bleeding in the first 20 wk of pregnancy without any passage of tissue or cervical dilatation. The cervical os is closed.
- Inevitable Abortion—Uterine bleeding in the first 20 wk of pregnancy without any passage of tissue but with a dilated cervical os.
- Incomplete Abortion—Uterine bleeding in the first 20 wk of pregnancy with a dilated cervical os and only partial expulsion of the products of conception.
- Complete Abortion—Uterine bleeding in the first 20 wk of pregnancy with complete expulsion of all the products of conception.
- Missed Abortion—Fetal death in utero at <20 wk gestation with the products of conception retained.
- Septic Abortion—This is an incomplete abortion in which infection has ascended into the uterus causing endometritis, parametritis and peritonitis.
- Blighted Ovum—An embryo that has failed to develop, although there is an identifiable gestational sac, can also lead to uterine bleeding.
- Molar Pregnancy/Gestational Trophoblastic Disease—Occurs in 1/2000 pregnancies; tumors that arise from proliferation of the placental trophoblast and occur in both a benign and malignant form. A fetus is generally absent in this process. These patients present with vaginal bleeding in the first half of pregnancy 90% of the time. They generally have higher than expected serum HCG levels and their pelvic ultrasound shows a classic snow storm appearance caused by intrauterine hydropic villi. OB/Gyn consultation is necessary for treatment.

Clinical Evaluation

- Obtain a detailed menstrual history from the patient.
- Information that is important to obtain in this history includes:
  - the date of the last normal menstrual period
  - the date of the last menstrual period
  - the number of pads used per hour can help to quantify the amount of vaginal bleeding that has occurred.
  - Past OB/GYN history should be obtained to include the number of prior pregnancies ,the number of term/preterm deliveries and the number of abortions both spontaneous and therapeutic
- An important step in the evaluation of vaginal bleeding in pregnancy is the complete pelvic examination which allows the localization and the quantification of the source of vaginal bleeding.
  - Bleeding from the cervical os this indicates a uterine source.
  - It should be noted whether or not the internal cervical os is open or closed and whether or not there are any products of conception present at the cervical os.
    - It can be determined whether or not the internal cervical os is open by gentle insertion of a finger into the os on bimanual examination. Ring forceps can also be carefully used to make this evaluation. Force should never be used in either of these techniques.
  - The uterus should be evaluated for tenderness, size, shape and consistency.
  - The patient's adnexa should be palpated for the presence of masses and/or tenderness.
  - The presence of cervical motion tenderness should be determined.

Diagnostic Evaluation

- Standard laboratory testing to be obtained in the setting of vaginal bleeding in the first half of pregnancy should include all of the following.
  - Complete blood count (CBC)
  - Quantitative serum HCG level (to help with interpretation of ultrasound results and to serve as a baseline)
  - Rh type
  - Urinalysis obtained by catheterization (to screen for urinary tract infection which is a common cause of abortion/threatened abortion.)
- Emergent pelvic ultrasound should be obtained to help with a definitive diagnosis_and to help guide subsequent treatment. Patients with a visualized intrauterine pregnancy with a closed cervical os can be considered to have a threatened abortion. Correlation of ultrasound results with the patient's history and physical findings will allow the type of abortion to be identified.
- Any products of conception (tissue) passed from the cervical os should be sent to pathology for evaluation.

Emergency Department Management

- The hemodynamically stable patient with a documented intrauterine pregnancy can usually be discharged home with a follow-up evaluation with their OB/Gyn physician in 48 h. They should be instructed to return to the Emergency Department if vaginal bleeding increases and/or returns, if they notice any passage of tissue, or if they develop significant pelvic pain or fever. The patient should be placed on pelvic rest which means no intercourse, no douching and physical activity. If the patient is limited Rh negative, they will need to receive Rh immune globulin(RhoGAM) to prevent isoimmunization.
- In the patient who is hemodynamically unstable and/or presenting with heavy persistent vaginal bleeding:

- intravenous access should immediately be obtained and volume resuscitation initiated with normal saline.
- A bedside hemacue and urine pregnancy test should be obtained while awaiting formal lab results.
- The patient should be placed on supplemental oxygen, cardiac monitor, automated blood pressure monitor and pulse oximetry.
- An emergent bedside ultrasound should be performed looking for an intrauterine pregnancy.
- Immediate OB/GYN consultation should be also be obtained.
- OB/GYN consultation should be obtained regarding patients with inevitable, incomplete, missed, septic abortions or molar pregnancies.

Vaginal Bleeding in the Second Half of Pregnancy

Vaginal bleeding after 20 wk of gestation can present a complicated clinical management situation with the lives of the mother and fetus often both in serious jeopardy.

Abruptio Placentae/Placental Abruption

- Definition—The complete or partial placental separation from the decidua basalis (uterine implantation site) after 20 wk of gestation. When this separation develops, blood vessels are ruptured leading to hematoma formation which leads to significant hemorrhaging and fetal hypoxia. DIC may also develop in this situation.

Incidence

- Abrubtio placentae occurs in approximately 1/100 pregnancies and is the cause of approximately 14% of all stillbirths in the United States.

Clinical Presentation

- Variability of clinical presentation is related to the quantity and location of hemorrhaging.
- Classical findings include painful vaginal bleeding with uterine tenderness and contractions. The patient may complain of back and/or abdominal pain, and the pain is usually relatively sudden in its onset and constant in nature.
- On exam the uterus may be firm and hypertonic.
- Vaginal bleeding is visible approximately 80% of the time and concealed 20% of the time.
- Fetal heart sounds and visualized fetal cardiac activity by ultrasound may be absent.
- In a small percentage of cases usually involving very small or marginal abruptions, the patient may not present with pain.
- Clinical evidence of DIC may be present as the abruption may activate the coagulation cascade. Early awareness of this possibility may become apparent if the patient is noted to have excessive hemorrhaging at venopuncture or intravenous access sites, mucosal/gingival hemorrhaging, easy bruising and/or hematuria.
- Pelvic examination should be delayed until a placenta previa has been ruled out.

Emergency Department Management

- Intravenous access should be immediately obtained.
- The patient should be placed on a cardiac monitor, blood pressure monitor and pulse oximeter. Supplemental oxygen should be initiated.
- CBC, PT/PTT, DIC panel, urinalysis, Rh and type and crossmatch.
  - The patient should be transfused in an attempt to keep the hematocrit >30 in the presence of active significant hemorrhage.

Table 7A.5. Causes of vaginal bleeding in the 2nd half of pregnancy

- Abruptio placentae/placental abruption
- Placenta previa
- Premature labor
- Premature rupture of membranes
- Lesions of the cervix and lower genital tract
- Uterine rupture
- Vasa previa

Table 7A.6. Risk factors for placental abruption

- Hypertension
- Preeclampsia
- History of prior placental abruption
- Trauma
- Cigarette smoking
- Increasing maternal age
- High multiparity
- Illicit drug use (cocaine abuse)
- Excessive alcohol consumption

- A Kleihauer-Betke test should also be obtained to detect fetal cells in the maternal circulation.
  - A positive Kleihauer- Betke test may be the only diagnostic finding in the presence of a very small abruption.
- The patient should be placed on a fetal monitoring unit as soon as it is feasible and fetal heart sounds should be checked frequently.
- An emergent bedside ultrasound should be performed.
  - Ultrasound may not always be able to identify placental abruption however visualization of the placenta will allow one to rule out a placenta previa. When abruption is visualized on ultrasound it appears as a hypoechoic area between the placenta and the uterine wall.
  - Magnetic resonance imaging (MRI)is able to detect placental abruption with a higher degree of accuracy than ultrasound and has been shown to be safe in pregnancy. The limitations of MRI are its lack of uniform availability as well as the inability to monitor hemodynamically unstable patients in the scanner.
- Obtain immediate OB/GYN consult while performing your evaluation.
- With mild placental abruption expectant management is indicated if the mother is stable. In the presence of more severe placental abruption, expedited vaginal delivery or emergent cesarean section may be necessary.

Placenta Previa

Definition

- Placenta previa describes a situation where any part of the placenta implants in the lower uterine segment and be associated with a high risk of significant serious maternal hemorrhaging. The implantation usually occurs below the fetal presenting part.
- Total Placenta Previa—The patient's internal cervical os is completely covered by the placenta.
- Partial Placenta Previa—The patient's internal cervical os is only partially covered by the placenta.

Table 7A.7. Risk factors for placenta previa

- Multiparity
- Multiple gestation pregnancies
- Previous cesarean section
- Prior uterine scar for any reason
- Increasing maternal age
- Previous abortion
- Prior placenta previa
- Diabetes mellitus
- Erythroblastosis fetalis
- Any process that increases placental size

- Marginal Placenta Previa—The placenta is located adjacent to the patient's internal os but is not covering it.

Incidence

- Placenta previa occurs in approximately 1/200 pregnancies in the United States.
- Approximately 90% of placenta previas resolve spontaneously prior to term.

Clinical Presentation

- The patient usually presents with **painless vaginal bleeding** in the 2nd half of pregnancy.
- Vaginal examination should be avoided as manipulation of the placenta during the examination may cause the tearing of blood vessels leading to potentially life threatening hemorrhage to the mother and/or the fetus.

Emergency Department Management

- When placenta previa is clinically suspected emergent obstetrical consultation should be obtained.
- Pelvic examination should initially be avoided.
- Intravenous access should be established with appropriate hemodynamic monitoring and pulse oximetry.
- Blood should be obtained for a CBC, Type/Cross and Rh.
- Ultrasound should be performed on an emergent basis.
- Fetal monitoring should be placed.
- Emergent cesarean section versus close observation may be the indicated treatments to be decided by the obstetrical consultant.

Hyperemesis Gravidarum

- It is normal to have some degree of nausea and vomiting for most patients during the course of pregnancy and these episodes are especially frequent in first 12 wk.
- The diagnosis of hyperemesis is defined as severe refractory nausea and vomiting with evidence of dehydration, weight loss, ketonuria and increased urine specific gravity.
- Hyperemesis gravidarum is a diagnosis of exclusion after no other cause for the patient's symptoms are found.

Diagnostic Evaluation

- Laboratory testing in hyperemesis gravidarum should include a CBC, electrolyte panel, urinalysis, checking for the presence of urine or serum ketones.
  - The finding of large amounts of ketones in the patient's urine or serum indicates that the patient may be obtaining much of her caloric requirement from lipolysis due to depleted glucose and glycogen stores.

Emergency Department Management

- The initial Emergency Department treatment of hyperemesis gravidarum includes intravenous rehydration with crystalloid solution (normal saline or lactated ringers).
- Subsequent intravenous rehydration continues with the use of either D5-lactated Ringers or D5-normal saline.
  - Urine ketones and the specific gravity should be monitored during the course of the patients rehydration for improvement.
- Parenteral antiemetics are frequently used for control of further nausea and vomiting. Frequently used antiemetics are promethazine (phenergan), prochlorperazine (compazine) and trimethobenzamide (tigan).
- Potassium may need to be added to the patient's intravenous fluids based on laboratory findings.
- Indications for admission in a patient with hyperemesis graviderum include:
  - uncertainty regarding the etiology of the patients severe nausea and vomiting;
  - electrolyte imbalances and/or ketosis that are not resolving with treatment;
  - persistent nausea and vomiting in spite of adequate treatment;
  - a >10% weight loss.
- The patient who shows significant improvement in symptomatology with Emergency Department treatment is able to tolerate PO intake; has no major electrolyte abnormalities and demonstrates resolving ketosis can be discharged home.
  - The patient should be instructed to eat small frequent meals and increase oral fluid intake.
  - The patient may be discharged home with a prescription for an antiemetic, usually in the form of a rectal suppository.
  - They should be instructed to be rechecked either in the Emergency Department or with their private obstetrician in 24 h.
  - The patient's obstetrician should also be contacted to inform them of the patient's visit to the ED.

Pregnancy Induced Hypertension

Definition

- Hypertension in pregnancy is defined as a blood pressure ≥140/90 in the second half of pregnancy in a previously normotensive patient.
  - Hypertension in pregnancy occurs in approximately 5% of pregnancies.
  - In most cases, the blood pressure usually returns to normal within 10 days after delivery.
  - When severe preeclampsia develops (>170/105), there is an increased risk of intracranial hemorrhage in the mother and also of placental abruption.
- Preeclampsia is defined as hypertension in pregnancy occurring after 20 week's gestation with associated proteinuria and generalized edema.
  - The most common risk factor for preeclampsia is being primigravida.
  - There is also a hereditary component to its occurrence as daughters and sisters of individuals with preeclampsia are at increased risk.
- Eclampsia is defined as the occurrence of seizures in a patient diagnosed with preeclampsia. This occurs during the third trimester or in the immediate postpartum period. There is a significant mortality rate for both the mother and the fetus with eclampsia.
- Chronic hypertension in pregnancy also occurs and involved the presence of hypertension before the onset of pregnancy that continues during and persists long after the completion of the pregnancy. Despite the chronic nature of this underlying condition, preeclampsia and eclampsia can both can occur in these patients.

Table 7A.8. Differential diagnosis for severe nausea and vomiting in early pregnancy

- Hyperememesis gravidarum
- Hepatitis
- Viral gastroenteritis
- Appendicitis
- Partial intestinal obstruction
- Diabetic ketoacidosis
- Molar pregnancy/gestational trophoblastic disease
- Urinary tract infection
- Multiple gestation pregnancy
- Migraine
- Gallbladder disease

- Transient hypertension is hypertension that develops in the latter half of the pregnancy. It is very mild hypertension that doesn't compromise the pregnancy and spontaneously regresses in the postpartum period.
- The HELLP syndrome is a clinical variant of preeclampsia. The letters stand for H-Hemolysis, EL-Elevated liver enzymes and LP-Low platelets.
  - This syndrome tends to occur more commonly in the multiparous patient.
  - Along with the usual signs and symptoms of preeclampsia, the patient with HELLP syndrome complains of epigastric or right upper quadrant abdominal pain.
  - Diagnosis of this syndrome can be made through obtaining an adequate history and physical along with supporting laboratory testing such as abnormal liver function test, decreased platelet count and hemolysis.
  - As a form of preeclampsia, the initial treatment of the HELLP syndrome is the same as for preeclampsia.

Clinical Presentation

- The diagnosis of preeclampsia is made in the presence of the following:
  - Sustained systolic blood pressure ≥140 or diastolic blood pressure ≥90 measured on two separate occasions ≥6 h apart

    AND EITHER
  - Significant proteinuria

    OR
  - Generalized edema or weight gain of at least 5 lb in 1 wk
- Although preeclampsia usually occurs after 20 wk gestation, it may be seen earlier in the presence of molar pregnancy.
- In cases of severe preeclampsia the patient may also complain of scotomas, severe headache and upper abdominal pain.

Diagnostic Evaluation

- Laboratory evaluation should include a CBC with a peripheral blood smear, platelet count, liver function tests, electrolytes, blood type and screen and urinalysis.
- If the fetal gestational age is <24 wk (nonviable), intermittent fetal heart rate assessment should be performed. If the fetal gestational age is >24 wk, then fetal monitoring and biophysical profile should be obtained.

Emergency Department Management

- When blood pressure is noted at this level it should be lowered with either intravenous hydralazine or labetalol with a goal of a systolic pressure of 140-150 and a diatolic pressure of 90-100 mm Hg.

Table 7A.9. Risk factors for the development of preeclampsia and eclampsia

- Chronic hypertension
- Primigravida
- Family history of preeclampsia / eclampsia
- Multiple gestation pregnancy
- Extremes of age
- Molar pregnancy
- Diabetes mellitus
- Chronic renal disease
- Infrequent or nonexistent prenatal medical care

- Magnesium sulfate is also administered to these patients to prevent seizures (eclampsia) from developing.
  - Magnesium sulfate is loaded by giving 4 g intravenously over 20 min followed by an infusion at 2 g/h.
  - Careful monitoring of the patient's respiratory status, deep tendon reflexes and urine output are necessary to avoid magnesium toxicity.
    - When symptomatic magnesium toxicity is suspected, it can be reversed by intravenous administration of 10 ml of 10% calcium gluconate over 10 min.
- Obstetrical consultation should also be obtained to further manage and admit the patient. Early or expedited delivery of the fetus may be indicated for both maternal and fetal well being.

7

Cardiopulmonary Arrest during Pregnancy

- The incidence of cardiac arrest is approximately 1 in every 30,000 pregnancies.
- The most common etiologies are pulmonary embolism, trauma, post-partum hemorrhage with hypovolemia, amniotic fluid embolism and congenital/acquired cardiac disease.
- In the treatment of cardiac arrest in pregnancy initial resuscitative measures and procedures should be employed as they would in the nonpregnant patient.
- Large bore intravenous access should be obtained at a site above the diaphragm so as to enhance the systemic circulation of resuscitation medications. Again, vena caval compression may decrease venous return and medications administered below the diaphragm may be delayed in their ability to reach the central circulation in a timely manner.
- The standard medications indicated in the Current ACLS Guidelines are considered safe for use in the pregnant cardiopulmonary arrest patient based on limited data.
- Electrical therapy which includes defibrillation or cardioversion are performed if clinically indicated in the same manner and with the same energy levels as in the non pregnant patient.
- Perimortem C-section may be clinically indicated. The circumstances regarding the arrest , gestational age, potential survival of the fetus and the time interval since the onset of the maternal cardiac arrest are all factors to be considered when making this decision.
  - If the postmortem C-section can be performed in <5 min from the onset of the maternal cardiac arrest there is a good prognosis for the survival of a viable infant.
  - The emergency physician must also be prepared to perform a neonatal resuscitation and have appropriate equipment and staff available should it be needed.
- Bedside ultrasound in the emergency department is extremely helpful in the decision making process regarding whether to initiate a perimortem C-section by providing information regarding fetal age and viability.
  - If gestational age is >20 wk and there is positive fetal cardiac activity perimortem, C-section may be indicated.

Part B: Selected Gynecologic Emergencies

Vulvovaginits

Clinical Presentation

Generally presents with vaginal discharge of varying types depending on the underlying etiologic agent.

- The color, odor and consistency of vaginal discharge should be noted during examination as should the presence of any erythema, ulcerations, blistering or edema.
- Historical items which should be elicited include any associated history of vaginal itching or irritation, any changes in contraception methods, any changes in sexual activity or partners, any recent use of antibiotics, any previous episodes of vaginitis and pregnancy /menstrual history.
- Vaginal cultures should be obtained including those for gonorrhea and Chlamydia. A wet mount and KOH preparation should be obtained.
- Examination should include identifying the presence of any cervical motion tenderness or adnexal/uterine tenderness or masses.
- A urinalysis and urine HCG should be obtained also as part of the evaluation.

Bacterial Vaginosis (BV)

- This is the most common cause of vaginitis and is caused by overgrowth of mixed bacterial flora causing a disturbance in the vaginal mucosal ecosystem. There is no single infectious agent that is responsible.
- Vulvar/vaginal pruritis or irritation is relatively rare in this type of vaginitis.
- The vaginal discharge is classically described as having a fishy odor and is usually gray in coloration. The discharge is thin and is noted to adhere to the vaginal mucosa. There is minimal inflammation of the vaginal mucosa.
- BV is diagnosed by noting an elevated vaginal pH >4.5. Also if a 10% solution of KOH is added to a sample of the discharge, it causes the release of a strong fishy odor. Microscopic examination of the discharge reveals the presence of clue cells. Clue cells consist of vaginal epithelial cells with clusters bacteria adhering to the cell membranes.

Treatment

- Topical
  - Metronidazole gel 0.75%, one applicator full applied twice a day X 5 days
  - Clindamycin cream 2%, one applicator full applied at bedtime X 7 days
- Oral
  - Metronidazole 500 mg PO twice a day X 7 days. Metronidazole can also administered as a one time dosage of 2.0 g which also delivers effective initial treatment but has a higher recurrence rate.

Candida Vaginitis

- This is the second most commonly diagnosed cause for vaginitis. The most frequent presenting complaint is vulvar/vaginal pruritis and discharge which, when it is present, is minimal. The discharge is classically described as cottage cheese-like in appearance, and there is no significant odor. The patient may complain of dyspareunia. The pelvic examination generally shows erythema and edema of the mucosal surfaces. The patient who is immunocompromised or diabetic is at increased risk.
- Candida vaginitis is diagnosed in large part based on the patient's history and physical examination. Obtaining a wet mount and KOH preparation will demonstrate the presence of hyphal elements and yeast cells confirming the diagnosis.

Treatment

- Topical
  - Clotrimazole 1% cream applied QHS for 7-14 days (OTC)
  - Miconazole 2% cream applied QHS for 7-14 days(OTC)
  - Butaconazole 2% cream applied QHS for 3 days
  - Miconazole 200 mg vaginal suppository inserted QHS for 3 days
  - Clotrimazole 500 mg tablet inserted vaginally QHS once
- Oral
  - Fluconazole 150 mg tablet PO once

Trichomonas Vaginitis

- The causative agent is the protozoan *Trichomonas vaginalis* and should be considered a sexually transmitted disease (STD).
- The presenting signs and symptoms include yellow/gray frothy discharge which may be malodorous. Dysuria and dyspareunia may be present. The vaginal mucosa is typically erythematous, and the patient typically complains of pruritis. Punctate hemorrhages occur on the cervix (strawberry cervix) <20% of the time.
- The diagnosis of Trichomonas vaginitis is confirmed by the visualization of pear-shaped flagellated trichomonads on wet mount. Leukocytes may also be visualized on microscopic examination. The pH of vaginal secretions is increased and is >4.5.
- The treatment of *Trichomonas vaginalis* consists of the oral administration of Metronidazole either as a single dose of 2.0 g or 500 mg PO bid for 7 days. It is important that the patient's sexual partner be treated also to prevent further reoccurrence and further spread. Topical metronidazole treatment is generally not recommended because of the inability to eradicate the organism from the urethra and skenes glands leading to reoccurrence.

7

Atrophic Vaginitis

- Usually occurs in postmenopausal women due to diminished levels of circulating estrogens. Presents with vaginal and vulvar itching and discomfort. Physical examination usually demonstrates a pale, thin vaginal mucosa that is often friable. Wet mount, KOH evaluation and vaginal cultures are all found to be negative. A pap smear should be obtained at the time of the examination.
- Appropriate emergency department treatment is to prescribe estrogen replacement therapy either orally or in the form of a vaginal cream and to arrange gynecological follow-up.

Contact Vulvovaginitis

- This occurs secondary to a localized allergic reaction or chemical irritation after exposure to various substances. Common etiologies are soaps, deodorants, douches, tampons, panty hose, toilet paper and underwear. The patient presents with vaginal irritation, itching and discomfort. The physical examination reveals erythema and edema of varying degrees. Infectious etiologies are eliminated by the wet mount, KOH and cultures. The treatment consists of eliminating further contact with the causative agent, the use of sitz baths, topical steroids, oral antihistamines and gynecologic follow-up.

Genital Herpes

- Symptoms of primary vaginal infection include vulvovaginal discomfort but may be accompanied by systemic symptoms such as fever and malaise.
- Genital herpes is considered an STD and occurs most frequently in the young adult and adolescent patient populations.
- Physical examination demonstrates fluid filled vesicles on the mucosal surfaces that may eventually progress on to painful ulcerations.

Table 7B.1. The differential diagnosis of acute PID

- Acute appendicitis
- Ectopic pregnancy
- Ovarian torsion
- Urinary tract infection (UTI)
- Ovarian cyst/ruptured ovarian cyst
- Endometriosis
- Urolithiasis
- Gastroenteritis
- Diverticulitis

- The initial episode lasts for approximately 2-4 wk. The herpes virus continues to live in the dorsal root ganglia after the initial episode has resolved and predisposes the patient to chronic recurrent episodes.
- The diagnosis is initially made by history and noting vesicular lesions/ulceration on physical examination. Confirmation of the diagnosis can be made by viral cultures and/or by finding multinucleated giant cells on a Tzanck smear.
- The treatment of genital herpes is not curative but is aimed at shortening the course of episodes and decreasing the frequency of reoccurrence. 7
  - Antiviral therapy for the initial episode consists of the use of acyclovir (Zovirax) 400 mg 5x/day for 7-10 days or valocyclovir (Valtrex) 1 g PO BID for 7-10 days.
  - Analgesic agents and sitz baths may also be used to help alleviate the patients discomfort.
  - Patients with severe systemic symptomatology may need to be admitted for intravenous antiviral therapy.

Vaginal Foreign Bodies

- Foreign bodies left in place either intentionally or accidentally for >24 h may lead to overgrowth of vaginal flora leading to foul smelling vaginal discharge. This frequently occurs in the pediatric population but may occur in adults when forgotten tampons or diaphragms are left in place.
- The treatment consists of removal of the foreign body.

Pelvic Inflammatory Disease

- This disease process represents an infection of the upper female reproductive tract that is sexually transmitted and starts as an ascending infection from the cervix and vagina.
- This disorder is the most common gynecological cause for hospitalization in reproductive age women.
- Most cases of PID are polymicrobial.
  - In approximately 50% of cases *Chlamydia trachomatis* and *N. gonorrhea* are isolated and represent the primary pathogens.
  - Pathogens that are also responsible less frequently include *Bacteroides, Peptostreptococcus, E. coli, Haemophilus influenzae* and *Gardnerella vaginalis.*
- Risk factors for PID include sexual activity with multiple sexual partners, adolescence or young adulthood, douching, presence of an IUD and a prior history of other STDs.

Clinical Presentation

- Initial presentation frequently includes a complaint of lower abdominal/pelvic pain but may include complaints of vaginal discharge, fever, malaise, nausea and vomiting.
- The minimal criteria for diagnosing PID per the CDC are lower abdominal tenderness, adnexal tenderness and cervical motion tenderness.

Table 7B.2. Current indications for inpatient therapy in PID

- Uncertain diagnosis
- Tuboovarian abscess
- Simultaneous pregnancy
- Adolescence
- The immunocompromised patient (HIV)
- Clinically toxic-appearing patient
- Inability to maintain adequate PO intake/ persistent nausea and vomiting
- Failure of attempted outpatient treatment
- Unreliable patient/ inability to provide adequate follow-up

    - Other more minor determinants that increase the clinical suspicion of PID are fever, abnormal cervical or vaginal discharge, systemic signs, an increased ESR or CRP, and dyspareunia.
- Fitz-Hugh-Curtis syndrome is one of the serious sequelae of PID and presents with right upper quadrant pain secondary to liver capsule inflammation and perihepatitis.
- Appropriate laboratory evaluation for PID should include CBC, qualitative HCG, urinalysis, GC and Chlamydia cultures or DNA probe and screening for other common STDs.
- Ultrasound evaluation is generally not indicated unless there is suspicion of a possible tuboovarian abscess or an associated pregnancy.

7

Emergency Department Management

- The diagnosis of PID is based on clinical evaluation as early treatment is necessary in order to minimize the possibility of serious complications.
- Patients with PID can be treated as inpatients (Table 7B.2) or outpatients depending on the severity of their symptoms.
- Male sexual partners of patients with PID should be evaluated for STDs and treated for GC and Chlamydia

Inpatient Therapy

- Cefotetan 2 g IV Q12H or Cefoxitin 2 g IV Q6H plus Doxycycline 100 mg IV Q12H
- Alternative Treatment—Clindamycin 900 mg IV Q8H plus Gentamycin 2 mg/kg IV load and then 1.5 mg/kg Q8H IV

Outpatient Therapy

- Ceftrioxone 250 mg IM plus Doxycycline 100 mg PO BID x 14 days
- Alternative Treatment-Ofloxin 400 mg PO BID x 14 days
  plus
  Metronidazole 500 mg PO BID x 14 days
- At discharge patients may also be given analgesics as indicated by the severity of their symptoms. They should be educated about the importance of preventing STD spread and reinfection by having their male partners evaluated and treated appropriately.

Tuboovarian Abscess (TOA)

- This occurs in approximately 5% of patients diagnosed with PID and the constellation of symptoms and etiology of TOA is similar to that discussed previously with PID.
- Lower abdominal pain, cervical motion tenderness, adnexal tenderness and a palpable adnexal mass are present.
- Ultrasound is the imaging tool of choice in confirming the diagnosis of TOA.
- Gynecologic consultation should be obtained.

- In most cases IV antibiotic therapy is the sole necessary treatment.
  - However 20-40% of cases do require surgical intervention for successful treatment.

Pediculosis Pubis

- Pediculosis pubis is a cutaneous infestation with the louse, *Phthirus pubis*. Found in the area of pubic hair after contact with an infected individual, it is frequently transmitted through sexual contact.
- The adult form is approximately 1-2 mm in length and the nits, small 0.5 mm ova, are found at the base of pubic hair shafts.
- Patients present with the complaint of severe itching in the pubic area and the diagnosis is confirmed by direct visualization of either the adult or nit form.
- The most effective treatment is Permethrin cream applied to the involved area.
  - Lindane shampoo can be used as an alternative treatment but is contraindicated in pregnancy or lactation.
- All clothing and bed linens should be cleaned to eliminate sources of reinfection.
- All recent sexual contacts should be informed and subsequently treated.

Pubic Scabies

- This represents a highly contagious infestation by the mite *Sarcopetes scabiei*. The female mite which is approximately 0.2-0.4 mm long burrows into the patients skin to deposit eggs.
- Transmission usually occurs from intimate contact with an infected individual or with infested clothing.
- The clinical presentation is that of severe pruritis in the pubic area.
- Physical examination may show the presence of burrows which are noted as small (<1 cm) raised threadlike structures.
  - Further confirmation of the diagnosis can be made by microscopic evaluation of skin scrapings to identify mites, eggs or fecal material.
- The most effective treatment is the topical application of Permethrin 5% cream from the neck down which is subsequently washed off 8 h later.
- Machine washing of all clothing and bed linens in hot water reduces the incidence of reinfection.
- Antihistamines may be given for control of pruritis and the patient should be warned that this pruritis may persist for several weeks after treatment because of residual skin inflammatory response.
  - Topical application of corticosteroids can help alleviate residual pruritis.

Bartholin Abscess

- A Bartholin abscess is a polymicrobial infection of a Bartholin duct cyst; *E. coli* being the organism found most frequently (*N. gonorrhea* can be found as the etiologic agent in <10% of cases).
- These abscesses occur most frequently in women of reproductive age and present with a painful lump on the labia.
- The abscess is palpable as a tender fluctuant mass over the vulva on the involved side and are usually unilateral.
- Incision and drainage of the abscess using local anesthetic is the treatment of choice.
  - The area of the abscess is locally infiltrated with 1% lidocaine and an incision is made on the mucosal surface of the abscess parallel to the hymenal ring.
  - The appearance of purulent drainage indicates successful penetration of the abscess wall.
  - The abscess should be irrigated with normal saline and a Word catheter should be inserted to allow further drainage.

- The patient should be discharged on antibiotic therapy and be given gynecologic follow-up in two days for packing removal and further evaluation.
  - Amoxicillin/clavulanic acid 875 mg PO BID for 5 days plus Metronidazole 500 mg PO BID for 5 days offers excellent coverage.
  - Alternatively Ciprofloxacin 500 mg PO BID plud Metronidazole 500 mg PO BID can be used.
- Most patients can be treated as outpatients. Patients that require admission are usually septic or have severe cellulitis/necrotizing fasciitis.

Suggested Reading

1. American College of Emergency Physicians. Clinical policy: Critical issues in the initial evaluation and management of patients presenting to the emergency department in early pregnancy. Ann Emerg Med 2002; 20:575.
2. Hicks JL, Rodgerson JD, Heegard WG et al. Vital signs fail to correlate with hemoperitoneum from ruptured ectopic pregnancy. Am J Emerg Med 2001; 19:448.
3. Sibai BM. Treatment of hypertension in pregnant women. New England J Med 1996; 335:257.
4. ACOG, Practice Bulletin, no 43. Management of preterm labor. Obstet Gynecol 2003; 101:1039.
5. Mattar F, Sibai BM. Eclampsia. VIII. Risk factors for maternal mobility. Am J Obstet Gynecol 2000; 182:307.
6. Sobel JD. Vaginitis. New Engl J Med 1997; 337:1896.
7. Mehta SD, Rothman RE, Kelen GD et al. Clinical aspects of the diagnosis of gonorrhea and chlamydia infection in an acute care setting. Clin Infect Dis 2000; 32:655.
8. BrennerPF: Differential diagnosis of abnormal uterine bleeding. Am J Obstet Gynecol 1996; 175:166.
9. Carter J, Fowler J, Carson L. How accurate is the pelvic examination as compared to transvaginal sonography? J Reprod Med 1994; 39:32.
10. MacKenna A, Fabres C, Alam V et al. Clinical management of functional ovarian cysts: A prospective and randomized study. Hum reprod 2000; 15:2567.
11. Mateer JR, Aiman EJ, Brown MH et al. Ultrasonographic examination by emergency physicians of patients at risk for ectopic pregnancy. Acad Emer Med 1995; 2:867.
12. Burgher SW, Tandy TK, Dawdy MR et al. Transvaginal ultrasonography by emergency physicians decreases decreases patient time in the emergency department. Acad Emerg Med 1998; 5:802.

CHAPTER 8

Orthopedic Emergencies

Susan Zapalac and Mark G. Richmond

Pelvis and Sacrum

Anatomy and Function

- The anatomy of the pelvis consists of the right and left innominate bones, the sacrum, and the coccyx.
  - The innominate bone consists of the ischium, pubis, and ilium.
  - The sacrum lies between the two innominate bones forming the SI joints posteriorly. The innominate bones join together anteriorly at the symphysis pubis.
  - The acetabulum is a deep, cup-shaped indentation in the lateral aspect of the innominate bone. It is made up of the ilium superiorly, the ischium inferolaterally and the pubis anterosuperiorly.
- Vascular supply consists of branches of the internal iliac artery, median sacral artery, and the superior rectal artery.
- The neurologic anatomy consists of the sacral nerve roots within the sacrum and the lumbosacral plexus surrounded by the pelvic ring.
- The function of the pelvis is to distribute weight of the trunk to the lower extremities while standing, and to the ischial tuberosity while seated.
- The pelvis aids in protection of the sigmoid colon and the genitourinary organs including the bladder, the distal ureters, the urethra, and the male and female reproductive organs.

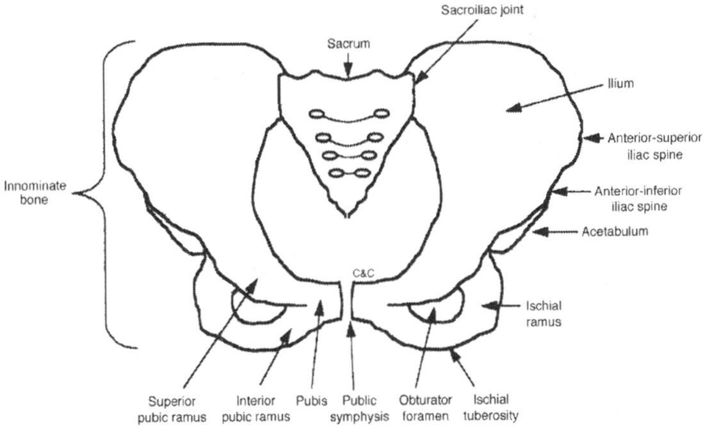

Figure 8.1. Anatomy of the pelvis. From Emed Clin North Am 2000; 18(1):2, ©2000 Elsevier, with permission.

Table 8.1. Tile classification system

Type A:	Stable pelvic ring Injury
	A1: Avulsion fracture of the innominate bone
	A2: Stable iliac wing fractures or stable minimally displaced ring fractures
	A3: Transverse fracture of coccyx and sacrum
Type B:	Partially stable pelvic ring injury (rotationally unstable, vertically stable)
	B1: Open book injury (unilateral)
	B2: Lateral compression injury
	B3: Bilateral type B injuries
Type C:	Unstable pelvic ring injury (vertical sheer, rotationally and vertically unstable)
	C1: Unilateral
	C2: Bilateral, one side type B, one side type C
	C3: Bilateral type C lesions

Trauma

- Pelvic trauma accounts for 3% of all orthopedic trauma.
- Mechanism of injury is either low or high energy in nature.
  - Low energy fractures include domestic falls, straddle injuries, avulsion fractures, and low velocity vehicular injuries. These mechanisms usually result in isolated fractures of individual bones and do not disrupt the pelvic ring.
  - High energy fractures result in pelvic ring disruptions. Mechanisms include motor vehicle accidents (57%), pedestrian vs. auto (18%), motorcycle (9%), falls from heights (9%) and crush injury (4%) (J Trauma 29:981-1002, 1989).

8

Classification System

- The ideal classification system would allow the clinician to identify associated injuries and their complications, help formulate a treatment plan and predict morbidity and mortality.
- Several classification systems exist for pelvic fractures, however the Orthopedic Trauma Association and the OA group uses the Tile classification.
  - The Tile system is based on the direction of force applied to the pelvis ring including lateral compression (LC), anteroposterior compression (APC), vertical sheer (VS), and combined mechanical injury (CM). It is further divided into radiographic evidence of stability or instability. Young expanded this classification by subdividing the LC and APC fractures based on the amount of ring disruption as viewed by three pelvic radiographic views.

Management

Prehospital Care

- If the patient is unstable or there are prolongued transport times, pneumatic antishock garments (PASG) should be applied if available. Spinal immobilization with c-collar and backboard are essential.

ABCs

- As with all patients, the patient's airway, breathing and circulation are first to be evaluated.
- Specific attention to hemodynamic stability is important since certain pelvic fractures can lead to hemorrhagic shock.
- IV sites in the lower extremity should be avoided if a pelvic fracture is suspected. In the event that the patient is unstable, crystalloid fluids (either normal saline or lactated Ringers) should be used.

- Use of blood products should be initiated early in the resuscitation phase if the patient is hypotensive and a severe pelvic fracture is suspected.

History

- When time permits, a detailed history from witnesses, the patient, or ambulance personnel must be obtained.
- Details of the mechanism of injury, direction of force, amount of force, and associated injuries are pertinent.
- Ask the patient if they have pain with movement, ambulation, sitting, standing, or defecating.
- Attention to associated injuries is crucial (is patient able to void, are they pregnant, do they have any motor or sensory deficits etc.).

Physical Exam

- As part of the secondary survey, inspect for swelling, ecchymosis, and tenderness over the hips, groin, and lower back.
- Note deformity or asymmetry in the lower extremities.
- Examine the anterior-superior iliac spine by pressing medially and then laterally to check for internal and external rotation of the pelvis respectively. Checking for vertical and rotational deformity should be performed only once and by the most experienced physician as this examination may dislodge blood clots and result in hemorrhage and rapid decompensation.
- Examine the skin to verify open versus closed fractures. It is essential to perform digital and perineal examinations. Assess for continuity of the anus and rectum, condition of the prostate and for fresh blood.
- Check for blood at the opening of the urethra and in the female for continuity and lacerations of the vagina.
- If the pelvis is unstable, detailed exams with an anoscope and speculum should be postponed until stabilization is possible.
- A detailed vascular and neurologic exam should be performed.

Radiography

- Plain films of the pelvis should be ordered when the patient is symptomatic or if the patient is not assessable. An AP view should be ordered early. Additional plain film views or CT scan of the pelvis, if the patient is stable, should be ordered to determine the type of fracture.
    - Plain films
        - AP pelvis view should be the initial film ordered for any pelvic trauma. Most significant fractures will be seen on this view; however it does not demonstrate the degree of bony displacement well. The pubic symphysis should be no more than 5 mm wide with <2 mm offset of left and right pubic rami. The SI joint should be no >2-4 mm wide.
        - Inlet view (30° caudal view) provides visualization of the posterior arch, widening of the SI joint, and displacement of the anterior arch.
        - Outlet view (30° cephalic view) allows visualization of cephalad or vertical displacement of the hemipelvis, nondisplaced sacral fractures and SI joint widening.
    - Computed tomography
        - CT is useful for delineating pelvic fractures that may be missed on plain radiographs and for elucidating associated complications including retroperitoneal hematomas. Studies have shown that CT scanning is superior for demonstrating acetabular fractures and posterior arch disruptions. It is important to remember that only the hemodynamically stable patient is a candidate for CT scanning.

- MRI
  - MRI may be useful for neurologic and vascular injury and to delineate genitourinary complications. Its role in initial evaluation of an acute fracture is limited due to prolonged time for evaluation and availability.

Labs

- Type and crossmatch, serial hemoglobin and hematocrit, platelets, and PT/PTT should be ordered on all patients with severe pelvic fractures.

Other Studies

- High energy fractures have a high association with nonorthopedic injuries. In one study of patients with high energy pelvic fractures, 47% had associated abdominal injury (J Trauma 23:535, 1983.).
- A diagnostic peritoneal lavage (DPL) or an ultrasound FAST exam should be performed.
  - If the DPL aspirate or the FAST exam is postive, a laparotomy is warranted.
  - If the DPL cell count is positive, an external fixator should be applied and the patient taken for laparotomy.
  - If the cell count is negative and the patient is a candidate for external fixation, then one should be applied.
  - If the patient remains hemodynamically unstable in any case, then angiography and selective embolization is necessary.

8

Complications

Hemorrhage

- Hemorrhage is the leading cause of death in patients with pelvic fractures.
- 20% of hemorrhages are due to disruption of the iliac and femoral vessels and the remaining occur from venous, marrow or small vessel disruption.
- The fractures that are most likely to place patients at risk for hemorrhage are Tile's Type B1 fractures (open book injury/sprung pelvis), any open fracture, or those that meet Cryer's criterion. Cryer et al noted that a gap or displacement of 0.5 cm or more at any fracture site in the pelvic ring or acetabulum as seen on the AP pelvis plain film view had a high correlation with significant hemorrhage.
- It is essential that the ED physician recognize and evaluate those fractures that place the patient at high risk for retroperitoneal hematomas and associated abdominal visceral injury.
- Resuscitation and stabilization is paramount until definitive therapy can be assured.

Genitourinary Complications

- Rupture of the bladder and posterior urethra occurs in approximately 5% of patients with pelvic injury.
- Blood at the meatus, a high riding prostate, laceration of the vagina, or inability to void are all correlated with GU injury.
- Fractures associated with GU injuries include Tile's Type B2 (subluxation of the pubic symphysis and bucket-handle fractures), Type B1 (open book injury), Type C (Malgaigne fracture), and straddle fractures.
- Microscopic hematuria alone does not mandate cystography or urethrography.
- Clinical suspicion of urethral injury or gross hematuria requires further evaluation including a combination of urethrogram, intravenous pyelography, cystography, and CT scan.

Neurologic Injury

- A complete neurologic exam, including motor, sensation and deep tendon reflexes, is critical when pelvic injury is suspected.
- Fractures associated with neurologic injury include sacral and acetabular fractures.

- Neurologic injury is found in 22% of vertical sacral fractures and horizontal sacral fractures at or above the S3 level.
- Acetabular fractures are associated with sciatic nerve injury.
- Stabilization of the fracture is essential.
- Neurosurgical or orthopedic consultation is required.

Orthopedic Injury

- Orthopedic consultation for surgical fixation is required in all patients that have a double break in the pelvic ring, who do not meet Cyer's Criterion, and who are hemodynamically stable.

Gynecologic Injury

- Vaginal bleeding or uterine bleeding may result from lacerations secondary to open fractures or, in the gravid uterus, abruptio placentae or uterine perforation.
- Fallopian tube or ovarian damage is less likely, but can occur.
- If any gynecological injury is suspected, gynecological consultation is required.

Uncomplicated Fractures

- These fractures are simple, involve only a single break in the pelvic ring, are not associated with secondary injuries, and do not require orthopedic repair.
- Fractures that are considered uncomplicated include Tile's Type A1, A2 and A3 fractures (ASIS avulsion, AIIS avulsion, ischial tuberosity avulsion, ischial body fractures, iliac wing fractures, and coccyx fractures).
- Uncomplicated fractures require bed rest and analgesia and can be managed as an outpatient.

Acetabulum

Anatomy and Function

- As mentioned previously, the acetabulum is a deep, cup-shaped indentation in the lateral aspect of the innominate bone.
- It is made up of the ilium superiorly, the ischium inferolaterally and the pubis anterosuperiorly.
- Functionally, the acetabulum is an inverted Y, with one limbs forming the anterior and posterior columns.
  - The anterior column extends from the iliac crest to the symphysis pubis and involves the anterior lip of the acetabulum.
  - The posterior column starts at the superior gluteal notch, through the acetabulum, the obturator foramen, and the inferior pubic ramus. It includes the posterior lip of the acetabulum.
  - The acetabular dome is the notch of the inverted Y. It provides the superior weight-bearing area and includes the anterior and posterior columns.

Management

Prehospital care, ABCs, and history and physical is similar to the pelvis and hip. On physical exam, vascular, visceral, and neurologic deficits need to be evaluated.

Radiography

- Plain Films
  - Acetabular fractures are difficult to identify on only an AP pelvis view. If suspicious, the posterior column and the anterior lip can be visualized with a 45 degree external view. The posterior lip and the anterior column are best visualized on the 45 degree internal view. Posterior oblique films are best used to visualize central acetabular fractures.

- CT is by far the best means to elucidate fine details of acetabular fractures.

Classification

- Classification of acetabular fractures is quite complex. Letounel breaks the classes into:
  - Type A: Partial articular one column fracture
  - Type B: Partial articular transverse oriented fracture
  - Type C: Complete articular, both column fracture

Treatment

- Treatment is based on the type of fracture, its displacement or dislocation, patient factors and fracture factors.
- In the ED, immobilization, early orthopedic consult and admission are required for acetabular fractures.

Complications

- Include osteoarthritis, traumatic arthritis, avascular necrosis, and sciatic nerve injury.
- Central dislocation of the femoral head through the acetabulum has a high incidence of significant blood loss.

Hip

Anatomy and Function

8 • The hip is a ball and socket joint comprised of the acetabulum and the proximal aspect of the femur ending 5 cm distal to the lesser tubercle.
- There is a strong fibrous capsule that surrounds the joint on all sides. The capsule extends from the acetabulum proximally to the femoral neck posteriorly and the intertrochanteric line anteriorly.
- The function of the hip is to distribute weight to the lower extremities and allow movement including flexion, extension, internal and external rotation, abduction and adduction.
- Vascular supply of the proximal femur is tenuous. It has essentially three main sources: the epiphyseal arteries, the metaphyseal arteries, and the foveal artery of the ligament teres.
- Disruption of the femoral neck or joint capsule can disrupt flow to the femoral head resulting in avascular necrosis.
- Sensory innervation to the hip joint is through the femoral, obturator, sciatic and superior gluteal nerves.

Trauma

Hip Dislocation

- In adults, significant force is required to dislocate the hip.
- Motor vehicle accidents are a common cause of hip dislocations, and they are often associated with multisystem trauma.
- Children and individuals with prosthesis require minimum force to cause hip dislocation.
- Hips can dislocate anterior, posterior or central.
- Posterior dislocation is most common, followed by anterior and central dislocations.
- It is important to look for associated hip fractures in both the high speed motor vehicle patient as well as the osteoporotic older patient.
- Children can have associated slipped capital femoral epiphysis.
- Hips should be relocated no later than 12 h after the event.
- Delay in reduction places the patient at increase risk for femoral head necrosis.

Presentation and Complications

Anterior Dislocation

- Anterior dislocations commonly occur after a fall or injury where the femur is abducted suddenly.
- There are three types of anterior dislocations including superior iliac, superior pubic and inferior (obturator) depending on the anatomic position of the femoral head after dislocation.
  - Superior iliac dislocation presents with the hip extended and externally rotated.
  - Superior pubic dislocation presents with the hips flexed, abducted, externally rotated.
  - Inferior dislocation presents markedly flexed, abducted and externally rotated. Complications of anterior dislocation include disruption of the femoral artery, nerve and vein, postrraumatic arthritis and pulmonary embolism.

Posterior Dislocation

- Posterior dislocations are commonly caused in motor vehicle accidents when the knees hit the dashboard.
- The affected hip presents shortened, flexed, and internally rotated.
- Complications include sciatic nerve injury, femoral head necrosis, posttraumatic arthritis, and pulmonary embolism.

Central Dislocation

- Central dislocation refers to the disruption of the acetabulum with the femoral head displaced into the pelvis. (See acetabular fractures.)
- Surgical reduction is required.

Management and Treatment

Initial Assessment

- As with all trauma, airway, breathing, and circulation should first be assessed and managed.
- Assess the patient's neurologic and vascular status of the affected extremity.
- In anterior dislocations, check the patient's femoral, popliteal, and pedal pulses. Diminished pulses, swelling of the thigh, weakness of the quadriceps, diminished patellar reflex, or decreased sensation of the anteromedial thigh should alert the ED physician to femoral nerve or vascular injury.
- If a posterior dislocation is suspected, check for sciatic nerve impairment: assess the posterolateral leg and sole of the foot for sensation, and flexors of the knee for strength.
- Careful documentation should be noted before and after reduction.
- Early analgesia or local anesthesia (in the elderly or unstable patient) is recommended.

Radiography

- Plain radiographs in two views are necessary to evaluate the dislocated hip.
- The AP view and the lateral view are the most common.
- Films must be evaluated for occult fractures of the acetabulum, femoral head, femoral neck and femoral shaft prior to reduction.
- Additional views including the judet, 15 degree oblique and the cross-table axiolateral view as well as CT may be needed to adequately assess these areas.
- To distinguish between anterior and posterior dislocation, assess the femoral head size and the position of the lesser trochanter.
- In anterior dislocations the femoral head appears larger than the opposite side and the lesser trochanter is easily visualized.
- In posterior dislocations the femoral head appears smaller and the lesser trochanter will be superimposed on the femoral shaft.

Reduction

- If a dislocation is associated with a fracture, reduction is controversial depending on the location and type of fracture.
- Orthopedic consultation is recommended.
- If no fracture is present, then conscious sedation and primary closed reduction is the treatment of choice:
  - Anterior dislocation is best reduced with the Allis maneuver. The patient is place supine, the pelvis is stabilized at the ASIS, and gentle traction is applied in the line of the deformity. The hip is slowly flexed to 90 degrees then slowly internally and externally rotated.
  - Posterior dislocation is best reduced with the Stimson maneuver. The patient is placed prone, and the pelvis is stabilized at the sacrum. The limb is allowed to hang over the table and positioned in flexion at the hip, knee, and ankle at 90 degrees. Gentle downward traction is placed just distal to the knee.
- If reduction is not obtained easily then orthopedic consult and open reduction must be considered.
- After all reductions, post-reduction films and neurovascular assessment is critical.

Disposition

- All hip dislocations require orthopedic consult. Most hip dislocations require admission.

Hip Fracture

Epidemiology

- Hip fractures account for over 250,000 hospital visits annually in the US. They are associated with a high morbidity and mortality.
- One-year mortality rates are as high as 14-36% in the elderly. (N Engl J Med 334:1519-25, 1996).
- The mechanism of injury is commonly minor trauma in the elderly compared to high energy trauma in the young adult.
- Fractures, dislocations, and fracture dislocations are an orthopedic emergency.

Management

- Hip fractures have a broad spectrum of presentation.
- They may present with an ambulatory elderly person complaining of groin, hip, thigh, or knee pain to the hypotensive multisystem penetrating or blunt trauma patient.
- Management is tailored to the presentation.

Prehospital Care

- The ABCs should be assessed and treated initially.
- After stabilization, a brief neurovascular exam of the extremity should be performed.
- Open wounds should be dressed with a sterile dressing. Bleeding should be stopped with direct pressure.
- The patient should be splinted and placed in a position of comfort.
- Traction devices such as a Hare splint may be placed as it can reduce pain and can tamponade further blood loss.
- Contraindications to traction include open fractures, suspected sciatic nerve injury, fracture of the ipsilateral pelvis or lower extremity, and fractures near the knee.

ABCs

- Patients' airway, breathing, and circulation must be stabilized initially.
- Once assessed and treated, the hip fracture should be evaluated during the secondary survey.

Physical Exam

- Signs and symptoms of hip fracture include tenderness, ecchymosis, deformity, and shortening.
- Examine the wound and determine if it is open or closed.
- Evaluate neurologic and vascular status. Check the deep tedon reflexes, sensation, and pedal pulses.
- If vascular injury is suspected, compare the blood pressures of the injured and noninjured extremity.
- Pay close attention to sciatic nerve injury (see hip dislocation).
- Evaluate the entire extremity for associated injuries (femoral shaft, knee, tibia and fibula).

Radiography

- Plain films should be performed as soon as possible.
- Splints should be removed to obtain adequate films.
- AP pelvis and a lateral hip film are standard.
- If necessary, an oblique hip with the femur internally rotated at 15-20 degrees will enhance visualization of the femoral neck.
- If a fracture is not identified on plain films but is highly suspected, a CT, bone scan or preferably a MRI must be performed to rule out fracture.

Treatment

- Adequate analgesia is pertinent.
- Opioid analgesia is recommended.
- In the unstable patient where narcotics use is not acceptable, a femoral nerve block can be used to reduce pain.
- Patients with open fractures should be started on antibiotics immediately.
- For a clean open wound <1 cm, a first-generation cephalosporin should be started.
- For wounds that appear contaminated or are >1 cm, antibiotics with Gram-negative and Gram-positive coverage should be used.
- Tetanus prophylaxis should be given.
- All fractures require orthopedic consultation and admission.

Classification and Associated Complications

There are multiple classification systems for hip fractures. However, hip fractures can easily be described by location. Presentation, treatment, and complications are different with each fracture.

- Intracapsular
    - Femoral head
        - Femoral head fractures are most commonly associated with dislocations.
        - The incidence of femoral head fractures in anterior dislocations is between 22-77% while the incidence of fractures in posterior dislocations is between 10-16%.
        - It is imperative to obtain post-reduction AP pelvis films and assess for femoral head fractures.
        - Fractures can be very subtle, and if suspected a CT or MRI should be obtained. Avascular necrosis and post-traumatic arthritis are the most common complications.
    - Femoral neck
        - Femoral neck fractures can be nondisplaced (Garden Type I-II) or displaced (Garden Type III-IV).
        - Femoral neck fractures occur most commonly in low energy fractures in the elderly.
        - In younger patients they may result from multisystem trauma.

        - Young patients will have an associated femoral shaft fracture 20% of the time, and the femoral neck fracture can be overlooked.
        - A high index of suspicion should be maintained and appropriate imaging modalities used.
        - Nondisplaced fractures or those with other injuries may not be visualized, and MRI or bone scanning is necessary.
        - Avascular necrosis is the most common complication of femoral neck fractures.
        - The incidence is higher in displaced fractures and increases with delay of definitive treatment.
        - The remaining complications are due to prolonged immobilization and nonunion.
- Extracapsular
    - Trochanteric
        - Trochanteric fractures are uncommon and usually seen in young or old patients after a direct fall.
        - Trochanteric fractures are either displaced or nondisplaced fractures of the lesser or greater trochanter.
        - Fractures of the lesser trochanter are commonly avulsion fractures.
        - Nondisplaced fractures can be treated with bed rest and analgesia.
        - In the young patient, displaced greater trochanteric fractures >1cm or lesser trochanteric fractures >2 cm require internal fixation.
        - Complications include associated muscle funtion loss, but generally outcome and function are good.

8

    - Intertrochanteric
        - The classic intertochanteric fracture consists of the fracture line between the greater and lesser trochanter.
        - It is seen most commonly in the elderly after falls.
        - Intertrochanteric fractures can be classified as stable or unstable.
        - Unstable fractures are those that are displaced, comminuted or with multiple fracture lines.
        - Definitve treatment is almost always open reduction and internal fixation.
        - Moribidity is high—49% of patients sustaining intertrochanteric fractures are unable to regain their original ability to ambulate (J Bone Joint Surg 60A:930, 1978).
        - Mortality is associated with complications of immobilization including pulmonary embolism, pneumonia, UTIs, decubitus ulcers etc.
    - Subtrochanteric
        - The subtrochanteric region is the area between the lesser trochanter to a point 5 cm distally.
        - In the elderly, fractures commonly occur secondary to low energy trauma and poor bone quality.
        - In the young, the fracture is secondary to high energy trauma.
        - Pathologic fractures are common at this site.
        - Definitive care is controversial but orthopedic consultation is always required.
        - Traction splinting is recommended.
        - Subtrochanteric fracture complications include significant blood loss, associated injuries of high energy trauma, and nonunion.

Disposition

- All fractures require orthopedic consultation and admission.
- Open reduction and internal fixation versus arthroplasty is dependent on the type of fracture, the age and condition of the patient, and the condition of the bone.

Femur

Anatomy and Function

- The femur is the strongest and longest bone in the body.
- The proximal extremity consists of the head, neck, greater and lesser trochanter.
- The shaft extends from an area 5 cm distal to the lesser trochanter to 6 cm proximal to the adductor tuberecle.
- The distal extremity is comprised of the medial and lateral condyles separated by the intercondylar fossa.
- The femur approximates with the acetabulum to form the hip joint proximally and with the tibia distally forming the knee joint.
- While the vascular supply of the proximal femur is tenuous, the femoral shaft is highly vascular.
- It obtains its blood supply from nutrient vessels originating from the profunda femoris artery.
- Due to the vast blood supply, fractures heal well with little complication; however, significant hemorrhage after injury is likely.
- The proximal and mid-femur are innervated by the femoral nerve and the sciatic nerve.
- The femur transmits weight from the body to the lower extremities.
- The shaft is surrounded by several strong muscles.
- These muscles are encased in fascia creating the anterior, medial and posterior compartments of the leg.
- By understanding the muscle attachment sites, the physician can predict the displacement of the femur after fracture.

8

Trauma

- **Femoral shaft fractures** are caused by high energy trauma including motor vehicle accidents, pedestrian vs. auto, motorcycle accidents, fall from heights and gunshot wounds.
  - Comminuted fractures occur more commonly from gunshot wounds and direct side impact of a motocycle accident.
  - Spiral fractures are more commonly seen after fall from heights.
  - Motor vehicle accidents and auto-pedestrian accidents commonly result in transverse fractures.
  - 70% of femoral fractures, specifically spiral fractures, in children <3 yr old are secondary to nonaccidental trauma.
  - Rarely, pathologic fractures of the femur occur.
- **Distal femur fractures** account for 4% of all femoral fractures.
  - These are also related to motor vehicle accidents and significant falls.
  - These fractures include the extra-articular supracondylar fractures and intra-articular fractures—the condylar, intercondylar and epiphyseal injury.
  - Patients with femur fractures often present with a swollen, painful, ecchymotic and shortened thigh. They usually are unable to move the knee or the hip.
  - It is important to remember that femur fractures rarely present without other associated injuries.

Classification

- There are multiple classification systems for femoral shaft fractures.
- For the ED physician, it is important to describe the location of the fracture (proximal, mid-shaft, or distal third), the geometric description (transverse, oblique, spiral, comminuted), whether it is open or closed, and whether it is displaced or angulated.
- Distal femur fractures can be divided into four types: supracondylar, intercondylar, condylar, and distal femoral epiphyseal fracture. These can be further subdivided depending if they are nondiplaced, displaced, comminuted or impacted.

Management

Prehospital Care

- The ABCs should be assessed and treated initially.
- After stabilization, a brief neurovascular exam of the extremity should be performed.
- Open wounds should be dressed with a sterile dressing.
- Bleeding should be stopped with direct pressure.
- The patient should be splinted and placed in a position of comfort.
- Traction devices such as a Hare splint may be placed as it can reduce pain and can tamponade further blood loss.
- Contraindications to traction include open fractures, suspected sciatic nerve injury, fracture of the ipsilateral pelvis or lower extremity, and fractures near the knee.

ABCs

- Femoral shaft and distal femoral fractures are associated with multisystem trauma.
- As such, the patient's airway, breathing, and circulation is to be assessed.
- Extremity injuries can be assessed during the secondary survey.
- It should be noted that significant blood loss, up to 1.5 L, can be associated with femoral shaft fractures.
- However, studies have shown that isolated femoral neck fractures are not associated with hypotension and the ED physician should not attribute hemodynamic compromise to such a fracture.

- A search for other sites of hemorrhage is mandatory.
- Early fluid resuscitation is mandated.

Physical Exam

- Femoral shaft and distal femur fractures are obvious.
- Swelling, deformity, tenderness, and ecchymosis are common.
- In performing a physical exam, the ED physician must assess the pelvis, ipsilateral hip, knee, and lower extremity.
- Neurovascular status is essential.
- In distal femoral fractures, it is important to evaluate the popliteal artery and vein as well as the tibial and peroneal nerves.
- Although compartment syndrome in the thigh is rare, the compartment should be evaluated and pressures obtained if elevated pressure is suspected.

Radiography

- Once the patient is stable, standard plain films including AP and lateral femur should be obtained.
- In one study, 13-31% of patients with associated femoral neck fractures went undetected.
- Be vigilant for other fractures or dislocations and obtain films to rule out these injuries.

Treatment

- Once stabilized, femoral shaft fractures can be splinted with a Hare traction or Sager splint.
- If there is a contraindication to splinting (see above), then a plaster or fiberglass-reinforced bulky dressing should be substituted.
- Early adequate analgesia is essential for the patient.
- If the fracture is open, broad spectrum antibiotics and tetanus prophylaxis should be administered.
- Early orthopedic consultation is critical.

Complications

- Complications in femoral shaft fractures include compartment syndrome, fat embolism syndrome, vascular injury, malunion, osteomyelitis, and myositis ossificans truamatica.

- The first four are critical for the ED physician to recognize.
  - Compartment syndrome occurs in the anterior, posterior, or medial compartments of the thigh after femoral fractures, crush injury, burns, PASG, and drug use. The condition is rare due to the fact that the compartments of the femur can expand substantially. Findings include tense compartments associated with pain out of proportion with passive stretching of the muscles involved. Paresthesias, diminished pulses, pallor, and poikiothermia may also be manifested. Absent pulses are a late finding, and often the damage is irreversible. Measurement of the compartments is necessary (>0 mm Hg is abnormal, and microcirculation is compromised when pressures reach 30 mm Hg). Emergent fasciotomy is recommended for compartment syndrome. Potassium, myoglobin, CK, and renal function should be assessed to rule out associated hyperkalemia, myoglobinuria, and rhabdomyolysis.
  - Fat embolism syndrome occurs in 2-23% of patients with femur fractures. Clincial findings are fever, tachycardia, respiratory distress, altered mental status and, although a late finding, a petechial rash over the chest. Management includes supportive care with vigilant airway management.
  - Vascular injury primarily occurs with penetrating trauma (GSW) although it can be associated with fractures and dislocations. Signs of vascular injury include pulsatile hematoma, hemorrhage, absent distal pulses, palpable thrill or audible bruit. More often than not, patients with penetrating trauma do not illicit these "hard" findings and have more subtle findings such a small nonexpanding hematoma, diminished pulses etc. Modrall, Weaver, and Yellin's algorithm for penetrating trauma is shown on Figure 8.2. If a patient has signs of vascular injury, then arteriogram and surgical repair is necessary. If soft signs are present, or the ABI is <1.00 arteriography is warranted. If the ABI is ≥1.00 observation is recommended.

8

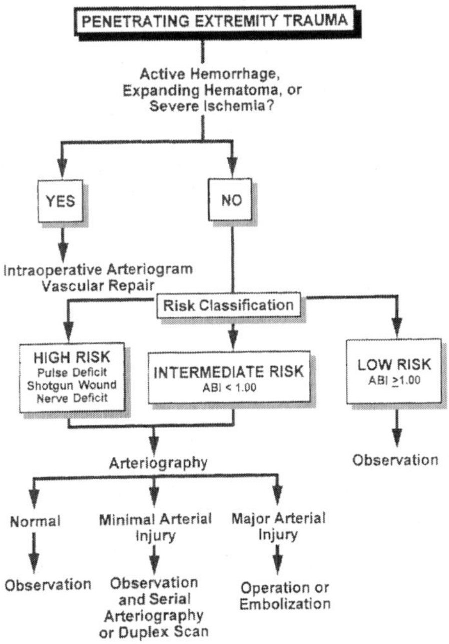

Figure 8.2. Management of penetrating trauma. Reprinted from Emed Clin North Am 2000; 18(1):61, ©2000 Elsevier, with permission.

The Knee Joint

Anatomy

- The knee is more complex than a simple hinge joint.
    - It is made of three articulations, the patellofemoral, the tibiofemoral, and the tibiofibular joints.
    - The ligamentous components of the joint are the lateral collateral ligament (LCL), medial collateral ligament (MCL), the anterior cruciate ligament (ACL), and the posterior cruciate ligament (PCL).
    - The extensor mechanism includes the quadriceps tendon, the patella, and the patellar ligament.
    - The important cartilaginous structures are the medial and lateral menisci.
    - The medial meniscus is connected with the MCL; the lateral menicus is not.

Management

Prehospital

- Prehospital care entails immobilization to prevent further neurovascular injury, elevation and ice.

History

- The emphasis is on mechanism.

- The following are classical injuries associated with common mechanisms of injury:

Mechanism	Injury
• Head on traffic collision	Posterior cruciate ligament injury
• Twisting injury (i.e., skier)	Anterior cruciate ligament tear
• Contact with lateral force (i.e., football player)	MCL tear, medial meniscus tear and ACL tear (aka O'Donaghue's triad)
• Hyperextension	ACL injury followed by PCL injury
• Turn with tibia rotated in opposite direction	Patellar dislocation

- In addition to mechanism, onset and quality of pain (static or dynamic) is important.
- Inquire about associated systemic symptoms or disease.

Physical Exam

- Neurovascular exam is crucial especially if knee dislocation is suspected.
- If dislocation has occurred, immediate reduction with sedation and analgesia is recommended.
- After reduction, neurovascular exam must be reevaluated.
- Inspect for areas of swelling, tenderness, and ecchymosis. The presence of an acute effusion suggests hemarthrosis and possible ACL tear.
- Aspiration of the knee may reveal clues to the underlying injury.
- Hemarthrosis is highly associated with cruciate ligament tears. Fat suggests the presence of fracture.
- All lacerations in the vicinity of the knee joint must be considered to involve the joint space until proven otherwise.
- Assess the stability of the knee joint by evaluating the ligaments and the menisci.
- The Lachman test is the most sensitive test for determining ACL disruption. It is performed in 15° to 30° of flexion. The proximal tibia is pulled anterior relative to the tibia. Excessive movement relative to the opposite side is considered positive. Anterior and posterior drawer tests in 90° flexion also help to diagnose ACL and PCL injuries respectively.
- The medial and lateral collateral ligaments can be tested by assessing the joint under valgus and varus stress respectively.

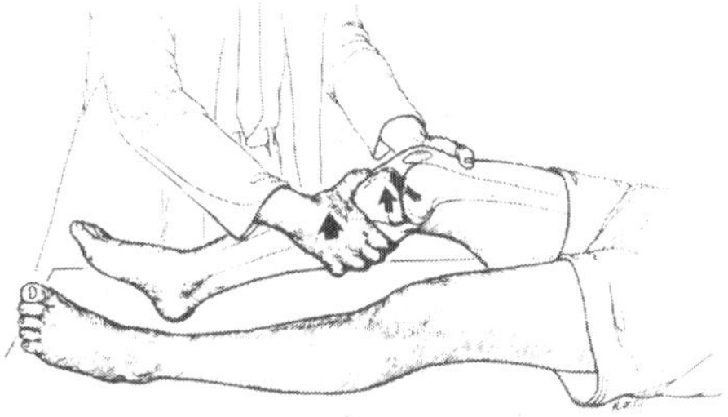

Figure 8.3. The Lachman test. (From Rockwood and Green, fort fourth ed., pg 2070)

- The medial and lateral menisci can be tested by performing the McMurray or grind test although the specificity and sensitivity are low.
- Examine the knee for range of motion, both active and passive. Always remember to identify other injuries by examining the femur, tibia, and patella.

Radiography

- Plain Films
  - The Ottawa Knee Rules were described in order to delineate which patients require plain films.
  - Standard views are the AP and lateral.
  - Other views may be helpful to elucidate individual injuries if suspected:
    - Cross-table lateral — May detect fat-fluid level, pathognomonic of fracture
    - Oblique views — Fracture or loose foreign body
    - Notch view — Osteochondral fracture
    - Sunrise view — Patellar injuries
    - Plateau view — Tibial plateau fracture
- CT and MRI are rarely used in emergency imaging of the knee.
  - MRI provides superior visualization of soft tissue structures including menisci and cruciate ligaments, but is usually performed by the primary care or orthopedist in a nonurgent setting.

A knee X-ray series is only required for knee injury patients with any of these findings:

1. age 55 yr or older, or
2. isolated tenderness of patella*, or
3. tenderness at head of fibula, or
4. inability to flex to 90°, or
5. inability to bear weight both immediately and in the emergency department (4 steps).**

* No bone tenderness of knee other than patella.
** Unable to transfer weight twice onto each lower limb regardless of limping.

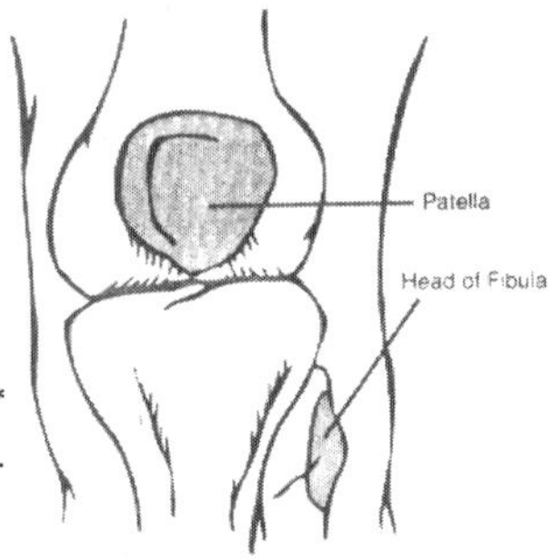

Figure 8.4. Ottawa Knee Rule for use of radiography in acute knee injuries. From Stiel et al. Implementation of the Ottawa Knee Rule for the use of radiography in acute knee injuries. JAMA 1997; 278:2075-2079.

8

Table 8.2. Classification of ligament sprains

STRETCH-GRADE I
A first-degree sprain is really a microscopic tear and can be treated with rest, ice, and protection with crutches and/or a splint.
PARTIAL-GRADE II
A second-degree sprain should be immobilized to prevent the ligament from tearing completely.
COMPLETE TEAR-GRADE III
Third-degree sprains may require surgery. Therapy is somewhat controversial.

From EM reports 1/6/1997

Classification, Treatment and Disposition

Soft Tissue Injury

Injury	Classification	Treatment	Disposition
Anterior Cruciate Ligament (ACL), Posterior Cruciate Ligament (PCL)	Grades I-III	Compressive dressing from midthigh to midcalf. Ice, Elevation	Urgent orthopedic referral for grade III. May include 24 h recheck
Collateral Ligament tears. (MCL and PCL)	Grades I-III	Knee immobilizer and crutches. Ice compression elevation and rest.	Grades I and II follow with primary care physician. Grade III requires orthopedic referral
Meniscus Injuries	NA	Reduction of "locked joint" knee immobilizer and crutches. Ice compression elevation and rest	Orthopedic referral.

Fractures and Dislocations

Injury	Classification	Treatment	Disposition
Distal Femur Fracture	Suprcondylar, Intercondylar, Condylar	Immobilization, Analgesia	Emergent orthopedic consultation
Tibial Plateau Fracture	Schatzker Class I-VI. (Class IV, V, VI suspect vascular injuries)	Immobilization, Analgesia	Emergent orthopedic consultation. If non-displaced lateral depression only (<8 mm) immobilization non weight-bearing and follow-up in 48hrs.
Knee Dislocation	Anterior, Posterior, Lateral, Medial, Rotatory	Immobilization. Exclusion of vascular injury usually requires angiography. Evaluation for compartment syndrome	Emergent orthopedic consultation
Patellar Dislocation	Medial and lateral	Reduction by knee extension simultaneous movement of patella. Knee immobilizer and crutches	Referral to orthopedics

Table 8.3. Compartments, lower leg

Fascial Compartment	Structures Contained	Clinical Findings
Anterior	Extensor muscles	Great toe extension, ankle dorsiflexion
	Anterior tibial artery and veins	Dorsalis pedis pulse
	Deep peroneal nerve	Sensation of great toe web space
Lateral	Peroneus longus muscle	Foot eversion
	Peroneus brevis muscle	
	Superficial peroneal nerve	Sensation to dorsum of foot
Deep Posterior	Deep flexor muscles	Great toe flexion
	Posterior tibial artery and veins	Posterior tibial pulse
	Tibial nerve	Sensation to plantar surface of foot
	Peroneal artery and veins	
Superficial Posterior	Superficial flexor muscles	Ankle plantar flexion
	Sural nerve	Sensation to lateral side of foot and distal calf

Tibia and Fibula

8

Anatomy and Function

- The tibia and fibula are joined together at the superior and inferior tibiofibular joint at each end and via the interosseous membrane along the length of the two bones.
- The tibia is the weight-bearing bone.
- The interosseous membrane is responsible for approximating the two bones inferiorly thereby stabilizing the ankle mortise.
- While the fibula is surrounded by muscle tissue except at its distal portion, the tibia is predominantly exposed over its anteromedial aspect with tenuous blood supply.
- Thus the tibia is predisposed to open fractures, osteomyelitis, delayed union, malunion, and nonunion.
- The lower extremitiy is divided into four compartments by its investing fascia: anterior, lateral, superficial posterior and deep posterior.
- Each compartment has various structures within.
- In a compartment syndrome, knowing these structures and their clinical significance can be helpful in elucidating which compartment is involved (Table 8.3).
- The vascular supply of the lower extremity is via the poplilteal artery. The artery divides inferiorly to the knee joint into three branches: the anterior tibial, the posterior tibial, and the peroneal artery.
- The anterior artery runs through the anterior compartment and can be evaluated by palpating the dorsalis pedis pulse.
- The posterior tibial artery travels through the deep posterior compartment and is evaluated by palpating the posterior tibial pulse.

Trauma

- Fractures of the tibia are the most common long bone fractures.
- They are caused by both high and low energy mechanisms.

Management

Prehospital Care

- Immobilization, ice, and elevation are important prior to transport to the hospital.
- Open wounds should be covered with normal saline sterile dressings.
- Assessment of neurovascular status is pertinent.

ABCs

- As with all patients, airway, breathing, and circulation are an ED physician's primary concern.
- If the patient has multisystem trauma, the extremity is a lower priority and can be managed during the secondary survey.
- If the extremity is pulseless or has gross deformity compromising the neurovascular status, the extremity should be reduced immediately with proper analgesia and sedation.

History and Physical Exam

- Knowing the mechanism and nature of the injury is always important in any fracture.
- A tibia fracture caused by a high-energy mechanism has a poorer prognosis than a fracture caused by a low-energy mechanism.
- If an open fracture is present it is important to know if it is a clean or contaminated wound.
- On physical exam, look for bony deformity, tenderness to palpation, and or swelling.
- Examine skin carefully for signs of an open fracture.
- Examine the knee and ankle joint for effusions, tenderness, and stability.
- Examining the compartments of the legs is critical.
- Neurovascular status should be assessed.

Radiography

8

- Plain films including AP and lateral views are necessary.
- Films should include the knee and ankle joint to rule out concomitant injury.

Classification, Treatment, and Disposition

- Tibial fractures can be divided into extra-articular proximal fractures and shaft fractures, and intra-articular proximal and distal fractures (see knee and ankle joint sections respectively).
- Fibular fractures are usually associated with tibia fractures although isolated injury does occur.

Extra-Articular Proximal Tibial Fractures

- Subcondylar tibial fractures are usually associated with tibial plateau fractures but can occur in isolation.
    1. Isolated, stable, nondisplaced transverse fractures are treated with analgesia, ice, long-leg posterior splint and referral to an orthopedic surgeon. The patient should remain nonweight-bearing.
    2. Comminuted fractures or those involving the joint require admission for open reduction and internal fixation.
- Tibial tuberosity fractures. The tibial tuberosity is the insertion site for the quadriceps tendon. Fracture of the tubercle is rare, occurring most commonly in younger individuals. The patients may present with swelling and pain at the site that is exacerbated with extension of the knee. A hemarthrosis may be present if the fracture expands into the joint space. There are three types of fractures.
    1. Incomplete avulsions require cast immobilization with the knee in extension and orthopedic follow-up. The patient should use crutches and remain nonweight-bearing until orthopedic evaluation.
    2. Extra-articular complete avulsions require closed reduction, and if adequate, cast immobilization with the knee in extension, and orthopedic follow-up.
    3. Intra-articular complete avulsions require open reduction and internal fixation.

Tibia Shaft Fractures

- There are multiple classification systems used to identify tibial shaft fractures and their treatment.

- Much controversy exists. Johner and Wruhs report a system by Muller based on the mechanism of injury, comminution, soft-tissue injury and displacement of the fracture.
- For simplicity, the ED physicians' treatment can be determined by displacement of the fracture and whether it is open or closed:
  - Closed non-displaced or minimally displaced tibial shaft
    - Adequate analgesia is imperative. If the fracture is displaced, closed reduction is required: the fracture should have no more than 10° angulation in AP plane, 5° in varus/valgus, and <1.5 cm of shortening.
    - Immobilize in a long leg posterior splint at 10-20° of knee flexion.
    - A circumferential cast should be avoided due to possibility of compartment syndrome.
    - The patient should remain nonweight-bearing.
    - Early orthopedic consult is required.
    - Admission is based on patient reliability, pain control, adequate reduction, articular involvement, and orthopedic follow-up.
  - Open tibial shaft
    - Adequate analgesia is imperative.
    - Cover the open area with a sterile dressing and splint similarly to a closed tibial shaft fracture.
    - The patient should receive tetanus prophylaxis as indicated by their immunization status.
    - In addition, IV antibiotics should be started immediately. First generation cephalosporins are recommended.
    - If the wound is >1 cm, heavily contaminated or has severe soft tissue injury, an aminoglycoside should be added.
    - Early orthopedic consultation and admission are required.

8

Proximal Fibula Fractures

- Isolated uncomplicated fibula fractures can be treated with ice, elevation and analgesia.
- If the patient is in considerable pain, a brace or short leg cast can be placed and immediate weight-bearing is allowed if tolerated.
- If there is minimal pain, an ace rap from toe to knee can be placed for patient comfort and the patient can weight bear as tolerated.
- Othopedic follow-up is recommended. It is imperative for the ED physician to recognize and rule out associated tibia fractures, i.e., a Maisonneuvre fracture. This is discussed in further detail in the following section on the ankle.
- Complications of an isolated proximal fibula fractures are very rare and include malunion.

Complications

- Tibia fractures, as mentioned previously, have a higher incidence of malunion, delayed union and nonumion.
- For the ED physician, one of the important concerns is compartment syndrome.
- The anterior compartment is the most common involved.
- Findings include tense compartments associated with pain out of proportion with passive stretching of the muscles involved.
- Paresthesias, diminished pulses, pallor, and poikilothermia may also be manifested.
- Absent pulses are a late finding and often the damage is irreversible.
- Measurement of the compartments is necessary (>0 mm Hg is abnormal and microcirculation is compromised when pressures reach 30 mm Hg).
- Emergent fasciotomy is recommended for compartment syndrome.
- Potassium, myoglobin, CK, and renal function should be assessed to rule out associated hyperkalemia, myoglobinuria, and rhabdomyolysis.

Ankle

Anatomy and Function

- The ankle consists of the articulations of the fibula, tibia, and talus supported by the syndesmotic, lateral collateral and medial collateral ligaments.
- The medial collateral ligaments consist of the deltoid ligament comprised of the posterior tibiotalar, tibiocalcaneal, anterior tibiotalar, and tibionavicular parts.
- The lateral ligaments are the calcaneofibular, anterior talofibular, and lateral talocalcaneal ligament.
- The anterior and posterior tibiofibular ligament create the syndesmotic ligmanets.
- The tibia, fibula, and the ligaments have been described as a ring-like structure surrounding the talus.
- One disruption in the ring may not cause instability; however if two or more elements are involved, instability ensues.
- Single disruptions are less common, and a second disruption should be sought carefully.
- The biomechanics of the ankle are complex with four axes of motion that allow dorsiflexion, plantarflexion, inversion and eversion.

Management

Prehospital Care

- Ankle injuries are commonly isolated injuries, but can be assosciated with multisystem trauma.
- As with all extremities with possible injury, the ankle should be immobilized in a position of comfort.
- The extremity should be elevated, and ice packs used to prevent further swelling.

ABCs

- As with all injuries, airway, breathing, and circulation should be assessed and stabilized first.
- If the extremity's neurovascular supply is compromised, there is gross deformity including skin tenting, then immediate reduction with adequate analgesia should be performed.
- Obtaining radiographic studies should not delay reduction.
- Be sure to reassess the neurovascular status after reduction.

History

- Understanding the mechanism and position of the ankle at the time of injury can help understand the fracture or sprain pattern.
- What position was the ankle at the time of injury?
- Was the pain acute or did it occur over time? Was the patient able to ambulate at the time of injury?
- Knowing if the patient heard a loud pop or snap may clue the physician to ligamentous injury
- Finally, ask if there is pain in other areas such as the foot or lower extremity.

Physical Exam

- Examination of the ankle must include the foot and proximal tibia and fibula.
- Look for tenderness, ecchymosis, and swelling.
- Palpate the proximal fibular head, the posterior aspect of the distal 8 cm of the tibia and fibula (medial and lateral malleolus respectively).
- Examine the navicular and 5th metatarsal for injury.
- Compare the range of motion of the injured ankle to the uninjured ankle.
- The anterior drawer test will assess the integrity of the anterior talofibular ligament and the squeeze test will test the integrity of the syndesmotic ligaments.

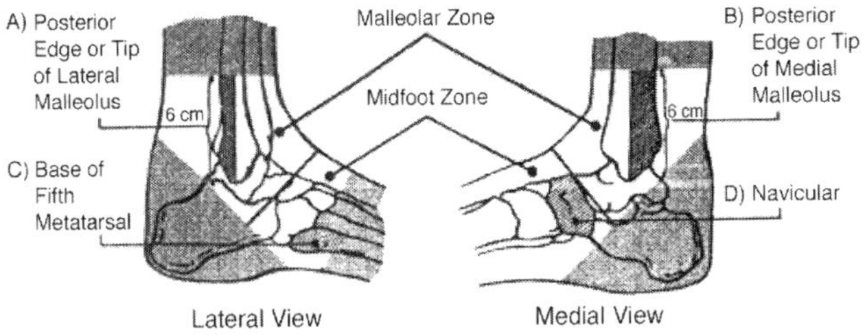

Figure 8.5.Ottawa ankle rules for foot and ankle radiographic series in acute ankle injury patients (adapted from Stiell et al).

8

Radiographs

- It has been estimated that $500 million is spent annually on ankle radiographs.
- The Ottowa Ankle Rules suggest that by adhering to a set of decision rules, unnecessary radiographs could be reduced without missing significant injuries. The study states that an ankle radiographic series should be performed if (1) there is bone tenderness at the posterior edge of the distal 6 cm of either the medial or lateral malleolus or (2) the patient is unable to bear weight both immediately after injury and in the ED. It further states, foot radiographic series is only required if (1) there is any pain at the navicular or the distal 5th metatarsal bone or (2) the patient is unable to bear weight both immediately after injury and in the ED. These rules apply for acute injuries and do not apply to pediatric patients.
- The standard ankle series includes the AP, lateral and mortise view. CT, MRI, and stress views are primarily used for more definitive work-ups by specialists and have little value in the ED.

Classification and Treatment

Ankle Fracture

- There are several classification systems for ankle fractures, however the Danis-Weber and the Lauge-Hansen are the most commonly used (Table 8.4).
- The former system is based on the location of the fibular fracture, and the latter is based on the position of the ankle at the time of injury (Table 8.4).
- Orthopedic consultation is recommended for all posterior malleolar, bimalleolar, trimalleolar, intra-articular fractures with step deformities, open fractures and pilon fractures (fracture of the distal tibial metaphysis resulting from high energy forces driving the talus into the tibial plafond).
- Treatment of unimalleolar fractures is more complex and is dependent on the location of the fracture and associated injuries.

Table 8.4. The two most commonly used classification systems for ankle fractures

Lauge-Hansen	Danis Weber	Treatment
Supination-adduction	**Type A: Fibula fracture below tibotalar joint**	
Stage I: Transverse fracture of the lateral malleolus at level below joint or lateral collateral ligament tear.	A1: isolated	Cast for 6-8 wk, 3 wk nonweight-bearing, close ortho follow-up
Stage II: Vertical fracture of medial malleolus.	A2: With fracture of medial malleolus	Ortho consult in ED
	A3: With posteriomedial fracture	Ortho consult in ED
Supination-eversion (lateral rotation)	**Type B: Fibula fracture at the level of tibiotalar joint**	Ortho consult in ED
Stage I: Rupture of anterior tibiofibular ligament.	B1: Isolated.	
Stage II: Spiral fracture of distal fibula near or at joint.	B2: With medial malleolus or medial ligamentous rupture.	
Stage III: Disruption of posterior tibiofibular ligament with or without avulsion from posterior malleolus	B3: With a medial lesion and fracture of the posterolateral tibia.	
Stage IV: Oblique fracture of medial malleolus or rupture of the deltoid ligament.		
Pronation-abduction		
Stage I: Transverse fracture of medial malleolus or deltoid ligament rupture.		
Stage II: Disruption of posterior and anterior tibiofibular ligaments with or without avulsion from the posterior malleolus.		
Stage III: Oblique fracture of distal fibula at level of joint.		
Pronation-eversion (lateral rotation)	**Type C: Fibula fracture above tibiotalar joint.**	Ortho consult in ED
Stage I: Transverse fracture of medial malleolus or deltoid ligament rupture.	C1: Diaphyseal fracture of the fibula, simple	
Stage II: Disruption of anterior tibiofibular ligament complex and interosseous membrane.	C2: Diaphyseal fracture of the fibula, complex	
Stage III: High fracture of fibula ≥6 cm above joint level.	C3: Proximal fibula fracture	
Stage IV: Disruption of posterior tibiofibular ligament with or without avulsion from posterior malleolus.		

- An isolated medial malleolar fracture may be treated with casting for 6-8 wk, nonweight-bearing for 3 wk and close orthopedic follow-up.
- It is imperative to remember that these fractures are rare (<1% of all ankle fractures) and concomitant injury should be sought.
- All other medial malleolar injuries require orthopedic consultatation in the ED.
- If a lateral malleolar fracture is below the tibiotalar joint line and there are no other associated injuries, the management consists of casting for 6-8 wk, at least 3 wk of nonweight-bearing and close orthopedic follow-up.

- If there is associated tenderness over the deltoid ligament orthopedic consult is recommended.
- Likewise, if the fibula fracture is above the level of the tibiotalar joint line (Weber Type C) orthopedic consult in the ED is required.
- Fractures of the lateral malleolus at the tibiotalar joint line are more controversial although orthopedic consult in the ED is recommended especially if there is a positive squeeze test or a widened medial joint space.

Ankle Sprains

- As discussed previously, the ankle is supported by the syndesmotic, lateral, and medial ligaments.
- Approximately 85% of ankle sprains involve the lateral ligaments with the anterior talofibular ligament being the most common.
- Injury usually occurs secondary to inversion.
- Ankle sprains can be divided into three grades based on function and pathology:
  - Grade I: Ligamentous stretching without gross joint instability. Patient is in mild pain and usually can ambulate.
  - Grade II: Partial tear of the ligament with moderate joint instability, often has substantial swelling and pain. Patient has moderate difficulty with walking.
  - Grade III: Complete tear of the ligament with gross instability, often has severe edema and ecchymosis. Patient is unable to ambulate.
- Treatment for ankle sprains starts with rest, ice, analgesia and elevation.
- Grade I and Grade II injuries are best treated with air splints, gel splints, or tape. Crutches are appropriate for a few days with early advancement as tolerated.
- Follow-up with a primary care giver is sufficient for Grade I Sprains.
- Severe grade II or grade III injuries require immobilization with a posterior ankle splint or sugar tong splint.
- Crutches are indicated and should be used until the patient is able to weight-bear as tolerated without significant pain.
- Orthopedic referral is recommended for all severe Grade II and III lesions.
- Full sport activities should not be started until the ankle is pain-free with running and turning.

Tendon Rupture

- Achilles tendon ruptures are commonly occur in middle-aged men during physical activity.
  - The patient may describe the onset as a loud pop or the feeling of being struck in the back of the ankle.
  - They may complain of difficulty walking with weakness in plantar flexion.
  - Physical exam may reveal a palpable defect over the tendon and a positive Thompson.
  - To perform the Thompson test, have the patient kneel on a chair or table with the knees at 90°. Squeezing the calf muscles should result in passive plantar flexion at the ankle. If the movement is decreased when compared to the opposite or ankle, or absent, then an Achilles tendon rupture is present.
  - Treatment by operative or nonoperative means is controversial and consultation with an orthopedist is recommended.

Foot

Anatomy and Function

- The foot is comprised of the hindpart (seven tarsal bones including the calcaneus and the talus), the midfoot (navicular, cuboid, and three cuneiform bones), and the forepart (five metatarsals, two phalanges of the great toe, and three phalanges of each of the lateral four toes).

8

- The articulations between the hindpart and the midfoot are called Chopart's joint or midtarsal joints.
- There are five tarsometatarsal joints between the midfoot and forefoot called the Lisfranc's joints.
- All together there are 28 bones and 57 articulations.
- The surrounding soft tissue including tendons, ligaments, muscles and fascia, create a complex system that plays a vital role in ambulation.
- Weakness or failure of any of these parts, places increased stress on the others, hindering both weight-bearing and nonweight-bearing function and with time deterioration of the other uninjured parts.
- Blood supply is from the anterior and posterior tibial arteries.
- Motor and sensory innervation is by the peroneal, posterior tibial, saphenous, and sural nerves.

Management

Prehospital Care

- Foot injuries are commonly isolated injuries, but can be associated with multisystem trauma.
- As with other extremity injuries, the foot should be immobilized in a position of comfort.
- The extremity should be elevated, and ice packs used to prevent further swelling.

ABCs

- As with all injuries, airway, breathing, and circulation should be assessed and stabilized first.
- If the extremity's neurovascular supply is compromised, there is gross deformity including skin tenting, then immediate reduction with adequate analgesia should be performed.
- Obtaining radiographic studies should not delay reduction.
- Be sure to reassess the neurovascular status after reduction.

History

- The history of a patient should include age, job, recent events, mechanics of the injury, nature of the pain, and functional loss.
- It is important to understand if the foot was punctured, crushed, lacerated, repeatedly stressed or had direct or indirect forces applied in order to narrow the diagnosis.
- Moreover, systemic medical problems such as peripheral vascular disease or diabetes should be explored.
- Neurovascular status is pertinent—patients with increasing pain and decreasing sensation are at high risk for compromise and are in need of immediate medical attention.

Physical Exam

- Always undress the foot completely including shoes and socks.
- Use the noninjured foot for comparison.
- Look for ecchymosis, swelling, wounds, deformity, pallor, cyanosis, points of tenderness and crepitus.
- Examine the foot through range of motion both passive and active.
- If pulses are not obtainable by palpation, Doppler should be used.
- Document the patient's ability to ambulate and whether there is an antalgic gait.

Radiographs

- Three basic plain film views are indicated for foot injury—anateroposterior, lateral and obliques.

Table 8.5. Hawkins classification of talar neck fractures

Classification	Anatomical Abnormality	Treatment
Type I	Nondisplaced No joint involvment	Short-leg cast for 8-12 wk
Type II	Displacement of the talar neck No ankle joint involvement with subluxation or dislocation of subtalar joint	Closed reduction followed by casting in equines in a short-leg nonweight-bearing cast
Type III	Displaced fractures with dislocation of the talus at the subtalar and ankle joint	Emergent ortho referral
Type IV	Displacement of the talar neck with subluxation or dislocation of subtalar joint No ankle joint involvement Talar head dislocation	Relocation if neurovasculary compromised Emergent ortho referral

- Because the foot has many small bones that frequently overlap, as well as numerous accessory bones, special views are indicated if certain fractures are suspected (discussed below).
- Bones scans can be useful for subtle fractures not seen on plain films including stress fractures.
- CT is the modality of choice for imaging complex anatomy including the subtalar joint, the calcaneus, and the Lisfranc joint complex.
- MRI has been recommended for soft tissue injury, including tendon ruptures, however it is rarely used emergently in the ED.

8

Classification. Treatment. Disposition. Complications

Talar Fractures

- The talus can be divided into the head, body and neck.
- Subtle fractures of the talar dome (osteochondral lesions) are commonly missed and should be looked for diligently. They result when the talus hits the distal tibia or fibula.
- All patients require orthopedic consultation.
- Talar head fractures can be displaced or nondisplaced. Nondisplaced fractures can be treated with a short-leg walking cast for 6 wk while displaced fractures require immediate reduction and emergent orthopedic consultation.
- Talar neck fractures have been described by Hawkins as indicated below.
- Complications include avascular necrosis, skin necrosis, arthritis, nonunion and malunion.
- Talar body fractures that are nondisplaced or are laterally displaced can be immobilized in a short-leg nonweight-bearing cast for 6-8 wk.
- Displaced or comminuted fractures require emergent orthopedic referral.

Calcaneal Fractures

- A twisting mechanism or a fall commonly results in calcaneal fractures.
- The key is identifying associated injuries (spine fractures, contralateral calcaneal fractures, compartment syndrome) and determining whether the fracture is extra-articular or intra-articular.
- Plain films should be examined closely (Bohler's angle, angle of Gissane) to elucidate subtle fractures.
- CT is usually necessary to classify fractures.
- Nondisplaced and extra-articular fractures
    - Nondisplaced and extra-articular fractures are treated with compressive dressings, nonweightbearing, ice, and elevation.

- These fractures are usually nonoperative.
- Intra-articular fractures require bed rest, elevation, a posterior splint or Jones dressing until operative management occurs.
- All calcaneal fractures require consultation with an orthopedic surgeon.
- Admission may be required for pain control or if the fracture is open or complicated with other injuries.

Midfoot Fractures

- Nondisplaced navicular and cuboid fractures can be treated with immobilization in a short leg walking cast for 6 wk.
- Displaced fractures require immediate orthopedic consultation.
- The Lisfranc injury, or injury of the articulation of the base of the first three metatarsals with the cuneiform and the fourth and fifth metatarsals with the cuboid, requires orthopedic evaluation in the ED.
- Injury to this area can be missed, and it is important to scrutinize radiographs carefully.
- Always check the alignment of the medial aspect of the metatarsals with the medial aspect of its corresponding tarsal bone.
- A fracture of the navicular and cuneiform bone, the base of the second metatarsal, or widening of the spaces between the first and second or second and third metatarsals suggest a Lisfranc injury.

Forefoot

- Displaced metatarsal fractures (>3 mm or 10 degrees angulation) require orthopedic consultation and open reduction.
- Nondisplaced first metatarsal fractures requires nonweight-bearing in a short-leg walking cast while the second through fourth requires a stiff shoe or casting as needed for comfort.
- The body and neck of all the metatarsals requires precise alignment with immobilization.
- Fractures of the base of the metatarsal are associated with a Lisfranc injury and should be analyzed closely.
- A Jones fracture is a fracture of the base of the fifth metatarsal between the diaphysis and metaphysis. It requires nonweight-bearing immobilization for 6-8 wk.
- A Dancer's fracture is an avulsion fracture of the tuberosity at the base of the fifth metatarsal.
- Orthopedic consult is required only if displacement is more than 2 mm or involves >30% of the articular surface.
- Nondisplaced phalangeal fractures can be buddy taped with the adjacent toe.
  - A short leg walking cast may be needed for adequate support of the great toe.
  - Open, comminuted fractures require orthopedic consultation.

The Shoulder Girdle

Anatomy and Function

- The shoulder girdle is a complex structure comprised of the clavicle, the scapula and the humerus articulating at the glenohumeral, acromioclavicular and the sternoclavicular joints.
- The glenohumeral joint is a ball and socket joint that allows for flexion, extension, abduction, adduction, circumduction and rotation. Because of its mobility, it sacrifices strength.
- The Skeletal Components—The glenoid of the scapula and the head of the humerus—are supported by an articular capsule, the catilagenous glenoid labrum, the glenohumeral and coracoacromial ligaments and tendons of the rotator cuff muscles (teres minor, teres major, supraspinatous and subscapularis muscles) and the long head of the biceps brachii.

- The sternoclavicular joint is a fibrocartilagenous joint between the sternum and the clavicle. Because the bony articulations are incongruous, the joint lacks stability. Support is provided by the joint capsule, an articular disc, the sternoclavicular ligament and the costoclavicular ligament. The sternocalvicular joint is highly mobile and moves along with upper extremity motion; it is capable of elevation, forward and backward movement and rotation.
- The acromioclavicular (AC) joint is the articulation of the acromion process of the scapula and the lateral aspect of the clavicle. Like the sternocalvicular joint, the AC joint is fibrocartilagenous. The coracoclavicular ligament (extracapsular), the acromioclavicular ligament (intracapsular), the joint capsule and the intra-articular disc help stabilize the joint.

Mangement

Prehospital Care

- Immobilization, ice and elevation are important prior to transport to the hospital. Open wounds should be covered with normal saline sterile dressings.
- Assessment of neurovascular status is pertinent.

ABCs

- As with all patients, airway, breathing and circulation are an ED physician's primary concern.
- If the patient has multisystem trauma, the shoulder girdle is a lower priority and can be managed during the secondary survey.
- If there is neurovascular compromise, then immediate reduction is indicated.
- Early immobilization is important.

8

History

Shoulder pain is a common complaint.

- It is important to delineate if the pain is acute, chronic or associated with trauma.
- Moreover, the quality and character of the pain should be noted.
- Inquire if the pain occurs at rest or with movement.
- Knowing the patient's occupation and hand dominance may be helpful.
- Always ask about other medical conditions.

Physical Exam

- If possible, examine the patient seated or standing up.
- Begin by evaluating the patient's neurovascular status: examine the sensory and motor components of the brachial plexus (C5-T1 dermatomes, strength of the trapezius, deltoid, biceps, thumb extensors, finger flexors and interossei muscles).
- Examine the sternoclavicular joint, the clavicle, the acromioclavicular joint, the humerus, the scapula and the glenohumeral joint.
- Look for deformities, ecchymosis, lacerations or swelling.
- Palpate for tenderness or crepitus. Observe range of motion, both active and passive.
- For soft tissue injuries, it may be necessary to inject 10 ml of 1% lidocaine into the joint, so range of motion can be adequately tested.
- Always compare to the contralateral side.

Radiography

- Glenohumeral: a three view series is recommended including a true AP (35 degree oblique), transcapular lateral and an axillary view. The axillary view may be difficult to obtain secondary to the patient's injury and limited range of motion. In this instance, a modified axillary or "y-view" may be useful.

- Sternoclavicular plain films including AP, oblique and 40 degree cephalic tilt views can be performed, although interpretation can be difficult. CT scan is considered the best technique to delineate injury to the joint.
- Acromioclavicular: plain films should include a standing or sitting AP of both AC joints (on a large cassette), a lateral axillary view of both shoulders and a 15 degree cephalic tilt. Rockwood recommends a stressed (person holding weight in hands) AP and lateral "shoulder-forward" view however, Bossart et al found this less efficacious and Rosen does not recommend them.

Classification Treatment Disposition Complications

Fractures

- Clavicle—Clavicle fractures are common and can be classified by location of the fracture: medial, middle and lateral third. The middle third is the most common fracture site. Open fractures, fractures resulting in skin tenting or neurovascular injury require immediate orthopedic consultation. ED treatment is as follows:
  - Medial—Immobilization with sling/swathe, ice, rest and analgesia. Follow-up with primary care physician.
  - Middle—Immobilization with sling/swathe, ice, rest and analgesia. Figure-of-eight splints and clavicular spica casts have been recommended for reduction of the clavicle. However, they are used less frequently in the ED for three reasons: (1) malunion is not commonly associated with a functional or cosmetic deficit; (2) closed reductions are difficult to maintain; and (3) they commonly cause increased discomfort for the patient. If fracture is displaced >20 mm, then orthopedic evaluation within 72 h is recommended. Otherwise, the patient should follow-up with their primary care physician.
  - Distal (lateral)—Further classified into three types:
    - Type I (coracoclavicular ligament intact): Immobilization with sling/swathe, ice, rest and analgesia. Patient may follow-up with primary care physician.
    - Type II (coracoclavicular ligament intact, proximal fracture fragment displaced superiorly): Immobilize with sling/swathe, ice, rest, analgesia. Follow-up with orthopedic surgery in 72 h.
    - Type III: (Intra-articular fracture involving the AC joint) Immobilization with sling/swathe, ice, rest and analgesia. Follow-up with primary care physician.
- Scapula—Although fractures to the scapula are rare, they are associated with a high incidence of injury to the ipsilateral chest wall, lung, shoulder, extremity, brachial plexus and underlying vascular structures. Always look for other injuries, particularly in Type II and Type III fractures listed below. There are multiple classification systems for scapula fractures. The following classification is by Thompson et al.:
  - Type I—Fractures of the acromium or coracoid process and small fractures of the body. Treat with sling immobilization, ice, rest and analgesia. Significantly displaced acromium or coracoid processes, those that impinge on the glenohumeral joint, or those associated with 3rd degree AC separation require orthopedic surgery.
  - Type II–Fractures of the glenoid and neck. Initial management is immoblization, ice and analgesia. Orthopedic consultation is indicated after associated injuries are ruled out.
  - Type III—Major fractures of the body. Initial management is immobilization, ice and analgesia. Orthopedic consultation is indicated after associated injuries ruled out.
- Proximal Humerus—Proximal humeral fractures are more common in the elderly or osteopenic patient whereas shaft fractures are more common in the young athletic patient (see Humerus section). Proximal humeral fractures involve either the anatomic neck, the surgical neck, the greater tuberosity or the lesser tuberosity. They can be considered as either displaced or nondisplaced.

- Nondisplaced fractures are those fracture fragments that are displaced <1 cm or angulated <45 degrees. Nondisplaced fractures can be treated by immobilization with sling and swathe, ice, rest and analgesia. Orthopedic follow-up should be arranged. Adhesive capsulitis is an associated complication.
- Displaced fractures are classified by the Neer classification system. The system describes the fracture fragments by the number of parts, involvement of the articular surface and dislocation if present. Impression fractures (those occurring with dislocation, i.e., Hill-Sachs deformity) are considered separately. Two-part, three-part, four-part fractures and those involving the articular surface (including fracture dislocations) require immediate orthopedic consultation. Many of these injuries require surgical repair. Complications include adhesive capsulitis, avascular necrosis of the humeral head, myositis ossificans and neurovascular injuries.

Dislocations/Separations

- Glenohumeral Dislocation—There are four types of glenohumeral dislocations: anterior, posterior, inferior (luxatio erecta) and superior. Greater than 95% of glenohumeral dislocations are anterior. Posterior dislocations are second most common; inferior and superior dislocations are rare. All dislocations require immediate neurovascular exam and radiographic studies to document the type of injury. Be mindful of associated injuries. Reduction is most successful when the patient has adequate analgesia and muscle relaxation. Conscious sedation is commonly used. Nerve blocks and intra-articular analgesia may be considered. Closed reduction for any dislocation is performed by reproducing the mechanism of injury through traction-countertraction. Multiple techniques exist (see Table 8.6). Post-reduction films are required and a post-reduction neurovascular exam must be documented.
- Sternoclavicular Dislocations—A high degree of clinical suspicion will allow the diagnosis of this uncommon dislocation. The most common cause of these injuries is motor vehicle accidents or sports injuries. Patients commonly will have extreme pain to palpation over the joint, swelling and/or palpable deformity. The patient will have pain with adduction or flexion of the arm. Dislocations occur either anterior or posterior. Patients <25 yr old may have a Salter-Harris Type I injury since the medial epiphysis of the clavicle has not fused (see Table 8.7).
- Acromioclavicicular Separation—Acromioclavicular injuries occur most commonly by a direct blow to the point of the shoulder with the arm in adduction. Less commonly, injury can occur by an upward or downward indirect force on the upper extremity. Tenderness and edema may be noted over the AC joint; a palpable deformity may be present. AC joint injuries are classified by the extent of ligament damage and clavicle dislocation. Table 8.8 is based on Rockwoods modified classification system of Allman and Tossy.

Soft Tissue Injury

- Impingement Syndrome—Impingement syndrome describes the chronic and progressive degeneration of the tendons of the rotator cuff and the long head of the biceps and the subacromial bursa secondary to mechanical trauma in the subacaromial space. The greater tuberosity of the humerus compresses the tendons against the lateral aspect of the acromion causing impingement and progressive deterioration. The classic test for impingement syndrome is performed by stabilizing scapular movement with one hand while passively flexing the upper extremity over head. Reproduction of pain in the shoulder suggests impingement. This test may also be positive in other conditions. A better test to distinguish impingement involves resolution of pain with injection of 10 ml of 1% lidocaine into the subacromial space.
- Neer classifies impingement into three stages. Presentation, treatment and disposition are discussed below:

Table 8.6. Glenohumeral dislocation

Glenohumeral Dislocation	Presentation	Treatment	Disposition	Complications
Anterior Subglenoid Subcoracoid Subclavicular	Arm in abduction and sligh external rotation. Squared off shoulder.	Closed reduction Immobilization. Post-reduction films and neurovascular assessment.	Analgesia, rest, ice. Uncomplicated primary dislocation—primary care or orthopedic follow-up. Complicated (associated fracture, rotator cuff tear, neurovascular injury)—orthopedic follow-up	Adhesive capsulitis, neurovascular injury, rotator cuff injury, Hill-Sach deformity (posterolateral corticle depression of humeral head), Bankhart's lesion (avulsed fragment of anterior glenoid labrum).
Posterior	Arm held in adduction and internal rotation. Unable to externally rotate. Coracoid process prominent.	Closed reduction—pt. supine, axial traction of humerus applied to adducted arm, with gentle lifting of the head into glenoid fossa. For locked posterior dislocations, lateral traction on the proximal humerus with gentle internal rotation may be added. Cast in neutral rotation, slight extension and 15-20° abduction (handshake cast). May require open reduction.	Analgesia, rest, ice. All reduced posterior dislocations require orthopedic referral. If closed reduction unsuccessful —immediate orthopedic consultation/admission.	Fracture of the posterior glenoid rim and proximal humerus.

continued on next page

Table 8.6. Continued

Glenohumeral Dislocation	Presentation	Treatment	Disposition	Complications
Inferior	Arm abducted, forearm resting on head	Closed reduction—pt. supine, in-line humeral shaft traction with countertraction across ipsilateral shoulder (using a sheet). Gentle abduction applied then gentle adducted after relocation. Immobilize. Post-reduction films, neurovascular assessment.	Ice, rest, analgesia. Orthopedic referral.	Neuropraxia, thrombosis of the axial artery. Rotator cuff tears occur with all inferior dislocations.
Superior	Arm shortened, adducted to side.	Immediate orthopedic consultation. Open reduction.	Admit.	Acromioclavicular, scapula, humerus fractures. Rotator cuff, biceps tendon tears. Neurovascular deficits.

Table 8.7. Sternoclavicular injury

Type	Description	Treatment	Disposition	Complications
Anterior	Clavicular head displaced anteriorly.			
Grade I	Mild sprain-ligaments intact, joint stable but painful.	Ice, analgesia, and immobilization in sling for 3-4 days.	Follow-up with PMD.	Rare
Grade II	Subluxation-ligaments partially disrupted or severely stretched.	Ice, analgesia, and immobilization with padded figure-of-eight splint and sling.	Follow-up with orthopedic surgery.	Rare
Grade III	Dislocation—Capsular ligament, intra-articular disc ligament disrupted.	Ice, analgesia. Orthopedic consultation. Closed reduction with conscious sedation; patient supine, 3-4 towels between scapulas posteriorly, press on shoulders anteriorly towards table while gently pressing clavicle head posteriorly. Immobilize with figure-of-eight splint.	Follow-up Orthopedic surgery.	Anterior dislocations are unstable injuries and usually dislocate despite immobilzation. Although there is cosmetic defect, joint function is not affected. Open reduction and internal fixation is NOT recommended.
Posterior	Clavicular head displaced posteriorly.	Immediate orthopedic consult. CT and aortogram. Closed reduction in the OR with anesthesia recommended. If airway compromised, then immediate reduction un ED; patient supine with 3-4 towels between scapulas, apply traction to extended and abducted arm. Manual traction on the clavicle may be required.	Ortho admit.	Intrathoracic and superior mediastinal injury. Airway and vascular life-threatening injuries.

Table 8.8. AC joint injury

AC Joint Injury	Description	Treatment	Disposition	Complications
Type I	Strain to AC ligament; AC joint stable. No widening or separation.	Ice,rest, sling for comfort and analgesia.	Follow-up with PMD.	AC arthritis (pain with 120-180 degrees abduction).
Type II	AC ligaments ruptured, coracoclavicular ligament intact, widening of AC joint.	Ice, rest, sling for comfort and analgesia.	Follow-up with PMD.	AC arthritis
Type III	AC and CC ligament disrupted, clavicle displaced upward, AC joint widened.	Ice, rest, sling for comfort and analgesia.	Ortho referral within 72 hours.	AC arthritis, cosmetic deformity.
Type IV	Posterior dislocation of the clavicle, into or through the trapezius	Ice, immobilization, analgesia, ortho consult.	Ortho admit.	Anterior sternoclavicular joint dislocation. AC arthritis. Neurovascular or muscle injury.
Type V	Severe type III with disruption of the trapezius and deltoid muscles.	Ice, immobilization, analgesia, ortho consult.	Ortho admit.	AC arthritis. Neurovascular or muscle injury.
Type VI	Complete ligament and muscle disruption. subcoracoid dislocation of the lateral clavicle.	Ice, immobilization, analgesia, ortho consult.	Ortho admit.	AC arthritis. Neurovascular or muscle injury.

  - Stage I: Edema; anterolateral shoulder pain described as a dull ache which increases with activity; usually seen in young adults. Rest, ice and anti-inflammatories recommended. Follow-up with PMD.
  - Stage II: Bursa fibrosis and thickening with tendonitis. Occurs in older adults (25-45 y/o). Pain occurs at night and with strenuous activity. Rest, ice, anti-inflammatories and physical therapy to strengthen rotator cuff muscles is recommended. Follow-up with PMD. If patients have failed nonsteroidal therapy, corticosteroid injection may improve pain. Orthopedic referral recommended for these patients.
  - Stage III: Rotator cuff tear, bicep tendon rupture, with changes of the acromion. Patients are older with long-standing history of tendonitis and bursitis. Decreased range of motion noted. Rest, ice and corticosteroid injection may improve pain. Orthopedic referral recommended.
- Biceps Tendinitis—Pain from biceps tendonitis is described as anterolateral shoulder pain that radiates down the ipsilateral arm. Usually worsens with overhead activity and may be worse at night. Diagnostic tests include reproducible anterolateral shoulder pain with either the Yergason's test (elbow flexed at 90 degrees, with forearm supinated against resistance) or the Speed's test (shoulder flexed at 90 degrees, elbow extended, forearm supinated against resistance). Treatment consists of rest, ice, anti-inflammatories and a sling for comfort. Early range of motion exercises is imperative. Orthopedic follow-up recommended.

- Subacromial bursitis is usually associated with impingement syndrome (see above). The bursa becomes inflamed in response to inflammation of the surrounding structures. The impingement sign is positive, and there is primarily lateral shoulder pain with abduction of the upper extremity between 70 and 100 degrees. It is difficult to delineate bursitis from tendonitis or impingement syndrome, but treatment is similar. Rest, ice and anti-inflammatories are recommended. If conservative therapy fails, a corticosteroid injection into the bursa may improve the pain. Plain films are usually not indicated. Other studies such as MRI may be ordered during follow-up with a PMD or orthopedist.
- Calcific tendonitis results from calcium deposition in the rotator cuff tendons (commonly supraspinatous). This deposition is thought to be due to repeat microtrauma and degenerative changes. The process is seen primarily in adults, 40-50 yr of age. Calcium deposits may first be an incidental finding on radiographs, and the patient may be asymptomatic. An inflammatory reaction usually occurs, causing severe pain. Patients may present with pain between 70 to 100 degrees of abduction and may have marked swelling and tenderness over the joint. Conservative therapy is indicated (rest, ice, analgesia, anti-inflammatories). Injection of analgesia into the joint may improve pain greatly. Orthopedic follow-up is recommended.
- Rotator cuff tears are seen most commonly in patients over 40 yr old; they present with intense pain with or complete inability to abduct the upper extremity. The examiner may hold the arm at 90 degrees abduction and see if the patient can maintain the position (Drop Test). With significant tears, they will not be able to hold abduction. Plain film radiographs may show narrowing of <6 mm between the superior aspect of the humerus and the undersurface of the acromion (normal 7-14 mm).
- Rotate cuff tears can be classified into acute or chronic tears. Acute tears are usually secondary to forced abduction and may be associated with glenohumeral dislocations (commonly inferior glenohumeral dislocations). Patients should receive ice, analgesia, sling immobilization and prompt orthopedic follow-up.
- Chronic tears occur over time, with gradually worsening shoulder pain and decreased range of motion. Patients should receive analgesia and orthopedic follow-up for physiotherapy and possibly surgical correction.

The Elbow Joint

Anatomy and Function

- The elbow joint is a hinge joint comprised of the distal humerus and the proximal ulna and radius. The elbow is capable of flexion and extension from 0 to 150° as well as supination and pronation. The humerus divides into medial and lateral columns distally. The condyles have both an articulating and nonarticulating surface. The trochlea of the medial condyle articulates with the ulna. The radius articulates with the capitellum, the articulating surface of the lateral condyle. The lateral epicondyle is the origin of extensor and supinator muscles, the medial epicondyle gives rise to the flexor muscles.
- When the elbow is flexed, the coronoid process of the ulna and the radial head are received anteriorly by the coronoid and radial fossae on the anterior aspect of the humerus. During extension, the olecranon fossa on the posterior aspect of the humerus receives the olecranon. The main function of the ulno-trochlear articulation is flexion-extension; the radio-capitellum joint is responsible for supination-pronation. The joint is stabilized by the lateral and medial collateral ligaments.
- Anteriorly, the median nerve and brachial artery run midline at the joint. The radial nerve runs posterolateral around the joint, coursing anteriorly as it moves distally down the radius. The ulnar nerve courses posterior to the medial condyle before moving anterior-lateral in the forearm.

Management

Prehospital

Prehospital care entails immobilization to prevent further neurovascualar injury, elevation and ice.

History

If the injury is traumatic, obtain the mechanism of injury—did the patient fall on an outstretched hand (FOOSH) or was there a direct blow to the elbow and if so, in which direction/position (anterior, lateral, flexed, etc.)? Always ask about numbness or weakness to assess the neurovascular status. Inquire about range of motion. In addition to mechanism, inquire about onset, quality and radiation of pain (static or dynamic). Ask about associated systemic symptoms or disease.

Physical Exam

Neurovascular exam is crucial especially if elbow dislocation is suspected. If dislocation has occurred, immediate reduction with sedation and analgesia is recommended. After reduction, neurovascular exam must be reevaluated.

Inspect for areas of swelling, tenderness and ecchymosis.

All lacerations in the vicinity of the elbow joint must be considered to involve the joint space until proven otherwise.

Examine the elbow for range of motion, both active and passive. Always remember to identify other injury by examining the clavicle, shoulder, humerus, forearm, wrist and hand.

Radiography

- Plain Films—AP and lateral films are necessary for proper evaluation of the elbow. The AP film is shot in full extension and a true lateral at a 90° of flexion with the thumb up. A true lateral is pertinent for correct interpretation. The radial head should be aligned with the capitellum in all views. In the lateral view, make special note of the posterior fat pad; even in occult fractures, the posterior fat pad will be

visible secondary to the displacement of the fat from the olecranon fossa by the effusion. False positives can occur if there is laxity of the triceps when the arm is not in true flexion. An anterior fat pad can be seen normally on the lateral film but will be more pronounced when an effusion or hemarthrosis is present.
- CT and MRI are rarely used in emergency imaging of the elbow and are usually performed by the primary care or orthopedist in a nonurgent setting.

Labs

In traumatic injury, routine labs are not necessary. If there is concern for an infected joint, aspirate should be sent for cell count, glucose, protein and culture.

Classification, Treatment, Disposition and Complications

Fractures and Dislocations

Injury	Classification	Description	Treatment	Disposition	Complications
Distal Humerus Fracture	Suprcondylar: Extension	Distal humerus displaced posteriorly			Volkmann's ischemic contracture (0.5%), cubitus varus (10%) neurologic deficits—motor and sensory (7%).
	Type I	Stable, nondisplaced.	Analgesia. Posterior splint; 90° flexion.	F/U ortho in 24hrs	
	Type II	Minimally displaced. Stable after reduction.	Analgesia. Reduction. Posterior splint; 90° flexion.	Determined by ortho.	
	Type III	Displaced	Analgesia, ortho consult	Determined by ortho.	
	Type IV	Displaced with neurovascular deficits.	Analgesia/ sedation with immediate closed reduction, emergent ortho oconsult	Determined by ortho.	
	Supracondylar Flexion	Distal humerus displaced anteriorly			Neuropraxia, most commonly to the ulnar nerve. Cubitus valgus or varus may also occur.
	Type I	Nondisplaced or minimally displaced.	Analgesia; posterior Splint at 90° flexion.	F/U ortho in 24hrs	
	Type II	Incomplete fracture; ant cortex intact.	Analgesia, ortho consult	Determined by ortho	
	Type III	Completely displaced. Distal fragment proximal and anterior.	Analgesia; ortho consult	Determined by ortho	
	Intercondylar	T or Y shaped fx with separation of the condyles.	Analgesia, immobilization	Immed. Ortho consult	Fibrosis or ankylosis from prolonged joint immobilization. Loss of joint funtion.

Injury	Classification	Description	Treatment	Disposition	Complications
	Transcondylar	Fracture line through both condyles within the joint capsule.	Analgesia, immobilization.	Immed. Ortho consult.	Poor union, loss of joint function. Commonly occurs in elderly osteoporotic patients which impairs healing process.
	Condylar	Fracture of the articular and nonarticular surface of either condyle:			Ulnar nerve impingement, nonunion, cubitus valgus or varus, arthritis.
		Undisplaced	Posterior splint; 90° flexion	F/U ortho in 24 h.	
		• Lateral	• In supination, wrist extended.		
		• Medial	• In pronation; wrist flexed.		
		Displaced >3 mm	Analgesia, immobilization, ortho consult.	Admit; ORIF	
	Articular	Capitellum	Analgesia, posterior splint Reduction if displaced.		Arthritis, decreased range of motion, avascular necrosis of the capitellum.
		Trochlea	Analgesia, posterior splint. ORIF if displaced.		
	Epicondylar	Nondisplaced.	Anagesia, posterior splint, elbow and wrist flexed with forearm pronated.	F/U ortho in 24 h.	Ulnar nerve entrapment and palsy with medial epicondylar fracture.
		Displaced.	Immediate ortho consult.		
Radial Head Fracture *Note: In all radial head fractures, look for additional ulna fractures.*		Type I: nondisplaced.	Sling support with early range of motion exercises (within 24 h). Consider aspiration for hemarthrosis.	F/U ortho	Malunion. Loss of joint function.
		Type II: marginal fractures with displacement	Aspiration, analgesia, immobilization. If mechanical block present-ortho consult.	Consult ortho	
		Type III: Comminuted fractures of head	Analgesia, immobilization, ortho consult.	Consult ortho	
		Type IV: any of the above with elbow dislocation	See elbow dislocation.	Consult ortho	

8

Injury	Classification	Description	Treatment	Disposition	Complications
Proximal Ulna Fracture	Olecranon	Nondisplaced	Analgesia, immbolization at 45-90° of flexion.	F/U ortho in 24 h.	Ulnar neuropraxia—sensory loss over hypothenar eminence or weakness of the interossei muscles.
		Displaced (>2 mm) or Fracture dislocations	Emergent orthopedic consult.	ORIF, admit	
Elbow Dislocation	Anterior; Posterior, Lateral, Medial		Immediate reduction with conscious sedation if neurovascularly compromised. After reduction, reexamine neurovascular status and check post-reduction films. If stable through ROM, then immobilize in posterior splint in flexion. Exclusion of vascular injury usually requires angiography. Evaluate for compartment syndrome.	Emergent orthopedic consultation. If joint is stable post-reduction and reliable pt. with good f/u, may d/c home (rest, elevate, ice, f/u ortho, return for sx of vascular compromise).	Anterior: High incidence of vascular deficits. Posterior elbow dislocation: Brachial artery compromise. Median nerve deficits.
Radial Head Subluxation (Nursemaids Elbow)		Annular ligament Subluxes between radial head and capitellum.	Reduce by palpating radial head while supinating and the flexing the forearm. Recheck function.	F/U with PMD prn	Rare
Epicondylitis (Tennis Elbow)		Angiofibroblastic tendinosis of the lateral >medial epicondyles. Exercise related.	Rest, elevate, ice, analgesia.	F/U PMD	Rare

Forearm

Anatomy

The bones of the forearm are the radius and ulna. The ulna is relatively fixed , the radius rotates around the ulna. The proximal radius is supported by the annular ligament. The articulation of the ulna and radius at the wrist is known as the distal radioulnar joint, (DRUJ). The two bones are connected along their length by fibrous tissue. Due to this complex interconnection, energy can be transmitted both above and below and injury site. Joints above and below a site of trauma should be included in radiographic evaluations of the forearm. There are two fascial compartments in the forearm, dorsal and volar. Compartment syndrome is a potential complication of all forearm injuries.

	Motor	Sensory
Ulnar nerve	Intrinsic hand muscles	Small finger and Ulnar side of ring finger
Median nerve	Finger flexion, innervation of thenar eminence	Most of palm
Radial nerve-proximal	Wrist, finger and thumb extension	No sensory
Radial nerve-distal (aka. Posterior interosseous nerve)	Pure sensory	Dorsum of hand

History and Physical Exam

Injuries to the forearm are most commonly related to either direct trauma or fall on outstretched hand or "FOOSH." Specific mechanisms will be discussed with each specific fracture type. The physical exam should note specific areas of point tenderness and lacerations. A detailed neurovascular exam is essential because deficits can help pinpoint specific injuries.

Radiography

Radiographic evaluation of the forearm is generally accomplished with AP and lateral X-rays. A widened radioulnar joint suggests disruption of this complex and subluxations at the wrist or elbow must be carefully noted. Fracture of the ulnar styloid is commonly seen with Colles fractures but can also be a clue to injury of the DRUJ and triangular fibrocartilage complex (TFC). The lateral projection of the distal radius usually demonstrates a volar tilt of 10-25 degrees. Loss of this volar tilt represents a fracure of the distal radius (Fig. 8.6). 8

Diagnosis, Treatment, Disposition and Complications

Diagnosis of forearm injuries in usually straightforward and based on the physical and radiographic findings. Due to the relatively fixed nature of the ulna, exacting reduction is necessary to retain function. Orthopedic correction is needed for angulation of >10 degrees or displacement of >50% of the diameter of the bone. Radius fractures are considered displaced when angulation is >20 degrees or >1 cm of shortening. Complications or all fractures include nonunion and compartment syndrome. Be sure to exclude epiphyseal injuries of the distal radius in children with wrist trauma because the carpal bones are cartilaginous and rarely injured.

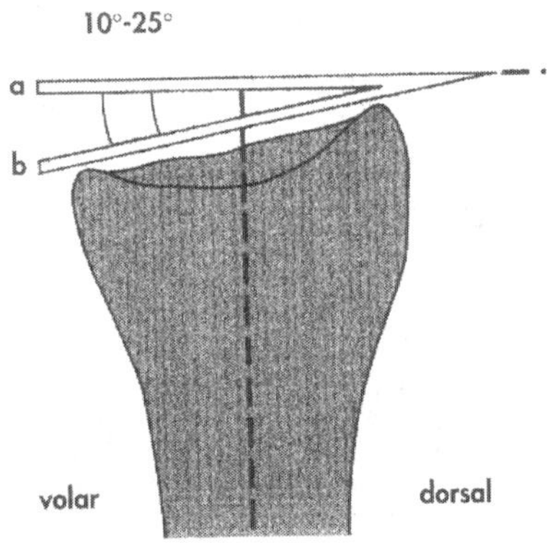

Figure 8.6. Normal lateral view of the wrist. The distal radius has a normal volar tilt of 10-25 degrees. Reprinted from Rosen's Emergency Medicine: Concepts and Clinical Practice, 5th ed. ©2002 Mosby/Elsevier, with permission.

Injury	Mechanism	Treatment and Dispostion	Complications
Ulnar shaft fracture Night stick fracture more	Direct trauma to forearm	Splint, referral to orthopedics for short arm cast and follow-up	Middle and proximal one-third are likely to progress to malunion
Ulna fracture with proximal radial head dislocation. Monteggia's fracture dislocation	Forced pronation of the forearm during "FOOSH" also direct trauma	Orthopecic consultation in the ED for surgical treatment	Radial nerve injury
Radius, distal fracture with dorsal angulation, Colles fracture	"FOOSH"	Orthopedic referral vs. consult in ED dependent upon angulation and preferance of orthopedist.	Median nerve injury. Missed fractures, this is more common with fracture of radial styloid
Radius, distal fracture with volar angulation Smith's fracture or "reverse Colles"	Direct trauma to dorsum of hand or "FOOSH"	Orthopedic referral vs. consult in ED dependent upon angulation and preference of orthopedist.	Median nerve injury
Radial styloid fracture Chauffeuror Hutchinson fracture	Twisting	Nondisplaced fracture can be splinted and referred to orthopedics. Displaced fractures require orthopedic consultation and possible surgical fixation	Fracture displaced by more than 3 mm; associated with scapholunate dissociation
Oblique intra-articular fracture of the rim of the radius with dorsal or volar carpal subluxation. Dorsal or volar Barton's fracture dislocation	High energy trauma to wrist joint	Orthopedic consultation for reduction in the ED. Usually requires surgical treatment	Wrist instability. The fracture can be subtle and missed.
Distal radius fracture with distal radioulnar joint dislocation; Galeazzi fracture dislocation	"FOOSH" with wrist in extension and forearm pronated	Orthopedic consult for surgical fixation	Ulnar styloid fracture may indicate disruption of DRUJ and TFC
Radius and ulna shaft fracture		Nondisplaced fractures splinted with prompt referral to orthopedics. Displaced fractures require orthopedic consultation and typically require ORIF	Compartment syndrome, nonunion
Dislocation of distal radioulnar joint DRUJ	"FOOSH" with hand in pronation or supination	Orthopedic consultation frequently requires ORIF	Delayed diagnosis can lead to severe chronic arthritis

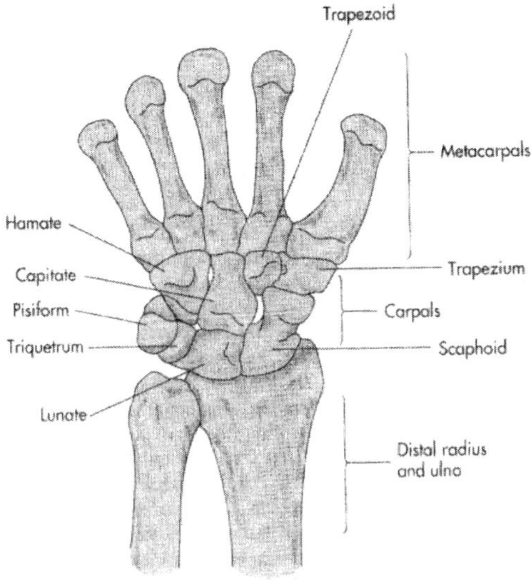

Figure 8.7. Bones of the wrist. The wrist joint includes the distal articular surfaces of the radius and ulna, as well as the proximal and distal carpal rows. Reprinted from Rosen's Emergency Medicine: Concepts and Clinical Practice, 5th ed. ©2002 Mosby/Elsevier, with permission.

8

Wrist

Anatomy

The wrist is a complex articulation involving eight carpal bones and their interface with the distal radius and ulna and the proximal metacarpals. Two sets of volar ligaments and one set of dorsal ligaments support the carpal bones. The Scaphoid and lunate articulate with the distal radius, the ulna articulates with a fibrocartilagenous structure known as the triangular fibrocartilage (abv. TFC). The blood supply to most carpal bones enters distally. The scaphoid, lunate and capitate bones are supplied by a single vessel which predisposes them to avascular necrosis especially with proximal fractures. The scaphoid by far the most commonly injured bone and is palpable within the anatomical snuff box which is bordered by the distal radius and the extensor policis longus and brevis tendons. A bony landmark called Listers tubercle is located on the dorsal aspect of the distal radius just ulnar to the extensor pollicis longus tendon (Fig. 8.7).

Prehospital

Immobilization to prevent further pain and further injury.

History

Most wrist injuries are secondary to a fall on outstretched hand or FOOSH mechanism. There are however some specific traumatic mechanisms that are associated with specific injuries. Please refer to the following table.

Physical Exam

As with all fractures, point tenderness over a bone suggests fracture or significant injury. There are, however, specific maneuvers which are associated with individual carpal injuries as follows.

Injury	History and Mechanism	Physical Exam Findings
Scaphoid fracture	Dorsal wrist pain after or on hyperextension	Tenderness with longitudinal thumb compression, Snuff box tenderness
Lunate fracture	Dorsal wrist pain post- or on hyperextension	Tenderness to dorsal wrist palpation with palmar flexion, tenderness with axial loading of third digit
Triquetrum fracture	Direct trauma to dorsum of wrist or hyperextension	Tender over dorsal wrist just distal to the ulnar styloid
Pisiform fracture	Direct trauma to hypothenar area	Paresthesia in ulnar nerve distribution, tenderness increased with ulnar deviation
Hamate fracture	Direct trauma to base of hypothenar with hypothenar tenderness	Paresthesia in ulnar nerve distribution, decreased grip strengh
Trapezium fracture	Direct trauma to adducted thumb	Tenderness with movement of thumb
Trapezoid fracture	Direct axial trauma to index finger	Tender to base of second metacarpal
Capitate fracture	Direct trauma to dorsal wrist	Dorsal wrist tenderness and swelling
Perilunate dislocation	Hyperextension of wrist	Wrist pain and deformity
Lunate dislocation	Similar to perilunate but typically more force	Wrist pain and deformity, neurpathy in median nerve distribution
Scapholunate dissociation	Forced hyperextension of wrist, clicking with wrist motion	Dorsal wrist tenderness, distal to Listers tubercle

8

Radiography

Plain Films

Standard views are the AP and lateral and oblique projections. Other views may be helpful to elucidate individual injuries if suspected, please refer to table. It is important to understand the normal radiographic anatomy when interpreting wrist X-rays. The radial styloid projects beyond the distal ulna and the distal radius has a volar tilt of 10-25 degrees in lateral projection (Fig. 8.6). The capitate lunate and distal radius should align on the lateral projection and the long axis of the scaphoid should intersect at 30 to 60 degrees (Fig. 8.8). The distance between carpal bones should be uniform and about 2 mm; any increase suggests ligamentous disruption.

Diagnosis Treatment and Disposition

Injuries to the wrist must be treated carefully. All fractures or suspected fractures should be immobilized with thumb spica splint in neutral position. Special radiographic views should be ordered if specific injuries are suspected. Ligamentous injuries can sometimes be diagnosed by a six-view motion study. Risk of nonunion and AVN is high with scaphoid lunate fractures. There is also significant risk of radiographically occult injuries, particularly of the scaphoid and lunate. Plain radiographs miss up to 15% of scaphoid fractures, for this reason, wrist injuries

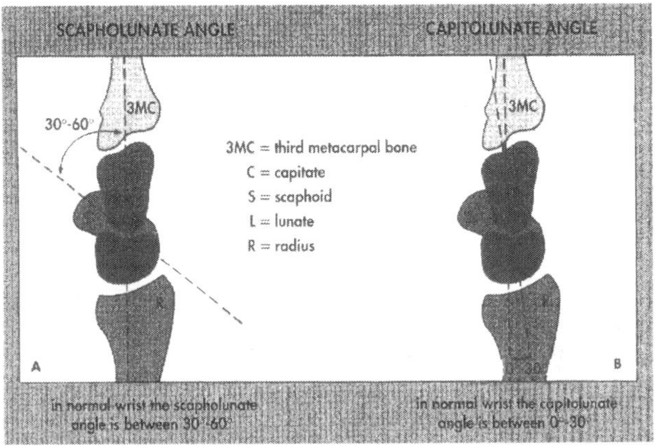

Figure 8.8. A) The normal scapholunate angle is formed by the intersection of the longitudinal axes of the scaphoid and lunate and normally measures 30-degrees. B) The normal capitolunate angle is formed by the intersection of the capitate and lunate central long axes and normally measures 0-30 degrees. Reprinted from Rosen's Emergency Medicine: Concepts and Clinical Practice, 5th ed, ©2002 Mosby/Elsevier, with permission.

8

without apparent radiographic abnormality should immobilized. This is typically done with thumb spica splint. Orhopedic follow-up is essential. Classically, repeat X-rays are obtained in 10-14 days. Bone scan, CT or MRI obtained at variable times post injury may be able to exclude injury more acutely. Any neurovascular injury or displaced fracture requires immediate orthopedic consultation.

Injury	X-Ray Views and Findings	Treatment and Disposition
Scaphoid fracture	Scaphoid views; displaced fracture is >1 mm	All fractures or suspected fractures require immobilization in neutral position with thumb spica splint. Displaced fracture requires immediate orthopedic consultation. Nondisplaced fracture requires urgent referral
Lunate fracture	Frequently difficult to visualize on plain radiographs	See scaphoid fracture
Triquetrum fracture	Standard lateral frequently shows a "chip", oblique pronation views may help	Thumb spica immobilization and orthopedic referral
Pisiform fracture	Carpal tunnel views and 30 degree semisupinated views	Thumb spica immobilization, if ulnar nerve compromise then immediate orthopedic consultation
Hamate fracture	Carpal tunnel views and 30 degree semisupinated views	Immobilization, if ulnar nerve compromise then immediate orthopedic consultation otherwise referral in 1-2 wk
Trapezium fracture	Pronated oblique view or an AP view of trapezium known as a "Roberts view"	Thumb spica immobilization and orthopedic referral

Injury	X-Ray Views and Findings	Treatment and Disposition
Trapezoid fracture	Oblique views or tomograms	Thumb spica immobilization and orthopedic referral
Capitate fracture	Standare PA and lateral X-rays are usually adequate	Thumb spica immobilization and orthopedic referral
Perilunate dislocation	Lateral view demonstrates normal position of lunate, dorsal displacement of capitate. Sometimes referred to as the"peanut out of the teacup"	Orthopedic consultation in the ED for reduction.
Lunate dislocation	PA view shows a triangular appearance or the lunate and on the lateral view the lunate is displace in a volar direction and has the apperance of a "spilled teacup"	Orthopedic consultation in the ED for reduction.
Scapholunate dissociation	Clenched fist view. A diastasis of 3 mm or more known as a "Terry Thomas sign" is pathognomonic. Scapholunate angle >60 degrees suggests injury	Orthopedic consultation for surgical repair

Complications

8

Scaphoid fractures have a 3% incidence of AVN, up to 10% result in nonunion. Lunate fractures have high incidence of AVN, known as Kienbock's disease. Hamate and pisiform fractures are associated with ulnar nerve injuries. Capitate fractures are also more predisposed to AVN. Carpal dislocations can result in median nerve injury and chronic instability.

The Hand

- Approximately 5-10% of all ED visits represent hand injuries.
- The general principal in the management of hand injuries is to preserve function.

Examination

- The examination of the hand should proceed in an orderly and deliberate way.
- Vascular integrity is measured by first assessing skin color and temperature. The next step is palpating for peripheral pulses and checking for normal capillary refill (<1-2 sec). The Allen test is a good way to test radial and ulnar blood flow to the hand. To perform this test, have the patient open and close their fist several times. With the fist in the closed position the examiner occludes both the radial and ulnar arteries by applying pressure at the wrist. As the patient opens their fist the pressure is released over one of the arteries. The hand should immediately flush. This test should be repeated with occlusion of the opposite artery to assess adequate blood flow from both the ulnar and radial arteries. Failure of the hand to immediately flush indicates partial or complete compromise of either the radial or ulnar arterial supply to the hand. A Doppler should be used in situations when the examiner is unable to palpate peripheral pulses.
- Neurologic testing of the hand and digits must assess both sensory and motor systems.
- Three Nerves—The median, ulnar, and radial nerves—provide sensory and motor innervation to the hand.
- Sensory Testing—Includes both gross assessment and two-point discrimination.
  - While there is considerable overlap of sensory innervation, each nerve has an autonomous zone of innervation.

- Gross sensory testing
- The radial zone is the first dorsal web space between the thumb and index finger.
- The ulnar zone is the volar tip of the 5th finger.
- The median zone is the volar tip of the index finger.
- Two-point discrimination of the digits—normally the fingertips should be able to distinguish two points that are 2-5 mm apart. If a sensory deficit is detected one should also suspect digital artery injury as the artery and nerve run closely together.

- Motor testing.
  - The radial nerve innervates the extrinsic dorsal muscles of the forearm and is responsible for wrist, thumb, and MP joint extension as well as abduction of the thumb. As the radial nerve enters the hand it becomes a pure sensory nerve and thus provides no motor innervation to the intrinsic muscles of the hand.
    - Testing radial nerve function is accomplished by checking abduction and extension of the thumb by having the patient give the thumbs up sign and by having the patient extend the wrist against resistance.
  - Compared to the radial nerve, the ulnar nerve innervates many of the intrinsic muscles of the hand. This nerve provides motor innervation to all seven interosseus muscles, the abductor pollicis, the hypothenar muscles, and the lumbrical muscles of the ring and 5th fingers.
    - Testing the ulnar nerve is accomplished by having the patient cross their index and middle finger.
    - Another test is to have the patient adduct and abduct their fingers against resistance
    - To test abductor pollicis strength have the patient tightly pinch a piece of paper between thumb and index finger. If the abductor muscles are weak the IP joint of the thumb will flex (**Froment's sign**).
  - The median nerve supplies innervation to the thenar muscle groups via a branch called the recurrent median nerve. This nerve also supplies motor innervation to the lumbrical muscles of the index and middle fingers.
    - Testing median nerve motor function is accomplished by having the patient oppose their thumb and index fingers or their thumb and 5th fingers. However this function is sometimes possible with loss of median nerve function due to ulnar nerve overlap. To isolate the median nerve have the patient flex the distal phalanx of their thumb against resistance.

      The median nerve enters the hand along with nine extrinsic flexor ligaments of the hand through the carpal tunnel and may become compressed causing **carpal tunnel syndrome (CTS)**. CTS can produce pain or paresthesia along the distribution of the median nerve. There are two useful tests for CTS. The first test is known as **Tinel's sign** and consists of paresthesia in the median nerve distribution with percussion over the flexor retinaculum. A second method of testing is **Phalen's test.** With the Phalen's test the patient is asked to flex their wrists completely for one minute. If symptoms are reproduced before one minute, the test is considered positive.

- Bones and Joints
  - Deformities should be noted and palpation should direct the examiner to localizing points of tenderness. Examination of the hand should include palpation of the anatomic snuffbox. Tenderness in this region suggests a scaphoid injury.
  - In order to test for rotational deformity of the fingers have the patient flex all of their fingers except the thumb. Normally, all tips should point to the region of the scaphoid.
  - Ligamentous stability should also be assessed during this phase of the examination.

- Tendons
  - It is important to note that tendon function may still be preserved with partial tendon lacerations. Therefore it is imperative to closely inspect all wounds for partial tendon lacerations during any hand laceration repair.
  - To test the common extensor tendons have the patient extend all their digits.
  - Testing the extensor indicis and extensor digiti minimi is accomplished by having the patient extend the index and 5th fingers respectively while making a fist. While repair of extensor tendon lacerations is safely performed by the ED physician in many emergency room facilities, flexor tendon laceration repair requires an orthopedic hand specialist.
  - To test the integrity of the palmaris longus, flexor carpi radialis, and flexor carpi ulnaris flexor tendons have the patient flex the wrist against resistance while palpating the individual tendons.
  - Having the patient bend the tip of the thumb against resistance tests the integrity of the flexor pollicis longus.
  - The flexor digitorum superficialis tendons are tested individually by having the patient flex the PIP joint while the other digits are held in extension. The profundus tendons are tested by having the patient flex each digit while the PIP joint of that digit is stabilized in extension.

Inflammatory Disorders of the Hand

8

- **Paronychia and eponychia** are common inflammatory conditions of the radial/ulnar lateral nail fold or basal nail fold respectively.
  - The most common causative organisms are staphylococcus and group A streptococcus.
  - The presentation is an erythematous, tender, edematous nail fold with pus often visible.
  - Treatment for most paronychia and eponychia is simple I & D via insertion of an 11 blade between the nail and nail fold after digital block.
  - Drainage should be followed by warm water soaks for 10-15 minutes three times per day.
  - If no fluctuance is identified conservative therapy with warm water soaks may be appropriate.
  - Antibiotics are generally only indicated for complicated cases.
- A **felon** is a collection of pus in the pulp space of the finger tip usually precipitated by a small and often trivial puncture wound.
  - The presentation of a felon differs from that of a paronychia/eponychia, in that the pad of the finger tip rather than the nail fold is affected.
  - Treatment is accomplished with digital block and I & D. Several different techniques for drainage are described. However, a unilateral longitudinal approach is probably the most widely used and also has the advantage of smaller incision that spares the sensate volar pad. A volar approach should be reserved for felons pointing to the volar surface.
  - After I & D a sterile wick should be left in the tissue space for 24-48 hrs to allow continued drainage.
  - Warm water soaks after the procedure are encouraged.
  - In contrast to paronychia and eponychia, felons tend to have more significant associated cellulitic changes and antistreptococcal/antistaphylococcal antibiotics are generally recommended for 7-10 days.
  - Adequate treatment of the felon is important because if left untreated this infection may progress to flexor tendon synovitis **(FTS)**.
- **FTS** is a surgical emergency requiring swift diagnosis and aggressive management.

- To accurately make the diagnosis one should be familiar with the four cardinal signs described by Kanavel:
  - Tenderness over the flexor tendon sheath with maximal tenderness usually located in the midpalmar area
  - Symmetric swelling of the digit sometimes referred to as "sausage digit"
  - Pain with passive extension of the digit
  - Finger usually held in flexed posture at rest
- These infections are often polymicrobial being caused by staphylococcus and streptococcus species as well as anaerobes and potentially *Neisseria gonorrheae.*
- Once the diagnosis has been made, emergent surgical consultation should be obtained and intravenous antibiotics (usually a β-lactamase inhibitor or first-generation cephalosporin and penicillin) should be initiated.

- **Closed-fist injury (CFI)** results from a human bite, typically after the patient's clenched fist striking another person's teeth.
  - CFI might initially appear as a small and benign looking wound to the dorsum of the hand in proximity to the MCP joint. However the potential for deep penetration by the tooth and the significant force involved with this injury facilitate the transfer of saliva into multiple planes.
  - The skin, extensor tendons, joint space, and bones can all be affected.
  - Fractures are commonly associated and therefore X-rays must be obtained.
  - As with FTS, emergent consultation with a hand surgeon should be obtained and broad-spectrum antibiotics initiated.
- **Herpetic whitlow** is a common viral infection usually affecting the distal digit.
  - The causative virus is the herpes simplex virus type 1 or 2. The virus is spread by direct contact, and therefore infection may result from occupational exposure. Healthcare professionals and cosmetologists have traditionally been at increased risk for this infection.
  - Edema, erythema, burning, and pruritis may be present along with the hallmark vesicles.
  - Distinguishing herpetic whitlow from a felon is sometimes difficult, but important to do as I & D of herpetic whitlow may lead to increased morbidity.
  - Herpetic whitlow is usually a self-limited infection that requires only immobilization, elevation, and pain medication.
  - Oral acyclovir may shorten the duration of infection but is reserved for more severe infections.

Soft Tissue Injuries to the Hand

- **Nail bed injuries** are relatively common and are usually the result of direct hand trauma.
  - Nail bed lacerations are often associated with significant subungual hematomas.
  - While open lacerations should be repaired with a 6-0 or 7-0 absorbable suture, the appropriate treatment of nail bed lacerations associated with significant subungual hematomas remains controversial. Traditional literature supports removal of the nail and repair of the nail bed for subungual hematomas occupying greater than 25-50% of the nail bed. However, some data suggests that leaving an intact nail plate to approximate the nail bed laceration is less traumatic and equally effective.
  - Nail bed injuries associated with greater than 25% subungual hematomas and pain should be trephinated with an 18 gauge needle or electrocautery device.
  - If a **distal phalanx fracture** accompanies a nail bed injury then the fracture should be reduced, the nail removed, and the nailbed copiously irrigated with normal saline.

- If the nail plate is not severely damaged it can be used as a splint for the fracture. The splint should be anchored by a nonabsorbable suture though both the splint and the area just proximal to the nail sulcus.
- If the nail is too damaged for use a silicon sheet cut in the shape of a nail should be used as a splint. Xeroform or Vaseline gauze should be placed under the nail fold to prevent adhesions of the eponychium to the matrix.
- Prophylactic antibiotics are controversial but are generally prescribed.

- **High pressure injection injuries to the hand** have become more common with the increased use of high pressure injection guns in the work place.
  - The severity of the injury is affected by the pounds per square inch of force, the agent injected, and time interval to treatment.
  - Solvent agents such as paints and stains produce a greater inflammatory response and necrosis.
  - These injuries require emergent surgical debridement and decompression.
  - Antibiotics should be started and tetanus updated while awaiting surgical consultation.
- **Gamekeeper's thumb** is an acute injury to thumb ulnar collateral ligament.
  - This injury is caused by forced extension and abduction of the thumb usually occurring with skiing or playing football.
  - The patient presents with tenderness and edema at the dorsoulnar aspect of the MCP joint.
  - Radiographs should be obtained to rule out a fracture.
  - Treatment involves placing the patient in a thumb SPICA splint and referral to a hand surgeon.

8

Fractures

- Bones of the hand are the most commonly fractured in the body and the fifth metacarpal is the most frequently fractured bone in the hand.
- Fractures of the metacarpal bones are divided into those involving the thumb metacarpal and those involving the remaining four metacarpal bones. Additionally, the ring and little finger metacarpals have between 15-25 degrees of AP mobility at their base while the index and long finger metacarpals are essentially immobile.
  - This difference in mobility is an important aspect in fracture management. In order to preserve function, anatomic reduction is generally required for fractures of the index and long finger metacarpals.
  - In contrast, anatomic reduction is not required for fractures of the ring and little finger metacarpals because their normal mobility allows for some compensation.
- Fractures of all but the thumb metacarpal are further classified as head, neck, shaft, or base fractures.
- **Metacarpal head fractures** tend to occur as the result of a direct blow, missile, or crush injury.
  - The hand should be immobilized in the "safe" position with the wrist in 20 degrees of extension, the MCP joints flexed to 90 degrees, and the DIP and PIP joints in full extension.
  - These fractures tend to heal poorly even with optimal therapy and referral to a hand surgeon is indicated.
  - If these fractures are complicated by a puncture wound or laceration then emergency surgical consultation is required. Broad-spectrum antibiotic should be initiated as well.
- **Metacarpal neck fractures** usually occur as the result of a direct impact.

- These fractures are typically unstable and reduction is difficulty to maintain secondary to deforming muscle forces. As a result, rotational deformity is a common complication.
- In general less than 15 degrees of angulation is tolerated in fractures of the index and long finger metacarpals whereas less than 35 and 45 degrees, respectively can be accepted in fractures of the ring and little fingers.
- Nondisplaced metacarpal neck fractures of the ring and little fingers are immobilized in an ulnar gutter splint.
- A radial gutter splint is used to immobilize nondisplaced metacarpal neck fractures of the index and long fingers.
- Displaced metacarpal neck fractures of the ring and little fingers should be reduced and splinted in an ulnar gutter splint with the hand in the "safe" position.
- Displaced metacarpal neck fractures of the index and long fingers usually require surgical fixation for anatomic reduction and surgical consultation is warranted.

- **Metacarpal shaft fractures** can be transverse, oblique, comminuted, or spiral.
  - A direct blow is usually responsible for the transverse, oblique, or comminuted type of fractures while a rotational force or indirect blow are responsible for the spiral type fracture.
  - The management of shaft fractures differs from that of neck fractures in that even less angulation is acceptable.
    - No angulation is acceptable in shaft fractures of the index and long fingers.
    - Only 10 to 20 degrees, respectively is acceptable in shaft fractures of the ring and little fingers.
  - As with neck fractures all rotational deformity must be corrected.
  - These fractures are immobilized in a gutter splint (up to but not including the MCP joints).
- **Metacarpal base fractures** are uncommon and usually result from an axial loading mechanism.
  - These fractures are typically stable, and immobilization with a volar splint along with surgical referral is acceptable treatment.
- **Thumb metacarpal fractures** are classified as either intra- or extra-articular.
  - Extra-articular fractures are more common and can typically tolerate 20-30 degrees of angulation. Closed reduction and treatment in a thumb spica splint is the preferred treatment.
  - The two most common types of intra-articular fractures are the Bennett's and Rolando's fractures.
  - The Bennett type fracture is an intra-articular metacarpal base fracture combined with dislocation or subluxation at the carpo-metacarpal joint.
  - The Rolando type fracture is typically an intra-articular Y or T shaped fracture of the metacarpal base.
  - Both fracture types require surgical reduction. Therefore, after initial splinting with a thumb splica splint, the patient should be referred to a hand surgeon.
  - The Rolando type fractures typically have a poor prognosis regardless of management as they are often complicated by osteoarthritis, malunion, and chronic joint stiffness.
- **Fractures of the proximal and middle phalanges**
  - For stable nondisplaced fractures, dynamic splinting with buddy tape to an adjacent normal digit and early range of motion within 3-5 days after injury is recommended.
  - Unstable or displaced fractures can typically be managed with closed reduction and immobilization either in an ulnar gutter splint for fractures involving the ring and little fingers or a radial gutter splint for fractures of the index and long fingers.
  - These fractures should be evaluated by an orthopedic surgeon in 7-10 days.

- **Distal phalanx fractures** are the most common phalangeal fractures of the hand and may be associated with concomitant soft tissue injury of the nail bed.
  - **Mallet finger** is a common type of distal phalanx fracture described as an avulsion of bone and extensor tendon from the tendon insertion site of the dorsal surface of the base of the phalanx.
  - Most closed distal phalanx fractures are treated with a short volar hairpin splint.
  - Irreducible angulated fractures may need K-wire fixation and should be referred to a hand surgeon.
  - Fractures that are associated with nail bed injury should be treated as open fractures as discussed in the section on nail bed injuries.

Suggested Reading

1. Coppola P, Coppola M. Emergency department evaluation and treatment of pelvic fractures. Emerg Med Clin North Am 2000; 18(1):1-29.
2. Geiderman J. General principals of orthopedic injuries. In: Rosen, ed. Emergency Medicine Concepts and Clinical Practice. 5th ed. 2002:467-493.
3. Gurr D, Gibbs M. Femur and hip. In: Rosen, ed. Emergency Medicine Concepts and Clinical Practice. 5th ed. 2002:643-673.
4. Keene J, Lange R. Diagnostic dilemmas in foot and ankle injuries. JAMA 1986; 256(2):247-251.
5. Neer CS, Rockwood CA. Fractures and dislocations of the shoulder. In: Rockwood CA, Green DP, eds. Fractures in Adults, 4th ed. 1996.
6. Ritchie J, Munter D. Emergency department evaluation and treatment of wrist injuries. Emerg Med Clin North Am 1999;17(4):823-841.
7. Roberts D, Stallard T. Emergency department evaluation and treatment of knee and leg injuries. Emerg Med Clin North Am 2000;18(1):29-66.
8. Steill I, McKnight D, Greenberg G et al. Implementation of the Ottawa ankle rules. JAMA 1994; 271(11):827-832.

Chapter 9

Endocrine and Electrolyte Emergencies

Jeffrey Denham and Wendy Denham

Part A: Endocrine Emergencies

Diabetic Ketoacidosis

Diabetic ketoacidosis (DKA) is a syndrome of hyperglycemia, acidosis, dehydration, and electrolyte depletion caused by a relative or absolute deficiency of insulin in the setting of increased "stress"/counter-regulatory hormones. It usually occurs in type I diabetics but may also occur in type II diabetics associated with major concurrent illness, which alters the balance between insulin and counter-regulatory hormones. In either case the end result is uncontrolled catabolism.

Pathophysiology

- DKA can be precipitated by any disorder that severely alters the balance between insulin and its counter-regulatory hormones in the diabetic patient. The counter-regulatory hormones include glucagon, epinephrine, growth hormone and cortisol. Of these, glucagon is the most influential, and its levels are elevated four- to five-fold in DKA. These hormones promote gluconeogenesis and lipolysis while inhibiting peripheral glucose utilization. Insulin blocks gluconeogenesis and lipolysis while promoting peripheral glucose utilization. The hormonal imbalance between insulin and its counter-regulatory hormones promotes hyperglycemia and leads to acidosis through ketogenesis.
- DKA precipitants include: insulin deficiency, infection, ischemia/infarction (cardiac and CNS), infant (pregnancy), and injury (trauma). Therefore, inadequate insulin availability, increased physiologic stress, or a combination of the two may result in DKA.
- Inadequate insulin usage in the type I diabetic is the most common cause of DKA. Infections are the second most common cause.
- Acidosis occurs secondary to increased lipolysis, which leads to an elevation of free fatty acids, the substrate of ketone formation. Elevations in ketones lead to anion gap acidosis.
- Dehydration and electrolyte depletion are the result of an osmotic diuresis caused by hyperglycemia. As the kidney's capacity to reabsorb glucose is surpassed, the excess glucose in the renal tubule is excreted drawing out water and with it sodium, potassium, magnesium, calcium and phosphorous.

Diagnosis and Evaluation

The diagnosis of DKA is often obvious on the basis of clinical signs and symptoms alone. In the setting of hyperglycemia, metabolic acidosis, and ketonemia/ketonuria, the diagnosis of DKA is easily made.

- The typical history is that of a type I diabetic with several days of polydipsia, polyuria, polyphagia, nausea, vomiting and malaise. A history of recent discontinuation of insulin or intercurrent illness is usually obtained.

Clinical Signs and Symptoms

- Indicators of dehydration are present including polyuria, polydipsia, weight loss, dry mucous membranes, skin tenting, sunken eyes, depressed fontanels (pediatric patient), tachycardia and, in late stages, hypotension and lack of urine production.
- Indicators of metabolic acidosis include tachypnea, Kussmaul breathing (hyperpnea), sensation of shortness of breath and a fruity odor of the breath.
- Nausea and vomiting is often severe and contributes to dehydration and hypokalemia.
- Abdominal pain is present in approximately 30% of patients, especially children. The etiology is unclear but is likely due to gastric distention or stretching of liver capsule and resolves with treatment. In adults with DKA, abdominal pain is more likely an indicator of true abdominal disease, (i.e., the precipitant of DKA as opposed to the result of it).

Laboratory/Studies

- Potassium is the most important laboratory value to determine in DKA. There is total body potassium depletion due to urinary losses and vomiting. The average deficit is 3-5 meq/kg. However, initial potassium levels obtained may be normal or high secondary to extracellular shifts in the setting of metabolic acidosis. For every 0.1$\uparrow$ in pH, K^+ will $\downarrow$ by 0.6.
- There is a total sodium deficit secondary to urinary losses. Also, pseudohyponatremia results from glucose drawing water into the intravascular space, thereby diluting the serum sodium concentration. To correct add 1.6 to the sodium value for every 100 mg/dl glucose above normal.
- Corrected Na = serum Na + 1.6 (serum glu – 100)/100

- There is total body depletion as serum phosphorous follows potassium into the urine. As a component of ATP and 2,3-DPG, a severe phosphorous deficiency affects respiration, myocardial contractility and tissue oxygenation. Treat levels <1.0. There is no evidence that routine supplementation is useful.
- There is depletion secondary to urinary losses and supplementation may be necessary.
- Prerenal azotemia secondary to dehydration is present in most cases. Underlying renal insufficiency is common secondary to diabetic nephropathy. Serum creatinine determinations may be spuriously elevated in the setting of DKA, as ketones interfere with laboratory analysis.
- An initial blood gas is recommended in moderate to severe DKA to determine the degree of metabolic acidosis and respiratory compensation. In some facilities, the blood gas has the added advantage of providing a rapid set of electrolyte results (i.e., potassium). Venous blood pH is not significantly different from arterial blood pH among patients with DKA and is a reasonable alternative, especially if multiple measurements are to be taken.
- CBC is standard. Leukocytosis is common, often >15,000-20,000 without infection, making bandemia the best clue to infection. The hematocrit may be elevated due to hemoconcentration from prolonged dehydration.
- Urinalysis is mandatory. Glucose and ketones are elevated. Look for signs of urinary tract infection, a common precipitant of DKA.
- Urine pregnancy test is essential in any female of reproductive age.
- EKG should be examined for signs of a precipitant (i.e., ischemia/infarction). Look for EKG evidence of hypo/hyperkalemia (peaked T-waves of hyperkalemia of flattened T-waves or U-waves of hypokalemia)
- CXR should be examined for signs of pneumonia or CHF.

ED Management

The treatment of DKA can be divided into four basic components: rehydration, potassium replacement, insulin administration, and a careful search for and treatment of acute precipitants or underlying disease.

Rehydration

- Fluid replacement is the most critical and effective therapy for DKA. Fluids simultaneously correct dehydration, acidosis and hyperglycemia.
- The average fluid deficit is 5-10 L.
- To restore intravascular volume use normal saline. The first 1-2 L of normal saline (NS) should run in over 30-60 min/L. More normal saline may be needed depending on the level of dehydration and hemodynamic stability.
- Once intravascular volume has been restored, switch to 0.45% NS for maintenance fluids (usually 150-250 ml/h). It is important to avoid too rapid or overhydration. Complications include a precipitous drop in K^+ if not supplemented early, fluid overload among patients with significant preexisting cardiac or renal disease, and cerebral edema (particularly in pediatric patient).
- Among pediatric patients, the average fluid deficit is 100-150 ml/kg. In the presence of hemodynamic instability an initial normal saline bolus of 20 ml/kg is given over 1 h. In the dehydrated, but hemodynamically stable pediatric patient an initial fluid bolus is not necessary and may increase the risks of cerebral edema. Instead, judicious administration of 1.5x maintenance fluids may be all that is needed during the emergency department course.

9

Potassium Repletion

- Beware of life-threatening hypokalemia once fluids and insulin are begun. K^+ shifts back into cells and urinary losses are temporarily increased as renal excretion returns to normal. Monitor the potassium hourly over the first few hours.
- An initial EKG is a quick, helpful window to the K^+ level before lab results are available.
- Potassium supplementation should be started when the potassium level is in the upper half of the normal range and the patient is making urine. Oral replacement, if the patient can tolerate it, is as effective and safer than intravenous routes. It may be given IV in concentrations of 20-40 meq/L at a rate of 10 meq/h (may approach 15-20 meq/h in severe hypokalemia).

Insulin Administration

- Insulin is not the first thing a person in DKA needs and can be lethal in the presence of hypokalemia. It is prudent to wait for a potassium level prior to insulin administration.
- Low dose insulin therapy is equally effective as high dose insulin therapy with fewer complications (i.e., hypokalemia, hypoglycemia).
- The frequent practice of administering a bolus of insulin is no longer recommended, as it increases risk without adding benefit.
- The half-life of insulin is 3-10 min; therefore continuous infusion provides a steady, reliable and easily titratable amount. The therapy of choice is 5-10 u/h continuous infusion or 0.1 u/kg/h drip. Before starting a drip, it is important to prime the tubing with 50 ml of the insulin infusion, as insulin will bind to the tubing.
- Once the serum glucose is ≤ 300 switch to a glucose containing solution. The optimal rate of glucose decline is 100 mg/dl/h keeping the glucose above 250 mg/dl during the first 5 h of treatment.
- The insulin infusion should be continued until the anion gap acidosis is corrected. It is important to remember that the goal is not euglycemia but normalization of the anion gap acidosis. Also, the anion gap may correct while the serum bicarbonate level

remains low. This is usually secondary to nonanion gap hyperchloremic metabolic acidosis, which may persist after overhydration with normal saline. Once the anion gap closes, and the patient has clinically improved and is able to tolerate POs, the insulin drip may be discontinued. It is important to give a dose of subcutaneous regular insulin 30 min prior to discontinuing the insulin drip to prevent a rebound of hyperglycemia and acidosis.

- A flow sheet provides an organized method of tracking a patient's progress (Table 9A.1).

Search/Treatment for Acute Precipitant

- Infections are more common in diabetics and their presentations are often masked in DKA. The typical indicators of infection, fever and leukocytosis, are not reliable in DKA. For reasons not well understood, patients in DKA generally do not have fever, in fact, mild hypothermia is frequent. Leukocytosis is also common in DKA secondary to "stress" demargination. Bandemia is more specific for infection. It is important to rule out diabetic infectious emergencies: necrotizing fasciitis, osteomyelitis, Fournier's gangrene, malignant otitis externa, rhinocerebral mucormycosis, emphysematous pyelonephritis and emphysematous cholecystitis.
- Diabetics are at risk for early cardiac events, which may precipitate DKA. Myocardial infarction or ischemia is often masked in adults with DKA. Pain may be atypical or absent in the setting of diabetes secondary to diabetes-related neuropathy ("silent MI").
- All female patients of reproductive age must be screened for pregnancy. Fetal mortality may be 50-90% after a single episode of DKA.

9

Other Interventions

- Phosphorus: Give 20 meq of potassium phosphate for a phosphorus level <1.
- Magnesium: If magnesium is low, give 0.35 meq/kg in fluids over first 3-5 h or 2.5-3 g for a seventy kg patient.
- Sodium Bicarbonate: The role of sodium bicarbonate is controversial. May be considered for a pH<6.9. However, there is no evidence that patients given sodium

Table 9A.1. Example flow sheet

	Time 0	1 h	2	3	4	6
Na^+	128	130	132		136	140
K^+	4.2	4.0	3.8		3.9	4.0
Cl^-	96	102	106		110	114
HCO_3^-	5	7	9		11	16
AG	27	21	17		15	10
glu	580	490	400	310	260	180
insulin	10u/kg/h	10u/kg/h	10u/kg/h	7u/kg/h	5u/kg/h	5u/kg/h
IVF	1 L bolus	1 L bolus	1/2 NS @ 250 ml/h	1/2 NS @ 250 ml/h	D5 1/2 NS 250 ml/h	D5 1/2 NS 250 ml/h
Mental Status	normal	normal	normal	normal	normal	normal

Time 0: Corrected for pseudohyponatremia Na = 136; AG = Na-(Cl +HCO_3). Potassium level is normal at 4.2 but expected to drop, therefore oral potassium is given at Time 0 if patient is able to make urine. Electrolytes should be obtained hourly for the first few hours, then every 2 h once a positive trend is established. Glucose should be monitored hourly (may alternate d-sticks with lab draws). Potassium may need to be added to IV maintenance fluids if oral replacement not possible. Time 6: Continue insulin infusion at 5u/kg/h, give 6u regular insulin SQ, and plan to discontinue infusion in approximately 30 min. Also notice that the anion gap has normalized @ 10, but HCO_3 is still low @ 16, due to a **non-anion** gap hyperchloremic acidosis from saline hydration.

bicarbonate do better. If it is used, do not correct to above a pH of 7.1 and do not bolus. Dilute 1 amp $NaHCO_3$ in 1 L 0.45 NS and infuse over 1 h. Complications of sodium bicarbonate therapy include hypokalemia, paradoxical CSF acidosis, cerebral edema, alkalosis, hypernatremia, and shifting of the oxygen-hemoglobin saturation curve to the left (decreasing peripheral tissue oxygenation).

Special Considerations

Euglycemic DKA

- A subset of patients with DKA will have normal or only slightly elevated blood glucose. Euglycemic DKA is defined as a blood glucose level <300 mg/dl. This may occur if insulin levels are sufficient to maintain a normal blood glucose but not sufficient to block lipolysis and ketogenesis. For example: a diabetic in early DKA precipitated by infection who has decreased carbohydrate intake secondary to nausea/vomiting and has increased their insulin dose in response to DKA.

Ketoalkalosis

- In the setting of severe vomiting (metabolic alkalosis), dehydration (contraction alkalosis), and hyperventilation (respiratory alkalosis), acidemia may not always be present.

Negative Ketone Assays

- Initially, patients in DKA may have negative tests for ketones in the serum and urine in the presence of significant ketoacidosis. The laboratory detects only acetoacetate, while the predominant ketone of early and untreated DKA is betahydroxybutyrate. As treatment is begun the levels of detectable ketones (acetoacetate) increase.

9

Pediatric DKA Considerations

- DKA is responsible for approximately 70% of diabetes-related deaths in children. It must be taken seriously with prompt diagnosis and treatment and early consultation of a pediatric critical care specialist and or pediatric endocrinologist. If hemodynamically unstable, give 20 ml/kg bolus of 0.9% NS. Otherwise, 0.45% NS run at 1.5x maintenance is generally sufficient and minimizes risks of rapid overhydration. During treatment the glucose level should decrease no faster than 50-100 mg/dl/h, and should be checked hourly. The insulin dosage is 0.05-0.1 u/kg/h IV infusion. When under the age of 5 yr, only mildly ill, or within 6 h of a subcutaneous dose use 0.05-0.08 u/kg/h. It is important to add 5% dextrose to IV fluids when glucose level is 250-300, or to slow the rate of glucose fall. Consider adding 10% dextrose when glucose falls below 200. The goal is to keep glucose approximately between 180-200 throughout the first 24 h of therapy.

Cerebral Edema

- Most commonly seen in children, but may be seen in adults. It involves an acute alteration in mental status usually 6-10 h after initiation of therapy. The mortality is approximately 90%. The exact etiology is unclear but may be due to “idiogenic osmoles” developed in brain cells in response to a hypertonic environment. This leads to intracellular swelling once fluids are administered. Overzealous hydration and rapid declines in serum glucose play a role. However, some evidence suggests that subclinical cerebral edema exists in most children with DKA. Treatment is mannitol 0.25-2 g/kg IV, which should be kept readily available and given at the first sign of neurologic deterioration.

Insulin Pump

- The use of insulin pumps is becoming more common, particularly in the pediatric population, secondary to the convenience and steady glucose control it provides. However, the insulin pump is associated with an increased incidence of DKA, usually secondary to mechanical problems with the tubing.

Admission Criteria

- Most patients with DKA should be admitted to the hospital. An ICU admission is recommended in the following situations:
  - extremes of age
  - severe concurrent illness
  - altered mental status
  - persistent severe acidosis
  - severe hypokalemia
- Patients with mild to moderate DKA can generally be sent to the regular ward. In some settings a 24 h "observation" unit is available. A small subgroup of patients may be discharged from the emergency department: those with only mild DKA and complete resolution of anion gap acidosis, no precipitating illness, not the first episode, able to tolerate oral fluids and self-administer insulin, and close follow available.

Nonketotic Hyperosmolar Syndrome

Nonketotic hyperosmolar syndrome (NKHS) consists of hyperglycemia, hyperosmolarity, severe dehydration and altered mental status without significant ketosis or acidosis.

Pathophysiology

- NKHS typically occurs in the elderly type II diabetic; although about half of the patients have no known prior history of diabetes. It may also occur in children (rare) and nondiabetics in special circumstances. The classic scenario is that of an elderly, type II diabetic who encounters a stressful event.

9

- As with DKA, the underlying mechanism of NKHS is a relative insulin deficiency in the setting of elevated "stress"/counterregulatory hormones. In contrast to DKA, insulin levels are sufficient to prevent significant ketoacidosis. The result is severe hyperglycemia, osmotic diuresis, profound dehydration, and electrolyte depletion.
- The mortality rate of NKHS is higher than that of DKA. This can be explained by three reasons: (1) more profound dehydration and electrolyte disturbances, (2) older demographics, (3) life-threatening precipitants and coexisting disease are more common. Coma and death are the end-result when left untreated.

Common Precipitants

- Infection (most common; usually Gram-negative pneumonia or sepsis)
- Ischemia/infarction (cardiac or CNS)
- Injury (trauma, burns)
- GI bleed
- Pancreatitis
- Pulmonary embolus
- Medications (β-blockers, phenytoin, steroids, thiazide diuretics)
- Peritoneal dialysis, hyperalimentation (may precipitate NKHS in nondiabetics)

Common Contributing Factors/Coexistent Disease

- Renal insufficiency (impaired elimination of glucose)
- Cardiac disease (CHF, A-fib, unstable angina or previous MI)
- Altered mental status/Altered thirst mechanism (Alzheimer's, CVA; unable to keep up with fluid losses)
- Physical debilitation (unable to reach water)
- Diuretic medication (exacerbates fluid losses and thiazides promote hyperglycemia)

Diagnosis and Evaluation

- There are four basic diagnostic criteria: (1) marked hyperglycemia (>600 mg/dl, often >1000 mg/dl); (2) hyperosmolarity (>320 mosm/L); (3) pH >7.3 (may be more acidotic secondary to coexistent illness; sepsis, lactic acidosis); (4) minimal or absent ketosis.

Clinical Signs and Symptoms

- Polyuria, polydipsia, weight loss, fatigue and weakness often begin days to weeks before presentation. The average length of symptoms is 12 days in NKHS as opposed to 3 days for DKA.
- Signs and symptoms of dehydration are present: decreased skin turgor and sweating, dry mucous membranes, tachycardia, and, in the late stages, orthostasis, hypotension and shock. The average fluid deficit in NKHS is 8-12 L as opposed to ≈ 6 L in DKA.
- Urinary output (UOP) is not a good indicator of hydration status in NKHS as the osmotic diuresis inappropriately maintains urinary output in the face of severe dehydration.
- Neurologic signs and symptoms are usually present and correlate with the elevation of osmolarity. Mental status may range from mild drowsiness to lethargy and frank coma. Focal neurologic deficits are often found including hemiparesis, hemianopsia, cranial nerve findings, aphasia and dysphagia, and focal seizures. The focal seizures are best treated with benzodiazepines.

Laboratories/Studies

- Hyperglycemia must be present for diagnosis. Serum glucose is usually >600, and often >1000.
- Hyperosmolarity is the cardinal laboratory indicator of NKHS. Osmolarity is always >320.

  Calculated osmolarity = 2 Na + glucose/18 + BUN/2.8

  Effective osmolarity/tonicity = 2 Na + glucose/18 (urea is a freely permeable, "inaffective" osmole).
- Mild anion gap acidosis often secondary to lactic acidosis from dehydration, infection, or cardiac failure is present.
- Potassium deficit secondary to osmotic diuresis occurs. The usual total body deficit is 5-10 meq/kg. Serum potassium levels may initially be normal or even high depending on extracellular shifts. Without significant acidosis, these shifts are less pronounced in comparison to DKA.
- Sodium deficit is present secondary to osmotic diuresis (5-10 meq/kg). Despite total body sodium depletion, the patient with NKHS is often hypernatremic (greater free water losses than Na losses). As in DKA, the laboratory value for sodium must be adjusted in the presence of hyperglycemia, as it will be factitiously lowered.

  Corrected Na = serum Na + 1.6 ([serum glucose – 100] / 100).
- Magnesium and phosphorus losses occur secondary to osmotic diuresis.
- Complete blood count commonly reveals leukocytosis secondary to "stress" demargination and or infection, as well as elevated hematocrit secondary to dehydration.
- Hyperamylasemia: associated pancreatitis is common; it may precipitate NKHS or be caused by NKHS.
- Hypertriglyceridemia is common.
- CXR should be evaluated for signs of occult pneumonia or cardiac failure.
- EKG should be evaluated for signs of precipitant ischemia/infarction, as well as indications of hypokalemia.
- Head CT may be necessary in the setting of new altered mental status, seizures, or focal neurologic findings, although all may be explained by the hyperosmolarity of NKHS.

ED Management

The treatment of NKHS can be divided into four basic categories: (1) fluid replacement, (2) electrolyte correction, (3) insulin administration, and (4) treatment of the precipitant.

Fluid Replacement

- The first treatment objective is to establish hemodynamic stability with 0.9% NS as necessary. Usually 1-2 L of 0.9% NS over the first hour is required. Once hemodynamic stability is restored, switch to 0.45% NS as 0.9% NS may lead to or exacerbate hypernatremia.
- The average fluid deficit is 8-10 L or 20-25% of total body water. The goal is to replace 50% of losses over the first 12 h, with the remainder over the next 24 h.
- Fluid management with NKHS can be difficult and may require invasive monitoring, since most patients are elderly with preexisting renal and cardiac impairment.

Electrolyte Correction

- Anticipate and treat hypokalemia early. Begin replacement once potassium is in the normal range and the patient is making urine. Oral replacement is adequate and safer if patient able to tolerate it. Otherwise, replace intravenously at a rate of 10-20 meq/h.
- Magnesium replacement may be needed for effective replacement of potassium.
- As in DKA, replace phosphorus if <1.0.

Insulin Administration

9

- In NKHS, insulin therapy is *less* important than in DKA. This seems counterintuitive, as glucose levels are higher with NKHS. However, in DKA the goal of insulin therapy is to stop lipolysis and, thereby, ketoacidosis. In NKHS there is already sufficient insulin to block lipolysis, and, therefore, the insulin requirement is generally lower.
- The attendant risks of insulin therapy are higher with NKHS than DKA. (1) Hypoglycemia: many previously undiagnosed diabetics and those on oral regimens may be very sensitive to insulin and require slower infusion rates. The goal is to slowly lower glucose levels at a rate of ≈100 mg/dl. (2) Hypotension: insulin should not be given until the patient is hemodynamically stable. Glucose osmotically maintains the intravascular compartment in the face of profound dehydration. An abrupt shift of glucose into the intracellular compartment post-insulin administration may cause sudden intravascular collapse. (3) Hypokalemia: insulin should not be given prior to obtaining a potassium result or if the potassium level is <3.5. Insulin should be given as an infusion of 5-10 u/h or 0.1 u/kg/h. Bolus insulin is potentially harmful.

Search for and Treat Acute Precipitants and Coexisting Disease

- Early diagnosis and treatment of infection with antibiotics, and appropriate management of cardiac and renal disease are extremely important in decreasing the morbidity and mortality of NKHS.

Admission Decision

- ICU admission is necessary for most patients. Very mild cases may be managed in a monitored observation setting. Admission decisions require evaluation of the severity of the NKHS, precipitating diseases, and concurrent conditions.

Alcoholic Ketoacidosis

Alcoholic ketoacidosis (AKA) is a syndrome of anion gap ketoacidosis, dehydration, nausea, vomiting, weakness and, commonly, abdominal pain 12-72 h following an episode of heavy alcohol consumption with abrupt cessation. It occurs in chronically malnourished alcoholics.

Pathophysiology

- The mechanism of increased ketone production is similar to starvation ketosis. Two factors may account for this: (1) decreased intake of food secondary to binge drinking, underlying abdominal pathology (gastritis, peptic ulcer disease, pancreatitis), or concurrent infection; (2) depleted liver glycogen stores secondary to chronic alcoholism. The end result is the mobilization of free fatty acids for ketone production.
- In AKA the predominant ketone is β-hydroxybutyrate (B-Hb). Alcohol causes an elevation of intracellular NAD/NADH which leads to preferential production of B-Hb. The ratio of B-Hb to acetoacetate in AKA is ≈ 12:1. In contrast, the ratio is ≈ 3:1 in DKA.

Diagnosis and Evaluation

- AKA is probably underdiagnosed. Nausea, vomiting, abdominal pain and dehydration are all common and nonspecific presentations of chronic alcohol abuse, making the diagnoses of AKA difficult.
- Furthermore, the term AKA is often a misnomer. AKA frequently involves neither A (alcohol), nor K (ketones), nor A (acidosis). Serum alcohol is usually low or absent, as the patient has stopped drinking 12 to 72 h previously. Ketones are often absent on initial tests for ketonemia or ketonuria, as B-Hb is nonreactive with the ketone assay. However, post-hydration ketones will be present, as B-Hb is converted to the reactive acetoacetate. Alkalemia may be more common than acidemia. Ketoalkalosis is a common finding in AKA secondary to vomiting and dehydration which lead to metabolic and contraction alkalosis respectively.

9

Signs and Symptoms

- Nausea and vomiting, often severe, is present in almost all cases.
- Abdominal pain is common and usually diffuse and nonspecific. As in DKA, the abdominal pain is often benign, of unclear etiology, and usually resolves as the ketosis clears. However, it is imperative to rule out other more serious pathology common in alcoholics such as gastritis, PUD, pancreatitis, alcoholic hepatitis, and subacute bacterial peritonitis. Surgical illness, such as appendicitis and acute cholecystitis, must also be considered.
- Manifestations of dehydration are often present including dry mucous membranes, dry skin with decreased turgor, increased thirst, tachycardia, decreased urine output, and in late stages hypotension.
- Tachypnea +/- Kussmaul respirations may be present in the setting of significant metabolic acidosis.
- Generalized weakness and drowsiness is a common finding. True alterations in mental status mandate a thorough search for other conditions. Common coexistent etiologies of altered mental status in the alcoholic population include hypoglycemia, alcohol withdrawal or delirium tremens, and occult head injury with a subdural hematoma.

Laboratory/Studies

- Mixed acid-base disturbances are common. Anion gap metabolic acidosis (ketoacidosis), metabolic alkalosis (vomiting), contraction alkalosis (dehydration), and respiratory alkalosis (compensatory) may all be present concurrently. Acidemia or alkalemia may predominate the clinical the picture.
- Hypoglycemia or hyperglycemia may be present. Hypoglycemia may be present secondary to poor dietary intake and low glycogen stores in the liver. However, a mildly elevated glucose is more common in AKA.
- Hypokalemia can reach life-threatening levels secondary to severe vomiting combined with poor dietary intake.
- Hypomagnesamia is commonly present in alcoholics.

- Serum and urine ketones initially may be negative becoming positive post-hydration. In contrast to DKA, the urinalysis shows ketonuria without glucosuria.
- Elevated BUN/Cr ratio secondary to dehydration (prerenal azotemia) may be present.
- Mild leukocytosis from stress demargination can occur, making leukocytosis a non-specific marker for infection.
- Anemia and thrombocytopenia, typical of chronic alcohol abuse, is often found concurrently.
- Liver function tests and pancreatic enzymes (amylase, lipase) should be sent in search of coexistent alcoholic hepatitis or pancreatitis.
- Stool for occult blood should be performed to rule out a GI bleed in any alcoholic with nausea, vomiting, and abdominal pain.
- EKG should be considered in middle age/elderly population with nonspecific abdominal pain to rule out ischemia/infarction. It may also provide an early clue to electrolyte abnormalities.
- Head CT is an important study in the alcoholic patient to rule out other etiologies of altered mental status.

ED Management

There are four basic components in the treatment of AKA: (1) fluid replacement, (2) glucose administration, (3) electrolyte replacement, and (4) treatment of a precipitant or coexisting illness.

- Fluid replacement: If hemodynamically unstable secondary to volume depletion, administer 0.9% NS as a bolus and repeat as necessary. Once stability is restored, begin maintenance fluid with D5 0.45% NS.

9

- Glucose administration: Glucose-containing IV solutions should be started early in the course of treatment, even in the euglycemic patient. The early use of glucose containing solutions decreases production of ketoacids and replenishes glycogen stores leading to a more rapid resolution of metabolic abnormalities than saline alone. Thiamine repletion is usually needed along with glucose administration for patients with AKA. Theoretically, thiamine should be given before or with glucose to avoid precipitation of Wernicke's encephalopathy.
- Electrolyte replacement: Hypokalemia should be anticipated and treated early. Administration of fluids alone will initially increase urinary losses of potassium. Replacement of magnesium is usually needed in the chronic alcoholic and will aide in the treatment of hypokalemia.
- Other interventions for AKA include antiemetics as needed and benzodiazepines for withdrawal symptoms.
- Admission criteria include persistent nausea and vomiting, persistent abdominal pain and abdominal pain of unclear etiology, significant electrolyte abnormalities, or significant comorbid illness.

Thyroid Storm

Thyroid storm represents an extreme and rare life-threatening form of thyrotoxicosis. It is defined as exaggerated thyrotoxicosis with an elevated temperature, tachycardia out of proportion to fever, altered mental status, and cardiovascular dysfunction. Hyperthyroidism, thyrotoxicosis, and thyroid storm represent a continuum of disease.

Pathophysiology

- Only 1-2% of patients with hyperthyroidism will progress to thyroid storm. Most patients with thyroid storm have previous symptoms of uncomplicated hyperthyroidism for an average duration of 6-8 mo.

- Thyrotoxicosis + an acute precipitant leads to thyroid storm.
- The exact mechanisms underlying the shift from thyrotoxicosis to thyroid storm are unclear. It is more than simply excess hormone production, as thyrotoxicosis cannot be distinguished from thyroid storm on the basis of hormone levels. An acute precipitant is usually the catalyst for converting thyrotoxicosis to thyroid storm.
- The etiology of thyroid storm includes all the etiologies of thyrotoxicosis. Grave's disease represents 85% of all thyrotoxicosis. Other etiologies include: toxic diffuse goiter, toxic multinodular goiter, thyroiditis, metastatic follicular carcinoma, struma ovarii, or medications (thyroid hormone, amiodarone, IV contrast).
- Common precipitants include infection (most common, especially pulmonary), ischemia (cardiac, CNS, mesenteric), trauma, surgery (especially thyroid), burns, or medications (thyroid hormone, radioactive iodine, contrast dye).
- Historically, the mortality of untreated thyroid storm approached 100%, but this is now decreasing.

Diagnosis and Evaluation

Signs and Symptoms

- Constitutional symptoms including weight loss, anorexia, weakness, nervousness, heat intolerance, and fever are common.
- Cardiovascular symptoms include palpitations, dyspnea, chest pain, tachycardia, widened pulse pressure, CHF (high output), arrhythmias (PVCs/PACs/A-fib/A-flutter), or shock in severe cases.
- CNS symptoms of tremor and restlessness are common. Altered mental status ranging from confusion to psychosis and coma distinguish thyroid storm from severe thyrotoxicosis.
- GI symptoms of diarrhea, abdominal pain, jaundice, and tender hepatomegaly (due to cardiac congestion) may be present.
- Musculoskeletal symptoms may include proximal muscle weakness (thyrotoxic myopathy).

Laboratories/Studies

- Thyroid function studies demonstrate a low TSH, elevated Total T4, and Free T4. There are no pathognomonic values, and these studies are often not available during the emergency department time course.
- Complete blood count may demonstrate an anemia of chronic disease and leftward shift of WBCs. The peripheral white count may be elevated from thyroid storm or from precipitating illness.
- Mild azotemia from dehydration may be present.
- Mild hyperglycemia secondary to a hypermetabolic state is common.
- LFT elevation (liver congestion secondary to high-output CHF) is occasionally found.
- EKG demonstrates sinus tachycardia, PVCs, PACs, A-fib, A-flutter or evidence of concurrent or precipitating ischemia.
- CXR may demonstrate CHF, cardiomegaly, or pulmonary infiltrates suggestive of concurrent disease or precipitating illnesses.
- Head CT scan to rule out other causes of altered mental status may be indicated.
- Other studies are indicated as needed in order to diagnose precipitating/concurrent illness.

ED Management

The treatment of thyroid storm can be divided into three components: (1) supportive care, (2) anti-thyroid therapy, and (3) treatment of acute precipitant and coexisting disease.

Supportive Care

- Fluid replacement is given as needed to compensate for dehydration from insensible and GI losses.
- Supplemental oxygen is indicated for the increased oxygen demand.
- Fever control with cooling measures and acetominophen is indicated. Avoid aspirin as it displaces T4 from thyroglobulin.
- Glucocorticoid replacement is often needed. Plasma cortisol levels are low in thyroid storm. Steroids have been reported to increase survival in doses equivalent to 300 mg of hydrocortisone. Dexamethasone is preferred; added benefit of decreasing peripheral conversion of T3 to T4.

Anti-Thyroid Therapy

- Medications are used to block the peripheral effects of thyroid hormone, to block the synthesis and release of thyroid hormone, and to block the peripheral conversion of thyroid hormone to its active state.
  - *Block peripheral effects*: β-blocking agents inhibit the end-organ effects of thyroid hormone. They have become the cornerstones of thyroid storm management, reducing both morbidity and mortality. Propranalol is the B-blocking agent of choice, as it also helps to block peripheral conversion of T4 to T3. The standard dose of propranalol is 1-2 mg titrated to tachycardia q 10-15 min to a maximum dose of 10 mg. High-output cardiac failure in thyroid storm will benefit from β-blockade, but β-blockers may worsen CHF among patients with preexisting heart failure.
  - *Block synthesis*: Propylthiouricil (PTU) 150-300 mg orally or per NG tube every 6 h inhibits thyroid peroxidase. PTU produces a more rapid effect than methimazole (within 1 h) and has the added advantage of decreasing peripheral conversion of Free T4 to T3.
  - *Block release*: Iodides (potassium iodide, sodium iodide, Lugol's solution) block release of T4 from the thyroid, but must be given 1-2 h after PTU in order to block the uptake of iodides for synthesis of thyroid hormone. Options include (1) SSKI: 5-10 gtts po q 12 h or (2) Lugol's solution: 30 gtts po q 12 h or (3) sodium iodide: 1 g IV.
  - *Block peripheral conversion*: Peripheral conversion of T4 accounts for 85% of circulating T3, a more physiologically active form. Dexamethasone, PTU, and propranalol all play a role in blocking peripheral conversion.
- The treatment of thyroid storm necessitates the management of any underlying precipitant.
- The patient in thyroid storm requires ICU admission.

Special Considerations

Apathetic Thyrotoxicosis

- Apathetic thyrotoxicosis is thyrotoxicosis without the usual hyperkinetic manifestations. It is a difficult diagnosis to make. The clinical picture is dominated by cardiac manifestations and altered mental status. CHF and A-fib are common. It is usually seen in the elderly with toxic multinodular goiter. These patients probably have an attenuated response to thyroid hormone and, therefore, do not display the typical symptoms.

Myxedema Coma

Myxedema coma is a rare, exaggerated, and life-threatening form of hypothyroidism. It is not well understood, and the exact diagnostic criteria are poorly defined.

Pathophysiology

- Myxedema coma occurs among patients with long-standing, untreated or inadequately treated hypothyroidism when subjected to stress. A precipitating event can be found in most cases of hypothyroidism.

9

- **Poorly controlled hypothyroidism + Acute precipitant ⇒ Myxedema Coma**
- Etiologies of hypothyroidism include primary, secondary, or tertiary causes.
  - Primary hypothyroidism (disease of the thyroid gland) accounts for 95% of hypothyroidism. Diseases include Hashimoto's thyoiditis most commonly, iatrogenic causes (post-radiation, post-surgical, anti-thyroid medications), congenital abnormalities, enzymatic defects, neoplasms, or infections
  - Secondary hypothyroidism (disease of the pituitary gland) accounts for 4% of hypothyroidism. Diseases include pituitary tumors, infiltrative diseases (sarcoid), infarction, hemorrhage, or trauma.
  - Tertiary hypothyroidism (disease of the hypothalmus) accounts for <1% of hypothyroidism. Etiologies include injury, infarction, infiltration or hemorrhage.
- Acute precipitants of myxedema coma are diverse. They include infection most commonly, environmental (prolonged cold exposure), cardiac (ischemia, CHF), pulmonary diseases, metabolic disease, trauma, surgery, or medications.
- The incidence is greater in women, the majority are elderly (half are between the ages of 50 and 70).
- Before the advent of thyroid hormone replacement, the mortality rate approached 100%. Even with treatment, the mortality may be as high as 50%.

Diagnosis and Evaluation

- Diagnosis is purely clinical and based on a constellation of signs and symptoms (Table 9A.2).

Laboratory/Studies

- Thyroid function tests demonstrate an elevated TSH, low total T4 and free T4. There are however no pathognomonic lab values, and these studies may not be available in the Emergency Department time course.
- Complete blood count is standard. A normal WBC count in the setting of infection is expected (left shift and bandemia more sensitive). Anemia of chronic disease can be seen.
- Hypoglycemia is usually mild but can be severe.
- Hyponatremia may be caused by SIADH secondary to hypothyroidism.
- Arterial blood gas may demonstrate hypoxemia, hypercapnia, and/or acidosis.

9

Table 9A.2. Signs and symptoms of hypothyroidism/myxedema coma

Hypothyroidism Symptoms	Hypothyroidism Signs	Myxedema Coma
Weakness, fatigue	Prolonged relaxation of DTRs	Exaggerated s/sx of hypothyroidism plus:
Cold intolerance	Non-pitting edema (myxedema)*	Hypothermia
Paresthesias	Dry, scaley skin	Bradycardia
Myalgias	Yellow skin**	Hypoventilation
Menorrhagia	Hypothermia	Hypotension
Hearing loss	Sparse pubic and axillary hair	Altered mental status/coma
Emotional lability	Thinning eyebrows	Pericardial/pleural/ peritoneal effusions
Forgetfulness	Mononeuropathy***	Ileus
	Hoarse voice*	Acute urinary retention
	Goiter (10% of cases)	
	Thyroidectomy scar	

*Non-pitting edema and hoarse voice are due to hyaluronic acid deposition in the dermis and vocal cords, respectively. **Yellowing of the skin is caused by decreased conversion of carotene to vitamin A. ***Mononeuropathies, such as carpal tunnel syndrome, are fairly common.

- Elevated enzymes including CPK, LDH, or LFTs may be present.
- Elevated cholesterol levels are often present
- CXR should be done to check for signs of pneumonia, CHF, or cardiomegaly (pericardial and pleural effusions in one-third of cases).
- EKG should be examined for signs of ischemia/infarction, bradycardia, conduction blocks, or low voltage (effusion).
- Head CT may be needed to rule out other explanations for altered mental status.

ED Management

The treatment of myxedema coma can be divided into three components: (1) initial supportive therapy; (2) thyroid hormone replacement; and (3) identification and treatment of precipitating illness.

Supportive Therapy

- Hypoxia and hypoventilation may be life-threatening. Administer supplemental oxygen and consider early intubation if necessary.
- Hypoglycemia needs immediate treatment. Obtain rapid D-stick and administer glucose if necessary.
- Hypotension should be treated with 0.9% NS as necessary for hemodynamic stability with cautious attention to underlying CHF. Pressor agents must be used with caution, as they may precipitate arrhythmias in the setting of myxedema coma. The administration of thyroid hormone will also augment the pressor effect.
- Treat hypothermia with gradual, passive rewarming. Rapid, active rewarming can be harmful.
- Correct hyponatremia gradually with fluid restriction. Hypertonic saline should be considered for Na <110, worsening mental status, or seizures.
- Avoid the use of sedative/hypnotics that may further depress respiratory, CNS and metabolic function.
- Administer stress doses of corticosteroids. Cortisol levels are low in the hypothyroid patient. Hydrocortisone 100 mg IV q 8 h should be given empirically.

Thyroid Hormone Replacement

- Prompt intravenous replacement of thyroid hormone has been shown to improve the survival of myxedema coma. L-thyroxine (T4) is the drug of choice. It has a more gradual onset with fewer cardiac complications as compared to triiodothyronine (T3). The exact dose is controversial, but usually 300-500 mg IV is given while on a cardiac monitor. This can be followed by 50-100 mg IV daily after admission.

Search for and Treat Precipitant Illness

- A thorough history and physical exam is required, as well as laboratory and radiologic investigation for infection/sepsis, metabolic abnormalities, cardiac, CNS, or renal disease. Treatment should be given as needed.
- All patients with myxedema coma should be admitted to the ICU.

Acute Adrenal Insufficiency

Acute adrenal insufficiency (AI) is a life-threatening condition, and the diagnosis is often missed in the emergency department. The signs and symptoms of acute AI are nonspecific, making diagnosis difficult.

Pathophysiology

- The adrenal gland is divided into the cortex and medulla.
- The adrenal cortex secretes glucocorticoids, mineralocorticoids, and androgenic hormones.

Table 9A.3. Etiology of Addison's disease

Type of Adrenal Insufficiency	Cause
Functional adrenal insufficiency (most common type) Suppression of the HPA due to exogenous steroids	Rapid withdrawal of exogenous steroids
Primary adrenal insufficiency (disease of the adrenal gland) ↓cortisol, ↑ACTH, ↑CRH	Autoimmune: Addison's disease* Infection: TB, cocci, histo, CMV, HIV** Infiltration: metastasis***, sarcoid, amyloid, lymphoma Post-adrenalectomy Hemorrhage: Waterhouse-Frederichson†
Secondary adrenal failure (disease of the pituitary gland) ↓cortisol,↓ACTH, ↑CRH	Infarction: Sheehan's syndrome†† Hemorrhage: Pituitary apoplexy††† Infiltration: neoplastic, granulomatous
Tertiary adrenal failure disease of the hypothalamus) (↓cortisol, ↓ACTH, ↓CRH	Infection Infarction Hemorrhage Infiltration

*Addison's disease: most common cause of AI. ** HIV patients are at increased risk for AI from adrenalitis due to TB, fungal infections, lymphoma and HIV itself. Also, medications, such as rifampin and ketoconazole inhibit steroid synthesis. *** Adrenal glands are a common site of metastasis (i.e. from lung, breast); however, clinically significant AI is rare since 90% of the gland must be destroyed. † Waterhouse-Frederichson: bilateral adrenal hemorrhage, usually seen in children, associated with septicemia (meningococcemia most common). †† Sheehan's syndrome: ischemic pituitary necrosis associated with sever postpartum hemorrhage. Approximately 30% of women suffering from postpartum hemorrhage may demonstrate some degree of anterior pituitary insufficiency. ††† Pituitary apoplexy: acute, intrapituitary hemorrhage. A life-threatening event manifested by severe headache, ocular paralysis, ptosis and collapse.

9

- The release of glucocorticoids (cortisol) is regulated by the hypothalamic-pituitary axis (HPA): Hypothalamus → corticotropin-releasing hormone (CRH) ⇒ pituitary → adrenocorticotropic hormone (ACTH) ⇒ adrenal cortex → cortisol ⇒ end organ sites.
- Mineralocorticoids (aldosterone) are primarily influenced by the renin-angiotensin system and will be secreted despite defects in the HPA axis.
- The adrenal medulla secretes catecholamines and is regulated by the sympathetic nervous system.
- AI will occur in three situations: rapid exogenous steroid withdrawal in a steroid-dependent patient (most common), chronic adrenal insufficiency subjected to severe stress (i.e., trauma, surgery, infection, pregnancy), or acute adrenal failure de novo (rare; i.e., bilateral adrenal hemorrhage) (Table 9A.3).

Diagnosis

The diagnosis of AI requires a high index of suspicion and is based on a constellation of signs and symptoms. Definitive diagnosis is often not possible in the emergency department (Table 9A.4).

Laboratory/Studies

- Low random cortisol level: <20 mg/dl in setting of severe stress suggests AI, but levels are often unavailable in the emergency department setting.
- Hypoglycemia occurs in two-thirds of cases and may reach life-threatening levels. Cortisol deficiency impairs gluconeogenesis and glycogenolysis. Treat with dextrose replacement.

Table 9A.4. Signs and symptoms of adrenal insufficiency

Symptoms	Signs
Weakness, fatigue	Hypotension***
Nausea, vomiting	Tachycardia
Malaise	Orthostasis
Anorexia	Hyperpigmentation†
Myalgias	Vitiligo††
Abdominal pain*	Altered mental status
Lethargy	Sparse axillary and pubic hair
Salt cravings**	Auricular cartilage calcification

*Abdominal pain of AI may resemble an acute abdomen, but it is important to rule out primary abdominal pathology as the precipitant of AI. ** Salt cravings seen in primary AI secondary to a deficiency of aldosterone (a salt-retaining hormone regulated by the renin-angiotensin system). Aldosterone production remains intact with functional (exogenous steroid-induced), secondary, and tertiary forms of AI. *** Signs of volume depletion will be more prominent in pituitary AI secondary to aldosterone deficiency. † Hyperpigmentation: In primary AI, ACTH is upregulated. Melanocyte stimulating hormone (MSH) is a byproduct of ACTH production. This does not occur with functional, secondary or tertiary forms of AI. †† Vitiligo: hypopigmentation is present in 10-20% of patients with Addison's disease; an autoimmune process.

9

- Hyperkalemia and hyponatremia: when seen together, consider AI. Both are usually mild, but may reach severe life-threatening levels, particularly with primary AI. Aldosterone acts on the renal tubules to conserve sodium in exchange for potassium and hydrogen. Aldosterone deficiency leads to hyponatremia, hyperkalemia, and acidosis.
- Hypercalcemia may be present. Mild elevations are commonly seen; the etiology is unclear.
- Azotemia and elevated hematocrit secondary to dehydration may be found.
- Mild metabolic acidosis secondary to decreased hydrogen ion excretion (aldosterone deficiency) is common.
- CXR/EKG are standard to rule out precipitant pulmonary and cardiac disease.
- Head CT may be needed to rule out CNS pathology in the setting of new altered mental status.
- ACTH stimulation test assures the diagnosis in unclear cases. This is not usually performed in the ED setting. Dexamethasone does not interfere with the ACTH stimulation test.

ED Management

There are five basic components of AI therapy: (1) fluid resuscitation; (2) correct hypoglycemia; (3) glucocorticoid replacement; (4) correct electrolyte abnormalities; and (5) search for and treat underlying disease.

Fluid Resuscitation

- Fluid resuscitation with 0.9% NS as necessary to restore hemodynamic stability is a treatment priority. D5NS at infusion rates of 500-1000 ml/h over the next 3-4 h may be needed to correct dehydration. Close attention must be given to underlying cardiac or renal disease during aggressive hydration.
- It is important to note, that refractory hypotension may be seen in the presence of normovolemia, as cortisol deficiency depresses myocardial contractility and vascular tone. If possible, avoid vasopressors, as they tend to be less effective and precipitate arrhythmias in the setting of AI.

Correct Hypoglycemia

- Treatment with D50 is given as necessary. It is recommended to add dextrose to the fluids (D5NS) over the early course of the disease.

Glucocorticoid Replacement

- Frequently, the diagnosis and treatment of AI must proceed simultaneously. There is little risk of giving a single dose of steroids when the diagnosis of AI is suspected but not confirmed. In this situation dexamethasone is preferred as it does not interfere with the ACTH stimulation test. Recommended dose: dexamethasone 4-10 mg IV q 6-8 h.
- If the diagnosis of AI is clear (i.e., previously diagnosed or history of abrupt withdrawal of steroids), hydrocortisone is preferred as it has a greater mineralocorticoid (salt-retaining) effect. Recommended dose: hydrocortisone 100 mg IV q 6-8 h.
- When dexamethasone is used Florinef 0.1 mg po (synthetic mineralocorticoid) may need to be added in the setting of primary AI.

Correct Electrolyte Abnormalities

- Electrolyte imbalances usually correct with the administration of fluids and glucocorticoids alone. However, hyperkalemia may be severe, particularly with primary adrenal insufficiency, and may require treatment with calcium, sodium bicarbonate, insulin/glucose, and kayexalate.

Search for and Treat Underlying Illness

- A thorough history and physical exam is needed, along with appropriate laboratory tests and investigations (CXR, EKG, head CT) as necessary to rule out underlying infection, cardiac, pulmonary, or CNS disease. Treatment should be initiated as necessary.

9

Admission Criteria

- Most patients with acute adrenal insufficiency require ICU admission. Some mild cases may be appropriate for regular ward.
- Steroid-dependent patients with chronic adrenal insufficiency may be managed as an outpatient for minor illnesses. This is done by doubling their daily steroid dose for ≈ 3 days in consultation with their primary physician. The use of prefilled dexamethasone 4 mg IM syringes at home and medical alert bracelets in case of emergency should be discussed with all AI patients.

Hypoglycemia

Hypoglycemia is a life threatening endocrine emergency that, once recognized, is easily treatable. Delay in diagnosis or treatment may lead to irreversible neurologic injury or death.

Pathophysiology

- Hypoglycemia is generally defined as a blood glucose level <50 mg/dl in adults/children, and <30 mg/dl in neonates.
- Glucose homeostasis involves a tightly regulated balance between insulin and its counterregulatory hormones (glucagon, epinephrine, cortisol, growth hormone). An excess of insulin or deficiency of counterregulatory hormones will tip the balance toward hypoglycemia.
- Hypoglycemia is most often seen as a complication of diabetes therapy but is also seen in nondiabetics, usually as a complication of other disease processes (Table 9A.5).
- In pediatric patients, hypoglycemia may be seen in the acutely ill or septic child due to lack of oral intake, hypermetabolic state, or secondary to accidental ingestions (alcohol, salicylates, oral hypoglycemics).

Table 9A.5. Etiologies of hypoglycemia

Increased insulin levels	Overdose of insulin or oral hypoglycemics
	Insulinoma
	Post-gastrectomy/gastrojejunostomy (due to rapid gastric emptying and exaggerated insulin response).
Underproduction of glucose	Liver disease
	Alcohol
	Sepsis
	Malnutrition
	Adrenal insufficiency
	Pituitary insufficiency
	Hypothyroidism
	Growth hormone deficiency
	Glucagon deficiency
	Chronic renal failure
	Medications (salicylates, β-blockers)
	Akee fruit
	Fasting*

*Normal adults can maintain blood glucose levels up to 72 h of fasting.

Diagnosis and Evaluation

Signs and Symptoms

- The exact level of glucose at which patients demonstrate the signs and symptoms of hypoglycemia varies among individuals based on age, weight, sex, activity level, and coexisting disease. Most adults will be symptomatic with levels <50 mg/dl, but the rate of fall also contributes to symptoms.
- It is conceptually useful to divide the signs and symptoms of hypoglycemic into two categories: neuroglycopenic and adrenergic.
- Neuroglycopenic symptoms represent the direct CNS affects of hypoglycemia. Signs and symptoms include dizziness, fatigue, inability to concentrate, confusion, psychosis, headache and focal neurologic findings.
- Adrenergic symptoms are produced by the counterregulatory surge (i.e., epinephrine) in response to hypoglycemia. Signs and symptoms include tremor, anxiety, diaphoresis, tachycardia, nausea, and hunger.
- A subset of patients with diabetes has "hypoglycemic unawareness" due to an impaired adrenergic response secondary to autonomic neuropathy.
- The neonate and the young infant may be asymptomatic or demonstrate only subtle, nonspecific signs (lethargy, tachycardia, seizures, or apnea).

Laboratories/Studies

- All patients with altered mental status require immediate finger stick glucose (accucheck, D-stick).
- Further laboratory tests (metabolic panel, CBC, cultures, CXR, EKG, or head CT) should be tailored to the patient's history and physical condition when ruling out precipitant or underlying illness.

ED Management

- For symptomatic hypoglycemia, intravenous dextrose should be given as 50 ml of D50W (1 amp) and may be repeated as necessary.

- Oral glucose can be given if the patient is awake and there is no risk of aspiration (milk, juice, fruit).
- Some patients may need continuous infusion of glucose for persistent hypoglycemia, particularly in oral hypoglycemic overdoses.
- Consider glucagon 1 mg IM/SC/intranasal if IV access is delayed. Glucagon has a more important role in the prehospital setting. The effect will be seen in 10-20 min after stimulation of hepatic glucose release and will not work in the setting of hepatic insufficiency. Patients at risk for hypoglycemia should have glucagon available at home and a family member instructed in proper use.
- When chronic alcohol abuse is suspected, thiamine should be given concurrently with glucose, as there is a theoretical risk of precipitating Wernicke's encephalopathy.
- Pediatric hypoglycemia is treated with lower concentrations of dextrose to avoid hyperosmolarity. Young children and infants: Give 25% dextrose, 2-4 ml/kg. Neonates: Give 10% dextrose 1-2 ml/kg.

Admission Criteria

- Patients need a monitored setting in the presence of unresolved neurological injury, severe hypoglycemia, recurrent hypoglycemia in the emergency department despite treatment, long-acting oral hypoglycemic overdoses, hypoglycemia at the extremes of age or with severe underlying illness.
- Patients may be discharged home if all of the following conditions are met: mild hypoglycemia only, complete resolution of symptoms, close primary physician follow-up, and the ability to administer insulin or oral hypoglycemic correctly.

Part B: Electrolyte Emergencies

Electrolyte abnormalities usually occur in conjunction with other, often life-threatening emergencies. Because severe derangements of sodium, calcium and potassium are common and can be life-threatening on their own, they are discussed individually below. After the recognition and initial treatment of severe electrolyte abnormalities, patients are generally admitted, although their location within the hospital is largely dependent on the underlying diagnoses.

Hyponatremia

Hyponatremia is defined as a serum sodium <136.

Pathophysiology

- Hyponatremia exerts its most prominent effects on the central nervous system. Low serum sodium causes water to move into brain cells through an osmotic gradient. The severity of the clinical syndromes depends on the magnitude as well as the rate of decline of the serum sodium.
- When hyponatremia occurs rapidly, there is less time for the brain to adapt, causing more marked cerebral edema and more marked symptoms.

Diagnosis and Evaluation

- Patients are generally asymptomatic at levels >120.
- Symptoms of hyponatremia include nausea, vomiting, lethargy and confusion, with extreme manifestations such as seizure and coma, at levels <115.
- Factitious hyponatremia occurs when the plasma osmolality is either normal or increased. In these cases, sodium is falsely decreased secondary to hyperosmolarity syndromes (such as extreme hyperglycemia, mannitol or glycerol therapy), severe hyperlipidemia or hyperproteinemia (e.g., multiple myeloma). In true hyponatremia, the plasma osmolality is decreased.

- True hyponatremia is classified according to the overall volume status of the patient; hypovolemic, euvolemic or hypervolemic. The treatment of hyponatremia will depend upon the clinical assessment of the patient's volume status.
- Consider causes of hyponatremia that may require other critical actions:

Causes	**Actions**
Hypothermia	Warming measures
Hypothyroidism	Administration of thyroid hormone
Adrenal crisis	Intravenous corticosteroids
Excessive fluid therapy	Turn off IV fluids
Excessive water intake	Fluid restriction
Head trauma or other insult	Consider CT Head

Laboratory/Studies

- Obtain serum electrolytes including STAT glucose; electrolytes should be followed during treatment every 1-2 h.
- Head CT is indicated for any alteration of mental status.
- In severely hyponatremic and critically ill patients, a Foley catheter is necessary to closely follow urine output and for urinary electrolytes.

ED Management

- A serum sodium >120 rarely requires emergent treatment. The urgency of treatment increases with lower levels and rapid decline in mental status.
- Decreased level of consciousness (LOC) may cause airway obstruction and may require airway adjuncts and/or intubation. If using RSI, caution must be used with succinylcholine (contraindicated if concomitant hyperkalemia present) and with barbiturates (if porphyria is cause of hyponatremia).

- If patient is hypovolemic or hypotensive, use normal saline (0.9%) boluses as the initial fluid of resuscitation. If patient is hypervolemic (e.g., signs of fluid overload are present), fluids may be restricted and loop diuretics administered intravenously (e.g., furosemide)
- Follow neurological status for decompensation, as change dictates more aggressive sodium repletion.
- Use normal saline (0.9%) at approx 50 ml/h (in adults),once euvolemic, for altered mental status. Keep 3% saline solution at bedside available for worsening neurological status.
- If seizures are present or there is rapid deterioration in LOC, give boluses of 50 ml of 3% normal saline until symptoms improve; watch for signs of fluid overload.
- If 3% saline is not available, a single ampule of sodium bicarbonate may be substituted for hypertonic saline solution (one ampule of 50 ml of 7.5% sodium bicarbonate is roughly equivalent to 100 ml of 3% saline)
- Important note: Focal weakness, hemiparesis, ataxia and abnormal plantar reflexes may be seen with hyponatremia itself **or** with central pontine myelinolysis from overaggressive correction. To avoid this complication, correction of hyponatremia should ideally not occur faster than 0.5 meq/L/h. More rapid correction appears to be safer when hyponatremia has developed more acutely and when diuretics are used during correction.

Hypernatremia

Hypernatremia is defined as a serum sodium >145.

Pathophysiology

- The vast majority of hypernatremia cases seen in the ED are a result of severe volume loss.

- As in hyponatremia, the predominant effects of hypernatremia are on the central nervous system. High plasma osmolality results in a shift of water out of cells, causing decreased brain cell volume.
- The overall decrease in brain volume may lead to intracranial hemorrhage, causing further worsening neurological findings.

Diagnosis and Evaluation

- Hypernatremia is manifested clinically with progressive neurological symptoms corresponding to increasing serum osmolality. Also contributing to the serum osmolality are glucose, urea and alcohols. Initial irritability is followed by tremulousness and ataxia, with extreme manifestations (seizure, coma) as osmolality increases.
- Consider causes of hypernatremia that may require other critical actions:

Causes	Actions
Severe dehydration	Volume resuscitation with NS
Central lesion causing DI	Head CT
Renal etiology of high sodium	Consider emergent dialysis
Excessive sodium intake	Discontinue diet, medications containing sodium
Lithium toxicity	Saline diuresis, consider dialysis

Laboratory/Studies

- Obtain serum electrolytes including STAT glucose, electrolytes should be followed during treatment every 1-2 h.
- Head CT is indicated for any alteration of mental status.
- A lithium level is indicated if any suspicion of its use exists.
- A Foley catheter is necessary to closely follow urine output.

9

ED Management

- Decreased level of consciousness (LOC) may cause airway obstruction and may require airway adjuncts and/or intubation.
- Hypovolemia and hypotension should be treated initially with normal saline (0.9%) boluses.
- Follow neurological status for deterioration—this dictates more aggressive normalization of sodium.
- Use 0.45% normal saline at approximately 50 ml/h once euvolemic for altered mental status. The rate of infusion may need to be increased or switched to free water if seizures or worsening mental status occur.
- Be mindful that overaggressive normalization of sodium, especially if it is chronic, may cause cerebral edema and death. Ideally one should not correct serum sodium faster than 0.5 meq/L/h. There is also a risk of causing pulmonary edema with rapid hypotonic fluid administration. The risk of cerebral edema is less of a concern in cases when hypernatremia develops acutely, before the brain has the opportunity to create idiogenic osmoles.
- For hypervolemic patients, one may consider the addition of loop diuretics (e.g., furosemide)
- Consider vasopressin or desmopressin if central diabetes insipidus (DI) is the suspected etiology
- Dialysis may be required if kidneys are unable to excrete sodium.

Hypocalcemia

Hypocalcemia is defined as a serum calcium <8.5 or ionized calcium <2.0.

Pathophysiology

- In the nervous system, low serum calcium levels cause increased membrane permeability to sodium and hence, neuronal excitability. This effect is counteracted by potassium and magnesium.
- In the heart, low ionized calcium reduces the strength of myocardial contraction.

Diagnosis and Evaluation

- As with other electrolytes, the severity of clinical manifestations are magnified with rapidity on onset.
- Clinical manifestations are both neurological and cardiovascular. Neurological features range from paresthesia and weakness to tetany, altered mental status and seizures. Cardiac features may include congestive heart failure, hypotension and dysrhythmias.
- Important physical examination findings suggestive of hypocalcemia include a anterior neck scar from thyroid surgery, hyperactive deep tendon reflexes and a positive Trousseau sign (carpal spasm resulting from inflation of a blood pressure cuff.
- Consider causes of hypocalcemia that may require other critical actions:

Causes	**Actions**
Magnesium depletion	STAT IV magnesium replacement
Renal dysfunction	Consider dialysis
Acute pancreatitis/sepsis	Aggressive treatment of underlying process
Rhabdomyolysis	Saline diuresis, sodium bicarbonate
Fat embolism	Mechanical ventilation, consider heparin
Exogenous drug administration	Discontinue agent
- phosphate containing agents	
- loop diuretics	
- corticosteroids	
- theophylline	
- heparin	
- sodium nitroprusside	
- cimetidine	
- phenytoin	

- Look for signs of focality on neurological examination and consider other causes of neurological deficits

Laboratory/Studies

- Measure STAT electrolytes, including Mg, PO_4 and albumin.
- Although hypoalbuminemia will decrease the measured total calcium, the clinically relevant ionized calcium will not be affected.
- A STAT EKG is indicated to rule out prolonged QT interval (normal width T wave with prolonged ST segment) as this predisposes the patient to lethal dysrhythmia
- Consider head CT to rule out intracranial pathology
- Electrolytes should be measured every 1-2 h during correction.

ED Management

- Beware of over-aggressive ventilation in the mechanically ventilated patient; respiratory alkalosis causes a decrease in the ionized fraction of calcium, with no change in the serum level (each 0.1 rise in pH lowers ionized Ca by about 5%)
- For symptomatic patients or those with a prolonged QT interval, use 10% calcium gluconate IV (20 ml over 10 min) followed by infusion of 60 ml in 500 ml of 5% dextrose in water (D5W) at 1 mg/kg/h.
- In cardiac arrest, use 10% calcium chloride (1 g) IV push. This dose may be repeated.

Hypercalcemia

Hypercalcemia is defined as a level >10.5.

Pathophysiology

- Hypercalcemia slows cardiac conduction and decreases automaticity.
- Due to the creation of a tubular defect, hypercalcemia impairs the concentrating ability in the kidney. This leads to profound dehydration.
- Prolonged increases in calcium, especially in conjunction with increased phosphorus, leads to deposition of the solid phase in tissues throughout the body. Deposition in the conduction system of the heart and the renal parenchyma may cause permanent organ dysfunction.

Diagnosis and Evaluation

- Symptoms are varied and span multiple systems. They include weakness, lethargy, polyuria, renal calculi, bony and abdominal pain and other gastrointestinal symptoms. As calcium continues to rise, life-threatening manifestations are cardiac (dysrhythmias) and neurological (seizures, coma). Cardiac arrest may occur at levels above 20.

Laboratory/Studies

- STAT EKG should be performed, looking for signs of cardiac toxicity such as shortened QT interval, flattened T waves, conduction delays and blocks.
- EKG changes are of special concern in the presence of digitalis as digitalis effects are amplified.
- Measure STAT electrolytes, initially including Mg, PO_4 and albumin and follow electrolytes every 1-2 h during correction.
- Consider head CT to rule out intracranial pathology

9

ED Management

- Decreased level of consciousness (LOC) may cause airway obstruction and may require airway adjuncts and/or intubation.
- Saline diuresis is the cornerstone of management in the ED, as most of these patients are significantly dehydrated. Fluid resuscitation with normal saline should commence prior to the initiation of loop diuretics.
- The clinician should not be misled by the presence of hypertension. Despite dehydration, hypercalcemic patients may be hypertensive secondary to arteriolar vasoconstriction.
- Thiazide diuretics are to be avoided as they raise calcium
- Dialysis may be necessary for renal patients
- Bisphosphanates, mithramycin, calcitonin and steroids are not usually necessary in the initial resuscitation, but may be considered after specialist consultation
- Hyperreflexia, fasciculations and alteration of mental status progressing to coma may be present. Look for signs of focality on neurological examination and consider other causes of neurological deficits.

Hypokalemia

Defined as a potassium level <3.5

Pathophysiology

- Hypokalemia arises through either a shift of potassium into cells or by increased losses, usually GI or renal. Its adverse effects result from abnormalities in membrane polarization. This occurs in virtually every organ system in the body.

Diagnosis and Evaluation

- Hypokalemia is generally symptomatic below 2.5 with weakness and fatigue. As levels decrease, the most serious manifestations are neurological and cardiac. There is progressive paralysis, hypoventilation and ultimately, cardiac dysrrhythmias.
- The neurologic findings in hypokalemia are nonspecific; consideration of the differential diagnosis must be made when neuromuscular findings are present. Weakness may be caused by various infectious, vascular, traumatic and toxicologic etiologies that require prompt identification.
- Consider coexistent conditions that may require other critical actions in the setting of hypokalemia:

Conditions	Actions
Hypomagnesemia	IV magnesium sulphate (2 g slow IV push)
Digitalis toxicity	Administer digoxin specific Fab fragments
Hyperglycemia/DKA	Hold insulin administration
Metabolic acidosis	Hold sodium bicarbonate/alkaline infusions
Drug related:	
- amphotericin B	Hold these infusions
- synthetic penicillins	
- aminoglycosides	
- theophylline	
- loop diuretics	

Laboratory/Studies

- A STAT 12-lead EKG should be performed looking for PR prolongation, flattened T waves, U waves, ST depression, decreased voltage and ventricular ectopy/ tachydysrrhymias
- STAT electrolytes should be performed, including Mg, PO_4, Ca and albumin and repeated every 1-2 h during potassium repletion
- Head CT and/or neurological consultation may be indicated if there is uncertainty regarding the etology of neurological signs.
- Measure digitalis level if its use is known or suspected.

ED Management

- The patient must be placed on telemetry to monitor for dysrrhythmias
- Potassium replacement is safest via oral route, where the GI tract serves as a homeostatic regulator of absorption.
- Intravenous potassium replacement, in those patients with life-threatening levels and signs, should not normally occur faster than 20 mmol/h with a maximum concentration of 40 mmol/L peripherally and 60 mmol/h via central vein—faster rates can be considered only with patient extremis.
- Mix KCl solutions in saline, not dextrose solutions (increasing glucose may exacerbate hypokalemia).

Hyperkalemia

Hyperkalemia is defined as a potassium level >5.0. Even when severe, it may present simply as a laboratory abnormality in an asymptomatic patient.

Pathophysiology

- Hyperkalemia arises through either a shift of potassium out of cells or failure of its excretion (renal). Like hypokalemia, it exerts its adverse effects by disturbing membrane polarization.

- The most important toxicity of hyperkalemia is cardiac. There may be a progressive series of warning signs on the 12-lead EKG and cardiac monitor, which if left untreated, lead to ventricular dysrhythmia and asystole.

Diagnosis and Evaluation

- Consider coexistent conditions that may require other critical actions in the setting of hyperkalemia:

Conditions	Actions
Renal failure	Evaluate and treat the cause If prerenal, saline fluid resuscitation If post-renal, relieve obstruction (drain urine) May require emergent dialysis
Adrenal insufficiency	Administer mineralocorticoids IV
Digitalis toxicity	Administer digoxin specific Fab fragments
Metabolic acidosis	Administer sodium bicarbonate
Drug related: - β-blockers - ACE Inhibitors - heparin	Consider holding dose/infusion

Laboratory/Studies

- It is imperative to repeat elevated serum potassium levels as they are frequently falsely elevated due to hemolysis in the specimen. The therapy for elevated levels may itself be dangerous if potassium is not, in fact, elevated—this underscores the importance of the STAT EKG.
- A STAT 12-lead EKG should be performed looking for the progressive signs of hyperkalemia. These begin with tall, tented T waves and QRS widening. As these changes become more exaggerated and P waves become flattened and disappear, the QRS complexes become difficult to distinguish from T waves. This creates the omninous "sine wave" pattern.
- The EKG abnormalities of hyperkalemia are a better indication of the degree of cardiac toxicity present than the serum level—hence the EKG findings should be the principal guide for the aggressiveness of therapy.
- It must be recognized that hyperkalemia can mimic the changes of acute MI on the EKG—its aggressive treatment and careful observation of the EKG will help resolve this dilemma.
- STAT electrolytes are essential, including Mg, PO_4, Ca and albumin. These will need to be repeated frequently (e.g., every hour) when severe derangements of potassium are involved.
- When an acid-base disturbance is also present, serial measurements of serum pH is also critical during management because of the effect of pH on serum potassium shifts.
- Measure digitalis level if its use is known or suspected.

ED Management

- When using RSI, succinylcholine must not be included in the drug regimen if hyperkalemia is known or even a reasonable possibility. This includes rhabdomyolysis, burns or crush injuries after 48-72 h, a history of renal failure, acute urinary retention, tumor lysis syndrome and other causes. Succinylcholine elevates potassium levels and may be fatal in this setting.
- Patient must be placed on telemetry to monitor for dysrhythmias.
- Administration of calcium chloride or gluconate (10 mL of a 10% solution given by slow IV push) is the most time-critical of interventions. It acts within minutes to

reverse EKG manifestations and risk of ventricular dysrhythmia. It may be repeated, directing therapy toward the EKG and cardiac monitor. Its effect is only temporary and steps to reduce potassium concentration must occur simultaneously.

- Potassium may be shifted into cells temporarily using a combination of insulin, glucose and bicarbonate. A typical regimen may include 10 units of insulin IV, 50 g of dextrose (one ampoule of D50) and one ampoule of sodium bicarbonate (use with caution in patients with potential volume overload). Inhaled β agonists, such as albuterol nebulizers, are also effective. These interventions take effect in a 15-45 min time frame.
- The next step in therapy is to promote excretion of potassium. This is achieved with the use of diuretics (e.g., furosemide 20-80 mg IV push in patients who produce urine) and oral/nasogastric cation exchange resins (e.g., sodium polystyrene sulfonate 25-50 g PO). These are effective in a hour or more after administration.
- Dialysis is the definitive method of removing excess potassium and may be necessary if the above techniques are ineffective.

Suggested Reading

1. Adams SL. Alcoholic ketoacidosis. Emerg Med Clin North Am 1990; 8(4):749-760
2. Adrogue HJ, Madias NE. "Hyponatremia". N Engl J Med 2000; 342(21):1581-9.
3. Burch HB, Wartofsky L. Life-threatening thyrotoxicosis. Thyroid storm. Endocrinol Metab Clin North Am 1993; 22:263-277.
4. Comi RJ. Approach to acute hypoglycemia. Endocrinol Metab Clin North Am 1993; 22(2):247-262.
5. Duffens K, Marx JA. Alcoholic ketoacidosis-A review. J Emerg Med 1987; 5(5):339-406.
6. Fleckman AM. Diabetic ketoacidosis. Endocrinol Metas Clin Noerth Am 1993; 22(2):181-207.
7. Gennari FJ. "Hypokalemia". N Engl J Med 1998; 339(7):451-8.
8. Gross P, Reimann D et al. "Treatment of se vere hyponatremia: Conventional and novel aspects". J Am Soc Nephrol 2001; 12(Suppl 17):S10-4.
9. Halperin ML, Goldstein MB. Fluid, Electrolyte and Acid-Base Physiology: A Problem-Based Approach. 3rd ed. Toronto: WB Saunders, 1999.
10. Jordan RM. Myxedema coma. Med Clin North Am 1995; 79:185-194.
11. Kitabchi AE, Wall BM. Diabetic ketoacidosis. Med Clin North Am 1995; 79(1):9-33.
12. Lee CT, Guo HR et al. "Hyponatremia in the emergency department". Am J Emerg Med 2000; 18(3):264-8.
13. Lorber D. Nonketotic hypertonicity in DM. Med Clin North Am 1995; 79(1):39-52.
14. Nicoloff JT, Lopresti JS. Myxedema coma. A form of decompensated hypothyroidism. Endocrinol Metab Clin North Am 1993; 22:279-290.
15. Oelhers W. Adrenal insufficiency. N Eng J Med 1996; 355:1206-1212.
16. Pope DW, Dansky D. Hyperosmolar hyperglycemic nonketotic coma. Emerg Med Clin North Am 1989; 7(4):849-857.
17. Rastegar A, Soleimani M et al. "Hypokalaemia and hyperkalaemia". Postgrad Med J 2001; 77(914):759-64.
18. Service SJ. Hypoglycemia. Med Clin North Am 1995; 79(1):1-6.
19. Service SJ. Hypoglycemia disorders. N Engl J Med 1995; 27:1144-1150.
20. Tietgens ST, Leinung MC. Thyroid storm. Med Clin North Am 1995; 79:169-184.
21. Umpierrez GE et al. Review: Diabetic ketoacidosos and hyperosmolar nonketotic syndrome. Am J Med Sci 1996; 311(5):225-33.
22. Wrenn KD, Slovis CM et al. The syndrome of alcoholic ketoacidosis. Am J Med 1991; 91(2):119-128.
23. Ziegler R. "Hypercalcemic crisis". J Am Soc Nephrol 2001; 12(Suppl 17):S3-9.

CHAPTER 10

Hematologic Emergencies

Alicia Haglund

Part A: Transfusion in the Emergency Department

Transfusion Components

Blood

- The indication for blood transfusion is to improve oxygen delivery to tissues, primarily the brain and heart.
- There is no absolute level of hemoglobin (Hg) that mandates transfusion although this is commonly done when Hg is <7 g/dL. Transfusion may be required at higher Hg levels depending upon the clinical setting (i.e., acute myocardial infarction).
- Several types of components containing red blood cells are available.
  - Whole blood
    - In addition to red blood cells (RBCs), whole blood contains platelets, white blood cells (WBCs) and plasma. However, platelet and WBC activity as well as various factor activity falls significantly over a period of hours to days.
    - Massive hemorrhage is the only potential ED indication for whole blood. However, whole blood is never truly necessary. RBCs, platelets and plasma components are all available separately and intravenous crystalloid is used with packed RBCs for volume replacement in the setting of major hemorrhage.
    - Disadvantages of whole blood include the higher number of antigenic components and the potential for volume overload.
  - Packed RBCs (PRBCs)
    - PRBCs are most commonly used to increase the Hg. Each unit contains approximately 300 ml of total volume, of which about two-thirds is RBCs.
    - One unit will raise the Hg by 1 g/dL.
    - RBCs can be frozen and stored for up to 3 yr. They are good for 24 h after thawing. Because frozen blood also lacks essentially all WBCs, platelets and plasma components, it is also useful in patients with a history of allergic or febrile transfusion reaction.
    - Patients scheduled for elective surgery may donate their own blood ahead of time for later transfusion. These units can be stored for up to 35 days.
  - Leukocyte-Poor PRBCs
    - This blood product has had plasma, WBCs and platelets removed via a process of filtration.
    - Leukocyte-poor PRBCs are indicated in patients with a history of allergic or febrile transfusion reaction. They also decrease the likelihood of alloimmunization to the various HLA antigens and, therefore, are indicated in patients who receive multiple transfusions.
    - Leukocyte-poor PRBCs carry a reduced risk of CMV transmission and should be used in immunosuppressed patients.

Platelets

- Platelet transfusion is indicated in patients with acute hemorrhage when thrombocytopenia is a suspected cause. It is also indicated as a prophylactic measure in severely thrombocytopenic patients and those who will undergo invasive procedures. The threshold for prophylactic platelet transfusion is generally considered to be a count <10-20,000/mm³. In patients with active hemorrhage or planned invasive procedures, 50,000/mm³ is the usual cutoff.
- Prophylactic platelets are not routinely indicated in patients receiving massive PRBC transfusion.
- Typically 6-8 units are transfused (250-300 ml total volume). Each unit is expected to raise the platelet count by 5-10,000/mm³.
- Those who have received multiple transfusions often develop alloimmunization. These patients usually have little or no increase in platelet counts after transfusion. Consider use of single-donor HLA-matched platelets or leukocyte-poor platelets in these patients.
- Platelet transfusion is not indicated in cases of immune or thrombotic thrombocytopenic purpura (TTP) except in rare clinical circumstances.

Fresh Frozen Plasma (FFP)

- FFP contains plasma, coagulation factors, proteins C and S and antithrombin.
- FFP is indicated for consumptive coagulopathy as seen with disseminated intravascular coagulation (DIC) or TTP. It is also used in hemophilia when specific factor replacement therapy is not available. Patients with coagulopathy secondary to liver disease or warfarin toxicity are treated with FFP if significant hemorrhage is present.
- Rarely, FFP is indicated in the setting of massive PRBC transfusion although patients generally require transfusion of more than one blood volume before coagulopathy develops. Even then, bleeding is unlikely with prothrombin time <1.5 times control.
- Dose ranges from 5-25 ml/kg.

Cryoprecipitate

- Cryoprecipitate contains factor VIII, fibrinogen, von Willebrand factor (vWF) and factor XIII.
- It is used in patients with congenital hypofibrinogenemia and hypofibrinogenemia secondary to consumptive processes such as DIC. It can be used in the emergent setting for patients with hemophilia A when specific factor concentrates are not available. Cryoprecipitate is also used for certain patients with von Willebrand disease when DDAVP is not effective and factor VIII concentrates containing vWF are not available.
- The usual dose is 1-2 bags for every 10 kg of bodyweight.

Transfusion Reactions

- Transfusion of any blood product carries some degree of risk whether from allergic reaction or infection.
- It is important to inform the patient of potential risks as well as benefits prior to transfusion.
- Always screen patients for suspected risks and minimize them when possible. Some of the most severe reactions occur with bedside errors such as mislabeling a specimen and transfusing a patient with the wrong blood.

Immune-Mediated

Acute Hemolytic Reaction

- Occurs when the patient has preformed antibodies that result in the intravascular lysis of donor RBCs.
- The most severe reactions involve the ABO system.

- Clinical presentation
  - Symptoms include chest pain, low back pain, dyspnea and pain at the IV site.
  - Examination is significant for hypotension, tachypnea, tachycardia and temperature elevation. Patients may have evidence of microvascular bleeding. The severity of the reaction is dependent upon the amount of transfused blood.
  - Laboratory studies show findings consistent with intravascular hemolysis. These include hemoglobinemia, hemoglobinuria, hyperbilirubinemia, elevated lactic dehydrogenase and decreased haptoglobin.

Treatment

- If a reaction is suspected, the transfusion must be stopped immediately.
- Check the identification of the patient and the donor blood since mislabeling is a common cause of acute hemolytic reactions.
- Various studies are indicated to assess the type and severity of reaction.
  - A blood specimen from the patient plus all of the unused donor blood must be sent to the blood bank for analysis.
  - Other studies are obtained looking for evidence of hemolysis (see above).
  - The Hg will remain the same or decrease after a transfusion reaction.
  - Coagulation studies and platelet count are necessary since these reactions sometimes lead to DIC.
- Hemodynamic support is the mainstay of therapy. Hypotension should be treated aggressively with intravenous fluids. Patients with continued hypotension despite fluid resuscitation are treated with dopamine.
- Renal failure is a serious complication of these reactions. The patient's renal function should be followed closely. It is important to maintain urine output using intravenous fluids ± loop diuretics.
- Depending upon the severity of the reaction, patients may have varying degrees of microvascular bleeding. This should be managed in a standard fashion.
- Note that hemolytic transfusion reaction secondary to mismatched blood (i.e., wrong ABO group) does not preclude subsequent transfusion with the appropriate specimen.

Delayed Hemolytic Reaction

- Occurs in patients previously sensitized to various RBC antigens (usually Rh, Kell, Kidd, or Duffy) who have a negative antibody screen secondary to low levels.
- Patients experience an anamnestic antibody response resulting in extravascular hemolysis occurring approximately 7 days after transfusion.
- Presentation is similar to an acute hemolytic reaction but is usually less severe. Subsequent renal failure and DIC are unusual. Patients are sometimes asymptomatic with the diagnosis based on the blood bank identification of new antibodies on a subsequent cross match.
- Treatment is supportive and follows the guidelines used for acute hemolytic reactions.

Febrile Nonhemolytic Reaction

- Febrile nonhemolytic reactions are characterized by fever and chills developing during or within 12 h of transfusion.
- They result from antibodies directed against donor leukocytes and platelets. They occur most frequently in multiparous women and those patients with a history of multiple transfusions.
- Clinical findings are usually mild and there is no hemolysis. However, it is sometimes difficult to differentiate these reactions from acute hemolytic reactions. When the type of tranfusion reaction remains in question or if the reaction is severe, transfusion should be held pending further investigation.
- The incidence and severity of these reactions are reduced by premedication with antipyretics or transfusion of leukocyte-poor RBCs.

Anaphylaxis

- This acute, severe allergic reaction occurs immediately after contact with the offending antigen in donor blood. Only a minute amount of donor specimen is necessary.
- This type of reaction is very rare occurring mainly in patients who are IgA-deficient.
- Findings include cardiovascular collapse, bronchospasm, laryngeal edema, vomiting and urticaria.

Treatment

- The transfusion must be stopped immediately.
- Specific therapy depends on the severity of the reaction.
  - Intravenous antihistamines, epinephrine and steroids are indicated for all severe cases.
  - Glucagon is administered to patients resistant to epinephrine secondary to β-blockade.
  - Bronchodilators are indicated with mild bronchospasm. However, more definitive airway management may be needed. Always be prepared for a surgical airway.
  - Hypotension is treated with intravenous fluids ± vasopressors.
- Milder allergic reactions can occur in response to donor plasma proteins. These reactions are treated with antihistamines. Transfusion can often be continued.

Graft-Versus-Host Disease (GVHD)

- GVHD results when donor lymphocytes recognize the host as foreign and attack the host's lymphatic tissues. The reaction occurs in patients who are already immunosuppressed.
- Clinical manifestations usually occur days after the transfusion. They include fever, rash, diarrhea and liver function abnormalities. Sepsis and death are the ultimate sequelae.
- The reaction is usually resistant to all therapy.
- The incidence of GVHD is minimized by use of irradiated, leukocyte-poor blood products.

Transfusion-Related Acute Lung Injury

- This rare reaction is thought to be secondary to transfusion of donor plasma with high titers of antibodies that bind to antigens on recipient WBCs.
- Patients develop noncardiogenic pulmonary edema and respiratory distress within hours of transfusion.
- Treatment is supportive.

Non-Immune Mediated

Infection

- All blood products can transmit viral diseases. The risk is small but persists despite the current use of routine screening that includes hepatitis B, hepatitis C, HIV, cytomegalovirus (CMV), and syphilis.
- The chance of transmitting hepatitis B is 1:63,000 units and hepatitis C 1:103,000 units. For HIV, it is approximately 1:450,000 to 1:660,000 units.
- CMV is commonly transmitted via transfusion. This is rarely a problem except in the patient who is immunosuppressed.

Hypothermia

- Blood components are either refrigerated or frozen. Rapid infusion, especially with large volumes, can result in hypothermia and its numerous complications.
- Use of an in-line warmer can prevent tranfusion-related hypothermia.

Electrolyte Abnormalities

- Most blood products contain citrate as an added anticoagulant. Citrate binds calcium and can lead to hypocalcemia. This tends to occur in patients receiving massive tranfusion over a short period of time. Sequelae are related to cardiac arrhythmias. Treatment is with intravenous calcium.

- Hyperkalemia is possible, especially in neonates and patients with renal failure. This is caused by leakage of potassium from RBCs during storage. Standard recommendations for hyperkalemia should be followed.

Bacterial Contamination

- Rarely, blood has bacterial contaminants. The likelihood of contamination is related to the duration of storage.
- Transfusion with contaminated blood can lead to septicemia, shock and death. Onset of symptoms is usually sudden and rapidly progressive unlike the more common febrile nonhemolytic reactions.
- If bacterial contamination is suspected, the transfusion should be stopped and the donor unit cultured. The recipient is started on empiric broad-spectrum antibiotics with supportive care as indicated.
- There is also a risk of bacterial contamination of platelets; this risk is increased with pooled platelet concentrates.

Part B: Thrombocytopenia

- Defined as <100,000 (100K) platelets.
- Platelets play a role in hemostasis via adhesion, aggregation and secretion of procoagulant mediators. With fewer than 50K platelets, a moderate bleeding risk exists. With fewer than 20K, there is significant potential for severe hemorrhage and spontaneous bleeding.
- Platelets are produced in the bone marrow. Two-thirds are distributed in the serum and one-third stored in the spleen. Normal life span is 7-10 days with a basal, daily demand of 7-10K. Platelets are destroyed in the spleen via phagocytosis.
- Regardless of cause, clinical findings are related to mucosal and cutaneous hemorrhage and include petechiae, purpura, epistaxis, gingival bleeding, hemoptysis, hematemesis and hematuria. Bleeding does not usually occur unless platelet counts are severely depressed.
- Thrombocytopenia results from three basic mechanisms—splenic sequestration, decreased bone marrow production and accelerated destruction.
- Work-up should focus on determining the etiology since treatment will vary depending upon the underlying pathology.

Sequestration

- Sequestration of platelets in the spleen and sometimes the liver can cause pooling of a certain fraction of circulating platelets although counts rarely fall below 20K.
- Sequestration is commonly caused by portal hypertension secondary to hepatic disease. Other causes include malignancy, congestive heart failure, sickle cell disease, sarcoidosis and systemic lupus erythematosus (SLE).

Decreased Production

- Impaired production occurs with both congenital and acquired disorders. Most of the congenital disorders are rare and seldom diagnosed in the Emergency Department (ED).
- Acquired disorders are more common and are usually the result of bone marrow insults from toxins, infection or malignancy. Specific etiologies include:
    - Alcohol—this toxin is a direct bone marrow suppressive agent and can lead to significant thrombocytopenia with platelet counts <20K. The effect is reversible with abstinence.
    - Many medications including thiazide diuretics and many chemotherapeutic agents.
    - Various infectious agents including numerous viruses and tuberculosis.

- Deficiencies of vitamin B_{12} and folate.
- Ionizing radiation.
- Neoplastic infiltration such as leukemia and lymphoma.
- Marrow fibrosis and myelodysplastic syndromes.

Accelerated Destruction

Immune-Mediated

- Immune-mediated thrombocytopenia results when platelets coated with antibody, immune complexes or complement are cleared by mononuclear phagocytes in the spleen or other tissues.
- Various drugs and the autoimmune disorder idiopathic thrombocytopenic purpura (ITP) are commonly implicated. Other associated processes include transfusion reactions, SLE, Grave's disease and the human immunodeficiency virus (HIV).
- Both therapeutic and nontherapeutic drugs cause immune-mediated destruction including quinine, quinidine, digoxin, sulfonamides, rifampin, indomethacin, valproic acid, H_2 antagonists, heparin, aspirin, furosemide, procainamide, heroin and cocaine.

ITP

- ITP is defined as isolated thrombocytopenia with normal bone marrow and no other identified etiology. ITP is a diagnosis of exclusion.
- Leukocyte count is normal. Anemia, if it occurs, is secondary to blood loss rather than immune-mediated destruction. Coagulation studies are normal. Bleeding time is not useful.
- There are two distinct clinical syndromes—acute and chronic.

Acute ITP

- Seen predominantly in children, it is characterized by rapid onset of severe thrombocytopenia following a viral illness or immunization. In most children, the diagnosis is made clinically. Rarely, a bone marrow examination is needed to rule out other causes such as aplastic anemia or leukemia although the latter will not usually present with isolated thrombocytopenia.
- Physical exam is remarkable for mucosal and/or cutaneous bleeding manifestations but lymphadenopathy and splenomegaly are usually absent. If present, these findings should lead the physician to suspect other diagnoses.
- In the vast majority, only supportive care is necessary—limiting the child's activity, avoidance of aspirin products and reassurance. Life-threatening and intracranial hemorrhage are exceedingly rare with this disorder. Outpatient care is appropriate as long as platelets are >20-30K and the patient has only cutaneous bleeding. In these patients it is imperative that the emergency physician (EP) verify that an adequate social network exists and also consult with the specialty physician who will be providing follow-up care.
- Rarely, patients will mandate specific intervention. Some sources say that this decision should not be based solely on the platelet count even if <20K. If possible, emergent consultation with a pediatric hematologist is indicated prior to initiating therapy. Various treatment options exist.
  - High-dose steroids (methylprednisolone 30 mg/kg/day) are indicated for patients with life-threatening or intracranial bleeding. Lower doses (prednisone 1-4 mg/kg/day) are often advocated for patients with platelet counts <10-20K.
  - Intravenous immunoglobulin (IVIG) in given a single dose of 0.8 g/kg. Note that there is a risk of viral transmission with IVIG.
  - Platelet transfusion is given for patients with life-threatening or intracranial bleeding.
  - A minority of children will progress to chronic ITP. These patients rarely have severe disease and may have delayed remission.

Chronic ITP

- Seen in adults, it affects women three times more often than men. Patients usually present with an insidious history of easy bruising or menometrorrhagia. Diagnosis is usually made clinically. The American Society of Hematology does recommend screening for HIV as well as a bone marrow biopsy in patients >60 yr of age.
- As with children, physical exam is notable for cutaneous and mucosal bleeding while other findings such as lymphadenopathy and splenomegaly are absent.

Treatment

- Adults with platelet counts >50K are managed expectantly as outpatients. Those with counts between 30-50K can also be managed as outpatients as long as they have only mild cutaneous bleeding. In these patients, prednisone is often recommended. Outpatient steroid therapy should be coordinated with the primary hematologist.
- Those with life-threatening bleeding and/or platelet counts <30K mandate specific therapy and inpatient care. Hematology consultation is recommended prior to initiating treatment. Options are the same as for acute ITP.
- Prednisone (1-2 mg/kg/day) will usually cause a decrease in bleeding within 24 h and a rise in the platelet count within 1-3 wk. There is no consensus on the duration of therapy although steroids are tapered in order to prevent recurrence.
- IVIG (1 g/kg/day for 3 days) is effective in raising the platelet count but is generally reserved for bleeding emergencies or prior to surgery.
- Platelet transfusions are rarely used since exogenous platelets are also rapidly destroyed. However, they are indicated for life-threatening hemorrhage.
- Immunosuppressive agents are used in those patients with refractory disease.
- Splenectomy is indicated for patients with chronic disease who have bleeding or when all other treatment options are unsuccessful.

Non-Immunologic Platelet Destruction

- Non-immunologic mechanisms involve peripheral destruction as a result of abnormal vessels, fibrin thrombi and intravascular prostheses.
- These patients are usually acutely ill. Etiologies include sepsis, thrombotic thrombocytopenic purpura (TTP), disseminated intravascular coagulation (DIC), preeclampsia, severe burns, mechanical heart valves and vasculitic disorders. Envenomation by snakes of the Crotalus species (rattlesnakes, copperheads and water moccasins) causes platelet destruction that can be halted with antivenin.

TTP

- TTP is an uncommon but severe syndrome that occurs as a consequence of widespread thrombosis secondary to disruption of the endothelium by a number of mechanisms. TTP is associated with neoplasm, various medications, pregnancy, vascular disease and infection including *E. coli* serotype 0157 and HIV. Pregnancy-associated TTP and the postpartum state accounts for 10-25% of cases. In these cases, onset is usually prior to 24 wk gestation.
- TTP is defined clinically by thrombocytopenia, microangiopathic hemolytic anemia, fever, neurologic involvement and renal dysfunction. The complete pentad is present in a minority of patients and is not necessary to make a diagnosis.
- Patients usually present with evidence of thrombocytopenia such as purpura or easy bruising. Fever is often present and may be quite high although this should make the EP also consider sepsis. Central nervous system findings typically fluctuate and are disproportionate to alterations in blood pressure or renal dysfunction. These findings include headache, altered mental status, focal deficits and seizures. Hematuria is often present, and patients have varying degrees of renal involvement. Jaundice may be noted

depending on the amount of hemolysis. Splenomegaly is often palpated. Gastrointestinal and cardiac signs and symptoms also occur as a result of local thrombosis.
- Complete blood count reveals anemia and thrombocytopenia (often <20K). Peripheral smear shows macroangiopathic changes such as schistocytes. Electrolytes are significant for mild to moderate elevation of blood urea nitrogen and creatinine (usually <3.0 mg/dl). Lactic dehydrogenase, reticulocyte count, and unconjugated bilirubin are elevated as a result of hemolysis. Urinalysis demonstrates hematuria that varies from microscopic to gross. DIC studies including fibrinogen, D-dimer, and coagulation parameters are usually normal. A pregnancy test should be obtained in all reproductive age females. Computed tomography of the brain is indicated to rule out intracranial hemorrhage.

Treatment

- Initial therapy must address the ABCs. Other supportive measures include benzodiazepines for seizure control and hemodialysis if necessary for renal failure.
- Definitive care is via plasma exchange therapy. This has substantially decreased mortality secondary to TTP although it is still as high as 20%. Fresh frozen plasma (FFP) is used in the interim if exchange therapy is not immediately available.
- Steroids are often used in addition to plasma exchange therapy although their effectiveness has not been definitively established.
- Platelet transfusions are not indicated as they are thought to worsen the thrombotic process.

Hemolytic Uremic Syndrome (HUS)

- HUS is a variant presentation of TTP. Typically, HUS has predominantly renal involvement and a better outcome although there is overlap, particularly in adults. It may not be possible to make a distinction between the two processes.
- HUS is most often associated with infectious diarrhea secondary to certain strains of *E. coli Shigella*, *Salmonella*, *Yersinia* and *Campylobacter* species. Characteristic clinical findings include acute renal insufficiency/failure, thrombocytopenia, microangiopathic hemolytic anemia and fever. A minority of cases follow upper respiratory infection.

Treatment

- Supportive measures are usually adequate in stable pediatric patients. Patients with limited symptoms and normal renal function may be candidates for outpatient care although this must be coordinated through the appropriate specialist. Some patients require therapy directed at renal failure, hypertension and their sequelae.
- Patients who are severely affected will require plasma exchange therapy as with TTP.
- Avoid platelet transfusion.

DIC

- DIC is a systemic activation of both the coagulation and fibrinolytic systems. Disruption of the balance between thrombus formation and clot lysis results in both hemorrhagic and thrombotic sequelae.
  - Microvascular thromboses in major organ systems lead to ischemic damage. In addition, low-flow states may result in macrovascular thromboses. Subsequently, thrombocytopenia occurs when platelets are trapped within the clots. Circulating red blood cells become damaged, causing a microangiopathic hemolytic anemia.
  - Systemic activation of plasmin results in hemorrhagic complications associated with DIC. Plasmin lyses fibrinogen and fibrin to form fibrinogen degradation products (FDPs). FDPs prevent fibrin monomers from polymerizing to form a clot and cause platelet dysfunction. Plasmin degrades various clotting factors and activates the complement system which results in red blood cell and platelet lysis.

- Any physiologic insult that results in endothelial damage, circulating antigen-antibody complexes, endotoxemia, tissue damage or platelet or red cell damage can predispose to DIC. It is frequently associated with obstetrical complications, malignancy, massive trauma, burns and sepsis.
- Clinical presentation depends both upon the underlying etiology and the degree of thrombosis versus hemorrhage. In acute DIC, clotting factors are used more rapidly than they can be replaced and hemorrhagic manifestations predominate. In chronic DIC, the body is able to keep up with the rate of clotting factor consumption. Hemorrhagic findings include oozing at venipuncture sites, petechiae, ecchymoses and mucosal bleeding as well as bleeding in other major organ systems. Thrombotic complications occur in any organ system and are often more difficult to manage.
- There are a wide range of laboratory findings. Decreased (<100K) platelets, decreased fibrinogen (<150 mg/dl), increased PT and PTT, increased D-dimer and increased FDPs are noted. The peripheral smear shows evidence of hemolysis. Patients often have other organ-specific abnormalities depending upon the underlying etiology and the degree of involvement.

Treatment

- The most important aspects are to identify and aggressively treat the underlying disease and to maximize cardiovascular and pulmonary support.
- Replacement therapy
  - FFP is indicated for an elevated PT. The dose is 10-15 ml/kg.
  - Cryoprecipitate is usually reserved for severe hypofibrinogenemia (<50 mg/dL) or active bleeding with a fibrinogen level of <100 mg/dL. The usual dose is 8 Units.
  - Platelet transfusion is indicated for patients with marked thrombocytopenia (<20K) or active bleeding. The dose is 1.0 U/10 kg of body weight.
- Use of heparin is controversial although possible indications include purpura fulminans (gangrene of digits/extremities), amniotic fluid embolism, malignancy with thrombotic complications, acute promyelocytic leukemia and macrovascular thromboembolic complications. Dosing regimens vary based on indication.
- Fibrinolytic inhibitors such as aminocaproic acid are sometimes used with severe life-threatening bleeding that has not responded to other measures. They are be used with extreme caution in only select patients as they may create a life-threatening thrombotic condition. Consultation with the admitting critical care physician is advised prior to use of heparin or fibrinolytic inhibitors.

Part C: Sickle Cell Anemia

- Sickle cell disease (SCD) is an inherited, autosomal recessive disorder caused by a substitution of valine for glutamine on the β-hemoglobin chain (S gene). Deoxygenation causes these β-hemoglobin chains to polymerize and take on a distorted sickle shape. These irregular cells result in microvascular occlusion with subsequent tissue ischemia and infarction. In addition, the abnormal erythrocytes are prematurely destroyed resulting in a chronic hemolytic anemia.
- There are many variants of SCD—all have in common the abnormal S gene and clinical evidence of disease.
  - Patients with homozygous SCD (SS) inherit an abnormal S gene from each parent. Disease is clinically severe.
  - In sickle C disease (SC), patients inherit an S gene from one parent and a C gene from another. The C gene carries a different amino acid substitution and also produces an abnormal hemoglobin molecule. Patients experience mild to moderate symptoms. Notably, patients tend to have ocular and obstetric complications.

- In sickle β-thalassemia, patients an S gene from one parent and a β-thalassemia gene from the other. Patients who carry a β-thalassemia gene produce decreased amounts of normal β-hemoglobin. Disease severity for both β-thalassemia and sickle β-thalassemia varies with the amount of β-hemoglobin that is produced. Patients with sickle β-thalassemia who produce some normal β-hemoglobin usually have the mildest form of SCD.
- Patients with sickle cell trait (one normal β-hemoglobin gene) have symptomatic disease only under conditions of extreme hypoxia. However, these patients often have painless hematuria and isosthenuria.
- SCD occurs in people of African descent. In the United States, approximately 8% of the black population are carriers of the S gene, and 0.15% of black children are homozygous for disease (SS).
- The clinical manifestations of SCD are diverse, and nearly any organ system may be affected.

Vasoocclusive Crisis

- Painful vasoocclusive crises are the most frequent presenting complaint among patients with SCD.
- Acute pain results from tissue ischemia when sickled red blood cells (RBCs) obstruct the microcirculation. Subsequent regional hypoxia and acidosis causes further sickling and tissue injury.
- Recurrent episodes cause irreversible tissue necrosis, often leading to chronic pain and organ dysfunction.
- Precipitating factors include infection, low oxygen tension, dehydration, acidosis, physical exertion, physical or psychological stress, alcohol, pregnancy and extremes of weather. In many patients, the exact cause is unknown. However, it is important to look for and treat reversible causes.

- Clinical presentation will vary depending upon the organ system involved (see Table 10C.1). Low-grade temperature is common in adult patients even in the absence of infection.

Diagnosis and Work-Up

- The history and physical should focus on identification of any precipitating factors, especially infection and dehydration.
- It is important to inquire about the patients regular pain medications including the last dose.
- Determine the nature and location of pain as well as the similarity to previous episodes. Typical pain crises tend to recur in the same pattern in an individual patient.
- Ask the patient about any previous complications and prior transfusions. If available, knowledge of the baseline and most recent hemoglobin (Hg) levels are helpful.

Laboratory Studies

- A complete blood count (CBC) is indicated in most patients with SCD who present to the Emergency Department (ED). The Hg should be compared with baseline levels which generally range from 6-10 g/dL. A major decrease (>2 g/dL) is indicative of a hematologic crisis. Infections may also result in a decreased Hg due to increased hemolysis. Most adults have a mild leukocytosis (12-15,000/mm^3) even in the absence of infection. Platelet count is often elevated. Other findings on peripheral smear include the typical sickled RBCs, nucleated RBCs and target cells.
- Reticulocyte count is normally elevated; a low count may indicate an aplastic crisis.
- A type and screen (or cross) should be obtained in patients who appear ill.
- A pregnancy test is recommended for all women of child-bearing age.

Table 10C.1. Clinical presentation of sickle cell disease (SS)

Organ System	Signs and Symptoms	Comments
Musculoskeletal	• Pain localized in the extremities, back, sternum and joints • Onset may be sudden • Chronic ulcerations and venous stasis changes often noted	• Patients with SCD are at risk for osteomyelitis. If warmth and erythema are present, suspect underlying infection. • *Salmonella* causes the majority of osteomyelitis. • In patients with hip pain, suspect avascular necrosis of the femoral head. • Always inquire about trauma.
Pulmonary (Chest Syndrome)	• Pleuritic chest pain, fever, cough and hypoxia • Possible infiltrates on chest radiograph • Baseline hypoxia and ↑ A-a gradient common	• Possible etiologies include infection, fat embolism and infarction secondary to thromboembolic disease.
Central Nervous System (CNS)	• Focal and/or generalized neurologic deficits including seizure, hemiplegia, visual disturbances and altered mental status	• Ischemic strokes more common in children; most occur before 10 yr of age. • With increasing age, the incidence of cerebral and subarachnoid hemorrhage rises • Most patients have complete recovery after the first event, but recurrences are common.
Genitourinary	• Microscopic and gross hematuria • Isosthenuria • Priapism • Chronic erectile dysfunction	• Papillary necrosis is common and may be asymptomatic.
Gastrointestinal & Hepatobiliary	• Diffuse abdominal pain ± nausea, vomiting and diarrhea • May mimic a surgical abdomen • Chronic hyperbilirubinemia common	• Rule out other causes before making diagnosis of pain crisis. • Gallstones are common due to the chronic hemolysis. A minority develop cholecystitis. • The majority of adults with SCD have a small, nonfunctional spleen due to infarction. Some also have hepatic infarction.

- For patients with abdominal pain, additional studies include urinalysis, blood urea nitrogen and creatinine. Bilirubin is often abnormal but if markedly elevated may be indicative of choledocholithiasis or hepatic infarction.
- Cultures are ordered based upon clinical presentation.

Radiographic Studies

- Chest radiograph (CXR) should be ordered in all patients with pulmonary symptoms as well as those with fever.
- Plain radiographs are indicated in patients with localized bone tenderness. However, plain radiographs are insensitive for both osteomyelitis and bone infarction especially with early disease. Bone scan can help differentiate between osteomyelitis and bone infarction.

- Diagnosis of PE is challenging. Perfusion scans are usually nondiagnostic. Angiography and contrast helical computed tomography (CT) are best avoided since contrast can worsen sickling.
- Obtain a head CT for all patients with neurologic symptoms.
- Patients with right upper quadrant abdominal pain warrant ultrasound examination.

Other Diagnostic Studies

- When joint effusion is present, aspiration is recommended to rule out infection and inflammatory arthropathy.
- Any patient with severe headache and/or persistent neurologic deficit with a normal head CT should undergo lumbar puncture (LP) to rule out subarachnoid hemorrhage.

General Treatment

- The mainstays of ED therapy consist of parenteral analgesia and treatment of any identified precipitant.
- Do not underestimate or dismiss the severity of patient discomfort. Also do not withhold narcotic analgesia because of concerns for addiction or "drug seeking" behavior. Tailor pain management to each individual patient. Generally, morphine is preferred over meperidine whose metabolite can cause seizures when large doses are used (>600 mg/day).
- Hydration is important. All vasoocclusive crises may be precipitated or exacerbated by dehydration. Oral therapy is appropriate unless the patient is severely dehydrated or vomiting. Patients with SCD often develop heart failure as a result of chronic hypoxia and anemia; avoid over-aggressive use of intravenous fluid.
- Supplemental oxygen is commonly advocated but is only beneficial if the patient is hypoxic.
- Transfusion of PRBCs may be needed for patients with severe anemia as occurs with splenic sequestration, increased hemolysis, blood loss or aplastic crisis. The emergency physician should remember that chronic transfusion therapy is associated with several potential complications including alloimmunization, transmission of infection and iron overload.
- Exchange transfusion entails removal of sickled RBCs followed by replacement with normal donor RBCs. The goal is to decrease Hg S to <30%. Possible indications include cerebrovascular accident (CVA), acute chest syndrome and priapism.

Organ-Specific Treatment

- Chest syndrome—broad-spectrum empiric antibiotics are recommended. In addition, simple and exchange transfusion are useful.

CVA

- Acutely, patients require a simple or exchange transfusion in order to reduce the overall burden of Hg S. Emergent hematology and neurology consultations are indicated.
- After the initial event, patients often receive preventative transfusion therapy.

Genitourinary

- Rest and hydration are usually sufficient management for hematuria. Severe or persistent bleeding may require transfusion. Antifibrinolytic amino acids such as epsilon-aminocaproic acid are sometimes used for refractory bleeding. However, these agents may result in persistent clots and should be used with extreme caution.
- Patients with priapism are initially treated with hydration and analgesia. When conservative treatment fails, other options include simple or exchange transfusion and aspiration of the corpora. Emergent and urologic consultations are indicated.

Infectious Crisis

- Infection is a leading cause of death among patients with SCD.

- The vast majority of patients develop functional asplenia by 5 yr of age. This predisposes them to life-threatening infection with organisms such as *Haemophilus influenzae*, *Streptococcus* species and *Salmonella* species.
- Common sites of infection include the lungs, kidneys and skeletal system since these are the organs most affected by microvascular occlusion.
- Fever and leukocytosis are more predictive of infection in the pediatric population. Fever in a pediatric patient <5 yr old indicates life-threatening infection until proven otherwise. In adults, presentation is often more subtle and usually associated with a pain crisis.
- History and physical should focus on identification of the source, if possible. CBC, urinalysis and CXR are routine in the setting of suspected infection. Consider LP for patients who appear toxic and those who have meningismus or altered mental status. Blood and urine cultures are indicated. Sputum cultures are added in those with pulmonary symptoms.
- Empiric broad-spectrum antibiotics are recommended.

Hematologic Crisis

- Hematologic crises present with a sudden worsening of the baseline anemia.
- Causes include splenic sequestration and decreased bone marrow production.

Sequestration

- Sequestration occurs primarily in children who have not developed functional asplenia.
- Circulating RBCs are trapped in the spleen (and less commonly the liver), resulting in a sudden drop in Hg.
  - Presentation
    - Dizziness, syncope and abdominal fullness are common presenting symptoms.
    - Examination reveals splenomegaly, pallor, hypotension and tachycardia.
    - Laboratory is remarkable for a unusually low Hg and an elevated reticulocyte count.
- Treatment consists of intravenous fluids to help mobilize trapped RBCs. Transfusion is sometimes necessary.

Aplastic Crisis

- In aplastic crisis, the bone marrow does not produce new RBCs. Bone marrow suppression is usually secondary to viral infection (parvovirus B19) and has also been linked to folate deficiency.
- It is more common in the pediatric population.
- Patients experience fatigue, dyspnea and dizziness. Examination is notable for marked pallor.
- Laboratory reveals a markedly low Hg as well as a low reticulocyte count.
- The condition is generally self-limited. Sometimes patients will require a simple transfusion. Supplemental folate is often recommended.

Special Pediatric Considerations

- The presence of a "focus" of infection does not change the importance of full work-up and parenteral antibiotics.
  - With administration of ceftriaxone, a well-appearing child with a normal work-up may be safely managed as an outpatient, provided reliable follow-up is available within 24 h.
  - Admission is recommended in children <1 yr of age.
- Splenic sequestration occurs in 10-30% of children with Hg SS, usually between the ages of 6 mos to 3 yr, frequently in patients with previously palpable spleens.

- Splenic sequestration may also occur in older children and adolescents.
- Precipitating factors include viral illnesses.
- In rapidly progressive cases exchange transfusion may be required.
- Transfusion should be initiated promptly.
- Surgical splenectomy following the first episode of splenic sequestration remains a controversy.

- Acute chest syndrome
  - The etiology is unclear but is believed to be infection secondary to pulmonary infarction.
  - Acute chest is a common postoperative complication of general anesthesia, possibly due to excessive hydration.
  - May be life-threatening.
  - The child with acute chest syndrome should be managed in a monitored setting.
    - Therapy includes oxygen to avoid hypoxemia, judicious use of fluids, and incentive spirometry to avoid atelectasis, and a broad spectrum antibiotic.
    - Proper management of acute chest syndrome is important as many patients with repeated episodes of acute chest syndrome go on to develop chronic lung complications such as restrictive lung disease, pulmonary hypertension, and cor pulmonale.

Part D: Oncologic Emergencies

Mechanical Complications

Superior Vena Cava (SVC) Syndrome

- SVC syndrome is the clinical manifestation of SVC obstruction resulting from direct tumor invasion, thrombus or external compression of the vessel by an adjacent mass. As a result, patients experience a severe reduction in venous return from the head, neck and upper extremities.
- Malignant tumors account for >90% of cases. Most notable are lung cancer and lymphoma. Less common nonmalignant sources are tuberculosis, aortic aneurysm, sarcoidosis, goiter and trauma. Thrombi may also occur as a result of central venous catheters.

10

Clinical Manifestations

- Patients usually complain of neck and facial swelling as well as facial flushing. They sometimes report that their shirt collars are tighter than normal. Other symptoms include fatigue, dyspnea, headache, dysphagia, chest pain, dizziness and syncope.
- The Emergency Physician (EP) will note dilatation of the veins of the neck, upper extremities and upper chest. Other signs include facial edema and plethora. Severe cases exhibit proptosis, glossal and laryngeal edema and mental obtundation.
- The severity of symptoms and clinical presentation depends upon the level of obstruction and the rate at which it develops. SVC syndrome occurring gradually is better tolerated since patients have compensatory collateral vessels.
- Death can occur from airway compromise, cerebral edema or cardiac compromise.

Diagnosis

- Diagnosis is based on the history and physical.
- For patients with a known malignancy, a detailed work-up is not necessary. For patients without history of malignancy, a work-up is required to rule-out nonneoplastic causes and to determine the specific tumor histology.
- A chest radiograph (CXR) may reveal a widened mediastinum or findings consistent with underlying neoplasm. The diagnosis can be confirmed by computed tomography (CT) of the chest.

Treatment

- Emergency Department (ED) treatment consists mainly of supportive measures. These include elevation of the head of the bed and airway support. The EP should anticipate difficulty if intubation is required.
- Definitive therapy is determined by the oncologist based upon on the type of tumor. Radiation treatment is useful for nonsmall cell and other metastatic solid tumors. Other patients may require chemotherapy or surgery.
- If SVC syndrome results from a catheter-related thrombosis, the catheter should be removed. Antithrombotic or fibrinolytic therapy is an option provided that there are no contraindications.

Airway Obstruction

- Airway obstruction occurs in the setting of malignancy via direct airway involvement or external compression by adjacent tumor. Common etiologies include laryngeal, thyroid and pulmonary neoplasm as well as lymphoma.
- Upper airway obstruction refers to obstruction at the level of or proximal to the mainstem bronchi. Signs and symptoms include stridor, respiratory difficulty and hoarseness. If time permits, otolaryngology consultation is advised, as these patients sometimes require placement of a surgical airway. If necessary, immediate airway control is best accomplished via direct laryngoscopy or fiberoptic laryngoscopy. The EP must always be prepared for cricothyrotomy.
- Patients with lower airway or endobronchial obstruction present with cough, dyspnea, wheezing and hemoptysis. They sometimes experience lobar collapse and infection distal to the obstructing lesion. Immediate airway control via endotracheal intubation is the mainstay of ED treatment.

Hemoptysis

10

- Hemoptysis results from friable neovascular tissue near the tumor or local erosion of tumor into a nearby blood vessel. Bronchogenic carcinoma is the most frequent neoplastic cause of hemoptysis.
- Death from exsanguination is uncommon, but serious complications can occur as a result of impaired gas exchange.
- The primary goals of ED management are airway control and cardiovascular support. In a patient with significant bleeding, early intubation is recommended. Endotracheal intubation with a large diameter tube (8 mm) is recommended in order to facilitate subsequent fiberoptic diagnosis. The treatment is discussed in detail elsewhere.

Pericardial Tamponade

- Detailed discussion in Cardiovascular section.

Spinal Cord Compression (SCC)

- SCC usually results from tumor extension from neighboring vertebrae. It is more frequent in the thoracic spine. Commonly implicated neoplasms are lung, breast, and prostate carcinoma as well as multiple myeloma and lymphoma.

Clinical Manifestations

- Back pain is the initial symptom in 95% of patients occurring at the site of tumor metastasis. Pain may be localized or radicular in nature and is worsened by percussion of the affected vertebral bodies. Pain is often worse at rest.
- Symptoms of more serious disease are extremity weakness or paresis, sensory deficits and bowel or bladder dysfunction.

Diagnosis

- The initial diagnostic study is plain radiographs of the spine. These are abnormal in 60-90% of cases. Note that SCC sometimes occurs via the intervertebral foramina without direct bony involvement. Furthermore, a certain degree of vertebral destruction is necessary before bony changes are seen on plain films. Other studies are indicated for suspected SCC in the setting of normal plain radiographs.
- Other diagnostic options are bone scanning, CT myelography, and magnetic resonance imaging (MRI) with MRI being the study of choice.

Treatment

- ED treatment consists of high-dose dexamethasone (25 mg every 6 h).
- Emergent oncology and neurosurgery consultations are necessary to plan for definitive therapy such as radiation or surgery.

Medical Complications

Tumor Lysis Syndrome (TLS)

- TLS is a group of metabolic derangements caused by tumor cell destruction.
- TLS typically occurs when there is rapid cell turnover or a large burden of cells. The result is hyperkalemia, hyperuricemia and hyperphosphatemia. Hyperphosphatemia causes a secondary hypocalcemia via precipitation of calcium phosphate. Complications include renal failure, cardiac arrhythmias and death.
- TLS usually occurs within 1-2 days of chemotherapy or radiation for rapidly growing tumors, such as leukemias or lymphomas. Other risk factors include poor renal function and decreased urine output. Spontaneous TLS is less common.

Clinical Manifestations

- Patients present with signs and symptoms of electrolyte disturbance and possibly renal failure. These include tetany, altered mental status, seizures, arrhythmias, flank pain, hematuria and decreased urine output.

Diagnosis

- Most patients under the care of an oncologist receive hydration and allopurinol prior to their chemotherapy or radiation. The EP will often see the cases that develop unexpectedly—patients with undiagnosed malignancies and spontaneous tumor necrosis, cancer patients given steroids for an unrelated problem or patients who receive outpatient chemotherapy for tumors not classically associated with TLS.
- The clinical picture in conjunction with electrolyte abnormalities will assist the EP in diagnosis. Laboratory studies demonstrate elevated potassium, elevated uric acid level, low calcium and elevated phosphorus. Blood urea nitrogen and creatinine are often elevated. EKG should be obtained looking for arrhythmias.

Treatment

- ED therapy is guided by the severity of electrolyte and renal abnormalities.
- If renal function is intact, initial treatment is aggressive hydration with normal saline.
- Urine alkalinization with sodium bicarbonate to a pH of >7.0 is recommended to increase uric acid secretion. It is important to monitor the calcium because alkalinization will aggravate the preexisting hypocalcemia.
- Allopurinol is also recommended. Initial ED dose is 300-900 mg.
- Treatment for hyperkalemia is administered as indicated by the potassium level and EKG.
- Hemodialysis is indicated in patients with renal failure and those who are refractory to standard therapy. It may also be necessary in cases with severe electrolyte abnormalities. Hemodialysis portends a poor prognosis.

Hyperviscosity Syndrome (HVS)

- HVS refers to the clinical sequelae of increased blood viscosity. It is seen primarily with dysproteinemias (Waldenstrom's macroglobulinemia and multiple myeloma) and the blastic phase of leukemias. It also occurs less commonly with polycythemia vera, connective tissue disorders and sickle cell disease.
- Increased blood viscosity results in decreased microvascular perfusion and vascular stasis.

Clinical Manifestations

- The most common signs and symptoms are related to mucous membrane bleeding, retinopathy and neurological disturbances. It is also common to have cardiovascular effects such as heart failure and acute myocardial infarction.
- The most common sites of bleeding are the oral cavity, nose and genitourinary tract. Ocular symptoms include diplopia and varying degrees of vision loss. Neurologic symptoms include headache, hearing loss, vertigo, paresthesias and gait disturbances.
- Examination reveals altered mental status, retinal hemorrhage, retinal detachment and papilledema.

Diagnosis

- Early recognition of HVS is critical in order to prevent irreversible sequelae.
- Diagnosis is made via the clinical presentation in association with direct measurement of serum or whole blood viscosity. The normal range is 1.4-1.8 units; HVS typically requires a viscosity above 5 units. In most EDs, clinical impression is particularly important since measurement of blood viscosity is not immediately available. The EP should suspect HVS if lab evaluation is hampered by analyzer dysfunction secondary to increased blood viscosity.

Laboratory Studies

- In patients with leukemia, leukocyte count will be markedly elevated.
- Coagulation studies are indicated to rule out an underlying clotting disorder, but these will be normal in patients with HVS.
- Anemia and renal dysfunction are commonly noted in patients with HVS.
- Hypercalcemia is often present in the setting of multiple myeloma.
- Serum and urine protein electrophoresis are not helpful for the EP but should be obtained after the patient is admitted.

The Use of Diagnostic Imaging Is Tailored to the Individual

- CXR may reveal findings consistent with heart failure.
- Head CT scan is indicated in all patients with altered mental status, seizures or focal neurologic deficits.
- Note that intravenous contrast dye increases the risk of renal failure.

Treatment

- Plasmapheresis is the treatment of choice for HVS. In extreme cases where plasmapheresis is not readily available, phlebotomy of 100-200 ml of blood may be helpful.
- Standard therapies should be employed for renal failure, heart failure and bleeding.

Hypercalcemia

- Hypercalcemia occurs in approximately 10-20% of known malignancies, making it the most common life-threatening metabolic disorder associated with cancer. It usually is seen in those with advanced disease but is occasionally the presenting feature.
- It is often the symptoms or signs of underlying malignancy that bring the hypercalcemic patient to seek medical attention.
- There are two primary mechanisms for malignancy-related hypercalcemia.

- Some neoplasms result in direct bone destruction via osteolytic skeletal metastasis. This is frequently the case in breast, lung, renal, thyroid, ovary and colon cancers.
- Cancers sometimes produce parathyroid hormone-related proteins or other substances that promote bone reabsorption.

- Detailed discussion found in the Electrolyte abnormality section.
  - The presence and severity of symptoms is related to the rate of rise of calcium but does not necessarily correlate with the absolute level. However, levels over 14 mg/dL are considered critical and are most often associated with severe signs and symptoms.
  - Hypercalcemia affects many organ systems and often has a nonspecific presentation.
    - Gastrointestinal effects include anorexia, nausea, vomiting, abdominal pain, constipation and ileus.
    - Neurologic symptoms include headache, fatigue, weakness, difficulty concentrating, confusion, irritability, lethargy, hyporeflexia and coma.
    - EKG manifestations include shortening of the QT interval, prolongation of the PR interval and widening of the QRS complex. At very high levels, bradyarrhythmias and bundle branch blocks can occur with progression to complete heart block and cardiac arrest. It should be noted that calcium accentuates the side effects of digoxin.
    - Renal findings include dehydration, polyuria, polydipsia and oliguric renal failure.

Diagnostic Studies

- Laboratory studies should include total and ionized serum calcium levels as well as an albumin level. Normal total calcium level is <10.5 mg/dL. It may be necessary to correct for protein-bound calcium. It is also possible to measure the level of free ionized calcium.
  - Correction formula for calcium:

Corrected calcium (mg/dL) = measured total calcium (mg/dL) + 0.8 x [4.0 – albumin (g/dL)]

- Renal function and electrolytes should be measured. Serum phosphate levels are usually low or normal.
- An EKG is indicated in all cases.
- Other studies are ordered based on clinical presentation.

Treatment

- Treatment is guided by the calcium level, symptom acuity and the underlying disease.
- Patients with mild to moderate elevations of calcium who are asymptomatic are appropriate for outpatient evaluation and treatment.
- Immediate treatment is necessary for severe hypercalcemia (corrected level >14 mg/dL) regardless of symptoms and for symptomatic hypercalcemia regardless of level. Cardiac monitoring is mandatory in these patients given the risks of arrhythmias.
  - Aggressive hydration is the mainstay of ED treatment. Hydration helps decrease calcium via a dilutional effect and via an increase in renal calcium clearance. The rate of fluid therapy is based on the degree of hypercalcemia, severity of dehydration and ability of the patient to tolerate fluid replacement.
  - Loop diuretics such as furosemide promote renal calcium excretion but should not be used until hypovolemia has been corrected and urine output is adequate. Avoid thiazide diuretics as they increase serum calcium levels. When diuretics are used, urine output should be followed closely. It is recommended that output be matched with intravenous fluid replacement in addition to maintenance fluids. Watch closely for signs of fluid overload, especially in elderly patients.
  - Patients with renal failure will require dialysis using a calcium-free dialysate.

- Other adjunctive medications include calcitonin, mithramycin, bisphosphonates, gallium nitrate and steroids. The drug regimen most appropriate for each individual depends on the underlying cause. Consultation with the admitting oncologist is recommended.

Syndrome of Inappropriate Antidiuretic Hormone (SIADH)

- SIADH is characterized by hyponatremia with an inappropriately elevated urine osmolality.
- Approximately 1-2% of cancer patients will develop SIADH. It is often associated with small cell lung cancer that produces ectopic ADH. It is also seen with primary tumors that have metastasized to the brain or lungs and can be precipitated by certain chemotherapeutic agents.

Clinical Manifestations

- Most patients are asymptomatic and are diagnosed only after routine labs are obtained for other reasons.
- Those who become symptomatic usually have a rapid drop in the serum sodium or an absolute value <115 mg/dL. Early symptoms include anorexia, nausea and myalgias. Later manifestations include confusion, psychosis, seizures and obtundation.
- The physical examination is likely to be unremarkable. Patients with SIADH should not have evidence of either hypovolemia or fluid overload unless other comorbid conditions are present.

Diagnosis

- The diagnosis is made via the identification of hyponatremia and inappropriately concentrated urine in the euvolemic patient (see Table 10D.1).
- Other laboratories are often normal unless they are affected by other comorbid conditions or medication use.
- Note that the following criteria only apply to those with normal renal, adrenal and thyroid function. In addition, use of diuretics will interfere with diagnosis.
- A CXR and/or CT scan of the head may help with diagnosis of the underlying lesion. CT scan of the head is also indicated in certain cases to rule-out other potential causes of acute neurological changes.
- Further discussion found in electrolyte abnormality section.

Neutropenic Fever

- Neutropenia (see Table 10D.2) may occur secondary to many chemotherapeutic agents as well as the patient's underlying neoplastic process.
- Fever in all neutropenic patients must be considered infectious in origin until proven otherwise.
- There are many other causes of neutropenia including aplastic anemia, sepsis, hypersplenism, human immunodeficiency virus and systemic lupus erythematosus. There are also numerous medications that can cause marrow suppression and neutropenia. These factors should be considered in the neutropenic patient.

Table 10D.1. Laboratory diagnosis of SIADH*

Serum sodium <135 mEq/L,
Serum osmolality <280 mOsm/L
Urine sodium >20 mEq/L
Urine osmolality >100 mOsm/L

* Euvolemic patient

Table 10D.2. Cell counts in neutropenia

Neutropenia = Neutrophil count* <1500 cells/mm^3
Severe neutropenia = Neutrophil count* <500 cells/ mm^3
* Absolute Neutrophil Count = WBC x (% Bands + % Neutrophils)

- Patients with neutropenia are prone to all types of infection, especially Gram-negative bacilli. *Staphylococcus* and *Streptococcus* species are also common pathogens, particularly in those with indwelling catheters.

Clinical Manifestations

- The diagnosis is made via identification of fever in a neutropenic patient. The definition of fever for those with documented neutropenia it is considered to be a single temperature >38.5° or a sustained temperature >38°.
- Chemotherapy patients are at highest risk for neutropenia approximately 7-12 days after their last treatment.
- ED evaluation should include a thorough history and physical exam, focusing on potential sources of infection. In the majority of neutropenic patients, a primary source will not be identified. These patients often do not have the usual signs of inflammation, thus the typical clinical findings are minimal or absent. Any indwelling catheters should be considered as a possible source.

Diagnostic Studies

- Laboratory studies should include a complete blood count with differential, electrolytes, blood cultures, urinalysis and urine culture. Most neutropenic patients will not have pyuria, so urine Gram stain and culture are especially important.

- A CXR is recommended even in the absence of symptoms although many patients with pneumonia will have an unremarkable CXR.
- Lumbar puncture should be performed only if clinically indicated.

Treatment

- Empiric antibiotics are recommended for neutropenic fever when the absolute neutrophil count is <1000/mm^3.
- There are several suggested broad-spectrum antibiotic regimens (see Table 10D.3).
- For chemotherapy patients, myeloid growth factors such as G-CSF and GM-CSF may be useful in shortening the duration of neutropenia.
- For selected low-risk patients, outpatient management with oral or intravenous antibiotics may be feasible. The EP should discuss outpatient therapy with the patient's oncologist prior to making a definitive treatment decision.

Table 10D.3. Antibiotic regimens for neutropenic fever

Combination therapy
Aminoglycoside + anti-Pseudomonal penicillin*
Aminoglycoside + ceftazidime
Single-drug therapy
Ceftazidime
Imipenem/cilastatin
Cefepime
Add vancomycin if indwelling line or suspected MRSA
*Includes ticarcillin, ticarcillin/clavulanate, piperacillin, piperacillin/tazobactam, mezlocillin

Part E: Congenital Bleeding Disorders

Von Willebrand's Disease

- von Willebrand's Disease (vWD) is the most common congenital bleeding disorder. Patients have an abnormality of von Willebrand factor (vWF), a plasma component that facilitates platelet adhesion. Additionally, vWF protects factor VIII from proteolytic degradation so that there may also be a decrease in circulating factor VIII levels.
- The majority of patients exhibit an autosomal dominant pattern, but certain subtypes have an autosomal recessive or X-linked pattern of inheritance. More than 21 distinct clinical subtypes have been described, and there are numerous variants. The type of disease is characterized by either quantitative or qualitative defects of vWF.
  - Type I
    - Structurally normal vWF
    - Mild to moderate decrease in vWF plasma levels
    - Proportionate decrease in factor VIII levels
    - Most common type (80%)
  - Type II
    - Multiple variants exist
    - Qualitative abnormalities
    - Most patients are missing the portion of the molecule that mediates platelet adhesion
  - Type III
    - Rare (<1%)
    - Little or no vWF present
    - Greatly reduced levels of factor VIII
    - Markedly defective hemostasis

Presentation

- Patients with vWD experience mucosal and cutaneous bleeding as would be expected in patients with defective platelet function or thrombocytopenia. In addition, those patients with type III vWD may have soft tissue and joint hemorrhage as a result of the associated factor VIII deficiency.
- The severity of bleeding depends upon the type of disease—patients with type I vWD have mild symptoms whereas those with type III may have life-threatening bleeding.

Diagnosis

- vWD should be suspected in patients with mucocutaneous bleeding and a family history.
- Complete blood count (CBC) and prothrombin time (PT) are normal. Partial thromboplastin time (PTT) is prolonged if there is an associated factor VIII deficiency.
- Bleeding time (BT) is prolonged in the majority of patients with vWD but does not necessarily correlate with the presence or degree of bleeding. Other patients may have symptoms in the setting of a normal BT. This test is sometimes inconsistent and is of little use for the Emergency Physician (EP).
- Other more specific tests such as ristocetin cofactor activity, quantitative measurement of vWF plasma antigen and qualitative structural assessments are helpful for the EP if that information is available from the patient or chart.

Treatment

- Specific therapies will be dictated by the type of disease and degree of bleeding.

- In all cases, it is vital to communicate with hematology and reference prior treatment records, if available.
- Always avoid the use of any antiplatelet medications in these patients.
- Several treatment options exist.

Desmopressin (DDAVP)

- DDAVP results in a 2-5 fold increase in plasma vWF and factor VIII levels and is the first line agent for type I vWD. Response is rapid and lasts up to 12 h.
- The usual dose is 0.3 μg/kg IV per day. Doses are repeated as needed depending upon the severity of bleeding. DDAVP can also be given subcutaneously or intranasally.
- DDAVP alone or in combination with antifibrinolytic agents (see below) is usually adequate treatment for patients with mild disease.
- DDAVP does not carry the risk of viral transmission as do the blood products. However, patient response to the drug diminishes with repeated use.
- DDAVP may be useful for some vWD III subtypes. It is contraindicated in certain vWD II subtypes. Consultation with a hematologist is strongly recommended prior to use of DDAVP unless previous response to therapy is known.

Antifibrinolytic Amino Acids (Episilon aminocaproic acid and tranexamic acid)

- These amino acids bind to plasminogen to interfere with clot lysis.
- They are used primarily for dental procedures, epistaxis, and menorrhagia. They should be avoided in cases of hemarthrosis and hematuria as the persistence of formed clots may lead to subsequent fibrosis.
- The dose of episilon aminocaproic acid is 50 mg/kg every 6 h and tranexamic acid 25 mg/kg every 8 h. The drugs can be given orally or via the intravenous route.
- Like DDAVP, these medications avoid the risk of viral transmission.

Transfusion Therapy (Generally required for all patients with severe disease and those who do not respond to DDAVP)

- Cryoprecipitate
  - Contains both factor VIII and vWF.
  - Dose is 1-2 bags/10 kg body weight every 12-24 h as dictated by the patient's clinical condition. Higher doses may be indicated depending upon the clinical situation.
  - The primary problem with cryoprecipitate is its risk of viral transmission. As a result, factor VIII concentrates containing vWF have become a mainstay of therapy.
- Factor VIII concentrates containing vWF are available. Note that most factor VIII products do not contain significant amounts of vWF so hematology consultation is recommended for specific product and dosing information.
- Fresh frozen plasma (FFP)
  - FFP can be used in emergent situations if cryoprecipitate or factor VIII concentrates are not available.
  - Large doses are needed since FFP contains relatively little vWF.

Hemophilia

- Hemophilia A is characterized by factor VIII deficiency or defective factor function. Hemophilia B has a clinical presentation identical to hemophilia A but is characterized by factor IX deficiency. The majority of both hemophilia A and B patients are X-lined recessive although sporadic cases do occur.
- The incidence of hemophilia is 1-2 in 10,000 males, with hemophilia A accounting for approximately 80-85% of cases.
- The disease is classified by severity.
  - Patients with >5% factor activity are considered to have mild disease. Bleeding usually occurs only in the setting of trauma or surgical procedures.

- Factor activity from 1-5% denotes moderate disease. These patients occasionally have spontaneous hemorrhages but most often bleed after trauma.
- Patients with <1% factor activity are considered to have severe disease with the greatest incidence of spontaneous hemorrhage.

Presentation

- Spontaneous bleeding into muscles and joints is the hallmark of severe disease. Larger joints are affected more often. Hemarthroses usually first occur in early childhood just after the child begins to ambulate. Recurrent joint bleeding results in permanent articular damage and contractures.
- Bleeding also occurs after even minor trauma and can present in a delayed fashion (days).
- Although joint and soft tissue hemorrhage is more common, bleeding can occur in any organ system.

Diagnosis

- In the vast majority of patients encountered in the Emergency Department (ED), the diagnosis has already been established. The disease type and severity can be obtained form the patient and/or old records.
- In younger male patients without a previous diagnosis, the disorder should be suspected in the setting of deep muscle bleeding or hemarthrosis. Also consider hemophilia when hemorrhage (intracranial or otherwise) occurs after seemingly trivial trauma.
- Laboratory studies reveal a prolonged PTT and normal PT. Platelet count and bleeding time are normal. Factor assays are necessary for definitive diagnosis.

Treatment and Disposition

- Early, aggressive therapy is paramount in order to minimize morbidity and mortality.
- It is important to remember that most hemophiliacs and their families have an excellent understanding of their disease. The physician should respect the patient's judgment about the presence of bleeding which is not clinically apparent.

10

Factor Replacement

- Except for the most trivial bleeding, treatment must include immediate factor replacement. The desired factor activity level is determined by the location and severity of hemorrhage (see Table 10E.1). Factor levels >25-30% for factor VIII or 15-30% for factor IX provide effective hemostasis.
- The desired factor activity level for patients who need invasive diagnostic or therapeutic studies is 100%.

Table 10E.1. Desired factor activity levels

Location of Bleeding	Desired Activity Level
Central nervous system	100%
Gastrointestinal	
Retroperitoneal/Iliopsoas	
Retropharyngeal	
Ocular	
Any major trauma	
Surgery	
Muscle	40-50% (100% if compartment syndrome suspected)
Oral or epistaxis	
Hematuria	
Deep lacerations	25-50%
Joint	

Table 10E.2. Dosage calculations for factor replacement therapy

Desired factor activity level (%) x weight (kg) x 0.5 = dose of factor VIII
Desired factor activity level (%) x weight (kg) x 1 = dose of factor IX

- For factor VIII, 1 unit/kg will increase the activity level by 2%. For factor IX, 1 unit/kg will raise the blood level by 1% (see Table 10E.2). For all patients, the EP should assume an initial factor activity level of 0%.
- The half-life is approximately 8-12 h for factor VIII and 24 h for factor IX. For major bleeds or surgery, doses are repeated until the underlying condition resolves or any surgical or traumatic wound is healed.
- There are many products available for both factors VIII and IX including those derived from purified plasma, monoclonal antibodies and recombinant technology. Most patients will be able to tell the EP what product they use. Prothrombin complex concentrates (PCCs) contain factor IX as well as variable amounts of X, VII and II. As a result, this product is associated with thrombotic complications, especially in patients with prolonged immobilization.
- About 10-20% of patients with hemophilia A develop antibodies to exogenous factor VIII. This should be suspected in any patient who does not respond to factor replacement. Patients with low levels of inhibitor respond to higher doses of factor replacement whereas others will mandate alternative therapy. In these cases, hematology consultation is imperative. Treatment options include porcine factor VIII, factor VIIa and PCCs. Another option is to decrease the amount of inhibitor using plasmapheresis. Factor IX inhibitor antibodies are less common (about 3% of patients). These patients can also be managed using PCCs.

Alternative and Adjunctive Treatments

- FFP contains all the clotting factors and is appropriate for both hemophilia A and B if specific factors are not immediately available. Cryoprecipitate contains factor VIII and vWF and can be used for hemophilia A. It is important to note that both FFP and cryoprecipitate are substandard treatments with higher complication rates. Arrangements should be made for emergent patient transfer to a capable center if factor specific treatment is not readily available.
- DDAVP is effective in patients with mild hemophilia A (see "vWD" section for dosing).
- Antifibrinolytic amino acids are indicated in hemophilia patients with anticipated dental procedures, oral bleeding, or epistaxis (see "vWD" section for dosing). They should not be used in conjunction with PCCs because of the increased thrombotic risk.

Organ-Specific Ancillary Measures (Refer to Tables 10E.1 and 10E.2 for factor dosing)

Central Nervous System

- No matter how trivial, all head and spinal trauma is considered significant. Maintain a very low threshold for diagnostic imaging.
- Factor replacement should be given immediately. Never delay for diagnostic studies.
- Patients with neurological signs or symptoms who have negative computed tomography (CT) of the head must be admitted for close observation and continued factor replacement (3-5 days).
- Patients with negative imaging and a normal exam can be discharged with strict head injury precautions as long as they have a reliable psychosocial support system.

Oral Cavity

- Always remove any existing friable clot.

- Give a one time dose of factor replacement therapy.
- Antifibrinolytic agents can be used after factor replacement in order to decrease the incidence of rebleeding. Duration of therapy is 5-7 days.
- Soft diet is recommended.

Epistaxis

- Initial intervention should include use of direct pressure and local vasoconstrictors.
- Avoid nasal packing as removal will cause recurrent hemorrhage.
- If bleeding does not resolve with local measures, administer a one time dose of factor replacement therapy.
- Saline drops, petrolatum jelly and/or antifibrinolytic agents are used to prevent rebleeding.

Joint

- Early symptoms of joint bleeding include pain, tingling and warmth. Later findings are swelling and decreased range of motion.
- Factor replacement should be given as soon as possible after symptom onset. An early bleed with minimal pain and swelling usually responds to a one-time dose. For severe cases, higher doses (50 U/kg) are warranted with 3-5 day duration of treatment. Symptomatic relief is achieved with rest, ice and elevation. Splint immobilization provides comfort but early range of motion is encouraged.
- Radiographic studies are not indicated unless history suggests possible fracture.
- Joint aspiration is usually not necessary. Symptoms should resolve promptly with adequate coagulation factor replacement.
- Hemorrhage into the hip joint in children has the potential to cause aseptic necrosis of the femoral head.

Soft Tissue

- Muscle hematomas are a common complication of hemophilia. Forearm and calf hemorrhage are particularly concerning because of the potential for compartment syndrome.
- Extensive hemorrhage and compartment syndrome requires higher doses of factor replacement (50 U/kg) with continued therapy until resolution. Minor hemorrhage is managed with a single dose of factor replacement.
- Extensive hemorrhage should be managed as an inpatient. Emergent orthopedic consultation is indicated if there is any concern about compartment syndrome.

Iliopsoas

- Common symptoms of iliopsoas hemorrhage include hip or groin pain that radiates to the back or thigh. It may also present as abdominal pain. The hip is held in slight flexion and internal rotation. Hip extension causes pain. Anteromedial thigh paresthesias suggest femoral nerve compression. On physical exam a mass may be palpable in the iliac fossa.
- It is often impossible to distinguish from hip joint hemorrhage based on clinical exam alone. CT scan or magnetic resonance imaging confirm the diagnosis. Ultrasound is sometimes helpful but has limited sensitivity with early bleeds.
- Initiate immediate treatment as soon as the diagnosis is suspected. High-dose factor replacement is indicated. Patients are admitted for further treatment until condition has resolved.

Neck/Retropharynx (RP)

- Any bleeding into the neck, posterior pharynx or sublingual region poses the threat of airway compromise. The EP should suspect RP hemorrhage in any patient complaining of neck swelling, difficulty swallowing, voice changes or respiratory difficulty.

- Immediate factor replacement is indicated. Patients are admitted for continued therapy until the condition has completely resolved. With prompt therapy, intubation is often avoided.
- Plain radiographs of the neck are sometimes diagnostic although diagnosis may require ultrasound or CT scan.

Hematuria

- Hematuria can originate from the upper or lower urinary tract and is often a benign condition.
- Steroids are recommended as first-line therapy. Dose is 2 mg/kg up to maximum of 60 mg/day. Duration of therapy is 2 days, then rapid taper. Maintenance of urinary output is critical.
- Persistent bleeding requires further treatment with factor replacement for several days.
- Antifibrinolytic agents are contraindicated, as clots can lead to permanent renal damage.
- A further urologic evaluation should be considered to rule out other causes of hematuria.

Trauma

- Severe trauma requires immediate, high-dose factor replacement that is continued until would healing is complete.
- Minor cuts and abrasions rarely require treatment other than standard first aid. Deeper lacerations require single-dose factor replacement.
- Wounds should be closely observed by the patient and repeat doses of factor given if needed.
- All fractures (even minor) require immediate treatment. Initial high-dose factor replacement is followed by moderate doses (25 U/kg) until fracture is stable (about 1 wk). Initial immobilization is with a splint or bivalved cast to allow for further swelling. A full cast can be placed after swelling has resolved.

10

Suggested Reading

1. Freifeld AG, Pizzo PA. The outpatient management of febrile neutropenia in cancer patients. Oncology 1996; 10(4):599-606.
2. Gin-Shaw S, Moore GP. Selected white cell disorders. Emerg Med Clin N Amer 1993; 11(2):495-516.
3. Pimentel L. Medical complications of oncologic disease. Emerg Med Clin N Amer 1993; 11(2):407-19.
4. Maguire WM. Mechanical complications of cancer. Emerg Med Clin N Amer 1993; 11(2):421-30.
5. Janz TG. Managing the Cancer-Related Emergency. Emergency Medicine March 1990; 22:6-58.
6. Bunn HF. Pathogenesis and treatment of sickle cell disease. N Engl J Med 1997; 337(11):762-9.
7. Wethers DL. Sickle cell disease in childhood: Part I. Laboratory diagnosis, pathophysiology and health maintenance. Am Fam Physician 2000; 62(5):1013-20, 1027-8.
8. Lane PA. Sickle cell disease. Ped Clin N Amer 1996; 43(3):639-64.
9. Steingart R. Management of patients with sickle cell disease. Med Clin N Amer 1992; 76(3):669-82.
10. Pollack Jr CV. Emergencies in sickle cell disease. Emerg Med Clin of N Amer 1993; 11(2):365-78.
11. Bolton-Maggs PH. Idiopathic thrombocytopenic purpura. Arch Dis Child 2000; 83(3):220-222.
12. Cines DB, Blanchette VS. Immune thrombocytopenic purpura. N Engl J Med 2002; 346(13):995-1008.
13. Elliott MA, Nichols WL. Thrombotic thrombocytopenic purpura and hemolytic uremic syndrome. Mayo Clin Proc 2001; 76(11):1154-62.

14. Furlan M, Robles R, Galbusera M et al. von Willebrand factor-cleaving protease in thrombotic thrombocytopenic purpura and the hemolytic-uremic syndrome. N Engl J Med 1998; 339(22):1578-84.
15. George JN, Raskob GE. Idiopathic thrombocytopenic purpura: Diagnosis and management. Am J Med Sci 1998; 316(2):87-93.
16. Goebel RA. Thrombocytopenia. Emerg Med Clin N Amer 1993; 11(2):445-464.
17. Ruggenenti P, Remuzzi G. The pathophysiology and management of thrombotic thrombocytopenic purpura. Eur J Haematol 1996; 56(4):191-207.
18. Ruggenenti P, Remuzzi G. Thrombotic thrombocytopenic purpura and related disorders. Hem-Onc Clin N Amer 1990; 4(1):219-241.
19. Goodnough LT, Brecher ME, Kanter MH et al. Transfusion medicine. N Engl J Med 1999; 340(6):438-47.
20. Labadie LL. Transfusion therapy in the emergency department. Emerg Med Clin N Amer 1993; 11(2):379-406.
21. Practice guidelines for blood component therapy: A report by the American Society of Anesthesiologists Task Force on blood component therapy. Anesthesiology 1996; 84(3):732-47.
22. Schreiber GB, Busch MP, Kleinman SH et al. The risk of transfusion-transmitted viral infections. N Engl J Med 1996; 334(26):1685-90.
23. Cohen AJ, Kessler CM. Treatment of inherited coagulation disorders. Am J Med 1995; 99(6):675-82.
24. Hoyer LW. Hemophilia A. N Engl J Med 1994; 330(1):38-47.
25. Mannucci PM. Treatment of von Willebrand's disease. J Intern Med Supplement 1997; 740:129-32.
26. Nichols WC, Ginsburg D. von Willebrand disease. Medicine 1997; 76(1):1-20.
27. Pfaff JA, Geninatti M. Hemophilia. Emerg Med Clin North Am 1993; 11(2):337-63.
28. Shaffer LG, Phillips MD. Successful treatment of acquired hemophilia with oral immunosuppressive therapy. Ann Intern Med 1997; 127(3):206-9.
29. Wald DA, Alleyne P. Congenital bleeding disorders: Diagnostic features and syndrome-specific management in the emergency department. Emergency Medicine Reports 2001; 22(8):81-92.

CHAPTER 11

Infectious Disease Emergencies

Ellen M. Slaven, Fred A. Lopez and James Rhorer

Infectious Diseases

Emergency physicians (EPs) are frequently confronted by a wide array of infections, from trivial upper respiratory viral infections to life threatening necrotizing soft tissue infections. Emergency department (ED) patients are also diverse, ranging from septic neonates to febrile travelers from foreign countries. Potentially life-threatening infections must be identified early and appropriate antimicrobial therapy initiated promptly. This section reviews specific infectious diseases issues in the ED. Certain individual diseases are covered in detail in other chapters.

Sepsis

Definition

- Sepsis, septic shock, and sepsis syndrome are terms that have been used interchangeably. Specific terminology has been adapted in both research and clinical settings to strictly define these conditions. (Table 11.1)

Epidemiology

- Despite improvements in antimicrobial therapy and critical care techniques, the mortality of patients with sepsis remains approximately 35%. The incidence of sepsis and septic shock, continues to rise due to increasing numbers of elderly and immunocompromised patients, more frequent use of invasive procedures and devices, and the development of antimicrobial resistance among common bacterial pathogens.

Table 11.1. Definitions

Systemic inflammatory response syndrome (SIRS)—two or more of the following:
Temperature >38° C or <36° C
Heart rate >90 beats/min
Respiratory rate >20 breaths/min or pCO2 <32 mm Hg
White blood cell count >12,000 cells/mm^3 or <4,000 cells/mm^3 or >10% immature neutrophils
Sepsis
SIRS with a confirmed infectious etiology
Severe Sepsis
Sepsis with hypotension or systemic manifestations of hypoperfusion (i.e., lactic acidosis, oliguria, altered mental status)
Septic Shock
Severe sepsis despite adequate fluid resuscitation

From: American College of Chest Physicians/Society of Critical Care Medicine Consensus Conference: Definitions for sepsis and organ failure and guidelines for the use of innovative therapies in sepsis. Crit Care Med 1992; 20(6):866, with permission.

Diagnosis and Evaluation

- Historical data should be obtained from the patient and/or any other available source. Important factors include recent hospitalizations, invasive procedures, antibiotic usage, recent travel, and vaccinations as well as past history of infections or immunosuppressive diseases or medications.
- A thorough physical examination should be performed with an emphasis on identification of a possible site of infection, evaluation for organ system dysfunction, and evidence of hypoperfusion.
- Laboratory analysis should include white blood cell count with differential, hemoglobin, platelet count, serum chemistry profile, protime/partial thromboplastin time, serum lactate, urine analysis, and arterial blood gases.
- Blood, urine, and sputum should be collected and cultured. Material for culture should ideally be obtained prior to antimicrobial therapy. However, the administration of antibiotics should never be delayed.
- A chest radiograph should be obtained.
- Additional studies are warranted as dictated by the clinical scenario, such as computed tomography (CT) of the abdomen and lumbar puncture.

ED Management

- If a patient is suspected of having severe sepsis or septic shock, initial treatment priorities are directed to reverse life-threatening abnormalities. The patient's airway, breathing and circulation must be quickly established and maintained.
- Endotracheal intubation
  - May be required to protect the airway, deliver increased oxygen concentrations, or reduce the work of breathing.
- Hypotension
  - Requires aggressive intravascular fluid resuscitation with crystalloids. Vasopressor therapy (dopamine, dobutamine and norepinephrine) must be considered for persistently low mean arterial blood pressures despite fluid resuscitation. Urine output should be closely monitored as a measure of adequate volume resuscitation.
- Antimicrobial therapy must be instituted early. Antibiotic selection in the ED must be empiric because it is often not possible to identify the responsible organism(s) with certainty. This selection depends upon the presumed site of infection, the results of Gram stains, resistance patterns of common pathogens, as well as patient factors such as allergies, renal/hepatic dysfunction, immune status, and history of recent hospitalization. General guidelines for severe sepsis without an identifiable source are defined in Table 11.2. If anaerobic pathogens are likely, then metronidazole or clindamycin should be added. Imipenem, a carbapenem with broad spectrum coverage, may be used as a single agent. None of the above antibiotic regimens is effective against methicillin-resistant *Staphylococcus aureus* (MRSA). If MRSA is suspected (e.g., the presence of an indwelling vascular catheter) vancomycin should be added.
- Steroids have not been demonstrated to improve outcome.

11

Table 11.2. Empiric antibiotics for severe sepsis without a source

Anti-pseudomonal cephalosporin (ceftazidime or cefepime)
OR
Anti-pseudomonal penicillin (ticarcillin/clavulanate or piperacillin/tazobactam)
WITH
Aminoglycoside (gentamycin or tobramycin)
OR
Fluoroquinolone

- Patients should be admitted to an appropriate level of care based on their clinical stability and ongoing therapeutic needs.

Suggested Reading

1. Rangel-Frausto MS. The epidemiology of bacterial sepsis. Infect Dis Clin N Amer 1999; 13:299-312.
2. Dellinger RP. Current therapy for sepsis. Infect Dis Clin N Amer 1999; 13:495-509.
3. A report from the National Nosocomial Infections Surveillance (NNIS) System. National Nosocomial Infections Surveillance (NNIS) Report, Date Summary from October 1986 April 1996, Issued May 1996. Am J Infect Control 1996; 24:380-8.

Unusual Infections

There are many infectious diseases that present diagnostic dilemmas. Vague constitutional symptoms are often associated with these illnesses making the diagnosis challenging. A careful and detailed history, incorporated with physical examination, will aid in developing a broad differential diagnosis that includes some of the more unusual infections. Following are examples of such illnesses.

Toxic Shock Syndrome

- In 1978 a multisystem illness caused by *Staphylococcus aureus* was described by Todd and coworkers. They reported on seven children aged 8-17 yr who had common clinical features of high fever, hypotension, diarrhea, erythroderma, mental confusion, and renal failure. This entity was named toxic shock syndrome.
- In the early 1980s reports of women with toxic shock syndrome associated with menses appeared. Menstrual toxic shock syndrome was believed to be related to hyperabsorbable tampons and since their removal from the market the incidence of disease has decreased.
- Nonmenstrual toxic shock syndrome has become more clearly recognized as cases of menstrual toxic shock syndrome decline. Nonmenstrual toxic shock syndrome may occur following *S. aureus* colonization of the vagina and vaginal infections, the use of contraceptives, postpartum states, or abortions. Men are also at risk following surgical procedure where wounds may become infected, from osteomyelitis, or from respiratory tract infections.
- The management of toxic shock syndrome requires aggressive intravenous fluid resuscitation for hypotension.

Table 11.3. Diagnostic criteria for toxic shock syndrome

Temperature >38.9° C
Systolic blood pressure <90 mm Hg
Rash with subsequent desquamation (especially palms and soles)
Involvement of ≥3 of the following organ systems
- Gastrointestinal: vomiting, profuse diarrhea
- Muscular: myalgia, or >5-fold increase in CPK
- Mucous membrane hyperemia: vagina, conjunctiva, or pharynx
- Renal insufficiency: at lease twice normal BUN or creatinine
- Hepatic: at least twice normal bilirubin, transaminases
- Blood: thrombocytopenia (<100,000 platelett/mm^3)
- Central nervous system: disorientation without focal neurologic signs

Negative serologic results for Rocky Mountain spotted fever, leptospirosis, and measles

CPK = creatine phosphokinase; BUN = blood urea nitrogen. From: MMWR Morb Mortal Wkly Rep 1990; 39(RR-13):1, with permission.

Table 11.4. Frequency of signs and symptoms of toxic shock syndrome

Clinical Signs and Symptoms	Frequency
Diarrhea	98%
Myalgia	96%
Vomiting	92%
Temperature ≥40°	87%
Headache	77%
Sore Throat	75%
Conjunctival hyperemia	57%
Decreased sensorium	40%
Vaginal hyperemia	33%
Vaginal discharge	28%
Rigors	25%

*Note: Rash and shock are not included because they are part of the definition of toxic shock syndrome. Adapted from: Shands KN, Schmid GP, Dan BB et al; Toxic shock syndrome in menstruating women—Association with tampon use and staphylococcus aureus and clinical features in 52 cases; N Engl J Med 303:1436-1442; ©1980 Massachusetts Medical Society, with permission.

- Vaginal examination should be performed early to remove any tampons and to collect cultures, including both cervical swabs and the tampon itself.
- Other potential sources of infection should be cultured and potentially infectious material removed, i.e., drainage of abscesses or wound debridement.
- Antibacterial therapy should be administered immediately with a β-lactamase resistant antistaphylococcal antibiotic, such as intravenous nafcillin or oxacillin (2 g q 4-6 h).
- Clindamycin (600-900 mg IV q 8 h) is a suggested addition to nafcillin/oxacillin due to its ability to decrease toxin production.
- With the incidence of methacillin-resistant *S. aureus* increasing in community acquired infections the addition of vancomycin (1 g IV q 12 h) may be prudent prior to culture and sensitivity testing of the pathogen.
- Patients with toxic shock syndrome should be admitted to an intensive care unit.

11

Suggested Reading

1. Todd J, Fishaut M et al. Toxic-shock syndrome associated with phage-group-1 Staphylococci. Lancet 1978; 2:1116-8.
2. Center for Disease Control and Prevention. Follow-up on toxic shock syndrome—United States. MMWR 1980; 29:279-9.
3. Waldvogel FA. *Staphylococcus aureus* (Including staphylococcal toxic shock syndrome) In: Mandell GL, Bennett JE, Dolin R, eds. Mandell, Douglas, and Bennett's Principles and Practice of Infectious Disease. 5th ed. Philidelphia: Churchill Livingstone, 2000:2089-92.

Syphilis

- *Treponema pallidum* is the pathogen responsible for causing syphilis.
- The clinical course following infection evolves through stages.
  - The primary stage is characterized by a painless and indurated ulcer, or chancre, at the site of innoculation.
  - Skin rash, mucocutaneous lesions, and lymphadenopathy are typical clinical findings associated with secondary infection. Latent infection is defined as the stage without clinical manifestations but positive serologic tests. Duration of infection of <1 yr is known as "early latent syphilis". "Late latent syphilis" refers to asymptomatic infection of >1 yr.

- Tertiary syphilis is a slowly progressive disease that manifests as central nervous system, (e.g., tabes dorsalis and general paresis), cardiovascular (e.g., aortitis), or gummatous disease.
- Congenital syphilis may cause rhinitis, rash, jaundice, notched and widespread incisors (Hutchinson's teeth), and bony abnormalities (saber shins).
- Dark field examination of exudate from a mucocutaneous lesion demonstrating spirochetes is the easiest and most rapid test for syphilis.
- The diagnosis is most commonly made by serologic testing using two general types of tests, nontreponemal tests (Venereal Disease Research Test, or VDRL, and the Rapid Plasma Reagin, or RPR), and treponemal tests (Fluorescent Treponemal Antibody Absorbed, or FTA-ABS, and *T. pallidum* Particle Agglutination, or TP-PA).
- A positive nontreponemal test is defined as a four-fold change in titer (equivalent to an increase in two dilutions, e.g., 1:4 to 1:16).
- Nontreponemal tests usually become nonreactive with time after treatment. False positives may be seen with viral infections, connective tissue disease, pregnancy, and malaria.
- The VDRL test is used to assess the cerebral spinal fluid for the presence of neurosyphilis.
- Treponemal tests are specific antibody tests used to confirm the positive reactions to VDRL or RPR. Most patients with a positive treponemal test will remain positive for life despite treatment.
- Penicillin G is the drug of choice for all stages of syphilis. Patients with penicillin allergy who are pregnant, who have neurosyphilis, or congenital syphilis require desensitization and treatment with penicillin,

Arthropod-Borne Infections

- Symptoms of malaise, myalgias, headache, and the signs of fever and rash are commonly associated with arthropod-borne infections. (Table 11.5) Ticks, mites, lice, and fleas are common vectors. These infections have a geographic distribution based upon the habitat of the animal reservoirs and the insects that transmit the diseases to humans.

Cat Scratch Disease

- The causative agent is *Bartonella henselae.*
- The disease mainly affects children and is typically benign, subsiding within several months.
- The hallmark is regional lymphadenopathy occurring proximal to the site of a cat scratch or bite.

Table 11.5. Arthropod-borne infections

Disease	Pathogen	Signs and Symptoms	Treatment
RMSF*	*Rickettsia rickettsii*	Fever, headache, rash	Doxycycline
Scrub typhus	*R. tsutsugamushi*	Fever, headache, rash	Doxycycline
Epidemic typhus	*R. prowazekii*	Fever, headache, rash	Tetracycline
Ehrlichiosis	*Ehrlichia chaffeensis*	Fever, rash, leukopenia	Doxycycline
Lyme Disease	*Borrelia burgdorferi*	Erythema chronicum migrans, joint pains, arthritis, headache	Doxycycline
Tularemia	*Francisella tularensis*	Lymphadenopathy, cough, pneumonitis, ulcerating lymph nodes, oculoglandular disorder	Streptomycin
Babesiosis	*Babesia microti*	Fever, shaking chills, arthralgias, headache, asplenic patients high risk	Clindamycin, Quinine

*Rocky Mountain spotted fever

Table 11.6. Nontuberculous mycobacteria

Agent	Signs and Symptoms	Treatment
Rapid Growing		
M. fortuitum group *M. chelonei/abscessus*	Cutaneous disease	Surgical, Resistant to anti-TB Rx Amikacin and cefoxitin
Slow Growing		
MAC (M. avium-intracellulare complex)	Pulmonary disease	Clarithromycin Rifampin Ethambutol +/- Surgical
	Disseminated disease (HIV +)	Clarithromycin Ethambutol +/- Rifabutin
M. kansasii	Pulmonary disease	Isoniazid Rifampin Ethambutol
	Disseminated disease (HIV +)	Same as pulmonary
Intermediate Growing		
M. marinum	Cutaneous disease	Clarithromycin **OR** Rifampin and Ethambutol

From: Mandell: Principles and Practice of Infectious Diseases. 5/e:2632. ©2000 Elsevier Inc., with permission.

- Atypical manifestations, occurring <5% of cases, include conjunctivitis with lymphadenopathy (oculoglandular fever), encephalitis, myelitis, peripheral neuopathy among others.
- Treatment is with azithromycin but is usually reserved for severe disease.
- The differential diagnosis of unilateral lymphadenopathy includes nontuberculosis mycobacterial infection, tularemia, brucellosis, syphilis, histoplasmosis, neoplasms, and others.

11

Nontuberculous Mycobacteria

- Mycobacteria species other than *M. tuberculosis, M. bovis, M. africanum*, and *M. leprae* are considered "atypical" mycobacteria. There are over 50 species that are frequently categorized by growth rates, rapid (<7 days) and slow (>7 days) growing mycobacteria (Table 11.6). 95% of all human mycobacteria infections are caused by seven species, *M. tuberculosis, M. leprea, M. avium-intracellulare* comples, *M. kansasii, M. fortuitum, M. chelonae*, and *M. abscessus*.

Suggested Reading

1. Manguina C, Gotuzzo E. Bartonellosis: Old and new. Infect Dis Clin N Amer 2000; 14:1-22.
2. Brown B, Wallace Jr RJ. Infections due to nontuberculous mycobacterium. In: Mandell GL, Bennett JE, Dolin R, eds. Mandell, Douglas, and Bennett's Principles and Practice of Infectious Diseases. 5th ed. Philadelphia: Churchill Livingstone, 2000:2630-6.

Human Immunodeficiency Virus

The appearance of the human immunodeficiency virus (HIV) approximately two decades ago has changed the face of medicine. HIV infected patients are at risk for opportunistic infections and many require additional prophylactic antimicrobial therapy. Each year numerous developments in the management of HIV infection, primarily new medications, are introduced. In 1986 zidovudine was the first antiretroviral approved for the treatment of HIV infection. Today 16 antiretroviral drugs are currently available. Patients infected with HIV are treated with complex regimens of antiretroviral "cocktails" including reverse transcriptase inhibitors, protease inhibitors, and nonnucleoside reverse transcriptase inhibitors. Each medication has numerous side effects, some of which are lethal. This chapter reviews the common problems faced by emergency physicians when treating HIV infected patients, opportunistic infections and reactions to highly active antiretroviral therapy (HAART).

Opportunistic Infections

- The precise opportunistic infection (OI) any given patient will develop depends upon the level of immunosuppression and the types of exposures the patient experiences.

Pneumocystis carinii Pneumonia (PCP)

- *P. carinii* is an organism of low virulence in the immunocompetent host and is found in the lungs of both humans and animals worldwide.
- PCP continues to be a common OI in North America and is a common presentation in those newly diagnosed with HIV infection. Despite prophylaxis with trimethoprim/sulfamethoxazole and the ensuing reduction in the incidence of PCP, this infection remains a substantial cause of morbidity and mortality.

Clinical Presentation

- PCP is an interstitial pneumonia with an insidious onset of shortness of breath, dry cough, and fever that develops over weeks to months, usually in HIV infected patients with CD4 count <200.
- Tachypnea and tachycardia are common findings, and auscultation of the lungs reveals crackles in approximately one-third of infected adults. Hypoxemia and an elevated arterial-alveolar gradient may be noted and are reflections of the severity of disease.
- The chest radiograph classically shows bilateral diffuse interstitial infiltrates. Unilateral infiltrates, cavities, pneumatoceles, lymphadenopathy, effusions and normal appearances on chest radiograph are atypical, but do occur.
- Serum lactic dehydrogenase (LDH) may also be elevated but is nonspecific.
- *P. carinii* is rarely found in expectorated sputum but may be detected by inducing sputum with inhaled saline. Fiberoptic bronchoscopy has the highest yield of >90%.

11

ED Management

- Many patients with mild to moderate disease are treated empirically based on clinical diagnosis.
- Trimethoprim/sulfamethoxazole remains the drug of choice and may be administered either orally or intravenously. The dose is 15-20 mg/kg/day trimethoprim and 75-100 mg/kg/day sulfamethoxazole in 3 or 4 divided doses. Clindamycin (900 mg IV q 8 h) plus primaquine (30 mg po q day) is an alternative therapy for patients who are allergic to sulfa medications. Other alternatives include atovaqone (750 mg po BID), pentamidine (4 mg/kg/d IV).
- Prednisone is indicated for patients who are hypoxic (pO_2 <70 mm Hg) and should be administered 15-30 min before the antibiotics are given. The dose is 40 mg po BID for 5 days, then 40 mg po q day for 5 days, and then 20 mg po q day for 11 days.

Toxoplasmosis

- *Taxoplasma gondii* is a coccidian parasite that infects animals and humans worldwide. Infection in immunocompetent individuals is rarely of clinical significance, but in immunocompromised patients infection commonly involves the brain, lungs, and eyes.

- Encephalitis is the most common presentation, and it may demonstrate a wide array of clinical manifestations including altered mental status, weakness, sensory abnormalities, meningismus, cerebellar signs, movement disorders, seizures, and neuropsychiatric disturbances.
- Presumptive diagnosis is made by detecting multiple bilateral cerebral lesions on computer tomography of the brain. The lesions are "ring-enhancing", or contrast enhancing in up to 80% of patients. A single lesion may be seen with Toxoplasma encephalitis in up to 40% of patients, but the suspicion for central nervous system (CNS) lymphoma is elevated with this finding.
- These patients should be admitted and treated with oral pyramethamine, folinic acid and either sulfadiazine or clindamycin for 2 wk.
- Maintenance therapy is required for life due to a relapse rate of 80%.

Cytomegalovirus (CMV)

- Retinitis is the most common manifestation of CMV disease and usually occurs when the patient's CD4 cell count is <50 cells/mm^3. CMV also causes esophagitis, enteritis, colitis, and pneumonia.
- The symptoms of CMV retinitis are nonspecific and include floaters, flashing lights, loss of visual field, or a vague sense of visual loss. Visual impairment caused by focal necrotizing retinitis can rapidly progress to blindness.
- The presumptive diagnosis is made by observing white fluffy retinal infiltrates occurring within areas of hemorrhage on fundoscopic exam.
- Treatment with intravenous gancyclovir or foscarnet is usually empiric due to the invasive nature of obtaining retinal or vitreous material for examination.

Cryptococcus

- *Cryptococcus neoformans* is a ubiquitous encapsulated organism.
- *Cryptococcus* fungus causes subacute meningitis, as well as pneumonia and skin lesions, in patients with T-cell defects. AIDS is the predisposing factor in approximately 90% of cases.
- The presenting symptoms of crytococcal meningitis are mild and nonspecific, such as nausea, headache, dizziness, irritability, somnolence, or confusion. Cranial nerve dysfunction and papilledema are noted in up to 20% and 30%, respectively. Seizures occur late in the course of infection. Fever is not always present and neurologic examination is often normal. Most patients have minimal or no nuchal rigidity.
- Diagnosis is made by identifying the yeast in the cerebrospinal fluid (CSF) with an India ink preparation, by cryptococcal antigen detection, or by culture. Over 90% of patients will test positive for serum cryptococcal antigen. The opening pressure of the CSF during lumbar puncture is often elevated, and the CSF glucose is decreased. Usually more than 20 white blood cells are noted. A normal CSF profile in the setting of suspected cryptococcal meningitis mandates fungal culture of the CSF.
- Routine laboratory testing, i.e., CBC and sedimentation rate, are usually normal.
- The treatment of choice is intravenous amphotericin B (0.5 0.8 mg/kg/d IV) plus 5 FC (25 mg/kg po q 6 h) for 2 wk or until the patient is clinically stable is. Oral fluconazole is required for lifelong suppressive therapy.

Suggested Reading

1. USPHS/IDSA Prevention of Opportunistic Infections Working Group. 1999
2. USPHS/IDSA Guidelines for the Prevention of Opportunistic Infections in Persons Infected with Human Immunodeficiency Virus. Clin Infect Dis 2000; 30:S29-65.

Table 11.7. Adverse reactions of nucleoside reverse transcriptase inhibitors

Zidovudine (AZT, Retrovir)	Didanosine (ddI, Videx)	Zalcitabine (ddC)
headache	diarrhea	peripheral neuropathy
nausea	peripheral neuropathy	stomatitis
bone marrow suppression	* pancreatitis	rash
myopathy	elevation of ALT/AST	
Stavudine (d4T,Zerit)	**Lamivudine (3TC, Epivir)**	**Abacavir (Zigen)**
peripheral neuropathy	nausea	* hypersensitivity reaction
gastrointestinal upset	headache	nausea/vomiting/ diarrhea
headache	fatigue	* severe rash
elevated ALT/AST	insomnia	fever

* Warrants discontinuation of all antiretrovirals

Highly Active Antiretroviral Therapy (HAART)

- The widespread use of combination therapy with multiple antiretroviral agents has reduced the morbidity and mortality due to HIV infection. Death rates due to HIV have declined as much as 65% in the United States. Despite these optimistic results, only one-half of patients achieve maximal viral suppression. Many factors prevent successful treatment including viral resistance, inadequate adsorption due to gastrointestinal disease, lack of adherence to complex drug regimens, adverse drug effects, expense, and coincident mental health and substance abuse problems.
- Emergency physicians should never discontinue HAART unless potentially lethal adverse effects develop. If this situation occurs, then all antiretrovirals should be stopped simultaneously to prevent the development of resistance.
- Many new antiretroviral agents are currently under development and emergency physicians must continue to be vigilant and prepare for the recognition and management of adverse effects.

11

Nucleoside Reverse Transcriptase Inhibitors

- The first class of antiretrovirals to be introduced was the nucleoside reverse transcriptase inhibitors (NRTIs). There are currently six NRTI's available and although their mechanism of action is similar, adverse reactions for each agent are unique.
- Common symptoms include fever, rash, and gastrointestinal complaints. Over 85% of cases began within the first 6 wk of therapy. A rare but deadly adverse reaction is lactic acidosis with severe hepatomegaly. Other common adverse reactions are listed in (Table 11.7).
- Adefovir dipiroxil is a novel nucleotide reverse transcriptase inhibitor. Renal toxicity is the most important adverse effect and is seen in over one-third of patients. Adefovir should be discontinued if renal function is impaired.

Nonnucleoside Reverse Transcriptase Inhibitors

- All NNRTIs are capable of causing a rash that may be mild and self-limited or severe and capable of progressing to Stevens-Johnson syndrome. Mild rashes are treated with oral antihistamines or topical steroid creams. Antiretrovirals may be continued and the patient promptly reevaluated by their primary physician. However, if the rash is severe or associated with fever, severe pruritis, ulcers, blisters, vomiting, diarrhea, mucosal involvement, or muscle or joint pain, then all antiretrovirals must be discontinued.
- Common adverse reactions of the NNRTIs are listed in (Table 11.8).

Table 11.8. Adverse reactions of non-nucleoside reverse transcriptase inhibitors

Nevirapine (Viramune)	Delavirdine (Rescriptor)	Efavirenz (Sustiva)
* Stevens-Johnson syndrome	* severe rash	CNS effects
* severe rash	diarrhea	* severe rash
fever	fatigue	elevated ALT/ALT
elevated ALT/AST	headache	
headache		

* Warrants discontinuation of all antiretrovirals

Table 11.9. Adverse effects of protease inhibitors

Saquinovir (Invirase)	Indinavir (Crixivan)	Ritonivir (Norvir)
gastrointestinal symptoms	nephrolithiasis	gastrointestinal_symptoms
elevated ALT/AST	hyperbilirubinemia	headache
	gastrointestinal symptom	elevated ALT/AST
		taste perversion
Nelfinavir (Viracept)	**Amprenavir (Angenerase)**	**Lopinavir/ritonivir (Kaletra)**
diarrhea	* severe rash	gastrointestinal symptoms
nausea	nausea/vomiting	headache
	headache	weakness

* Warrants discontinuation of all antiretrovirals

Protease Inhibitors

- Adverse effects are common (Table 11.9). Other complications include hyperlipidemia, hyperglycemia, diabetes, and fat redistribution.

Suggested Reading

1. Max B, Sherer R. Management of the Adverse Effects of Antiretrovial Therapy and Medication Adherence. Clin Infect Dis 2000; 30(Suppl 2):S96-116.

Prevention and Prophylaxis

Isolation

Preventing the transmission of infectious microorganisms from infected or colonized patients to other patients and health care workers must be a priority in the ED. The Centers for Disease Control and Prevention and Hospital Control Practices Advisory Committee's recommendations for isolation are reviewed below (Table 11.10).

- **Standard Precautions** have replaced "universal precautions" and apply to all patients. Standard precautions stipulate that gloves should be worn when touching blood, bloody fluids, secretion (other than sweat), mucous membranes, or nonintact skin. Hands should always be washed after touching a patient and after removing gloves. (Of note are observational studies demonstrating compliance with hand washing by health care workers to be <50%.) Chorhexidine and isopropyl alcohol are superior to soap and water in removing transient flora from the hands. Chlorhexidine additionally provides a residual antibacterial effect. Gowns, face shields, and/or eye protection are recommended for procedures or activities that are likely to generate a splash of blood, bloody fluids, secretions, or excretions.

11

Table 11.10. Isolation guidelines for selected conditions

Standard Precautions	To be used for all patients
Airborne Precautions	Tuberculosis—pulmonary, laryngeal (suspected or confirmed)
	Varicella/Zoster virus—chickenpox
	Herpes zoster—disseminated
	Herpes zoster—immunocompromised patient
	Rubeola (measles)
Droplet Precautions	Adenovirus
	Epiglottitis (*H. influenzae* in children)
	Diphtheria
	Influenza
	Meningitis (suspected bacterial)
	Menigococcal disease
	Mumps
	Mycoplasma pneumoniae
	Parvovirus B 19
	Pertussis
	Plague (pneumonic)
	Rubella
	Streptococcal (group A) pneumonia, pharyngitis, scarlet fever
Contact Precautions	Abscesses
	Encephalitis
	Gatroenteritis (including *C. difficile*)
	Herpes simplex
	Impetigo
	Lice or scabies
	Multidrug resistant organism (including VRE*)
	Respiratory syncytial virus
	Rubeola
	Varicella/Zoster virus—Herpes zoster—localized
	Wound infections

* Vancomycin resistant Enterococci

- **Airborne precautions**, in addition to standard precautions, are required for all patients known or suspected to be infected with a pathogen that is transmitted by the airborne route (Table 14). Droplet nuclei (particles ≤5 µm) or dust particles containing the infectious agent are produced when an infected patient coughs, sneezes, or speaks. These particles may remain suspended in the air and travel long distances. Airborne precautions mandate a private room with negative pressure and a closed door. All visitors must wear an approved respirator mask (N95) that is fit tested. Transporting the patient out of a negative pressure room should be minimized and the patients must wear a standard surgical mask when outside of the room. Patients infected with the human immunodeficiency virus (HIV) presenting with pneumonia, i.e., cough, fever, and pulmonary infiltrates on chest radiographs, should be placed in respiratory isolation until pulmonary tuberculosis can safely be excluded.
- **Droplet precautions**, in addition to standard precautions, are implemented for all patients known or suspected of having an infectious illness transmitted by droplets, i.e., large particles that do not remain suspended in the air (Table 14). Droplets are produced when infected patients talk, cough, or sneeze. The droplets generally travel no farther than 3 feet. These patients require a separate room but do not require negative pressure or air

filtering. A regular surgical mask should be worn by visitors or health care workers when they enter the patient's room and by the patient when transported outside of the room.

- **Contact precautions** are required for patients known or suspected of being infected or colonized by organisms that are spread by direct contact, or by contact with contaminated surfaces. Patients requiring contact isolation also require standard precautions. These patients require a private room with strict disinfection of all equipment after use. Barrier precautions, such as gloves and gowns, are used to prevent contamination of exposed skin and clothing.

Suggested Reading

1. Edmond MB, Wenzel RP. Isolation. In: Mandell GL, Bennett JE, Dolin R, eds. Mandell, Douglas, and Bennett's Principles and Practice of Infectious Diseases. 5th ed. Philadelphia: Churchill Livingstone, 2000:2991-5.

Vaccinations

Vaccines have proven to be one of the most important means for preventing morbidity and mortality from infectious diseases worldwide. In the ED, the prevention of infectious complications of wounds or bites is a high priority. In particular, tetanus toxoid and rabies vaccination are routinely administered. Guidelines for their use are summarized below. (Tables 11.11 and 11.12)

Table 11.11. Guidelines for the routine use of tetanus prophylaxis in the U.S.

	Simple and Uncomplicated Wounds	All Other Wounds *
Vaccinated	**Td 0.5 ml IM q 10 yr	Td 0.5 ml IM q 5 yr
Unvaccinated (unknown vaccination or <3 doses)	Td 0.5 ml IM now and 2 later doses	Td 0.5 ml IM now and 2 later doses TIG 250 units IM

* Including, but not limited to, burns, puncture wounds, avulsions, crush injuries, and wounds contaminated with soil, saliva or feces. ** If 3 doses of "fluid toxoid" were given, then a fourth dose of tetanus toxoid should be given, preferably the "absorbed toxoid".

Table 11.12. Guidelines for post-exposure treatment for rabies

Animal Type	Recommendations
Dog, cat, ferret	Observe for 10 days if animal is healthy -hold PET* Suspect rabid animal -begin PET Unknown (animal escaped) -contact public health officials
Raccoon, bat, skunk, fox, coyote, other carnivores	Suspect rabid animal -begin PET (unless animal is available and tests negative)
Large rodents (beavers and woodchucks), Small rodents (chipmunks, squirrels, etc), livestock, and other mammals	Consider each case individually -contact public health officers

* Post-exposure treatment

11

Tetanus

- A primary vaccination series includes three doses providing protective antibodies in more than 95% of recipients. It is recommended for children over 7 yr of age and adults. For children <7 yr the vaccination series begins at 2 mo of age and includes 5 doses. The diphtheria toxoid, tetanus toxoid, and either whole-cell or acellular pertussis vaccine adsorbed (DTP or DTaP) are recommended. Patients older than 7 yr should receive the adsorbed tetanus and diphtheria toxoids (Td).

Rabies

- Rabies is a viral illness transmitted by the contaminated saliva of an infected animal. It causes progressive and inevitably fatal encephalitis. Fortunately, it is uncommon in the United States. Only 36 human cases of rabies have been reported between 1980 and 1996.
- Rabies post-exposure treatment is recommended for persons exposed (i.e., bites, scratches or mucous membrane contact) to bats unless the bat is available for immediate testing and is found to be negative for rabies. Coyotes, foxes, raccoons, skunks, and large rodents (beavers and woodchucks) are often infected with rabies and post-exposure treatment is recommended for exposures associated with these animals. Small rodents (squirrels, hamsters, guinea pigs, gerbils, chipmunks, rats and mice) and lagomorphs (hares and rabbits) are rarely found infected with rabies and have not been known to transmit the disease to humans. Domestic dogs, cats, and ferrets are infrequently infected with rabies and the likelihood varies by geographic region. If the animal is healthy it may be confined and observed for 10 days. If the animal becomes ill it must be euthanized, its neural tissue examined for infection, and if present, post-exposure treatment begun. If the animal remains well, then no treatment is warranted. State and local health departments should be consulted for uncertain cases, those involving small rodents, and cases where treatment is begun.
- All wounds potentially exposed to rabies must be aggressively and thoroughly cleaned and irrigated.

- Patients suspected of being exposed to a rabid animal should begin post-exposure treatment immediately. If a patient delays seeking evaluation, treatment should not be withheld, because incubation periods >1 yr have been reported.
- The recommendation for previously unvaccinated individuals is a regimen of rabies immune globulin (RIG) 20 IU/kg and a series of rabies vaccines, 1.0 ml IM in the deltoid area, to be given on day 0, 3, 7, 14, and 28. If anatomically feasible, the full dose of RIG should be infiltrated around the wound, with the remainder given IM at another anatomical site distant from the wound and the vaccine site. If the patient is previously vaccinated, then RIG is not given, and the passive vaccine is administered, 1.0 ml IM in the deltoid area, on days 0 and 3 only.

Prophylactic Antimicrobials in Wound Care

- Prophylactic antimicrobials for low risk lacerations have not been shown to reduce the incidence of infection. However, the American College of Emergency Physicians' (ACEP) "Clinical Policy for the Initial Approach to Patients Presenting with Penetrating Extremity Trauma" recommends antibiotics for lacerations at high risk. (Table 11.13)
- Antibiotics should provide coverage against pathogens most likely to infect skin and soft tissues, i.e., *Staphylococcus aureus* and *Streptococcus* species. First generation cephalosporins (cephalexin orally or cefazolin parenterally) are adequate. Erythromycin may be substituted in the penicillin allergic patient with a history of IgE mediated hypersensitivity (i.e., urticaria, angioedema, or anaphylaxis). Wounds that are contaminated with saliva, soil, or feces require antibiotic prophylaxis that includes additional activity against anaerobic and Gram-negative bacteria; amoxicillin-clavulanic acid is recommended in this setting.

Table 11.13. Wounds at high risk for infection

Located on hand or lower extremity
Devitalized tissue
Contamination
Involvement of joints, tendons or bone
Human or animal bites
Impaired host immune response (e.g., diabetes or HIV infection)

- Bite wounds are common injuries treated in the Emergency Department. Although these wounds may initially appear as trivial superficial wounds, the infectious complications can include tenosynovitis, local abscesses, septic arthritis, and osteomyelitise. It is estimated that 15-20% of dog bites and over 50% of cat bites become infected. Most authorities recommend prophylactic antibiotic therapy for 3-5 days in all but the most trivial superficial bite wounds. Amoxicillin/clavulanate is the antibiotic of choice due to its spectrum of activity that includes the most common pathogens found in infected bite wounds, i.e., *S. aureus, Eikinella corrodens, Pasturella multocida*, streptococci, and anaerobes. Ciprofloxacin with clindamycin may be given penicillin allergic patients.

Suggested Reading

1. Fesmire FM, Dalsey WC, Howell JM et al. American College of Emergency Physicians: Clinical Policy for the Initial Approach to Patients Presenting with Penetrating Extremity Trauma. Ann Emerg Med 1999; 33.
2. Talan DA, Citron DM, Abrahamiam FM et al. Bacteriologic analysis of infected dog and cat bites. N Engl J Med 1999; 340:85-92.

Post-Exposure Prophylaxis

Neisseria meningitidis *Exposure*

- Antimicrobial chemoprophylaxis is recommended for "close contacts" of patients with known disease caused by *Neisseria meningitidis*. (Table 11.14) The risk of sporadic meningococcal disease is up to 800 times greater for "close contacts" of patients. 11
- Prophylaxis should be administered as soon as possible, ideally within 24 h, after the case patient is identified. Antimicrobial agents recommended by the Center for Disease Control and Prevention (CDC) are listed in (Table 11.15).

Blood and Body Fluid Exposure

- The initial management includes cleaning the exposure site and appropriate wound care. There is no evidence to support squeezing of the site, use of caustic materials, or injection of any substance.
- Post-exposure issues include prophylaxis and/or testing for infection with HIV, hepatitis B, and hepatitis C. (Table 11.16)

Table 11.14. Definition of "close contacts" of patients with N. meningitidis disease

a. Household members
b. Day care center contacts
c. Anyone directly exposed to the patient's oral secretions
- kissing
- mouth-to-mouth resuscitation
- endotracheal intubation/ airway management

- Assessment of the risk of HIV includes the evaluation of the source material, the type of exposure, degree of exposure, and the source patient. (Tables 11.17-11.19)
- If the HCW is vaccinated against hepatitis B, then serum antibody titers for HBsAb are measured following an exposure. When the levels are adequate for protection (i.e., >10 mIU ml) no post-exposure prophylaxis is warranted. If the HCW levels fall below 10 mIU/ml, then post-exposure prophylaxis is recommended. (Table 11.20)
- There are no recommendations for post-exposure prophylaxis for hepatitis C. Routine testing includes anti-HCV of the source patient, and both anti-HCV and ALT at baseline and after 6 mo for the exposed HCW.

Table 11.15. Antimicrobial prophylaxis against meningococcal disease

Drug	Age	Dose
Ciprofloxacin*	>18 yr	500 mg po x 1 dose
Ceftriaxone	Adults	250 mg IM x 1 dose
Ceftriaxone	<15 yr	125 mg IM x 1 dose
Rifampin	<1 mo	5 mg/kg po q 12 h x 4 doses
Rifampin	Children	10 mg/kg po q 12 h x 4 doses
Rifampin*	Adults	600 mg po q 12 h x 4 doses

*Not recommended for pregnant women.

Table 11.16. Risk of infection following percutaneous injury

HIV	0.3%
Hepatitis C	1.8%
Hepatitis B	2-40%

Table 11.17. Exposures at risk for the development of HIV

Source Material:	Blood, bloody fluids, or other potentially infectious materials*
Type of Exposure:	Mucous membranes, non-intact skin, or percutaneous injury
Degree of Exposure:	Large volume, prolonged contact, or deep penetration
Source Patient:	HIV positive, high viral load

*Semen, vaginal secretions, pericardial, cerebrospinal, peritoneal, synovial and amniotic fluids.

Table 11.18. Recommendations for post-exposure prophylaxis for HIV

1. PEP is not recommended—brief exposure on intact skin
2. PEP may not be warranted—low titer exposure, low volume, non-intact skin or mucous membranes
3. Consider basic regimen—low titer exposure with large volume and/or long duration exposure or less severe percutaneous injury (scratch or solid needle)
4. Recommend basic regimen—high titer exposure, low volume, non-intact skin or mucous membrane
5. Recommend expanded regimen—more severe percutaneous injury (large bore needle, deep injury, visible blood or needle from vessel) regardless of titer: high titer exposure, large volume and/or long duration; or high titer, less severe percutaneous injury (scratch or solid needle)

Table 11.19. Post-exposure prophylaxis regimens for HIV

Basic	zidovudine (AZT) 300 mg po BID x 4 wk **and** lamivudine (3TC) 150 mg po BID 4 wk
Extended	BASIC plus indinavir (Crixivan) 800 mg po q 8 h x 4 wk **or** nelfinavir (Viracept) 750 mg po TID x 4 wk

Table 11.20. Recommended post-exposure prophylaxis—Hepatitis B

Vaccinated HCW*:		
	Serum HbsAb titers >10 mIU/ml	No treatment
	<10 mIU/ml	HBIG** 5.0 ml IM **and** HB*** vaccine x 1
Unvaccinated HCW		HBIG 5.0 ml IM **and** HB vaccine x 3 (at 0,1 and 6 mo)

*HCW (Health care worker). **HBIG (Hepatitis B immunoglobulin). ***HB (Hepatitis B)

Suggested Reading

1. Centers for Disease Control and Prevention. Control and prevention of meningococcal disease: Recommendations of the Advisory Committee on Immunization Practices. MMWR 1997; 46 (RR-5):1-9.
2. Centers for Disease Control and Prevention. Public health service guidelines for the management of health-care worker exposure to HIV and recommendation for postexposure prophylaxis. MMWR 1998; (RR-7):1-33.
3. Centers for Disease Control and Prevention. Recommendations for prevention and control of hepatitis C virus (HCV) infection and HCV-related diseases. MMWR 1998; 47(RR-19):1-38.

CHAPTER 12

General Management of the Poisoned Patient

Kathryn Challoner

Part A: Initial Management

As with all emergency situations, initial management must immediately assess and address the ABCs of emergency management.

A. Airway

- The clinican must guarantee an adequate airway and a decreased level of consciousness, absence of a gag and inability to control secretions should result in a decision to intubate.
- Many poisoned patients with altered mental status have been involved in trauma (motor vehicle accidents, falls, altercations) so intubation should proceed with in line stabilization of the neck.
- C-spine precautions should be observed.

B. Breathing

- A pulse oximetry reading should be immediately obtained and the adequacy of tidal volume assessed.
- There are two types of pulse oximeters—fractional and functional and only the fractional may detect significant qualities of carboxyhemoglobin and methemoglobin.

C1. Circulation

- A perfusing blood pressure is essential to maintain adequate cerebral and coronary circulation.
- A word of caution—if a poison is severe enough to depress the blood pressure—it may also depress left ventricular function (especially the classes of sedative-hypnotics, calcium channel blockers and β-blockers).
- To give the patient large volumes of intravenous saline in an attempt to treat hypotension may result in putting the patient into fluid overload.
- Fluids must be administered cautiously and usually in conjunction with a central venous pressure line.

C2. Cerebral Metabolic Need

- A patient with altered mental status may have undetected hypoglycemia.
- Several drugs may lower the blood glucose level including pentamidine, ethanol and β-blockers.
- A fifth vital sign should be an immediate measurement of the blood glucose level on any patient with altered mental status.

D. Decontamination

- If a poison is absorbable through the skin or mucous membranes, emergency staff must first don appropriate barrier gear to prevent cross contamination.

Emergency Medicine, edited by Sean Henderson. ©2006 Landes Bioscience.

Table 12A.1. Absorbents used

Toxin	Gastrointestinal Agent
Paraquat	Fuller's earth
Lithium	Kayexalate
Iodine	Starch
Thallium	Prussian blue
Phosphorus	Mineral oil, potassium permanganate solution
Iron	Sodium bicarbonate
Fluoride	Lime water (calcium carbonate)

- Skin should be flushed with copious amounts of water and eyes irrigated.
- Water **should not** be used in cases of heavy metal exposure (phosphorus, sodium, and calcium oxide).
- As a general principle in gastrointestinal decontamination, if a poison is organic it will bind to charcoal and if it is inorganic, a specific precipitating, binding or oxidizing agent that could be used (Table 12A.1).

  There are certain categories of poisons where gastric emptying is contraindicated or controversial:

  1. Alkaline caustic ingestions
     - These ingestions cause liquefactive necrosis of the esophagus resulting in deep burns and perforation.
     - Care must be taken not to damage or manipulate the esophagus further so emesis or lavage is contraindicated.
  2. Petroleum distillate ingestions
     - The decision to perform gastric lavage in these ingestions must be made on a case by case basis.
     - In general if the ingestion is an aliphatic hydrocarbon only, gastric emptying is not performed as these hydrocarbons carry a low risk of systemic toxicity but a high risk of pulmonary toxicity if aspirated.
     - Difficulty is encountered if the hydrocarbon is a vehicle for another substance of high toxicity—in which case it is wise to obtain the opinion of a toxicologist as to the risk benefit ratio for the patient.

12

Gastric Emptying—Use of Activated Charcoal Plus Cathartic

- The decision to perform gastric emptying is made on a case by case basis; there is no blanket rule.
- The clinician must consider the toxicity of the substance, coingestions, bioavailability, and pharmacokinetics of the substance when the decision to empty the stomach is made.
- Some substances can be recovered 24 h or more after ingestion (e.g., Lomotil).

Ipecac

- There are several serious concerns in using ipecac, and it is now rarely used.
- Ipecac cannot be used if there is any possibility that the patient's level of consciousness will decrease or develop seizures.

Charcoal Lavage

- The most effective means of gastric decontamination is charcoal lavage—i.e., instill 50 g of activated charcoal into the patient's stomach—lavage until clear and then reinstill 50 g of activated charcoal plus cathartic.
- Often an awake patient more than one hour from ingestion may just be given oral charcoal mixed with cathartic to drink.

Cathartics

- There are two types of cathartics—the saccharides (sorbitol or mannitol) and the salines (magnesium citrate, magnesium sulfate and sodium sulfate).
- The concern with the saccharides is that if the gut is not moving, water might be drawn into the gut causing distension and increased transluminal pressure across the gut wall resulting in ischemic necrosis of the bowel.
- Cathartics should not be used if the patient has an ileus or already has severe diarrhea.

Whole Bowel Irrigation

- In whole bowel irrigation, electrolyte balanced solutions are used to flush the toxin from the bowel.
- This can be very effective in heavy metal (iron) or lithium poisoning.

E. Enhanced Elimination (Toxicokinetics)

There are four basic ways the body can be assisted in eliminating the toxin:

1. Pulsed Serial Activated Charcoal
    - This modality can be used if the toxin has an active toxic metabolite that undergoes enterohepatic or enteroenteric recirculation and if no ileus is present.
2. Forced Diuresis
    - This modality has decreased in usage because the risk usually outweighs the benefit.
    - To perform this modality the toxin must be eliminated by the kidneys and usually it is necessary to give the toxin a ionic charge so that it will be trapped in the renal tubules.
    - Forced alkaline duresis is still sometimes employed in moderate salicylate and phenobarbital overdoses but there are many concerns about this maneuver.
3. Hemodialysis
    - To consider this modality the clinician needs to know the degree of protein binding, whether a toxin is water soluble and the molecular weight of the toxin.
    - The higher the protein binding, the less dialyzable the toxin is.
    - The most common dialyzable toxins are salicylate, lithium and the alcohols.
4. Hemoperfusion
    - The principle of hemoperfusion is that anticoagulated blood is passed through a column containing either charcoal or resin absorbents.
    - The clinician should know the volume of distribution of the toxin.
    - After hemoperfusion the toxin will undergo redistribution from the peripheral tissues creating a rebound effect in the patient.
    - This is an excellent form of treatment in most of the sedative hypnotics, lipid soluble drugs, and theophylline intoxication.

F. Antidote

Once the clinician has addressed decontamination and enhanced elimination, the next step is the administration of an antidote (Table 12A.2).

Work-Up

- It is important to determine the identity of the ingestion as some poisons may manifest their toxicity at a later stage.
- Sometimes the poison can be deduced by the toxidrome that the patient manifests or by certain breath, excretia, skin colors or laboratory abnormalities.
- The toxicology requests should always include a request for serum acetaminophen level as this may be difficult to detect clinically. Acetaminophen will manifest its toxicity later and has a definitive and effective antidote.
- Other lab tests may include blood gases to detect acid base abnormalities or unusual hemoglobins and liver and kidney function tests.

Table 12A.2. Antidotes

Poison	Antidote
Acetaminophen	N-acetylcysteine
Arsenic	BAL (British Antilewisite Agent), d-penicillamine
Carbon monoxide	Oxygen
Cyanide	Sodium nitrite and sodium thiosulfate
Digitalis, Cardiac glycosides	Digibind
Ethylene glycol	Ethyl alcohol
Fluoride	Calcium and magnesium
Gold	BAL
Iron	Deferoxamine
Isoniazid	Pyridoxine
Lead	EDTA (edetate calcium disodium) and BAL, d-penicillamine
Mercury	BAL, d-penicillamine
Methyl alcohol	Ethyl alcohol
Nitrites	Methylene blue
Opiates	Naloxone
Organophosphates	Atropine and pralidoxime
Thallium	Prussian blue (ferric ferrocyanide).

- An EKG looking for conduction abnormalities.
- A KUB radiograph can be used to determine the presence of radioopaque tablets.

Dispositon On-Going Care and Prevention

- Often hospital observation is necessary because the toxin may have a latent phase where the patient seems to recover then decompensates.
- The complications of the poisoning will require further treatment and management.
- Especially important is the prevention factor—the clinician needs to identify who is at risk for future ingestion
- Psychiatric or social work consultation may be necessary.
- Education for poison prevention must be given.

12

Part B: Individual Toxins

Acetaminophen

Pharmacology and Pathophysiology

- This product is well absorbed from the gastrointestinal tract and has a pKa of 9.5.
- It has a peak serum level of o 1/2 to 2 h following a therapeutic oral dose.
- Its effects are observed at 30 min and last 4 h, while sustained-release formulations last 6-8 h.
- The plasma binding (e.g., to albumin) is low—this drug does not displace other drugs from binding sites.
- The drug is eliminated by hepatic metabolism.
- 90% is converted to inactive glucuronide and sulfate conjugates which are excreted.
- The major metabolite in adults is APAP-glucuronide and in neonates, APAP-sulfate.
- Plasma half-life is 2 h after a therapeutic dose.
- In normal doses, only 2-4% is processed down the P450 pathway to form the toxic intermediary NAPBQI (N-acetyl-para-benzoquinoneimine).[1]
- Minimal toxic doses are estimated to be 140 mg/kg in children and 7.5 g in adults.

- Significant toxicity may occur after a 250 mg/kg acute overdose.
- The elimination T/2 can be related to the extent of liver damage: at half lives >4 h some hepatic necrosis is likely and half lives >10-12 h, fatal hepatic encephalopathy may occur.
- Toxic doses of APAP depletes glutathione, saturates conjugation pathways and depletes sulfate stores.
- The capacity of the liver to detoxify NAPBQI is exceeded and NAPBQI covalently binds to the hepatocyte proteins causing hepatic necrosis.
- Toxicity is increased with:
  a. Drugs that induce P450 enzymes
  b. Ethanol/INH
  c. Fasting
  d. Depleted reserves of hepatic glutathione (preexisting liver disease)
- In chronic poisoning the data is very unclear.
- Adults and children should not exceed 4 g daily.
- Currently, treatment is recommended if the ingested estimated dose is greater or equal to 6-8 g/day for two or more days or if patient has elevated liver function tests.
- Currently, there is available over the counter sustained released caplets consisting of 325 mg of immediate release acetaminophen on one side and 325 mg in a matrix formulation designed for slow release on the other are available.
- In a therapeutic dose, 95% is released in 5 h, but studies have shown a decreased rate of dissolution with increasing tablets.
- One case report with a coingestion demonstrated a second toxic peak at 14 h when the 6 and 10 h levels were nontoxic.[2]

Clinical Features

There are classically four stages of acetaminophen toxicity.[1]

1. Stage one occurs from 1/2 to 24 h and may be asymptomatic or consist of mild gastrointestinal complaints.
2. Stage two lasts from 24-48 h and consists of right upper quadrant pain, elevated liver function tests with 25% of patients showing some renal dysfunction.
3. Stage three occurs at 72-96 h and consists of hepatic and renal failure, coagulation defects, myocardial pathology, pancreatitis, and encephalopathy.

12

4. Stage four occurs at 4 days to 2 wk and consists of recovery.
   - Liver biopsy at this point will reveal centrilobular necrosis.
   - Death can occur at 3-5 days.

Work-Up and Laboratory

- Blood levels should be sent for serum acetaminophen levels at 4 h or more after an acute ingestion. If acetaminophen levels have not peaked, interpretation of results could be misleading.
- Liver and renal function tests should be done.

Treatment

- Gastric decontamination should be done if within 4 h of ingestion.
- Co-ingestion may alter the kinetics of absorption.
- The acetaminophen levels after 4 h can then be plotted on an acetaminophen toxicity nomogram (Prescott or Rumack-Matthew).
- Any level above the safety (potentially toxic) line on the nomogram receives the full course of treatment: N-acetylcysteine (Mucomyst) (NAC) load at 140 mg/kg, then 70 mg kg orally every 4 h for 17 doses.
- If the patient vomits, try administration by nasogastric tube with perhaps intravenous Reglan or an anti-emetic.
- If vomiting persists, there is a 48 h intravenous protocal.[3,4]

- The later treatment is not approved by the FDA as the solution is not pyrogen free—and anaphylactoid reactions have been reported in 2-10% of the intravenous administrations.
- The *British Medical Journal* in 1991 reported that the use of intravenous acetylcyteine with conventional liver care improves survival compared with conventional liver care alone in patients with fulminant hepatic failure after acetaminophen overdose.[5]
- If in doubt, it is safer to treat.

Salicylates (Aspirin)

Pharmacology and Pathophysiology

- Absorption of enteric coated or large amounts of aspirin tablets and capsules can be erratic as concretions can be formed in the stomach.
- With liquid preparations, absorption is rapid with peak levels occurring in 2 h.
- Once absorbed, salicylates are rapidly hydrolyzed to salicylic acid.
- In therapeutic doses, salicylates are 90% bound to albumin.
- Salicylic acid and its derivatives are conjugated with glycine or glucuronide and excreted in the urine.
- The renal excretion is pH dependent and is increased if the urinary pH is above 6.0.
- At therapeutic concentration, elimination kinetics are zero order.
- With large intoxications, elimination switches to first order kinetics as metabolic pathways become saturated.
- Salicylates stimulate the sensitivity of the medullary respiratory center resulting in hyperventilation and respiratory alkalosis.
- Salicylates also uncouple oxidative phosphorylation increasing metabolic rate, stimulating gluconeogenesis, tissue glycolysis, lipid metabolism producing lactic acid and ketone bodies and metabolic acidosis.[1]
- Salicylates also cause decreased platelet aggregation, prolongation of bleeding time and gastritis resulting in gastrointestinal bleeding.

Clinical Features

- The main clinical features are CNS excitation and depression, hyperventilation, tinnitis, warm flushed skin, vomiting and abdominal pain, fever and dehydration.
- Severe intoxications can cause pulmonary edema, GI bleeding, seizures, acute renal failure and cerebral edema.

Work-Up and Laboratory

- A chemistry seven panel, PT, CBC, stool guaiac, arterial blood gases and a serum salicylate level should be sent.
- The Done Nomogram is unreliable and should not be used if the patient has ingested sustained released, enteric-coated products or has had multiple small ingestions.
- The clinicians should treat primarily on the clinical signs and symptoms of the patient.

Treatment

- Charcoal lavage should be considered and the patient should then be given activated charcoal plus a saline cathartic.
- Fluid and electrolyte abnormalities and hypoglycemia should be corrected and the patient monitored for GI bleeding.
- Alkaline diuresis can be considered in moderate poisoning in preselected individuals.
- Severe poisoning should be treated with hemodialysis.

Nonsteroidal Inflammatory Agents

Pharmacology and Pathophysiology

- The primary mechanism of action of NSAIDs is inhibition of prostaglandin synthesis.

- They also inhibit platelet function and disrupt the gastric mucosal barrier to cause gastrointestinal bleeding.
- They are well absorbed and extensively protein bound (98%).
- Elimination is by hepatic biotransformation to metabolites, which are mostly conjugated with glucuronic acid and are excreted in the urine.

Clinical Features

- The patient may exhibit some mild GI and CNS disturbances.
- The course of intoxication is usually benign.
- Renal dysfunction has been reported after massive overdose in patients with underlying volume depletion, or renal disease or in chronic long-term use.
- Seizures have been reported with the fenamate class.[6]

Work-Up and Laboratory

- Baseline electrolytes, renal function and serum acetaminophen levels should be drawn.

Treatment

- Care is supportive and symptomatic.
- Activated charcoal should be given if within 4 h of ingestion.
- The patient should be monitored for gastrointestinal bleeding.
- The patient should be monitored for 6-12 h after acute overdose.

Calcium Channel Blockers

Pharmacology and Pathophysiology

- Calcium channel blockers relax arterial smooth muscle, but they have little effect on most venous beds.
- They cause coronary and peripheral vasodilatation.
- In the heart, membrane depolarization in atria and ventricles occurs as a result of two inward currents, one carried by sodium through the fast channel and one carried by calcium through the slow channel.
- Blockage of the slow channel by calcium channel blockers can result in a negative inotropic effect, weakening the force of muscular contraction.

12

- Verapamil and diltiazem depress the SA nodal activity and slow AV conduction.
- Verapamil delays the rate of recovery of the slow calcium channel and therefore affects AV conduction and the rate of the sinus node pacemaker. It undergoes hepatic metabolism yielding one active metabolite, norverapamil, with 20% of the original activity. Verapamil blocks calcium entry into pancreatic islet β cells resulting in an impairment of insulin release.[7,8]
- Nifedipine has its greatest effects against vascular spasm and has no AV node activity. It has no active metabolites and reaches peak drug level between 2-6 h post ingestion.[7,8]
- Diltiazem acts on the AV node slowing conduction and prolonging the AV node functional refractory period. It has a large first pass effect and undergoes hepatic metabolism with one active metabolite desacetyl diltiazeam.[8]
- The three calcium channel blockers represent three different chemical classes, bind to different receptors and so have different overdose toxicities.[7-9]
- Calcium channel blockers are rapidly absorbed with large first pass hepatic metabolism and minimal renal excretion.
- There is a large volume of distribution and high degree of protein binding (98%).
- Peak effects of overdose may be delayed with coingestions or if sustained release preparations are ingested. Sustained release preparations tend to form large concretions in the stomach.[10]

Clinical Features

- The initial symptoms are hypotension, bradyarrhythmias, myocardial depression, peripheral dilatation, hyperglycemia, congestive heart failure and lethargy.[9]

Work-Up and Laboratory

- Electrolytes, glucose, Ca^{++}, Mg^{++}, and EKG need to be obtained.
- It is always wise to do a serum drug screen to check for the presence of acetaminophen.
- Serum levels of calcium channel blockers do not correlate with the degree of toxicity and are not obtained.
- An abdominal flat plate may demonstrate the presence of a concretion in the stomach.
- Since many of these drugs are class C, a pregnancy test should be obtained on all women of child-bearing age.

Treatment

- An intravenous line needs to be placed and the patient placed on continuous cardiac monitoring.
- Gastric decontamination is performed by instilling activated charcoal plus a cathartic if there is no lieus.
- In massive acute ingestions gastric lavage may be done and charcoal instilled at the end of the lavage.[11]
- Whole bowel irrigation is performed in the presence of a concretion.
- With bradydysrythmias, a menu approach is used: if one pharmacological agent fails, move to the next agent.
  1. Atropine—0.5 to 1 mg to a total of 3 mg for an adult and 0.02 mg/kg for a child.
  2. Calcium chloride or calcium gluconate—10 ml of a 10% solution slow IV push
  3. Glucagon—1-10 mg slow intravenously
  4. External or temporary pacing
  5. Beta-1 agonists used with phosphodiesterase inhibitors have been successful—a case of successful treatment with amrinone with isoproterenol—has been reported.[12]
  6. Vasopressors—Dopamine and Norepinephrine have been used—the patient must be continuously monitored to ensure that congestive heart failure is not developing. If therapy is not successful, pulmonary arterial catheters and arterial catheters are needed to titrate treatment.
  7. Large volumes of intravenous fluid may precipitate congestive heart failure.
  8. Because of hepatic metabolism, high protein binding and large volume of distribution, calcium channel blockers are not removed by hemodialysis or hemoperfusion.

12

Admission or Disposition

- Unless the patient ingested a sustained release preparation or a coingestion, totally asymptomatic patients with normal EKGs may be discharged after 6 h of monitoring.
- All other patients should be admitted to a monitored bed and psychiatric consultation should be obtained as appropriate.

Beta-Blockers

Pharmacology and Pathophysiology

- Beta-blockers inhibit the effect of catecholamines at the β receptor site, inhibiting the formation of adenyl cyclase and reducing the cellular level of cyclic AMP.
- They exhibit a membrane stabilizing (quinidine-like) effect. This decreases myocardial contractility, decreases cardiac conduction and decreases heart rate. (B1 block).
- Block of the $\beta 2$ receptors results in vasoconstriction, bronchospasm, and insulin release.
- Beta-blockers are well absorbed with a large first pass effect and undergo hepatic metabolism.

- The exceptions are atenolol and nadolol which have little hepatic metabolism and are eliminated by the kidneys.[1]
- Generally volumes of distribution is large in these drugs, but protein binding is variable depending on the β-blocker.

Clinical Features

- The main clinical features are bradydysrhythmias, conduction blocks, hypotension, and heart failure.
- The patient may also have hypoglycemia, CNS depression and seizures.[1,13,14]

Work-Up and Laboratory

- Electrolytes, glucose, renal function, and EKG are required.
- Serums levels of β-blockers are not helpful.
- It is always wise to get serum toxicology to rule out the presence of a coingestion such as acetaminophen.
- Since many of these drugs are Class C and atenolol is Class D, a pregnancy test should be done on all women of child-bearing and age.

Treatment

- An intravenous line is established and the patient placed on continuous cardiac monitoring.
- If the patient exhibits altered level of consciousness, an immediate bedside accucheck must be performed and glucose administered as necessary.
- For significant bradydysrhythmias and conduction blocks and hypotension, again a menu is followed, titrated to patient response.[11]
  1. Atropine—0.5 mg to a total of 3 mg in an adult.
  2. Glucagon—1-10 mg slow intravenously—acts by activating adenyl cyclase and has positive inotropic effect even with β blockade.
  3. External or temporary pacing
  4. Isoproterenol 1 mg IV followed by infusion, 2-10 μg/min in adults titrated to effect.
  5. The use of vasopressors without providing additional inotropic cardiac support may precipitate cardiac failure. Agents such as dobutamine or agents with both α and β properties (e.g., epinephrine) may be helpful.
  6. Phosphodiesterase inhibitors—such as amrinone—do not appear to be superior to glucagon.
  7. Large volumes of fluid will precipitate congestive heart failure.
  8. Hemodialysis and hemoperfusion may be helpful in those β-blockers that have low protein binding, a small volume of distribution and renal excretion (e.g., atenolol, nadolol, sotalol).[15]

Admission or Disposition

- Unless a sustained release preparation or a coingestion was taken, totally asymptomatic patients with normal EKGs may be discharged after 6 h.
- All other patients must be admitted to a monitored bed and a psychiatric consultation obtained as appropriate.

Digitalis

Pharmacology and Pathophysiology

- Digitalis binds to a specific receptor site on the cardiac cell membrane inactivating the sodium-potassium ATPase pump. This pump maintains sodium extracellulary and potassium intracellularly—hence when the pump is inhibited, sodium is exchanged for calcium. This increases the concentration of calcium within the sarcoplasmic reticulum and results in increased inotropic action of cardiac muscle.[16,17]

- Digitalis also increases vagal tone and decreases the rate of conduction through the AV node.
- Digitalis is rapidly absorbed and is eliminated by renal excretion. It has a large volume of distribution and a long half-life.

Clinical Features

- The patient will have GI symptoms such as nausea, vomiting.
- CNS symptoms such as dizziness, syncope, seizures, confusion, disorientation.
- Cardiovascular symptoms such as tachydysrhythmias with AV block or bradydysrythmias.
- Patients sometimes complain of yellow halos around objects.

Work-Up and Laboratory

- Chemistry panel to include electrolytes, renal and hepatic function, serum acetaminophen level, digitalis level and an EKG should immediately be obtained.
- If the patient is a female of child-bearing age, a pregnancy test should be requested.

Treatment

- The patient should be placed on a cardiac monitor and an intravenous line established.
- Perform gastric decontamination by administering oral activated charcoal. Consider charcoal lavage with large ingestions within the 2 h of ingestion.
- Repeat activated charcoal 1/2 mg/kg every 4-6 h if no ileus is present.
- Treat hyperkalemia only with drugs that will cause an intracellular shift: insulin and glucose and bicarbonate. Avoid kayexalate and **calcium is contraindicated**.
- Ventricular irritability and arrhythmia may be treated with the Class 11 antidysrhythmics—phenytoin and lidocaine. **Class 1 antiarrhythmics are contraindicated (Procainamide and Quinidine)**.
- Treat bradyarrythmias with atropine or pacing.
- Administer magnesium sulphate 2 g intravenously.
- For bradydysrhythmias unresponsive to therapy, ventricular dysrhythmias, $K^+ >5.5$, deterioration of clinical status, severe renal insufficiency. Administer digoxin specific antibody fragment.[18,19]
- The amount can either be calculated:

$$\#\text{vitals of FAB} = \frac{\text{serum digoxin level} \times \text{patient's weight in kg}}{100}$$

- Or 5-10 vials given empirically if amount ingested is unknown.
- The administration of FAB fragments will interfere with the serum digitalis determination for up to a week.

Disposition

- Asymptomatic patients with normal EKGs and nonelevated digitalis levels after 12 h observation may be discharged home.
- All other patients should be admitted to a monitored bed.

Tricyclic Antidepressants

Pharmacology and Pathophysiology

- The tricyclic antidepressants are well absorbed from the gastrointestinal tract.
- Absorption may be prolonged in overdose secondary to their anticholinergic effect.
- They have extensive first pass metabolism, large volumes of distribution and large degrees of protein binding.
- The four major effects are TCA's are:[1]

  1. Inhibition of reuptake of norepinehrine and serotonin

2. Sodium channel blockade resulting in a quinidine like membrane effect
3. An anticholinergic effect
4. Alpha-adrenergic blockade.

Clinical Features

- Serious toxicity is characterized by CNS excitation and seizures or coma, cardiac dysrhythmias and conduction blocks, hyperthermia, ileus, urinary retention, hypotension.

Work-Up and Laboratory

- The most important test to obtain is an EKG, to look for prolongation of the QRS, conduction blocks and rightward terminal 40-msec deviation of the QRS axis especially in lead AVR.[20]

Treatment

- The patient should be immediately placed on continuous cardiac monitoring.
- Charcoal lavage should be performed and activated charcoal instilled.
- Early intubation with hyperventilation is recommended.
- If the patient has conduction blocks, bradyarrhythmias, or a prolonged QRS begin serum alkalinizxation to a pH of 7.5.
- Hypotension, unresponsive to intravenous saline, may be treated with vasopressors under hemodynamic monitoring.
- Seizures are treated with benzodiazepines.
- Beta-blockers, class 1A antiarrhythmics and flumazenil are **contraindicated** in a TCA overdose.
- Asymptomatic patients with normal EKGs and good gastric decontamination may be discharged after psychiatric consultation and 6 h of cardiac monitoring.
- All other patients should be admitted to a monitored unit.
- Pulsed serial activated charcoal (if no ileus is present) and charcoal hemoperfusion may be beneficial in serious overdoses.

Serotonin Reuptake Inhibitors

Pharmacology and Pathophysiology

12

- This class of drugs have the ability to inhibit the presynaptic reuptake of serotonin.
- They also may have the ability to block the 5-HT2 receptor.
- Most are well absorbed from the gastrointestinal tract and are extensively bound to plasma proteins.
- They are highly lipophilic and have large volumes of distribution.
- They are primarily eliminated by hepatic metabolism.

Clinical Features

- The most common features of overdose are CNS agitation or seizures, tachycardia and gastrointestinal symptoms of cramping and diarrhea.

Serotonin Syndrome

- This syndrome may develop if a serotonergic agent is added to a patient's regimen or if the dose of a serotonergic agent is increased.
- Sternback published several diagnostic criteria for this syndrome that include symptoms from the neuromuscular, cardiovascular and gastrointestinal systems.[21]
- Providing that other etiologies have been ruled out and the patient is not on other neuroleptics, the syndrome can be diagnosed with the presence of three of the following: agitation, ataxia, diaphoresis, diarrhea, hyperreflexia, hyperthermia, mental status changes, myoclonus, shivering tremor.

Work-Up and Laboratory

- A basic chemistry panel should be obtained as well as a CPK and uric acid to rule out rhabdomyolysis.
- A urine toxicology panel and EKG should be obtained to rule out the presence of other sympathomimetic drugs of abuse.

Treatment

- The treatment of this overdose is supportive and symptomatic.
- Activated charcoal should be given.
- Benzodiazepines can be used to control tremors, agitation and seizures if they develop.
- If the patient develops severe muscular rigidity and hyperthermia, the use of dantrolene should be considered (see section on Neuroleptic Malignant Syndrome).

Neuroleptics

Pharmacology and Pathophysiology[22]

- These drugs are well absorbed with peak concentrations seen 2-6 h after ingestion.
- They have large volumes of distribution and are highly lipophillic.
- They are highly protein bound (>95%) and undergo metabolism in the liver with inactive and active metabolites that may undergo enterohepatic circulation.
- While there are five classes of neuroleptics (phenothiazines, thioxanthines, butyrophenones, dibenzoxazepines, dihydroindole), they all act primarily blocking the dopamine receptors.
- Some agents also have a peripheral α-adrenergic blockade causing hypotension
- May lower the seizure threshold.
- They block the muscarinic receptor sites producing paralytic ileus and urinary retention.
- They demonstrate quinidine-like cardiotoxicity by prolonging both the QRS and QT interval and interventricular conduction delay.
- This puts the patient at risk for a torsade de pointes dysrhythmia.
- They interfere with the body's ability to thermoregulate and may cause hypothermia or hyperthermia.
- Due to a suppression of the chemoreceptor trigger zone, most phenothiazines are anti-emetics.

12

Clinical Features

- The clinical triad of overdose consists of CNS depression, hypotension and seizures.
- QRS and QTC prolongation may be seen.
- Some anticholingergic physical signs may be present including ileus, urinary distension, dry skin and mucous membranes, and dilated pupils.

Work-Up and Laboratory

- Blood phenothiazine levels do not correlate well with toxicity.
- Urine phenothiazine screens (Forrest test) confirm recent ingestion only.
- An abdominal radiograph may demonstrate radioopaque tablets, which would indicate the need for further decontamination measures.
- An EKG should be obtained and if there is a conduction block present, the patient should be hospitalized on a monitor.

Treatment

- A charcoal lavage should be performed and then charcoal mixed with cathartic administered.
- If there is enterohepatic recirculation of an active metabolite, pulsed serial activated charcoal may be given at 1.2 g/kg every 6 h provided there is no ileus.
- A Foley catheter should be inserted for urinary retention.

- Treatment in general is supportive and symptomatic.
- Seizures are treated with benzodiazepines.
- Prolonged QTC can be treated with 2 g of magnesium sulfate intravenously.
- Hypotension is treated with fluids and adrenergic vasopressor therapy.
- Class 1A agents (quinidine, procainamide) and physostigmine are **contraindicated**.
- If the patient is a woman of child-bearing age, a pregnancy test should be performed as these agents are class C (teratogenic).
- Because of the pharmacokinetics, forced diuresis, hemodialysis and hemoperfusion are unlikely to be of value.

Dopamine Deficiency Syndromes Secondary to Neuroleptic Adminstration

- A mild dopamine deficiency, clinically manifested as a dystonia or extrapyramidal reaction may be treated with a centrally acting anticholinergic agent such as diphenhydramine.
- The patient may require benztropine treatment for on-going neurochemical imbalance.
- A patient with moderate dopamine deficiency will clinically present with a catatonic face and posturing and rigid muscles but normal vital signs.
- This requires the administration of amantadine or bromocriptine to increase dopamine concentrations centrally to resolve the syndrome.
- The patient should be monitored for dehydration and rhabdomyolysis and the neuroleptic discontinued.
- Since other entities can mimic this syndrome, if there is no history of neuroleptic use or failure to respond to therapy a further work-up including a CAT scan, lumbar puncture, metabolic panel and full toxicological analysis is indicated.
- A severe lack of dopamine can result in the **neuroleptic malignant syndrome**.
- This is a medical emergency with a high mortality rate.
- The patient typically will present with muscle rigidity, altered level of consciousness, autonomic instability with a high temperature.
- Cooling measures and hydration must be initiated immediately.
- Dantrolene sodium should be administered for the severe hyperthermia in a dose of 1-2 mg/kg to a maximum dose of 10 mg/kg.[22]
- Amantadine and/or bromocriptine should then be started and the patient admitted to the intensive care unit.[23,24]
- The patient needs to be aggressively monitored for sepsis, rhabdomyolysis and renal failure.

Lithium

Pharmacology and Pathophysiology

- The exact mechanism of lithium action is unknown but is probably due to the substitutions of cations across cellular membranes.
- Lithium has a very narrow toxic therapeutic index.
- The drug is well absorbed with peak levels appearing 30 min to 2 h after ingestion with regular tablets, while delayed release tablets can give peak levels 4-12 h later.
- There is no protein binding or hepatic metabolism.
- The volume of distribution is that of body water (Vd 0.6 L/Kg).
- Clearance is by the kidneys and therefore dependent on renal function, glomerular filtration rate, sodium and volume status.

Clinical Features

- Clinical presentations centers around CNS effects and GI disturbance.
- The patient may exhibit tremors, fasciculation, hyperreflexia, muscle rigidity, agitation, confusion, stupor, seizures, coma, nausea, vomiting, diarrhea, conduction blocks and QT prolongation.[25]

- Lithium will produce urine sodium and water loss secondary to nephrogenic diabetes insipidus.
- Toxicity may be precipitated by drug interactions, an acute febrile illness and is exacerbated by dehydration.

Work-Up and Laboratory

- A full metabolic panel (including electrolytes and renal function), serial lithium levels, EKG, and toxicological panel looking for coingestions needs to be obtained.

Treatment

- Activated charcoal does not bind lithium but might be of benefit if coingestions are involved.
- Kayexalate (sodium polystyrene sulfonate) may be given orally although clinical efficacy has not been proven in humans.[26]
- Hemodialysis is the treatment of choice and is indicated if there is a deterioration in renal function, if the lithium level is 2-3 mEq/1 and the patient is symptomatic or exhibits a deteriorating clinical state and if there are severe fluid and electrolyte disturbances unresponsive to supportive care.
- The patient should be rehydrated with normal saline.
- Diuretics are contraindicated. Haldol is contraindicated in lithium toxicity.

Alcohols

Ethanol

Pharmacology and Pathophysiology

- Ethanol is a primary alcohol with a rapid absorption from the GI tract.
- It is metabolized by alcohol dehydrogenase in the liver and follows zero order kinetics.
- 5-10% is excreted in sweat, urine and respirations.
- When the level of ethanol is very high, as with overdose, the enzymatic system of the liver becomes overloaded, and elimination may switch to first order kinetics. This condition may be fatal.[23]
- The volume of distribution of ethanol is 0.6 L/kg distributed in the body water of the patient.

12

Clinical Features

- Besides for the features of CNS depression, respiratory depression, ataxia, the patient may exhibit vomiting, gastrointestinal bleeding, pancreatitis, acid base and electrolyte imbalance.

Work-Up and Laboratory

- A full metabolic panel including serum electrolytes and liver function should be obtained.
- A toxicological panel including serum alcohol level is also needed, looking especially for the coingestion of other toxic alcohols.
- A stool hemoccult should be documented as well as a CXR, EKG and pulse oximetry.
- A patient with altered mental status requires a CAT scan and an aggressive search for infection as part of the work-up.

Treatment

- The patient should receive vitamins, especially thiamine, and magnesium in the intravenous fluid. Generally the treatment is supportive and symptomatic.
- If the patient should present in deep coma with a history of massive alcohol intake and there is no other etiology for the coma (i.e., hypoglycemia, infection, head trauma), then the patient requires immediate hemodialysis.[23]

Methanol

Pharmacology and Pathophysiology

- Methanol is absorbed rapidly from the GI tract.
- It is metabolized in the liver by alcohol dehydrogenase to formaldehyde, which is then metabolized to formic acid.
- Formic acid is metabolized further to carbon dioxide by a folate dependent pathway.
- Both formaldehyde and formic acid are very toxic.
- Methanol increases GABA inhibition to a degree and is very irritating to the GI tract.

Clinical Features

- While the patient may appear mildly intoxicated and complain of GI discomfort, serious symptoms generally do not appear until 24 to 48 h after ingestion.
- The patient at that time may have serious abdominal pain and visual disturbances and/or visual loss.

Work-Up and Laboratory

- A serum methanol level, full toxicological panel, metabolic panel including pancreatic enzymes, stool hemocult and EKG need to be obtained.
- A visual acuity also needs to be performed.

Treatment

- Gastric lavage with activated charcoal should immediately be performed if within 2 h of ingestion.
- An alcohol infusion should be started to stop the conversion of methanol to formaldehyde by competitively occupying the liver enzyme alcohol dehydrogenase.
- The loading dose of 10% ethanol infusion is 10 mg/kg load followed by a maintenance dose of ~1.6 ml/kg/h to maintain a serum ethanol level of 125 mg %.
- In patients who are heavy alcohol consumers, a higher maintenance infusion rate is needed.
- Sodium bicarbonate should be given intravenously to obtain a normal pH as well as folic acid 50 mg q 4 h for 6 doses intravenously.
- If the patient is already acidotic, the active principle of folic acid, leucovorin, 1 mg/kg should be given intravenously as the initial dose.

12

- Indications for hemodialysis include: serum methanol levels >40 mg% or if the blood pH shows metabolic acidosis with a history of methanol ingestion and normal lactate levels.[1]

Isopropyl Alcohol

Pharmacology and Pathophysiology

- Isopropyl alcohol is a secondary chain alcohol, also known as rubbing alcohol.
- It is rapidly absorbed from the GI tract and is metabolized in the liver to acetone.
- It is also excreted by the kidneys and follows primarily first order of kinetics.
- It has a volume of distribution of 0.6 L/kg.

Clinical Features

- Isopropyl alcohol is approximately twice as potent as ethanol in causing CNS depression and is highly irritating to the GI tract.
- The symptoms are CNS depression, hypotension, hemorrhagic gastritis with nausea, hematemesis and abdominal pain.

Work-Up and Laboratory

- A full chemistry 7 panel, serum and urine toxicology panel, serum acetone and ketones and arterial blood gases should be run.
- A hemoccult and EKG should be obtained.

Treatment

- Gastric decontamination after 2 h, unless there is a coingestion, will not be beneficial.
- Treatment is supportive and symptomatic.
- If the patient is in deep coma with autonomic instability, hemodialysis is necessary.

Ethylene Glycol

Pharmacology and Pathophysiology[1]

- Ethylene glycol is a polyalcohol and a class six poison.
- It is a major component of antifreeze to which a fluorescent dye is added to make the solution Kelly green or orange.
- It has a volume of distribution of 0.8 L/Kg and follows zero order kinetics.
- It is rapidly absorbed from the GI tract and is metabolized in the liver by alcohol dehydrogenase to glycoaldehyde.
- Glycoaldehyde is further metabolized by aldehyde dehydrogenase to glycolic acid and glyoxalic acid by oxidation of the aldehyde groups.
- Glyoxalic acid forms oxalate which binds with calcium to form monohydrate and dehydrate crystals of calcium oxalate, which are excreted in the urine.
- The organic acids and other substances that are formed interfere with the Krebs cycle and glucose utilization and will result in multisystem organ failure and death if metabolism proceeds unchecked without appropriate therapy.

Clinical Features

- Symptoms may be delayed for 12 h or more and consist of nausea, vomiting, acidosis and seizures. In the next 12 h coma, cardiopulmonary problems such as hypotension, congestive heart failure and cardiovascular collapse occur.
- The third phase, which occurs 24-72 h after ingestions, consists of nephrotoxicity, flank pain, and oliguric renal failure.
- Hypocalcemia may develop secondary to precipitation of calcium as calcium oxalate and may cause prolongation of the QTc interval.
- Myositis and myalgias that occur may be accompanied by elevation in uric acid and creatinine phosphokinase.

12

Work-Up and Laboratory

- Most hospitals cannot perform ethylene glycol serum concentrations so while a level may be drawn and sent to contract laboratory, it will not be used in making a definitive diagnosis.
- A full chemistry panel including renal and liver function tests, CPK, stat calcium level, serum and urine toxicology, serum osomality and arterial blood gases must be immediately obtained.
- The urine can be examined for calcium oxalate crystals to substantiate the diagnosis.
- The urine can be checked under a Wood's lamp to see if the urine will fluoresce from the fluorescent dye in the antifreeze.[27]

Treatment

- Gastric decontamination with activated charcoal should be performed.
- The patient with a pH of 7.2 or less should be given intravenous sodium bicarbonate.
- Begin IV alcohol infusion (load 10 mg/kg of 10% ethanol infusion) to reach a blood alcohol level of 120-130 mg% then follow with a maintenance infusion.[1]
- Alternatively fomepizole (15 mg/kg followed by 10mg/kg q 12 h x 4 doses then 15 mg/kg q 12 h thereafter) may be used to inhibit the alcohol dehydrogenase enzyme.[28]
- Pyridoxine 100 mg and thiamine 100 mg should be given intramuscularly and magnesium sulfate 2 g intravenously.

- If the calcium level is depleted, 10 mL 10% calcium gluconate should be given intravenously.
- Hemodialysis should be implemented as quickly as possible.

Barbiturates

Pharmacology and Pathophysiology[23]

- Barbiturates augment the GABA (γ-aminobutyric acid) inhibition.
- When the barbiturate binding sites in the neuronal system are occupied, the chloride ion channel stays open longer and the increased flow of chloride ion from neurons over the membranes of effector cells decreases neuronal excitability.
- Barbiturates are largely undissociated at a pH of 7.4 and are highly lipid soluble.
- The short-acting barbiturates have variable elimination half-life since their elimination depends on the enzymatic cleavage of a side chain in the liver.
- They are not significantly excreted in the urine in active form.
- Long-acting barbiturates (phenobarbital) have limited liver metabolism and are mainly excreted by the kidney.
- Phenobarbital is also excreted or secreted into the lumen of the intestinal tract and exhibits an enterohepatic recirculation.

Clinical Features

- The patient will demonstrate CNS depression, respiratory depression hypothermia, ileus, muscle flaccidity and may have barbiturate burns over pressure points.
- Complications include aspiration pneumonia and rhabdomyolysis.

Work-Up and Laboratory

- A full metabolic and toxicologic panel must be obtained plus CPK and uric acid to rule out the presence of rhabdomyolysis.
- A CXR, EKG and arterial blood gases or pulse oximetry need to be documented.
- The level of short-acting barbiturate corresponds to the level of coma in the patient.
- A level of coma inconsistent with the measured level of short-acting barbiturate should prompt the clinician to search for other etiologies.[29]

12

Treatment

- With short-acting barbiturates: gastric decontamination, charcoal and cathartic (no sorbitol) and supportive care.
- With long-acting barbiturates: gastric decontamination, charcoal and cathartic (no sorbitol) and pulsed serial activated charcoal every 6 h if no ileus is present.
- Alkaline diuresis with moderate overdose and stable patient may be performed, but great care must be taken to avoid fluid-overloading the patient.
- Patients in deep coma require charcoal or Amberlitre resin hemoperfusion.

Benzodiazepines

Pharmacology and Pathophysiology

- There is a wide variation in the duration and action of the different classes of benzodiazepines.
- In general they are well absorbed from the GI tract and are metabolized in the liver by two major pathways—demethylation and conjugation with gluronide.
- They are highly protein bound.
- Some are metabolized to pharmacologically active compounds with longer half-lives than the parent compounds.
- Benzodiazepines bind to benzodiazepine binding sites in the central nervous system.

- This potentiates the inhibitory action of GABA by increasing the frequency of opening of the chloride channel, allowing an increase in the flow of chloride ion over effector membranes and inhibiting excitation.[23]

Clinical Features

- The clinical presentation is that of CNS depression and respiratory depression.
- Most deep comas are associated with coingestions.

Treatment

- Treatment should consist of gastric decontamination and supportive care.
- The use of the antidote, flumazinil, is not recommended as it may precipitate seizures.

Other Sedative Hypnotics

Pharmacology and Pathophysiology

- In general, all sedative hypnotics act by augmenting GABA inhibition.
- Many sedative hypnotics have enterohepatic or enteroenteric recirculation of active metabolites and large volumes of distribution.

Clinical Features

- Sedative hypnotic overdoses are characterized by CNS depression, hypothermia, hypotension, respiratory depression and ileus.

Work-Up and Laboratory

- All overdoses require a CXR, EKG, pulse oximetry, full metabolic panel and serum and urine toxicology.
- Complications of these overdoses are aspiration pneumonia and rhabdomyolysis.

Treatment

- Gastric decontamination should be performed; sorbitol in the cathartic is not recommended.
- If the toxin has an enterohepatic or enteroenteric recirculation of an active metabolite, the patient should receive pulsed serial activated charcoal if there is no ileus.
- Mild to moderate overdoses in young healthy patients require supportive care.
- Severe overdoses or moderate overdoses in elderly or debilitated patients require hemoperfusion.

Narcotics

Pharmacology and Pathophysiology

- Opioid receptors are located throughout the central nervous system and consist of primarily three types.
  1. The "mu" receptors mediate euphoria and respiratory analgesia.
  2. The "kappa" receptors produce miosis and sedation.
  3. The "sigma" receptors cause dysphoria, hallucinations, respiratory and vasomotor stimulation.
- The emergency medicine physician will be dealing primarily with: heroin, demerol, and lomotil in the course of clinical practice. All will give the narcotic picture: CNS depression, miosis, respiratory depression and hypotension in overdose.[23,30]

Heroin

Pharmacology and Pathophysiology

- Heroin is a semisynthetic morphine derivative.

- It is deacetylated by plasma esterases to monoacetylmorphine in the liver, hydrolyzed to morphine, then conjugated with glurconide to be excreted in the urine.
- Heroin is a drug of abuse and is bought on the street as black tar, brown tar, cut with alduterants, sold in bags or balloons often contaminated with alduterants or bacteria.[23]

Work-Up and Laboratory

- Pulse oximetry and CXR needs to be documented.
- The chart must record the pertinent physical negatives that would rule out bacterial endocarditis, sepsis or infection, neurological findings, hepatitis, and tetanus.
- A pregnancy test should be performed on all women of child-bearing age.

Treatment

- The antidote is naloxone (Narcan) intravenously.
- If an intravenous line cannot be administered, Narcan may be injected sublingually or down an endotracheal tube.
- The half-life of narcan is 20-40 min—therefore the patient may require a rebolus or a Narcan drip.
- Heroin pulmonary edema is a noncardiogenic pulmonary edema. The treatment is intubation and PEEP.[23]

Demerol

Pharmacology and Pathophysiology

- Meperidine is metabolized to normeperidine, which is 2-3 times more toxic than the parent compound and has a longer serum half-life.
- This metabolite accumulates in renal failure and in chronic dosing.
- Fatal reactions can occur if meperidine is given to a patient who has received MAO inhibitors within 2 wk.

Clinical Features

- The patient may present with the narcotic syndrome (respiratory depression, CNS depression and hypotension) or with CNS agitation, seizures, hypertensions, tachycardia and hyperpyrexia.[23]

12

Treatment

- The treatment is the same as for heroin.

Lomotil

Pharmacology and Pathophysiology

- Lomotil consist of diphenoxylate – 2.5 mg and atropine sulfate – 0.025 mg.
- Diphenoxylate is a long acting narcotic that is metabolized to diphenoxylic acid which is five times more potent as the parent compound and has a serum half-life that is twice as long.[30]

Clinical Features

- Ingestions will present with respiratory depression, CNS depression, flaccidity, hypotension and may be accompanied by the anticholinergic signs of mild facial flushing, pupillary dilatation and urinary retention.[30]

Work-Up and Laboratory

- Pulse oximetry and a chest radiograph should be obtained as well as a full toxicological screen looking for coingestions.

Management

- Gastric decontamination should be carried out regardless of the time of ingestion.
- Because of the strong narcotic/anticholinergic over-ride of this medication, pills may be recovered many hours after ingestion.
- The patient may be treated with Narcan intravenously for respiratory depression.
- All patients then need to be admitted as the active metabolite of diphenoxylate—difenoxine—will accumulate causing a second period of respiratory depression 13-24 h after acute ingestion.[30]
- Pulsing the gastrointestinal tract with activated charcoal is not possible due to the ileus that develops after acute ingestion.

Stimulant Drugs of Abuse

Cocaine

Pharmacology and Pathophysiology[1]

- Cocaine acts as an α-2 antagonist and as a sympathomimetic.
- It is well absorbed from the gastrointestinal tract and is hydrolyzed by plasma esterases.
- The serum half-life is short and in the range of 40-60 min.
- The metabolites are eliminated in the urine for up to 8 days.
- Cocaine also has an anesthetic effect on the skin and mucous membranes and a quinidine-like effect on the myocardium.

Clinical Features

- The major features of this intoxication are cerebral agitation, seizures, coma, elevated vital signs including hyperthermia, tachdysrhythmias, pulmonary edema, and cardiac arrest.
- Complications include myocardial ischemia and infarction, cerebral infarction, abruptio placenta, renal infraction, pneumomediastinum, pneumothorax, cardiac dysrhythmias and rhabdomyolysis can cause renal failure.

Work-Up and Laboratory

- A chemistry panel, CPK, uric acid and urinalysis should be sent.
- A CXR and EKG should be obtained.
- If the patient complains of chest pain, serial EKG and troponins need to be ordered.

12

Treatment

- Gastric decontamination after ingestion with activated charcoal plus cathartic is appropriate.
- Hydration should begin and the patient monitored for rhabdomyolysis.[31]
- If the EKG shows a prolonged QT_C, 2 g of magnesium should be given intravenously.
- A patient complaining of chest pain needs to be ruled out for myocardial ischemia or infraction and admitted to a monitored unit.
- A patient with altered mental status or a seizure requires a CT scan.
- A patient who has a viable pregnancy requires an immediate pelvic ultrasound and fetal monitoring.
- Hyperthermia may be treated with cooling measures.
- Cerebral agitation and seizures treated with benzodiazepines.
- A patient complaining of eye symptoms needs an eye examination with fluorescein staining to rule out crack keratopathy.
- Severe cases of cocaine ingestion with uncontrolled hypertension may require intravenous nitroprusside or phentolamine for blood pressure control.
- It is important that a β-blocker NOT be administered until an α-blocker is already in place and the patient's blood pressure is under control.

12

Admission and Follow-Up

- Asymptomatic patients with normal CPK, uric acid levels and vital signs may be discharged after 6 h.

Methamphetamine

Pharmacology and Pathophysiology

- Methamphetamine displaces catecholamines from their storage sites on neuronal terminals increasing the amount of norepinephrine the effector cell (α-1 stimulation) and blocks the reuptake of norepinephrine (α-2 inhibition).
- It is well absorbed, partly metabolized by the liver and excreted by the kidneys.
- The elimination half-life is decreased in an acid urine.[1]

Clinical Features

- The clinical features are those of elevated vital signs: hypertension, tachycardia, hyperthermia, tachypnea.
- The patient will exhibit cerebral excitation, seizures, intracerebral and subarachnoid hemorrhage and rhabdomyolysis.
- Rhabdomyolysis may lead to acute renal failure.
- The patient may be very hard to sedate and injury to the extremities such as compartment syndromes have been reported.
- Chronic methamphetamine use is associated with vasculitis.

Work-Up and Laboratory

- A full chemistry panel, CPK, uric acid and urinalysis should be sent.
- If the patient exhibits altered mental status, a CT scan should be obtained.

Treatment

- If appropriate, gastric decontamination should consist of activated charcoal plus a cathartic.
- An intravenous line should be started for hydration, and the patient must be adequately sedated.
- Sedation may be quite difficult and require the use of multiple agents: benzodiazepines and haldol may be used.
- Hyperthermia may be addressed with cooling measures and the clinician should examine for muscular rigidity that could point to the neuroleptic malignant syndrome.
- An α-1 blocking agent such as phentolamine should be used for severe hypertension.
- A β-blocker should NOT be given unless an α-block is already in place.
- Urine acidification is not recommended.

Admission and Follow-Up

- These patients for the most part need to be admitted and carefully monitored for rhabdomyolysis and the other complications of this intoxication.
- Patients who are alert and oriented with normal vital signs and normal chemistries may be safely discharged home.

Phencyclidine (PCP)

Pharmacology and Pathophysiology

- This drug of abuse has multiple sites of action—antogonism at the dopamine-2 site, release of catecholamines from the raine neurons, interference with the serotonin pathways, also cholinergic and anticholinergic activity.[32,33]
- In addition, it is a powerful dissociative anesthetic, which makes the clinical examination of a patient difficult to interpret.

- It is well absorbed from the gastrointestinal tract, has a large volume of distribution and is primarily metabolized by oxidative metabolism in the liver.
- It exhibits gastroenteric recycling and is excreted in the urine.
- Elimination of PCP in the urine is pH dependent in that acid urine traps PCP in its ionized form in the renal tubules and inhibits reabsorption.

Clinical Features

- The clinical patterns have been divided into the major and minor syndromes.[32,33]
- Minor patterns may exhibit lethargy, bizarre behavior, euphoria or dysphoria.
- Major patterns demonstrate coma, catatonia, toxic psychosis and an acute brain syndrome.
- Other clinical signs are alterations in behavior, motor disturbances, abnormal vital signs, seizures, nystagmus, hypersalvation, diaphoresis, bronchorrhea, bronchospasm, and urinary retention.
- These patients may have hypoglycemia, rhabdomyolysis, renal failure, malignant hyperthermia, urinary retention, and associated trauma that will be hard to detect because of their abnormal perception to pain.

Work-Up and Laboratory

- A full chemistry panel, CPK, uric acid and urinalysis should be obtained.
- Women of child-bearing age should have a pregnancy test performed.

Treatment

- Any patient with altered mental status should have an immediate accucheck or glucose determination performed.
- A thorough search for trauma should be performed and as the patient's perception of pain may be altered, radiographic clearance and repeated examinations are recommended.
- Serious cases of intoxication should be treated with gastric decontamination and repeat doses of activated charcoal orally.
- An intravenous line should be inserted for hydration and if the patient has difficulty urinating, a Foley should be placed.
- The patient should be sedated with benzodiazepines and monitored for rhabdomyolysis.
- Urinary acidification, hemodialysis and charcoal hemoperfusion are ineffective.

Admission or Follow-Up

- Minor patterns of intoxications with normal vital signs and laboratory values may be safely discharged home in 6 h.
- Major patterns of intoxication and patients with abnormal labs should be admitted.

Psychedelic Drugs of Abuse

LSD

Pharmacology and Pathophysiology

- LSD is a 4 substituted dimethyltryptamine derivative that effects serotonergic and dopaminergic pathways in the CNS.
- The drug is usually ingested from a sugar cube or a blotter and is rapidly absorbed.
- The onset of action is 15 min to 1 h and the volume of distribution is 0.27 L/kg with over 80-protein binding.[1,23]
- LSD is extensively metabolized in the liver and excreted in the urine and bile.

Clinical Features

- The common manifestation of this ingestion are hallucinations, panic attacks and synesthesia.
- The patient may exhibit a serotonin flush across the neck and face.

- The action inhibition of serotonin on the chromaffin cells of the GI tract may lead to nausea, vomiting abdominal cramps and diarrhea.
- Severe intoxications may demonstrate coagulopathies.
- Autonomic instability is rare.

Hydrocarbons

Hydrocarbons are divided chemically and clinically into three categories:

1. aliphatic
2. aromatic
3. chlorinated

Aliphatic Hydrocarbons—The Petroleum Distillates

Pharmacology and Pathophysiology

- The petroleum distillates are produced from the fractional distillation of crude petroleum.
- Aliphatic hydrocarbons are straight chain hydrocarbons and consist of kerosene, gasoline, naphtha, and mineral seal oil.
- The major toxicity from these products comes from aspiration into the lungs causing bronchopneumonitis.
- The tendency for aspiration is reflected in the SSU (Saybolt Seconds Universal Units) of the product—the lower the SSU, the higher the volatility and the lower the viscosity of the compound.[1]

Clinical Features

- The clinical symptoms fall into three categories—GI, pulmonary and CNS.
  1. GI symptoms are common and consist of nausea, vomiting and abdominal cramping.
  2. CNS findings of lethargy, seizures and coma are usually a reflection of pulmonary hypoxia.[34]
  3. The pulmonary findings are the most serious and consist of coughing, choking, tachypnea, retractions and cyanosis. Physical examination may be normal or may reveal rales, rhonchi and wheezing.

Work-Up and Laboratory

12

- Pulse oximetry, arterial blood gases and a CXR must be performed.
- An increased A-a gradient in the blood gas is predictive of pulmonary toxicity.
- Common CXR findings include unilateral and bilateral basilar infiltrates, densities, pneumatoceles, and atelectasis.
- Pneumothorax has been reported.
- Symptoms and CXR findings may be delayed in presentation.
- An EKG should be obtained looking for dysrhythmias and evidence of myocardial ischemia.

Treatment

- There is a great deal of controversy on the subject of gastric emptying in these ingestions.
- There is general agreement that if the hydrocarbon contains a toxic additive, then gastric decontamination should be considered on a case by case basis.
- Any attempt at GI decontamination should include a consideration to protect the airway.
- Activated charcoal is indicated only if the aliphatic hydrocarbon contained a toxic additive.
- All patients with pulmonary symptoms should be given oxygen and placed on a cardiac monitor and pulse oximetry and admitted to an intensive care setting.
- Respiratory failure is treated with intubation plus PEEP, and there are case reports of treatment with extracorporeal membrane oxygenation as an alternative therapy.[35]
- Antibiotics in the absence of infection and steroids have not been shown to be beneficial.

Admission

- Asymptomatic patients with normal blood gases and CXR may be discharged after 6 h of observation.
- Any other patient must be admitted to an intensive care setting.

Aromatic Hydrocarbons

Pharmacology and Pathophysiology

- Aromatic hydrocarbons have a benzene ring consisting of benzene, toluene, xylene and vinyl chloride.
- They are extremely toxic to multiple organ systems, and exposure in the workplace is regulated by state and federal regulations.

Treatment

- All aromatic hydrocarbon exposures require decontamination, admission and consultation with a toxicologist or physician trained in occupational medicine.
- There is no antidote.
- Treatment is symptomatic and supportive.

Halogenated Hydrocarbons

Pharmacology and Pathophysiology

- These agents are both highly volatile and lipophilic. The six compounds most commonly encountered are:
  - Carbon tetrachloride
  - Chloroform
  - Methylene chloride
  - Trichloroethylene (TCE)
  - Trichloroethane (TCA)
  - Tetrachloroethylene
- These compounds are well absorbed through the lungs and the gastrointestinal tract.
- They are excreted through the lungs and in the urine.
- Their toxicity is mainly to the CNS system, liver and kidneys.
- Carbon tetrachloride especially is a powerful hepatotoxin that causes fatty degeneration of the liver and hepatic necrosis.

Clinical Features

- Classically acute exposure presents with CNS depression, nausea and occasionally vomiting.
- Both carbon tetrachloride and methylene chloride are strong skin irritants and defatting agents that may produce burns.
- Liver and kidney abnormalities may develop from 12 h on, with hepatic and renal failure in 1 wk.

Work-Up and Laboratory

- The patient requires extensive work-up including arterial blood gases if the toxin has been inhaled, a full metabolic panel to include hepatic and kidney function tests, EKG, CXR, and urinalysis.

Treatment

- There is no antidote, and these toxins are treated with symptomatic and supportive care.
- Decontamination should occur—the skin should be washed and one dose of activated charcoal should be given.
- Liver and renal function test results should be addressed.

Admission

- These patients need to be admitted for follow-up evidence of end-organ damage.

Organophosphate Pesticides

Pharmacology and Pathophysiology

- Organophosphate pesticides are powerful inhibitors of the cholinesterase enzymes.
- This results in an accumulation of acetylcholine at the neuronal synapses resulting in first stimulation and then paralyzing transmission.
- The organophosphate-cholinesterase bond is permanent, and after a period of time (24-48 h) of binding the cholinesterase enzyme is destroyed.
- The cambamate-cholinesterase bond reverses spontaneously in 4-8 h.[36]
- These compounds are readily absorbed through the skin and all mucous membranes and metabolized in the liver.

Clinical Features

- The patient will exhibit both the muscarinic and nicotinic signs of poisoning.
- The muscarinic signs include the SLUDGE toxidrome—salvation, lacrimation, urination, defecation, emesis plus bradycardia and miosis.
- The nicotinic toxidrome includes muscular twitching, fasiculations, weakness and paralysis, tachycardia, hypertension and CNS symptoms of anxiety, restlessness, confusion, seizures and coma.[36]

Work-Up and Laboratory

- An EKG, CXR, full metabolic panel, serum acetaminophen level and serum and RBC cholinesterase level (before giving pralidoxime) should be obtained.

Treatment

- Decontamination—full decontamination needs to be performed and health care providers should don protective gear. The skin and eyes if exposed should be irrigated and the water collected for safe disposal. Charcoal lavage should be performed after the airway is secured.
- Secure the airway and immediately place the patient on oxygen, cardiac monitor, pulse oximeter and establish intravenous access.
- Administer 1.0 mg of atropine in adults and observe effects. The ability to resist atropinization is one of the hallmarks of organophosphate poisoning.
- If muscarinic symptoms continue, administer IV atropine 2 mg every 30 min until secretions dry. (This may take a lot of atropine.)
- Insert a urinary catheter.
- Unless you know this is a carbamate pesticide, administer pralidoxime 1 g slowly in 20 ml 5% D5W. May repeat one-half in 30 min if muscular weakness persists.
- Seizures are treated with benzodiazepines.

Disposition

- Admit to a monitored bed.
- Poisoned workers should not return to work until their plasma and RBC cholinesterase levels return to normal.
- All pesticide exposures must be reported to local public health authorities.

Anticholinergics

Pharmacology and Pathophysiology

- The blockade of the cholinergic receptors—both peripheral and central—result in the anticholingeric toxidrome.

- Atropine overdoses are very rare so the main poisoning that the clinician in California will encounter is that of Jimsonweed intoxication.

Clinical Features

- The classic presentation is characterized as:
  Hot as Hades
  Blind as a Bat
  Dry as a Bone
  Red as a Beet
  Mad as a Hatter
- The patient in severe intoxications is confused, hallucinating, agitation, hot dry, flushed, with mydriasis, ileus, and urinary retention.

Work-Up and Laboratory

- Chem 7, serum and urine toxicology and an EKG can be obtained.

Treatment

- Place the patient on oxygen, cardiac monitor and establish intravenous access.
- Charcoal gastric lavage should be performed and since there is gastric retention—should be considered up to 24 h post ingestion. Cathartics can be administered only when bowel sounds return.
- The antidote is reserved for severe poisonings with autonomic instability ONLY. After adequate oxygenation has been achieved, slowly administered 1-2 mg of physostigmine in adults in 0.5 mg increments and not to exceed 1 mg/min. Physostigmine is CONTRAINDICATED in mild anticholingergic toxicity and in TCA overdoses.
- Treat seizures with benzodiazepines.
- Insert a urinary catheter.
- Conventual cooling measures are used for hyperthermia.

Disposition

- Admit to a monitored bed for observation.

Caustics

Pharmacology and Pathophysiology

- Acid ingestions result in a coagulative type of necrosis.
- The esophagus, distal stomach (antrum and pylorus) and small bowel are usually affected.
- Esophageal perforation tends not to occur—instead the patient may experience small bowel perforation resulting in acid peritonitis 2-3 days after the original ingestion.[37,38]
- Systemic toxicity may occur in large ingestions including acidosis and hemolysis.

Clinical Features

- The patient will complain of oral pain, dysphagia, inability in control secretions, chest and abdominal pain, nausea, vomiting and hematemesis.
- The patient usually (but not always) will have oral burns, and pain on abdominal palpation.
- Aspiration of acid may result in laryngeal edema, dyspnea, tachypnea and severe aspiration pneumonitis.

Work-Up and Laboratory

- The patient requires a CBC, chem 7 panel, renal function, blood type and crossmatch, stool guiac, and if acidosis is present—a DIC panel.
- A CXR is required to detect abnormal perforation or an inhalation injury.

Treatment

- Airway stabilization should take priority with oxygen and definitive airway management in the presence of laryngeal edema and stridor.
- Intravenous lines should be started.
- Aspiration and cool water lavage with a soft N/G tube within 1 h of ingestion may be beneficial.
- Emesis, charcoal and cathartic are contraindicated.
- The patient should undergo endoscopy for assessment of the degree of injury approximately 24 h after the ingestion. There is no role for the use of steroids.

Admission and Follow-Up

- Patients should be admitted and kept NPO until endoscopy can be performed.
- Patients with no evidence of injury on endoscopy and who are asymptomatic can be discharged home.
- Otherwise the patient needs to be admitted and monitored for delayed perforation of the stomach and small bowel.
- After recovery the patient may develop pyloric stenosis.

Hydrofluoric Acid

Pharmacology and Pathophysiology

- Hydrofluoric acid burns are unique in that the fluoride ion will penetrate tissue deeply to bind calcium and magnesium to form insoluble salts.
- Profound systematic hypocalcemia and hypomagnesemia may result leading to respiratory arrest, cardiac arrhythmias and arrest and death.

Clinical Features

- Fluoride burns are characterized by excruciating pain that develops slowly over several hours.
- The skin is usually ertythematous with blisters and later develops a white or gray blanched appearance.
- The fingernails may appear white.
- The direct corrosive effect causes burns and bleeding from the GI tract.
- In cases of severe ingestions, coma, cardiac arrhythmias and arrest will occur.

12

Work-Up and Laboratory

- If the area of the burn is large, an immediate EKG needs to be obtained to look for prolongation of the QTc that might alert the clinician to systemic hypocalcemia.
- Stat calcium and magnesium levels should be obtained as well as a chemistry 7 panel type and cross.
- Stool guicac, and coagulation panel.
- A CXR should be obtained.

Treatment (Skin Burns)

- The area of burn to be flushed with copious amounts of water and then washed with soap and water.
- The burned area needs to be immersed into a cool slushy bath of magnesium salts or calcium.
- A calcium gluconate gel should be applied to the affected area, rubbed in well and covered with a latex glove.
- If the pain is relieved, the treatment could be continued for 45-60 min, then discontinued, the gel washed off and the patient observed for recurrence of pain.
- If the pain does not recur after an hour or so, the patient may be sent home with the gel and a pair of gloves.

- If the pain returns, the patient should be instructed to reapply the gel and return to the hospital.
- If the pain persists, intradermal injection of calcium gluconate needs to be given.
- A 10% calcium gluconate solution can be injected intradermally using a 26 gauge needle. (Do not use calcium chloride, as it will cause tissue necrosis).
- The clinician can map out areas of pain and then use a local anesthetic or conscious sedation.
- Blistered or necrotic tissues should be debrided and if the nails are discolored, they should be removed and the subungal tissues injected.
- The clinician also needs to consider antibiotics and tetanus prophylaxis.
- For extensive burns, the clinician has two approaches. Intra-arterial regional perfusion of calcium gluconate or chloride may be given or intravenous regional perfusion.[39,40]

Alkali

Pharmacology and Pathophysiology

- Alkali corrosives produce a necrotizing liquefactive type of burn particularly at the three anatomical points of narrowing of the esophagus.
- Perforation of the esophagus is the greatest hazard of this ingestion and should be suspected if the patient presents in shock.

Clinical Features

- The patient will complain of severe pain of the mouth, chest and epigastric area with dysphagia and nausea.

Work-Up and Laboratory

- A large bore intravenous line should be started and the patient typed and crossed.
- A CXR should be obtained looking for perforation.
- A stool guaiac should be obtained.

Treatment

- The patient should be kept NPO, and there should be no attempt at gastric emptying or decontamination.
- The patient should be given pain relief and admitted.
- A flexible endoscopy will be done ~24 h after admission looking for the degree of burn to the esophagus and stomach.
- The role of broad spectrum antibiotics is questionable.

Admission and Follow-Up

- The patient should be admitted to Intensive Care and carefully monitored for perforation and gastrointestinal bleeding.
- After endoscopy, a stent can be placed and there is some evidence that steroids might be helpful in partial thickness burns to the esophagus.[41]
- The sequelae of this ingestion is esophageal stricture.

Iron

Pharmacology and Pathophysiology

- When excessive iron is ingested, the body's transport and storage proteins become saturated resulting in circulating free iron.
- This is highly toxic to cellular mechanisms and causes damage to the cardiovascular, liver, CNS and hematopoietic systems.
- In addition iron is a directly corrosive agent to the gastrointestinal tract.

Clinical Features

- The clinical presentation of iron toxicity has classically been divided into four stages:
  1. Stage 1—30 min to 6 h after ingestion. The patient may complain of gastrointestinal symptoms such as nausea. Vomiting, diarrhea and abdominal pain.
  2. Stage 2—6-24 h after ingestion—The patient may appear well and improved.
  3. Stage 3—12-48 h post ingestion—The patient will demonstrate evidence of multisystem organ damage and failure such as abnormal blood tests, liver, kidney and cardiac failure, coma, GI bleeding, respiratory distress syndrome. Death may occur.
  4. Stage 4—4-6 wk after ingestion—recovery, but the patient may have gastric outlet obstruction.

Work-Up and Laboratory

- Serum iron and total iron binding capacity should be measured at 4 h as well as a chemistry panel including liver and kidney function tests, CBC, stool guaiac.
- The patient should be typed and crossed.
- A KUB should be obtained looking for iron concretions in the stomach and tablets in the gastrointestinal tract.
- If the number and type of tablets is known, the mg/kg of elemental iron ingested may be calculated by the formula: # tablets x % elemental iron in mg per tablet / kg body weight.[42-44]

Treatment

- Predictors of iron toxicity are:
  - Iron level >400 μg/dL
  - Iron level >TIBC
  - WBC >15,000/mm^3
  - Serum glucose >150 mg/dL
  - Significant symptoms (CNS, acidosis, vomiting)
- The patient should receive a gastric lavage and a KUB should be taken.
- If the KUB is positive for iron tablets, then whole bowel irrigation with GoLYTELY should be performed.
- If a iron bezoar remains after decontamination, a gastroenterologist should be consulted for removal.
- The decision to administer deferoxime is made on the estimate of iron ingested, if the patient is symptomatic and the serum iron and TIBC levels at 4 h. The starting dose is 15 mg/kg/h.
- Recent literature argues that the presence or absence of leukocytosis and hyperglycemia do not appear to be reliable predictors of the severity of iron overdose in adults.[45]

12

Deferoxamine Challenge Test

This test is sometimes performed in cases where toxicity is not clear and laboratory results are still pending (1st 6 h). The patient is given deferoxamine 50 mg/kg up to 1 g IM, and all the urine is collected. If the urine turns a vin rose color, the test is positive and chelation is continued intravenously. However it is important to note that *the urine color change is an insensitive marker and that a negative challenge test does not rule out the presence of iron toxicity.* For that reason, many toxicologists no longer recommend this test.

Isoniazid (INH)

Pharmacology and Pathophysiology

- Isoniazid blocks the action of metal-pyridoxine enzyme-substrate complexes producing acute pyridoxine deficiency.

- Pyridoxine deficiency seizures are believed to be related to blockage of decarboxylation and transaminase reactions involving glutamic acid and γ aminobutyric acid.
- Isoniazid is rapidly absorbed in the small intestine and is excreted by the kidney with no protein binding.
- Patients may be either slow acetylators (55% Afro-Americans and Caucasians) or fast acetylators (90% Asians).
- Fast acetylators will metabolize isoniazid rapidly increasing the amount of toxin produced by the liver, causing liver necrosis.[46]

Clinical Features

- The hallmarks of this intoxication are intractable seizures, metabolic acidosis, coma, hyperglycemia and respiratory distress.

Work-Up and Laboratory

- Electrolytes, glucose, renal function, liver function, pulse oximetry, arterial blood gasses, EKG, CXR, all need to be obtained.
- A serum toxicology screen to rule out the presence of coingestions such as acetaminophen is necessary.

Treatment

- Immediately establish an airway and administer oxygen.
- Establish an intravenous line and give 2-4 mg of ativan IV or 10 mg of diazepam IV.
- Give pyridoxine intravenously—1 g for each gram of isoniazid taken.
- If amount of isoniazid taken is unknown, give 5 g pyridoxine stat followed in 30 min by an additional 5 g.
- Administer IV sodium bicarbonate to correct acidosis.[46-48]
- If patient does not improve, perform hemodialysis.
- Admit to intensive care.
- Follow INH levels and acidosis.
- Follow liver function tests for hepatotoxicity for at least 3 days.
- Monitor for medical concerns during prolonged seizure activity such as aspiration pneumonia and cardiac ischemia.
- Obtain a pregnancy test in any woman of child-bearing age.
- Obtain a psychiatric consultation as appropriate.

Dilantin

Pharmacology and Pathophysiology[1]

- Absorption of phenytoin is slow and variable as the pKa is 8.3 and tends to precipitate in the acid of the stomach.
- It is 90% bound to protein and has a large volume of distribution.
- 95% of the drug is metabolized in the liver; inactive metabolites are excreted in the bile and then in the urine.
- The metabolism is dose-dependent—at therapeutic serum concentrations elimination follows first order kinetics, and at high concentrations, elimination is by zero order kinetics.
- The major manifestations of clinical toxicity is drowsiness, ataxia and nystagmus.
- With a massive overdose, nausea, vomiting, hypotension, cardiac bradydysrhythmias and conducted blocks may occur.

Work-Up and Laboratory

- A dilantin level and an EKG should be obtained.

Treatment

- Activated charcoal plus a cathartic is given.
- Treatment is supportive and symptomatic. Patients with levels >30 mg/ml, neurologic or cardiovascular findings should be admitted.
- Adults admitted for oral phenytoin overdose in whom the serum level is below 75 mcg/ml do not require admission to a telemetry unit.[49]

Heavy Metals

- It is unusual for the emergency physician to treat an acute ingestion of the heavy metals arsenic, mercury and lead.
- All lead to multisystem organ failure and all are cellular poisons.
- Generally in symptomatic ingestions, chelation must be started as quickly as possible.
- The chelating agent for arsenic, lead and mercury is BAL and is given as an IM injection at 3-6 mg/kg per dose to a maximum of 300 mg on a variable and tapering dosing schedule.
- BAL may produce several adverse side effects.
- BAL is in a peanut oil vehicle and can produce hemolysis in patients with a glucose-6-phosphate dehydrogenase deficiency.
- With lead ingestions, a second chelator is also added—calcium disodium ethylene diamine tetraacetic acid (CaNa2EDTA) at 1-2 g daily as an continuous intravenous infusion for up to 5 days.

Cyanide

Pharmacology And Pathophysiology

- Cyanide produces histotoxic hypoxia by binding with the ferric ion of mitochondrial cytochrome oxidase thus inhibiting the ability of the cells to utilize oxygen in oxidative phosphorylation.
- It is rapidly absorbed and excreted in the urine.
- Cyanide binding to the cytochrome oxidase is a reversible process.
- The enzyme rhodanese will catalyze cyanide and complex it with sulfur to form thiocyanate, which can be excreted in the urine.

Clinical Features

- The patient will present serious intoxications with coma, convulsions, respiratory failure, pulmonary edema and cardiac arrest.
- Early symptoms in milder intoxications can consist of headache, anxiety, paresthesiae, nausea and hyperventilation.

Work-Up and Laboratory

- A chemistry panel, lactate level, CXR and EKG should be obtained.
- An increased oxygen level in a venous gas or a narrowing of the O_2 difference in the arterial-venous gas suggests a cellular poison that is inhibiting cellular respiration.

Treatment

- The patient should immediately be placed on high flow oxygen and a monitor.
- The antidote is the Lilly Cyanide antidote kit.
- Amyl nitrite ampules should only be used until intravenous access is established.
- Intravenous sodium nitrite should be administered (300 mg slow push) followed by 12.5 g of sodium thiosulfate intravenously; second doses of sodium nitrite and sodium thiosulfate at one half the initial amounts may be administered 30 min after the first doses if the patient is not improving clinically.
- Hyperbaric oxygen therapy is also recommended in severe poisoning.[50]

Admission and Follow-Up

- Asymptomatic patients may be discharged in 6 h unless the substance was a nitrite compound where symptoms may be delayed.
- All symptomatic patients should be admitted to an intensive care setting.

References

1. Ellenhorn M, Barceloux D. Medical Toxicology. Diagnosis and Treatment of Human Poisoning. New York: Elsevier, 1988.
2. Bizovi K, Aks S, Paloucek F et al. Late Increase in acetaminophen concentration after overdose of tylenol extended relief. Ann Emerg Med 1996; 28:549-551.
3. Smilkstein M, Bronstein A, Linden C et al. Acetaminophen overdose: 48 hour intravenous protocol. Ann Emerg Med 1991; 20:1058.
4. Prescott LF. Treatment of severe acetaminophen poisoning with intravenous acetylcysteine. Arch Intern Med 1981; 141:386-399.
5. Keays R, Harrison PM, Wendon JA et al. Intravenous acetylcysteine in paracetamol induced fulminant hepatic failure: A prospective controlled trial. BMJ 1991; 303:1026.
6. Balali-Mood M, Critchley J, Proudfoot AT et al. Mefanamic acid overdose. Lancet 1981; 1:1354.
7. Mitchell LB, Schrooeder JS, Mason JW. Comparative clinical electrophysiologic effects of Diltiazem, Verapamil, and Nifedipine: A review. Am J Cardiol 1982; 49:629-35.
8. Erickson FC, Ling LJ, Grande GA et al. Diltiazem overdose: Case report and review of the literature. JEM 1991; 9:357-366.
9. Ramosak EA, Spiller HA, Myers A. Calcium channel blocker toxicity. Ann Emerg Med 1993; 19:649.
10. Sporer KA, Manning JJ. Masive ingestion of sustained-released Verapamil with a concentration and bowel infarction. Ann Emerg Med 1993;22:603.
11. Cox J, Wang R. Critical consequences of common drugs: manifestations and management of calcium channel blocker and B-adrenergic antagonist overdose. Emerg Med Rep 1994; 15:83-90.
12. Goenen M, Col J, Compere A et al. Treatment of severe verapamil poisoning with combined amirone-isoproterenol therapy. Am J Cardiol 1986; 58:1142.
13. Weinstein RS. Recognition and management of poisoning with β adrenergic blocking agents. Ann Emerg Med 1984; 13:1123.
14. Weinstein RS, Cole S, Knaster HB et al. Beta blocker overdose with Propranolol and Atenolol. Ann Emerg Med 1984; 13:1123.
15. Weinstein RS, Cole S, Knaster HB et al. Beta blocker overdose with Propranolol and Atenolol. Ann Emerg Med 1985; 14:161.
16. Kelly RA, Smith TW. Recognition and management of digitalis toxicity. Am J Card 1992; 69:108.
17. Sharff J, Bayer M. Acute and chronic digitalis toxicity presentation and treatment. Ann Emerg Med 1982;11:327-330.
18. Hickey AR, Wenger TL, Carpenter VP et al. Digoxin immune FAB therapy in the management of digitialis intoxication: Safety and efficacy results of an observational surveillance study. Am J Card 1991; 17:590.
19. Antman EM, Wenger TL, Butler VP et al. Treatment of 150 cases of life threatening digitalis intoxication with digoxin-specific FAB antibody fragments. Final report of a multicenter study. Circulation 1990; 81:1744-52.
20. Wolfe TR, Caravati EM, Rollins DE. Terminal 30- millisecond frontal plane QRS axis as a marker for tricyclic antidepressant cardiotoxicity. Ann Emerg Med 1989; 18:348.
21. Sternbach H. The serotonin syndrome. Am J Psychiatry 1991; 148:705-713.
22. Challoner K, Newton E. Toxicity Neuroleptic Agents. E-medicine on line @http://www.emedicine.com, 2002.
23. McCarron M. Diagnosis and treatment of drug overdose. In: Maronde RF, ed. Topics in Clinical Pharmacology and Therapeutics. New York: Springer-Verlag, 1986.

24. McCarron MM, Boettger ML, Peck JJ. A case of neuroleptic a nantadine. J Clin Psy 1982; 43:381.
25. Groleau G, Barish R, Tso E et al. Lithium intoxication. Am J Emerg Med 1987; 5:527-532.
26. Roberge R, Martin T, Schneider S. Use of sodium polystyrene sulfonate in a lithium overdose. Ann Emerg Med 1993; 22:1911-1915.
27. Winter ML, Ellis MD, Snodgrass WR. Urine fluorescence using a wood's lamp to detect the antifreeze additive sodium fluorescein: A qualitative adjunctive test in suspected ethylene glycol ingestions. Ann Emerg Med 1990; 19:663-667.
28. Jacobsen D, Sebastian CS, Barron SK et al. Effects of 4-methylpyrazole, methanol/ethylene glycol antidote, in healthy humans. JEM 1990; 8:455-461.
29. McCarron MM, Schultz BE, Walberg CB et al. Short-acting barbiturate overdosage – correlation of intoxication score with serum barbiturate concentration. JAMA 1982; 248:55-61.
30. McCarron M, Challoner K, Thompson G. Dipenoxylate—Atropine (Lomotil) overdose in children: an update report of eight cases and review of the literature. Pediatrics 1991; 87:694-700.
31. Curry S, Chang D, Conner D. Drug and toxin induced Rhabdomyolysis. Ann Emerg Med 1989; 18:1068-1084.
32. McCarron M, Schulze B, Thompson G et al. Acute phencyclidine intoxication: Clinical patterns, complications and treatment. Ann Emerg Med 1981; 10(6):290-7.
33. McCarron MM, Schulze BW, Thompson GA et al. Acute phencyclidine intoxication: incidence of clinical findings in 1,000 cases. Ann Emerg Med 1981; 10:237.
34. Beamon R, Siegel C, Landers G et al. Hydrocarbon Ingestion in children; a six year retrospective study. JACEP 1975; 5:771
35. Scalzo AJ, Weber TR, Jager RW et al. Extracorporeal membrane oxygenation for hydrocarbon aspiration. Am J DIS Child 1990; 144:867-71.
36. McCarron MM. Anticholinesterase poisoning. Drug Bulletin, County of Los Angeles Department of health Services. 1978 Vol 7.
37. Penner GE. Acid ingestion: Toxicology and treatment. Ann Emerg Med 1980; 9:374-379.
38. Zargar SA, Kochhar R, Nagi B et al. Ingestion of corrosive acids. Gastroenterology 1989; 97:702-7.
39. Vance MV, Curry SC, Kunkel DB et al. Digital hydrofluoric acid burns: Treatment with intra-arterial calcium infusion. Ann of Emerg Med 1986; 15:890.
40. Henry JA. Intravenous regional calcium gluconate perfusion for hydrofluoric acid burns. Clin Toxicol 1992; 30:203.
41. Howell JM. Alkallne ingestions. Ann Emerg Med 1986; 15:820-5.
42. Skinkinson C, Rumach BH, Sullivan JB. The dilemmas of acute iron intoxication. Emerg Med Rep 1989; 10:91-98.
43. Robotham JL, Lietman PS. Acute iron poisoning-a review. Am J Dis Child 1980; 134;875-879.
44. Schauben JL, Augenstein L, Cox J et al. Iron Poisoning: report of three cases and a review of therapeutic intervention. J Emerg Med 1990; 8:309-319.
45. Palantnick W, Tenenbein M. Leucocytosis, hyperglycemia, vomiting and positive X-rays are not indicators of severity of iron overdose in adults. Am J Emer Med 1996; 14(5):454.
46. McCarron MM. Isoniazid intoxlcation. Drug Bulletin, County of Los Angles Department of Health Services. 1978 Vol 5.
47. Yarbrough BE, Wood JP. Isoniazid overdose treated with high dose pyridoxine. Ann Emerg Med 1983; 12:303-5.
48. Wason S, Lacouture P, Lovejoy F. Single high-dose pyridoxine treatment for isoniazid overdose JAMA 1981; 246:1102-4.
49. Evers ML et al. Heart lung 1997; 26(4):325.
50. Hall A, Rumack B. Clinical toxicology of cyanide. Ann Emerg Med 1986; 15:1067.

CHAPTER 13

Selected Environmental Emergencies

Paul Silka

Hypothermia

Definition

Any patient with a core temperature of 35°C (95°F) or less is hypothermic. Hypothermia is further delineated into the following categories based on temperature:

- Mild hypothermia = 32-35°C (89.6-95°F)
- Moderate hypothermia = 28-32°C (86-89.5°F)
- Severe hypothermia = <28°C (86°F)

These categories of hypothermia are based on pathophysiologic changes that are routinely found at given temperature ranges and are important in treatment decisions.

Hypothermia can be further divided into primary and secondary;

- Primary hypothermia refers to patients who are hypothermic due to environmental exposure and are not found to have medical conditions, which lead to loss of normal physiologic temperature regulation.
- Secondary hypothermia results when normal thermogenesis is disrupted as is seen with patients with hypoglcemia or intoxication.

Finally, hypothermia may be **accidental** such as that seen is prolonged exposure to cold or **intentional** as is created during some surgical procedures.

The diagnosis of hypothermia is dependent on proper temperature measurement.

- A thermometer or thermocouple capable of measuring as low as 28°C should be used.
- The core temperature is best measured with electronic thermometers with flexible probes placed in the deep rectal position (15 cm) or the esophagus.
- Use proper equipment when more significant hypothermia is suspected.
  - Standard clinical thermometers only measure to 34.4°C (94°F) so diligence is necessary in pursuing this diagnosis.

Etiology

- Hypothermia is due to many causes although an accidental environmental (**primary**) source is most common. Environmental hypothermia may be immersion or nonimmersion in origin.
- Causes of secondary hypothermia include:
  - Metabolic
  - Drug-intoxication
  - Sepsis
  - Hypothalamic or CNS dysfunction
  - Dermal disease
  - Severe medical illness
  - Iatrogenic, can be considered a primary cause and is seen with surgical interventions and fluid resuscitation

Pathophysiology

Four terms are important in understanding temperature regulation.

1. *Conduction* is the transfer of heat by direct contact down a temperature gradient. Water is notorious for transferring heat away from victims of environmental hypothermia causing a loss of heat at 25 times the rate heat is lost to air.
2. *Convection* is the transfer of heat by movement. Wind plays a significant role in heat loss for many individuals with environmental hypothermia.
3. *Radiation* is responsible for heat exchange down a temperature gradient for those individuals who lack insulation.
4. *Evaporation* of water is also a major component of heat loss.

The hypothalamus is responsible for conservation and generation of body heat.

- Actions to conserve body heat are:
  - Vasoconstriction
  - Behavioral
- Trauma, intoxication, and psychosis may impair this normal response.

Thermogenesis can occur by

- Shivering
- An increase in the metabolic rate as adrenal and thyroid glands are stimulated.

Clinical

Patients with hypothermia may have a history that is obvious, a prolonged exposure to a cold environment. However, some cases may be less clear, and significant investigation must be undertaken to identify an accurate history that points to the causes of hypothermia. Be aware that the ambient environment need not be significantly cold for patient to develop hypothermia. Dry wind or wet clothes, and prolonged exposure even in relatively warm climates can still lead to significant hypothermia, particularly if an underlying medical illness or intoxication is present.

Predictable physical findings occur as core temperature decreases. (Table 13.1)

Table 13.1. Physical findings associated with specific core temperature ranges

Mild hypothermia:
- Lethargy, although excitability may be seen
- Shivering
- Confusion
- Loss of fine motor coordination

Moderate hypothermia:
- Delirium
- Slowed reflexes
- Bradycardia
- Shivering ceases below 31°C

Severe hypothermia:
- Rigidity
- Areflexia
- Cold skin
- Fixed pupils
- Coma
- Pulseless
- Apnea

Diagnostic Evaluation

- Electrocardiogram:
  Well-described changes in EKG tracings occur:
  - T-wave inversion
  - PR, QRS, QT prolongation
  - Muscle tremor artifact
  - Osborn J wave—a slow positive deflection at the end an of the QRS complex

  Dysrhythmias are also seen:
  - Sinus bradycardia
  - Atrial fibrillation
  - Nodal rhythms
  - AV block
  - PVC's
  - Ventricular fibrillation
  - Asystole

ED Management

Airway, breathing, and circulation are the first priorities, with caveats for the hypothermic patient:

- *Gentle handling* of these patients is indicated as sudden or vigorous movements may precipitate ventricular fibrillation, a risk as core temperature drops and the myocardium becomes more irritable.
  - In the prehospital setting, care and planning for movement of patients when possible may reduce unnecessary movements.
- *Airway management* can be complicated by muscle rigidity occurring in moderate and severe hypothermia.
  - Neuromuscular blockers may not be effective at temperatures <30° and should be avoided.
  - Intubation alone is unlikely to induce dysrhythmias. Risk may be minimized with adequate preoxygenation.
- *Oxygen*, warmed, and humidified, should be given to these patients as soon as possible.
- *Circulatory support*
  - Hypothermic patients are volume depleted.
  - Intravenous fluid must be warmed to 40°-42° C in a commercial warmer.
  - Lactated Ringer's solution to be avoide as the liver will be unable to metabolize lactate in moderate to severe hypothermia.
- Controversy exists over the initiation of CPR. Severely hypothermic patients may have a perfusing rhythm; however vital signs are hard to detect. CPR could induce fibrillation in the patients who actually may have a rhythm. Conversely, not initiating chest compressions will insure an unsuccessful outcome in the nonperfusing patient. Prehospital providers should take extra time to establish absence of vital signs before chest compressions are initiated; 30-60 seconds of assessment is essential. A cardiac monitor may add to the confusion, as nonarrest rhythms may be interpreted as pulseless electrical activity. In the ED, a bedside ultrasound can be used to determine cardiac activity.
- Pulse oximetry may be inaccurate when vasoconstriction is present.
- Central venous line must be placed with care as contact with the irritable myocardium may initiate dysrhythmias.
- Dysrhthmias are often refractory to standard treatment, rewarming is the best therapy;
  - Bradycardia is resistant to atropine in hypothermia.
  - Amiodarone has been determined to be the drug of choice for ventricular rhythms including ventricular fibrillation in the setting of hypothermia from animal studies.

  - Ventricular fibrillation should be treated with three attempts at defibrillation. CPR should be initiated if such intervention fails and the patient should be rapidly rewarmed.
- Glucose and thiamine should be given when hypoglycemia is documented by bedside glucose test.
- Alcoholics are predisposed to hypothermia as alcohol intoxication will decrease their ability to seek proper protection from cold injury.
- Antibiotics must be given if bacterial infection is thought to be present.
- Steroids should be given when adrenal suppression is suspected.

Rewarming

Passive

- Removal from cold environment
- Insulate with dry blankets

  Passive rewarming allows patients to use endogenous heat production correct body temperature.
  - Allows for steady correction without rapid changes that may lead to instability.
    - Patients must have an intact thermoregulatory system for passive rewarming to be successful.
    - Unstable patients who require rapid rewarming must also have active measures initiated.

Active External Rewarming

  - Warm water immersion
  - Heating blankets
  - Heated objects (e.g., IV bags)
  - Radiant heat
  - Forced air
- Active rewarming is indicated for patients who fail passive methods or for patients who have secondary causes for hypothermia.
  - Several disadvantages of active external rewarming must be considered;
    - Of minimal value in the setting of profound vasoconstriction or cardiac arrest.
    - External heat may lead to vasodilatation with subsequent hypovolemia, so-called "rewarming shock."
    - Rewarming acidosis occur as lactic acid from the peripheral circulation is mobilized centrally after vasodilatation.
    - "Core temperature afterdrop" is a phenomenon of continued core temperature decline after active external rewarming is initiated. This is probably due to continued conduction of heat from the relatively warm core to the periphery, although it had been previously ascribed to blood from the peripheral circulation "rushing" to the core after active external rewarming.

Active Core Rewarming

  - Inhalation rewarming
  - Heated IV fluids
  - GI tract lavage
  - Bladder lavage
  - Peritoneal lavage
  - Pleural lavage
  - Extracorporeal rewarming
    - Hemodyalysis
    - Arteriovenous
    - Venovenous
    - Cardiopulmonary bypass

- Mediastinal irrigation via thoracotomy
- Rewarming the core has the advantage of warming internal organs directly.
  - This improves cardiac function and decreases myocardial irritability.
  - Peripheral vasodilitation is avoided.
- These methods are invasive and considerable risks exist depending on the procedure chosen.
- Extracorporeal rewarming should be used when rapid correction of hypothermia is indicated.
- Thoracotomy with mediastinal irrigation should be reserved for patients who have arrested as the procedure is associated with considerably risk even in the hands of a qualified surgeon.
  - A recent study has shown that ED thoracotomy with mediastinal irrigation facilitated successful resuscitation in a majority of patients with severe hypothermia. Allows for rapid active rewarming, approximating that of cardiac bypass at 8°C/h.

Heat Emergencies

Heat illnesses encompass a spectrum of clinical scenarios which are mild and do not require medical care such as heat edema to serious disease with significant morbidity that requires the full armamentarium of an emergency department, from definitive airway control to pharmacologic intervention and critical care monitoring. Early recognition and treatment of these illnesses will decrease illness severity and mortality.

Hundreds of deaths occur annually in the U.S. due to heat-related illness. All victims of these disease entities have one or several risk factor that predisposes victims to heat illness. The following are common factors predisposing individuals to heat illness:

- Exertion in heat
- Dehydration
- Febrile illnesses
- Drug use, e.g., cocaine and amphetamines
- Medications, e.g., antipsychotics and anticholinergics
- Obesity and poor physical fitness
- Cardiovascular disease
- Hyperthyroidism
- Poor socioeconomic conditions
- Extremes of age

The body systems responsible for cooling;

- Skin. Skin surface acts as a cooling unit as vasodilatation allows core heat to be delivered to the periphery, and sweating permits significant evaporation.
- Cardiovascular. Cardiac output increases to meet the demands of vasodilatation and pushes blood carrying heat from the core to the periphery.
- Respiratory. A small degree of evaporative cooling occurs via respiration.

Pathophysiology

Body temperature is dependent on balance between heat production from metabolism and heat loss.

- Heat exchange is mediated through radiation, convection, conduction, and evaporation. Radiation is efficient in transferring heat away from the body when ambient temperatures are roughly less than that of the body, or 95° F.
- Evaporation is the most efficient mechanism the human body has to dissipate heat as ambient temperatures increase above 95° F. Heat energy is carried by liquid and converted into a gaseous phase allowing for substantial cooling. Sweating allows

considerable cooling as 1 L of sweat consumes 600 kcal of hea.. When high humidity is present this mechanism for cooling also will fail.

During heat stress, the hypothalamus stimulates the autonomic nervous system, leading to peripheral vasodilatation and splanchnic vasoconstriction. Cardiac output must rise to meet increasing demand. Stroke volume is diminished therefore leading to an increase in heart rate. Parasympathetic fibers stimulate sweating. As heat increases both the rate of sweating as well as the number of sweat glands activated increase.

As temperatures increase, splanchnic vasoconstriction fails and blood flow increases to the core and decreases to the periphery. As core temperature increases, cerebral dysfunction occurs due to:

- Cerebral edema
- Cerebral vascular congestion
- Increase intracranial pressure

Febrile illnesses cause increases in the body's production of heat. This is due to circulating pyrogens that are released by an activated immune system. These pyrogens "reset" the hypothalamus to increase heat production. This concept is important as cooling measures used for other mild to moderate heat illnesses are not useful in treating this etiology. If the patient is actively cooled, such as by enhancing evaporative mechanism with a cold bath, the body's response will be to shiver to maintain the core temperature at the indicated set-point. Antipyretics must be given to counteract the molecular mechanism which led to the hypothalamic stimulation.

Clinical Manifestation and Treatment of Heat Illnesses

- *Heat edema*—this entity is due to vasodilatation and orthostatic pooling of interstitial fluid.
  - Patients complain of swelling of hands and feet that usually occurs within the first few days of exposure to a hot environment.
  - No specific treatment is indicated.
- *Heat rash*—similar to heat edema in that it occurs with exposure to warm/hot environment and requires no specific intervention as it is a physical discomfort but has no significant sequelae.
  - Also called prickly heat, miliaria rubra, lichen tropicus, this is an inflammatory process involving the sweat ducts after the pores are obstructed by macerated stratum corneum.
  - Clinically, superficial vesicles on a red base are seen. This rash is often puritic.
  - Treatment is best accomplished with nondrying soap and loose-fitting clothes. Secondary bacterial infections occur, and initial treatment with topical bacteracidal agents is indicated.
- *Heat cramps*—manifest by involuntary painful contractions of skeletal muscle usually in the lower extremity.
  - Usually occurs after sustained heat stress and sweating, often in individuals who replace fluid losses with hypotonic solutions.
  - Best treated with oral balanced electrolyte solution or intravenous fluid replacement.
- *Heat exhaustion*—occurs when there is depletion of salt and water due to sweating. Venous return is compromised by a combination of heavy exertion and maximal cutaneous vasodilatation. Core temperature can be normal or elevated up to 104° F.
  - Symptoms are varied and include dizziness, weakness, light-headedness, fatigue, nausea, vomiting, headache, muscle cramps, syncope and myalgias.
  - Physical findings include orthostatic hypotension, sinus tachycardia, tachypnea, diaphoresis, and hyperthermia. Mental status is normal. As no findings are

pathonogmonic, other causes such neurologic, cardiovascular, or infectious disease should be considered.

- Treatment includes:
  - moving patient to a cool environment
  - administration of intravenous fluids, particularly if attempts at oral rehydration with a balanced salt solution has failed.
    - normal saline is the initial intravenous fluid of choice with subsequent fluids being chosen after review of serum chemistry.

- *Heat stroke*—occurs when thermoregulatory mechanisms are overwhelmed due to exogenous heat stress. The core temperature will rise to 104.9° F or greater. CNS dysfunction and anhidrosis are hallmarks of heat stroke. There may be a continuum of disease between heat exhaustion and heat stroke: therefore patients with heat exhaustion who develop end-organ dysfunction, particularly altered mental status must be assumed to have heat stroke.
  - Classified as *Classic* or *Exertional.*
    - *Classic* heat stroke occurs in high ambient temperatures and humidity.
      - Affects the extremes of age, in particular the elderly with cardiovascular disease.
      - Medication with anticholinergic properties, and lack of air conditioning as seen with lower socioeconomic groups are often contributing factors.
      - Core temperature rises slowly as mechanisms to disperse heat are overtaxed and fail.
      - Dehydration and electrolyte disorders are present.
    - *Exertional* heat stroke occurs rapidly, often in healthy individuals who have severe heat stress due to excess heat production, which occurs during exercise or physical labor in hot, humid conditions.

ED Management of Heat Stroke

Regardless of the type, treatment is the same and should be done while further work-up is undertaken;

- Rapid reduction of core temperature to 104° F should be attempted, after establishing and maintaining the ABCs.
- High flow oxygen—sicker patients with hyperthermia will require intubation.
- Vigorous fluid management with normal saline or lactated Ringer's solution is appropriate.
- A Foley catheter will help determine response to fluid replacement.
- Central venous monitoring or pulmonary artery occlusion pressure monitoring may be appropriate.
- Continuous temperature monitoring should also be performed with a thermister probe inserted rectally or in-line with the Foley catheter.
- Routine laboratory work-up including creatine phosphokinase and clotting studies are appropriate.
  - End-organs systems damaged by heat-illness are neurologic, cardiovascular, liver, and renal, appropriate studies include;
  - Computed tomography of the brain
  - Electrocardiogram and serum cardiac markers or enzymes
  - Serum hepatic functions
  - Serum creatinine and BUN

Rapid cooling is best accomplished by evaporation and should be started in the prehospital setting;

- Patient removed from heat stress.
- Clothing should be removed.

- Placing the patient in a cooled vehicle, applying a tepid bath, and surrounding with circulating air will enhance evaporation.
- Ice packs to the groin and axilla are appropriates however care should be taken to not induce vasoconstriction by packing the patient in ice.

In the emergency department a tepid water mist generated by a handheld spray bottle should be applied to exposed skin. A fan or fans should be directed at the patient to allow evaporative cooling to occur. Commercial cooling blankets are also available. A reasonable goal is to drop the patient's core temperature 0.2° C/ min. Remember, in severe heat illness, such as heat stroke, anitpyretics are not efficacious. Further care may include;

- Neuroleptics such as chlorpromazine can be used to inhibit shivering induced by rapid cooling. This will minimize endogenous heat production created during shivering. Caution, this class of medications may be additive with anitcholinergics responsible for the heat stroke episode. May also exacerbate hypotension.
- Benzodiazepine for treatment of seizure
- Diuretics, such as mannitol, enhance renal blood flow thereby decreasing renal injury. Such injury is common in heat stroke that is exertional in nature causing rhabdomyolysis with myoglobinuria.

ED Disposition

- *Heat exhaustion*: these patients can generally be cooled, hydrated, and have electrolytes corrected in the ED with subsequent discharge home.
- *Heat stroke*: this is a complex illness, and disposition and outcome are dependent on age and comorbitiy. Admission is indicated for all patients diagnosed with heat stroke.
    - Younger patients with the exertional variety have a good prognosis when treated early.
    - Older patient's or those with comorbid medical conditions who have classic heat stroke are more likely to have delayed complications ranging from clinically significant electrolyte disorder to cerebral edema or cardiopulmonary collapse and therefore will require admission to an intensive care setting.

Illnesses Due to Altitude

The partial pressure of oxygen decreases with increasing altitude. Resulting hypoxia can lead to a varied constellation of symptoms and disease for patients venturing to elevations >8,000 feet.

- *Moderate* altitude is >8000 feet, and even though only mild hypoxia occurs at this level, rapid assent, particularly in those with preexisting medical conditions may develop illness.
- *High* altitude is between 10,000 and 18,000 feet. Oxygen saturation falls below 90% at this elevation and most serious altitude illnesses develop here.
- *Extreme* altitude is elevation above 18,000 feet. Prolonged visits at this elevation will lead to deterioration even for those acclimatized to moderate and high altitude.

Acclimatization

A gradual, systemic process in which organ systems adapt to elevation and alter function allowing for increasing oxygenation of tissues.

Factors Promoting Acclimatization—Pulmonary

- Increase respiratory rate—hypoxic ventilatory response (HVR) occurs when the carotid bodies detect decrease pO_2 and feeds back to the central respiratory center of the medulla to increase ventilation.
- The ensuing respiratory alkalosis from hyperventilation will act on the respiratory center to slow respiration.

- Excess bicarbonate is excreted via the kidneys to maintain homeostasis allowing for an appropriate increase in ventilation.
- Central chemoreceptors reset to lower pCO_2, i.e., inspiration is triggered more frequently.
  - Acetazolomide causes bicarbonate diuresis thereby facilitating acclimatization.
- This process is referred to as respiratory or ventilatory acclimatization and takes 4 to 7 days. People who have insufficient HVR may be at risk for other high altitude illnesses such as acute mountain sickness (AMS) and high altitude pulmonary edema (HAPE).
- Pulmonary hypertension occurs as a result of hypoxia. This response appears to be accentuated for patients who develop HAPE.
- Ultimately, failure to acclimatize along with fluid shifts from intra- to extracelluar spaces lead to HAPE and high-altitude cerebral edema (HACE).

Blood

- Increased red cell mass occurs over days to weeks resulting in chronic mountain polycythemia.

Fluid Balance

- Diuresis follows bicarbonate build-up from hyperventilation with subsequent hemoconcetration.
- Fluid retention with shift of fluid into the extracellular spaces is seen and becomes clinically apparent as AMS and HAPE when pronounced.

Cardiovascular

- Decreasing pO_2 will lead to increase sympathetic tone.
- Heart rate increases and stroke volume decreases.
- The net result is decrease cardiac reserve and decrease exercise tolerance.

Cerebral Vascular

- Cerebral blood flow will increase at higher altitude as pO_2 decreases even in light of hypocapnia due to hyperventilation.

Clinical Syndromes

Acute Mountain Sickness (AMS)

Due to mild cerebral edema induced by hypoxia. This constellation of symptoms may be the early end of the spectrum of disease that progresses to HACE or HAPE.

- Complaints are often nonspecific and vague, particularly early.
- Headache is the predominate symptom. Headache along with the complaint of a gastrointestinal disturbance, such as anorexia, or a complaint of fatigue, dizziness, or sleep disturbance is necessary to confirm the diagnosis.
- Most commonly seen following rapid ascent to moderate or high altitude.
- Physical examination will often reveal a patient who appears ill but is otherwise unremarkable. Crackles may be heard upon auscultation of the lungs. Ataxia is a harbinger of onset of HACE.

Differential Diagnosis

- Dehaydration
- Hypothermia
- Intoxication
- Carbon monoxide
- Central nervous system infection, injury or mass

Treatment

- For most this disease is self-limited and will resolve in days. Occasionally some may fail to acclimatize and may have symptoms for weeks.

- If ataxia or mental status changes occur, treatment should be instituted rapidly.
- Descent or low flow oxygen will cure mild AMS rapidly.
- Further treatment includes:
  - Analgesics for headache. Acetominophen is appropriate.
  - Anti-emetics with prochlorperizine being first choice as this agent increases HVR.
  - Acetazolamide, 125-250 mg PO BID, can be used prophylactically or to treat symptoms. This agent works by stimulating a sodium and bicarbonate diuresis which prevents fluid retention.
  - Dexamethasone can also be used to diminish symptoms.

Prevention

- Avoid rapid ascents
- Acetozolamide 1-2 days prior to ascent and dexamethasone have been efficacious in preventing AMS.
- Ginko biloba has been shown to work prophylactically.
- High-carbohydrate diet
- Avoid alcohol

High-Altitude Pulmonary Edema (HAPE)

- This is a form of noncardiogenic pulmonary edema. Its onset can be rapid and is the most common cause of death from altitude illness worldwide.
- Cold exposure and exertion at elevation are predisposing factors to HAPE.
- Children and those with pulmonary hypertension are at increased risk.
- Key complaints include dyspnea at rest. Symptoms of AMS may also be present.
- A cough with frothy sputum occurs late and suggests worsening disease.
- Physical findings include crackles, first heard in the right middle lobe.

Differential Diagnosis

- Cardiogenic pulmonary edema
- Pneumonia

Treatment

- Keep warm and minimize exertion.
- Descent
- Oxygen, guided by pulse oximetry
- Hyperbaric oxygen. Portable hyperbaric bags are often carried by climbers on expeditions.

- Medications that have been used include furosemide, morphine, and nifedipine. Mild cases can be sent to base camp or hotel room with home oxygen, nifedipine and close observation. Inhaled nitric oxide has also been shown to be effective.
- Persistent hypoxia requires more aggressive treatment.
- Patients traveling by air after recovery should either delay their travel or arrange for portable oxygen.

Prevention

- Gradual ascent
- Nifedipine, 20 mg TID, 1 day before and 3 days after ascent
- Acetazolamide

High-Altitude Cerebral Edema (HACE)

- HACE is presumed when patients with AMS develop ataxia or altered mental status.
  - commonly associated with HAPE
  - progresses to coma in hours to days. Onset of coma is associated with mortality of >60%.

Differential Diagnosis
- Dehydration
- Hypothermia
- Intoxication
- Carbon monoxide
- Central nervous system infection, injury or mass

Treatment
- Oxygen
- Descent
- Dexamethasone
- Hyperbaric oxygen
- Intubation with optimal ventilation for patients with coma.

Prevention

As with AMS, and HAPE. Most importantly, gradual ascent.

Submersion Injury (SI)

Definition
- This is injury following prolonged immersion in a liquid environment.
- Near-drowning is the term that is used to describe survival, for at least 24 h, following a submersion.
- Drowning is defined as death resulting from suffocation within 24 h of a submersion in liquid. Death is usually a complication of pulmonary and/or neurologic injury.
- Immersion syndrome is sudden death seen following submersion in very cold water. This is probably due to dysrhythmias which are vagally mediated.
- "Wet" drowning or near-drowning refers to submersion followed by aspiration.
- "Dry" drowning or near-drowning refers to submersion without aspiration. This occurs in 10-20% of submersion injuries.

Epidemiology
- Submersion injury is a significant cause of morbidity in the U.S. resulting in over 8000 deaths annually.
- Two ages are most affected:
  - Toddlers most commonly injured or killed when adult supervision is interrupted. These injuries are commonly seen in bathtubs, swimming pools, and industrial buckets.
  - Adolescents most commonly injured when attempting to swim distances or in environments that are dangerous, or require more skill than the victim possesses. Alcohol is frequently a factor in submersion injury in adolescent and adult submersion injury.

Pathophysiology
- A well-defined sequence of physiologic events has been outlined;
- Breath-holding → voluntary apnea → panic → air hunger → involuntary gasp → aspiration → larnygospasm.
- Hypoxia → unconsciousness → active aspiration → convulsions → death follows if no rescue is performed.
- Previous work has focused on the difference between fresh and saltwater submersion injuries.
  - Fresh water submersion injury can lead to hemodilution and red cell lysis.

    - Salt-water submersion injury has been shown to cause a decrease in blood volume and an increase in serum electrolyte concentration.
    - The clinical significance of fluid and electrolyte changes is minimal as most SI victims typically aspirate a small amount of water, 4 ml/kg. Both salt and fresh water lead to washout of pulmonary surfactant leading to pulmonary injury.
- Pulmonary injury, frequently manifest as noncardiogenic pulmonary edema, and anoxic brain injury with subsequent cerebral edema are the insults that lead to morbidity and mortality following submersion.

Outcome

- The diving reflex may improve survival in infants and small children. This mammalian reflex occurs when cold water, <68° F, contacts the victims face. Peripheral and splanchnic circulation is diminished. Apnea and bradycardia ensue with subsequent neurologic circulation preservation.
- Cold water submersions induce rapid hypothermia. In this setting, good neurologic outcomes following prolonged submersion in water 10° C.
- Predictors of good outcome:
    - Victims who are awake and alert in the ED following submersion.
    - Those with decreased consciousness but are arousable have a survival of 95% without neurologic insult.
    - Cold water submersion.
- Predictors of poor outcome:
    - Prolonged submersion
    - Delay in CPR following rescue
    - Glasgow Coma Score <5
    - Severe acidosis
    - Hyperglycemia detected at the time or shortly after resuscitation
    - Fixed and dilated pupils
    - Asystole

  Despite the correlation of the above findings with poor outcome, victims have presented with each of the above findings and yet gone on to have a good neurologic outcome. This makes it very difficult for emergency providers to not provide full resuscitative efforts unless obvious signs of death are present.

History

- Often obvious in cases of submersion injury. Can be more obscure when a victim is found with altered mental status near a body of water.
    - Clues to submersion include choking, coughing, and vomiting.
- Events or comorbidity leading to the submersion event are important in treatment.
    - All the following must be considered:
        - Drug and alcohol intoxication
        - Head injury
        - Cervical spine injury
        - Cardiac disease or arrest
        - Seizure
        - Diabetes
        - Suicidality

Physical

- A careful cardiopulmonary examination must be done.
    - Stabilization with definitive airway and cervical spine protection must be accomplished concurrently.

- Any pulmonary finding such as tachypnea, rhonchi, crackles, or wheeze suggest some degree of injury.
- A careful and thorough neurologic examination must be done to determine level of consciousness, brain function, and peripheral nerve function.
- A search for accompanying traumatic injury, particularly spinal injury, should be made.

Treatment

Prehospital

- Often inadequate information exists on scene to terminate rescue and resuscitative efforts (the same may be true for the ED). Aggressive rescue measures appear to have positive effects on outcome.
- Mouth-to-mouth should be initiated early, even with the victim still in the water.
- The cervical spine should be protected during rescue.
- Subdiaphragmatic thrust should be performed if foreign material is found to obstruct the airway. This need not be done prophylactically at the start of basic life support.
- CPR performed early after rescue is a predictor of positive outcome.
- Attention to factors coinciding with submersion need to be addressed, such as,
  - Traumatic injury
  - Ischemic heart disease
  - Reactive airway disease
  - Hypoglycemia
  - Hypothermia

Emergency Department Managment

- Airway with cervical spine precautions must be paramount.
- Pulse oximetry should be used liberally, and a blood gas analysis should be done if oxygen saturation is abnormal or physical examination reveals altered mental status, rhonchi, wheeze, or crackles.
- Humidified oxygen should be given to all victims early and titrated to an oxygen saturation of 95% if possible.
- Albuterol can be used to treat bronchospam that may develop.
- Early intubation is indicated for patients who are altered after a submersion, particularly if pulmonary findings or hypoxia are present.
- Patients with respiratory failure will need positive pressure ventilation. Positive end-expiratory presure is usually indicated for patients with hypoxia requiring intubation. Diuretics are not efficacious.
- Hemodynamic and cardiac monitoring should be instituted.
  - These patients may have acidosis predisposing to myocardial dysfunction and dysrhythmia.
  - Sodium bicarbonate can be used to correct acidosis, but only after ventilatory management and correction of hypovolemia, which is frequently present in submersion injury.
  - Fluids must be used judiciously as cerebral edema complicates many severe submersion injuries.
  - This having been stated, the best way to reduce risk of secondary neurologic injury is to ensure optimal cerebral perfusion.
- Rewarming is essential. Some centers use extracorporeal circulation techniques to reduce the risk of dysrhythmias seen in other rewarming protocols.
- Routine laboratory plus toxicology and anti-convulsant drug levels should be obtained.
- CT imaging of the brain is only indicated at this stage if traumatic injury is suspected.

Disposition

- Patients who do not have a history of significant submersion and a normal examination and pulse oximetry may be discharged home.
- Patients who have a history of a significant submersion but are asymptomatic or have only mild symptoms in the ED and have a negative work-up may be discharged home after a period of observation.
  - These patients should have clear instructions to return for dyspnea and have an adequate method to return if the need arises.
  - The observation period should be at least 4 h.
  - Patients with preexisting cardiopulmonary disease may need more conservative management and therefore have a lower threshold for admission.
- Patients who have mild hypoxia that are readily corrected with oxygen should be admitted, given supportive care, and discharged when hypoxia resolves.
- Finally, patients with impending or actual respiratory failure or neurologic insult need admission to an intensive care setting in a center with resources to manage such a patient.

Dysbaric Injuries

Background

Those at risk:

- Scuba
- Commercial divers
- Caisson workers
- Aviators

Physical Principles

- Pressure is the force exerted on a given unit of area. The weight of air at sea level is 14.7 lbs/in^2 (psi) or 1 atmosphere (atm).
- Water is, of course, denser than air so large changes in pressure occur as an individual descends underwater. Each 33 feet of seawater (fsw) adds another atmosphere absolute (ATA). Therefore, at 99 fsw a diver will experience 3 ATA of pressure.
- Barotrauma refers to the injury caused by the mechanical effects of pressure.

Three basic physics concepts are the key to understanding these types of injuries.

13

- When pressure is exerted on a gas its volume will decrease proportionally. Boyle's law states that if pressure is doubled the volume will be halved. The inverse is also true, less pressure the greater the volume.
- The pressure of a mixture of gas is the sum of the partial pressure of each gas. This is Dalton's law and helps explain the cause of decompression sickness (DCS) and nitrogen narcosis.
- The amount of gas that will dissolve in a liquid is proportional to the partial pressure of that gas in contact with the liquid. This is Henry's law and tells us that the deeper a scuba diver goes, the more gas that will dissolve in the body tissues.

Pathophysiology of Dysbaric Injuries

The basis for disease is the damage that results from the expansion or contraction of gas spaces that occur in the body as a result of pressure.

History

Accurate history of the patients dive is essential. This "dive profile" should elucidate the following:

- The length and depth of each dive

- Rate of ascent
- Time of onset of symptoms
- Change in the type or intensity of the symptoms

Clinical Manifestations

Dysbaric injuries can be separated into those occurring during descent or ascent.

Descent Injuries

Squeeze. If a pressure differential develops in gas-, usually air, filled body compartments, pain and injury can occur. Individuals who descend underwater can experience a variety of these injuries commonly called "squeeze."

- Middle ear squeeze (barotitis media) occurs when the noncollapsible middle ear does not equilibrate with ambient water pressure during descent.
  - External water pressure exerts force on the tympanic membrane (TM). Patient's experience ear pain and fullness.
  - Both the TM and the middle ear mucosa will develop inflammatory changes with edema and hemorrhage.
  - TM rupture can occur leading to an influx of water to the middle ear space.
  - This will induce vertigo due to the caloric vestibular stimulation. Symptoms can be dramatic with severe vomiting and dysequillibrium.
  - This can be prevented by gently forcing air into the middle ear via the eustashian tube by performing a valsalva or Frenzel maneuver prior to and during the descent as needed. Individuals with disease preventing equilibration include an upper respiratory infection, allergic rhinitis, and otits media, all of which can cause eustashian tube dysfunction.
  - Ascent should occur at first sign of a diver being unable to equilibrate.
  - Treatment includes the following:
    - Analgesic may be needed, and topical aural anesthetics can be used with an intact TM.
    - Oral and nasal spray decongestants are appropriate with an added antihistamine if there is an allergic component.
  - Patients must abstain from diving until symptoms resolve.
  - Patients with perforated TMs should not use aural drops or dive until healing occurs.

External ear squeeze (barotitis externa) occurs when a portion of the external canal is occluded by cerumen, tight-fitting wet suit hood, or an ear-plug during descent.

- A negative pressure gradient develops within this space leading to tympanic membrance bulging, and is exerted on the soft tissue structure of the external canal causing edema and hemorrhage with subsequent pain.
- Treament is ascent and cortosporin otic suspension.

Inner ear barotrauma (IEBT) is far less frequent, however more serious than the other ear squeeze's.

- IEBT results from damage to the cochleovestibular apparatus.
- Pressure from middle ear is transmitted to the oval window by the ossicles during descent if the diver fails to equilibrate.
- Inward movement creates a pressure wave within the perilymph of the cochlea leading to outward distention of the round window into the middle ear.
- Ultimately, if a sudden equilibration occurs through vigorous valsalva, the round window may rupture.
- Even without rupture, the inner ear may be injured due to hemorrhage or tearing of the labyrinthine (Reissner's) membrane.

- Clinical findings include:
    - History is usually that of a difficult descent with onset not far from the surface.
    - Severe vertigo
    - Tinitus
    - Nystagmus
    - Fullness in the affected ear
    - Sensorineural hearing loss
- IEBT should be differentiated from alternobaric vertigo (ABV). ABV is due to unilateral pressure differential between the middle and inner ear. ABV tends to occur on deeper dives, often during ascent, and is transitory.
- Decompression sickness localized to inner ear structures may also mimic IEBT but again would occur after ascent.
- Treatment of IEBT includes medication to reduce vertigo such as, droperidol, meclizine, and diazepam.
- Due to damage to inner ear structure and the hearing loss, referral to an ENT surgeon is warranted.

Ascent Injuries

Decompression Sickness (DCS)

With increased ambient pressure, inspired compressed air dissolves in tissue. Oxygen is metabolized but nitrogen is an inert component of compressed air and must be released into the blood for elimination via the lungs. If too rapid an ascent occurs, nitrogen will be rapidly expelled into the interstitial, intralymphatic, or intravascular spaces. At this point nitrogen bubbles can block flow or distort tissue. These bubbles may cause indirect injury by causing alterations in coagulation, protein denaturation, and endothelial damage.

Clinical

- Decompression illness will manifest in many different forms depending on the amount of inert gas dissolved in the tissue and the location of the bubbles that develop as ambient pressure decreases.
- Accurate history of the patient's dive and the onset, location, and quality of pain is imperative in diagnosing DCS.

13

Type I Decompression Sickness

This grouping of DCS is manifest as either musculoskeletal pain, dermal complications, or constitutional symptoms.

- Musculoskeletal pain
    - Aching pain of the limbs, often in or about joints is the most common manifestation of DCS.
    - This manifestation is the classic DCS pain first described to occur in workers who were breathing compressed air in the subterranean shafts where the Brooklyn Bridge pylons were being placed. Workers emerged from these depths and within hours, developed pain forcing them into a bent-at-the-waist position. From this came the lay description of the disorder, "the bends."
    - DCS pain is described as deep, aching, dull, and hard to localize. Pain may improve with direct pressure as inert gas bubbles trapped in tissue causing discomfort are compressed.
    - Occasionally, the pain will be rapid in onset.
    - The sensation of fleeting aches or odd feelings may be present. This is referred to the "niggles." This may be a premonitory to an episode of true limb bends.
- Dermal DCS

- Commonly manifest as pruritis.
- The prurits is normally confined to the trunk and can be associated with reddening due to vasodilatation.
- Painless swelling with a "peau d'orange" can be seen suggestive of lymphatic obstruction due to bubbles.

- Consitutional DCS
  - Symptoms of DCS include, minor fatigue and malaise. Profound exhustion on the other hand may be a prodrome to more serious DCS.

Type II DCS

These groups of injuries are manifest as neurologic, cardiorespiratory, or vestibular symptoms. Type II DCS injuries are at risk for permanent disability or death must be treated rapidly.

- Pulmonary DCS
  - This relatively rare injury occurs when a sudden and massive venous gas bubble obstructs the pulmonary artery. This is typically seen after rapid ascent minutes after reaching the surface.
  - The diver will complain of pleuritic substernal pain and cough.
  - Deep breathing will exacerbate the discomfort and can provoke paroxysms of nonproductive coughing. This manifestation led to this entity being labeled the "chokes."
  - Cynosis, right-heart failure, and cardiovascular collapse can follow and immediate recompression is indicated.
- Neurologic DCS
  - This injury will manifest as a crescendo of symptoms starting with tingling or constricting about the trunk and progression to numbness, ascending motor weakness with loss of bowel and bladder control.
  - Onset is from minutes to hours after a dive but may also be sudden and devastating with paraplegia resistant to treatment.
  - Coma has also been reported.
  - The pathophysiology of the injury is uncertain. It is clear however that this injury is prone to occur at the spinal cord and may be due to arterial gas embolism, epidural venous thrombosis, or local tissue bubble formation.
- Vestibular DCS
  - Sudden onset of vertigo, nausea, vomiting, nystagmus, tinnitus, and hearing loss can be seen when inert gas damages membranes of the semicircular canals and cochlea.
  - The impressive presentation of this form of DCS has led to the classical labeling by divers as the "staggers."
  - Must be differentiated from middle ear barotrauma, which is more common. Vestibular DCS can be permanent while middle ear barotrauma generally does not cause prolonged disability.

Treatment

- Advanced airway, positive pressure ventilation and crystaloid support of blood pressure may be necessary for those divers severely injured.
- All patients with DCS, type I and type II, should be treated with hyperbaric oxygen.
- Early consultation with an experienced dive physician is mandatory, as delays in recompression are associated with significant sequelae and death.

Pulmonary Barotrauma and Arterial Gas Embolism (AGE)

This injury with associated AGE is the leading cause of death during diving after drowning. It is typically seen after breath-holding during a rapid ascent and is most common among novice divers who are most likely to undergo a panic ascent and/or be unfamiliar with their equipment.

Pathophysiology

- Pulmonary barotrauma is due to over-inflation that occurs during ascent when expanding gases cannot exit the lung.
- This is most commonly the result of breath-holding during ascent but can also be due to trapped gases from local pulmonary obstruction such as bronchospasm.
- Intrinsic lung abnormalities may also result in pulmonary barotrauma.
- As over inflation occurs the central vascular bed, subclavian veins, the aorta, heart, and carotid arteries are filled with gas.

Clinical Presentation

Symptoms and signs are manifest within minutes of surfacing.

- A catastrophic presentation in which the victim develops apnea, unconsciousness, and/or cardiac arrest is seen in 5% of cases. A lethal arrhythmia secondary to coronary artery embolism or cerebral embolism or complete filling of the central vascular bed is the etiology of death.
  - These individuals are resistant to advanced cardiopulmonary resuscitation.
  - Presentation is more likely to consist of varying degrees of neurologic injury.
- AGE most commonly produces the following findings:
  - Stupor and confusion or loss of consciousness
  - Hemiparesis
  - Seizure
  - Vertigo visual disturbance
  - Sensory changes
  - Headache

Treatment

- Advanced airway, positive pressure ventilation and crystaloid support of blood pressure may be necessary for those divers severely injured.
- Positioning of patients with AGE has been the source of some controversy.
  - Traditional teaching that a patient should be placed in the head-down position, with the head below the heart. This was thought to prevent any remaining bubbles from embolizing to the cerebral circulation. Further study fails to support this theory in clinical practice and the supine position is most appropriate.
  - Supine positioning is less likely to promote cerebral edema that can occur following an AGE and which would be made worse with head-down positioning.
- Recompression in a hyperbaric chamber is imperative.
  - Early consultation with an experienced dive physician is mandatory, as delays in recompression are associated with significant sequelae and death.
  - Location of treatment facilities and physicians experienced with these injuries can be achieved by contacting the Divers Alert Network (DAN).

Electrical Injuries

Electrical injuries account for approximately 3% to 4% of admissions to burn centers. They are seen most commonly in adolescent males and industrial injuries. Small children also have a predisposition to low-voltage injuries due to the relationship of their stature and electrical cords and outlets.

Factors Determining Severity of Electrical Injury

Types of Current:

- Direct current (DC) tends to cause injury after a brief encounter with current and can throw the victim causing secondary blunt trauma.

- Alternating current (AC) is more dangerous than DC at the same voltage. Tetany of flexor muscles occurs when a victim makes contact with a current of 6 to 9 mA. This draws the victim closer to the current and prolongs contact.

Resistance of Tissues

- Each tissue has a specific resistance to flow of electrical current.
  - Nerves, muscle and blood vessels have low resistance and are good conductors of electrical current.
  - Bone, tendon, and fat have a high resistance and heat rapidly causing a coagulation necrosis.
  - Skin is relatively resistant to electrical current. Resistance is diminished with the presence of moisture (e.g., water or sweat).
  - In high-voltage (high-tension) injuries the entire cross section of the body will act as a resistor, and individual resistance of each tissue is of little importance.

Amperage

- This is a measure of energy and reflects the flow of electrons per second through an object.
- Different levels will have differing physiologic consequences:
  - 0.2-0.4 mA will cause tingling sensation
  - 6-9 mA is known as the "let-go" current at which individuals will not be able to let go of the energy source causing prolonged contact with the injurious energy source
  - 20-50 mA will cause tetany of the thoracic muscles and subsequent respiratory arrest
  - 60-120 mA will induce ventricular fibrillation

Voltage

- Current flows when there is a potential difference between two points. This potential is measured in volts.
- High voltage is generally considered to be above 1000 V.
  - A 600 V current however can cause significant tissue damage and death. Due to the public's accessibility to lower voltage nearly half of electrocutions are due to low-voltage sources.
  - Even the 120 V used in household currents has caused death. This occurs when environmental circumstances enhance energy by decreasing resistance.

Environmental Factors

- Moisture such as water or sweating lowers the resistance significantly; lower voltage can cause more injury.
- Water immersion is a classic precipitant of electrocution as a household current of 120 V can become lethal.

Current Pathway

- Injury will depend on the path current takes through the body.
- "Source" and "ground" describe the entry and exit of current.
- Source and ground site may be most dramatic of all findings on physical examination; however the tissue that the current passes through will have the most extensive damage.
  - These findings have been referred to as the "tip of the iceberg" phenomenon, in that the soft tissue injury seen during examination of the patient with electrical injury is only a small portion of the damage, particularly as voltage increases.
  - Current passing through the heart can lead to myocardial injury and dysrhythmia.
  - Current passing through the brain can lead to seizures, paralysis, and respiratory arrest.

Duration

- The longer the duration of contact the higher the tissue damage.
- Carbonization of tissue occurs as tissue damage reaches it's peak.

Pathophysiology

- Contact burns at source and ground sites will be evident.
  - The more extensive these lesions the greater the injury to deeper tissue.
  - The converse is not true however, as lack of cutaneous injury does not mean that there is no injury. Extensive tissue damage may occur even without significant source and ground lesions.
  - Moisture may lower resistance such that no external trauma is present.
- "Kissing pattern" is the name for burns caused from current arcing across the flexor creases of the knees, elbows, and axilla.
- Current passing along arteries can lead to spasm, thrombosis, and aneurysm formation.
- Extensive vascular damage results in myonecrosis and compartment syndrome.
- Bone is a poor conductor of electricity and necroses from extensive high-voltage injury along with muscle and nervous tissue.
- An electrical arc is a high-voltage current between two potentials.
  - Injuries from arc contact include:
    - Thermal burns
    - Blunt musculoskeletal injury from fall following contact with the source.

Clinical

History

- The following are important:
  - Voltage
  - Type of current
  - Duration of contact
  - Pathway of current
  - Factors modifying skin resistance, i.e., water
  - Comorbitity, such as diabetes or coronary artery disease.

Physical

- Primary survey must be done to stabilize airway, breathing, and circulation.
- Secondary survey should focus on delineating cutaneous burns and secondary trauma.
- A careful neurovascular assessment must be done.

Laboratory and Ancillary

- EKG and a period of cardiac monitoring is indicated for:
  - High-voltage exposure (>600 V)
  - Low-voltage if loss of consciousness, amnesia, altered mental status, neurologic findings, palpatations, ectopy, or chest pain present
- Extensive cutaneous burns or high-voltage electrical injuries require:
  - Blood count
  - Electrolytes, BUN, creatinine, and Ca^{++}
  - Creatine kinase
  - Serum myoglobin
  - Urinalysis, if dip heme positive but no RBC's suspect rhabdomyolyis
  - Liver functions/amylase if abdominal injury suspected
- Cervical spine radiographs and computed tomography of the brain must be performed if patient is altered or has neurologic findings

Treatment

Prehospital

- Rescuers must secure a safe scene by disconnecting the power supply before attempting a rescue.
- Airway must be secured, breathing established, and circulation restored with ACLS protocol with the following caveats:
  - Maintain cervical spine precautions.
  - Aggressive fluid resuscitation should be initiated in the field.

Specific Treatment Depends on Specific Injuries

- Cardiovascular—sudden cardiac death from asystole due most commonly from DC current and high-tension AC current, and ventricular tachycardia, ventricular fibrillation due most commonly low-tension AC current as seen in household accidents, must be treated by ACLS protocol.
  - Myocardial dysfunction and infarction have been described and require intensive care admission. Seen when current travels in a vertical pathway through the body.
    - Myocardial damage is normally nontransmural and ST-T changes may be absent.
    - Coronary artery vasospasm can also occur but is less frequent.
    - Common dysrhythmias include; sinus tachycardia, nonspecific ST-T changes, atrial and ventricular ectopy, first and second-degree heart blocks, atrial fibrillation, bundle branch blocks, and QT prolongation.
    - Cardiac enzymes may fail to delineate cardiac injury and cardiac markers have not been studied in this senario. Echocardiogram may be helpful.
  - Crystalloids should be used to treat hypotension initially with pressors added in a stepwise fashion. Significant burns must be resuscitated with normal saline or lactated Ringer's to maintain urine output at 1 to 1.5 ml/kg/h.
- Pulmonary—respiratory arrest can occur.
  - Aggressive airway management and positive pressure ventilation are indicated if respiratory effort and hypoxia is present.
  - Pulmonary edema and contusion have been described.
- Neurologic—coma must be treated with intubation and optimal cerebral oxygenation.
  - Seizure can occur and should be treated in standard fashion.
- Cutaneous—electrothermal contact injuries, flash burns, and secondary burns from clothing and accessories such as belts or rings.
  - Burn dressings with sulfadiazine or mafenide acetate should be applied.
  - Tetanus immunization must be insured, as electrical injuries are "tetanus prone."
  - Transfer to a burn center is appropriate for significant burn injuries.
- Vascular—injury can result in compartment syndrome.
  - Partial-thickness cutaneous burns can convert to full-thickness when vessels thrombose.
- Musculoskeletal—vertebral compression, dislocations, and long bone fractures are common with high-voltage following falls as patients are thrown from the energy source.
  - Perform reductions and splinting and obtain appropriate consultation.
  - Compartment syndrome requires fasciotomy and extensive soft-tissue and vascular damage may lead to amputation.
- Oral injuries—children can chew through the electrical cords resulting in lip injury. Current may arc resulting in high temperatures (up to 2500° C).
  - Injury often involves the commissure, the tongue, and alveolar ridge with or without injury to the labial artery.
    - Bleeding will be delayed as initial thrombosis and eschar will prevent bleeding.
    - Delayed hemorrhage occurs at approximately 5 days when the eschar separates.

- Inpatient observation has been the traditional recommended management although outpatient management with close follow-up and parent education has been described.
- Surgical and dental consultation is necessary for repair of these injuries.

Disposition

- High-voltage exposure (>600V), no obvious injuries = observation
- High-voltage exposure with injuries should be transferred to a burn center.
- Monitoring is indicated for all patients with high-voltage injury whose first EKG is abnormal.
- Low-voltage exposure that is asymptotic and normal EKG may be discharged home.

Bites and Stings

For the purposes of this work, bites and stings from many sources will be discussed together, as many of the treatments in the ED will be similar regardless of the source of injury.

Bites can be with and without envenomation. Mammals, arthropods and reptiles are all sources of common mechanical bite wounds.

Pathophysiology

Mechanical bite wound without envenomation can lead to extensive soft-tissue injury. As the size of the animal inflicting injury increases the amount of force applied will increase. In addition to soft-tissue damage, vascular, nerve, and bone injury must be considered. This becomes particularly important with dog bites as the force increases to 150 - 450 psi.

Infection is the common complication of all wounds. Several factors will increase the risk for infection:

1. Puncture type wound as seen with cat bites and foreign bodies
2. anatomic location with hand and feet being the site of greatest infection risk
3. delay in treatment
4. patient condition, including age and comorbid medical conditions

The animal and organisms commonly causing infection are grouped below:

- Dog—*Stapholococcus aureus*, *Pasturella multicida*, and *Capnocytophaga canimorsus* causing severe sepsis in immunocompromised individuals
- Cat—*Pasturella multicida*, also *Bartonella henselae* seen in cat-scratch disease, which results in rash and regional lymphadenopathy following exposure to a cat (not necessarily a bite)
- Monkey—*Neisseria*, α-hemolytic streptococci, *Hemophilus parainfluenza*, and *Eikinella corrodens*. *Herpesvirus simiae* must also be considered.
- Large feline carnivores (lions, tigers)—*Pasteurella*
- Alligator—*Aeromonas hydrophila*
- Iguana—generally innocuous, although moderate soft tissue damage can occur.
- Rat—leptospirosis

Illness following bites from venomous creatures will vary depending on the venom.

Snake Bites

Most North American snakebites involve pit vipers (family *Crotalidae*).

- Venom from these animals is designed to kill prey rapidly and causes direct tissue injury with capillary leakage, coagulopathy, and neurotoxicity.
- 2-20% of these bites are "dry," as no venom is inoculated from the bite to the victim.
- Clinical findings include tenderness at regional lymph nodes, nausea, metallic taste, muscle fasiculation, bleeding from puncture site, and occasionally hypotension and shock.

- These bites are painful with significant soft tissue swelling, often leading to swollen, tense extremities, ecchymosis and hemorrhagic bullae. Vascular compromise can occur, but compartment syndrome is rare.
- Tissue necrosis can occur at the wound site.
- The venom acts on blood and diminishes its ability to clot.
  - This syndrome resembles disseminated intravascular coagulation (DIC) as a generalized fibrinogenolysis.
  - True DIC with intravascular thrombosis, microangiopathic hemolytic anemia, and thrombocytopenia is rare.

Treatment

"Do no harm" is the first most important concept of snakebite treatment. Incision into the wound followed by suction, tourniquets, and ice should not be used.

- Prehospital care should include:
  - ABCs
  - Minimize patient activity
  - Remove restrictive clothing or jewelry
  - Splints
  - A venous constriction band—this is not a tourniquet, it should be placed to impede venous and lymphatic return. This should not decrease arterial flow.
- In the ED:
  - ABCs must be insured. Although rare, a generalized reaction can occur resulting in swelling of the lips, face and airway proper. These patients should be endotracheally intubated rapidly before swelling occludes the airway. A surgical airway may be indicated if significant swelling has already occurred.
  - Early intravenous access must be assured.
  - Opioid analgesia is commonly indicated
  - Tetanus status must be assessed and updated
  - Splinitng should be done, particularly for bites below the elbow
  - Antibiotics are often initiated but need only be given if an infection exists.
  - Baseline laboratory studies are indicated: complete blood count including platelet count, prothrombin time, fibrinogen, and electrolytes.
  - Antivenin Crotalid Polyvalent (ACP) is available which neutralizes venom from North, Central, and South American crotalids.
  - A new, polyvalent, ovine-derived, fragment antibody (Fab), crotalid antivenin has been approved for clinical use.
    - This antivenin may be associated with improved efficacy and fewer side effects then ACP.
  - Consult regional poison center for dosing instructions.
  - Indicated for use in the following snakebite patient:
    - Mild crotilid envenomations with clinical progression.
    - Moderate and severe crotilid envenomation.
    - This can be broken down into the following;
      - Local injury—swelling and ecchymosis
        - Progression is present when circumfrential measurements done serially are found to increase 0.5 cm/h.
      - Coagulopathy—thrombocytopnia, hypo-prothrombinemia, prolongation of INR
      - Systemic—hypotension or altered mental status
  - ACP use is complicated by a high rate of adverse reactions.
  - The Wyeth ACP is derived from horse serum. The following complications are commonly seen.

- Anaphylactic reaction (type I hypersensitivity)
- Anaphylactoid reactions
- Serum sickness (type III hypersensitivity)
- Skin testing is indicated before use.

Prognosis

Mortality rate has been reported at 1.4%. This may be high as many insignificant bites go unreported.

- Characteristics of fatal crotalid bites include:
  - Proximal site of bite
  - Minimal or no antivenin administered
  - Nonaccidental
- Most sequelae occur as a result of upper extremity or face wounds and include:
  - Decreased range of motion
  - Weakness
  - Hypesthesia
  - Paresthesia
  - Anesthesia
  - Skin discoloration
  - Rarely amputation

Arthropods

Bites caused by this family of organisms are often complicated by envenomation. Arthropods are animals with segmented bodies and jointed appendages. Medically important animals include the following classes:

- Arahnida
  - Spiders
- Hymenoptera
  - Bees
  - Wasps

Spiders

*Brown Recluse or Missouri Fiddle Back (*Loxosceles reclusa*)*

13 While many spiders may cause bites, spider-bites from either *loxosceles* or *latrodectus* are most commonly complicated requiring medical care.

The brown recluse spider (*Loxosceles reclusa*) produces the clinical condition of loxoscelism, occasionally referred to as "necrotic arachnidism."

The Spider

- Nonaggressive
- Prefer quiet, warm and dry environment, e.g., wood pile, garage, clothes piles
- Nocturnal and most active from April to October
- Only bite when trapped against the skin
- *L. reclusa* is found throughout U.S. but mostly in South Central United States

Pathophysiology

- *Loxoceles* venom causes enzymatic digestion of prey.
- Venom consists of a phospholipase, sphingomyelinase D, which promotes hemolysis, aggregation of platelets with release of serotonin, and vascular thrombosis and polymorphnuclear (PMN) leukocyte chemotaxis.
  - Other proteins in the venom include hyaluronidase, esterases, proteases, and others.
- The result of envenomation is activation of complement leading to inflammation and hemolysis. A dermal necrotic lesion is seen clinically.

Clinical

Dermal

- The bite itself can be mild or severe. Mild to moderate pain followed by itching is experienced.
- Within 24 h a central blister with surrounding erythema and edema develop. This is tender.
- Lymphangitis can occur.
- Local vasospasm leads to ischemia, pain and continued itching.
- A white ring of vasospasm surrounds the central violaceous center of the wound. This complex is surrounded by erythema and is the classic red, white, and blue of dermonecrosis.
- The erythema can progress and central serous or hemorrhagic bullae may develop at the wound center between 24 and 72 h.
- At 4-7 days an eschar forms and can remain in place days to weeks before sloughing off. An ulcer involving skin and adipose tissue remains. The muscle beneath the wound is spared.

Systemic

Systemic reactions from *Loxosceles* are infrequent and generally occur in 24-96 h and resolve in one week. The following signs and symptoms may be seen:

- Fever
- Chills
- Arthralgias
- Malaise
- Rash
- Nausea
- Vomiting
- Leukocytosis
- Thrombocytopenia, hemolytic anemia, jaundice, with progression to renal failure, DIC and shock has been described.
- Anaphlactic or anaphylactoid reaction may be seen.
- Other rare complications include endotoxemic-like effect of the venom, and seizures.

Treatment

- ABCs are generally not compromised unless anaphylaxis occurs or soft-tissue swelling in the face and neck region results from a bite to the neck or face region.
- Tetanus toxoid should be given if patients have not been recently updated.
- Local wound care should include thorough cleaning and elevation.
- Antihistamine may be of limited value for pruritis.
- Antibiotics should be administered only if a secondary infection occurs.
- Dapsone has been recommended by some experts in treating *Loxosceles* envenomation; however this remains controversial:
  - Dapsone, a sulfone antibiotic, limits PMN migration theoretically limiting lesion progression.
  - Dapsone is associated with considerable complications which outweigh the benefits which way be experienced by its use.
  - These complications include: hypersensitivity syndrome, methemoglobinemia, and hemolytic anemia in patients with G6PD deficiency.
  - Dapsone may be indicated if a *Loxosceles* bite occurs in a cosmetically sensitive area and results in progressive ischemic necrosis within 24 h.
- Hyperbaric oxygen has been advocated by some; however no clinical improvement of *Loxosceles* wounds has been consistently demonstrated.

- Early surgical excision is contrandicated as this can lead to further necrosis, delayed wound healing, and scarring. Only after the area of necrosis is clearly demarcated, 4-8 wk after the bite, should surgical alternatives be explored.
- Patients with significant systemic reactions should receive supportive care, common interventions include:
  - Inpatient management.
  - Hydration and monitoring of renal function when hemoglobinuria is present. Dialysis is indicated for renal failure.
  - A complete blood count with a coagulation profile should be obtained. Type and cross should be ordered.
- Antivenin for *Loxosceles* exists; however its use is usually clinically impractical as most patients are not aware of the envenomation until significant tissue destruction has occurred.

Prognosis

Serious complications are rare with most bites healing with only supportive care. Most severe systemic reactions are seen in children.

Black Widow (Latrodectus)

The bite of the black widow is diagnosed infrequently. Yet, the web of this spider seems to be spun specifically to attract children, as the webs are large and described as graceful and geometric on bushes and in grass.

The Spider

- Found throughout North America
- Aggressive when disturbed, as they are territorial, prefer warm, dry, quiet, locations such as garages or basements.
- Bites occur when their web is disturbed.

Pathophysiology

- Black widow venom lacks cytotoxic agents so no local wound or pain is experienced.
- The venom is a neurotoxin which acts at the presynaptic membrane of the neuromuscular junction causing acetylcholine release with decreased reuptake.
- Excess acetylcholine leads to muscle contraction/cramping/spasm.

Clinical

13

- The bite itself is nonpainful, perhaps a pricking sensation.
- Within 30 min there can be a sensation in the bitten extremity described as uncomfortable. Regional lymph node tenderness has been described.
- A "halo" lesion may appear at the bite site, which is specific for black widow bites but fades within 12 h.
- Significant symptoms are commonly delayed 6 to 12 h.
- Proximal muscle cramping is seen and progresses to pain in the back, chest, or abdomen depending on the site of the envenomation.
- Pain can wax and wane.
- Autonomic nervous system dysfunction may include nausea, vomiting, sweating, hypertension, tachycardia. Priapism has also been described.
- Patients have described feeling "not quite right" for 1 to 2 wk after the bite.
- Patients frequently appear quite ill, especially children.

Treatment

- ABCs are generally not compromised.
- Tetnus toxoid should be given if patients have not been recently updated.
- Analgesia is a priority for these patients:

- Intravenous opioids are indicated and may be sufficient to manage the patient.
- Benzodiazepines should be used as an adjunct to opioids.
- Some authors have advocated calcium gluconate over benzodiazepines for treatment of the cramping pain.
- Antivenin is available for patients with refractory pain.
  - *Latrodectus* antivenin is horse serum-derived and is noted to be very efficacious. However, acute hypersensitivity reactions are seen, and death has been described.
  - The indications for antivenin should be limited to pain not controlled by analgesia or life-threatening hypertension and tachycardia uncontrolled with supportive care.

Prognosis

- If pain is controlled and the patient tolerates oral medication, outpatient management can be considered.
- If pain is not readily controlled, admission for hydration and analgesia is indicated.
- Patients with significant hypertension or refractory to analgesia should be admitted to a critical care unit.
- Death is rare.

Hymenoptera

Stinging insects are members of the order Hymenoptera. Over 40 people in the U.S die annually from hymenoptera stings making this an important injury for emergency physicians to understand.

Subgroups

- Vespids
  - Wasp
  - Yellowjacket
  - Hornet
    - these insects can deliver multiple stings
- Apids
  - Bumblebee and honeybee
    - barbed stinger, which bee cannot retract after stinging these insects can only deliver one sting and then are eviscerated when removed from stinger

Venom

- Biogenic amines—histamine, dopamine, serotonin, acetylcholine, noradrenaline
- Enzymes—phospholipase, hyaluronidase
- Peptides—are found in wasps and hornets

Reactions

Local reactions at the site of sting are most common.

- Pain
- Swelling
- Pruritis

Local reaction can become extensive within the first few days and last upward of 1 wk. Reactions are difficult to distinguish from cellulitis, but these stings rarely become infected. Lymphangitis and adenopathy are indicators of infection.

Treatment

- Ice and analgesia are sufficient for majority of these injuries.
- Diphenhydramine—indicated for more extensive local reaction
- Prednisone—can be given for more extensive and bothersome local reaction

Even people with large local reaction have <5% chance of anaphylaxis following future stings.

Toxic Reactions

- Occur following multiple stings, hundred or thousands, as can be seen when Africanized or "killer" bees are the attacking insects.
  The large volume of venom mediates these reactions.
  Africanized bees have been found in many regions of the southwestern U.S. as they have migrated from Brazil after Southern African bees were mated with Europeanized bees in 1956 in an attempt to create a species, which were thought, might produce more honey. The result however created a species that has a pack mentality with great ferocity. Victims of attacks of these bees will have thousands of stings.

Anaphylaxis

Anaphylaxis is an acute, IgE mediated, life-threatening allergic reaction with manifestations that may begin with rash and pruritis, but may quickly progress to airway edema and cardiovascular collapse. Common precipitants include bee stings, peanuts, shellfish and prescription medications.

Primary Survey

Airway

• Attention must immediately be directed to the airway for signs of acute compromise such as drooling, stridor and dysphonia. The presence of these signs dictates the most aggressive treatment, even when there is a concern about cardiac compromise with use of parenteral epinephrine.

• Airway occlusion is a grave concern in anaphylaxis and will usually respond to IV epiniephrine administration. If not, emergent, definitive airway control is indicated. If orotracheal intubation is attempted, it should be done with smaller tubes and with the simultaneous preparation for a surgical airway such as a cricothyroidotomy. Airway swelling may make endotracheal intubation impossible.

• In certain situations, such as angioedema secondary to a drug reaction, swelling may be limited to the anterior airway and nasotracheal intubation in the breathing patient is a possible rescue maneuver before a surgical airway is necessary.

13

Breathing

• Inhaled β agonists such as albuterol (5 mg increments by inline nebulization) should be used to reduce bronchospasm, including after a definitive airway is established.

• Inhaled agents should not be delivered at the expense of maximal supplemental O_2. Oxygen by nonrebreather mask (or an FiO_2 of 100% during mechanical ventilation) should be used initially.
Methylprednisone 125 mg IVP should be given.

• A type 1 histamine blocker such as diphenhydramine (25-75 mg IVP) and a histamine type 2 blocker such as famotidine (20 mg IVP) should be considered.

• IV aminophylline or IV terbutaline may be considered for refractory bronchospasm.

Circulation

In the presence of hypotension, intravenous normal saline should be initiated immediately as a bolus. 1-2 L may be delivered along with epinephrine, with frequent reassessments for the development of pulmonary edema. Trendelenburg position should be utilized initially for shock.
Epinephrine is the mainstay of treatment for circulatory compromise. It should be administered subcutaneously until intravenous access is available.

The initial adult dosages of epinephrine are as follows:

IV: 0.1 mg initial dose (1 ml of 1:10,000 solution)

SQ: 0.3-0.5 mg initial dose (0.3-0.5 ml of 1:1,000 solution)

These may be repeated and titrated to clinical effect, and once stabilized, epinephrine may need to be delivered as a continuous infusion.

Important note: Although epinephrine is indicated for any circulatory compromise such as frank hypotension or other clinical signs of shock, caution should be exercised in patients with known coronary artery disease and in older patients. In these circumstances, the doses of epinephrine used should be reduced accordingly and titrated until there is a clinical response or signs of cardiac ischemia or dysrhythmia appear.

Disability: Alternative diagnoses should be considered when altered mental status does not improve with oxygenation and restoration of adequate perfusion. This includes a head CT.

Exposure: A diffuse rash, urticaria or the presence of a medical alert bracelet or identification card may be important clues to the diagnosis of anaphylaxis in undifferentiated shock or airway obstruction.

Resuscitation Phase

Critical Questions: Clues to an anaphylactic etiology of shock may be obtained by searching for a history of allergy, new medications or exposures.

Critical Investigations: Chest X-ray may reveal critical information, such as a pneumothorax.

Lateral soft tissue neck films may assist with evaluating the airway for alternative diagnoses of airway obstruction, such as foreign bodies, hematomas and abscesses—but these should not be done in lieu of stabilizing a compromised airway.

Suggested Reading

1. Anderson PC. Spiders bites in the United States. Dermatologic Clinics 1997; 15(2):307-12.
2. Bond GR. Snake, spider, and scorpion envenomation in North America. Pediatr Rev 1999; 20(5):147-50.
3. Dart RC, McNally J. Efficacy, safety, and use of snake anti-venom in the United States. Ann Emerg Med 2001; 37:168-174.
4. Jerrard DA. ED management of insect stings. Am J Emerg Med 1996; 14(4).
5. McKinney PE. Out-of-hospital and interhospital management of crotaline snakebite. Ann Emerg Med 2001; 37:181-188.
6. Walter FG, Bilden EF, Gibly RL. Envenomations. Crit Care Clin 1999; 15(2):353-86.
7. Weber EJ, Callaham M. Animal Bites and Rabies. In: Rosen et al, eds. Emergency Medicine: Concepts and Clinical Practice. St Louis: Mosby, 1998.
8. Brunette DD, McVaney K. Hypothermic cardiac arrest: An 11 year review of ED management and outcome. Am J Emerg Med 2000; 18(4):418-22.
9. Danzel DF. Accidental Hypothermia. In: Rosen et al, eds. Emergency Medicine: Concepts and Clinical Practice. St Louis: Mosby, 1998.
10. Decker W, Li J. Hypothermia. In: Danzel DF, ed. eMedicine/Emergency Medicine. Website: http://www.emedicine.com/EMERG/topic279.htm 2001.
11. Hanania NA, Zimmerman JL. Accidental hypothermia. Crit Care Clin 1999; 15(2):235-49.
12. Kunihiro A, Foster J. Heat Exhustion and Heat Stroke. In: Danzel DF, ed. eMedicine/Emergency Medicine. Website:http://www.emedicine.com/EMERG/topic236.htm. 2001.

13. Schaffer, James A. Body Temperature Regulation. In: Johnson LR, ed. Essential Medical Physiology. 2nd ed. Philadelphia: Lippincott-Raven, 1998:815-824.
14. Yarbrough B, Bradham A. Heat Illness. In: Rosen et al, eds. Emergency Medicine: Concepts and Clinical Practice. St Louis: Mosby, 1998.
15. Dietz TE. Altitude Illness Pulmonary Syndromes. In: Keim SM, ed. eMedicine/Emergency Medicine/Environmental. Website: http://www.emedicine.com/emerg/topic795.htm.
16. Dietz TE. Altitude Illness—Cerebral Syndromes. In: Danzel D, ed. eMedicine Emergency Medicine/Environmental. http://www.emedicine.com/emerg/topic22.htm.
17. Hackett PH, Rabold MB. High Altitude Medical Problems. In: Tintinalli, ed. Emergency Medicine: A Comprehensive Study Guide. New York: McGraw-Hill, 2000.
18. Klocke DL, Decker WW, Stepanaek J. Altitude-Related illnesses. Mayo Clin Proc 1998; 73:988-993.
19. Mardirosoff C, Tramer MR. Efficacy and harm of pharmacological prevention of acute mountain sickness: Quantitative systemic review. Br Med Jour 2000; 321:267-272.
20. Yaron M, Honigman B. High-Altitude Illness. In: Rosen et al, eds. Emergency Medicine: Concepts and Clinical Practice. St Louis: Mosby, 1998.
21. Causey AL, Tilelli JA, Swanson ME. Predicting discharge in uncomplicated near-Drdrowng. Am J Emerg Med 2000; 18(1).
22. Hayes BE. Near Drowning. In: Tintinalli, ed. Emergency Medicine: A Comprehensive Study Guide. New York: McGraw-Hill, 2000.
23. Kallas HJ, Marcin JP, Pollack MM. Prognostic indicators for near-Drdrowning patients admitted to the pediatric intensive care unitabstract). Crit Care Med 1999; 27(12):A120.
23. Rumbak MJ. The etiology of pulmonary edema in fresh water near-drowning. Am J Emerg Med 1996; 14:1761-69.
24. Weir E. Drowning in Canada. Can Med Assoc Jour 2000;162:1867-8.
25. Carvalho MD, Shockley LW. Diving Injuries. In: Rosen, ed. Emergency Medicine: Concepts and Clinical Practice. St Louis: Mosby, 1998.
26. Dutka AJ, Francis TJ. Pathophysiology of Decompression Sickness. In: Bove, ed. Diving Medicine. Philadelphia: Saunders, 1997.
27. Neuman TS. Pulmonary Barotrauma. In: Bove, ed. Diving Medicine. Philadelphia: Saunders, 1997.
28. Cooper MA. Electrical and Lightning Injuries. In: Rosen, ed. Emergency Medicine: Concepts and Clinical Practice. St Louis: Mosby, 1998.
29. Jain S, Bandi V. Electrical and Lightning Injuries. Crit Care Med 1999;15(2).
30. Rai J, Jeschle MG, Barrow RE et al. Electrical Injuries: A 30-Year Review. J Trauma 1999; 46(5):933-36.

CHAPTER 14

ENT, Maxillofacial and Dental Emergencies

Kirsten Calder and Jason Ruben

Nasal/Sinus Emergencies

Basic Examination

- Appropriate equipment including adequate illumination and suction is required.
- Topical vasoconstrictors and anesthetics facilitate examination (Table 14.1). The medication is applied to a cotton pledget, inserted into the naris and left in place for 5-10 min. An alternative is insufflation of aerosolized medication.

Epistaxis

The vascular system of the nasal structures has multiple sources including terminal branches of the internal and external carotid arteries. Kiesselbach's plexus is a rich venous network located at the anterior septum and is the source of bleeding in the majority of cases.

Etiology/Risk Factors

- Children and persons exposed to cold, dry air conditions are often affected. Minor nasal trauma is a common cause, but the possible etiologies are numerous (Table 14.2).
- Hypertension contributes to the severity and duration of bleeding. However, there is no clear evidence to support elevated blood pressure as a primary etiology.

Diagnosis

- Remove existing blood clots prior to examination and pretreat the patient with local anesthetics and/or vasoconstrictors (see "Basic Examination").
- Determine whether bleeding is anterior or posterior, which chamber in involved and localize the exact site if possible. More than 90% of cases are anterior. In general, posterior bleeding is very brisk and difficult to control. These patients may also have hemodynamic instability, bilateral nares involvement, isolated oropharyngeal bleeding and continued bleeding after placement of adequate anterior packing.
- If the patient has severe bleeding, hemodynamic instability or significant underlying disease or risk factors such as coagulopathy, appropriate laboratories should be obtained.

Table 14.1. Vasoconstrictors/anesthetics (V/A)

4% Lidocaine (A)
4% Cocaine (V/A)
0.5% Phenylephrine (V)
Epinephrine 1:1000 (V)

Emergency Medicine, edited by Sean Henderson. ©2006 Landes Bioscience.

Table 14.2. Etiology of epistaxis

Idiopathic	
Iatrogenic	Surgical procedures, nasal foreign body removal, nasogastric/ nasotracheal tubes
Trauma	Nose-picking, fractures, foreign bodies
Local irritants	Cocaine, nasal sprays, cigarette smoke, toxic gases
Inflammatory	Rhinitis, sinusitis, granulomatous disease
Mass lesions	Nasal/sinus tumors, carotid artery aneurysm
Medications	Anti-platelet agents, NSAID's, warfarin, heparin
Systemic disease	Liver/renal failure, DIC, thrombocytopenia
Congenital	Hemophilia, von Willebrand disease, Rendu-Osler-Weber disease, sickle cell disease

NSAID: Non-steroidal anti-inflammatory drug
DIC: Disseminated intravascular coagulation

Treatment

Many methods exist. In addition, attention to any premorbid conditions and/or hemodynamic instability is mandatory.

- **Direct pressure**: Manual pressure applied to both nasal alae in a pinching fashion for 10-15 min is often effective and should be used during prehospital care and during preparation for more definitive treatment.
- **Cautery**: An excellent method of treatment when an obvious site of bleeding is identified.
  - Both silver nitrate applicator sticks and electric devices are available. Silver nitrate cautery is usually ineffective in cases of profuse bleeding.
  - Risks include increased bleeding and septal perforation. These risks are minimized by use of silver nitrate vs. electrical. Avoid use of cautery on both sides of the septum and do not apply for more than 5-10 seconds. Use with caution in pediatric patients.
- **Nasal packs**: Placement of nasal packs is very effective in controlling epistaxis especially when the bleeding site cannot be adequately visualized or is not accessible to cautery. Unfortunately, packing is uncomfortable, unaesthetics and carries risks of both sinusitis and toxic shock syndrome. When placing nasal packs, take care not to over-distend the cavity or alae as this may lead to pressure necrosis. Numerous materials are available.
  - **Gauze strips**: Either petrolatum or iodoform gauze is appropriate. The gauze is layered using bayonet forceps from the posterior base of the nasal cavity upward and outward. Covering the gauze with bacitracin antibiotic ointment prior to placement is recommended to decrease risk of toxic shock syndrome.
  - **Expandable sponges/cotton pledgets**: This includes Merocel and the popular Rhinorocket. Lubricate with antibiotic ointment prior to placement. Once in place, adequate expansion is achieved via injection of 3-10 ml of saline or otic antibiotic solution into the body of the sponge.
  - **Absorbable hemostatic material**: Includes Gelfoam sponges and Surgicel. These may be used as packs and also work by forming a clot upon contact with the bleeding site. Ideal for patients with coagulopathy because they don't require removal.
  - **Posterior packs**: For posterior bleeding, balloons (see below) provide more rapid control. If balloons are not available, a posterior pack can be made by rolling several 4 x 4 gauze pads and securing with suture material. They are placed in a retrograde fashion by using a nasally inserted Foley catheter as a guide. Posterior packing is also achieved by placement of a Foley catheter through the naris followed by slow balloon inflation with 10-15 ml of saline once the catheter tip is in the nasopharynx.

- **Epistaxis balloons**: These are easier to place and better tolerated than conventional packing. Once in place, the balloons are inflated with enough water to control bleeding. Inflation should not exceed the maximum volume specified on the device. The posterior balloon is inflated first. If only anterior control is needed, the posterior balloon remains empty.
- **Other**: Cautery and packing sometimes fail to control severe bleeding especially if posterior. In these cases, emergent ENT consultation should be obtained. Endoscopic cautery, arterial ligation and arterial embolization are other therapeutic options.

Disposition and Discharge Planning

- Patients with packing or balloons are at risk for developing sinusitis secondary to occlusion of the sinus ostia so appropriate antibiotic coverage is warranted. Analgesics are also recommended. Packs and balloons are left in place for 2-5 days prior to removal.
- Instruct patients to avoid straining, nose blowing, alcohol intake and use of antiplatelet medications. Appropriate follow-up should be arranged.
- Admission for continuous pulse oximetry and supplemental oxygen is recommended for patients with posterior packs or balloons. In addition, consider admission for patients with anterior packs if they are elderly or have underlying cardiovascular disease.

Nasal Foreign Bodies

Risk Factors

- Nasal foreign bodies (FBs) are seen most commonly in children.
- Others at risk include psychiatric patients and people with nasal piercings.

Clinical Presentation and Diagnosis

- A detailed history is often available by the caretaker who witnessed placement of the FB.
- Symptoms include unilateral nasal discharge, difficult nasal respiration and epistaxis. Purulent drainage and sinusitis are seen with FBs that have been present for an extended period of time. A complete head and neck exam is indicated to rule out additional FBs.
- The object is often visible on physical exam. Nasal radiographs are useful for radiopaque FBs. Order soft tissue neck and chest films if aspiration is suspected.
- Differential diagnosis includes nasal polyps, nasal tumors, unilateral choanal atresia and sinusitis.

Treatment

- Have cooperative patients try forceful nasal exhalation with manual occlusion of the uninvolved naris.
- Depending upon the shape and size, FBs are removed with a small suction catheter, ear loop or alligator forceps. Alternatives include cyanoacrylate glue applied to the tip of a wooden applicator or Fogarty catheter placement just posterior to the object followed by gentle outward pressure after balloon inflation.
- FB removal in uncooperative pediatric patients is often difficult. Repeated attempts traumatize the mucosa with subsequent epistaxis and edema limiting visualization. Removal by ENT under general anesthesia may be indicated in these cases. If the object is not visible or not easily removed, ENT referral is necessary.

Disposition

- Patients who have had uncomplicated removal of nasal FBs are safely discharged without additional treatment or follow-up.
- Most patients who require ENT evaluation for difficult to remove items can be seen within the following 24 h. Emergent evaluation is indicated for corrosive FBs, infants with bilateral nares involvement, organic FBs and those with respiratory difficulty.

Sinusitis

Risk Factors

- Sinusitis results from obstruction of the sinus ostia. In the majority of cases, this is secondary to viral upper respiratory infection. Other causes include allergic rhinitis, trauma, polyps, septal deviation, nasogastric/nasotracheal intubation and nasal packing.
- Patients with cystic fibrosis, diabetes mellitus and immunocompromise are at increased risk.
- The sinuses grow in stages. Aeration of the maxillary and ethmoid sinuses is present at birth or shortly thereafter. In contrast, the frontal and sphenoid sinuses do not begin to aerate until early childhood.

Etiology

- Acute sinusitis lasts <3-4 wk. The most common pathogens include *S. pneumoniae*, nontypable *H. influenzae* and respiratory viruses.
- Subacute disease lasts longer than 4 wk but resolves before 12 wk.
- Chronic sinusitis denotes persistence of signs and symptoms >12 wk and is usually a polymicrobial infection with organisms that are more likely resistant to β-lactams.
- Patients with diabetes and neutropenia are at risk for fungal sinusitis such as Mucormycosis.
- Severely ill, hospitalized patients and those with indwelling nasogastric tube or nasotracheal tubes have disease caused by Gram-negative bacilli including Pseudomonas.

Clinical Presentation and Diagnosis

- Symptoms strongly indicative of acute sinusitis include facial pain (Table 14.3), purulent rhinorrhea, dental pain, altered sense of smell and fever. Pain secondary to sinusitis increases with head movement. Nonspecific symptoms include headache, fatigue, otalgia and malaise. Sore throat and cough are common as a result of sinus drainage into the oropharynx. Sphenoid disease often presents as an isolated headache.
- Chronic disease has a similar presentation with low-grade symptoms that are persistent or that recur intermittently over time.
- Pertinent physical examination findings include tenderness with palpation of the involved sinus, visualization of purulent nasal secretions, fever and abnormal sinus transillumination.
- Viral rhinitis is often mistakenly diagnosed as sinusitis. Patients with bacterial sinusitis are more likely to have a poor response to decongestants, abnormal sinus transillumination, dental pain and purulent nasal discharge. Other differential diagnoses include tension headache, vascular headache, brain abscess, meningitis and intracerebral mass lesions/hemorrhage.
- Diagnosis in the ED is via clinical presentation.
    - Plain radiographs are not necessary for the diagnosis and are not recommended for routine use. If films are obtained, pertinent findings include an air-fluid level, sinus opacification and mucosal thickening but none of these findings reliably differentiates between bacterial and viral disease.

14

Table 14.3. Symptoms of sinusitis

Sinus	Location of Symptoms
Maxillary	Cheek/zygoma, teeth
Ethmoid	Medial aspect of eye/orbit
Frontal	Forehead
Sphenoid	Retro orbital, occiput

- CT scan provides visualization of all four paired sinuses. Indications for CT scan include: severe disease, refractory symptoms, preoperative planning and suspected involvement of contiguous structures or invasive disease.

Treatment

- Antimicrobials
  - For acute disease, a 10-day course is recommended for patients with continued facial pain and purulent nasal discharge after a 7-day course of decongestants. Appropriate first-line agents include amoxicillin, trimethoprim/sulfamethoxazole, and doxycycline. Other, more broad-spectrum agents such as fluoroquinolones, macrolides, and second and third generation cephalosporins should be reserved for treatment failures. The majority of patients will eventually have symptom improvement and no complications even without antibiotics.
  - Chronic sinusitis is not usually responsive to antibiotics. Otolaryngology follow-up is indicated for patients with chronic disease.
  - Diabetic/neutropenic patients with fungal disease require hospitalization in a monitored setting and treatment with IV antifungals.
- Ancillary treatment consists of mucolytics, systemic vasoconstrictors such as pseudoephedrine, and *brief* use of topical vasoconstrictors. Avoid antihistamines which dry and thicken secretions. In addition, oral fluids and saline nasal sprays promote moisture and mobilization of secretions. Steroids have no proven efficacy in acute disease. Patients with sinusitis secondary to obstructing nasal foreign body such as packing should have prompt removal of the obstructing device if possible.
- Surgical indications include: acute symptoms unresponsive to antibiotics, chronic disease, invasive/intracranial disease, nasal polyposis and suspected underlying tumor or mucocele.

Disposition

- Most patients are discharged home with ENT follow-up.
- Admission is indicated for toxic patients, failed outpatient therapy and those in whom invasive disease in suspected. This includes those patients with facial abscesses, orbital swelling, visual disturbances, cranial nerve and/or focal neurological deficits.

Otic Emergencies

Basic Anatomy

The ear is a complex structure that is beyond the scope of discussion. However, there are salient points to keep in mind.

- The middle ear has direct communication with the nasopharynx via the eustachian tube whose patency helps keep the middle ear pressure equalized with the atmosphere. There is also direct communication with the middle cranial fossa via the mastoid antrum and air cells.
- Multiple nerves supply the ear including cranial nerves V, VII, IX, X, and XI as well as cervical nerves C2 and C3. As a result, ear pain can occur with nonotic pathology especially dental and oropharyngeal disease. Patients with complaints of otalgia and an unremarkable ear exam need to undergo evaluation of the teeth, oral cavity, pharynx, TMJ and neck. Disease of the larynx and upper esophagus should also be considered.

Otitis Media

Risk Factors

- Eustachian tube dysfunction and obstruction: This leads to negative pressure within the middle ear and subsequent fluid collection. Children have a higher incidence because their Eustachian tube has less cartilaginous support. Other common etiologies

of tube obstruction are adenoid hypertrophy and edema secondary to viral upper respiratory infection.
- Adults with OM and middle ear effusion who lack obvious risk factors need to be evaluated for possible mass lesions such as nasopharyngeal carcinoma.

Etiology
- A significant percentage of OM is secondary to respiratory viruses, most commonly RSV.
- The most common bacterial etiologies are *S. pneumoniae, H. influenzae* and *M. catarrhalis.*

Clinical Presentation and Diagnosis
- Older children and adults present with otalgia and ear fullness. Infants and younger children who are unable to verbalize sometimes have ear pulling and nonspecific symptoms such as irritability, decreased PO intake, vomiting and diarrhea. Fever may be present; however, the majority of children with OM have only moderate temperature elevation.
- The most sensitive indicator of OM on physical examination is immobility of the TM with air insufflation. Other physical findings include: air-fluid levels or bubbles behind the TM, retraction or bulging of the TM, blurring of osseous middle ear landmarks and TM color change. Mild erythema of the TM occurs with crying and fever and should be considered indicative of OM only when other abnormalities are present.

Treatment
- Although most practitioners still prescribe antibiotic therapy, it has been shown that a majority of pediatric OM will have spontaneous cure without antibiotics. The American Academy of Pediatrics Subcommittee on the Management of Acute Otitis Media now recommends a treatment approach that takes into account the patient's age, the severity of disease, and the certainty of diagnosis. Observation alone is considered as an appropriate treatment option for children greater than 2 yrs of age with non-severe illness. For all others, the duration of antibiotic therapy is 10 days unless azithromycin is used. When making a treatment selection consider ease of dosing, side-effect profile, probability of bacterial resistance, cost and presence of any allergies.
  - High-dose amoxicillin (80 mg/kg/d)is still the recommended agent for most patients with OM
  - Alternate medications include macrolides, amoxicillin/clavulanate, various cephalosporins, trimethoprim/sulfamethoxazole, erythromycin/sulfisoxazole, and clindamycin.
- Ancillary treatment should include antipyretics and analgesics as needed.
- Suspect resistant pathogens or an alternate diagnosis in patients with persistent symptoms after 48-72 h of treatment. These patients as well as those with recurrent disease should be treated with amoxicillin/clavulanate or one of the second or third generation cephalosporins.
- Recurrent and persistent disease with middle ear effusion places pediatric patients at risk for deafness. These children should be referred to ENT for placement of pressure equalization (PE) tubes and consideration of tonsillectomy and adenoidectomy.
- Complications of OM include chronic otitis, meningitis, intracranial abscess, mastoiditis, labyrinthitis and invasion of neighboring structures including bone and facial nerve.

Disposition
- Most OM patients are discharged home with ENT follow-up.
- Admission is indicated for febrile neonates, toxic patients (usually infants) and any patient in whom serious complication is present.

Mastoiditis

This infection of the mastoid air cells is rare since the development of modern antibiotic therapy. It typically occurs in the setting of inadequately treated OM but can also be the initial presentation of OM. Despite concerns to the contrary, an initial period of observation alone for OM has not resulted in a significant increase in mastoiditis cases.

- Symptoms include otalgia, fever, malaise and purulent otorrhea.
- Physical examination reveals painful swelling and fluctuance over the mastoid process. The TM sometimes has signs of concurrent OM. Complications include hearing loss, facial nerve paralysis and intracranial infection. Look for corresponding physical signs.
- Emergent ENT consultation and admission for broad-spectrum IV antibiotics is necessary. Head and temporal bone CT should be obtained if complications are present.

Otitis Externa

Risk Factors

- Otitis externa (OE) occurs with auditory canal trauma and maceration of the canal skin.
- Trauma is usually secondary to foreign bodies such as Q-tips and hearing aids. Maceration occurs after prolonged exposure to water or in areas of high temperature and humidity.

Etiology

- Most cases of OE are causes by *Pseudomonas, Proteus or S. aureus.*

Clinical Presentation and Diagnosis

- Symptoms include ear pain, hearing loss, itching and otorrhea. There is often a history of recent swimming, minor ear trauma or attempted cerumen removal using irrigation.
- On examination, the skin of the auditory canal will have varying degrees of edema and exudates that may be so severe as to prevent visualization of the TM. Exudates should be gently suctioned in order to allow for thorough evaluation. The TM itself may have moderate erythema. Palpation of the tragus and other external structures produces pain.
- Diagnosis of OE is usually obvious but the differential includes both Herpes zoster and necrotizing OE. With Herpes zoster, patients present with burning pain followed by a vesicular rash to the external ear and auditory canal.

Treatment

- Therapy consists of topical polymyxin/neosporin/hydrocortisone solution. Consider topical ciprofloxacin for diabetic patients and suspected *Pseudomonas.* If edema is severe, an ear wick should be placed so that medication will be delivered the entire length of the canal.
- Systemic antibiotics are necessary only if there is cellulitic involvement of the external structures or for patients with diabetes/immunosuppression. Some practitioners also prescribe systemic antibiotics for OM when the TM cannot be adequately visualized. An alternative is to reevaluate these patients for concurrent OM in 24 h after canal edema has lessened.
- Instruct patients to follow dry ear precautions and avoid placement of Q-tips and other foreign bodies into the canal.

Disposition

With few exceptions, patients with OE are discharged home. However, follow-up is indicated to ensure that infection is resolving. Patients may also require repeat cleansing of the auditory canal and removal of ear wick if placed.

Necrotizing Otitis Externa (NOE) AKA Malignant OE

- NOE is associated with diabetes mellitus and other immunosuppressive states.
- Etiology is usually *P. aeruginosa.*
- Infection starts in the external ear and rapidly spreads to contiguous structures including bone, soft tissue, nerves and mastoid air cells. The facial nerve is commonly affected but other cranial nerves are sometimes involved as well.
- Pertinent examination findings include granulation tissue in the auditory canal, cellulitic changes of the external ear, ipsilateral facial nerve palsy and possibly other cranial nerve deficits. Patients describe deep, severe ear pain. They may be febrile and toxic.
- Patients with suspected NOE require CT scan to define extent of infection. Obtain emergent ENT consultation and initiate intravenous anti-pseudomonal antibiotics as soon as possible. These patients often require prolonged duration of therapy and surgical debridement by ENT.

Perichondritis

- This soft tissue infection of the external ear follows trauma, burns/exposure and ear piercing.
- Symptoms are pain and swelling involving the entire external ear. Examination reveals erythema and tenderness.
- *S. aureus* and *P. aeruginosa* are most commonly involved.
- Mild disease is treated with oral ciprofloxacin. More severe cases require intravenous medications. Early ENT is encouraged as progression of infection could lead to chondritis and permanent cartilaginous deformity.

Relapsing Polychondritis

Relapsing polychondritis is a rare inflammatory condition easily confused with perichondritis. Accurate diagnosis is important, as complications and treatment differ.

- Multiple cartilaginous sites are involved including the ear, larynx, epiglottis, joints, etc.
- Ear involvement is common and may spare the lobule. Other symptoms such as hoarseness occur as a result of widespread cartilage involvement.
- Treatment is with steroids. ENT should be involved in patient care.

Otic Foreign Bodies

Clinical Presentation and Diagnosis

- Otic FBs are seen most frequently in pediatric patients who will place any imaginable object into the auditory canal. In adults, insects and cotton Q-tips are commonly encountered.
- Patients present with ear pain, ear fullness and hearing loss. Purulent discharge is seen with long-standing FBs. Vertigo and facial nerve paralysis occur when the middle ear has been traumatized.
- Examination usually reveals the offending object although associated canal edema or bleeding will make visualization difficult. Attempt should be made to visualize the TM in order to rule out associated perforation.

Treatment

- Removal of otic FBs can be accomplished with a variety of methods:
  - Irrigation using lukewarm water via a plastic 18 g angiocath attached to a 30 ml plastic syringe (avoid if TM perforation is a possibility).
  - Depending upon the shape and size, FBs are also removed with a small suction catheter, ear loop or alligator forceps. An alternative is Fogarty catheter placement just posterior to the object followed by gentle outward pressure after balloon inflation.
- Immobilize live insects prior to attempted removal by placing 2% lidocaine in the auditory canal. This reduces patient discomfort and maximizes cooperation.

- When the FB is lodged firmly in the auditory canal or cannot be easily removed, the patient should be referred to ENT. Repeated attempts at removal cause bleeding and edema and make the situation worse. Pediatric patients are often uncooperative; it may be necessary for ENT to remove the object under general anesthesia or to employ procedural sedation in the ED.
- Complications of FB removal are auditory canal laceration, TM perforation and ossicle disruption. Make sure to reexamine the ear after FB removal.

Disposition

- Patients are discharged home after successful FB removal. Antibiotics are indicated only if there is associated infection. Follow-up is not necessary unless complications are present.
- Most patients with a difficult to remove FB can be referred nonemergently to ENT within 1-2 days. Emergent evaluation is warranted for organic, animate and corrosive FBs.

Cerumen Impaction

- Cerumen is cleared naturally via lateral migration of the underlying skin. Certain individuals are more prone to impaction based upon amount/consistency of cerumen produced as well as effectiveness of the migration process. People increase the likelihood of impaction by attempting to clean their ears with Q-tips or other similar objects.
- Impacted cerumen is easily cleared using lukewarm water via plastic angiocath attached to a syringe. Avoid irrigation if there is history of TM perforation. Another method is to soften the impacted material with triethanolamine (Cerumenex) by filling the ear canal and leaving the agent in place for 15-30 min. Cerumen is then removed with an ear loop.
- Prior to discharge, instruct patients to avoid future Q-tip use.

Tympanic Membrane Perforation

Etiology

- Blunt and penetrating trauma: includes temporal bone fractures, direct blows and attempted otic hygiene with Q-tips and iatrogenic injury from removal of cerumen impaction or FBs.
- Pressure changes from scuba diving, loud noise and lightning.

Clinical Presentation and Diagnosis

- Symptoms include ear pain, bloody otorrhea and mild hearing loss. Suspect injury to the ossicles or labyrinth if patients have vertigo, significant hearing loss or facial nerve deficits.
- Diagnosis is via direct visualization with an otoscope. Assess hearing if middle or inner ear damage is suspected.

Treatment

- Most uncomplicated TM perforations heal without specific therapy. Smaller and more centrally located perforations tend to heal faster than larger, peripheral lesions. Persistent perforations sometimes require surgical treatment.
- Antibiotics have not been shown to improve healing and are rarely indicated. The exception to this rule is when there is concurrent OE or OM. In cases of OE, treat patients with topical antibiotic suspension (vs. solution) which is less likely to penetrate into the middle ear space.

Disposition

All cases of TM perforation mandate ENT follow-up. This is done on a routine basis unless middle or inner ear injury is suspected. Instruct patients to follow dry ear precautions.

Hearing Loss

- Hearing loss is either conductive or sensorineural. Conductive hearing loss is caused by occlusion of the auditory canal, middle ear fluid or dysfunction of the TM and ossicles (Table 14.4). The etiology of sensorineural hearing loss varies (Table 14.5).
- Sudden sensorineural hearing loss (SSHL) is that occurring within the preceding 3 days.
- Patients with hearing loss need a thorough ear exam to rule out obvious causes such as otic FB, otitis media/externa and cerumen impaction.
- If the ear exam is unremarkable, consider other etiologies as noted above. Obtain a thorough history and perform a complete physical assessment including neurologic exam. Laboratories and CT scan are indicated if metabolic or structural causes are a concern.

Table 14.4. Conductive hearing loss

External ear	Congenital atresia Cerumen impaction, foreign bodies Otitis externa Obstructing lesions—osteomas, exostoses Trauma
Middle ear	Congenital ossicular abnormalities Otitis media, middle ear effusion Cholesteatoma, otosclerosis TM perforation

Table 14.5. Sensorineural hearing loss

Infectious	Syphilis Viral—MMR, VZV, CMV, others Meningitis Encephalitis
Medications	Aminoglycosides Loop diuretics Antineoplastic agents Salicylates, NSAIDs
Vascular	Sickle cell disease Cerebrovascular accident Coagulopathy
Endocrine	Diabetes mellitus Hypothyroidism
Traumatic	Noise Barotrauma Temporal bone fracture
Neoplastic	Hematologic malignancy Acoustic neuroma
Misc.	Ménière's disease Multiple sclerosis Autoimmune Congenital

MMR—measles, mumps, rubella; VZV—varicella Zoster virus. CMV—cytomegalovirus

- The role of the EP in cases of hearing loss is to rule out life-threatening etiologies and to provide appropriate referral. SSHL is an indication for emergent ENT evaluation. Prompt initiation of care improves chances for hearing recovery. Therapy sometimes includes empiric high-dose steroids as well as treatment of the underlying disease process.

Oropharyngeal and Laryngeal Emergencies

Basic Anatomy

- The oral cavity extends from the lips to the oropharynx. The roof consists of the hard and soft palates, and the cheeks form the lateral borders.
- The pharynx consists of the nasopharynx, oropharynx and hypopharynx. The nasopharynx is located just posterior to the nasal choanae. The oropharynx extends from the soft palate to the hyoid bone at the level of the third cervical vertebrae. The hypopharynx is the distal continuation of the oropharynx extending from the third to the sixth cervical vertebrae at which point the esophagus begins.
- The larynx, the uppermost part of the tracheobronchial system, also runs the length of the third through sixth cervical vertebrae just anterior to the hypopharynx. The prominent thyroid and cricoid cartilages are usually palpable and are excellent references for structures of the neck. The thyroid cartilage is present at the level of the fourth to fifth cervical vertebrae while the cricoid is at the sixth cervical vertebrae. The vocal cords are located at the level of the midpoint of the anterior thyroid cartilage.
- The pharynx, larynx, deep neck spaces, carotid artery, jugular vein, mediastinum and prevertebral soft tissue are in close contact with one another. Thus, disease in one area can have widespread consequences.
- Innervation of the face, oropharynx and larynx is complex and includes cranial nerves V, VII, IX, X, XI, XII. Referred pain is common and thorough examination of all surrounding structures is warranted.

Common Presenting Symptoms

Dysphagia

- Difficulty swallowing (not painful swallowing).
- Most common etiologies: neurologic and musculoskeletal disorders.
- Other causes: collagen vascular disorders, myopathies, motility disorders, obstructive lesions, and extrinsic compression secondary to thoracic/neck masses.

Odynophagia

- Painful swallowing.
- Etiology: infectious and irritant lesions of the oropharynx and esophagus.

14

Hoarseness

- Change in phonation.
- Secondary to laryngeal, vocal cord, or recurrent laryngeal nerve pathology.
- Most common etiology: viral inflammation ("laryngitis") and voice abuse.
- Other causes: vocal cord nodules, neoplasm, allergies, local irritants, granulomatous/ automimmune disease, endocrine disease, and trauma.

Pharyngitis

- This common condition is of concern because of the possible sequelae. Streptococcal disease may lead to rheumatic fever, glomerulonephritis, scarlet fever and toxic shock syndrome. Infectious mononucleosis (IM) has numerous potential complications including hemolytic anemia, hepatitis, splenic rupture, myocarditis, pericarditis and neurologic dysfunction. Other complications of pharyngitis include deep space infections, peritonsillar abscess, OM and rarely sepsis or meningitis.

- Differential diagnosis includes deep space infections, peritonsillar abscess and epiglottitis. Symptoms of pharyngitis are also seen in irritant and thermal injuries, gastroesophageal reflux, pemphigus, hematologic malignancy and trauma.

Etiology

- Overall, respiratory viruses are the most common etiology of pharyngitis.
- The most common bacterial agent, Group A β-hemolytic streptococcus (GAS), occurs primarily in the 5-15 yr age group. Other bacterial etiologies include *C. hemolyticum, H. influenzae, M. pneumoniae, S. pneumoniae, C. diphtheriae, Chlamydia*, syphilis, anaerobes and gonococcus.
- Infectious mononucleosis (IM) caused by Epstein-Barr virus often presents as pharyngitis in young adults and college students. Other viral causes are CMV and HSV.

Clinical Presentation and Diagnosis

- The primary symptom is odynophagia. Cough, upper respiratory symptoms, myalgias and headache are classically associated with viral disease but are also seen in bacterial disease especially *Chlamydia* and *Mycoplasma.*
- Physical Examination Signs
  - Pharyngeal edema and erythema is present and may extend to the uvula and soft palate. Fever, purulent exudates and ulcerations occur in both viral and bacterial disease and do not pinpoint the source. When ulcerations are present, consider anaerobes, herpes (I and II), syphilis, and herpangina.
  - An oropharyngeal membrane is seen with diphtheria and anaerobes.
  - Tender cervical adenopathy occurs with both bacterial and viral pharyngitis. Posterior cervical adenopathy is classically associated with IM.
  - Patients with IM can have a generalized maculopapular rash especially if they have received amoxicillin or ampicillin. Also consider scarlet fever.
  - Extrapharyngeal findings are often found in IM including hepatomegaly and splenomegaly.
- Laboratory Evaluation
  - Throat culture is the gold standard for diagnosing GAS and is 90-95% sensitive. Results take about 48 h to finalize.
  - Rapid streptococcal antigen tests are readily available and provide results in as little as 30 min. Sensitivity of these tests is approximately 60-90%.
  - A positive monospot test, leukocytosis with atypical forms and abnormal liver function tests support the diagnosis of IM.
  - Serologies and special culture media exist (i.e., for HSV, gonococcus and diphtheria) although use is not frequently indicated and results are not readily available in the ED.

Treatment

Antibiotics reduce the incidence of both suppurative and nonsuppurative complications of bacterial pharyngitis. The decision to initiate antibiotic treatment depends upon patient presentation, results of rapid strep screening as well as regional incidence of streptococcal disease and rheumatic fever. In general, empiric antibiotic therapy for pharyngitis in adults is not considered cost-effective and should be reserved for those patients with a high likelihood of GAS.

- Bacterial Pharyngitis
  - GAS: The recommended therapy remains penicillin VK (or a single dose of benzathine penicillin IM if noncompliance is likely). Alternatives include macrolides and second and third generation cephalosporins. Azithromycin is the only agent approved for a five-day course. Ten days of therapy is recommended with all other

agents in order to prevent rheumatic fever. Patients with chronic or recurrent disease should receive clindamycin or amoxicillin/clavulanate.
  - *Mycoplasma* and *Chlamydia*: erythromycin or tetracycline.
  - Diphtheria: antitoxin plus erythromycin or penicillin.
  - Gonococcus: IM ceftriaxone and other approved regimens such as oral fluoroquinolones.
- Viral Pharyngitis
  - IM does not require specific therapy although a tapered course of steroids can provide symptomatic relief. Immunocompromised people with HSV are treated with acyclovir. Otherwise, viral disease is treated symptomatically.
- Ancillary treatment includes analgesics, oral hygiene and intravenous fluids for both viral and bacterial disease.

Disposition

- Most patients with pharyngitis are safely discharged. Patients with recurrent and chronic symptoms should be referred to ENT for possible tonsillectomy.
- Patients with IM need discharge information detailing the possibility of multi-organ involvement. All potential contact sports should be avoided to reduce risk of splenic rupture.
- Admission is warranted for dehydrated patients or if any question exists regarding airway patency or ability to handle secretions.

Peritonsillar Abscess

Risk Factors/Etiology

- This occurs most commonly in young adults. It is unusual in children.
- Patients with previous episodes are at increased risk for recurrent abscess development.
- The process is usually seen in the setting of tonsillitis and can occur even if appropriate treatment has been initiated.
- Infection originates from either the tonsillar crypts or the surrounding minor salivary glands. As a result, peritonsillar abscess is possible even if the patient has had a tonsillectomy.
- Infections are typically polymicrobial involving both aerobes and anaerobes.

Clinical Presentation and Diagnosis

- Patients complain of unilateral odynophagia, dysphagia, malaise and muffling of the voice. There is often a history of peritonsillar abscess or recent pharyngitis.
- Examination reveals localized erythema and unilateral swelling at the peritonsillar area. If present, trismus makes exam difficult. The uvula is often displaced medially. If the abscess has spontaneously ruptured, gross purulent material will be visible. Patients usually have temperature elevation and clinical evidence of dehydration. Respiratory difficulty is less common but occurs when there is inferior extension of purulent material.
- Diagnosis of peritonsillar abscess is made clinically.
- Complications: dehydration; airway obstruction; aspiration of purulent drainage; cavernous sinus thrombosis; spread to contiguous areas including the deep spaces of the neck and intracranial structures; and spontaneous carotid artery erosion or trauma during treatment.
- Peritonsillar abscess is commonly confused with peritonsillar cellulitis. Cellulitis tends to be bilateral whereas abscess is unilateral in over 90% of cases. Differentiation is made via needle aspiration. The differential diagnoses also include neoplasm, carotid artery aneurysm, infectious mononucleosis, parapharyngeal space infection and pharyngitis.

Treatment

- Drainage is required and, in most cases, is accomplished by aspiration using an 18 gauge spinal needle after application of topical anesthesia. Treating physicians must be aware of the inferior and lateral location of the carotid artery. Children and patients with significant trismus often require treatment under general anesthesia.
- In certain cases incision is required for adequate decompression. These include patients with recurrent or continued abscess after aspiration.
- Ideally, decompression is done by an ENT specialist unless the EP has the necessary expertise. If ENT is not available and the patient has impending airway obstruction, the EP should perform needle aspiration.

Disposition

- Patients with uncomplicated abscess drainage are sent home with antibiotics and pain medications provided that they have appropriate follow-up.
- Admission is necessary for toxemia, presence of severe underlying disease, significant dehydration, airway compromise or if the patient is unable to tolerate PO fluids.

Ludwig's Angina

- Ludwig's angina is infection of the submandibular space which includes both the sublingual space and the deeper submaxillary space located beneath the mylohyoid muscle. Infection begins in the submaxillary space and spreads upward and posteriorly.
- Typically, this is rapidly progressive bilateral cellulitis rather than a purulent fluid collection.
- Spread is via direct invasion rather than lymphatics or hematogenous route.
- Airway obstruction from tongue displacement is the main concern and can occur acutely.

Risk Factors

- The primary risk factor is dental infection. The second and third mandibular molars are most commonly involved; the roots of these teeth are located beneath the mylohyoid muscle in the submaxillary space.
- Other risk factors are penetrating trauma to the floor of the mouth and mandible fractures.

Etiology

- Infections are usually polymicrobial with both aerobes and anaerobes.

Clinical Presentation and Diagnosis

- There is often a history of dental disease or recent dental work. Symptoms include odynophagia and pain to the floor of the mouth and anterior neck. Difficulty swallowing and speaking occur because of tongue displacement.
- Examination reveals an indurated edema of the floor of the mouth and anterior neck. Serosanguinous discharge is present but there is usually no frank pus or fluctuance. When significant edema is present, the tongue will be elevated and posteriorly displaced. Drooling, muffled voice, trismus, high fever and toxicity are common. Ludwig's angina may progress to involve adjacent structures of the head and neck as well as the mediastinum and thorax. Look for corresponding signs and symptoms.
- Diagnosis is made clinically.
    - Both CT scan and MRI delineate the extent of infection and should be obtained if abscess is suspected. Only stable patients with mild disease or those with a secure airway should leave the ED for special radiographic studies.
    - Soft tissue films of the neck help define the extent of tissue edema and reveal localized gas collections.
    - CXR is indicated in patients with suspected thoracic or mediastinal spread.
    - Blood cultures and blood count should be ordered but are not helpful acutely.
- The primary differential diagnoses are deep space and submandibular abscess.

Treatment

- Airway control: airway compromise can occur rapidly and is the primary cause of mortality in Ludwig's angina.
  - Keep the patient in an upright position in order to minimize respiratory distress.
  - Orotracheal intubation is likely to be difficult or impossible. The best methods of securing the airway are fiberoptic nasotracheal intubation or primary tracheostomy; emergent ENT consultation is imperative. If ENT is not immediately available, the EP must be prepared for cricothyrotomy.
- Intravenous antibiotics: agents of choice are high-dose penicillin plus metronidazole or cefoxitin. An alternative for the penicillin-allergic patient is clindamycin. Other effective single-drug therapies are piperacillin/tazobactam, ampicillin/sulbactam and ticarcillin/clavulanate.
- Surgical treatment of this condition is indicated in cases where antibiotic therapy has failed or if there is evidence of an obvious purulent fluid collection.

Disposition

All patients with Ludwig's angina should be admitted to a monitored bed setting in anticipation of sudden airway deterioration.

Adult Epiglottitis

- Epiglottitis, also referred to as supraglottitis, is usually considered a disease of young children but occurs in adults as well. It often involves other supraglottic structures including the tongue base, arytenoid cartilages and aryepiglottic folds. Adult disease tends to have a more insidious onset and benign course although this is not always the case. Even the larger adult airway can rapidly lose patency.

Etiology

- Common bacterial pathogens include *H. influenza, Streptococcus sp.* and *S. aureus.* Also consider viral pathogens, namely herpes simplex. Epiglottitis has also been reported secondary to thermal injury after smoking crack cocaine. In the majority of cases, however, the etiology is unknown.

Clinical Presentation and Diagnosis

- Symptoms include severe odynophagia, dysphagia, anterior neck pain and dyspnea.
- Examination:
  - Tongue blade examination in the stable adult is acceptable and will be notable for a relative absence of obvious pathology. This lack of oropharyngeal disease should lead the EP to suspect epiglottitis.
  - Patients often have fever, muffled voice, difficulty in handling secretions, tender cervical adenopathy and tenderness to palpation of the anterior neck. When present, stridor and respiratory distress indicate impending airway obstruction.
- Epiglottitis is a clinical diagnosis made via history and indirect laryngoscopy. Stable adult patients can undergo bedside radiographs of the lateral soft tissue of the neck. An edematous epiglottis and narrowed pharyngeal airway support the diagnosis. As with pediatric patients, take care to avoid any agitation in adult patients with respiratory distress.
- The most common mistake is to diagnose the patient with pharyngitis. Other conditions to consider are deep space infection, allergic reaction, foreign bodies, inhalation and trauma.

Treatment

- Airway control is the priority. Patients with respiratory distress should undergo immediate airway control. Adult patients with an intact airway do not require prophylactic intubation. However, the EP should have airway equipment present at the bedside in anticipation of worsening symptoms.

- The ideal mode of airway control is either orotracheal or nasotracheal intubation under direct visualization by ENT and Anesthesiology in the operating room setting. Blind nasotracheal intubation is contraindicated.
- If the patient requires an immediate airway by the EP, orotracheal intubation can be attempted but the physician must be prepared to perform cricothyrotomy.
- Additional treatment includes humidified oxygen and antibiotics. Recommended antibiotic regimens include cefotaxime, ceftriaxone and ampicillin/sulbactam. Steroids are commonly administered although their effectiveness has not been proven.

Disposition

- All patients with epiglottitis, irregardless of stability, should be admitted to an ICU in case of precipitous airway obstruction. ENT and Anesthesiology should be involved early in the event that airway control becomes necessary.

Deep Space Infections

This term includes infection and abscess located in the retropharyngeal (RP) and parapharyngeal spaces. Both retropharyngeal and parapharyngeal abscesses represent an ENT emergency because of the likelihood of airway involvement. In addition, infection in these areas spreads contiguously or via fascial planes of the neck to involve the mediastinum, prevertebral space, vertebral bodies, spinal cord and great vessels of the neck.

Risk Factors

- RP abscess in childhood usually develops secondary to regional adenopathy with suppuration. After approximately 5 yr of age, these nodes atrophy. In adults, this condition occurs in association with trauma to the posterior pharynx as seen with ingested foreign bodies, endoscopic procedures and intubation. RP abscess also occurs secondary to direct extension from other contiguous areas of the head and neck including the vertebrae.
- Parapharyngeal abscess is usually secondary to dental or pharyngeal infections. Cervical vertebral osteomyelitis and other head and neck infections also predispose to this condition.

Etiology

- Infections are typically polymicrobial. The predominant organism depends upon the source. Abscesses secondary to local spread from cervical vertebral osteomyelitis are more likely to be caused by *S. aureus* while those with a dental source are usually mixed aerobic and anaerobic.

14

Clinical Presentation and Diagnosis

- Symptoms include dysphagia, voice changes, odynophagia and neck/jaw pain which is increased by movement. Patients may also give a history of IV drug use or recent head, neck or dental infection.
- Physical Examination Signs
  - Common findings with both RP and parapharyngeal abscess: meningismus, respiratory difficulty/stridor, trismus, cervical lymphadenopathy, dysphonia, drooling, and fever. Infants and small children may have nonspecific signs such as irritability.
  - Parapharyngeal abscess: submandibular induration and edema as well as swelling of the lateral pharyngeal wall on examination of the oral cavity.
  - RP abscess: pediatric patients will often prefer to lay supine and hold their neck in extension as this minimizes respiratory difficulty.
- Ancillary Evaluation
  - Lateral soft tissue neck radiographs: retropharyngeal disease will manifest as prevertebral air-fluid levels or abnormal widening of the prevertebral soft tissue.

In children, this would be anything >5-7 mm at the level of the second cervical vertebrae and 14 mm at the sixth cervical vertebrae (22 mm in adults). Ideally, neck films should be performed with the neck extended and the patient at end-inspiration since respiratory expiration and neck flexion cause prevertebral soft tissue prominence and mimic disease. Also look for loss of normal cervical lordosis. Swelling of the pharyngeal soft tissue is an indicator of possible parapharyngeal infection.
- Chest radiographs are recommended in order to rule out mediastinal spread.
- CT scan of the neck defines the extent of infection and delineates areas of abscess and cellulitis but is rarely necessary to make the diagnosis.
- Radiographs should only be done in patients who are completely stable or after the airway has been secured. A physician should be in attendance.

Treatment

- The first goal of therapy is to ensure a patent airway. Significant abscess formation compresses the airway and makes intubation difficult. In addition, intubation attempts may cause abscess rupture and subsequent aspiration. ENT should be consulted for emergent tracheostomy if the patient has respiratory distress. The EP should be prepared to establish an airway via cricothyrotomy if necessary. Avoid blind nasotracheal intubation.
- Once the diagnosis is suspected, intravenous antibiotics should be initiated. Appropriate agents include those listed under "Ludwig's Angina".
- All patients with deep space abscess require admission to a monitored bed.

Foreign Bodies of the Upper Airway and GI Tract

- Elderly patients are most often affected secondary to underlying altered mental status, dysphagia and use of dental appliances. Children are also frequently involved because of their propensity to put objects in their mouth, their immature teeth and inadequate chewing.
- Differential diagnosis includes peritonsillar abscess, deep space abscess, epiglottitis, croup, local trauma, allergic reaction and functional disorders.

Clinical Presentation and Diagnosis (distal obstructions discussed elsewhere)

- Hypopharyngeal/upper esophageal FBs:
  - Older children and adults present with dysphagia, odynophagia and a foreign body sensation at their throat or neck. The patient is often able to pinpoint the area of involvement. If complete or near-complete obstruction is present, patients will be unable to swallow their own secretions.
  - Hypopharyngeal and esophageal FBs, if they are very large, can cause airway compromise in pediatric patients or in older patients.
- Laryngeal and tracheal FBs
  - Partially obstructing upper airway FBs cause stridor, hoarseness, coughing and varying degrees of respiratory difficulty with a prolonged inspiratory phase.
  - Complete obstruction results in cyanosis, inability to phonate/cough and abrupt respiratory failure.
- Symptoms that are initially mild sometimes become more severe in the case of organic FBs secondary to progressive moisture absorption and swelling.
- Sharp objects can cause rapid perforation. Blunt objects also cause perforation via pressure necrosis.
- Diagnosis in older children and adults is straightforward based upon history and presentation. In younger children, suspect FBs in anyone who presents with the above mentioned signs and symptoms especially if there is sudden onset and no history of airway disease. Various methods exist for localizing the object:

- Direct visualization of the oral cavity and pharynx with a tongue blade.
- Direct and indirect laryngoscopy: both require a cooperative patient and, thus, may be difficult to perform in children. Appropriate topical anesthesia prevents gagging.
- Fiberoptic laryngoscopy.
- Radiographs of the neck and chest sometimes contribute to diagnosis if the offending FB is radiopaque. Many fish bones and aluminum pull-tabs are often not visible.
- Take care during examination not to further impact the FB and worsen obstruction.

Treatment

- Upper Airway
  - The priority is to maintain airway patency.
  - Complete obstruction: immediate Heimlich maneuver should be done in the prehospital care setting or if airway equipment is not available (back blows for infants). In the ED, removal of the obstructing FB should be attempted by direct laryngoscopy. If this is unsuccessful, it will be necessary to either advance the FB (usually to the right side) using the endotracheal tube or to place a surgical airway.
  - Patients with partial airway obstruction who are able to maintain oxygenation can sometimes clear the object on their own with coughing. If not, FB removal in the controlled setting of an operating room by ENT is suggested. In cooperative patients, laryngeal FBs can also be removed by the EP via direct laryngoscopy.
- Pharynx
  - Oropharyngeal FBs are easily removed using Magill forceps under direct visualization.
  - It is often possible for the EP to remove hypopharyngeal FBs using either indirect or direct laryngoscopy. If visualization is inadequate or if the patient cannot tolerate attempts at removal, direct laryngoscopy by ENT under general anesthesia is indicated. An alternative is to consider procedural sedation in centers that have appropriately trained staff.
  - Upper esophageal lesions warrant ENT evaluation for rigid esophagoscopy.
- Patients with mild FB sensation upon swallowing who are able to tolerate PO intake without difficulty may have symptoms secondary to minor mucosal trauma. These patients are safely discharged home with prompt ENT follow-up.

Post-Tonsillectomy Hemorrhage

- Bleeding after tonsillectomy typically occurs in the immediate postoperative period (<24 h) or 7-10 days after surgery. Significant hemorrhage places the patient at risk for aspiration, airway compromise and hemodynamic instability. If necessary, immediate care should be directed toward providing an airway and treating life-threatening hemorrhage.
- Adequate visualization of the area is imperative for control of bleeding. Remove all clots via suction or forceps prior to definitive care. Once the bleeding site has been identified, hemostasis is attempted through various means:
  - Generalized oozing or bleeding is treated with hemostatic material or direct pressure to the area using ring forceps and gauze that has been soaked with lidocaine and 1:1000 epinephrine.
  - More circumscribed bleeding is often controlled with silver nitrate cautery or local injection of lidocaine with epinephrine.
  - Patients with brisk bleeding unresponsive to above measures should have emergent ENT evaluation. In these cases, hemostasis is best achieved in the operating room via electric cautery or suture ligation. Pediatric patients should also be treated by ENT in a controlled setting if possible unless the bleeding is very minor and the child is able to cooperate.

Salivary Gland Emergencies

Basic Anatomy

- The three pairs of major salivary glands are the parotid, the submandibular and the sublingual glands. In addition, there are many smaller minor salivary glands spread throughout the oral and oropharyngeal mucosa.
- When evaluating and treating salivary gland disease, it is important to be familiar with the local anatomy in order to prevent injury and rule out involvement of vital surrounding structures.
  - The parotid gland lies just anterior and inferior to the ear. It is drained by Stensen's duct that empties into the oral cavity adjacent to the second maxillary molar. The gland is in close proximity to the lateral pharyngeal space, and its investing fascia is contiguous with the deep fascia of the neck. The facial nerve courses through the gland.
  - The submandibular gland is medial to the mandible adjacent to the mylohyoid muscle and above the hypoglossal nerve. The mandibular branch of the facial nerve is in close proximity to the gland. The submandibular duct (Wharton's duct) runs along the base of the mouth and opens at the floor of the mouth lateral to the frenulum of the tongue.
  - The smaller sublingual glands are located in the anterior floor of the mouth just beneath the oral mucosa. These glands are drained by numerous ducts that open into the floor of the mouth. One or more ducts also open into the submandibular duct.

Sialoadenitis

Salivary gland inflammation has numerous etiologies including infection and conditions that lead to salivary stasis or decreased saliva production (Table 14.6).

Mumps

- Mumps occurs most frequently in persons <15 yr of age but is also seen in older adolescents and adults. Younger patients tend to have a more benign course.
- The parotid glands are most commonly affected.

Table 14.6. Etiology of sialoadenitis

Viral	Mumps (most common viral etiology), CMV, Coxsackie, ECHO, influenza
Bacterial	*S. aureus* (most common bacterial etiology), *S. pneumoniae, E. coli,* anaerobes
Granulomatous	Tuberculosis, other mycobacteria Actinomycosis Cat-scratch disease Sarcoidosis
Autoimmune	Sjögren's syndrome
Obstruction	Sialolithiasis Neoplasm Ductal strictures
Decreased saliva Production	Dehydration/malnutrition/dehabilitation NPO/postoperative status Radiation Systemic: endocrine disorders, uremia, CHF Medications*

CMV—cytomegalovirus; ECHO—enteric cytopathic human orphan virus; NPO—nothing by mouth; CHF—congestive heart failure; * Includes analgesics, antihistamines, phenothiazines, anticholinergics, etc.

14

- Patients experience a prodrome of fever, headache and myalgias followed by unilateral or bilateral parotid pain and swelling. Parotid swelling may not appear until after other prodromal symptoms have completely resolved. Examination reveals tender and enlarged glands. In contrast to bacterial disease, there are no purulent secretions and the saliva will be clear. Patients are infective for up to a week after resolution of symptoms.
- Complications include pancreatitis, meningitis, encephalitis, sensorineural deafness and orchitis. Gonadal involvement in prepubertal patients is rare. However, up to 30% of older males have orchitis. Mumps orchitis is usually unilateral and sometimes occurs in the absence of parotid disease. Infertility is rare even in cases of subsequent testicular atrophy.
- Treatment is supportive with hydration and analgesics.
- Having mumps is thought to confer immunity for future episodes although rare recurrences have been documented.

Bacterial Sialoadenitis

- As with mumps, bacterial infection usually involves the parotid gland. The other salivary glands are thought to produce saliva with greater bacteriostatic activity.
- Most patients are debilitated elderly with dehydration. A significant number of patients (up to 40%) are postoperative.
- The most common causative organism is *S. aureus*. Disease is believed to be secondary to salivary stasis with movement of oral flora into the gland.
- Patients complain of acute onset of gland swelling and pain. Bilateral involvement occurs in up to 20% of cases.
- Examination reveals a tender, warm, enlarged gland. Patients are febrile and possibly septic. Purulent secretions are expressed from the involved duct unless it is totally obstructed.
- Initial antibiotic therapy should cover penicillin-resistant *S. aureus* and anaerobes. Treatment is then tailored as necessary based on results of Gram stain and culture of secretions. Additional treatment includes IV hydration, analgesics, warm compresses and massage of the involved gland. Surgical therapy by ENT should be considered for persistent symptoms.

Sialolithiasis

- Salivary gland calculi occur most frequently in the submandibular gland presumably secondary to higher mucus content of the saliva and the long, upwardly directed duct. Parotid stones are less common, and sublingual stones are rare.
- Most cases are idiopathic although gout is a known cause.
- The majority of patients are middle-aged adults with men being affected more often.

14

- Patients present with sudden onset of pain and enlargement of the involved gland that begins after food intake. The gland is tender to palpation, and it is sometimes possible to palpate a calculus along the course of Stensen's or Wharton's duct. If the obstruction is subacute or chronic, evidence of secondary infection is also noted.
- Diagnosis is made clinically. The majority of submandibular gland calculi are radiopaque. Soft tissue radiographs provide diagnostic confirmation and help localize the calculus. Most parotid calculi are radiolucent. More specialized radiology studies such as contrast sialography, ultrasound or CT scan are available but are rarely indicated in acute cases of uncomplicated disease.
- Treatment consists of analgesics and sialogogues such as lemon drops. When secondary infection is suspected, begin empiric coverage for *S. aureus*. If the calculus is located in the distal portion of the duct, it may be possible to manually express it from the ductal orifice. Patients with sialolithiasis require prompt ENT follow-up. Surgical excision of the gland is required for recurrent cases or persistent proximally located stones.

Dental Emergencies

Basic Anatomy

The center of the tooth is the pulp which contains the neurovascular supply for the surrounding dental structures. The pulp is covered by dentin, a thick homogenous material that makes up the majority of the tooth and provides cushioning during mastication. The crown is the visible white portion of the tooth. It is covered in enamel, the hardest substance in the body. The non-visible anchoring portion of the tooth, the root, is covered in softer cementum and extends downward into the alveolar bone of the mandible and maxilla. The periodontal ligament surrounds the root and firmly attaches the cementum and alveolar bone. Gums, or gingiva, consist of keratinized squamous epithelium and normally are firmly attached to the tooth at the junction of the enamel and dentin.

The adult dentition consists of 32 permanent teeth: 8 incisors, 4 canines, 8 premolars and 12 molars. The permanent teeth begin to erupt at 6-7 yr and are complete by young adulthood. The four third molars, or wisdom teeth, may erupt later or not at all if they become impacted. The pediatric mouth is occupied by 20 primary or deciduous teeth: 8 incisors, 4 canines, and 8 molars. The first teeth to present are usually central incisors at 6-8 mo and the pattern is complete by the age of two.

Several numbering systems exist for identifying tooth location. The Universal System for adults numbers the teeth 1-16 starting with the right to left upper third molars and 17-32 from left to right lower third molars. There are two other systems, and all three have plans for both adult and deciduous teeth. For this reason, the best means of identification is description of the tooth by name and location (e.g., right maxillary lateral incisor).

Dental Pain

- Odontalgia, or dental pain, is usually secondary to underlying dental caries. Other intraoral pathology may also cause pain including fractures, osteitis, gingivitis and periodontal abscess.
- Local extraoral and systemic processes may cause pain referred to the dentition (see below). These diagnoses should be considered in cases where the history and exam is atypical for caries or other oral etiology.

Dental Caries

- Pain from caries is severe and throbbing. It is exacerbated by hot/cold/sweet/sour extremes as well as head movement. In some cases, patients are even sensitive to air. Although symptoms are usually poorly localized, patients are often able to identify the involved tooth.
- Caries are often the result of enamel destruction by bacteria contained in overlying plaques. Destruction progresses through the dentin and causes eventual pulpitis and/or pulp necrosis. Note that pulpitis and pulp necrosis also occur in the absence of obvious enamel destruction. An example is the patient who sustains enamel microfractures from bruxism (grinding). In these cases, diagnosis is made by history and reproduction of pain with percussion using a tongue blade.
- Treatment involves analgesia with oral medications and referral to a general dentist. The EP may also administer a nerve block if experienced in the procedure.

Postoperative Dental Pain and Osteitis

- Postprocedural pain is common after extraction and instrumentation of the teeth. Patients may also have painful, postoperative trismus. If infection can be ruled out, analgesics and referral back to the dentist is appropriate.

- Postextraction alveolar osteitis or dry socket results from loss of the blood clot from the socket and subsequent localized osteomyelitis in exposed alveolar bone. Patients present with severe oral pain and halitosis 2-3 days postextraction. Treatment of dry socket includes local anesthesia, irrigation with saline and packing the socket with iodoform gauze soaked in eugenol (other preparations can also be used). Dental follow-up for recheck and dressing change within 24 h is suggested. Antibiotics are commonly prescribed for moderate to severe cases.

Periodontal Infections and Dental Trauma

- Discussed below.

Maxillary Sinusitis

- Causes throbbing pain and pressure sensation referred to the upper teeth and the eye. Purulent nasal discharge, localized percussion tenderness and poor sinus illumination are also present and direct the EP towards this diagnosis.
- Treatment is with appropriate antibiotics and outpatient referral to ENT (see "Sinusitis").

Temporomandibular Joint Disorder (TMJ)

- Arises from a multitude of factors and their interaction including local trauma and bruxism. Most cases are considered idiopathic. TMJ is much more common in women.
- TMJ pain is usually unilateral and located at the temporal or preauricular area. It is often referred to the ear, neck and shoulder. The pain is dull in nature and worsens with mandibular movement. Patients also complain of locking or clicking of the jaw. Trismus occurs secondary to spasms of the temporal and masseter muscles.
- The majority of patients improve with symptomatic therapy that includes warm compresses, NSAIDs, soft diet, nighttime muscle relaxant and follow up with a specialist.

Trigeminal Neuralgia

- An idiopathic disorder that presents as paroxysmal sharp, piercing pain along the branch of cranial nerve V. In most cases, the second and/or third divisions are involved. Stimulation of "trigger points" in the face precipitate the pain. Neurologic examination and work-up are normal.
- Patients should be referred to a neurologist or ENT and are most often treated with carbamazepine.

Other Diagnoses Not To Be Missed

- Acute myocardial infarction (AMI) and temporal arteritis: both should be easily ruled-out by an accurate history, physical examination, review of systems and diagnostic studies.

Dental Trauma

- Dental trauma is a common complaint in the emergency department. School age children and adolescents are especially at risk secondary to playground activities, falls, athletic events and fights.
- Associated injuries including intracerebral and cervical spine pathology must always be considered in any patient presenting with dental trauma. In addition, airway assessment and stabilization (as indicated) is the main priority.
- General work-up of dental injuries should include: (1) careful inspection of oral cavity including dental occlusion and (2) evaluation of sensory, motor, vascular and glandular function.
- Pertinent points in the history include tetanus status and time of incident. Time is critical when dealing with fractured and avulsed adult teeth in order to prevent necrosis of exposed pulp. Also inquire about the location of any missing teeth or fracture fragments—always consider the risk of aspiration especially in young children.

Table 14.7. Ellis classification system for dental fractures

Classification	Findings	Treatment
Ellis I	Involves only enamel Minimal symptoms	Routine dental referral No emergent Rx required
Ellis II	Involves exposure of dentin (will appear yellow) Tooth sensitive to hot and cold	Urgent dental referral (within 24 h) ED Rx: coverage of exposed dentin with calcium hydroxide or other dental base Oral analgesics
Ellis III	Dental emergency Involves exposure of pulp (will appear pink) Tooth usually extremely sensitive and painful	Immediate dental referral ED Rx: coverage of exposed dentin with calcium hydroxide or other dental base Oral analgesics

Dental Fractures

Clinical Presentation and Diagnosis

- History is straightforward with dental pain and loose/missing teeth after facial trauma.
- Examination findings: Teeth may be loose, fractured, or completely avulsed. Look for associated fractures and soft tissue injuries. Pain can usually be elicited through percussion of the involved tooth although structures may be insensate if neurovascular injury is present.
- Diagnosis is often confirmed by clinical findings. X-ray evaluation by panograph, facial series and/or mandible series is useful in cases where associated fractures are suspected. Consider CXR and/or KUB to rule-out aspiration if a fragment is missing.
- The **Ellis classification system** is a commonly used method of describing fractures to the anterior teeth. Treatment depends upon the type of fracture present (Table 14.7). Types II and III lead to pulpal necrosis and require prompt stabilization and referral.

Disposition

- Patients are discharged home as long as there are not associated injuries that require inpatient treatment. Confirm dental follow-up for Ellis Class II and III patients.

Avulsed Teeth

- There are two main priorities when teeth have been completely removed from their socket.
  - Prompt replacement of the avulsed adult tooth—failure to do so precludes viability of the pulp and subsequent successful replantation. Ideally, the avulsed tooth should be replaced within the first hour after injury. Note that primary teeth are not replaced.
  - Location of the missing tooth—aspiration, ingestion and soft tissue impaction must be diagnosed by appropriate X-ray evaluation (panograph, CXR, KUB).

14

Treatment

- Prehospital care consists of placing the tooth in a moist, protective environment. Several options exist: (1) under patient's tongue or in the buccal pouch (recommended only in adults), (2) in the parent's mouth if the patient is a child or (3) in milk. A commercially available transport medium, "Hank's solution" (Save-A-Tooth), is optimal but often not available in everyday situations.
- Emergency department care:
  - The avulsed tooth should be irrigated with Hank's solution or normal saline. The goal is to remove debris. Do NOT scrub the tooth as this may damage the periodontal ligament fibers that are essential for replantation.

- Gently suction remaining blood clot and irrigate the socket.
- Replace the tooth into the socket being careful to handle by the crown only.
- Ideally, the tooth is stabilized by a dentist or oral surgeon. As a temporizing measure, the EP can use one of several available resins that basically splint, or glue, the replaced tooth to adjacent teeth.
- Analgesics as well as antibiotics with coverage for oral flora should be prescribed.

Disposition

- Discharge home if there are no other injuries requiring hospitalization
- Prompt follow-up with an oral surgeon or dentist is mandatory (within 24 h).

Oral Hemorrhage

Etiology

- The most common source of spontaneous oral bleeding is gingivitis. Spontaneous oral hemorrhage may also be a result of systemic disease or medication. Examples include excessive iatrogenic anticoagulation (heparin, coumadin), leukemia, thrombocytopenia, blood dyscrasia, HIV, and liver disease.
- Other causes include recent extraction or other surgical procedure.

Clinical Presentation and Diagnosis

- With gingivitis, patients describe bleeding that occurs after brushing or flossing. Ask if there is a history of recent dental extraction or coagulopathy.
- Examination is straightforward. Identify the site of bleeding and look for evidence of underlying inflammation or recent dental procedure.
- If necessary, do a complete examination looking for other evidence of underlying disease i.e., petechiae, spider angiomata, etc.

Treatment

- Simple gingival bleeding is usually controlled by direct pressure using folded gauze.
- Post-extraction bleeding can often be managed by evacuating residual clot and then having the patient bite down on folded gauze. If this fails, the EP can try local infiltration of lidocaine with epinephrine or placement of hemostatic material within the socket. Hemorrhage that is unresponsive to all measures mandates oral surgery or ENT consultation.
- Patients with underlying disease or coagulopathy also require directed therapy to correct the defect.

Disposition

- Patients with minor bleeding secondary to gingivitis may be discharged home with referral to a general dentist and instructions on proper dental hygiene.
- Those patients with postextraction bleeding may also be discharged as long as bleeding has been controlled. Appropriate follow-up should be given. Instruct them to avoid activities such as smoking which can cause negative pressure within the mouth and dislodge existing clot.
- Those patients who have bleeding secondary to systemic disease or iatrogenic coagulopathy should be treated accordingly.

Periodontal Infections

Gingivitis

- Gingivitis is an inflammatory response to plaque build-up at the gingival margins. It manifests as painless gum redness, swelling and bleeding.

- Treatment and prevention are via good oral hygiene including brushing and flossing as well as routine scaling in the dentist's office. Antibiotics are not necessary. Untreated, gingivitis may progress to periodontitis.
- Other conditions can lead to gingival swelling and inflammation including pregnancy, leukemia, granulomatous diseases and multiple medications.

Periodontitis

- Periodontitis is a result of gingivitis that has gone unchecked. This leads to separation of the gingiva from the tooth. The end result is further deposition of plaques and calculus in these periodontal pockets followed by destruction of the periodontal fibers that attach the tooth to the alveolar bone. Periodontitis is the main cause of tooth loss and decay.
- Patients may report halitosis and loose teeth. The condition is typically painless.
- Definitive treatment is done in the dentist's office. Patients should seek timely follow-up.
- Periodontal abscess occurs when purulent debris becomes trapped within pockets created by separation of the gingiva from the tooth. It presents as localized swelling and, in contrast to periodontitis, is painful. Patients should be started on appropriate antibiotics (i.e., penicillin) and saline washes. Dental follow-up is encouraged. Occasionally, drainage is required.

Acute Necrotizing Ulcerative Gingivitis (ANUG)

- AKA trench mouth and Vincent's angina, ANUG is caused by bacterial invasion of the gingival tissue.

Risk Factors

- The primary risk factor is immunocompromise especially HIV.
- Other contributing factors include poor nutrition, stress, poor oral hygiene, and alcohol/tobacco use.

Etiology

- Anaerobes and spirochetes.

Clinical Presentation and Diagnosis

- Patients complain of painful, swollen gums and a foul metallic taste in their mouth. They may give a history of immunosuppression or other risk factor.
- Examination: The characteristic finding is ulcerated, swollen gingiva at the interdental areas (papillae). There is associated gingival bleeding and halitosis. Patients may also have pseudomembrane formation, lymphadenopathy and fever.

Treatment

- Oral hygiene including hydrogen peroxide oral rinses.
- Antibiotics: appropriate agents include penicillin, clindamycin, and metronidazole.
- Analgesics

Disposition

- The patient must follow-up with a general dentist for scaling, debridement and continued care.

Maxillofacial Trauma

- Facial trauma presents a unique challenge to the EP. There is often a high level of concern by both the patient and practitioner because of the potential for permanent disfigurement.
- Maxillofacial trauma is often associated with other injuries especially to the head, neck and cervical spine. Pediatric patients have a higher rate of associated injuries than do adults. The dramatic nature of facial injuries should never distract the

treating physician from the ABCs of resuscitation. In addition, delayed airway compromise must be anticipated.
- Even in the absence of life-threatening injuries, patients may have involvement of critical structures such as the globe, facial nerve and parotid duct. Examination must be methodical in order to determine the extent of injury.

General Treatment

- Most facial fractures do not cause airway compromise. However, patients with significant injury such as Le Fort II and III fractures should have initial care directed toward airway stabilization.
- CSF rhinorrhea and basilar skull fractures: Conservative measures to promote dural closure include bed rest, stool softeners, and head elevation. Prophylactic antibiotics are not indicated
- Hemorrhage: Except in children, life-threatening bleeding from facial trauma is rare. Control hemorrhage with direct pressure rather than clamping or suture ligation.
- Epistaxis: Usually stops spontaneously or with anterior packing. However, posterior epistaxis can occur and will require appropriate therapy (see "Nasal Emergencies").
- Broad-spectrum antibiotics are routine for facial fractures that extend into the tooth-bearing region. The use of prophylactic antibiotics for sinus fractures is controversial. It is recommended to coordinate care with the specialist providing definitive care.
- The majority of patients are discharged from the ED unless they have associated injuries that require admission. Prior to discharge, maxillofacial surgery should evaluate the patient in order to assess the stability of the injury as well as arrange for definitive care and follow-up.
- Patients should be given a prescription for analgesics prior to discharge.

Nasal Fractures

- The nose is frequently fractured secondary to its anterior location on the face.
- Isolated nasal fractures are of little clinical consequence unless there is deformity or septal deviation that limits airflow through the nares.

Clinical Presentation and Diagnosis

- Patients present with nasal swelling, pain and epistaxis. They often complain of a nasal deformity. There will be a history of recent trauma.
- Examination: There is usually soft tissue swelling and ecchymosis unless the trauma is very recent. Patients may also have deformity/asymmetry, crepitus and overlying laceration. If there is a deformity, it is important to inquire whether this is old or new. The EP should check the patency of both nares and look intranasally for septal hematoma and foreign body. When present, epistaxis is usually mild and will resolve spontaneously. Posttraumatic clear nasal discharge is suspicious for cerebrospinal fluid rhinorrhea secondary to fracture of the cribriform plate.
- Diagnosis is made clinically. Nasal X-rays add little to the management and disposition.

Specific Treatment

- Persistent epistaxis is treated with nasal packing. If there is a significant laceration of the nasal mucosa, the patient is referred to ENT for consideration of suture repair.
- Any septal hematoma must be drained (see below).
- Indications for reduction of nasal fractures include deformity and naris obstruction. Fractures should be reduced by an ENT specialist. Most prefer to do the reduction after soft tissue swelling has resolved (3-7 days). The EP needs to arrange appropriate follow-up. Patients should be instructed to apply ice intermittently to reduce swelling. Prescribe analgesics as well as antibiotics if nasal packing has been placed.

- Septal hematoma: This manifests as asymmetry or widening of the nasal septum. There may be local discoloration although this unreliable as a means of diagnosis. If not adequately drained, the hematoma can become infected and cause subsequent necrosis and deformity of the septum. Hematomas are drained either via needle aspiration or incision. It is important to make sure that all blood has been expressed from the hematoma. After drainage, both nares are packed so that blood does not reaccumulate. These patients need antibiotics and close follow-up.

Naso-Orbital Ethmoid (NOE) Complex Fractures

- NOE fractures are secondary to posterior forces directed at the midface and may involve the adjacent frontal bone and maxillary bone as well as the medial orbital wall, cribriform plate and anterior cranial fossa.
- These may be isolated injuries but are more likely to be associated with other trauma.

Clinical Presentation and Diagnosis

- Patients have findings as seen with nasal fracture (see above). Patients may also have CSF rhinorrhea, ocular injuries and telecanthus or widening of the space between the eyes secondary to laceration of the medial canthal tendon.
- CT is indicated in cases of suspected NOE fracture.

Frontal Sinus Fractures

These injuries are a result of direct blunt trauma to the forehead. A significant amount of force is necessary to cause a frontal sinus fracture; always consider associated intracranial injury.

Clinical Presentation and Diagnosis

- Patients will have a history of direct trauma, often a motor vehicle accident in which their head struck the steering wheel or dash.
- Examination: Look for local swelling, deformity and overlying laceration. Patients may also have CSF rhinorrhea and palpable step-offs or crepitus. Not all fractures are clinically obvious.
- Diagnosis is made by CT scan.

Specific Treatment

- All frontal sinus fractures warrant evaluation by an appropriate specialist. In particular, patients with posterior wall fractures should be seen by a neurosurgeon.

Mandible Fracture

Mandible fractures are common injuries encountered by the EP. Many cases (>50%) involve fractures at more than one site.

Clinical Presentation and Diagnosis

- Patients complain of pain, dental malocclusion and difficulty opening or closing the mouth.
- Examination: Inspect for local tenderness, swelling and deformity. Patients may have facial asymmetry, lower lip anesthesia and difficulty opening and closing the mouth. Intra-oral examination often reveals lacerations, sublingual hematoma, dental trauma and obvious fracture with separation between the teeth. Subtle dental malocclusion is identified by the absence of a firm bite using a tongue blade placed between the teeth.
- Plain radiographs, panograph or the standard mandible series, confirm the diagnosis.

Specific Treatment

- Bilateral mandible fractures can result in airway obstruction secondary to tongue displacement. The first priority is always to provide an adequate airway if necessary.
- Prompt fixation is desired. Make sure that close oral surgery follow-up is arranged.

Orbital Fractures

See section "Ocular Trauma" in *Ophthalmologic Emergencies.*

Zygoma Fractures

- The zygoma articulates with multiple other bones and forms a significant part of both the orbit and the maxillary sinus.
- Most zygoma fractures are complex and involve these adjacent structures.

Clinical Presentation and Diagnosis

- Patients complain of difficulty and pain with mouth opening. Diplopia is possible with fractures involving the orbit and extraocular muscle entrapment.
- Examination: Findings include facial deformity, edema, ecchymosis, epistaxis, infraorbital/upper lip anesthesia, subcutaneous emphysema and subconjunctival hemorrhage of the lateral aspect of the eye. Always evaluate for associated globe injury.
- Suspected complex fractures should be evaluated with CT scan. Isolated arch fractures are visualized by a submentovertex radiograph.

Le Fort Fractures

The Le Fort classification consists of three fracture patterns (Table 14.8) and was developed to describe midface and maxilla fractures. However, the majority of these fractures are actually much more complex and do not conform to any one type.

Clinical Presentation and Diagnosis

- These fractures present with significant facial swelling that may obscure underlying deformity. Other possible findings include epistaxis, CSF rhinorrhea, dental malocclusion, infraorbital/upper lip anesthesia and globe injury.
- Le Fort II and III fractures can cause airway compromise and are often associated with intracranial injury.
- Diagnosis is via CT scan.

Mandibular Dislocation

Temporomandibular joint (TMJ) dislocation is most often bilateral. Dislocation occurs when the mandibular condyle is displaced anterior to the articular surface; muscle spasm often precludes spontaneous reduction.

Etiology

- Includes blunt trauma, seizure, extrapyramidal reaction and wide opening of the mouth as occurs with yawning. Many patients have underlying hypermobility of the joint.

14

Table 14.8. Le Fort classification for midface fractures

Classification	Fracture Pattern
Le Fort I	Horizontal fracture across the maxilla above the hard palate. Mobile maxilla and palate on exam (nasal bridge remains stable).
Le Fort II	AKA pyramidal fracture Fracture through the maxilla extending to the medial orbit and over the nasal bridge. Mobility of the maxilla and nasal bridge as a unit. May have obvious nasal flattening.
Le Fort III	AKA craniofacial disjunction or "dishface" Fracture through the nasal bridge, orbits and frontozygomatic suture. Total mobility of the face including the zygoma and orbits.

Clinical Presentation and Diagnosis

- Patients present with significant pain and may describe trauma or other preceding event that involves wide mouth opening.
- Examination: The mouth is held open and can't be closed. In unilateral dislocations, the jaw will be deviated towards the contralateral side (in contrast to condylar fractures). Patients will have difficulty speaking and swallowing.
- Diagnosis may be made clinically in most cases. However, X-ray evaluation is necessary in cases of trauma in order to rule out associated fracture.

Treatment

- The ease of reduction depends upon when the dislocation occurred. Dislocations present for hours or more often require general anesthesia for reduction.
- Reduction is facilitated with IV muscle relaxants and/or local anesthesia.
- The physician's hands are gloved and the thumbs placed inside the patient's mouth on the posterior molars or alveolar ridge. The thumbs should be wrapped in gauze to protect from bite injury. The rest of the fingers wrap around the mandibular symphysis.
- Downward pressure is applied to unlock the condyle and the condyle then relocated with posterior pressure.
- In bilateral dislocations, reduction of one side at a time may be easier.
- The mandible is then stabilized with an Ace bandage wrapped around the jaw and head.

Disposition

- Pain is usually well controlled with NSAIDs.
- Patients should be given follow-up with an oral surgeon and instructions for a soft diet.

Soft Tissue Injuries/Lacerations

Many of the same principles for wound care on the body also apply to the face (i.e., wound preparation, tetanus update). However, because of the aesthetic importance of this area, these principles must be stringently followed. In addition, there are some differences and special circumstances to consider when dealing with soft tissue injuries of the face and scalp. Specific anatomical areas are discussed below.

- Simple facial lacerations are closed with 6-0 nonabsorbable interrupted sutures after proper wound preparation. For complex wounds and wounds under tension, a layered repair with absorbable sutures may be necessary. Suture removal is in 3-5 days.
- The face is very vascular and, as a result, has rapid healing and less chance of wound infection. Clean facial wounds may be closed up to 24 hours after injury. In addition, debridement is kept to a minimum. Even questionable skin flaps often have adequate blood supply and acceptable healing. 14
- Lidocaine 1% with 1:100,000 epinephrine is appropriate in most cases and will assist in hemostasis. However, avoid use of epinephrine containing solutions on the ears, nose and tarsal plate. Consider regional nerve blocks for the tongue, ear, large wounds or when distortion of the tissues must be avoided (i.e., lips).
- Examination must assess those structures that have critical functions such as the facial nerve, parotid duct and muscles of facial expression.
- Prophylactic antibiotics haven't been proven to decrease infection in routine wounds and aren't indicated. Obvious exceptions include significant time elapsed since injury (>24 h), presence of foreign body, bite injuries, etc.
- NEVER hesitate to involve plastic surgery or ENT for repair. Cases in which this is indicated include:

- Lacerations that involve critical structures such as the facial nerve.
- Avulsion injuries or those with soft tissue defects.
- Complicated anatomical location such as the ear, auditory canal, nasal alae, etc.
- Complex lacerations with potential for poor aesthetic outcome.

Scalp

- When exploring scalp wounds, take notice of any bony step-offs or galeal tears.
- Galeal defects must be repaired in order to avoid subgaleal hematomas and a depressed scar. Use 4-0 absorbable suture material.
- Skin is approximated with staples or nylon. Leave long tails on the suture to facilitate removal from within the hair.
- Consider a pressure dressing in large lacerations to reduce hematoma formation
- Remove sutures in 7-10 days.

Forehead

- Maintain important landmarks including the scalp line, expressive wrinkles and eyebrows. Make sure that these are the first sutures placed during the superficial skin closure.
- U-shaped flap lacerations, especially with the base of the flap superiorly positioned, have a tendency to form a "trap-door" scar (the flap forms a prominent raised area). Careful deep tissue approximation is required and a compression dressing should be applied. Patients should be warned that future scar revision might be necessary.

Eyebrow

- Never shave the eyebrow. It provides alignment and the shaved area may not grow back completely.
- If tissue excision is necessary, excise at an angle parallel to the shaft of the hair to avoid creating bald areas.

Eyelid Lacerations

- See *Ophthalmologic Emergencies*.

Ear

- Simple lacerations of the ear are closed with 6-0 nonabsorbable suture.
- Wounds with exposed cartilage require detailed attention. Cartilage must be completely covered by skin in order to prevent subsequent infection and deformity. Cartilage may be conservatively trimmed to allow proper closure. In addition, devitalized cartilage should be debrided.
- Cartilage lacerations do not require reapproximation unless the fragments are unstable or displaced. In these cases, use 5-0 absorbable material and include the perichondrium in the repair.

- After repair, place the ear in a compression dressing with posterior and interior support.
- Otohematoma: Like septal hematomas of the nose, these require drainage in order to prevent necrosis of the underlying cartilage and subsequent deformity. Drainage is accomplished either with needle aspiration or incision. Afterwards, a compression dressing is placed to prevent reaccumulation of fluid. Close follow-up is mandatory.

Nose

- Simple wounds can be closed with either steri-strips or 6-0 nonabsorbable interrupted sutures.
- Complex wounds include those that involve cartilage and/or mucosa. Cartilage does not require suture repair unless there is displacement or instability of the fragments. In these cases, repair is done with fine absorbable suture. Mucosa is approximated with 5-0 absorbable suture with the knotted ends facing into the cavity of the naris. ENT

repair of these complex lacerations is recommended. Also consider ENT repair for those wounds that involve the alae and free rim of the nostril since precise approximation is required.

- Examine for septal hematomas in all nasal trauma.

Lips

- All lip lacerations require a search for embedded dental foreign bodies as these retained elements greatly increase the risk of wound infection.
- The vermilion border is the landmark for repair of lip lacerations. Any wound involving the vermilion border requires meticulous placement of the first percutaneous stitch at that point in order to maintain proper alignment.
- Through-and-through lacerations of the lip are repaired in layers. The oral mucosa is closed first with 4-0 or 5-0 absorbable suture in order to minimize flow of saliva to the remainder of the wound. Subsequently, the deeper muscle layer is closed with 4-0 or 5-0 absorbable then the skin with 6-0 nonabsorbable material. Absorbable material may also be used for percutaneous lip closure and will avoid the trauma of suture removal that might occur in certain patients including children.
- Up to 30% of tissue loss is tolerated without significant resultant deformity.

Oral Cavity and Tongue

- Most isolated oral mucosal lesions heal well without repair. The exceptions are deep wounds that may trap food and wounds that are caught between the teeth during chewing. Note the contrast to through-and-through wounds where the intraoral component must be repaired in order to minimize salivary contamination of the extraoral area.
- Repair of tongue lacerations can be controversial. Some general rules apply.
  - Linear, superficial lacerations of the central tongue heal well without suturing.
  - Lacerations which involve the edge, form flaps, bisect the tongue, or bleed excessively require suturing. An inadequate or ignored repair mayresult in loss of function and food trapping.
  - When repair is required, use 4-0 or 5-0 absorbable and close all layers in a single stitch. Through-and-through lacerations are closed on both sides placing the suture through half the thickness of the wound. Sutures should be loose since the tongue may exhibit significant swelling after injury.
  - Saline or hydrogen peroxide mouth rinses for intra-oral lacerations should be encouraged.

Suggested Reading

1. Diagnosis and Management of Acute Otitis Media. Clinical Practice Guideline, American Academy of Pediatrics Subcommittee on Management of Acute Otitis Media. Pediatrics 2004; 113:1451-1465.
2. Wurman LH, Sack LG, Flannery JV et al. The management of epistaxis. Amer J Oto 1992; 13:193-209.
5. Snow V, Mottur-Pilson C, Hickner JM et al. Principles of appropriate antibiotic use for acute sinusitis in adults. Ann Intern Med 2001; 134:495-497.
6. Hickner JM, Bartlett JG, Besser RE et al. Principles of appropriate antibiotic use for acute sinusitis in adults: background. Ann Intern Med 2001; 134:498-505.
7. Vukmir RB. Adult and pediatric pharyngitis: A review. J Emerg Med 1992; 10:607-611.
8. Bisno AL. Acute pharyngitis. N Engl J Med 2001; 344:205-211.
9. Marra S, Hotaling AJ. Deep neck infections. Amer J Oto 1996; 17:287-98.
10. Villalobos T, Arango C, Kubilis P et al. Antibiotic prophylaxis after basilar skull fractures: a meta-analysis. CID 1998; 27:364-369
11. Ellis E 3rd, Scott K. Assessment of patients with facial fractures. Emerg Med Clin N Amer 2000; 18:411-448.

CHAPTER 15

Ophthalmologic Emergencies

Kirsten Calder and Jason Ruben

Basic Anatomy

A detailed understanding of the anatomy of the eye is essential for emergency physicians in order to properly diagnose and treat ophthalmologic emergencies. Furthermore, this understanding allows accurate communication with consultants.

- **Eyelids**: The eyelids protect the globe from mechanical injury and dehydration. The **tarsal plate** is horizontally oriented connective tissue that supports the eyelid. It contains small **Meibomian glands** that secrete oily, lubricating fluid. The tarsal plate and neighboring fascia form the **orbital septum** that separates the eyelid from the underlying orbit. The upper and lower lids meet at the **medial and lateral canthi**. Several muscles are responsible for opening and closing the eyelid (Table 15.1).
- **Lacrimal gland and nasolacrimal duct**: The lacrimal gland is located in the superior temporal border of the upper eyelid. Tears produced by the gland flow over the eye and then drain into the nasolacrimal duct via small openings at the medial aspect of the eyelid called the **lacrimal puncta**. From the puncta, tears flow into the **canaliculi** to the **nasolacrimal duct** that empties into the nose.
- **Conjunctiva**: A vascular mucous membrane that covers the sclera (bulbar) and the inner aspect of the eyelids (palpebral). The bulbar and palpebral conjunctiva meet at the fornices.
- **Cornea**: The cornea is avascular and normally transparent. It receives nutrients from tears, aqueous humor and blood vessels in the neighboring conjunctiva. Corneal shape varies between individuals. The cornea is largely responsible for light refraction onto the retina.
- **Sclera**: "The white of the eye." This tough tissue covers the posterior part of the eye and is continuous with the dural sheath of the optic nerve. Anteriorly, the sclera joins the cornea at the limbus. The extraocular muscles attach to the sclera.
- **Iris**: This pigmented structure gives people their eye color. Muscles in the iris control the size of the centrally located **pupil**.
- **Lens**: A biconvex body located just posterior to the iris. It is also normally transparent. The neighboring ciliary muscles change the thickness of the lens giving the ability to focus and accommodate.

Table 15.1. Muscles responsible for opening and closing the eyelid

Muscle	Innervation	Function
Orbicularis oculi	Cranial nerve VII	Closes the eyelid
Levator palpebrae	Cranial nerve III	Opens the eyelid*
Superior tarsal	Sympathetics	Opens the eyelid*

* Loss of the superior tarsal muscle will result in mild ptosis whereas loss of the levator palpebrae muscle results in significant ptosis and inability to voluntarily open the eye.

- **Choroid and ciliary body**: The choroid is a layer of tissue located between the retina and sclera. It contains numerous blood vessels that provide nutrients to the retina. Anteriorly, the choroid is continuous with the ciliary body which produces aqueous humor and contains the ciliary muscles.
- **Anterior chamber**: The space between the cornea and iris.
- **Posterior chamber**: The space between the lens and iris. Aqueous humor circulates from the posterior to anterior chamber via the pupil and drains via the **trabecular network** into the **canal of Schlemm**.
- **Fundus**: Structures posterior to the lens including the retina and optic disc. The space posterior to the lens is filled with gelatinous **vitreous humor** that maintains the shape of the eye.
- **Retina**: The retina has two layers: an inner neural layer consisting of sensory cells and an outer pigmented layer. The retinal **fovea** is responsible for fine, central vision. The area of surrounding the fovea is the **macula** and is devoid of blood vessels.

Examination

There are many components to the eye examination. In addition, many etiologies of ocular pathology are associated with other local and systemic signs. A thorough head, neck and systemic exam is often required for adequate assessment.

- **Visual acuity (VA)**
  - The "vital sign" of the eye. VA must be measured on all patients with eye-related complaints.
  - If a patient wears corrective lenses, they should be worn when VA is assessed. Pinhole evaluation is necessary if the patient doesn't have their lenses. Failure of VA deficits to correct with pinhole occurs in the setting of refractive errors.
  - Patients unable to read the eye chart may have VA grossly assessed via finger counting or light/dark perception.
- **Pupils**
  - Exam allows assessment of the anterior visual pathway.
  - The EP should evaluate pupil size, symmetry and reactivity.
- **Inspection and slit lamp examination (SLE)**
  - Ophthalmologic exams should be done in a systematic and organized method with progression from anterior to posterior and peripheral to medial structures. Evaluation should also include the periorbital skin and soft tissue.
  - The slit lamp allows for closer visualization of the conjunctiva, cornea, anterior chamber (AC), iris, lens and vitreous. Specialized light functions are present to assess both the surface and layers of the eye and to look for corneal uptake after fluoroscein staining. Various lamps differ; be familiar with the equipment available.
- **Extraocular muscles**: Evaluate all cardinal positions of gaze.
- **Fundoscopy**: Examination of the posterior chamber, retina and optic disc is essential. This is difficult in the emergency department but should be done with as much accuracy as possible.
- **Intraocular pressure (IOP)**: In the ED, IOP is measured with tonopen or applanation using the slit-lamp. Normal IOP is <21.

Ocular Medications

Numerous topical ocular medications are available. Appropriate use of these agents can provide necessary treatment, decrease patient discomfort and facilitate examination (Table 15.2). Always ask about drug allergies and underlying medical conditions before prescribing or using.

Table 15.2. Ocular medications commonly used by EPs

Antibiotics (S/O)	Ciprofloxacin (S/O) Erythromycin (O) Gentamicin (S/O) Sulfacetamide (S/O) Tobramycin (S/O)
Cycloplegics (Duration of action)	Atropine (5-10 days) Cyclopentolate (6-24 h) Homatropine (2-3 days) Tropicamide (6 h)
Mydriatic (no cycloplegia)	Phenylephrine
Topical Anesthetics	Proparacaine Tetracaine
Glaucoma Agents	β-blockers (i.e., timolol) Cholinergic agents (i.e., Pilocarpine) Carbonic Anhydrase Inhibitors (See text below for other treatment)

S: solution; O: ointment

- **Antibiotics**: Specific agents will be discussed in the following sections. Delivery systems include ointment and solution. Ointments have greater contact time with the eye but can cause blurred vision.
- **Topical anesthetics**: Use of one or two drops relieves patient discomfort and allows for tonopen measurements and fluoroscein evaluation. These agents should not be prescribed for repetitive use as they retard healing and mask the pain from a worsening ocular condition.
- **Cycloplegics**: Ciliary muscle spasm causes significant pain in patients with corneal abrasions, iritis and other conditions. Cycloplegics provide pain relief via paralysis of the ciliary muscle. They also cause mydriasis; patients should be warned about this side effect. Duration of action varies between agents.
- **Mydriatics**: As do cycloplegics, these agents cause pupil dilation. However, mydriatics do not necessarily produce cycloplegia. The main use is to allow for adequate fundoscopic examination. Use in angle closure glaucoma and ruptured globe is contraindicated.
- **Glaucoma agents**: See discussion below.
- **Corticosteroids**: Steroids may worsen certain conditions and should be prescribed only in conjunction with a consulting ophthalmologist (with the possible exception of iritis).
- **Other**: Includes topical antivirals, antifungals, lubricants, decongestants and anti-inflammatories.

Disorders of Extraocular Movement (EOM)

15

- Eye movement is controlled by six extraocular muscles. Assessment of EOM provides information regarding the function of the muscles themselves as well as the integrity of the cranial nerves (CN) that provide innervation.
  - Medial rectus (CN III): Adducts the eye (medial gaze).
  - Lateral rectus (CN VI): Abducts the eye (lateral gaze).
  - Superior rectus (CN III): Elevates the eye with maximum action in lateral gaze.
  - Inferior rectus (CN III): Depresses the eye with maximum action in lateral gaze.
  - Superior oblique (CN IV): Depresses the eye with maximum action in medial gaze; also abducts and medially rotates (eye down and out).
  - Inferior oblique (CN III): Elevates the eye with maximum action in medial gaze; also abducts and laterally rotates (eye up and out).

Table 15.3. Etiology of extraocular movement disorders

Mechanical	Aneurysm, intracranial neoplasm, arteriovenous malformation, cavernous sinus disease, intracranial hemorrhage, orbital fracture with muscle entrapment, retrobulbar hematoma, abscess
Systemic	Diabetes, hypertension, thyroid disease, Wernicke's syndrome, giant cell arteritis, sarcoidosis
Neurologic	Myasthenia gravis, multiple sclerosis, ischemia, internuclear ophthalmoplegia, meningeal inflammation

- Abnormal EOM is typically due to pathology of the cranial nerves arising either from systemic processes or direct trauma to one or more nerve(s) (Table 15.3). Trauma or entrapment of extraocular muscles may also be a cause.

Clinical Presentation Diagnosis and Treatment

- Cranial nerve III palsy
  - Presentation: The primary symptoms are eyelid droop and diplopia. Note that patients sometimes do not notice significant vision defects because of their ptosis.
  - Examination: The affected eye is deviated laterally (due to unopposed action of the lateral rectus) and the patient has weak or absent eye elevation, adduction and depression. Other signs include pupil dilation and ptosis with the patient unable to voluntarily raise the eyelid. Patients can have incomplete palsies with sparing of the pupil.
  - *Patients with complete CN III palsy should be considered to have either posterior communicating artery (PCA) aneurysm or uncal herniation until proven otherwise.*
  - Patients with a pupil-sparing palsy likely have microvascular disease secondary to diabetes or hypertension.
- Cranial nerve IV palsy
  - Patients have vertical diplopia that increases in downward gaze.
  - Examination: There is weakness of eye depression when attempting to look nasally. Patients hold their head tilted to relieve diplopia.
  - Isolated CN IV palsy is unusual. Look for other neurologic and cranial nerve defects.
- Cranial nerve VI palsy
  - The primary symptom is horizontal diplopia which becomes more pronounced when the patient attempts gaze to the ipsilateral side.
  - Examination: the patient is unable to abduct the eye past the midline. There is no ptosis and the pupil is unaffected.
- Depending upon the etiology, patients may also have other cranial nerve defects as well as neurologic and systemic findings.
- With all cranial nerve palsies, diplopia is binocular and will be suppressed with patching of either eye (versus monocular diplopia).
- Work-up and treatment depend upon the underlying etiology. Patching is acceptable symptomatic therapy except in children because of risk of subsequent amblyopia.

Anisocoria

Etiology

- Numerous conditions cause anisocoria. These range from benign, local ocular pathology to severe life-threatening systemic and neurologic disease (see below).
- Normal variant: up to 20% of the population may have slight differences in pupil size. In these cases, the pupils are round and briskly reactive to both light and accommodation.
- Pupils normally dilate in the dark and constrict in the light. In general, the abnormal pupil is the one that is smaller in dim conditions or bigger in light conditions.

Clinical Presentation and Diagnosis

- Medications (both topical and systemic)
  - Cholinergic agents produce miosis via constriction of the smooth muscle fibers of the iris.
  - Sympathomimetics and anticholinergics cause mydriasis via relaxation of these fibers.
  - Patients who have received topical anticholinergic medications have pupil dilation that is unresponsive to pilocarpine.
- Local trauma and inflammation: Includes iritis, iris lacerations, lens dislocation and previous eye surgery
- Adie's tonic pupil
  - Usually occurs in young women.
  - Patients have anisocoria that is increased in light conditions. Pupils have sluggish reaction to convergence and minimal reaction to light. These patients have a characteristic supersensitivity to pilocarpine 0.125% with the affected pupil having marked constriction versus the normal pupil.
  - Treatment: Ophthalmology referral for maintenance cholinergic therapy.
- Argyll-Robertson pupil:
  - Most commonly caused by tertiary syphilis.
  - Patients have a small, irregular pupils that react to convergence but not light. Visual acuity is normal.
  - Work-up should include evaluation for syphilis.
- Horner's syndrome:
  - Etiology: Lesions of the sympathetic chain. Potential causes include brainstem cerebrovascular accident, spinal cord lesion, carotid dissection/trauma, cervical rib, lymphoma and Pancoast tumor.
  - Anisocoria is increased in dim/dark conditions. The affected pupil is miotic. Patients have mild ptosis although voluntary eye opening is preserved. Ipsilateral anhidrosis may be present.
  - Work-up depends upon acuity:
    - Acute disease mandates evaluation for underlying lesion. Obtain chest X-ray, chest CT, brain MRI, carotid duplex, etc as indicated.
    - Chronic disease is more likely to be benign and may be followed as an outpatient.
- Cranial nerve III palsy (see above) and intracranial pathology: Anisocoria secondary to impending uncal herniation will be associated with other neurologic findings and altered mental status.

The Red Eye

This is a frequent ambulatory complaint encountered in the ED. While many cases are secondary to benign conjunctival irritation, the EP needs to be aware of vision threatening and systemic causes of "red eye" that demand immediate treatment and further work-up.

Conjunctivitis

- Conjunctival inflammation has numerous etiologies with classification based on cause: bacterial, viral, fungal, allergic or chemical.
- Symptoms vary depending upon the etiology. In most cases, visual acuity is unaffected. Suspect other etiologies in those patients who complain of vision loss.

Bacterial Conjunctivitis

- Etiology: *Streptococcus pneumoniae* and *Staphylococcus aureus* are the most common causative organisms. Gram-negative bacilli such as *H. influenzae* are also implicated. Consider Pseudomonas in patients who wear contact lenses. *N. gonorrhoeae* and

Chlamydia are special concerns in neonates and patients at risk for sexually transmitted disease (see discussion below).
- Presentation: Patients complain of foreign body sensation, eye redness and discharge.
- Examination: Conjunctival injection and purulent eye discharge are noted. Chemosis and eyelid edema are sometimes present.
- Diagnosis is clinical. Patients with recurrent, persistent or severe disease should have swabs sent for Gram stain and culture.
- Treatment: Broad-spectrum topical antibiotic such as sulfacetamide, a flouroquinolone or bacitracin-polymyxin compound. Drops are used for 5-7 days. Do not patch or prescribe steroids. Follow-up with the ED or ophthalmology is encouraged. The patient should notice improvement in symptoms within 2-3 days; continued problems indicate persistent disease or an allergic reaction to the prescribed medication.

Gonococcal Conjunctivitis

- Risk factors: Occurs in individuals after direct contact with others who have gonococcal cervicitis/urethritis and neonates within the first 5 days of life.
- Presentation: Patients have hyperacute onset of severe bilateral eye redness, eyelid swelling and copious discharge. Adults often give a history of risk factors for sexually transmitted disease.
- Examination: Significant for severe eyelid edema, chemosis and conjunctival injection. Discharge is purulent and copious. Patients may have rapidly progressive corneal involvement including ulcerations and perforation; fluoroscein staining is indicated.
- Diagnosis is often clinical but Gram stain and culture of fluid are indicated.
- Treatment:
  - Systemic antibiotics are necessary for gonococcal disease. Less severe cases are managed with IM ceftriaxone. Neonates and adult patients with severe disease and/or corneal involvement should be hospitalized for IV antibiotics. It is also recommended to treat with doxycycline or azithromycin for concomitant Chlamydia. Treat all sexual partners as well as parents of infected infants.
  - Topical antibiotics are necessary for those patients with corneal involvement.
  - Saline irrigation four times a day is continued until eye discharge has resolved.
  - Ophthalmology consultation should be obtained emergently.
  - Children and adolescents with gonococcal conjunctivitis need evaluation for possible sexual assault.

Chlamydia Conjunctivitis

- Risk factors: Sexual contact with individuals who have Chlamydia cervicitis/urethritis. Greater than 50% of patients have concomitant genital infection. Neonates are also at risk, but disease occurs later than gonococcus at 5 days to 5 wk of life.
- Presentation: Symptoms are similar to viral conjunctivitis (see below). Patients may also give a history of recent genital symptoms consistent with cervicitis/urethritis. Neonates and infants may also have respiratory symptoms from concomitant Chlamydia pneumonia.
- Examination: Findings are similar to viral conjunctivitis although discharge is typically mucopurulent.
- Diagnosis is usually based upon history and clinical examination. For definitive diagnosis, swabs are sent for flourescent antibody testing, Giemsa staining or culture.
- Treatment: Consists of systemic doxycycline or other appropriate antibiotic. Sexual contacts also need to be treated. Children and adolescents need evaluation for possible sexual assault.

Viral Conjunctivitis

- Etiology: Respiratory viruses, esp. adenovirus.

- Presentation: Patients have eye redness, itching and watery discharge. They often have a concurrent or recent upper respiratory tract infection.
- Examination: Findings include conjunctival injection, tearing, conjunctival follicles, eyelid edema, watery to mucoid discharge and preauricular lymphadenopathy.
- Treatment is symptomatic with artificial tears, cool compresses and topical antihistamines and vasoconstrictors. Topical antiviral medications are not effective.
- Viral conjunctivitis can spread rapidly from the affected to unaffected eye and to other individuals. Patients need to be excused from work and be instructed to avoid touching their eyes and sharing items with others.
- **Epidemic keratoconjunctivitis (EKC)**: This is very contagious disease also caused by adenovirus. Common settings include dormitories and military housing. In contrast to typical viral disease, the patients will not have any systemic or respiratory symptoms. Both the conjunctiva and cornea may be involved. Treatment is symptomatic. Patient contact with others should be limited. Symptoms sometimes last up to 4 wk.

Allergic Conjunctivitis

- Presentation: Symptoms include a characteristic intense bilateral itching and tearing. Patients give a history of allergies, atopy and seasonal symptoms.
- Examination: The conjunctiva may have marked chemosis with either hyperemia or pallor. Discharge is watery.
- Treatment includes cool compresses, topical/oral antihistamines, topical vasoconstrictors and removal of the allergen if possible. Topical steroids are sometimes prescribed in conjunction with ophthalmology.

Pingueculum and Pterygium

- A pingueculum is a band of fibrous tissue occurring secondary to conjunctival degeneration. It begins near the palpebral fissure and extends in the direction of, but does not involve, the cornea. A pterygium is a continuation of this process with the fibrous tissue extending onto the cornea. Both occur more frequently on the medial aspect of the eye.
- Risk factors: Chronic sun/dust exposure and irritation.
- Presentation: Most patients are asymptomatic. Rarely, the tissue becomes inflamed and causes eye redness and discomfort. Patients with large pterygia can also experience vision loss if the tissue extends into the visual axis.
- Examination: The primary finding is fibrous tissue extending across the conjunctiva from the palpebral area with/without corneal involvement. This tissue may be whitish to yellow or may be injected and erythematous if inflamed.
- Treatment: Asymptomatic lesions do not require treatment. Local inflammation is treated with topical vasoconstrictors and artificial tears. Elective surgical removal is indicated for persistent inflammation and large pterygia that interfere with vision or use of contact lenses.

Corneal Abrasions

- See Section "Ocular Trauma".

Corneal Ulcer

- Etiology: Staphylococcus, Streptococcus, Pseudomonas, Herpes simplex and fungi.
- Risk factors: The primary cause is chronic use of soft contact lenses. Trauma secondary to organic material such as tree branches is a risk for fungal disease.
- Presentation: Pain, tearing, photophobia, FB sensation and eye redness.
- Examination: Ulcerations are visible as a whitish infiltrate of the cornea that will stain with fluoroscein. Other findings include conjunctival injection, eyelid edema, chemosis and AC cells and flare.

- Treatment: Corneal ulcerations are an ophthalmologic emergency. Rigorous topical antibiotic therapy is necessary with agents that provide coverage of both Staphylococcus and Pseudomonas such as ciprofloxacin. Eye patching and steroids are contraindicated.

Herpes Simplex Keratitis

- Etiology: Herpes simplex virus (HSV), either primary or reactivation. HSV causes many pathologic conditions of the eye including periorbital ulcerations, conjunctivitis, corneal ulcers and uveitis. Probably the most often-mentioned eye complication is keratitis.
- HSV keratitis presents as mild eye pain, FB sensation, eye redness, tearing and photophobia. Patients may give a history of herpes infection or immunocompromise.
- Examination: The classic finding is a dendritic pattern of fluoroscein uptake. Associated skin ulcerations are sometimes present. In contrast to zoster, these lesions are not in a dermatomal pattern and cross the midline.
- Treatment: Topical antivirals such as trifluridine or vidarabine are necessary if there is corneal involvement. Milder disease such as conjunctivitis is treated with symptomatic measures. Oral antivirals are beneficial for patients with immunocompromise. Steroids are recommended for certain HSV manifestations but should be administered only by the ophthalmologist. All patients with HSV keratitis must have an emergent ophthalmology evaluation.

Herpes Zoster Keratitis

- Etiology: Herpes zoster virus (HZV). Ocular and periocular disease is secondary to reactivation of the virus that is dormant in the trigeminal nerve.
- Presentation: The classic presentation is a painful rash on one side of the forehead. The rash is sometimes preceded by a prodrome of malaise and local pain. HZV also causes numerous ocular pathologies including conjunctivitis, keratitis, uveitis, scleritis and optic neuritis. Any patient who presents with the typical HZV rash involving the nasal tip needs to be evaluated for keratitis and other ocular manifestations.
- Examination: The rash of HZV is vesicular and located in a dermatomal pattern on the forehead. The rash does not cross the midline except in the rare bilateral case. Ocular findings vary depending upon the extent of involvement. These patients often have punctate corneal lesions. Dendritic corneal lesions are mentioned in association with HZV keratitis but are actually mucous deposits that stain poorly with fluoroscein. They are not epithelial defects and can be removed with a Q-tip.
- Treatment: HSV dermatitis is treated with systemic acyclovir or other antiviral agent. Topical antivirals are not indicated. Severely ill patients require admission and intravenous medications. HZV ocular disease is often managed with steroids in conjunction with ophthalmology. Ancillary treatment should include analgesics.

Uveitis

- Inflammation of the uvea which includes the iris, ciliary body and choroid. Uveitis affects anterior structures, posterior structures or both. Clear determination of involvement is sometimes difficult.
- Etiology: Causes include trauma, recent eye surgery, autoimmune disease, infection, malignancy and sarcoidosis. Some cases are idiopathic.
- Presentation: Symptoms include deep, throbbing eye pain; photophobia; eye redness and tearing. There is often a history of symptoms related to other organ systems.
- Examination: Patients have mild anisocoria with the affected pupil having sluggish reaction to light. Conjunctival injection is pronounced especially at the limbus. SLE reveals the presence of cells and flare in the AC and possibly an associated hypopyon or layering of white blood cells in the AC. IOP varies. VA is usually normal. When

present, vision deficits are mild. The patient with nontraumatic iritis also needs a thorough systemic evaluation to identify signs of associated disease.

- Diagnosis is clinical although laboratories and other ancillary studies are often indicated depending upon patient presentation and history.
- Treatment: Uveitis is often managed with cycloplegics and steroids although this is not always the case. Additional treatment with systemic antibiotics, antivirals or immunosuppressants is sometimes necessary. Care should be coordinated with both ophthalmology and internal medicine.

Episcleritis

- Etiology: Inflammation of the episcleral vessels occurs most frequently in young women with the majority of cases being idiopathic. Other risk factors include collagen vascular disease, autoimmune disease, inflammatory bowel disease, gout and infection.
- Presentation: Patients complain of eye redness, tearing and dull pain. Vision is usually unaffected.
- Examination: Injection is typically located in segments or patches. The primary differential diagnoses are conjunctivitis and scleritis (see below). The episcleral vessels are large and arrayed in a radial fashion. They will react to topical phenylephrine and are mobile when lightly manipulated with a Q-tip applied to the surface of the eye.
- Treatment is symptomatic with artificial tears and systemic anti-inflammatory medications.

Scleritis

- Etiology: Inflammation of the sclera is often caused by underlying systemic disease, most commonly connective tissue disease, vasculitis and infection. Many cases are idiopathic.
- Presentation: Patients complain of severe, deep eye pain. They also report eye redness, tearing and photophobia. Vision is normal or mildly affected. Depending upon the etiology, other systemic symptoms are also reported.
- Examination: There is prominent eye redness, and the globe is tender to palpation. Injection secondary to scleritis will not diminish with topical phenylephrine as opposed to conjunctival and episcleral inflammation. In addition, inflamed scleral vessels will not move when a Q-tip is lightly applied to the surface of the eye.
- Patients with scleritis need emergent ophthalmology evaluation and further work-up to determine underlying etiology. Treatment varies depending upon the extent of involvement and includes systemic steroids, immunosuppressants and nonsteroidal anti-inflammatories.

Glaucoma

- Glaucoma is a progressive optic neuropathy as a result of increased IOP.
- There are numerous causes of glaucoma. For the purpose of the EP, glaucoma can be classified into primary vs. secondary and angle closure vs. open angle disease.
  - Primary vs. secondary: Is the disease idiopathic or is there an underlying secondary cause such as trauma, recent eye surgery, hyphema, inflammation, diabetes etc?
  - Angle closure vs. open angle: angle closure glaucoma occurs in far-sighted persons with an abnormally shallow anterior chamber or flat iris who are predisposed to blockage of aqueous humor flow from the posterior to anterior chamber. The outflow tract is normal in patients with open angle disease.
- The primary concern of the EP is acute angle closure glaucoma which can lead to rapid vision loss. Open angle disease is slowly progressive and patients remain asymptomatic until late in the course. The majority of glaucoma is open-angle.
- The treatment for glaucoma is generally the same no matter what the cause. Secondary glaucoma also requires attention to the underlying disorder.

Primary Acute Angle Closure Glaucoma

- Etiology: In predisposed patients, pupil dilation obstructs aqueous humor flow and causes acute symptom development. Causes include pharmacological dilation, stress and recent attendance in an area with dim lighting. It is important to obtain medication history—is the patient taking any anticholinergic or sympathomimetic agents?
- Presentation: Symptoms include sudden onset of blurred vision, light halos, photophobia, eye pain, nausea, vomiting and headache.
- Examination: Patients have a mid-range fixed pupil(s), marked conjunctival injection and corneal clouding secondary to edema. On SLE, the AC is shallow. IOP is increased.
- Treatment: Prompt initiation of therapy is necessary as permanent vision loss may occur within a period of hours. The goal of treatment is to increase aqueous humor flow, decrease production and decrease volume. Obviously, those patients with higher pressures and more pronounced vision loss mandate more aggressive treatment.
  - Decreased aqueous humor production:
    - Carbonic anhydrase inhibitors such as acetazolamide, brinzolamide and dorzolamide. Acetazolamide is given IV or by mouth at a dose of 500 mg. The other two agents are topical and used for maintenance therapy. *Avoid in patients with sulfa allergy.*
    - Topical β-blockers: The most commonly used agent in the ED is 0.25% or 0.5% timolol with one drop applied to the affected eye. Use with caution in patients who have diabetes, pulmonary disease and heart failure.
  - Decrease volume of aqueous humor: Achieved with mannitol or glycerol. Glycerol is p.o. and often poorly tolerated. Dose of mannitol is 1-2 g/kg IV over 30 min.
  - Increased aqueous humor flow: The most commonly used agent in the ED is topical pilocarpine 2% with one drop applied to both the affected and unaffected eyes. The dose is repeated every 15 min in the affected eye until the pupil constricts.

Chemical Injuries

Etiology

Exposure to any number of possible agents including acids, bases, detergents, pepper spray and super glue.

- Alkali
  - Sources include lime, ammonia, cement, plaster and some household cleaners.
  - Alkalis cause extensive tissue destruction via liquefaction necrosis and rapidly penetrate to a depth sufficient to cause severe damage to deep ocular structures.
- Acids
  - Sources include hydrofluoric acid (glass cleaning agents and rust removers) and car batteries.
  - Acids cause coagulation necrosis which tends to limit the depth of penetration.
- Super glue (cyanoacrylate adhesive): Exposure seals the eyelids shut. Corneal abrasions occur as a result of hardened glue rubbing against the epithelium.

Presentation

- Depends upon the duration of exposure and the pH and concentration of the offending agent. Signs and symptoms range from mild conjunctival injection, chemosis, pain and photophobia to severe burns with complete corneal opacification, corneal perforation, and vision loss. Patients may also have burns of surrounding periorbital tissue.

Examination

- The entire eye needs to be examined and the lid margins completely everted. Any particulate matter and necrotic tissue should be gently removed with a Q-tip in order to minimize continued chemical exposure.

- Fluoroscein staining: corneal defects are common. However, if all of the corneal epithelium has been denuded, there will be minimal or no fluoroscein uptake.
- IOP: Sometimes elevated. Assess with tonopen or tonometer.
- pH: Easily measured with litmus paper placed in the fornix. Check pH both before and several times after treatment.

Treatment

- Irrigation
  - No matter what the offending chemical, immediate irrigation is mandatory. If it is possible to direct prehospital care, instruct patients to begin irrigation with tap water as soon as possible after exposure.
  - In the ED, patients should receive topical tetracaine and then irrigation with normal saline or lactated ringers via a Morgan lens. Irrigation is continued until a neutral pH is achieved. When checking pH after irrigation, wait 10-15 min in order to avoid measuring the pH of the irrigating solution. Note that normal pH can vary slightly between individuals and the measured pH does not necessarily reflect the pH of the AC and deeper structures.
  - In general, all significant exposures will require irrigation with at least 2 L of fluid. With strong acids and bases, irrigation is continued over a period of hours.
  - **Hydrofluoric acid (HF)**: Treatment for ocular HF exposure is the same as for other ocular chemical injuries. While calcium gluconate is used for skin exposures, topical ocular therapy with this agent is not yet standard of care.
  - **Super glue (cyanoacrylate adhesive)**: Patients should be referred to ophthalmology. Do not attempt to pry the eyelids open or dissolve glue with other substances.
- Topical cycloplegics, eye patching and parenteral analgesics are helpful for the patient with severe photophobia and pain.
- Increased IOP requires treatment with appropriate agents: β-blockers, pilocarpine, etc.
- Patients should be given either antibiotic solution or ointment.
- Never attempt to neutralize acids or bases. This will cause a heat-producing reaction and further tissue damage.
- Ophthalmologic consultation is warranted for all significant exposures.
- Complications: perforation, glaucoma, cataracts, retinopathy, adhesions and neovascularization of the cornea.

Blunt and Penetrating Trauma

- Remember to first address airway, breathing, circulation and neurologic status when treating trauma patients. Ocular injuries are often alarming but not usually life-threatening.
- As always, a systematic approach should be taken when evaluating ocular trauma (see above). Examination should start with the surrounding facial structures (bony and soft tissue) and then progress to the eye itself.
- Because periorbital tissue is loose and areolar, even minor trauma may result in significant ecchymosis and swelling. This should not prevent a thorough examination of the globe, orbit and surrounding facial structures. Special eyelid retractors are often available to facilitate examination and to minimize pressure on the globe. Folded paper clips may also be used for this purpose.
- It is important to determine VA if possible as this has prognostic value.
- When indicated, adhere to basic principles of wound care including tetanus updates.

Periorbital Ecchymosis and Contusion

- Extraocular movements, VA and SLE are normal.
- Treatment for isolated soft-tissue eyelid injuries is cold compresses and analgesics.

- The EP must rule out fractures and other associated globe injuries prior to patient disposition.

Subconjunctival Hemorrhage

- Ruptured conjunctival blood vessels.
- Etiology: Valsalva, hypertension, zygomatic arch fracture and coagulopathy are known causes. Some cases are idiopathic.
- Presentation: Superficial, well-demarcated areas of conjunctival bleeding without chemosis. VA is unaffected.
- Treatment: Reassurance. For recurrent nontraumatic hemorrhage, a routine work-up for underlying coagulopathy is recommended.
- Patients with bloody chemosis, decreased VA or other ocular abnormality need closer evaluation for possible ruptured globe.

Corneal Abrasions

- Presentation: Foreign body (FB) sensation, tearing and photophobia. Decreased VA occurs if the abrasion involves the visual axis.
- Examination: Patients have conjunctival injection and fluoroscein uptake on SLE or Wood's lamp evaluation. Lid edges should be everted and the eye thoroughly visualized in order to rule out FB.
- Treatment:
  - Antibiotics: The majority of practitioners probably prescribe prophylactic antibiotics although routine use of these agents for simple abrasions is controversial. If antibiotics are used, sulfacetamide or trimethoprim-polymyxin are acceptable for patients who don't wear contact lenses. Patients using lenses do require antibiotic therapy and this should provide Pseudomonas coverage. Appropriate agents include tobramycin or ciprofloxacin.
  - Symptomatic care: Cycloplegics, topical anti-inflammatories and oral analgesics.
  - Patching: Previously advocated as routine treatment. It is not necessary although some patients prefer to be patched in order to minimize discomfort. Avoid in patients who use contact lenses or if the abrasion is secondary to organic material such as tree branches.
  - Steroids are not indicated as they may retard healing and predispose to infection.
- Patients should have 24 h recheck. Ophthalmology referral is indicated for persistent defects or complications such as hypopyon or corneal ulcer.
- Differential diagnosis includes corneal ulcers, corneal FB and herpes keratitis.

Ultraviolet Keratitis

- Commonly seen in welders, snow skiers/snowboarders and persons using tanning salons
- Presentation: Tearing, FB sensation, photophobia and blurred vision.
- Examination: Conjunctival injection, decreased VA and punctate areas of fluoroscein uptake on SLE or Wood's lamp evaluation.
- Treatment: Short-acting cycloplegics and topical antibiotics. Oral analgesics are also helpful. Patients should be instructed about the use of protective eye goggles or shield.

Conjunctival, Corneal and Intraocular Foreign Bodies (FB)

- Presentation: Pain, photophobia, tearing and FB sensation. Patients usually give a history of grinding, sanding or hammering. Suspect intraocular FBs with exposure to metal striking metal.
- Examination: The FB or a surrounding rust ring is often visible. Patients have conjunctival injection and possibly corneal abrasions. Multiple abrasions in a linear pattern are indicative of a FB embedded in the interior aspect of the eyelid. Hypopyon and corneal infiltrates are seen in association with corneal FBs present longer than 24 h.

- Diagnosis: Usually straightforward for conjunctival and corneal FBs. Intraocular FBs are localized with plain radiographs, CT scan or ultrasound.
- Treatment:
  - FB removal: After application of tetracaine, superficial conjunctival and corneal FBs are removed using irrigation, Q-tip or the tip of a plastic catheter. If available, eye spuds are useful for embedded FBs and rust rings.
  - Medications: After removal of corneal FBs, patients will have a resultant corneal defect. Treat with appropriate topical antibiotics, cycloplegics and analgesics.
- Disposition: Patients should have a 24-48 h recheck to verify healing of the corneal defect. Immediate referral to ophthalmology is indicated for intraocular FBs, complicating hypopyon and FBs not easily removed by the EP.

Traumatic Mydriasis and Miosis

- Etiology: Local irritation and muscle tears of the iris.
- Examination: Anisocoria and slight scalloping of the pupil margins.
- No specific treatment is necessary besides reassurance. However, the EP must exclude more significant pathology such as intracranial lesions or ruptured globe.

Iridodialysis

- Separation of the iris from the ciliary body.
- Examination: Patients appear to have two pupils. Also look for associated hyphema
- In the absence of hyphema or other injury, treatment is supportive with referral to ophthalmology.

Traumatic Iritis (Iridocyclitis)

- Ciliary spasm and intraocular inflammation secondary to local trauma to the ciliary body.
- Presentation: Deep throbbing pain, photophobia and tearing. Symptoms occur within one to several days after local trauma.
- Examination: Patients may have anisocoria with the affected pupil having sluggish reaction to light. **In addition, patients may complain of pain with light in both the affected and consensual eyes**. Conjunctival injection is pronounced and is increased at the limbic region. SLE shows the presence of cells and flare in the AC.
- Treatment: Topical cycloplegics with long-acting agents recommended. Steroids are also acceptable but should be given only after consultation with an ophthalmologist.
- Differential diagnosis: systemic diseases may cause uveitis—Suspect systemic pathology if there is no history of trauma.

Hyphema

- Blood located in the AC secondary to disruption of local vasculature.
- Symptoms: Pain, photophobia and decreased VA.
- Examination: Patients have varying amounts of blood present in the AC. Severity can range from microhyphemas that are visible only with SLE to partial hyphemas that present as a red meniscus to complete hyphemas resulting in a completely black or red AC. Poor VA, delayed presentation, increased OP and large hyphema size portend a worse prognosis. Always consider associated globe injury.
- Treatment:
  - Supportive care: Perforated metallic shields protect the eye and allow the patient to recognize acute changes in VA that are predictive of rebleeding. The head of bed should be elevated bed 30-45 degrees and patients instructed to limit eye movement.
  - Medications: Antiemetics and analgesics are useful but avoid aspirin and other platelet inhibitors. Treat increased IOP if present.

- Complications: Acutely, patients may have rebleeding that usually occurs in 2-5 days. Other complications include acute or chronic glaucoma and permanent corneal staining.
- Disposition: all patients require emergent ophthalmology evaluation. Certain patients may be managed as an outpatient but this should be done only in conjunction with the ophthalmologist. If the patient is sent home, he/she should be instructed to avoid ocular activity (i.e., reading and watching television).
- The differential diagnosis includes nontraumatic hyphemas. Spontaneous bleeding may occur with coagulopathy, local neovascularization, leukemia, retinoblastoma and sickle cell disease. Emergent referral for appropriate work-up is indicated.

Lens Subluxation and Dislocation

- Disruption of the zonular fibers which connect the ciliary muscles to the lens. Partial tears result in subluxation; complete tears in dislocation.
- Presentation: Patients have monocular diplopia; symptoms are unchanged despite covering of the involved eye.
- Examination: There may be obvious dislocation of the lens either anteriorly into the AC or posteriorly into the vitreous. Two other findings to note are a shimmering movement of the iris (iridodonesis) and similar movement of the lens (phacodonesis). If the lens is obstructing the pupil, increased IOP may be present.
- Treatment depends upon the degree of subluxation. Ophthalmology referral is indicated. Patients without a history of trauma should be referred for further workup.
- Differential diagnosis: suspect genetic disorders (i.e., Ehlers-Danlos) if there is no history of trauma.

Orbital Wall Fractures

- Caused by blunt trauma to either the bony orbital rim or the globe. Fractures are most common at the weakest points of the orbit: the floor, medial wall and suture lines.
- Presentation: Symptoms include binocular diplopia (secondary to entrapment of EOM), localized pain and swelling.
- Examination: Signs include limited EOM especially upward gaze, ptosis, subcutaneous emphysema, point tenderness/step-offs, endophthalmos and anesthesia of the upper cheek and lip. It is mandatory to rule out injury such as ruptured globe and lens dislocation.
- Disposition: Patients are usually discharged after consultation with ophthalmology. Instruct patients to avoid Valsalva and nose blowing. Close follow-up with ophthalmology is necessary; delayed surgical repair is indicated for patients with persistent diplopia, significant cosmetic defect/endophthalmos and large or displaced fractures. Consider prophylactic broad-spectrum antibiotics in cases with sinus involvement.
- **Orbital blowout fracture**: This specific type of orbital injury is caused by direct trauma to only the globe rather than the bony elements. All of the energy is transmitted to the globe which is then pushed through the orbital floor. Orbital blowout fractures are more likely to be associated with EOM entrapment and herniation of orbital contents into the maxillary sinus. Treatment and disposition are the same.

Retrobulbar Hemorrhage

- Bleeding posterior to the globe secondary to disruption of local vasculature. It is usually traumatic but also occurs after orbit surgery. Retrobulbar hemorrhage causes an acute rise in intraocular pressure that is transmitted to both the globe and optic nerve.
- Examination: findings include proptosis, chemosis, subconjunctival hemorrhage, decreased VA, increased IOP and limited EOM.
- CT scan confirms diagnosis although patients often require emergent treatment based upon clinical examination.

- Treatment: Patients with increased IOP are at high risk for vision loss and require emergent therapy including lateral canthotomy and standard medications as used for glaucoma. Obtain emergent ophthalmology consultation on all cases.

Ruptured Globe

- Occurs after penetrating trauma or a sudden increase in intraocular pressure secondary to blunt trauma.
- Presentation: Patients complain of eye pain and decreased VA. Acuity changes vary from mild to complete vision loss.
- Examination: Some patients have obvious globe rupture with extrusion of orbital contents. In other cases look for a peaked/irregular pupil, bloody chemosis, hyphema, lens dislocation and/or vitreous hemorrhage. Measurement of IOP should be avoided but, if done, will usually be decreased.
- Treatment:
  - Emergent ophthalmology consultation is necessary as soon as the diagnosis is suspected.
  - Preoperative care: patients need IV placement, broad-spectrum antibiotics, tetanus update and analgesics. Keep patients NPO and place a protective eye shield. Inform patients to avoid straining and treat with prophylactic antiemetics.
- Differential diagnosis includes those causes of nontraumatic bloody chemosis such as coagulopathy and cavernous sinus disease.

Optic Nerve Injury

- The optic nerve may be lacerated by bone fragments or FBs or injured indirectly via compression, shearing forces or loss of blood supply.
- Presentation: Decreased VA, visual field cuts and, possibly, abnormal color vision.
- Examination: Patients have visual field defects and an afferent pupil defect. Thorough history should be obtained to rule out preexisting optic neuropathy.
- Evaluation: CT scan of the orbits localizes penetrating FBs and identifies any bony injury.
- Treatment: Emergent ophthalmology consultation is necessary. High dose steroids are advocated for patients with significant VA changes.

Eyelid Lacerations

- The patient with an eyelid laceration needs to have a thorough examination in order to rule out injury to the cornea, globe and other important structures.
- Simple lacerations are repaired by the EP using 6-0 or 7-0 nonabsorbable sutures.
- Ophthalmology repair is indicated for lacerations involving the lid margin, canalicular system, lacrimal apparatus, canthal tendons, levator muscles and orbital septum. Suspect septum involvement if there is fat visible in the wound. Complex lacerations should also be referred for plastics repair.

Lacerations of the Conjunctiva Cornea and Sclera

- All patients need a thorough examination to rule out FB and to determine the extent of involvement (is there a ruptured globe?).
- Conjunctiva: Small lacerations of the conjunctiva heal rapidly without repair. Large lacerations (>1 cm) should be referred for suturing. Patients with minor injury sent home from the ED need topical broad-spectrum antibiotics and 24 h follow-up.
- All corneal and scleral lacerations mandate emergent ophthalmology referral and should be treated as a ruptured globe by the EP.

Vision Loss

- History is essential for diagnosis and disposition of the patient with vision loss.

- Are vision changes sudden or gradual in onset? Sudden loss of vision generally implies more severe etiology and need for emergent consultation.
- Does vision loss involve one or both eyes? Most cases of sudden vision loss are monocular. Sudden binocular vision loss should raise suspicion for conversion disorder.
- Does the patient have associated pain or systemic symptoms?
- Does the patient have a history of visual acuity changes?
- Is there a history of diabetes mellitus, hypertension, recent infection or collagen-vascular disease? Many comorbid illnesses predispose to ocular pathology.

- Physical examination also provides necessary information regarding underlying pathology.
  - Many disorders of the anterior structures of the eye such as glaucoma and keratitis can cause fairly sudden onset of vision loss. Examination findings will assist with diagnosis.
  - Superficial examination of the eye may be completely normal. In these cases, the EP should suspect vitreous, retinal or optic nerve pathology. Fundoscopy is mandatory.
- The EP always needs to consider those causes that are the most vision threatening.

Central Retinal Artery Occlusion (CRAO)

- Etiology: Most cases are secondary to emboli from the carotid artery or heart valves. Other causes include trauma, increased IOP, collagen-vascular disease, hypercoagulation, migraine, sickle cell disease and giant cell arteritis.
- Presentation: Classic complaint is unilateral painless, sudden vision loss or field defects.
- Examination: Classic fundoscopic findings include diffuse pallor of the retina with a "cherry red spot" at the macula and "box-carring" or segmentation of the retinal vessels. Other signs include an afferent pupil defect and decreased VA. The majority of patients retain some light perception and ability to finger count. Patients with complete vision loss or minimal light perception likely have occlusion of the more proximal ophthalmic artery.
- Treatment: Early intervention by the EP is vital with the goal to relieve the occlusion or move it distally and spare more retina.
  - Ocular massage with gentle digital pressure applied to the closed eyelid off and on for 5-10 seconds at a time.
  - Inhalation of carbogen and paper bag rebreathing dilate the retinal artery.
  - Acetazolamide and topical timolol are given to reduce IOP.
  - Emergent ophthalmologic consultation is warranted. Paracentesis of the anterior chamber may be necessary.
  - Additional treatment is tailored to the underlying etiology.

Central Retinal Vein Occlusion (CRVO)

- Risk factors: Include increased IOP, diabetes, hypertension, hematologic disease/hypercoagulable state and vasculitis.
- Presentation: Unilateral painless vision loss or field defects of variable acuity and severity. Bilateral disease occurs but is rare.
- Examination: The severity of vision loss and fundoscopic abnormalities depends upon the etiology and whether disease is ischemic or nonischemic. Patients with ischemic CRVO typically have very poor VA and significant fundoscopic findings such as retinal hemorrhages, dilated retinal veins, cotton wool spots and exudates. Afferent pupil defect may also be present. Vision loss and fundoscopic abnormalities are usually less severe in nonischemic disease.
- Treatment: Appropriate agents are given if necessary for increased IOP. Otherwise, ED treatment is not necessary. However, patients do require close ophthalmology follow-up and further work-up by an internist to determine etiology.

Amarurosis Fugax

- Etiology: Most cases are secondary to emboli from the ipsilateral carotid artery or heart. Other risk factors include dysrrhythmias, hypoperfusion, vasospastic disease and hypercoagulable state.
- Presentation: Transient painless monocular vision loss that typically resolves after a few minutes. Rarely, vision loss lasts one to several hours. Recovery of normal vision can be abrupt or stuttering in nature. Patients sometimes experience recurring episodes and give a history of additional neurologic symptoms consistent with transient ischemic attack.
- Examination: fundoscopic and SLE are often normal or significant only for mild changes of the retina and retinal vessels. Carotid bruits are auscultated in a minority of patients.
- Treatment: no specific ED treatment is necessary although patients need cardiac and carotid work-up. Patients with recurrent symptoms should be admitted for immediate evaluation.
- Differential diagnosis: impending CRAO/CRVO, glaucoma and temporal arteritis must be considered. Additional aspects of the history and examination should provide adequate clues for diagnosis.

Retinal Detachment (RD)

- Etiology: There are three types of retinal detachment with differing risk factors and etiologies:
  - Rhegmatogenous retinal detachment (RRD): A break in the retina causes the sensory and pigmented layers to become physically separated by vitreous fluid. Risk factors include aging, trauma and eye surgery.
  - Exudative retinal detachment (ERD): Fluid leaks from the retinal vessels separate the two retinal layers but the retina remains intact. Etiology includes choroid neoplasm, local inflammatory disease and hypertension.
  - Tractional retinal detachment (TRD): Fibrous bands form in the retina then contract and pull the retina inward. Bands are secondary to previous vitreous hemorrhage or retinopathy from various causes such as diabetes and sickle cell disease.
- Presentation: Classic symptoms include flashing lights, floaters and a "curtain" floating down the visual field. Patients may also report visual field defects and loss of central vision.
- Examination: Fundoscopic abnormalities include: obvious breaks in the retina; areas of detached/folded retina; and vitreous hemorrhage or subretinal fluid/hemorrhage. However, the fundoscopic exam is often inadequate for complete visualization of the retina. Indirect ophthalmoscopy by a specialist is necessary.
- Treatment: Emergent ophthalmology consultation should be obtained. Treatment is surgical.

Vitreous Hemorrhage

- Etiology: Disruption of the retinal vessels. This occurs with blunt trauma or spontaneously in patients with neovascularization (i.e., diabetes and sickle cell disease) and retinal detachment. **Vitreous hemorrhage is also associated with macular degeneration, ocular neoplasm, subarachnoid hemorrhage, infection and connective tissue disease.**
- Presentation: Patients complain of acute onset of "floaters", flashing lights and VA changes that improve when the patient is erect and blood settles.
- Examination: Depends upon the severity of hemorrhage. Vitreous bleeding may obscure part or all of the fundus. With significant hemorrhage, the red reflex may be completely lost.

- Treatment: ED therapy consists of elevation of the head of bed and avoidance of antiplatelet and anticoagulant medications. These patients should be evaluated emergently by ophthalmology. Depending upon the etiology, severity and chronicity, definitive treatment includes retinal repair or vitrectomy.

Optic Neuritis

- Etiology: Optic nerve demyelinization. It is sometimes associated with multiple sclerosis (MS). However, there are other causes, and many cases are idiopathic.
- Presentation: Monocular (occasionally binocular) vision loss occurring over days to hours. Other symptoms include pain associated with eye movement and altered color vision. Patients are usually <50 yr of age.
- Examination: An afferent papillary defect is the rule. The disc and retina appear completely normal if neuritis is retrobulbar. On the other hand, some patients have optic disc edema.
- Diagnosis is usually clinical. Patients should also undergo MRI scanning to look for focal areas of demyelinization.
- Treatment: Most patients have slow recovery of vision over a period of weeks without treatment. High-dose steroids are given to MS patients in order to hasten vision recovery and prevent recurrence. Steroids should be given only in conjunction with ophthalmology. Patients should undergo MRI and additional work-up as indicated.

Temporal Arteritis or Giant Cell Arteritis (GCA)

- Etiology: Vasculitis involving medium and large arteries. Vision changes occur secondary to subsequent optic nerve ischemia.
- Presentation: Acute onset of marked vision loss that is usually monocular but can rapidly become binocular. Other symptoms include bitemporal headache, scalp tenderness (manifested as pain with hair brushing), painful chewing, fever, anorexia and weight loss. Patients often have polymyalgia rheumatica and have a history of shoulder and pelvic girdle pain. The disorder is uncommon before the sixth decade.
- Examination: Findings include an afferent pupil defect, a palpable and tender temporal artery, diminished VA and visual field defects. Fundoscopic examination is often unremarkable.
- Ancillary studies: Erythrocyte sedimentation rate (ESR) will be markedly elevated in nearly all patients with GCA. Definitive diagnosis is made by biopsy of the temporal artery.
- Treatment: Immediate steroid therapy is necessary and should be initiated when the diagnosis is first suspected. Do not wait for results of the ESR or ophthalmology consultation. High dose IV methylprednisolone is recommended in cases of acute vision loss. Otherwise, the patient is given prednisone 1-2 mg/kg/day orally. Prompt steroid treatment reduces the likelihood of consensual eye involvement and increases the chances for vision recovery in the involved eye. Patients need immediate ophthalmology examination.

Lids Lashes and Soft Tissue Disorders

Blepharitis

- A recurrent, bilateral inflammation of the lid margins.
- Etiology: Commonly caused by seborrheic dermatitis and Staphylococcal infection.
- Presentation: Patients complain of local eye irritation and crusting of the eyelids. Symptoms are usually increased in the mornings just after awakening.
- Examination: Patients have erythema and scaling noted at the eyelid margins. Mild conjunctivitis may also be present.

- Treatment consists of warm compresses and twice daily cleaning of the lid margins with baby shampoo. Consider topical antibiotic ointment such as erythromycin or bacitracin for severe or persistent cases.

Hordeolum

- Hordeolum, AKA stye, is an acute infection of the eyelid that arises from obstruction of a local sebaceous gland.
- Etiology: Causative organism is usually *S. aureus.*
- Presentation: Patients with hordeolum complain of acute pain, swelling and redness of the eyelid. On examination, the patient will have an erythematous and tender eyelid nodule that may point either towards the skin or the inside of the eye. Spontaneous rupture can occur.
- Treatment includes warm compresses and topical antibiotics. Incision and drainage is indicated for persistent cases and for extremely large or painful lesions. Systemic antibiotics are not necessary unless there is associated cellulitis.

Chalazion

- Etiology: Persistent obstruction of a Meibomian gland.
- Presentation: Nontender, firm nodule of the eyelid margin.
- Treatment: Warm compresses may help initially. Marsupialization is necessary for persistent lesions.
- Differential diagnosis: In older patients with chronic lesions, also consider carcinoma.

Dacrocystitis

- Etiology: Acute infection of the lacrimal sac usually secondary to Staphylococcus or Streptococcus.
- Risk factors include obstruction of the nasolacrimal duct, nasal/sinus surgery, local trauma, ductal stones and ductal neoplasm. Some cases are idiopathic.
- Presentation: Patients complain of acute onset of pain, swelling and redness located at the inferior and medial aspect of the eye.
- Examination: Findings include erythema and fluctuance over the lacrimal sac. The area is tender to palpation and it is often possible to express purulent material from the puncta. Patients may also have temperature elevation.
- Treatment includes warm compresses, anti-Staphylococcal antibiotics and analgesics. Ill-appearing patients should be hospitalized for intravenous antibiotics. Some patients may require incision and drainage by ophthalmology.

Periorbital and Orbital Cellulitis

Infection of the periorbital and orbital soft tissues occur more commonly in children and younger adults and require meticulous evaluation and treatment. Periorbital disease is limited to the eyelid and periorbital soft tissue whereas orbital cellulitis involves the globe and possibly deeper ocular and intracranial structures. The etiologies and presentation are similar. However, the treatment and complications differ so it is necessary to make an accurate diagnosis.

- Etiology: Common causative organisms include Staphylococcus, Streptococcus, *H. influenzae* (esp. in children) and anaerobes.
- Risk factors: Facial fractures and sinusitis predominate. Other contributing factors include orbital/sinus surgery, dental infections and bacteremia (less common).
- Presentation: Eyelid edema and redness are present with both periorbital and orbital cellulitis. In ocular cellulitis, possible findings include decreased VA, pain with extraocular movement, limited ocular motility, increased IOP and proptosis. Patients with periorbital cellulitis have disease limited to the tissue anterior to the orbital septum and, therefore, do not have these signs and symptoms. Both can cause fever.

- Diagnosis of periorbital cellulitis is clinical although emergent CT scan of the head and orbits is necessary if there is any question of orbital involvement.
- Treatment:
  - Periorbital cellulitis: Mild disease in an older, nontoxic patient is treated with outpatient broad-spectrum antibiotics with amoxicillin/clavulanate recommended. Admission for infants and young children is suggested.
  - Orbital cellulitis: Inpatient treatment with IV antibiotics is mandatory. Recommended agents include ampicillin/sulbactam or a third-generation cephalosporin or penicillinase-resistant penicillins. Emergent ophthalmology and/or ENT evaluation is necessary.
- Disposition: Anyone with periorbital cellulitis treated as an outpatient needs close follow-up.

Suggested Reading

1. Le Sage N, Verreault R, Rochette L. Efficacy of eye patching for traumatic corneal abrasions: A controlled clinical trial. Ann Emerg Med 2001; 38:129-134.
2. Leibowitz HM. Primary care: the red eye. N Engl J Med 2000; 343:345-351.
3. Shingleton BJ. Eye injuries. N Engl J Med 1991; 325:408-413.
4. Morgan A, Hemphill RR. Acute visual change. Emerg Med Clin N Amer 1998; 16:825-843.

CHAPTER 16

Trauma

Stephen D. Docherty and Carla Valentine

Introduction

- In trauma, assessment and resuscitation must often be performed simultaneously. The initial priority is to detect and treat rapidly fatal injuries.
- Trauma is the leading cause of death up to 44 yr of age with an estimated 57 million Americans injured each year and more than 150,000 people dying annually from those injuries.
- Intervention during the so-called "golden hour" is critical to patient survival since 60% of hospital deaths from trauma occur during this time.

Assessment

Primary Survey

- **Goals**—The goal of the primary survey is to rapidly identify treatable life-threatening emergencies during the first few minutes of the initial evaluation.
  - **A,B,C,D,E**—The primary survey is approached in a systematic fashion using the "ABCs":
    - A—airway maintenance with C-spine stabilization
    - B—breathing and ventilation
    - C—circulation and hemorrhage control
    - D—disability/neurological status
    - E—exposure/environmental control preventing hypothermia

Airway

- The airway should be opened using the chin lift or jaw-thrust maneuver.
  - The head tilt-chin lift maneuver should NEVER be performed in trauma patients with known or suspected cervical spine injury since it may cause permanent neurological injury.
- In an obtunded patient, an oral or nasopharyngeal airway may facilitate ventilation and airway patency.
- If the above measures fail to allow proper airway management endotracheal intubation, cricothyroidotomy, or tracheostomy should be considered.

Cervical Spine Immobilization

- All trauma patients should have proper cervical spine stabilization using a hard collar or sandbags along with total body immobilization using a long board as appropriate.

Breathing

- Breathing is assessed by noting the rise and fall of the chest wall, listening for breath sounds and/or feeling for the patient's breath.
  - If breath sounds are unequal or absent, suspect possible hemothorax or pneumothorax.
  - These should be immediately treated with a chest tube or needle thoracostomy, respectively.

Table 16.1. Glasgow Coma Scale

Best Motor Response	Best Verbal Response	Eye Opening
6—follows commands	5—oriented	4—spontaneously
5—localizes pain	4—confused	3—to voice
4—withdraws from pain	3—inappropriate words	2—to pain
3—decorticates	2—incomprehensible sounds	1—none
2—decerebrates	1—none	
1—flaccid/none		

- Palpate the trachea for deviation and the chest wall for fractures or subcutaneous emphysema.

Circulation
- The third step in the primary survey is to assess circulation and control any hemorrhage.
- Observe the patient's skin for color (e.g., pale, cyanotic) and palpate for temperature.
- Peripheral pulses may be used to estimate cardiac output.
  - In trauma, tachycardia (heart rate >120 beats per min in adults) is always suggestive of hypovolemic shock, often secondary to uncontrolled hemorrhage.
  - A normal heart rate does not exclude the diagnosis of hypovolemic shock. Patients taking various medications (e.g., digoxin, β-blockers or calcium channel blockers) may have a normal heart rate despite being in hypovolemic shock.
- Two large-bore intravenous lines should be placed to allow for large volume fluid resuscitation.

Disability
- The fourth step in the primary survey is to assess the patient's disability and neurological status.
  - A rapid neurological assessment is performed in order to evaluate potential intracranial injuries that might necessitate immediate surgical intervention.
  - The Glasgow Coma Scale may be used to assess the patient's level of consciousness (Table 16.1).
  - Pupillary responses should be documented.

Exposure/Environment
- The final step in the primary survey is exposure and environmental control.
  - All of the patient's clothing should be removed to allow thorough examination of all areas for potential injury.
  - The patient should be examined paying special attention to areas where injuries are easily missed including: the head, axillae, perineum and back.
  - The patient should be kept warm using blankets and warm intravenous fluids as hypothermia can induce arrhythmias, aggravate acidosis and impair platelet function.

Radiological Studies
- The three most important radiographs to obtain during the primary survey are chest, pelvis and cervical spine radiographs.

Chest X-Ray
- Allows diagnosis of a pneumothorax, hemothorax or great vessel injury.

Pelvic X-Ray
- Allows diagnosis of pelvic fractures which may be associated with significant blood loss.
- An anterior-posterior view should be obtained, and, if there is any abnormality, inlet and outlet films should be obtained.

- Inlet views delineate injuries to the sacroiliac joints or posterior displacement of the hemipelvis.
- The outlet view allows for good visualization of the sacral foramina or cephalad displacement of the hemipelvis.

Cervical Spine X-Ray

- Up to 90% of all significant injuries can be detected on a lateral cervical spine film which must include the C7-T1 junction.
  - If any abnormality is detected an oblique view should be ordered. This view is useful in evaluating the pedicles, intervertebral foramina, facet joints, and laminae.
  - The anterior-posterior view is useful in evaluating the transverse processes and rotational injuries.
    - An increased distance between spinous processes suggests a flexion injury.
  - The odontoid (open-mouth) view is useful to evaluate the lateral processes of C1 and the odontoid.

Life-Threatening Injuries

- During the primary survey several life-threatening injuries require immediate recognition since they are rapidly fatal if left untreated.

Cardiac Tamponade

- Occurs as a result of penetrating injuries to the chest wall (i.e., stab wounds) which results in the accumulation of fluid in the pericardial space restricting proper function of the atria and ventricles.
  - As little as 150 cm^3 of fluid is capable of causing tamponade.
- Approximately 33% of patients with this type of injury display the classic Beck's triad of hypotension, distended neck veins and muffled heart sounds.
  - Tachycardia and pulsus paradoxus (>10 mm Hg drop in systolic blood pressure with inspiration) may also be seen.
- If this type of injury is suspected, immediate thoracotomy or pericardiocentesis should be performed.

Tension Pneumothorax

- Occurs when air escapes into the pleural space.
- Clinical findings commonly associated with this type of injury include: dyspnea, respiratory distress, diminished breath sounds on the affected side, hyperresonance to percussion on the affected side, tracheal deviation away from the side of the lesion, tachycardia, hypotension and jugular venous distension.
- Immediate needle decompression should be performed in patients suspected of having this type of injury.
  - A 14 gauge intravenous catheter is inserted in the second intercostal space in the midclavicular line.
  - The definitive treatment is tube thoracostomy.

16

Open Chest Wounds ("sucking chest wound")

- Occurs as a result of a penetrating injury to the chest wall.
- This type of wound occurs when the chest wall communicates with the pleural space, and each breath allows air to progressively enter the pleural space.
- A three-sided dressing may be placed over the wound to prohibit air from entering and allow air to escape while obtaining supplies for definitive treatment.
  - **Never** place an occlusive dressing over the wound as this will essentially convert an open chest wound into a tension pneumothorax.
- Immediate definitive treatment includes tube thoracostomy followed by an occlusive dressing.

Flail Chest

- Occurs due to chest wall trauma secondary to tremendous force.
- There are various criteria for a flail chest injury including:
  - Two or more ribs fractured in two or more places
  - Costal cartilage disarticulation associated with more than one rib fracture
  - Disarticulation of the costal cartilage on both sides of the sternum
- Clinically, dyspnea and paradoxical chest wall movements may be noted.
- Treatment for flail chest is mainly supportive with supplemental oxygen and analgesics as needed.
  - The clinician must assess the patient for underlying lung injuries, and intubation may be required if signs of respiratory failure become apparent.

Exsanguinating Hemorrhage

- The most common anatomic spaces for life-threatening hemorrhage are intrathoracic (hemothorax), intrabdominal (hemoperitoneum), retroperitoneal, extremity compartments (especially thigh compartments).
- External hemorrhage should be controlled with immediate direct pressure and wound management.
- Internal hemorrhage is treated with fluid and blood resuscitation and appropriate surgical intervention.

Secondary Survey

- The secondary survey should only be performed after the primary survey has been completed and resuscitation measures have been initiated. It should include a complete focused history and head-to-toe examination. The vital signs (blood pressure, heart rate, respiratory rate and temperature) should be reviewed and a focused history should be obtained.
- The pneumonic "AMPLE" can focus the questioning in order to obtain the most pertinent facts.
  - A—allergies
  - M—medications
  - P—past medical history and illnesses
  - L—last meal
  - E—events surrounding the injury
- Finally, a complete head-to-toe examination should include: head, ears, eyes, nose, throat, chest, abdomen, back, rectal/vaginal, musculoskeletal, integument, and neurological status.

Laboratory Studies

Type and Cross-Match

- A type and cross should be sent immediately in all trauma patients as it takes up to 30 min to perform.
- If a patient requires an immediate blood transfusion, O/Rh negative blood can be used.
- Type-specific blood (typed but not cross-matched) can usually be ready in about 10 min.

Complete Blood Count

- A complete blood count should be sent.
- The hemoglobin and hematocrit are useful indices to determine the oxygen-carrying capacity of the blood.
- A hemoglobin of 7 mg/dL (hematocrit 21%) is usually well tolerated in trauma patients.

- In the elderly or those with cardiovascular compromise, a hemoglobin of 10 mg/dL (hematocrit 30%) may be a more appropriate goal.
- Abnormal platelet counts may herald future problems with hemostasis.

Urinalysis

- A urinalysis is a quick and easy study that can demonstrate gross or microscopic hematuria that may be indicative of a urinary tract injury.

Coagulation Studies (prothrombin time (PT)/activated partial thromboplastin time (aPTT))

- The PT and aPTT can help determine the presence of an underlying coagulopathy that can inhibit the patient's ability to stop bleeding.
- It is important to remember that normothermia (>35 degrees Celsius) is crucial to correcting coagulopathies in trauma patients.

Spine/Spinal Cord Injuries

- Annually, there are approximately 10,000 people who sustain spinal cord injuries and 5% of patients with severe head injuries have an associated spinal cord injury.
- The three most common mechanisms of injury are: (1) blunt trauma secondary to motor vehicle accidents (90%), (2) falls and (3) gunshot wounds.

Cervical Spine Injuries

- Cervical spine injuries are uncommon but potentially lethal/debilitating.
- The NEXUS study has described the epidemiology of cervical spine injuries.
  - 2.4% of over 34,000 patients enrolled in this study had cervical spine injuries.
  - Patients are considered low risk for cervical spine injury and therefore do not need radiographic studies of the c-spine if: they have no evidence of intoxication, no posterior midline tenderness, no altered level of alertness, no distracting painful injury and normal neurologic function.

Evaluation

- Clinical assessment includes a complete neurologic assessment and evaluation of the presence/absence of NEXUS low risk criteria.
- Radiological evaluation should include three views of the c-spine.
  - The lateral view needs to include from C1 to the C7/T1 interface. This view will demonstrate approximately 90% of all c-spine injuries.
  - Complete three-view series (AP, lateral, and odontoid) will demonstrate 98% of injuries.
  - CT scans or MRI should be used as needed to evaluate patients with continued tenderness despite normal X-rays.

Unstable Cervical Spine Injuries

- All cervical spine fractures are important, but special emphasis needs to be placed on unstable injuries.
  - Odontoid fractures
    - Type 2 and 3 odontoid fractures (base of the odontoid process and involving the body of C2)
  - Jefferson's fracture
    - Vertical compression injury
    - Lateral displacement of the lateral masses of C2
    - Widening of the predental space seen on the lateral view. This space is normally <3 mm in adults and <5 mm in children.
  - Bifacet injuries

  - Flexion injury
- Anterior displacement more than 50% of the AP diameter of the vertebral body above the level of the injury
- Hangman's injuries
  - Hyperextension injury
  - Fracture of the bilateral pedicles of C2
- Flexion teardrop fracture
  - Hyperflexion injury
  - Wedge-shaped fracture of the antero-inferior portion of the vertebral body.
- Any fracture dislocation

Anterior Cord Syndrome

Etiology

- This syndrome occurs as a result of severe flexion with injury to the spinothalamic and corticospinal tracts with sparing of the posterior columns.

Symptoms

- Generally, there is variable loss of motor function, pain and temperature sensation below the level of the lesion with preservation of the posterior columns (proprioception, vibration and light touch).
  - Bilateral spastic paresis (lateral corticospinal tract)
  - Bilateral loss of pain and temperature sensation (lateral spinothalamic tract)
  - Bilateral flaccid paralysis (anterior horn).
  - Bilateral Horner's syndrome (hypothalamospinal tract T2 or above)

Prognosis

- Of incomplete spinal cord injuries, this syndrome carries the poorest prognosis.
  - Patients may have partial recovery of sensory function; however, only 10-20% regains functional motor control.

Central Cord Syndrome

Etiology

- This syndrome usually occurs due to hyperextension or hyperflexion injuries in patients with preexisting cervical spine disease.
- The central spinal cord is most affected and often involves the corticospinal and spinothalamic pathways.

Symptoms

- Motor weakness is most pronounced in the upper extremities and in the distal portion of the extremities.
- There may be a variable sensory loss (pain/temperature more than proprioception/vibration).
- Dysesthesias (e.g., a burning sensation in the hands or arms) are common.
- Bowel and bladder control is usually intact with sacral sensory sparing as the neural pathways controlling these functions are distributed laterally.

Prognosis

- Generally, the prognosis for central cord syndrome is good.
  - For patients sustaining spinal cord contusion without hematomyelia, 50% recover enough to allow independent walking.
  - However, patients have variable recovery of upper extremity function typically with poor fine motor control.

Brown-Sequard Syndrome

Etiology

- This syndrome is characterized by transverse hemisection of the spinal cord.
- The lesion is typically in the cervical region and may occur as a result of either a penetrating injury or blunt trauma.
- Other causes include: spinal cord tumors (unilateral cord compression), ischemia, herniated disk, hemorrhage or infectious or inflammatory diseases.

Symptoms

- Ipsilateral loss of proprioception, vibratory and tactile discrimination (posterior columns)
- Contralateral loss of temperature and pain sensation (lateral spinothalamic tract)
- Ipsilateral Babinski and spastic paralysis (lateral corticospinal tract)

Prognosis

- Of incomplete spinal cord injuries, this syndrome carries the best prognosis with 90% of patients recovering bowel and bladder control as well as the ability to independently ambulate.

Cauda Equina Syndrome

Etiology

- The cauda equina is made up of lumbar, sacral and coccygeal nerve roots.
- This syndrome is characterized by lumbosacral nerve root injury often as a result of a central lumbar disc herniation.

Symptoms

- Patients may experience a variable loss of motor and sensory function in the lower limbs; affected limbs are areflexic due to nerve root injury.
- The patient may also have areflexic bowel and/or bladder and saddle anesthesia (loss of pain sensation over the perineal area).

Prognosis

- This syndrome carries a much better prognosis than spinal cord injuries since peripheral nerves (i.e., lower motor neurons) have a greater regenerative capacity.

Conus Medullaris Syndrome

Etiology

- This syndrome occurs as a result of sacral cord injury with or without lumbar nerve root involvement.

Symptoms

- Patients typically experience variable loss of motor and sensory function in the lower limbs with areflexic bowel and/or bladder.

Prognosis

- In severe lesions, patients may experience pronounced bowel and bladder dysfunction.

Spinal Shock

Etiology

- Spinal shock is a transient condition appearing soon after injury with recovery typically seen in the first 24 h.
- This condition occurs after complete or partial injury at spinal cord level T6 or above.

Symptoms

- There are no reflexes or voluntary movements below the level of the injury.
- Patients may also experience sensory loss or loss of rectal tone.
- Autonomic dysfunction (hypotension, bradycardia and vasodilatation) is often seen.

SCIWORA

- Spinal cord injury without radiographic abnormality (SCIWORA) is typically seen in children and occurs in approximately 66% of youngsters with spinal cord injuries.
- Due to the increased flexibility of the immature spine and spinal column.

Radiographic Studies for Potential Spinal Injuries

- AP, lateral and odontoid cervical films
  - The lateral film is useful as a screening tool since it identifies approximately 90% of cervical injuries
    - It must include the C7-T1 junction in order to be an adequate study.
- CT scan
  - A CT scan is useful if:
    - There is any abnormality on plain film,
    - If the patient has a neurological deficit with normal plain films
    - Or if adequate visualization cannot be obtained with repeated plain films.
- MRI
  - MRI is an increasingly important modality for visualizing ligamentous injuries and evaluating the spinal cord and canal.
  - It is also useful in detecting soft tissue, muscular or neurologic damage, but is less sensitive than CT for detecting bony injuries.

Treatment of Spinal Cord Injuries

- All spinal injuries are treated with strict maintenance of spinal immobilization.
- The National Acute Spinal Cord Injury Study (NASCIS II and III) trials demonstrated significant improvement in both motor and sensory function in patients with spinal cord injury when treated with methylprednisolone within 8 h of the injury.
- Orthopedic and neurosurgical consultation should be obtained as needed (Table 16.2).
- **Intravenous antibiotics** should be given in the emergency department to those patients with penetrating spinal cord injuries.

Table 16.2. Treatment priorities for patients with spinal cord injuries

Prehospital—spine stabilization and immobilization

Primary and Secondary Survey

- **ABCs**
  - Indications for intubation of patients with spinal cord injury:
    - Decreased mental status (Glasgow Coma Scale <8)
    - Acute respiratory failure (e.g., fatigue, PCO_2 >50 etc.)
- **Supplemental oxygen**

High dose steroids within 8 h of the injury for all patients with spinal cord injuries:

- Methylprednisolone 30 mg/kg over 15 min then...
  - **If 0-3 h after the injury**...45 min after the bolus, start a methylprednisolone infusion at 5.4 mg/kg per h for 24 h

    OR

  - **If 3-8 h after the injury**...45 min after the bolus, start a methylprednisolone infusion at 5.4 mg/kg per h for 48 h

16

Head Trauma

- 50% of trauma deaths involve head injury.
- There are 2 million cases of traumatic brain injury (TBI) per year in the US, many of which require in-hospital treatment including either surgical intervention or close observation.
- TBI is classified as mild, moderate or severe based on the Glasgow Coma Score (GCS).
    - Severe TBI, accounts for 10% of all TBI, is defined as a GCS of <9 and has a mortality rate of close to 40%.
    - Moderate TBI accounts for approximately 10% of all TBI, and is defined as a GCS from 9 to 13.
    - Mild TBI is defined as a patient with a GCS equal to or >14.

Pathophysiology

- Acceleration/deceleration injuries
    - Serious intracranial hemorrhage may occur without any direct blows or penetrating injuries to the head.
    - In these injuries the brain moves within the cranial vault causing a shearing and/or stretching effect on intracranial vessels which can possibly lead to hemorrhage.
    - The brain may also strike against the inner wall of the skull causing further brain injury (a contracoup injury).
- The initial traumatic injury to brain tissue is defined as the primary brain injury.
- Secondary brain injury occurs at a time after the initial mechanical trauma.
    - Common causes of secondary brain injury may include hypoxia, hypotension, increased intracranial pressure (ICP) or severe anemia from exsanguination.
        - ICP
            - The intracranial space is essentially a nonexpandible space, and if there is expansion of one or more of the components of this space, such as occurs with intracerebral hemorrhage or cerebral edema, the ICP will rise above the normal of 15 mm Hg.
            - Limited compensatory mechanisms exist to help control this rise in ICP.
                - The most common mechanism is displacement of cerebral spinal fluid (CSF) from the brain to the spinal canal.
            - If the ICP becomes equal to or greater than the systemic arterial blood pressure, cerebral perfusion will stop and death of brain tissue will occur secondary to ischemia.
            - The elevated ICP may also lead to brain herniation.
                - Brain herniation occurs when brain tissue is displaced through the opening in the tentorium. This herniating brain tissue eventually causes compression of the brain stem leading to death.
                - A dilated and fixed pupil usually occurs on the ipsilateral side of the herniation due to compression of the oculomotor nerve.

Skull Fractures

16
- Occur when the applied trauma causes disruption of the normal anatomy of the skull.
- Categorized as linear or depressed, open (scalp violated) or closed.
    - Comminuted fractures may also occur and usually indicate an extremely forceful impact to the cranium.
- Because a significant amount of traumatic force is required to produce any kind of skull fracture, a high index of suspicion for additional associated head injuries such as traumatic brain injury should be maintained.
- Noncontrast CT scan of the head with bone windows is the diagnostic modality of choice.

- Basilar skull fractures are a special type of linear skull fracture where the fractures extend through the base of the skull.
  - The most common location is the temporal bone.
  - Clinical findings in basilar skull fracture may include:
    - Hemotympanum caused by hemorrhage into the middle ear.
    - Battle's sign described as ecchymosis occurring behind the patient's ear
    - Raccoon eyes (periorbital ecchymosis)
  - CSF rhinorrhea or otorrhea also occur frequently with basilar skull fracture.

Epidural Hematoma

- An epidural hematoma is a collection of blood between the dura and the inner table of the skull.
- In approximately 80% of cases there is associated skull fracture.
  - Commonly occurs across the location of the middle meningeal artery.
- Arterial hemorrhage ensues causing the dura to be separated from the inner table of the skull as the hematoma expands.
  - Tears of the dural sinuses also may lead to epidural hematoma formation.
- The rapidly expanding hematoma compresses underlying brain tissue which may lead to herniation.
- The classic presentation of the patient with an epidural hematoma often includes a transient loss or decrease in the patient's level of consciousness followed by an interval of normal or improvement in mental status (lucid interval). The patient may also present with nonspecific complaints commonly seen with head injuries such as headache, vertigo, nausea and vomiting.
  - Subsequent to this lucid interval the patient develops a second episode of decrease in mental status.
    - The time frames of these various phases of change in mental status are variable depending on the size, location and rapidity of development of the epidural hematoma.
    - Only 30% of patients with the diagnosis of an epidural hematoma will present classically.
- Statistically epidural hematomas are diagnosed in only 0.5% of head injury patients with a mortality rate of 15-25%.
- The diagnostic test of choice is a noncontrast CT scan of the head.
  - The CT scan will reveal a hyperdensity that is lenticular in appearance on the surface of the brain.
- Immediate neurosurgical consultation is necessary with many cases needing operative intervention.
- If an increase in ICP is suspected interval treatment aimed at decreasing ICP is indicated pending neurosurgical intervention to prevent herniation.

Subdural Hematoma

- A subdural hematoma is decribed as bleeding that occurs between the brain and dura.
  - This type of intracranial hemorrhage occurs more commonly than the previously discussed epidural hematoma. (approximately 30% of patients presenting with a serious head injury) Subdural hematomas have a venous source for their hemorrhage and therefore accumulate at a much slower rate than epidural hematomas.
- The most common mechanism is a tearing of bridging veins secondary to shearing force (e.g., acceleration/deceleration)
- The presence of brain atrophy places an individual at increased risk for this type of injury.
  - e.g., the elderly patient or chronic alcoholics.
- An acute subdural hematoma is defined as one that causes symptoms within the first 24 h.

- A subacute subdural hematoma becomes symptomatic in the time frame of 24 h to 2 wk after a traumatic event.
- A chronic subdural hematoma is seen in those patients who become symptomatic beyond the 2 wk time period.
- A lucid interval, although classically associated with epidural hematomas, may occur with any intracranial hemorrhage including subdural hematomas.
- Patient complaints on presentation to the emergency department may include headache, vertigo, altered mental status, focal neurological deficits, nausea and vomiting.
- CT scan makes the diagnosis and neurosurgical consultation should be obtained.

Traumatic Subarachnoid Hemorrhage

- Describes a condition of blood in the subarachnoid space secondary to injury to subarachnoid vessels.
- Commonly seen in patients with severe head injury (40% of patients).
- The patient who is not altered typically complains of photophobia and cephalgia.
- The CT scan demonstrates blood in the sulci and basal cisterns.
  - In a small number of patients the amount of blood present is not sufficient to be detected by CT.
- One of the major complications of any type of subarachnoid hemorrhage is the development of secondary vasospasm which may result in secondary injury.
  - Normally occurs in a time frame ranging from 2 days to 2 wk after the initial injury.
  - The use of the calcium channel blocker nimodipine may be helpful in reducing the occurence of this vasospasm.
- Neurosurgical consultation and admission to a monitored bed are required.

Cerebral Contusion/Intracerebral Hemorrhage

- A cerebral contusion is a traumatic brain injury that demonstrates areas of edema and hemorrhage.
- Usually secondary to rapid deceleration/acceleration.
  - Other common mechanisms include direct blunt trauma to the head, motor vehicle accidents and falls from a significant height.
- The most common locations for contusions to occur are the frontal and temporal lobes.
- The majority of patients present with some degree of alteration in mental status.
- CT scan of the head demonstrates small punctate areas of intraparenchymal hemorrhage.
- Hospital admission and neurosurgical admission are warranted.
- Contusions to the brain can occur at the direct site of trauma or may occur on the opposite side of the patient's brain (a contrecoup injury).
- In patients with a history of previous injury, larger intracerebral hemorrhages may occur.

Concussion

- Blunt head injury with associated transient loss of consciousness.
- The patient will usually have a nonfocal neurologic examination.
- Amnesia and occasional cognitive impairment are classic findings.
- Most patients make a complete recovery; however some may develop a postconcussion syndrome.
- Post-concussion syndrome consists of headache, vertigo, insomnia, anxiety and memory. impairment that may last for many weeks to months after the initial injury.
- This syndrome is considered to be a mild form of diffuse axonal injury.
- CT scan of the head in these cases is normal.

Diffuse Axonal Injury (DAI)

- DAI is a common cause of prolonged coma after head blunt head injury.
- Mortality rate is approximately 30%.

- The injury itself is a shearing type injury causing microscopic neuronal injury diffusely throughout the brain.
- The patient is found to be comatose with an otherwise nonfocal neurologic examination.
- CT scan of the head will reveal nonspecific findings such as evidence of edema, indistinct gray/white matter interface and loss of cortical sulci. The ventricles may also be compressed.
- All of these findings are secondary to the diffuse axonal tissue injury and edema formation.
- Autonomic dysfunction is commonly seen in these patients.

Clinical Evaluation of the Traumatic Head Injury Patient

- Primary survey and secondary surveys as previously discussed.
  - It is of paramount importance that patients with a head injury not become hypoxic or hypotensive both of which can lead to secondary brain injury.
  - All patients with a GCS of 8 or less should be endotracheally intubated using rapid sequence technique.
  - Patients who have higher GCS scores should be intubated if there is any potential risk of aspiration or airway compromise.
  - Intravenous access should be obtained immediately and volume resuscitation initiated with crystalloid (NS or LR) as clinically indicated.
    - If the patient is not volume depleted, avoid the administration of excess intravenous fluid which can exacerbate the development of cerebral edema.
    - In the patient who is volume depleted, after initial volume administration has stabilized the patient, be judicious with additional fluid administration.
    - Hypotonic fluid should be avoided for volume resuscitation.
- Laboratory studies should include:
  - CBC, basic metabolic panel and coagulation studies should be obtained.
  - Bedside glucose determination (Accucheck) and hemoglobin (Hemacue) should be obtained.
  - Other laboratory tests may be indicated based on physical findings, history and the possibility of additional traumatic injuries being present.
- Radiographic evaluation should include:
  - A CT scan of the head should be obtained as soon as possible.
  - Additional radiographic studies are indicated as dictated by the clinical situation.
  - Have a low threshold for obtaining cervical spine radiographs in the head-injured patient.

Treatment of the Head-Injured Patient

- If any evidence of an increased ICP is present aggressive management to lower the ICP should be initiated.
  - The head of the patient's bed should be elevated 30 degrees assuming that there is no contraindication such as hypotension or spinal injury.
  - In the intubated patient it was previously thought that hyperventilation was useful to decrease the ICP. Recent studies however show that routine hyperventilation can actually lead to worsening of neurological outcome. Although hyperventilation does cause a decrease in cerebral blood flow, in many cases it does so to the point of causing cerebral ischemia. Current guidelines only recommend hyperventilation in the setting of impending herniation.
  - Osmotically active agents such as mannitol can be helpful in lowering ICP. The dosage for mannitol is 0.25-1.0 g/kg intravenously. The loop diuretic furosemide can be given along with mannitol to potentiate its effects in lowering ICP.
  - The administration of hypertonic saline has shown some promise in lowering of ICP, but at this time further clinical studies are needed to clearly delineate its potential usefulness.

- Blood transfusion should be given in the setting of severe anemia to maximize cerebral oxygen delivery.
- There are no clinical benefits obtained by the administration of steroids to the patient with traumatic brain injury.
- Appropriate intravenous sedation and analgesia is extremely important in the endotracheally intubated patient.
  - The restless and agitated patient will develop increased ICP and by carefully sedating the patient ICP increases can be limited.
  - It is important not to use neuromuscular blocking drugs for paralysis in patients who are not adequately sedated, or significant increases in ICP will occur.
  - The patient who is not adequately sedated and is pharmacologically paralyzed will be noted to have an increased heart rate, blood pressure and ICP.
- Prophylactic anticonvulsant medication is recommended for high risk patients with severe traumatic brain injury.
  - The most commonly used medication for this is phenytoin (Dilantin).
  - Acute seizure activity should be treated immediately with lorezepam or diazepam both of which are benzodiazepines.
  - Seizure activity itself can cause a transient rise in intracranial pressure.
- Neurosurgical consultation should be arranged as soon as possible. This may require transfer to specialized facilities.
- For the patients with mild traumatic brain injury if their GCS is 15, normal neurologic exam and their CT scan is normal they may be discharged home with appropriate head injury instructions and follow-up.

Neck Trauma

- Approximately 5-10% of traumatic injuries involve the neck.
- The most critical and immediately life-threatening problem is airway obstruction and massive hemorrhage.
- Approximately 3,500 people die each year from neck trauma due to suicides, hangings and accidents.
- Most blunt trauma is caused by motor vehicle accidents (driver hitting the steering wheel or dashboard) commonly resulting in laryngotracheal injuries.

Critical Landmarks

- A determination should be made as to whether or not the **platysma muscle** has been violated. Platysmal violation is often a clue to damage and injury to deeper structures.
- Zones of the neck (Table 16.3)
  - In order to properly evaluate, diagnose and prioritize neck injuries, the neck is divided into three anatomic zones.
  - **Zone I** is bounded by the clavicles inferiorly and the cricoid cartilage superiorly and contains many crucial structures including: the lung apices, trachea, aortic arch, the great vessels, esophagus, cervical spine and spinal cord.
    - Injuries in this zone are often difficult to access and repair and carry the highest morbidity and mortality.
  - **Zone II** is bounded by the cricoid cartilage inferiorly and the angle of the mandible superiorly.
    - Crucial structures in this zone include: the trachea, larynx, esophagus, carotid arteries, jugular veins, vertebral vessels, cervical spine and cord.
    - This zone is easily surgically accessible, and an oblique neck incision can often be used with minimal morbidity.
  - **Zone III** is bounded by the angle of the mandible inferiorly and the base of the skull superiorly.

Table 16.3. Summary—Zones of the neck

Zones	Boundaries	Critical Structures	Key Points
III	Base of skull superiorly Angle of mandible inferiorly	trachea, esophagus, pharynx, vertebral arteries, salivary/parotid glands, distal internal carotid arteries, jugular veins, cranial nerves (9-12)	• Surgically difficult to access due to problems with obtaining proper exposure (may necessitate disarticulating the mandible)
II	Angle of mandible superiorly Cricoid cartilage inferiorly	trachea, esophagus, larynx, jugular veins, common carotid arteries (internal and external branches), vertebral vessels, cervical spine, cervical spinal cord	• Low morbidity, best prognosis • Easily accessible structures • Low threshold for surgical exploration • Internal jugular vein and common internal carotid arteries are the most common vascular injuries (9 and 7%, respectively)
I	Cricoid cartilage superiorly Clavicles inferiorly	trachea, esophagus, lung apices, aortic arch, great vessels, proximal carotid arteries, thoracic duct, major cervical nerve trunks, vertebral arteries, cervical spine, cervical spinal cord	• Highest morbidity and mortality • Injuries difficult to access and repair

- Structures in this zone include: the trachea, pharynx, esophagus, vertebral arteries, internal carotid arteries and cranial nerves.
- Injuries in this region are surgically difficult secondary to problems with obtaining proper exposure.

Evaluation of Neck Trauma

- **Early intubation** is preferred for proper airway management.
- An expanding hematoma or laryngeal edema can occur quickly causing airway compromise or occlusion leading to significant morbidity.
- **Neck wounds should NOT be explored in the emergency department due to the risk of dislodging a clot and disrupting hemostasis.**
- In patients with neck trauma, evaluation for arterial injury should be performed. The majority of injuries due to penetrating neck trauma are vascular injuries.
  - The internal jugular vein and common and internal carotid artery are the vessels most frequently injured with frequencies of 9% and 6.7%, respectively.
  - Vertebral artery injury is rare. Typically, this lesion is clinically unsuspected and incidentally identified on angiography. In hemodynamically stable patients, supportive and expectant management is advocated. Definitive treatment is required in patients with persistent bleeding, arteriovenous fistula formation or pseudoaneurysm.
  - Clues to arterial injury include: expanding hematoma, pulsatile bleeding, shock unresponsive to fluids, presence of a new bruit or thrill, diminished distal pulses.
    - Patients with the above findings often require immediate surgical intervention.
- A high index of suspicion should be maintained for esophageal injury secondary to blunt trauma.

Diagnostic Studies

- Most patients require a three-view cervical spine series.
- Plain films can demonstrate subcutaneous emphysema, fractures, tracheal deviation, and foreign bodies (e.g., bullet fragments).
- A **chest X-ray** is especially important to evaluate Zone I injuries since this region includes the lung apices. A pneumothorax, hemothorax, subcutaneous emphysema (due to an associated pneumothorax or injury to the larynx, trachea or esophagus), widened mediastinum (due to injury to a major mediastinal vessel) or foreign body may be visualized.
- **Angiography** is especially useful in evaluating Zone I and III injuries in hemodynamically stable patients with platysmal violation.
- The use of color flow doppler is increasing since it is noninvasive and relatively inexpensive. However, this technique is operator dependent and its role in assessing vascular injuries is still unclear.
- The **CT scan** is an important tool for diagnosing laryngeal injuries.
- The role of MRI in penetrating neck injuries is still being evaluated.
- Endoscopic evaluation of the trachea and/or esophagus should be performed in patients at risk for injuries to these structures. Esophagography (70-80% sensitivity) is important for evaluating esophageal injuries; there is a 17% mortality rate after a 12 h delay in diagnosis of esophageal injuries.

ED Management

- **Hemodynamic stability**: Hemodynamically stable patients may undergo a diagnostic evaluation depending on the zone involved (see below) and may not require surgical evaluation. However, hemodynamically unstable patients with neck injuries in any zone require immediate surgical intervention.
- **Zone I**: Injuries in this zone are often evaluated using angiography, esophagography or endoscopy (laryngoscopy or bronchoscopy). If the result of any of the above studies is negative the patient is observed and managed medically. If a study result is positive, surgical evaluation should be obtained.
- **Zone II**: There are two alternatives for evaluating penetrating Zone II injuries in a hemodynamically stable patient. In the past, any injury with platysmal violation necessitated mandatory surgical intervention. However, this approach is losing favor due to the high negative exploration rate; many centers now favor selective management involving endoscopy, esophagography and angiography as indicated to determine the need for surgical intervention.
- **Zone III**: Injuries in this zone are most commonly evaluated by a thorough oropharyngeal examination, as well as laryngoscopy and angiography as indicated.

Strangulation/Hanging

- **Pathophysiology**: There are various mechanisms by which death due to hanging or strangulation occurs:
  - Cervical spine fracture and transection of the spinal cord (person drops a large distance with feet not touching the floor)
  - Complete airway and arterial occlusion (compression of crucial structures with loss of consciousness)
  - Cardiac arrest (increased vagal tone and carotid sinus activation)
- **Signs/symptoms**: There are many signs and symptoms of strangulation such as ecchymoses, lacerations, abrasions, odynophagia, hoarseness and stridor. Petechial hemorrhages, known as Tardieu's spots, may be present especially on the skin and subconjunctivae.
- **Diagnosis**: Cervical spine injury is rare in nonjudicial hangings and diagnosis will often be made on the basis of the history or mechanism of injury. A lateral cervical

spine film may also demonstrate a Hangman's fracture (bilateral C2 pedicle fracture with anterior displacement of the C2 vertebral body).

- **Treatment**: The main priority of treatment is directed toward maintaining and, if needed, securing the airway. Cervical spine immobilization should be performed when indicated. Cardiopulmonary monitoring should be performed and patients should typically be observed for a minimum of 24 h. Finally, mental health providers should be involved to address the psychiatric component in cases of attempted suicide.

Maxillofacial Trauma

- Maxillofacial trauma typically occurs as a result of blunt injury to the face with motor vehicle accidents causing 50-70% of maxillofacial fractures.
- It is important to follow a systematic approach when assessing these patients since traumatic facial injuries can often distract from identifying other potentially life-threatening injuries.
- Airway management is critical in these patients.

Mandibular Fracture

- Mandibular fractures account for 10-25% of facial fractures.
- Over half of patients with mandible fractures have fractures at more than one site on the mandible.
- The mechanism of injury is often due to falls or altercations.
- On physical examination assess the patient for pain with jaw movement, normal teeth approximation, external or intraoral lacerations, ecchymosis under the tongue, trismus, numbness of the lower lip or positive tongue blade test.
    - The tongue blade test is performed by having the patient bite down on a tongue blade. The physician then tries twisting the tongue blade in an attempt to break it. A patient with a mandibular fracture will immediately open their mouth when you try to twist the blade. This constitutes a positive tongue blade test.
- Plain films (panorama view) should be obtained in patients with positive findings on physical examination or in those with a high index of suspicion for mandibular fracture.
- In patients with negative radiographs but high clinical suspicion, a CT scan should be considered and is especially useful for diagnosing condylar fractures.
- Patients with missing teeth should also receive a chest X-ray to exclude aspiration.
- Definitive treatment includes immobilization and establishing proper occlusion using open or closed reduction.
- These fractures are often contaminated with oral flora and patients should be treated with prophylactic antibiotics (penicillin G or clindamycin if pen-allergic) and tetanus immunization.

Zygomatic Fracture

- **Arch**—Fractures of the arch often occur at multiple sites.
    - Patients may present with a variety of symptoms including trismus (due to impingement of the temporalis muscle or masseter muscle injury) or cheek pain.
    - Depression of the malar eminence can also be seen although edema can often mask this finding.
- **Tripod**—A tripod fracture is the most common zygoma fracture and is found along three margins: the zygomaticofrontal suture, the zygomaticomaxillary suture, and the zygomatic arch.
    - This type of injury occurs as a result of a direct blow to the cheek often extending through the orbital floor.
    - Sensory deficits are common and include anesthesia or paresthesia of the anterior cheek, upper lip, or lateral aspect of the nose due to infraorbital nerve involvement.

- Marked periorbital edema and ecchymosis, loss of cheek prominence and subconjunctival and scleral hemorrhages are common findings of tripod fractures. With inferior displacement of the zygoma, depression of the lateral canthus may be noted.
- **Body**—Fractures of the body result from extreme force.
  - This type of fracture clinically mimics tripod fractures and has many of the same signs and symptoms as noted above.
- Radiologic Evaluation
  - Thin section axial and coronal CT scan has become the gold standard for evaluating zygomatic injuries. In facilities without CT scanning ability, plain films may be obtained.
    - For suspected arch fractures the underexposed submental view (bucket handle view) is best.
    - Tripod fractures can be evaluated with the Waters view which allows visualization of the inferior orbital rims, maxillary sinus, and the maxillary portion of the zygoma.

Frontal Sinus/Bone Fractures

- Frontal sinus fractures are the third most common facial fracture and can occur due to trauma.
- The clinical presentation is varied and can include:
  - Edema and/or hematoma of the glabellar region
  - CSF rhinorrhea (due to posterior wall fracture of the frontal sinus with dural penetration)
  - Supraorbital region depression
  - Lacerations
  - Supraorbital and/or supratrochlear nerve involvement
  - Down and forward globe proptosis
- The imaging technique of choice to evaluate this type of injury is thin section axial and coronal CT scan.
  - An intracranial pneumocele is evidence of dural violation and neurosurgical consultation should be obtained.
- Treatment for this type of injury ranges from observation to open reduction (e.g., nasofrontal duct involvement or displaced fractures) (Table 16.4).

Orbital Fractures

- The maxilla and ethmoid bone comprise the inferomedial aspect of the orbit. This region is often the weakest and thus most prone to orbital blow-out fractures.
- Approximately one-third of orbital blow-out fractures have an associated eye injury.
- The imaging technique of choice is an orbital CT scan.

Maxillary Fractures

- **LeFort I**—The fracture line extends horizontally above the teeth separating the lower maxilla from the upper face. The fracture line typically transects the maxillary sinuses.
  - This type of fracture does not typically cause hypesthesia since the fracture line is well below the infraorbital nerve.
  - Physical examination findings often include: facial swelling, mobile hard palate, epistaxis, fractured/avulsed teeth or malocclusion.
- **LeFort II (nasomaxillary fracture)**—The fracture line separates the nasomaxillary complex from the upper face and passes through the lacrimal bones, inferomedial orbital walls, and posterolateral maxillary sinuses.
  - This type of fracture often extends through the zygoma; hypesthesia is common due to infraorbital nerve involvement.

Table 16.4. Management of frontal sinus fractures

Fracture	Disposition	Treatment Required
Nondisplaced anterior wall fracture	Discharge (Follow-up in 1-2 wk for surgical evaluation and repeat films)	1. **ANTIBIOTICS** 2. avoid valsalva maneuvers or further trauma
Displaced anterior wall or sinus floor fracture	Admission	1. **ANTIBIOTICS** 2. surgical evaluation exploration
Posterior wall fractures	**ADMISSION**	1. **ANTIBIOTICS** 2. immediate neurosurgical evaluation
Maxillary fracture (inferior wall)	Inferior rectus muscle entrapment	Upward gaze dysfunction
Ethmoid fracture (medial wall)	Medial rectus muscle entrapment	Lateral gaze dysfunction

- Physical examination findings may include: facial edema, telecanthus (>45 mm), epistaxis, bilateral subconjunctival hemorrhages or CSF rhinorrhea.

- **LeFort III**—This fracture results in craniofacial separation extending through the nasofrontal and frontomaxillary sutures including a fracture line through the floor of the orbits.
  - Physical examination findings often include: "dishface" deformity (face appears elongated and flattened), anterior open bite (due to posterior maxillary displacement) or CSF rhinorrhea.
- All LeFort fractures commonly have blood in the maxillary sinuses.
- Radiologic evaluation—Coronal CT scan has replaced plain films as the study of choice for evaluating LeFort fractures. The CT scan is superior in delineating these injuries since patients rarely have an isolated LeFort fracture and are more likely to have multiple fractures. If CT scan is not available, plain films should be used.
- Treatment
  - The primary survey (ABCs) should be the focus of the initial evaluation of facial injuries.
  - Early intubation is critical in order to secure the airway since overwhelming edema can distort the anatomy necessitating cricothyroidotomy.
  - IV antibiotics are standard
  - Definitive care involves ENT, OMF, and plastic surgeons for repair of these complex injuries.

Nasal/Nasoethmoid Fractures

- Nasal bone fractures are typically diagnosed based on history and physical alone.
- Plain films with special nasal views can be obtained although they are rarely useful. Plain films are often useful to delineate injury or fractures to **adjacent** sites.
- Less severe injuries can be managed by closed reduction with nasal packing to provide internal support and a splint for external support.
- Patients with a simple nasal fracture may be discharged home with follow-up in 5-10 days for further evaluation when the swelling has improved.
- Nasoethmoid fractures should be suspected if the history and physical support a nasal bone fracture with additional evidence of ethmoidal bone involvement (e.g., CSF rhinorrhea or telecanthus).

- Most CSF leaks resolve in 24-48 h and are self-limiting.
- CSF leaks occur as a result of a cribriform plate fracture with dural penetration.
- Patients with this type of injury typically require admission for observation secondary to the risk of a brain abscess or meningitis.
- The use of antibiotics is controversial.

- Thin section axial and coronal CT scan is the radiographic study of choice to evaluate nasoethmoid fractures. Plain films are rarely useful.
- **Septal hematoma:** This manifests as asymmetry or widening of the nasal septum. There may be local discoloration although this unreliable as a means of diagnosis. If not adequately drained, the hematoma can become infected and cause subsequent necrosis and deformity of the septum.
  - Hematomas are drained either via needle aspiration or incision. It is important to make sure that all blood has been expressed from the hematoma.
  - After drainage, both nares are packed so that blood does not reaccumulate.
  - These patients need antibiotics and close follow-up.

TMJ Dislocations

- A temporomandibilar joint dislocation can happen for many different reasons including jaw trauma or excessive opening of the mouth.
- In this type of dislocation, the jaw typically deviates away from the side of the lesion since the condyle of the mandible is trapped anterior to the articular eminence. This is often due to a muscular spasm.
- Unless there is a high index of suspicion for possible mandibular fracture, X-rays are not usually indicated.
- Manual mandibular reduction can be performed in the emergency department.
  - Benzodiazepines or local anesthetic may facilitate the success of the reduction.
- The patient should be placed in a Barton bandage and instructed to consume a liquid diet for several days and follow-up with an oral surgeon.

Thoracic Trauma

- Thoracic trauma accounts for approximately 20% of all deaths secondary to trauma. Many of these deaths are preventable if they are recognized and treated early on in the trauma evaluation process.
- Approximately 10-15% of patients with thoracic trauma require operative exploration and intervention.
- The most common causes of death in the immediate post-injury period are usually related to a major vascular injury or a major airway obstruction.

Rib Fractures

- Simple rib fractures represent a relatively common occurrence in the setting of blunt thoracic trauma with an incidence of approximately 50%.
- The most common sites for simple rib fractures to occur is at the site of impact or at the posterior angle of the rib, the area of the ribs' greatest structural weakness.
- Ribs 4-9 are the ones most often fractured.
- The presence of fractures involving ribs 9-12 should alert one to the possibility of an associated intra-abdominal injury due to their location over the liver and spleen.
- Fractures of the upper ribs 1-2 usually indicate a significant amount of blunt force has been applied to the patient's thorax as these ribs are well protected by other structures, and the presence of these fractures increase the chance of significant intrathoracic injury.
- Rib fractures occur more commonly in the adult patient population because of the relative inelasticity of the rib cage in adults when compared to those of pediatric patients.

Clinical Evaluation

- The major goal is to detect the presence of underlying thoracic injuries such as pneumothorax, hemothorax, pulmonary contusion or significant vascular trauma.
- Clinical findings typically noted with rib fractures are point tenderness at the fracture site, increased pain with respiration or truncal motion, crepitus at the point of tenderness, ecchymosis, splinting and muscle spasm over the affected area of the chest wall.
  - When the involved rib is compressed at another anterior or posterior location distal to the fracture site pain is illicited at the site of the rib fracture.
- An upright PA chest radiograph and pulse oximetry reading should be obtained on every patient with suspected rib fractures. A "rib series" is generally not indicated as it provides little additional information.
  - Approximately 50% of rib fractures are not visualized on the initial chest radiograph and trauma to the cartilaginous areas of the thoracic cage is generally not visualized thus making the identification of rib fractures a purely clinical diagnosis in a large percentage of cases

Treatment

- The major goal in the treatment of simple rib fractures is adequate pain control. Inadequately treated pain may interfere with the patient's ability to have normal ventilatory function. This decrease in ventilation can lead to atelectasis and possible pneumonia.
- The use of rib belts and various types of chest wall binding devices is not routinely recommended. While providing symptomatic relief, these devices restrict normal chest wall motion and ventilation putting the patient at further risk for developing atelectasis.
- In some cases for relief of severe pain an intercostal nerve block using a long-acting local anesthetic agent is indicated to assist with pain control.
- Hospital admission is generally reserved for those patients at risk for respiratory compromise such as elderly patients and those with significant underlying pulmonary disease.
- In the vast majority of patients rib fractures heal in a 3 to 6 wk period without complications.

Flail Chest/Pulmonary Contusion

- Flail chest is the fracture of three or more adjacent ribs at two or more locations creating a segment of the chest wall that has no bony continuity with the rest of the patient's chest wall.
- This free floating segment of chest wall allows paradoxical chest wall motion of this segment compared to the remainder of the chest wall which in turn leads to inadequate ventilation.
- The major complication is direct trauma to the underlying lung parenchyma, which leads to the development of a pulmonary contusion.

Clinical Evaluation

- Flail chest is not always readily apparent due to chest wall splinting.
- As the splinting begins to gradually resolve secondary to progressive muscle fatigue, the paradoxical chest wall motion may be noted.
- The use of a tangential light source or palpation of the patients chest wall may be useful in helping to detect the early subtle presence of paradoxical chest wall motion.
- The patient may also be noted to have chest wall tenderness, crepitus and ecchymosis.
- Dyspnea may also be present.
- Chest radiograph may demonstrate the presence of multiple rib fractures in a pattern suggestive of the presence of a flail chest.
- Evidence of a pulmonary contusion may also be radiographically noted although this finding often is not present initially.

- An arterial blood gas demonstrating evidence of hypoxia may also give contributory evidence to the presence of a flail chest and/or a pulmonary contusion.

Treatment
- Initial therapy should be directed towards the ABCs of trauma resuscitation.
- Administration of supplemental oxygen, pulse oximetry and cardiac monitoring are necessary.
- Adequate ventilation and oxygenation must be maintained.
  - Intubation may be required.
- The placement of a thoracostomy tube is indicated in the presence of an associated pneumothorax or hemothorax.
- If hypotension or signs of hypovolemic shock are not present, it is prudent to avoid the overly aggressive use of crystalloid intravenous solutions which may lead to overhydration with subsequent worsening of any pulmonary contusion that may be present.
- It is extremely important to continually reassess the patient with flail chest or pulmonary contusion looking for any early signs of respiratory compromise or hypoxia and initiating rapid aggressive treatment.
- Appropriate use of parenteral analgesics is indicated for adequate pain control.

Sternal Fractures /Blunt Myocardial Injury (Myocardial Contusion)
- The fracture of the sternum requires a significant amount of energy to be directly applied to the anterior thoracic wall.
  - The most common mechanism for this to occur is in a motor vehicle accident when the patient's anterior chest contacts the steering wheel or other structures including the seat belt at a high rate of speed.
- The major concern regarding sternal fractures is not the fracture itself but the possibility of injury to mediastinal structures such as the myocardium and the aorta.
- Blunt myocardial injury/myocardial contusion, which represents myocardial injury secondary to the blunt force applied to the anterior chest, may occur with or without associated sternal fracture.
- Fortunately blunt myocardial injury occurs relatively infrequently.
- Recent study data indicates that sternal fractures can occur in up to 3% of motor vehicle accidents, and the associated morbidity and mortality rates are much lower than previously believed. The mortality rates for an isolated sternal fracture is approximately 0.7%.

Clinical Evaluation
- The patient typically complains of pain over the sternal area that increases with respiration or truncal motion.
- Point tenderness is usually illicited over the area of sternal injury.
- There may be in some cases signs of external trauma such as ecchymosis, crepitus and abrasions.
- A lateral chest radiograph is the best way for detecting a sternal fracture as they are generally difficult to visualze on a standard radiograph.
- The fracture may also be seen on a CT scan of the chest.
- Evaluation of underlying cardiac contusion includes an EKG and cardiac enzymes.
  - The presence of a normal EKG and cardiac markers has been demonstrated to indicate patients at low risk for cardiac contusion.
  - Continuous cardiac monitoring of high risk patients is indicated as cardiac arrythmias are the major complication of blunt cardiac injury.

Management
- The patient should receive adequate analgesia during the course of their emergency department evaluation.

- In selected cases after short term monitoring and observation patients may be discharged home with oral analgesic agents and appropriate close follow-up.
- Patients with dysrhythmias and EKG changes should be admitted for observation and further cardiac evaluation including echocardiogram and cardiac enzymes for possible blunt myocardial injury/myocardial contusion.
- Some of the dysrhythmias that may be seen in the setting of blunt myocardial injury include unexplained sinus tachycardia, premature ventricular complexes, atrial fibrillation and bundle branch blocks. Acute ST segment EKG changes can also occur.
  - Dysrhythmias are treated only if the patient becomes unstable.

Pneumothorax

- Pneumothorax is an abnormal accumulation of air in the pleural space that can occur secondary to blunt or penetrating trauma and can even occur spontaneously.
- This is one of the most common serious thoracic traumatic injuries.
- The presence of a pneumothorax will cause varying degrees of lung collapse on the involved hemithorax and has the potential for interfering with adequate oxygenation and ventilation.
- An **open pneumothorax**, also know as a communicating pneumothorax, is caused by penetrating trauma that violates the integrity of the patients chest wall and the pleura causing air to directly enter the pleural space leading to lung collapse.
- A **simple, closed** or **noncommunicating pneumothorax**, which may occur secondary to blunt trauma presents with an intact chest wall.
- A **tension pneumothorax** occurs in the presence of a one-way tissue valve that can occur with a traumatic penetrating injury of the chest wall or lung parenchyma.
  - During the process of respiration, air is forced into the pleural space but subsequently has no way to escape. As this process continues the intrapleural pressure on the involved side becomes progressively higher causing compression of the lung. The mediastinal structures are compressed and displaced to the opposite side of the patient's chest. This mediastinal compression decreases venous return to the heart and subsequently cardiac output falls. The opposite lung also becomes partially compressed, further interfering with adequate oxygenation and ventilation. This situation can lead to rapid development of cardiopulmonary arrest if not identified and treated immediately.

Clinical Evaluation

- The patient's clinical presentation may vary from asymptomatic to severe dyspnea depending on the amount of lung collapse or tissue injury.
- A decrease in breath sounds and hyperresonance to percussion on the involved side of the chest can be present.
- A tension pneumothorax presents with tracheal deviation, jugular venous distention, respiratory distress, tachycardia, hypotension,absence of breath sounds on the involved side of the chest, hyperresonance to percussion and cyanosis.
- All patients suspected of having any type of pneumothorax should be placed on supplemental oxygen, pulse oximetry, cardiac monitor and have a stat chest radiograph.
- Intravenous access should be obtained.

Treatment

- A tension pneumothorax should always be a clinical diagnosis as any delay in treatment can be life threatening. This includes any delay to obtain a chest radiograph.
- The initial treatment for a tension pneumothorax is to rapidly decompress the involved side of the chest by placing a 14 gauge angiocath in the second intercostal space in the midclavicular line.
  - Successful placement is met with a rush of air from the needle with improvement in the patient's symptoms.

- Subsequently, a thoracostomy tube (chest tube) should be placed and connected to a waterseal system for definitive treatment.

- In the case of an open or communicating pneumothorax, initial treatment should be directed at placing a sterile occlusive type dressing over the tissue defect on the patient's chest wall.
  - This occlusive dressing should be taped down on only 3 of its 4 sides serving a valve like function allowing air to escape during the exhalation phase of respiration and preventing air from entering during inspiration.
  - Subsequently, a thoracostomy tube should be placed along with surgical repair of the chest wall defect.
- A simple or noncommumicating pneumothorax is treated by thoracostomy tube in most cases.
- A small apical pneumothorax can be treated in some cases by hospitalization and subsequent observation without thoracostomy tube placement.

Hemothorax

- Hemothorax is an accumulation of blood in the pleural space.
- The traumatic event causes injury to the internal organs of the patient's chest and /or injury to vascular structures in the chest wall leading to the accumulation of blood in the pleural space.
- Lacerations of the lung parenchyma represent the most common intrathoracic organ injury that serves as the source of the hemothorax.
- Chest wall injuries most frequently involve the intercostal blood vessels or internal mammary arteries and these often serve as a source of persistent bleeding.
- Varying degrees of hemorrhaging may occur, and if the size of the hemothorax is significant enough compromise of respiratory function may develop along with hypovolemic shock.
- Pneumothorax frequently occurs simultaneously with a hemothorax.

Clinical Evaluation

- The clinical presentation of the patient with an acute traumatic hemothorax is often directly related to its size.
- The patient may present with clinical evidence of hypovolemia.
- Signs and symptoms of respiratory distress may be seen such as dyspnea, tachypnea, cyanosis and the use of accessory muscles of respiration.
- There may be a decrease in breath sounds on the involved side of the thorax, along with dullness to percussion.
- In some cases a decrease in the patient's oxygen saturation by pulse oximetry can be observed.
- The diagnosis will be verified by obtaining a chest film.
- It requires 200-300 ml of blood in an upright chest film to be able to visualize the presence of a hemothorax.
  - In a supine film a much larger volume of blood is usually required for visualization of the hemothorax, often close to 1000 ml.
- A bedside hemacue should be performed to allow rapid evaluation of the patient's hemoglobin level prior to the availability of the CBC results.

16

Treatment

- The presence of a small hemothorax may not require any specific treatment other than close observation.
- In most cases the presence of a significant hemothorax requires a thoracostomy tube (chest tube) be placed and connected to a waterseal system for drainage.

- Approximately 5% or less of all patients presenting with a hemothorax require a thoracotomy for control of intrathoracic hemorrhage.
  - Operative thoracotomy is indicated for persistant significant intrathoracic hemorrhage.
  - The presence of 1500 ml of blood with the initial thoracostomy tube placement, or continuing blood loss at 200-300 ml per h, are indications for a thoracotomy.

Traumatic Asphyxia

- Traumatic asphyxia is a relatively rare injury that occurs secondary to a severe crush injury to the patient's chest.
- Classically a purplish discoloration and/or petechiae are noted to the patient's upper thorax, neck and facial area secondary to retrograde venous blood flow caused by compression of the superior vena cava.
- Facial edema and subconjunctival hemorrhages also can be observed.
- The amount of force applied to the chest places this patient at risk for many of the intrathoracic injuries discussed elsewhere in this section. These injuries when identified should be treated accordingly.
- A CT scan of the chest is routinely indicated to rule out intrathoracic injuries.
- A CT scan of the head is indicated in the presence of any neurological deficits although intracranial hemorrhage is a rare occurrence in these patients.
- Transient loss of consciousness may occur in up to 30% of patients.

Pericardial Tamponade

- The pericardial space is very small and is extremely limited in its capacity to accumulate fluid.
- Abnormal accumulation of fluid in the pericardial space leads to the restriction of proper function of both the atria and ventricles. As the myocardium begins to fill with blood it significantly increases intrapericardial pressure thereby compressing the heart which limits both atrial and ventricular filling. This compression subsequently leads to decreased cardiac output.
- A volume of 150 ml in the pericardial space can create a pericardial tamponade.
- Pericardial tamponade can be secondary to either blunt or penetrating trauma.

Clinical Evaluation

- Pericardial tamponade should always be considered in the differential diagnosis for any patient presenting with thoracic trauma especially penetrating trauma.
- Clinical manifestations of pericardial tamponade can best be described by Beck's triad.
  - This triad consists of hypotension, jugular venous distention (JVD) and muffled heart tones.
- Dyspnea and tachycardia are usually also present.
- Pulsus paradoxus can also be a manifestation of pericardial tamponade.
- Bedside ultrasound performed by the emergency physician is rapidly becoming a popular diagnostic tool for making a quick diagnosis.
  - This ultrasound is normally performed as part of the routine trauma Fast Scan.
- A chest film should be obtained but is rarely diagnostic in the setting of an acute pericardial tamponade.
- It should be noted that tension pneumothorax presents with many of the same diagnostic features as pericardial tamponade and also occurs much more frequently.

Treatment

- Definitive treatment is an emergent thoracotomy. Evacuation of the constricting blood clot followed by repair of the underlying injury is necessary.
- Pericardiocentesis should be performed only as a temporizing measure until thoracotomy is performed.

Traumatic Aortic Injuries

- The most common cause of traumatic aortic injuries is sudden deceleration usually secondary to motor vehicle accidents or falls from a significant height.
- In many cases if the aorta is traumatically ruptured death occurs immediately at the scene.
- Those patients that survive to reach the emergency department have a chance for survival if the injury is rapidly identified and appropriate treatment is initiated.
- The location where these aortic injuries most commonly occur is in the descending aorta below the level of the left subclavian artery in the area of the ligamentum arteriosum.
- Many of those patients that survive the initial injury have partial tears of the wall of the aorta with only adventitia maintaining the integrity of the aortic wall.
- Blood in some cases can leak into the mediastinum forming a contained hematoma.
- Other serious intrathoracic injuries may occur simultaneously further complicating the patient's management.

Clinical Evaluation

- The mechanism of injury with regards to a trauma patient should raise our suspicions about the presence of a possible aortic injury. This is especially true in the presence of a history of sudden severe decelerating forces.
- Specific clinical findings are often difficult to identify because of other associated thoracic injuries occurring at the same time.
- Hypertension can initially be seen in some cases of aortic injury because of activation of stretch receptors at the area of the aortic isthmus secondary to aortic wall stretching secondary to the sheering injury.
- The presence of hypotension is an ominous sign indicating some degree of aortic rupture with hemorrhage.
- The patient may complain of chest pain or intrascapular back pain.
- The patient may have a loud systolic murmer heard throughout the precordium.
- A decrease in lower extremity pulses such as the femoral can be noted along with ischemic pain to the extremities.
- Paraplegia and ischemic stroke can occur in some cases.
- Dyspnea, dysphagia, voice hoarseness and stridor may occur secondary to compression of other intrathoracic structures by the aorta or periaortic hematoma.
- Obtaining a stat portable chest radiograph is extremely valuable in helping to make the diagnosis.
    - One of the most common findings suggestive of aortic injury is a widened mediastinum of >6 cm in an upright PA film and >8 cm ina supine AP film.
    - Although there are many other traumatic and nontraumatic causes of mediastinal widening, given the correct history and mechanism of injury a high level of suspicion is warranted until traumatic aortic injury is ruled out by further radiological evaluation. Further radiographic findings are listed in Table 16.5.
- If further radiographic evaluation is needed, a spiral CT scan of the chest is the next appropriate step and has in many cases eliminated the need for aortography.
- Unfortunately, spiral CT scans may be difficult to obtain in the extremely unstable patient.
- One effective option for use in the unstable trauma patient that brings diagnostic capabilities to the patient's bedside is the transesophageal echo (TEE).
    - A TEE is performed at the bedside in a manner similar to an endoscopic procedure.
    - The major contraindication to performing a TEE is the presence or suspicion of an esophageal injury.
    - The patient will require intravenous sedation, which in selected patients may require associated endotracheal intubation for airway protection.

Table 16.5. Chest radiographic findings associated with traumatic aortic injury

- Mediastinal widening—(most common finding)
- Deviation of the trachea to the right
- Obscured aortic knob/irregular aortic knob contour
- Depression of the left main stem bronchus
- Deviation of the nasogastric tube / esophagus to the right
- Mediastinal width/chest width ratio of greater than 0.25
- Hemothorax
- Left apical pleural cap
- Disappearance of the clear space between the aorta and pulmonary artery
- Right mainstem bronchus elevation

- Aortography is another option in the stable patient, but is being used with declining frequency since the spiral CT scan has become widely available.

Treatment

- Once the diagnosis is suspected or made of traumatic aortic injury immediate surgical consultation should be obtained.
- The patient should receive continuous cardiac, blood pressure and pulse oximetry monitoring.
- Two large bore intravenous accesses should be obtained should intravascular volume resuscitation be needed.
- The patient should be maintained on supplemental oxygen, and the airway and breathing should be managed as dictated by the individual clinical presentation.
- Routine lab work should be obtained along with serial hemoglobins.
- The patient should be typed and crossmatched for multiple units (6) of PRBC should massive hemorrhage occur.
- Pending operative repair of a diagnosed aortic injury, the patients blood pressure should be controlled between 100-120 mm Hg systolic.
  - By avoiding excessive blood pressure elevations, a decrease in shearing forces that are applied to the aorta during systole occurs.
  - Pharmacologic agents that are useful for blood pressure control include intravenous labetalol and esmolol because of their β-blocking properties.
  - Beta-blockers further decrease the shearing forces on the damaged aortic wall.
- Surgical consultation is necessary for definitive management with admission being directly to theoperating room or the SICU.

Esophageal Injuries

- The most common cause of esophageal injuries is penetrating trauma.
  - The most common traumatic cause of penetrating esophageal injuries is iatrogenic secondary to various procedures such as endoscopy.
- Blunt trauma is a relatively rare cause of esophageal injuries. In blunt trauma patients the usual mechanism of injury is that the gastric contents are forced into the esophagus in an explosive manner secondary to a severe sudden force applied to the abdomen.
- The esophagus is located in the central mediastinum protecting it from an isolated injury. Esophageal injuries therefore usually occur with significant injuries to neighboring structures.
- Any delay in recognizing and treating traumatic esophageal injuries causes a significant increase in a patient's morbidity and mortality.
- Injuries of other structures near the esophagus may cause the patient to present so dramatically that an esophageal injury can potentially be overlooked during the initial evaluation.

Clinical Evaluation

- Pain is present in virtually all cases of esophageal trauma.
- Many of the other clinical findings are nonspecific such as dyspnea, tachycardia, dysphagia, and odynophagia. Pneumothorax and pneumomediastinum may also occur. Subcutaneous emphysema may be noted on the upper chest and neck areas.
- Chest X-ray findings include: evidence of pneumothorax, left-sided pleural effusion, widening of the mediastinum and pneumomediastinum.
- A contrast esophagram using water soluble contrast such as Gastrografin is helpful in identifying esophageal injuries.
- Endoscopy is another diagnostic option in this situation and if available should be attempted first saving the esophagram to clarify equivocal findings noted on endoscopy.
- The use of CT scan in this situation is of limited diagnostic value usually only demonstrating indirect evidence in some cases of esophageal trauma. CT scan does however help evaluate other possible intrathoracic injuries that may have occurred simultaneously.

Treatment

- The treatment of patients with esophageal trauma involves appropriate airway management and volume resuscitation as needed.
- Evaluation for other injuries must be thorough and meticulous.
- Appropriate trauma laboratory evaluation should be obtained keeping in mind that there is in general no specific laboratory testing that identifies esophageal trauma.
- The presence of any associated pneumothorax or hemothorax should be treated in the usual manner.
- The patient must be kept NPO and prophylactic broad-spectrum antibiotic therapy should be initiated early.
- Emergent surgical consultation should be obtained and the patient admitted.

Abdominal Trauma

- When evaluating a patient for abdominal trauma, it is necessary to think of the abdomen as three separate areas, which consist of the peritoneal cavity, pelvis, and the retroperitoneum. There are a wide variety of complex structures in each of these areas each with their own unique problems when injured.
- The organs found in each of these areas can be broken down into two basic types, solid organs and hollow organs.
  - The liver, spleen, pancreas and kidneys are solid organs.
  - Hollow organs are small intestine, large intestine, stomach and bladder.
- The diaphragm must also be considered when evaluating abdominal trauma as subtle diaphragmatic injuries can often be missed.
  - It should be remembered that the diaphragm in expiration may rise as high as the fourth intercostal space.
  - The presence of intra-abdominal contents located under the lower thoracic cage means that clinical evidence of lower chest trauma places the patient at risk for injury to intra-abdominal structures in the upper abdomen.

Penetrating Abdominal Trauma

- Penetrating intra-abdominal injuries may be caused by direct contact of the injured structure(s) with the penetrating object or by blunt/sheering forces created by cavitation that occurs along the path of a high kinetic energy penetrating object (bullet).
- In general with penetrating abdominal trauma, the amount of tissue injury is directly related to the amount of kinetic energy that the penetrating object transmits to the patient's abdomen. The stab wound has much lower kinetic energy and generally only cause tissue injury to structures directly in its path.

- Stab wounds to the abdomen only require exploratory laporatomy in approximately 25% of patients.
- Gunshot wounds because of their much higher kinetic energy result in exploratory laporatomy in approximately 80-90% of abdominal trauma cases. The gunshot wound will not only injure structures directly in its path but because of the explosive cavitation effect may injure by secondary blunt force far from the bullets path. A bullet with a high level of kinetic energy may cause intraperitoneal tissue injury without ever having entered the peritoneum. Gunshot wounds therefore have much higher morbidity and mortality rates.
- Shotgun wounds represent a special case in the discussion of gunshot wounds.
  - Due to the rapid decrease in velocity of the shotgun pellets and the scattering of the pellets over a wide area, less tissue injury may occur. If the victim is more than 7 yd away from the shotgun, the pellets can in most cases only penetrate fascia and subcutaneous tissue. Certainly the closer the victim is to the shotgun at the time of discharge the greater the kinetic energy delivered per pellet and more pellets will make contact with the victim increasing the potential for tissue injury. A shotgun injury at <3 yd range will create tremendous amounts of tissue injury.
  - Explosions with fragmentation/shrapnel injuries behave similarly to shotgun injuries.
- Knowledge regarding the number of shots fired or stab wounds inflicted is useful.
- Looking at the relationships between entrance and exit wounds can give the emergency physician some estimate of the trajectory of the penetrating object which in many cases is suggestive of structures which may have been injured. This relationship is not 100% reliable as often bullets ricochet or fragment off various bony structures changing their direction of flight.

Blunt Abdominal Trauma

- The most common causes of serious blunt abdominal trauma in the United States are motor vehicle accidents and falls.
- Blunt abdominal trauma can occur through several mechanisms, which include direct blows or sudden rapid compression of the patient's abdomen. Other mechanisms include sheering forces caused by sudden rapid deceleration such as occurs in motor vehicle accidents or falls from significant heights.
- Solid organs are injured more frequently than hollow organs in blunt abdominal trauma.
  - Solid organs sustain burst type lacerations of their parenchyma secondary to the blunt force mechanisms described. The lacerations lead to hemorrhage with development of subsequent tachycardia, hypotension and other signs of hypovolemic shock.
  - The most commonly injured organ in blunt trauma is the spleen (solid organ).
- Hollow organ injury occurs secondary to rupture caused by compressive forces. The rupture of hollow intra-abdominal organs causes hemorrhage and also the contamination of the peritoneum with their contents.
- Many cases of trauma demonstrate simultaneous solid and hollow abdominal organ injuries.
- Vascular attachments may be torn or avulsed leading to further hemorrhage and possible ischemic parenchymal injury.
- It should be noted that the elderly and alcoholic patient populations are at increased risk for intra-abdominal injuries because of the decreased abdominal wall tone.

Clinical Evaluation

- Historical information should be obtained regarding the events and mechanism related to the trauma from the patient, family, friends, police, paramedics and any other available resource. This information is often helpful in evaluating the potential for certain types of traumatic injuries.

- One of the key decisions to be made rapidly in the abdominal trauma evaluation is whether this patient needs an emergent operative intervention such as an exploratory laparotomy.
- Clinical signs of shock may be present during the initial evaluation or can be a delayed finding. The importance of continual reassessment of the patient cannot be emphasized enough.
- The patient's abdomen should be carefully examined by inspection, auscultation and palpation.
  - Acute abdominal tenderness is a serious sign.
  - Look carefully for any evidence of ecchymosis, abrasions or penetrating traumatic wounds.
  - The presence of a seat belt sign described as a linear area of ecchymosis and/or abrasion located at the previous location of the seat belt/shoulder restraint indicates the application of a significant amount of force to the patient's abdomen. The presence of a positive seat belt sign should increase the suspicion of an underlying intra-abdominal injury.
  - Ecchymosis over the patient's flank area referred to as Gray-Turner sign suggests possible retroperitoneal hemorrhage.
  - Periumbilical ecchymosis (Cullen's sign) also is suggestive of retroperitoneal hemorrhage.
- The patient should be carefully log-rolled as a unit maintaining full cervical spine immobilization and there back carefully inspected and palpated for any evidence of trauma.
- The rectal examination should be performed while the patient is in this lateral position checking for the presence of blood, sphincter tone and in male patients whether or not their prostate is high riding.
- The presence of normal bowel sounds does not exclude the possibility of intra-abdominal trauma. Shock also can cause a secondary ileus leading to a decrease in bowel sounds even in the absence of any abdominal traumatic injury.
- It should be emphasized that physical findings may be difficult to identify in the patient with altered mental status secondary to illicit drugs or alcohol or in the patient with a spinal cord injury which can affect pain perception.
- Physical findings and mechanism of injury should direct laboratory and radiographic studies that are obtained.
- A rapid bedside hemoglobin should be obtained initially and serially to help evaluate the amount of blood loss and help guide any necessary blood transfusion.
- Basic abdominal trauma labs should consist of a CBC, Basic Metabolic Panel, urinalysis, PT/PTT, liver function tests, type and crossmatch, lipase and pregnancy test in females of child-bearing age.
- A bedside ultrasound (FAST scan) should be one of the first radiological evaluations obtained in an attempt to identify free fluid in the abdomen.
- Plain radiographs are of limited value in the evaluation of abdominal trauma.
  - Plain abdominal films are helpful in identifying the presence or location of an intra-abdominal foreign body.
  - A chest radiograph is helpful with regards to abdominal trauma in identifying free air secondary to a hollow organ injury or the possible herniation of abdominal structures into the thorax secondary to a diaphragmatic injury.
  - Pelvic x-ray may demonstrate pelvic fractures, which should increase the clinical suspicion for retroperitoneal injury.
- Diagnostic peritoneal lavage (DPL) is an option in the absence of the availability of bedside ultrasound to evaluate for intraperitoneal injury. DPL should be reserved for the hemodynamically unstable patient in the absence of bedside ultrasound. It serves as an alternative for these patients who by virtue of their injuries are too unstable to be

taken to obtain a spiral CT scan of their abdomen and pelvis. A positive DPL or FAST scan mandates operative exploration in most cases.

- The hemodynamically stable patient with suspicion of intra-abdominal injury should obtain a CT scan of the abdomen and pelvis in an attempt to identify any traumatic injuries.
  - Often these patients also require CT scans of multiple other body areas simultaneously as clinically indicated.
  - The advantage of spiral CT scan is its ability to clearly delineate specific injuries than either ultrasound or DPL. This greater ability at localizing and identifying specific intra-abdominal injuries gives the surgeon a preview of what to expect during surgery.
  - Also the CT scan allows evaluation of the retroperitoneal area which is not evaluated by either ultrasound or DPL.

Treatment

- The primary and secondary surveys along with appropriate resuscitation should be initiated as discussed previously in this chapter.
- The patient should be placed on cardiac, blood pressure and pulse oximetry monitors.
- Supplemental oxygen should be administered to the patient, and two large bore (14 or 16 gauge) intravenous lines should be established.
- In the patient manifesting signs of hypovolemic shock, initial volume replacement with crystalloid (normal saline or ringers lactate) intravenous solutions should be initiated.
- Blood, either O negative or type-specific, should be readily available for emergent transfusion when necessary.
- When necessary, surgical consultation should be obtained as early as possible.
- Indications for laparotomy include inability to stabilize a patient with signs of continued blood loss, evidence of peritonitis, evisceration, radiological evidence of significant organ injury.
- Other treatment modalities include angiographic embolization of bleeding sites. This is especially useful in pelvic and retroperitoneal hemorrhage.

Special Considerations

Diaphragmatic Injuries

- These injuries can be caused by either blunt or penetrating trauma with penetrating trauma being the most common cause.
- The most common location of injury is the posteriorlateral area of the left hemidiaphragm.
- Diaphragmatic injury secondary to blunt trauma is most commonly secondary to motor vehicle accidents.
- The most common radiographic findings on plain films are visible abdominal contents in the thorax on a chest film, indistinct diaphragmatic border, nasogastric tube in the lower left chest and focal atelectasis.
- Spiral CT scan and ultrasound have low sensitivity for detecting diaphragmatic injuries.
- In suspicious cases laparoscopy can be a useful diagnostic tool.
- In many cases diaphragmatic injuries may go undetected for years until the patient presents with delayed complications such as bowel obstruction.

Retroperitoneal Injuries

- Clinical signs and symptoms may be subtle in the presence of retroperitoneal injuries.
- Structures found in the retroperitoneum include the pancreas, kidneys, aorta, vena cava and some segments of the colon and duodenum.
- The area is difficult to evaluate by physical examination and peritoneal lavage or ultrasound does not evaluate this area.

- Spiral CT scan is the most effective method for evaluating the retroperitoneum in the stable patient.
- The mechanism of injury should increase our suspicion about possible retroperitoneal injuries.
- Duodenal hematomas or rupture are examples of possible retroperitoneal injuries that are often difficult to detect on initial evaluation and may take several hours before symptoms develop enough to arouse clinical suspicion.
- Because of the difficulty isolating retroperitoneal structures, treatment with angiographic embolization is the first treatment option.

GU/Pelvic Trauma

- Approximately 3-10% of all trauma patients have injuries involving the genitourinary system (GU).

Renal Injuries

- Of all GU injuries, renal injuries comprise the vast majority and are usually the result of blunt force trauma such as motor vehicle accidents.
  - **Renal contusions** are the vast majority (90%) of renal injuries.
    - Renal contusions maintain an intact renal capsule and range from subcapsular hematomas, small lacerations, to minimal parenchymal ecchymosis.
    - An intravenous pyelogram would be normal in such an injury and these injuries are typically minor.
  - **Renal lacerations** make up approximately 5% of renal injuries.
    - Renal lacerations are divided into two categories: minor and major.
      - Minor renal lacerations involve disruption of the renal capsule while sparing injury to the corticomedullary or collecting system.
      - Major renal lacerations involve disruption of the renal capsule including injury to the corticomedullary or collecting system.
  - **Renal pedicle injuries** account for 2% of renal injuries.
    - Renal pedicle injuries involve damage to the main renal vessels or their branches.
  - **Renal pelvis rupture** is rare and involves the collecting system resulting in urinary extravasation into the retroperitoneal space.
  - **Renal rupture** ("shattered kidney") accounts for 1% of renal injuries.
    - These patients often become hemodynamically unstable due to uncontrolled hemorrhage.

Radiographic Studies

- The imaging modality of choice in patients with suspected renal injuries is an abdominal/pelvis CT scan.
  - For most injuries, the CT scan has a higher sensitivity and specificity than intravenous pyelogram and carries the added benefit of being able to identify other intra-abdominal injuries.
  - Common indications for scanning a patient include: penetrating injuries, gross hematuria, microscopic hematuria with hemodynamic instability, hemodynamic instability or persistent hematuria.
    - Microscopic hematuria alone is not an indication for a CT scan since the study is often low yield in this setting.
- Intravenous pyelogram is still the study of choice for suspected ureteral injuries and is also useful to diagnose rupture of the renal pelvis.

16

Ureteral Injuries

- Ureteral injuries are uncommon and comprise approximately 6% of GU injuries. These often occur due to penetrating injuries and, in fact, are often iatrogenic, occurring

Table 16.6. Management guidelines for renal injuries*

Grade	Criteria	Comments
GRADE I	Renal contusion Microscopic or gross hematuria Subcapsular hematoma (nonexpanding) No parenchymal laceration	No intervention is required Supportive care Conservative management (bedrest, hydration, serial hematocrits, serial urinalyses, monitoring) 70% of renal injuries
Grade II	Parenchymal laceration involving the superficial cortex (<1 cm deep) No expanding hematoma No urinary extravasation	No intervention is required (usually have spontaneous resolution) Supportive care Conservative management 20% of renal injuries
Grade III	Parenchymal laceration >1 cm deep No involvement of the collecting system No urinary extravasation	Require admission +/- operative management**
Grade IV	Parenchymal laceration extending to the collecting system Urinary extravasation present Main renal vascular injury	Require admission +/- operative management**
Grade V	Pedicle/hilum avulsion Shattered kidney	Surgical intervention often requiring nephrectomy to control life-threatening hemorrhage

*Guidelines developed by the American Association for the Surgery of Trauma (J Trauma, 1989). **Absolute indications for surgical intervention include: Grade V injuries, hemodynamic instability, uncontrolled hemorrhage or expanding/pulsatile retroperitoneal hematoma.

during surgical procedures. The diagnosis of ureteral injuries is clinically difficult and the clinician must have a high index of suspicion. CT scan or retrograde pyelography are useful studies to detect ureteral injuries and surgical repair is required.

Bladder Injuries

- Approximately 80% of bladder injuries occur with pelvic fractures as a result of blunt trauma.

Bladder Contusions

- Nearly 100% of bladder contusions ("bladder bruise") are associated with gross hematuria without disruption of the bladder wall or urinary extravasation.
- These lesions typically require nonoperative management and may include catheter drainage for 7-10 days.

Bladder Rupture

- The classic triad for bladder rupture includes: inability to void, gross hematuria and abdominal pain/tenderness.
- There are two different types of bladder rupture: intraperitoneal (IP) and extraperitoneal (EP).
  - IP bladder rupture results in urinary extravasation into the peritoneal cavity after injury to the dome of the bladder; this can often lead to peritonitis.
    - This type of injury often occurs in patients experiencing trauma with a full bladder.
    - Surgical intervention is often required for this type of injury.
  - EP bladder rupture is more common than IP and results in urinary extravasation after injury to the lateral wall or base of the bladder.

- For small lesions, nonoperative management is possible with 7-10 days of catheter drainage with antibiotic prophylaxis.
- Retrograde cystography or a CT cystogram is useful to evaluate bladder injuries.

Testicular Injuries

- Testicular injuries often occur as a result of penetrating trauma (gunshot or stab wound) or blunt injury (kick or direct blow).
- In injuries involving penetrating trauma, surgical exploration is often required.
- Clinical features resulting from blunt trauma include: ecchymosis, pain, testicular tenderness or a scrotal mass (hematocele).
- Testicular ultrasound is useful for identifying testicular rupture (75% specific), hematocele or other testicular lesions.
- Appropriate urologic consultation is required.

Penile Fracture

- Penile fractures are rare and often occur as a result of trauma to an erect penis.
  - The majority of cases occurs during sexual intercourse and are associated with urethral injuries in approximately 23% of cases.
- A retrograde urethrogram is useful for evaluating suspected urethral injury.
- Symptoms of penile fracture may include: immediate pain followed by flaccidity, swelling or angulation of the penis with deviation away from the injured side.
- The fracture occurs due to injury involving Buck's fascia and the corpus cavernosum.
- Approximately 90% of penile fractures resolve spontaneously with conservative management (pain control and refraining from sexual activity).

Special Considerations

Pediatric Trauma

Introduction

- Traumatic injuries are the number one cause of mortality and morbidity in the pediatric patient population.
- The most common etiology of traumatic injuries in the pediatric patient is secondary to motor vehicle accidents either as a vehicle passenger, pedestrian or bicycle rider. The incidence of motor vehicle accidents as the cause of traumatic injuries in pediatrics is approximately 35% followed by trauma secondary to falls at 26%. Other causes of pediatric trauma that are commonly seen include drowning/near drowning, residential fires and assault with either nonpenetrating or penetrating trauma.
- The most frequent age group for pediatric trauma based on data collected by the National Pediatric Trauma Registry is between the ages of 5-9 years.
- The highest incidence of traumatic injuries is during the summer and spring seasons usually during daylight hours.
- It cannot be over-emphasized that the best way to treat pediatric trauma is through an active community education and prevention program focused on reducing the risks of trauma.
- In the evaluation and management of the pediatric trauma patient, it is important to remember that theses patients do not just represent small versions of the adult trauma patient. There are sufficient differences between the adult and pediatric trauma patient that the one algorithm concept of trauma management is in many cases ineffective and can even be hazardous. The purpose of this pediatric trauma section is to elucidate the major differences in the primary and secondary survey and management criteria (Table 16.7).

Table 16.7. Pediatric resuscitation formulas

- Patient <1 yr of age Weight in Kg = $\frac{\text{age in months}}{2} + 4$
- Patient >1 yr of age Weight in Kg = (2 times age in years) + 10
- Minimal Systolic Blood Pressure = (2 times the age in years) + 70
- Endotracheal tube size = Age in Years divided by 4 + 4
- Endotracheal tube placement depth in cm = 3 times ETT size
- NG tube size = 2 times ETT size
- Foley catheter size = 2 times ETT size
- Chest tube size = 4 times ETT size
- Initial crystalloid intravenous fluid bolus = 20 ml/kg
- Initial blood replacement 10 ml/kg
- Alternative to above is the use of Breslow Tape which will also supply commonly used resuscitation drug dosages

- The primary and secondary survey for the adult patient is discussed in detail elsewhere in this chapter. In the following section the significant differences with the pediatric trauma patient will be highlighted.

Primary Survey

Airway

- The pediatric airway has some important anatomical differences when compared to the adult which can at times make their airway management more challenging.
- The pediatric patient has a tongue and tonsillar tissue that are disproportionately larger compared to the size of their oropharynx making visualization of the vocal cords during endotracheal intubation much more difficult.
- The smaller mouth and orophayngeal size reduces the work area available thus limiting visualization and airway access.
- The larynx is in a much more anterior position and smaller making visualization of the vocal cords and subsequent ETT placement more challenging.
- The cricoid cartilage is the narrowest part of the pediatric airway, unlike the vocal cords in the adult. This cricoid narrowing serves as a functional endotracheal tube cuff in patients under the age of eight; therefore in this age group endotracheal tubes are uncuffed.
- Because of the small internal diameter of the pediatric airway, airway adjuncts and endotracheal tubes are more easily occluded with secretions.
- The pediatric epiglottis is larger and can be floppier interfering with visualization of the glottic opening (vocal cords).
- Until approximately 6 mo of age children generally are obligate nose breathers meaning that any nasal obstruction whether from secretions or otherwise must be treated to prevent subsequent development of hypoxia.
- The head size in the pediatric patient population is proportionally larger placing the supine patient naturally into a sniffing position.
- In the pediatric patient population, it is generally easier to perform endotracheal intubation using a straight laryngoscope blade because of their anterior larynx.
- It should be kept in mind that should a surgical airway be necessary in the pediatric patient, because of the small size of the cricothyroid membrane, a needle cricothyrotomy only should be performed in the patient <10 yr of age.

- The short length of the pediatric trachea makes it much easier to accidently intubate the patient's main stem bronchus.
- The use of uncuffed ETT in patients under age eight also predisposes the ETT to be more easily displaced.
- **It is important to emphasize that during all airway manipulations and procedures when clinically indicated cervical spine immobilization must be maintained.**

Breathing

- It is important to be able to recognize early on the pediatric patient who is in respiratory distress or failure or has the potential based on history and physical of developing these entities. The best way to treat a pediatric respiratory or cardiopulmonary arrest is prevention.
- Clinical evidence of the development of respiratory distress includes any or all of the following: tachypnea, the use of accessory musculature, retractions, nasal flaring, splinting and cyanosis.
- The child with an abnormally slow respiratory rate in the setting of trauma may be manifesting respiratory muscle fatigue and should be considered to be in respiratory failure. This must be considered a prearrest state and BVM ventilation commenced immediately with subsequent endotracheal intubation.
- Head bobbing is also frequently associated with respiratory muscle fatigue.
- Lethargy and/or alteration of mental status may be an indication of cerebral hypoxia either secondary to respiratory failure or to the development of a shock state.
- Supplemental oxygen should routinely be administered to pediatric trauma patients and pulse oximetry continuously monitored.
- Early intervention with assisted ventilation (BVM) and endotracheal intubation when clinically indicated is extremely important.
  - Infants require an assisted ventilatory rate of approximately 40 breaths per min and children 20/min.
  - The appropriate tidal volumes range from 7-10 ml/kg for both infants and children. Care must be taken to limit the amount of pressure generated in the pediatric airway during assisted ventilation. Excessively high ventilatory pressures can injury the fragile tracheobronchial tree leading to barotrauma such as pneumothorax.
- Hesitation and delay in the recognition and management of any compromise with the pediatric patient's airway or breathing can lead to a disastrous outcome. Early recognition and effective management are the keys to a successful outcome.

Circulation

- The pediatric patient has significant cardiovascular reserves and is able to sustain considerable blood loss before there is a significant change in the patient's vital signs.
  - The patient may actually be in a shock state and have a normal blood pressure which is referred to as compensated shock.
  - Uncompensated shock is when hypotension also occurs.
  - The blood pressure is usually the last vital sign to become abnormal in the pediatric patient in shock. Hypotension in the pediatric trauma patient is a late finding and usually indicates significant blood loss and/or inadequate volume resuscitation.
  - Tachycardia is one of the earliest signs of shock. The patient's pulse should also be evaluated for its strength and regularity.
  - The patient's capillary refill should also be assessed and normally is <2 seconds. In order to accurately measure a patient's capillary refill, the extremity to be tested should be elevated slightly above the level of the heart to avoid refilling of the capillary bed by venous backflow. Also capillary refill is only accurate in a warm environment. In a cold environment peripheral vasoconstriction occurs to maintain body core temperature which will abnormally prolong capillary refill.

    - Another indicator of perfusion is the comparison of distal to proximal pulses in the same extremity. The absence or decrease of more distal pulses in the presence of proximal pulses indicates severe peripheral vasoconstriction as the body attempts to maintain perfusion of vital organs is a sign of shock.
    - In shock the skin may become cool, pale and clammy as a result of vasoconstriction. The skin may also appear cyanotic in some cases that can be directly related to the shock state itself and/or respiratory failure with its resulting hypoxia.
    - The pediatric patient in shock may become lethargic, irritable or confused secondary to a decrease in CNS perfusion although we must keep in mind the many other traumatic and nontraumatic causes.
    - A decrease in urinary output may also be seen secondary to a decrease in renal perfusion.
- All of the above clinical manifestations of shock share one thing in common and that is that they are all indicators of a decrease in the patient's end organ perfusion. It is extremely important that shock is detected in the earliest stage possible and appropriate intervention is initiated immediately.
- One of the goals of initial therapy is to aggressively treat the patient's volume deficit and shock state and prevent the compensated shock from progressing to uncompensated shock.
- When shock is identified intravenous access should be obtained as rapidly as possible. At least two large bore IVs should be obtained (IV catheter size varies with patient).
- Intraosseus access is an excellent alternative should peripheral access become difficult. Intraosseus access used to be restricted to the patient <6 yr of age, but this is no longer true; there is currently no upper age limit for intraosseus access.
    - It is important that intraosseus access not be placed into an injured extremity.
    - Intraosseus access should be considered a temporary form of intravenous access, and once the patient has been stabilized and peripheral access obtained the intraosseus line should be discontinued.
    - The location of choice for intraosseus access is on the medial aspect of the proximal tibia just inferior to the level of the tibial tuberosity.
    - The complication rate of this procedure is low. Complications include cellulitis and osteomyelitis.
    - Central line intravenous access and venous cutdowns are other alternatives.
- The initial resuscitation fluid of choice is a crystalloid solution consisting of either normal saline or Ringers lactate.
    - The initial volume of resuscitation of crystalloids is 20 ml/kg as a bolus, which is approximately equivalent to 20-25% of the normal circulating pediatric blood volume.
    - This fluid bolus can be repeated as necessary based on the continuing reassessment of the patient.
        - If possible, warmed resuscitation fluids should be used to decrease the chance of developing hypothermia.
- If blood infusion becomes necessary, infused O negative or type-specific blood should be at 10 ml/kg.
- Any external hemorrhage should be controlled by applying direct pressure to the wound.

Disability

- Use the AVPU method

  A—Alert
  V—Response to verbal stimuli
  P—Response to painful stimuli
  U—Unresponsive

Exposure

- The patient should be completely undressed to receive a complete examination and evaluation.
- The pediatric patient is at higher risk for developing hypothermia than the adult during this process because they have an increased total body surface area and less subcutaneous fat stores.
  - Warm blankets and intravenous solution warmers should be available.
  - The development of iatrogenic hypothermia can worsen preexisting acidosis and also can induce a coagulopathic state through inactivation of various coagulation factors.

Resuscitation

- When performing the primary survey in the pediatric trauma patient, as in the adult, it is important to remember that the primary survey and the resuscitation phase are to be performed simultaneously. When compromise of airway, breathing or circulation is encountered, immediate resuscitative intervention should occur even if the entire survey is not complete.

Secondary Survey

- The secondary survey commences when the primary survey is completed and initial resuscitation is initiated. The secondary survey consists of a detailed history with a head-to-toe physical examination of the pediatric trauma patient.
- The patient's complete vital signs are obtained at this time.
- Appropriate radiological and laboratory studies are obtained.
  - A bedside ultrasound (FAST scan) can be performed to evaluate for evidence of intra-abdominal free fluid/hemoperitoneum or pericardial effusion.
- A nasogastric tube and Foley catheter should be placed as clinically indicated.
  - The nasogastric tube is important for gastric decompression.
  - In the pediatric patient population, aerophagia may occur with any degree of respiratory distress leading to gastric distention which limits diaphragmatic excursion further aggravating the patient's respiratory distress.
- A more detailed neurological exam including Glasgow Coma Scale should be performed at this time.
- Always consider transfer to a trauma center for further management of the pediatric trauma patient. It is helpful to have prearranged transfer agreements in place with trauma centers and tertiary care centers to help expedite the transfer process.

Elderly Trauma

- The Emergency Medicine healthcare provider must have a clear understanding of the physiologic differences that occur in the elderly trauma patient and how these changes affect resuscitation and stabilization in the Emergency Department.
- Elderly trauma patients account for only a small percentage (10%) of the total population of trauma patients yet they account for approximately 25% of all trauma-related healthcare expenditures in this country.
- The increased cost of care is related to many of the physiologic changes that occur with aging along with the increased incidence of various disease processes and comorbidities that exist in this population.
- The most common cause of trauma in the elderly patient is falls; approximately 40% of traumatic events in the elderly (Table 16.8).
- Burns are the third leading cause of trauma in the older patient population.
- Fractures and head injuries account for the largest percentage of injuries in this age group.

Table 16.8. Etiology of falls in the elderly

- Cardiac dysrhythmias
- Valvular heart disease
- Vasovagal syncope
- Myocardial ischemia
- Orthostatic hypotension
- Anemia
- Hypovolemia
- CVA/TIA
- Hypoglycemia
- Hypoxemia
- Medication side effects

- Penetrating trauma in the elderly occurs but to a lesser extent than blunt trauma but still has a disproportionately high morbidity and mortality rates compared to the general population.
- Elder abuse should always be considered as a potential source of geriatric trauma. Often under-recognized by healthcare providers when compared to spousal or child abuse, estimates of potential elder abuse range between 500,000 to 2,500,000 incidents per year in the United States.

Special Considerations

- Medications for more chronic medical problems may complicate the management of trauma in the elderly patient.
  - Beta-blockers and calcium channel blockers may blunt the hemodynamic response to hypovolemia.
  - Coumadin and antiplatelet drugs can increase the risk of hemorrhage and can make it more difficult to control.

Head Trauma

- Geriatric brain atrophy can lead to several important problems:
  - Cognitive/memory impairment, which may make it difficult to obtain an accurate history.
  - The risk of intracranial hemorrhage secondary to stretching of the bridging veins from the brain to the dural sinuses. This leads to subdural hemorrhage.
- Therefore CT should be used liberally in this population.

Spinal Injuries

- The presence of degenerative joint disease (DJD) and osteoporosis in the elderly increases their risk of sustaining spinal injuries.
- The more frequent incidence of spinal stenosis means a higher incidence of spinal cord injury with or without fracture or subluxation.
- Compression fractures can occur more often in the thoracic and lumbar spine than in the general population.
- The elderly population is more predisposed to the development of spinal cord contusions which can lead to cord syndromes especially central cord syndrome.
- In a symptomatic patient, much lower threshold must be maintained to obtain flexion-extension cervical spine radiographs and CT scans of the cervical spine in the elderly even in the presence of normal cervical spine radiographs.

Chest Trauma

- The chest wall in the elderly patient is much less flexible than in the younger patient, and these patients are more likely to sustain fractures of the bony structures of the thorax.
- Pneumothorax and hemothorax occur more frequently because of the increased incidence of rib fractures in this population.
- Delayed complications such as atelectasis, pneumonia and ARDS are also increased in frequency.
  - In some cases early endotracheal intubation should be considered at the first signs of respiratory insufficiency.

Abdominal Trauma

- The physical examination of the abdomen and pelvis is less reliable in the elderly population which should lead the clinician to more liberal use of radiographic imaging studies.
- Repeat serial examinations of the abdomen are important in all trauma patients but especially in this population.
- If intravenous contrast agents are to be used in imaging studies, the patient should be kept well hydrated to reduce the chances of the renal insufficiency.
- Overhydration should be avoided as it may exacerbate any cerebral edema or pulmonary contusions that are present.
- In the case of cardiac comorbidity, excess fluid may lead to respiratory compromise.

Treatment

- The primary and secondary surveys should be performed as discussed previously.
- Due to their general inability to tolerate physiologic stress as effectively as younger individuals, the elderly patient may decompensate from their traumatic injury much more readily.
- Transfusion should be performed in the face of anemia or dropping hemoglobin early on to maximize the oxygen carrying capacity of the blood.
  - This is necessary to avoid exacerbation of any ongoing ischemic processes in the elderly such as myocardial ischemia and strokes.
- Early and liberal consultation with a trauma surgeon and any other necessary consultants should be obtained.
- Hospital admission is necessary in most cases.

Trauma in Pregnancy

- In the evaluation and treatment of the pregnant trauma patient it is important to remember that fetal well being is dependent on effective resuscitation of the mother.
- The normal physiologic response to maternal trauma in pregnancy is self-preservation of the mother at the expense of the fetus.
  - Blood flow is shunted away from the uterus to maintain flow to the mother's vital organs in the presence of hypotension. The presence of this shunting can mask signs of clinical shock until maternal blood loss is approximately 30-35%.

- Initial treatment priorities for the pregnant trauma patient remain essentially the same as for the nonpregnant trauma patient with a few modifications which take into account the physiologic changes seen in pregnancy:
- An approximately 50% increase in circulating blood volume occurs during pregnancy, but there is a lesser increase in red blood cell mass leading to the physiologic anemia of pregnancy.
- Heart rate will also increase 10-15 beats/min during the course of pregnancy.
- Blood pressure decreases by 5-15 mm Hg in both systolic and diastolic readings usually starting in the second trimester.

- Supine hypotension may be seen in pregnancy secondary to the gravid uterus compressing the inferior vena cava thereby decreasing venous return.
  - Placing the patient in the left lateral recumbent position or manually displacing the uterus usually will correct supine hypotension in pregnancy.
- Elevated progesterone levels in pregnancy lead to a decrease in both gastric and intestinal motility leading to an increase in the risk of gastric aspiration.
- Early use of nasogatric tube decompression can help to reduce this risk of aspiration.
- The diaphragm becomes elevated during the course of pregnancy secondary to displacement by the enlarging uterus. This results in a decrease in functional residual capacity and residual volume, which leads to a lower oxygen reserve. The pregnant patient compensates for the decrease in functional residual capacity by increasing tidal volume by up to 40%, resulting in hypocapnea and a mild respiratory alkalosis. A normal pCO_2 therefore, may represent an early sign of respiratory compromise in these patients.
- The pregnant patient's symphysis pubis widens by approximately 28 wk gestation, as do the sacroiliac joint spaces.

Clinical Presentation

Blunt Trauma

- Motor vehicle accidents represent the most common source of blunt trauma in pregnancy with falls and assaults, including domestic violence, as other potential causes.
- The most common serious injury seen with blunt trauma is placental abruption which occurs in approximately 50% of major abdominal traumatic injuries and in 5% of minor trauma events.
- Fetal death may occur in the presence of only minor maternal injuries.
- Uterine rupture may also occur but is a relatively uncommon occurrence (0.6%); however, if it does occur, fetal mortality is close to 100%.
- The clinical presentation of uterine rupture is highly variable.
- Common findings are the loss of the normal uterine contour and the palpation and/or visualization of fetal parts outside of the uterus by ultrasound.
- Maternal hypotension and shock usually rapidly ensue with an associated nonreassuring fetal heart rate pattern.
- Premature labor and/or premature rupture of membranes may also occur secondary to blunt trauma.

Penetrating Trauma

- During the course of a developing pregnancy as uterine size begins to increase, the other abdominal and pelvic organs become increasingly protected from penetrating trauma.
- The enlarging uterus, however, becomes progressively more susceptible to penetrating trauma.
- Penetrating abdominal or pelvic trauma in pregnancy have a relatively high fetal mortality rate, in some studies approaching 66%.
- The maternal mortality rate (5%) for penetrating abdominal/pelvic trauma is surprisingly low due to the shielding effect of the enlarging uterus.

Emergency Department Treatment of Pregnant Trauma Patients

- The initial evaluation of the pregnant trauma patient does not differ significantly from the initial management of the nonpregnant trauma patient.
- The patient's airway patency should be maintained while simultaneously protecting the cervical spine.
- The chin lift or jaw thrust techniques can be used to open the patient's airway and allow maintenance of cervical spine immobilization.

- Supplemental oxygen should be administered to maintain adequate oxygenation of both the mother and fetus. Other airway adjuncts such as nasal pharyngeal and oral pharyngeal airways may be useful in selected patients in maintaining airway patency. Endotracheal intubation may be necessary to maintain airway protection along with adequate oxygenation and ventilation in some cases.
- Adequate circulating blood volume must be maintained at all times. The presence of the increased blood volume seen in pregnancy means that in the setting of trauma the pregnant patient can lose a large percentage of her circulating blood volume prior to the development of tachycardia, hypotension and other signs of hypovolemic shock occur.
- This is important because decreased placental perfusion with associated fetal hypoxia may be occurring when the maternal vital signs appear stable.
- The pregnant patient with significant trauma should have two large bore intravenous catheters (14-16 gauge) placed as soon as possible for intravascular volume replacement and resuscitation.
- Crystalloid either as lactated Ringers or normal saline are the resuscitation solutions of choice for initial resuscitation of the pregnant patient.
- In the supine patient the uterus can compress the vena cava which leads to decreased venous return and subsequently a decrease in cardiac output. This decrease in cardiac output can cause or exacerbate a preexisting shock state. The uterus can be manually deflected to the left and the patient can be placed on her left side by placing a wedge under her right hip or backboard, which will alleviate the vena caval compression.
- The patient should be placed on a monitor and pulse oximetry. If the fetal gestational age is at least 20-24 wk continuous fetal monitoring should be performed.
- A thorough secondary survey should be performed consisting of a complete head-to-toe physical examination to include a pelvic examination unless there is a contraindication such as a suspected placenta previa.
- A nasogastric tube and Foley catheter may be indicated.
- Laboratory evaluation should consist of CBC, urinalysis, type and screen, PT/PTT, and HCG Quantitative. In the case of abdominal trauma, LFTs and lipase should be added.
- If the mother is Rh negative, a Kleihauer-Betke test should be obtained to detect the presence of fetal RBCs in the maternal blood smear. Regardless of the results or availability of the Kleihauer-Betke test, every Rh negative patient with a history of abdominal trauma should be given Rh immunoglobulin (Rhogam) to avoid sensitization. This should occur, regardless of the gestational age of the fetus.
    - In the setting of a pregnancy of <13 wk, mini-Rhogam may be used
- Radiographic evaluation should be obtained as clinically indicated by physical findings during the secondary survey.
    - Bedside ultrasound is an extremely useful tool in evaluating the pregnant trauma patient. It can be used for initial evaluation of fetal age and viability pending monitoring. A Focused Abdominal Sonogram for Trauma (FAST) can also be performed looking for signs of free fluid (hemoperitoneum) in the abdomen and pelvis.

- Obstetrical and/or surgical consultation is indicated depending on the nature and severity of the mother's injuries. Patients with abdominal/pelvic pain, vaginal bleeding, ruptured membranes or uterine contractions should be admitted with obstetrical consultation for further evaluation and continuous fetal monitoring.
- Perimortem c-section in the emergency department in the setting of maternal hypovolemic cardiac arrest can occasionally be successful, but in many cases the fetus has already been exposed to a period of prolonged hypoxia prior to the maternal cardiac arrest. This contributes to a generally poor outcome in many of these cases. If perimortem c-section is to have any chance of being successful, it must be initiated within 4-5 min of the onset of the maternal cardiac arrest.

Suggested Reading

1. Van der Sluis CK, Ten Duis HJ, Geertzen JH. Multiple injuries: An overview of the outcome. J Trauma 1995; 38:681.
2. Hirshberg A et al. Causes and patterns of missed injuries in trauma. Am J Surg 1994; 168:299.
3. Branney SW et al. Critical analysis of two decades of experience with postinjury emergency department thoracotomy in a regional trauma center. J Trauma 1998; 45:187.
4. Hoffman JR et al. Validity of a set of clinical criteria to rule out injury to the cervical spine in patients with blunt trauma. N Engl J Med 2000; 343:94.
5. McKenny MG et al. 1000 consecutive ultrasounds for blunt abdominal trauma. J Trauma 1996; 40:607.
6. Richardson JD, Wilson ME, Miller FB. The widened mediastinum: Diagnostic and theraputic priorities. Ann Surg 1990; 211:731.
7. Schafermeyer RW. Pediatric trauma. Emerg Med Clin North Am 1993; 11:187.
8. Perclet MH et al. Patterns of Injury in children. J Pediatr Surg 1990; 25:85.
9. Santoro TA, Schinco MA, Troskin SZ. Management of trauma in the elderly patient. Surg Clin North Am 1994; 74:163.
10. Kannus P, Parkkari J, Koskinen S. Fall-induced injuries and deaths among older adults. JAMA 1999; 281:1895.
11. Schiller WR, Knox R, Chleborad W. A five-year experience with severe injuries in elderly patients. Accid Anal Prev 1995; 27:167.
12. Jager TE et al. Traumatic brain injuries evaluated in US emergency departments, 1992-1994. Acad Emerg Med 2000; 7:134-140.
13. Teasdale G, Jennett B. Assessment of coma and impaired consciousness: A practical scale. Lancet 1974; 2:81-84.
14. LoCicero J, Mattox KL. Epidemiology of chest trauma. Surg Clin North Am 1989; 69:15.
15. Demetriades D et al. Selective nonoperative management of gunshot wounds of the anterior abdomen. Arch Surg 1997; 132:178-183.

CHAPTER 17

Psychiatric Emergencies

Carrie S. Korn, Adam J. Trenton and Glenn W. Currier

Introduction

A psychiatric emergency is defined as a disturbance in thoughts, feelings or actions for which immediate therapeutic intervention is needed. For a variety of reasons, from substance abuse to violent behavior, the number of psychiatric patients in the Emergency Department is on the rise. Patients with psychiatric disorders account for as many as 15% of all emergency department visits thus creating a need for an increase in psychiatric awareness in medical personnel.

Medical Clearance

The term medical clearance of psychiatric patients has become widespread in emergency departments and signifies an initial medical evaluation of all patients whose symptoms may or may not be psychiatric in origin. Currently, emergency departments are mandated by COBRA and EMTALA to medically screen all patients presenting for treatment by a licensed personnel. The level and comprehensiveness of the initial medical screening must be consistent with the patient's complaint. Regardless of illness or injury severity, medical screening must be uniformly addressed with each patient and the appropriate response initiated. Recent studies, however, have found the need for comprehensive medical screening of patients with psychiatric chief complaints coupled with a psychiatric history are both labor intensive and financially unsound. Therefore, prior to evaluation of patients with psychiatric complaints, the individual guidelines for each institution must be consulted.

Basic Medical Screening May Include

Initial triage by licensed personnel documenting:

- Age and sex
- Chief complaint
- Vital signs (including temperature)
- Past medical and psychiatric history
- Current medications
- Physical appearance

Medical work-up:

- Physical examination
- Mental status exam
- Laboratory—glucose, electrolytes, complete blood count, toxicology Screen and prescription drug levels.
- Pregnancy screen

Aggression and Violence

The emergency physician frequently encounters acutely aggressive and potentially violent patients. Aggression is defined as any forceful or assaultive verbal or

Emergency Medicine, edited by Sean Henderson. ©2006 Landes Bioscience.

Table 17.1. Common medical causes of psychiatric symptoms

Medical Cause	Symptoms
Alcohol and drug withdrawal	Hallucinations, delirium tremens, paranoia, agitation, confusion, anxiety, delusions, psychosis
CNS trauma/tumors	Confusion, mood or perceptual disturbances, hallucinations, depression, personality changes, mania, panic attacks, amnesia
Encephalitis	ALOC, disorientation, olfactory and gustatory hallucinations, psychotic behavior, personality changes.
Hyperglycemia or hypoglycemia	ALOC, agitation, anxiety, confusion, belligerence, fatigue
Hypertension or hypotension	Dementia, anxiety, agitation
Hyperthermia or hypothermia	ALOC
Hypoxia	ALOC, confusion, agitation
Intintoxication	Euphoria, depression, agitation, aggression
Intracranial hemorrhage	Flat or inappropriate behavior
Tachycardia	Agitation
Thyroid disease	Dementia, anxiety
Seizure disorder with medication toxicity	Depression, lethargy, confusion
Sepsis	Dementia, agitation, confusion

physical action either towards oneself, inanimate objects or others. These acts may be unprovoked or result from mild external or internal disturbances.

Incidence

- Violence and aggression are more common among patients receiving psychiatric services than the general public. A recent NIMH Epidemiologic Catchment Area survey indicated that 55.5% of violent individuals met the criteria for a psychiatric diagnosis, compared to 19.6% of nonviolent individuals.
- As many as 17% of patients seen in psychiatric emergency services could be classified as homicidal.
- The most consistent predictor of future violence is a history of violence.
- Violence is most commonly associated with diagnoses of schizophrenia, substance abuse, affective disorders, and personality disorders.
- Command hallucinations can induce dangerous behavior in some cases. One report indicated voice recognition and less dangerous commands were associated with a higher level of compliance.

Clinical Presentation

When an acutely agitated, potentially violent, patient presents, the focus should be on short-term (24-h) risk assessment. The first priority of the staff throughout the assessment process should be to maintain personal safety.

Table 17.2. Medications causing psychiatric symptoms

Medication Category	Medications	Psychiatric Symptoms
Analgesics	Salicylates	Delirium, anxiety
	Narcotics	Euphoria, dysphoria
Antiarrhythmics	Quinidine, Lidocaine, Tocainide, Mexiletine	Delirium, excitement, agitation
	procainamide	Delusions, depression, panic
Antibiotics	Aminoglycocides	Toxic psychosis
	Cephalothin	Delirium, paranoia
	Penicillin	Psychosis
	Sulfonamides	Delirium, anorexia
	Trimethoprim	Psychosis, depression, insomnia, mutism
Antichlolinergics	Benztropine diphenhydramine	Confusion, memory impairment,
	Meperidine	Agitation, delirium, hallucinations,
	Tricyclic antidepressants	Severe anxiety
Anticonvulsants	Carbamazepine	Restlessness, anxiety
	Phenytoin	Irritability, depression, visual hallucinations, agitation
	Phenobarbital	Depression, confusion, disinhibition
Antihypertensives	Beta-blockers	Depression, insomnia, psychosis
	Calcium channel blockers	Agitation, depression, panic
Antiinflammatory agents	Indomethacin	Delirium, depression, hallucinations
	NSAIDs	Depression, anxiety, confusion
Diuretics	Furosemide	Apathy, confusion, delirium
Dopaminergics	Neuroleptics	Neuroleptic malignant syndrome
	Sinemet	Confusion, paranoia, anxiety
Narcotics	Alcohol, barbiturates, benzodiazepines, narcotics	Sedation, impaired cognition
Serotonergic agents	SSRIs, Tricyclic antidepressants, Lithium, MAOIs	Serotonin syndrome
Steroids	Anabolic steroids	Aggression, paranoia, mood disorders
	Corticosteroids	Mood change, mania, agitation
	Oral contraceptives	Depression, anxiety
Stimulants	Amphetamine, cocaine, caffeine, Theophylline	Anxiety, agitation, Paranoid psychosis, insomnia, confusion
Sympathomimetics	Pseudoephedrine, Albuterol	Anxiety, agitation, psychosis, delirium
Miscellaneous drugs	Cimetidine	Hallucinations, confusion, delirium, depression
	Cyclobenzaprine	Mania, psychosis
	Digitalis	Confusion, psychosis, depression
	Hypoglycemic agents	Anxiety

17

Psychiatric Assessment

Generally, a brief assessment leading to a general diagnostic category is sufficient. A comprehensive assessment leading to a diagnosis is appropriate, but may not always be practical.

- Presenting psychopathology
  - Patients presenting with psychosis, particularly command hallucinations, may represent a particularly high violence risk.
- Rate of onset of symptoms
  - Acute onset of symptoms in a previously well-functioning individual are unlikely to be the result of schizophrenia or a major affective disorder.
- Mental status examination
  - Mental status examination may be necessary for some patients particularly those presenting with behavioral abnormalities, drug ingestion, multiple trauma, or metabolic disturbances.
  - Mental status examination may be necessary to determine a patient's competence to refuse emergency care.
  - Elements that should be incorporated into the mental status exam include:
    - Level of consciousness
    - Orientation to time, place, and person
    - Memory functioning
    - Concentration
- Abnormal speech can be an indicator of etiology.
  - Intoxication: slurred speech
  - Mania: rapid, pressured, or loud speech
  - Psychosis: abnormal thought content
  - Neurological: aphasic speech
- Acute suicide risk
- Psychiatric history:
  - Prior psychiatric diagnoses
  - Personal as well as family history of violence
  - Substance abuse history
- Examination of external stressors
- Deterrents of violence or suicide should be delineated (e.g., religious beliefs, fear of legal problems).

Physical Examination

- It is important to identify medical conditions that may cause or exacerbate psychiatric symptoms.
  - Delirium is indicative of a medical disorder impairing brain function.
  - Potential of head trauma should always be explored.
  - CNS infection should be considered with elevated temperature and altered mental status.
- Vital signs
- Urine toxicological screen
- Visual examination
  - Including systematic neurological examination
  - Complete physical exam may be indicated depending on the nature of the problem and resources of the staff.
- Cognitive examination (e.g., MMSE)
- Pregnancy test for women of child-bearing age

Treatment

First-Line Treatment of Agitated or Violent Behavior

Management of a violent/aggressive patient is somewhat dependent on the etiology of the behavior. The least invasive measures should be attempted first in an effort to facilitate a positive patient-provider relationship and foster long-term compliance.

- Staff should initially attempt to calm the patient by verbal means.
- A show of force by staff may be an effective means of dissuading acts of aggression.
- If pharmacological management is indicated, oral preparations should be used whenever possible.
- An oral benzodiazepine, such as lorazepam, 0.5-2 mg, may be used first unless the patient is clearly psychotic.
- If the patient is psychotic an oral antipsychotic, such as risperidone or olanzapine, may be administered alone or in conjunction with a benzodiazepine. Orally dissolving medications, available for olanzapine, or liquid concentrates, available for risperidone, offer a faster onset of action than tablets.
  - Risperidone, 2 mg, tablet or liquid concentrate, is regarded by some as the oral atypical antipsychotic of choice. Risperidone, 2 mg liquid concentrate, used in conjunction with oral lorazepam, 2 mg, has proven comparable to a combination of intramuscular haloperidol, 5 mg, and intramuscular lorazepam, 2 mg, in the management of agitated psychosis.
  - Olanzapine, 5-10 mg p.o., has displayed efficacy in treating psychotic agitation, and is regarded by some as an atypical antipsychotic of choice.
  - The widespread utility of quetiapine and ziprasidone in PES has not yet been established.

When Emergency Management Is Appropriate

If less invasive forms of therapy are ineffective or impractical, emergency intervention may become necessary. Emergency interventions, in the form of physical restraints or intramuscular medications are indicated in certain situations. Refusal to cooperate, intense staring, motor restlessness, purposeless movements, affective lability and loud speech alone are not always cause for emergency intervention. However, these behaviors in association with the following often necessitate involuntary care.

- Emergency intervention is always appropriate if the patient is directly threatening or assaultive or is an acute danger to other patients, bystanders, staff, or self.
- Emergency intervention is usually appropriate: if the patient displays aggression and demeaning or hostile verbal behavior or if the patient displays irritability or intimidating behavior.

Choice of Medication

If oral medications are refused, intramuscular preparations of benzodiazepines or traditional antipsychotics may be necessary. In 2001, the expert consensus guidelines listed lorazepam, haloperidol, and droperidol as first-line intramuscular options for the treatment of agitation and violence. In the near future, IM preparations of novel antipsychotics such as olanzapine and ziprasidone may become available. In some emergency departments, intravenous preparations of lorazepam, haloperidol, or droperidol are used as an alternative to intramuscular medications.

- Lorazepam, 0.5-2 mg IM, is well-absorbed intramuscularly and is characterized by a short half-life and lack of active metabolites. Repeat doses may be given after approximately 60 min if necessary. Intravenous lorazepam at a dose of 0.5-4 mg is also effective in management of acute agitation.

Table 17.3. Side effects commonly associated with drugs used in the treatment of agitation and violence

Drug	Side Effects
Haloperidol	Extrapyramidal symptoms, akathisia, dystonia, hyperprolactinemia, hypotension, sexual dysfunction, tachycardia, tardive dyskinesia
Lorazepam	Apnea, dizziness, disorientation
Olanzapine	Anticholinergic effects, diabetes mellitus, hypotension, sinus tachycardia, sedation, weight gain
Quetiapine	Dizziness, hypotension, sedation, sinus tachycardia, weight gain
Risperidone	Hypotension, prolactin elevation, sedation, sinus tachycardia
Ziprasidone	Hypotension, QTc prolongation, respiratory disorder, sedation

- Haloperidol, generally the typical neuroleptic of choice, may be administered at doses of 1-10 mg IM, which may be repeated after approximately 60 min to effect. Intravenous haloperidol is also effective for the management of acute agitation and may be given at an initial dose of 2-10 mg. Additional boluses may be given as needed, every 15-60 min. Intravenous haloperidol produces a clinical response within 5-30 min and has a superior effect on psychotic agitation compared to oral preparations during the first 3 h of treatment.
- Intramuscular ziprasidone, 5-20 mg, is currently being evaluated for clinical use in agitated psychotic patients. Ziprasidone has proven to be effective in management of acute agitation.
- Intramuscular olanzapine, 2.5-10 mg 1-4 injections/day, has shown promise in the management of acute agitation associated with schizophrenia and bipolar mania.

Delirium

Delirium is defined as a transient disorder characterized by impaired attention, memory impairment, disorientation or language and perceptual disturbances. This is the manifestation of an underlying medical condition such as infection, coronary ischemia, hypoxemia, or metabolic derangement. Delirium usually has a rapid onset (hours to days) and is episodic within this time range.

Incidence

- Delirium is one of the most common mental disorders found in both the ED and in-patient populations. In medically ill hospitalized patients approximately 10-30% of patients experience delirium.
- Delirium is particularly common among elderly individuals, especially those with pre-existing cognitive impairments such as dementia. Approximately 40% of hospitalized demented patients are delirious.
- Risk factors include: low serum albumin, multiple medical problems, dementia or cognitive impairment, polypharmacy, metabolic disturbances, few social interactions, advanced age, infection, fractures, visual impairment, fever or hypothermia, psychoactive drug use.
- Delirium has been associated with increased mortality, though specific mortality risk depends on a variety of factors.

Clinical Presentation

Diagnostic Methods

- If delirium is suspected, the priority should be to identify contributory medical problems (some causes of delirium are identified in Table 17.4).

Table 17.4. Medical conditions that can cause delirium

CNS Disorders	Metabolic Disorders	Cardiopulmonary Disorders	Systemic Illness
Head trauma	Renal failure	Myocardial infarction	Sub intoxication/ withdrawal
Seizures	Hepatic failure	Congestive heart failure	Infection
Postictal state	Anemia	Cardiac arrhythmia	Neoplasm
Vascular disease	Hypoxia	Shock	Severe trauma
Degenerative disease	Hypoglycemia	Respiratory failure	Sensory deprivation
	Thiamine deficiency		Temperature dysregulation
	Endocrinopathy		Postoperative state
	Fluid or electrolyte imbalance		Cancer
	Acid-base imbalance		
	Low serum albumin		

- Evaluation should include a comprehensive physical examination incorporating medical history, neurological examination, vital signs, and anesthesia record (if postoperative).
- Mental status should be assessed periodically. Cognitive tests such as clock face, digit span, and trailmaking tests, may be useful in identifying symptoms of delirium.
- The patient's medications should be reviewed, as delirium is frequently associated with medication initiation or withdrawal (some of the substances that have been associated with delirium are listed in Table 17.5).
- Routine laboratory tests may be helpful in determining the etiology of delirium and may include: complete blood count, blood chemistries (electrolytes, BUN and serum creatinine, thyroid, glucose, calcium, albumin, liver function studies [SGOT, SGPT, bilirubin], alkaline phosphatase, magnesium, PO_4) urinalysis, electrocardiography, chest X-ray, and measurement of arterial blood gases or oxygen saturation.
- Other laboratory tests may be indicated depending on the patient's clinical condition. These tests may include urine culture and sensitivity, urine drug screen, blood tests (venereal disease research laboratory (VDRL), heavy metal screen, B12 and folate

Table 17.5. Drugs that can cause delirium

Drugs of Abuse	Medications	Toxins
Alcohol	Anesthetics	Anticholinesterase
Amphetamines	Analgesics	Organophosphate insecticides
Cannabis	Antiasthmatic agents	Carbon monoxide
Cocaine	Anticholinergics	Carbon dioxide
Hallucinogens	Anticonvulsants	Volatile substances such as fuel or organic solvents
Inhalants	Antihistamines	
Opioids	Antihypertensive and cardiovascular medications	
Phencyclidine	Antimicrobials	
Sedatives	Antiparkinsonian medications	
Hypnotics	Corticosteroids	
Other	Diuretics	
	Gastrointestinal medications	
	Muscle relaxants	
	Immunosuppressive agents	
	Lithium and psychotropic medications with anticholinergic properties	

levels, lupus erythematosus (LE) prep, antinuclear antibody (ANA), urinary porphyrins, ammonia, human immunodeficiency virus (HIV)), blood cultures, measurement of serum levels of medications, lumbar puncture, brain computerized tomography (CT), or magnetic resonance imaging (MRI), or electroencephalogram.

- Electroencephalograms (EEG) of patients experiencing delirium commonly reflect generalized slowing and in some types of delirium may show low-voltage fast-activity (e.g., alcohol or sedative withdrawal). This escalated activity is often associated with agitated behavior.
- Various nursing scales may have clinical utility in detecting symptoms of delirium: NEECHAM Confusion Scale, Confusion Rating Scale (CRS), Clinical Assessment of Confusion (CAC-A), and the MCV Nursing Delirium Rating Scale (MCV-NDRS).
- The Delirium Symptoms Interview (DSI) is an interview schedule used to guide the diagnosis of delirium.
- Checklist, analog, and algorithm methods used to detect symptoms of delirium include: Confusion Assessment Method (CAM), Delirium Scale (D-Scale), Global Assessment Rating Scale (GARS), Organic Brain Syndrome Scale (OBS), and Saskatoon Delirium Checklist (SDC).
- Other scales are useful for assessing symptom severity among patients already diagnosed with delirium including the Delirium Rating Scale and the Memorial Delirium Assessment Scale (MDAS).

Common Symptoms

- DSM-IV describes general classifications of delirium
  - Delirium due to a general medical condition
  - Substance-induced delirium
  - Delirium due to multiple etiologies
  - Not otherwise specified
- Cognitive/neurological symptoms may be the result of diffuse cerebral dysfunction and can include: impaired recall and short-term memory, abnormalities of thought process, language alterations (dysgraphia [considered to be a sensitive indicator of delirium] dysarthria, dysnomia, aphasia) visuoconstructional deficits.
- Physical symptoms include autonomic changes (tachycardia, dilated pupils, and sweating).
- Perceptual symptoms include: illusions, visual (most common), auditory, gustatory, olfactory, and tactile hallucinations.
- Behavioral/psychiatric symptoms include: anxiety, irritability and agitation, increased or decreased psychomotor behavior, delusions (often persecutory), sleep-wake cycle disturbances, altered or labile affect, depression, euphoria, apathy, hallucinations (visual and/or auditory).
- Before developing overt delirium the patient may display symptoms such as restlessness, anxiety, irritability, drowsiness, and insomnia.
- Two subtypes of delirium have been described based on psychomotor activity.
  - Hyperactive subtype: patient is agitated, hyperalert
  - Hypoactive subtype: patient is lethargic, hypoalert (may result from an effort to reduce stimulus overload)
- Mental status impairment tends to fluctuate during the 24 h period, between periods of quiet reserve and overt agitation) with increased impairment usually seen in the evening hours. This is referred to as "sundowning".

Differential Diagnosis

- Special attention should be taken as to the sudden versus progressive onset of symptoms, as rapid onset is associated with delirium, while gradual onset may signify dementia.

- Cognitive disturbances associated with delirium tend to be reversible, while those associated with dementia generally are not.

Treatment

- The first priority in delirium management is to address any underlying physical causes. Consequently, medical clearance guidelines should be followed to ensure all possible contributory medical conditions are addressed.
- Environmental regulation is important in the management of a patient experiencing delirium.
  - Patient should be provided with a safe quiet environment.
  - Family members may be of help to keep the patient calm and secure.
  - Sensory impairments should be reduced.
  - Environmental cues should be used to facilitate orientation (e.g., clocks, calendars, lighting cues).
- Cognitive-emotional support can be helpful in strengthening any retained adaptive cognitive functioning.
- The use of restraints should be avoided as this could increase agitation and carry risks for injury.
- Pharmacological intervention is indicated for behavioral control and subjective distress. Neuroleptics of the butyrophenone class (haloperidol and droperidol) are favored over phenothiazines, which cause more sedation and which may exacerbate delirium.
  1. Historically traditional neuroleptics have been the treatment of choice for delirium.
     - Haloperidol (doses of 1-2 mg every 2-4 h as needed) either orally or parenterally, may be repeated until agitation is controlled. Haloperidol is the treatment of choice for patients in whom oral administration is impractical. For patients experiencing severe delirium, intravenous administration may be helpful. An initial haloperidol bolus of 10 mg IV followed by continuous intravenous infusion of 5-10 mg/h) offers rapid onset and continued efficacy. Elderly patients should initially receive low doses (0.25-0.5 mg orally or 0.125-0.25 mg parenterally). The use of haloperidol is associated with side effects including extrapyramidal symptoms and hypotension.
  2. Atypical antipsychotic agents have shown efficacy in treating delirium.
     - Risperidone is regarded as the preferred atypical antipsychotic for the treatment of delirium. Risperidone can be started at 0.25-0.5 mg twice daily, and increased to 4 mg/day if symptoms initially fail to clear. Maintenance doses of 0.25-0.5 mg may be given every 4 h for persistent agitation or decreased delirium. In rare cases risperidone has been associated with the onset of delirium, but this has not occurred in the absence of other risk factors (e.g., coprescribed medications, old age, or medical conditions).
     - Initial findings indicate that olanzapine (5-15 mg/day) has comparable efficacy to haloperidol (1.5-10 mg/day) in treating delirium without substantial risk of EPS. An initial dose of 2.5-5 mg at bedtime is a reasonable starting dose, and may be increased to 20 mg/day if symptoms fail to respond. Supplemental doses may be given, but this practice has not consistently produced greater efficacy. The side effect most commonly associated with olanzapine has been sedation.
     - Initial findings indicate that quetiapine (25-750 mg/day) is equally efficacious as haloperidol (1-10 mg/day) in treating delirium. The effective dose of quetiapine is somewhat variable, but the initial dose should be in the range of 25-50 mg twice daily. If it is well tolerated, it can be increased every 1-2 days to 100 mg twice daily. Additional doses of 25-50 mg may be given every 4 h for agitated or delirium symptoms. One case report indicated that quetiapine use was temporally related to the onset of delirium, though the mechanism of this was not understood.

  - A case report indicated that ziprasidone 40-100 mg/day may be effective in treating delirium. In a case of extreme overdose (4020 mg) ziprasidone was associated with the onset of delirium.
3. Benzodiazepine monotherapy is usually ineffective for most types of delirium.
  - Benzodiazepines are generally reserved for cases of suspected alcohol or other substance withdrawal.
  - When benzodiazepines are used, a relatively short-acting medication with no active metabolites (e.g., lorazepam) should be selected.
  - Combined haloperidol and lorazepam therapy can be started with 3 mg IV of haloperidol followed immediately by 0.5-1.0 mg IV of lorazepam and then modified according to the patient's degree of improvement.
  - The use of lorazepam for treatment of delirium has been associated with ataxia, oversedation, disinhibition, and increased confusion.

- Delirium associated with specific etiologies requires specific pharmacological interventions:
  - For delirium caused by anticholinergics, cholinergic drugs may be helpful. Within this class of drugs, physostigmine, a cholinesterase inhibitor has been used most commonly. Physostigmine may be used in intramusuclar or intravenous doses of 0.16 to 2.00 mg, or as 3 mg/h continuous intravenous infusions. Side effects associated with cholinesterase inhibitors include bradycardia, nausea, vomiting, salivation, and increased gastrointestinal acid.
  - For delirium associated with hypercatabolic conditions, paralytic sedation and mechanical ventilation may be required.
  - For delirium related with alcohol withdrawal, folate and thiamine should be administered.
  - Palliative treatment, involving morphine or other opiates, may be effective for patients in whom pain is an aggravating factor. However, opiates are known to have anticholinergic effects which can exacerbate delirium.
  - Multivitamins might be helpful for malnourished patients who might be experiencing delirium as a result of B vitamin deficiencies.
- The prognosis for delirious patients is generally positive.
  - Elderly patients, however, often do not recover fully, and persistent cognitive deficits are common.

Dementia

Dementia is defined as a gradual alteration in mental and cognitive capabilities. It can manifest in the form of language, memory or behavioral changes signaling a decline from the previous level of function.

Incidence

- The prevalence of dementia in the population greatly increases with age. While only 1% of the population experiences dementia by age 60, by age 85 at least 50% of the population is affected.
- Currently, an estimated 1.5 million people have severe dementia and an additional 1-5 million persons have mild to moderate dementia .
- The majority of dementia cases, 60-70%, are attributed to Alzheimer's disease (AD).
- The second most common form is vascular dementia, which accounts for 10-20% of dementia patients .
- Less common causes of dementia include Lewy body dementia, Pick's disease, Creutzfeldt-Jakob disease, hydrocephalus, Parkinson's disease, brain tumors, and metabolic disorders.

Table 17.6. Common symptoms of dementia

Type of Deficit	Common Presentations
Cognitive	Impairments in language, praxis, judgment, visuospatial function, related mental activities
Extrapyramidal	Rigidity, bradykinesia, movement disorders, abnormalities of gait
Functional	Loss of ability to perform personal care tasks, changes in social functioning
Neurologic	Visual field deficits, hemiparesis, hemisensory loss, asymmetric deep tendon reflexes, unilateral extensor plantar response
Personality	Indifference, regression, impulsiveness

Clinical Presentations

- Dementia is characterized by gradual and progressive memory impairment.
- Memory loss, naming problems, forgetting items, and visuospatial confusion characterize the early stage of dementia.
- The middle stage is characterized by loss of reading ability, decreased performance in social situations, increased difficulty in finding words and names, intermittent disorientation to time, inability to recognize familiar persons, behavioral problems, and losing directions.
- Late-stage symptoms include extreme disorientation, inability to dress and perform self-care, increasing delusions, hallucinations, and progressive loss of other activities of daily living and personality change.
- Rapid deterioration of mildly demented individuals is sometimes prompted by urinary tract infection, congestive heart failure, hypothyroidism, or delirium.
- Emergency treatment of dementia may be necessary if aggression, psychosis, or activity disturbances are present.
- Risk factors for dementia include advanced age, family history, and abnormal apolipoprotein status.

Diagnostic Evaluation

- Generally, patients should be evaluated for dementia if any of the following are present:
  - Memory or cognitive complaints with or without functional impairment
  - Questions of competency in elderly patients
  - Depression or anxiety in patients with cognitive complaints, or physician suspicion of cognitive impairment during a clinical interview.
- If suspicion of dementia exists, reversible causes such as subdural hematoma, normal pressure hydrocephalus, hypothyroidism, and the dementia syndrome of depression must be eliminated to make a definitive diagnosis.
- Lab tests commonly used in the assessment of dementia include complete blood count, serum electrolytes, calcium, glucose, BUN, creatinine, liver function tests, serum B12,

Table 17.7. Diagnostic criteria of dementia according to WHO

17

Diagnostic Criteria for Dementia

1. Decline in verbal and nonverbal memory, significant decrease in ability to learn new information, present for at least 6 mo
2. Decrease from premorbid levels in cognitive abilities such as planning and organizing and general processing of information
3. Preserved awareness of environment, delirium is absent
4. Decline in emotional control or motivation or a change in social behavior

Table 17.8. DSM IV diagnostic criteria for AD

Diagnostic Criteria for AD

1. Development of multiple cognitive deficits manifested by memory impairment and one or more of the following: aphasia, apraxia, agnosia, disturbance in executive functioning
2. Cognitive deficits cause a reduction in functioning from premorbid state
3. Course is characterized by gradual onset and progressive cognitive decline
4. Cognitive deficits are not caused by other central nervous system disturbances that lead to memory and cognition deficits or systemic conditions known to cause dementia
5. Deficits do not occur exclusively during delirium
6. Disturbance not better accounted for by an axis 1 disorder

and serology for syphilis. In some cases sedimentation rate, serum folate level, HIV testing, chest X-ray, and urinalysis should be performed.

- Mental status tests examine orientation, recent and remote memory, language, praxis, visuospatial relations, calculations, and judgment.
- Neuroimaging is helpful in identifying potentially treatable conditions that can otherwise be missed such as tumors, subdural hematoma, hydrocephalus, and strokes.
- Neuropsychological testing is commonly used in cases of borderline or suspicious dementia.
- Cognitive screening should examine memory, ability to calculate language, visuospatial skills, and degree of alertness. The MMSE is commonly used to detect cognitive impairment.
- Scanning techniques such as PET and SPECT examine cerebral function.
- EEG is not routinely performed, but it can be used to identify toxic or metabolic disorders, partial complex seizures, or Creutzfeldt-Jakob disease.

Alzheimer's Disease (AD)

- The onset of AD is characterized by the impairment of memory and orientation, while speech and motor abilities are preserved.
- Other clinical features include depression, anxiety, behavioral disorders and speech difficulties.
- The ability to perform everyday activities may be hampered by impaired visuospatial processing.
- In the early stages of AD most neurologic and extrapyramidal functions are preserved in typical forms of the disease.
- Once a positive diagnosis of AD can be made, survival ranges from 8-10 yr.

Vascular Dementia

- The symptoms of vascular dementia largely parallel those of AD. However, diagnosis requires not only cognitive dysfunction, but also signs of cerebrovascular disease upon neurologic exam.
- Common physical findings include exaggerated or asymmetric deep tendon reflexes, gait abnormalities, weakness of an extremity, hemiparesis, a unilateral extensor plantar response, or visual field deficits.
- The presence of extrapyramidal signs in conjunction with gait abnormalities indicates Parkinsonism, progressive supranuclear palsy, or AD.

Frontal Lobe Dementia

- Pick's disease often presents with language impairments such as logorrhea, echolalia, and palilalia. Behavioral impairments are often present as well.

Table 17.9. Diagnostic criteria for vascular dementia according to WHO

Diagnostic Criteria for Vascular Dementia
1. Dementia of a specified level of severity
2. Unequal distribution of deficits in cognitive function
3. If focal damage is evident it will manifest as one of the following: unilateral spastic weakness of limbs, unilateral increased tendon reflexes, an extensor plantar response, and pseudobulbar palsy
4. History, examination, or tests disclose severe cerebrovascular disease which may be judged to be related to the dementia

- Creutzfeldt-Jakob disease is a rare, rapidly progressing form of dementia. It generally presents with vague initial symptoms such as irritability and somatic sensations. Motor signs like myoclonus, Parkinsonism, and motor neuron dysfunction may also be present.
- Normal pressure hydrocephalus is a poorly understood cause of dementia characterized by a triad of gait disorder, urinary incontinence, and cognitive decline.

Treatment

- Nonpharmacologic treatment is often employed for certain manifestations of dementia such as circadian rhythm disturbances, catastrophic reactions, and wandering.
- Pharmacologic treatment becomes necessary when agitation, physical outbursts, or significant delusions or hallucinations are present.
  - Antipsychotics are generally effective for psychotic symptoms and nonpsychotic agitated behavior.
  - Stronger neuroleptics like haloperidol have better side effect profiles than low potency agents such as thioridazine and chlorpromazine.
  - Benzodiazepines can be used if neuroleptics are contraindicated. It is generally best to use short-acting agents like lorazepam, temazepam, and oxazepam.

Alzheimer's Disease

- Acetylcholinesterase inhibitors are commonly used to slow the breakdown of acetylcholine, an essential neurotransmitter in cognitive functioning. These drugs have not proven to stop or reverse the progression of AD.
- The two most commonly prescribed drugs for AD are donepezil and tacrine.
  - Donepezil is regarded as the first-line treatment for dementia as it is more selective, longer-acting, and has fewer side effects than tacrine.
  - Tacrine is commonly prescribed as an alternative to donepezil. However, up to 20% of patients cannot tolerate tacrine's cholinergic side effects.
  - Other AChE inhibitors that have shown efficacy but have not yet been approved in the United States include rivastigmine, metrifonate, and galantamine.
- Besides AChE inhibitors there are alternative treatments that are sometimes used for the treatment of AD. Commonly prescribed agents include:
  - Ibuprofen (400 mg, 2-3 times/day)
  - Vitamin E (800-2000 IU/day)
  - Conjugated estrogens. Estrogen shows promise as a treatment for cognition, mood, behavior, and motor disturbances associated with dementia.
  - Another class of drugs that has been examined is anti-inflammatory drugs. Anti-inflammatory drugs change the cerebral inflammatory response to amyloid protein deposits, thereby reducing the risk of developing AD or slowing the progression of symptoms.

Table 17.10. Nonpharmacologic interventions for dementia

Presentation	Intervention
Verbal outbursts/ mild delusions	Reassurance Ignoring behavior Diversionary strategies
Circadian rhythm disturbance	Avoid night time fluids Treatment of painful conditions that may be disrupting sleep Encourage daytime exercise Discourage daytime napping
Catastrophic reactions	Approach patient in calm manner Avoid excessive external stimuli Limit demands

Vascular Dementia

- The treatment of vascular dementia is focused on the treatment of risk factors. As hypertension is one of the more common risk factors for vascular dementia, antihypertensive drugs are commonly prescribed.
- Systolic blood pressure should be kept below 150-160 mm Hg and diastolic blood pressure should be in the range of 85-95 mm Hg. Because of cerebral artherosclerosis, treatment that lowers diastolic blood pressure below 85-95 mm Hg may worsen cognitive impairment.
- Enteric-coated aspirin is sometimes prescribed in the range of 81-325 mg daily.
- Vitamin E, 800-2000 IU daily, is also prescribed in some cases.

Neuroleptic Malignant Syndrome

Neuroleptic malignant syndrome (NMS), although rare, is a potentially life-threatening side effect of neuroleptic agents. NMS is considered an idiosyncratic reaction that may occur at any time during the course of treatment or when dopamine antagonistic therapy is discontinued. Incidence of NMS is not believed to be dose-related.

Incidence

- Among patients treated with neuroleptics, 0.02-3.23% are diagnosed with neuroleptic malignant syndrome. This wide discrepancy is due to many factors including lack of agreement on diagnostic criteria and difficulty estimating the size of the population at risk.
- The mortality rate associated with NMS is approximately 11.6%. Patients receiving various pharmacological interventions including dantrolene, bromocriptine, and amantadine have displayed lower mortality rates than those treated only with supportive measures.
- There are two peak age groups for the incidence of NMS, including individuals younger than 40 and over 70. The majority of cases, 80%, affect people under the age of 40. Patients displaying NMS in the younger group are commonly people taking neuroleptics for psychotic disorders, while those in the older group commonly develop NMS after discontinuation of dopaminergic agents.
- NMS is approximately twice as common in men.

Clinical Presentations

Onset of NMS

- The onset of NMS generally occurs 3-9 days after the initiation of neuroleptic treatment or the addition of a second medication. However, NMS may occur after a single dose or prolonged treatment.

Table 17.11. Risk factors associated with NMS

- Male gender
- Parenteral administration of neuroleptics
- Withdrawal of anti-parkinsonian medication
- Prolonged use of restraints
- Dehydration
- Agitation
- Alcoholism

- The symptoms of NMS typically develop over the course of 1-3 days.
- The types of drugs most commonly associated with NMS are neuroleptic agents such as phenothiazines, butyrophenones, and thioxanthenes. Higher potency neuroleptics may be more likely to cause this syndrome.
- Newer atypical antipsychotics have been implicated in causing NMS in rare cases. A 2002 review cited isolated reports of NMS associated with clozapine, risperidone, olanzapine and possibly quetiapine, but indicated that these agents carry a lower risk for NMS than traditional neuroleptics.
- NMS may also occur when individuals cease taking dopaminergic agents such as L-dopa and carbidopa, which are used to treat Parkinson's disease.

Diagnosis of NMS

- NMS presents with a relatively consistent clinical picture.
  - Autonomic dysfunction is often the earliest clinical finding in NMS. Common manifestations of autonomic instability include massive peripheral hyperadrenergic state, hypertension, postural hypotension, tachycardia, tachypnea, hypoventilation, pallor, flushing, vasoconstriction, incontinence, excessive perspiration and diaphoresis.
  - Hyperthermia, generally in the range of 39-42°C (in rare cases fever is not present) is believed to be a result of excessive heat production (from muscle contraction) when heat dissipating mechanisms are overwhelmed (from vasoconstriction).
  - Altered mental status is standard and ranges from mild confusion, agitation, and lethargy to stupor and coma.
  - Rigidity of the musculature may range from mild hypertonicity to lead-pipe rigidity.
    - Other extrapyramidal symptoms that are common in NMS patients include tremors, cogwheeling, masked face, dystonia, dyskinesia, opisthotonos, dysphagia, sialorrhea, choreiform movements, trismus, opsoclonus, oculogyric process, retrocollis, festinating gait, and flexor-extensor posturing.
- Though no single laboratory test can produce a definitive diagnosis of NMS, certain laboratory abnormalities are commonly associated with this syndrome.
  - Above normal levels of creatinine phosphokinase (CPK), (>1000 units/L) are commonly present. Levels as high as 100,000 units/L have been reported.
  - Another indicator of NMS is elevated WBC counts in the range of 10,000-40,000/mm.
  - Elevated CPK and WBC levels are the result of prolonged muscle contraction and hyperthermia.
  - Toxicological screening should be performed to determine the presence of any causative or exacerbating substances, such as MDMA (ecstasy) or cocaine.
  - Other complications related to NMS that may be detectable in lab tests include: myoglobinuria, metabolic acidosis, renal insufficiencies, and elevated hepatic transaminases.
- The differential diagnosis of NMS is relatively straightforward. Infectious, metabolic, environmental, and toxicological etiologies should be ruled out first.

    - CNS infections mimicking NMS can generally be distinguished by lumbar puncture and encephalogram. NMS lacks changes in glucose, white cells, and protein expected in central nervous system infections. Encephalitis and sepsis are two problems that may be mistaken for NMS.
    - Metabolic problems that can resemble NMS include hyperthyroidism, hypothyroidism, and hypomagnesemic tetany.
    - People taking neuroleptics may be increasingly susceptible to heat stroke, as these agents are believed to suppress central heat loss mechanisms. However, presentation with neuroleptic-induced heat stroke is associated with hot, dry skin rather than the diaphoresis associated with NMS. Furthermore, unlike those with NMS, patients who had neuroleptic-induced heat strokes may present with seizures, absence of EPS, absence of sweating, and history of exposure to high temperature.
    - Allergic drug reactions may produce fever and autonomic instability, but not rigidity. Furthermore, neuroleptics may produce extrapyramidal symptoms in the absence of fever, leukocytosis, and autonomic disturbances. Serotonin syndrome and atropinism (overdose with anticholinergic drugs) may also be confused with NMS. Intoxication with MDMA (ecstasy) or cocaine may mimic the symptoms of NMS.
- Otherwise only a few disorders are similar in appearance to NMS. Two diseases commonly mistaken for NMS are malignant hyperthermia and lethal catatonia, both of which present with fever and muscle rigidity.
    - Malignant hyperthermia describes a genetically determined defect in calcium transport which manifests as an abnormal contraction response of muscle tissue. This condition can lead to lethal hyperthermia and its onset is most frequently associated with exposure to anesthetics. This can be diagnosed by exposing biopsied muscle tissue to caffeine or halothane in vitro which results in a hypercontractile response when compared with normal muscle.
    - Lethal catatonia is a syndrome characterized by mutism, motor hyperactivity, altered consciousness, and fever that may progress to severe autonomic disturbances, coma, and death. Unlike NMS, this disease can present with severe anxiety and agitation as well as choreiform stereotypy.

Treatment

- Patients believed to have NMS should be transferred to an intensive care unit.
- Discontinuation of the offending neuroleptic agent is the first priority in the treatment of NMS. If prescribed, lithium should be discontinued as should anticholinergic drugs, in the case of severe extrapyramidal symptoms. Alternatively, if NMS was caused by discontinuation of dopamine agonists, these agents should be reinstated.
- The foundation of NMS treatment generally includes rapid cooling, fluid and electrolyte repletion, critical care monitoring, and supportive antipyretics.
- The pharmacologic treatment of NMS is based on dopamine agonists.
    - Bromocriptine is regarded as the drug of choice, with most patients responding within one day. Bromocriptine should be started at a dose of 2.5 mg orally, 2-3 times/day, and can be increased by 2.5-7.5 mg daily to a maximum dose of 30-45 mg/day.
    - Amantadine, an indirect dopamine agonist, can be given orally at a dose of 100-200 mg twice daily.
    - Levodopa alone, or in combination with carbidopa (a dopadecarboxylase inhibitor), may be effective for treating patients who are dopamine depleted.
- Dantrolene, a nonspecific skeletal muscle relaxer, can be used when muscle rigidity is severe and body temperature is difficult to control. Dantrolene is given orally in daily dosages of 50-600 mg or intravenously at doses of 1-10 mg/kg every 6 h. Dantrolene can produce a rare but fatal hepatocellular injury and should not be used in patients with preexisting liver disease.

- Otherwise, the treatment of NMS is based on symptomatology. Depending on the nature and severity of the patient's condition, other interventions include short-term antihypertensives (such as nifedipine), vasodilators (such as minoxidil or nitroprusside), heparin (to prevent deep venous thrombosis and pulmonary embolism), oxygen, and intubation.
- Most patients recover from the acute complications of NMS within 2-14 days without any cognitive impairment. In some instances a residual catatonic state persists for weeks or months after acute hyperthermic symptoms subside. Prolonged dysfunction is usually related to high fever, hypoxia, or other complications. Some complications that may be associated with NMS include renal failure, electrolyte abnormalities, dysrhythmias, aspiration pneumonia, sepsis, and pulmonary embolism secondary to the formation of deep vein thromboses. NMS caused by depot injections also tends to take longer to resolve.
- If treatment is deemed successful, a minimum of 2 wk should be allowed before the reintroduction of neuroleptics. Approximately, one-third of patients in whom neuroleptics are rechallenged experience recurrence of symptoms.
- In some cases death results from NMS or associated life-threatening complications. Mortality among NMS patients is usually attributed to respiratory failure resulting from thromboembolism or pneumonia.

Serotonin Syndrome

Serotonin syndrome is a rare and potentially lethal disorder that can be caused by any drug with serotonergic activity.

Incidence

- The incidence of serotonin syndrome, while believed to be rare, is unknown. Serotonin syndrome is difficult to study as it cannot be prospectively examined and retrospective studies are limited by underreporting due to unrecognized symptoms and confusion with other pathologies.
- Serotonin syndrome is often attributed to newer antidepressants as a result of their serotonergic activity. Prescription event-monitoring studies have indicated an incidence of 0.5-1 per 1000 patient months of treatment with the drugs: fluoxetine, sertraline, paroxetine, moclobemide, nefazodone, and venlafaxine.
- Individuals with serotonin syndrome commonly have underlying illnesses. Predictably, depression is common among those with serotonin syndrome, as depressive disorders are often treated with one or more serotonergic drugs.
- The occurrence of death with serotonin syndrome is also unknown. Among 38 cases reported between 1982-1991 and 41 reported cases between 1995-2000, two patients died, both of whom had complicated presentations involving several drugs.

Clinical Features

Onset of Serotonin Syndrome

- The onset of serotonin syndrome may occur from minutes to days after ingestion of serotonergic drugs.
- Serotonin syndrome usually occurs in conjunction with the increase in dose of a serotonergic drug, or the combination of serotonergic drugs. Combining drugs with different mechanisms of increasing serotonergic activity may be particularly dangerous (see Table 17.12). Serotonin syndrome caused by a single agents is often associated with over ingestion.
    - Antidepressants including MAOIs, TCAs, and SSRIs have been associated with serotonin syndrome when used in monotherapy or with other serotonergic agents.
    - Elicit drugs including MDMA "ecstasy" and cocaine have been implicated in causing serotonin syndrome.

Table 17.12. Drug combinations commonly associated with serotonin syndrome

Primary Drug	Potentially Dangerous Combinations
SSRIs	L-Trp, MAOIs, tricycle antidepressants, buspirone, lithium, trazodone, selegiline, moclobemide, and dextromethorphan
MAOIs	L-Trp, 3,4 metachlorophenylpiperazine (mCPP), trazodone, lithium, meperidine, MDMA, selegiline, dextromethorphan, SSRIs, tricyclic antidepressants
TCAs	3,4 meta-chlorophenylpiperazine (mCPP), nefazodone, trazodone, venlafaxine
Reversible inhibitors of MAO-A (such as moclobemide)	Venlafaxine, lithium, tricyclic antidepressants, SSRIs

Diagnosis of Serotonin Syndrome

- The diagnosis of serotonin syndrome is made clinically based on strong suspicion or known exposure to serotonergic agents, demonstration of specific signs and symptoms of the disorder, and exclusion of other medical and psychiatric symptoms.
- Serotonin syndrome may not be an all-or-nothing phenomenon, rather there appears to be a continuum of serotonin-induced hyperactivity. Thus, patients may present with one or more symptoms, but not meet complete diagnostic criteria.
- The clinical presentation of serotonin syndrome is characterized by a triad of cognitive/behavioral changes, autonomic instability, and neuromuscular changes.
  - Cognitive and behavioral symptoms may include confusion, disorientation, hypomania, agitation, and coma in severe cases.
  - Autonomic problems include fever, dilated pupils, shivering, diaphoresis, tachycardia, tachypnea, nausea, and diarrhea.
  - Neuromuscular dysfunction manifests as clonus, ocular clonus, restlessness, hyperreflexia, myoclonus, tremor, incoordination, rigidity, trismus, bilateral Babinski signs, rhabdomyolysis, nystagmus, seizures, and ataxia.
- Definitive diagnostic criteria for serotonin syndrome are lacking, though the diagnostic criteria proposed by Sternbach are commonly referenced (see Table 17.13).
- As Sternbach's diagnostic criteria do not consider the severity of symptoms a Serotonin Syndrome Scale (SSS) has been developed which measures 9 factors on a scale of 0 (absent) to 3 (severe). The symptoms assessed are agitation, disorientation, myoclonus, hyperreflexia, tremor, dizziness, hyperthermia, sweating, and diarrhea. A score >6 is said to be indicative of serotonin syndrome. While this scale has not been rigorously validated, it may have value as a guideline for assessing the possibility of serotonin syndrome.

Differential Diagnosis

- Disorders with similar presentations to serotonin syndrome include sepsis, stiff-man syndrome, heat stroke, delirium tremens, poisonings, and neuroleptic malignant syndrome (NMS).

Table 17.13. Sternbach's diagnostic criteria for serotonin syndrome

Diagnostic Criteria
1. Recent increased dose or addition of serotonergic agent
2. Other causes excluded
3. Patient should not have had a recent increase of a neuroleptic agent
4. Three of the following must be present: altered mental status, agitation, tremor, shivering, diarrhea, hyperreflexia, myoclonus, ataxia, or fever

- Serotonin syndrome is commonly confused with NMS. Relative to serotonin syndrome, NMS is more likely to be associated with high fevers, rhabdomyolysis, and mortality. The symptoms of NMS are more likely to include lead-pipe rigidity and EPS, whereas the symptoms of serotonin syndrome are more likely to include myoclonus, hyperreflexia, or dilated pupils. Serotonin syndrome is also sometimes associated with unusual ocular movement and preferential increased tonicity of lower limbs that is not seen in NMS.

Treatment

- If serotonin syndrome is suspected, serotonergic drugs should be immediately discontinued. In general, if the offending agent is withdrawn the syndrome tends to resolve on its own within 1-3 days, though recovery may take longer depending on the severity of symptoms.
- Supportive care measures should be initiated and may include: cooling blankets for hyperthermia, hydration, and maintenance of cardiac and renal function. In some cases. In some cases paralysis with neuromuscular blocking agents and mechanical ventilation may be required.
- Benzodiazepines may be effective in decreasing anxiety and agitation and treating seizures that may develop as a result of serotonin syndrome. However, benzodiazepines are not regarded as a first-line treatment for serotonin syndrome.
- High urine outputs should be maintained to prevent myoglobinuria from causing renal injury. Sodium bicarbonate has proven efficacious in preventing renal impairment.
- Throughout the course of the treatment, autonomic functioning should be closely monitored and precautions against seizures should be taken.
- For patients in whom supportive therapy alone is insufficient, treatment with an antiserotonergic drug may be given (see Table 17.14). It is unclear whether these drugs shorten the duration of serotonin syndrome, but they do appear to provide symptomatic relief.
  - Currently, cyproheptadine is regarded as the antiserotonergic drug of choice for serotonin syndrome. Cyproheptadine is generally effective at doses of 2-8 mg orally, 3-4 times/day.
  - Chlorpromazine may also be effective in the treatment of serotonin syndrome and has the advantage that it is available in parenteral form. It can be given in an initial parenteral dose of 12.5 mg. Due to its action as a dopamine (D_2) antagonist, chlorpromazine should not be used if NMS is suspected. Chlorpromazine has also been associated with hypotension, dystonic reactions, NMS, and possibly reduction of seizure threshold.
  - Methysergide and propranolol may also be effective in treating serotonin syndrome.
- As a result of the antagonistic effect of certain atypical antipsychotics on certain serotonin receptors, these drugs may someday have a role in the treatment of serotonin syndrome. Though research in this area is virtually nonexistent, risperidone has displayed a prophylactic effect in preventing serotonin syndrome in an animal study.

Table 17.14. Mechanisms of drugs commonly prescribed for serotonin syndrome

Drug	Mechanism
Cyproheptadine	5-HT_{1A} and 5-HT_2 antagonism as well as additional antimuscarinic properties
Chlorpromazine	Antagonistic properties at 5-HT_{1A} and 5-HT_2 receptors, blocks D2 α-adrenergic receptors, and has antimuscarinic effects
Methysergide	Non-selective 5-HT antagonist
Propanolol	Beta-blocker with antagonistic properties at 5-HT_{1A} receptors

- Supportive treatment is essential as untreated patients are at a risk of developing rhabdomyolysis, renal failure, hepatic dysfunction, disseminated intravascular coagulation, and in massive overingestion, cardiovascular collapse and death.

Suicide

The most unpredictable threat to health care is the suicide-prone patient. This patient population is vulnerable to the uncontrolled environment of an emergency department thereby creating a volatile situation in which the outcomes can be severe.

Incidence

- In 1998, nearly 1 million people worldwide committed suicide, including more than 30,000 Americans.
- Approximately 90% of suicide victims meet the criteria for at least one psychiatric illness.
- Depression accounts for 80-90% of suicide victims.
- Bipolar disorder is a particularly common diagnosis.

Clinical Presentation

- The patient may present with initial complaints of lethargy, depression, feelings of isolation or withdrawn.
- Signs of previous suicide attempts such as hesitation marks or history of drug overdose.
- There are several risk factors commonly associated with suicide, including depression, schizophrenia, and bipolar disorder.
- The most important indicators of suicide include affective disorders, schizophrenia, and alcoholism.

Diagnostic Evaluation

- The first component of the suicide assessment is the suicide inquiry, in which the physician asks about suicide ideations, plans, and attempts. This gives an indication of the current severity of suicidality.
  Sample questions:
  1. "You said you are depressed, what is that like for you?"
  2. "Are there times when you feel like crying?"
  3. "When you are feeling that way, what sort of thoughts go through your head?"
  4. "Have you ever gone so far as to think of taking your own life?"
  5. "Have you made a plan?"
  6. "Do you have the means to carry out this plan?"
- Special attention should be given to personal and family characteristics. Potentiating factors can include genetic traits as well as environmental influences. Often potentiating factors present in the form of disorder-based or personality disorders.

Table 17.15. Common risk factors for suicide

Common Risk Factors	
• Acute depression with turmoil	• Mixed episodes
• Suicidal ideation	• Rapid cycling
• Hopelessness	• Substance abuse
• Global insomnia	• Aggression
• Anhedonia	• Impulsivity
• Poor Concentration	• Low serotonergic activity
• Anxiety	• Hypothalamic-pituitary-adrenal axis activation
• Panic	

Treatment

- The patient should not leave the ED or be left alone even if the suspicion level is low.
- A sitter should be assigned to watch the patient closely.
- If a sitter is not available, physical restraints can be used to assure the patient is unable to leave the ED.
- In the event medical intervention is needed, i.e., laceration repair, gastric lavage and charcoal for a drug overdose, the patient should have these completed prior to a psychiatric consultation.
- Psychiatric services must be consulted prior to disposition.

Suggested Reading

1. Rice MM, Moore GP. Management of the violent patient therapeutic and legal considerations. Emerg Med Clin North Am 1991; 9:13-30.
2. Tesar GE. The agitated patient, Part I: Evaluation and behavioral management. Hosp Community Psychiatry 1993; 44:329-331.
3. Kaufman DM, Zun L. A quantifiable, brief mental status examination for emergency patients. J Emerg Med 1995; 13:449-456.
4. Allen MH, Currier GW, Hughes DH et al. The expert consensus guideline series. Treatment of behavioral emergencies. Postgrad Med 2001:1-88.
5. Currier GW, Simpson GM. Risperidone liquid concentrate and oral lorazepam versus intramuscular haloperidol and intramuscular lorazepam for the treatment of psychotic agitation. J Clin Psychiatry 2001; 62:153-157.
6. Trzepacz PT. Delirium. Advances in diagnosis, pathophysiology and treatment. Psychiatr Clin North Am 1996; 19:429-448.
7. American Psychiatric Association. Practice Guideline for the treatment of patients with delirium. Am J Psychiatry 1999; 156(suppl 5):1-20.
8. Roberts BL. Managing delirium in adult intensive care patients. Crit Care Nurse 2001; 21:48-55.
9. Francis J, Martin D, Kapoor WN. A prospective study of delirium in hospitalized elderly. JAMA 1990; 263:1097-1101.
10. Fernandez F, Levy JK, Mansell PWA. Management of delirium in terminally ill AIDS patients. Int J Psychiatry Med 1989; 19:165-172.
11. Levin T, Petrides G, Weiner J et al. Intractable delirium associated with ziconotide successfully treated with electroconvulsive therapy. Psychosomatics 2002; 43:63-66.
12. Dudley WHC, Williams JG. Electroconvulsive therapy in delirium tremens. Compr Psychiatry 1972; 13:357-360.
13. Figiel GS, Hassen MA, Zorumski C et al. ECT-Induced delirium in depressed patients with parkinson's disease. J Neuropsychiatry Clin Neurosci 1991; 3:405-411.
14. Eisendrath SJ, Goldman B, Douglas J et al. Meperidine-Induced delirium. Am J Psychiatry 1987; 144:1063-1065.
15. Goff DC, Garber HJ, Jenike MA. Partial resolution of Ranitidine-Associated delirium with physostigmine: Case report. J Clin Psychiatry 1985; 46:400-401.
16. Gurrera RJ. Sympathoadrenal hyperactivity and the etiology of neuroleptic malignant syndrome. Am J Psychiatry 1999; 156(2):169-80.
17. Balzan MV. The neuroleptic malignant syndrome: A logical approach to the patient with temperature and rigidity. Postgrad Med J 1998; 74(868):72-6.
18. Schneider SM. Neuroleptic malignant syndrome: Controversies in treatment. Am J Emerg Med 1991; 9(4):360-2.
19. Harrigan RA, Brady WJ. Antipsychotics. In: Tintinalli JE, Kelen GD, Stapczynski JS, eds. Emerg Med (textbook) 2000:1086-7.
20. Ty EB, Rothner AD. Neuroleptic malignant syndrome in children and adolescents. J Child Neurol 2001; 16:157-163.
21. Caroff SN, Mann SC, Keck Jr PE et al. Residual catatonic state following neuroleptic malignant syndrome. JClin Psychopharmacol 2000; 20(2):257-9.

Chapter 18

Special Imaging Studies for the Emergency Department: Angiography MRI V/Q and Sestamibi

Matt Hendrickson

Angiography

Summary

- The emergence of multiple alternative noninvasive studies (CT, MRI and U/S) as well as a stronger reliance on clinical exam in certain injuries (penetrating neck and extremity trauma) has led to a decrease in the use of emergent angiography.
- On the other hand, angiography remains the gold-standard for many vascular emergencies, and the improvement in technology (nonionic contrast, safer microcatheters) and technique has considerably lowered the risk of complications.
- All contraindications are relative and most can be overcome either with nonionic contrast and antiallergic pretreatment, and correction of coagulopathy.
- The indications for angiography can be divided into four groups: (1) first-line diagnostic test; (2) first-line diagnostic test and therapeutic modality; (3) therapeutic modality; (4) not first-line diagnostic test.
- First-line diagnostic test:
  - Penetrating extremity trauma—due to the high sensitivity of physical exam, the self-limited nature of most vascular injuries discovered on angiography, and the good outcome of delayed repair, angiography is no longer commonly used for patients without hard signs of vascular injury.
  - Penetrating neck trauma—the high sensitivity of color-flow doppler and clinical observation has reduced the use of angiography for zone 1 and 3 injuries.
  - Pulmonary embolus—patients with nondiagnostic V/Q scans and negative CT scans are appropriate candidates for angiography, unless pretest clinical suspicion is not high and the patient has negative serial lower extremity duplex scans.
  - Any patient with strongly suspected mesenteric ischemia requires emergent angiography to confirm the disease and guide the vascular surgeon's approach.
  - Nontraumatic limb ischemia—Although angiography is the gold-standard for evaluating peripheral vascular disease, the diagnosis of acute limb ischemia is usually made reliably by clinical exam.
- First-line diagnostic test and therapeutic modality:
  - Subarachnoid hemorrhage—emergency cerebral angiography is required for all acute SAH patients to prepare surgical approach or embolectomy before rebleeding occurs.
  - Upper GI bleeding requires emergency angiography and embolization or vasopressin infusion to control massive bleeding in patients who do not respond to endoscopic treatment.
  - Heavy lower GI bleeding (>3 units/day) may require angiography to localize and control bleeding and will sometimes precede colonoscopy.
- Therapeutic modality:

Emergency Medicine, edited by Sean Henderson. ©2006 Landes Bioscience.

  - Pelvic fracture hemorrhage—although there is little evidence determining whether angiography improves outcome, this is the preferred therapy for pelvic fracture patients with persistent hypotension after other causes for hypotension are excluded and after application of an external fixator.
- No longer first-line diagnostic test:
  - Aortic dissection—although angiography is still used by many vascular surgeons preoperatively, MRI and TEE are both considered more accurate tests without the associated risks, while CT has equivalent accuracy with much greater convenience.
  - Blunt aortic injury—angiography is the gold standard but CT is recommended as the initial test with angio reserved for indeterminate cases.
  - Abdominal aortic aneurysm—angiography has been replaced by CT, ultrasound (U/S) and MRI.
  - Renal trauma—CT is first-line with angiography used for indeterminate vascular injuries.

Indications

Extremity Trauma

- Angiography remains the gold standard for evaluating potential arterial injuries in penetrating extremity trauma (PET), but the indications for its use have narrowed over the last ten years.
  - Most arterial injuries will present with classic "hard" signs: pulsatile hemorrhage, pulsatile hematoma, overt distal ischemia, audible bruit, palpable thrill.
    - These signs require immediate operative management, and generally angiography will only be performed intraoperatively if at all.
  - Soft signs such as small hematoma, transient hypotension, absence of hemorrhage, fracture, and nerve injury are no longer treated differently than an absence of hard signs because studies show they have no clinical correlation with vascular injury.
- The management of injuries in proximity to a vascular structure is more controversial but a trend toward noninvasive studies (duplex or doppler ABIs), observation, or discharge home has replaced surgical exploration and angiograms.
- Angiography not only has a 2.6-5% complication rate), but it has a 1.9% false positive rate, prompting unnecessary surgical exploration. Using physical exam with the hard signs listed above reduces the false positive rate to 0%.

Duplex U/S

- Many centers use duplex U/S to assess vascular injury, but there are no studies with long-term follow-up to confirm its safety. At least one study has demonstrated its lack of reliability.
  - Even if U/S had comparable sensitivity to angiography, since several studies show that there is no morbidity when patients are only repaired if they manifest hard signs, the U/S result (either positive or negative) may not change management.
- Indications for arteriography in extremity trauma include:
  - Hemodynamic instability
  - Blunt trauma with signs of vascular injury (Fig. 18.1)
  - Intraoperative or postoperative evaluation
  - Delayed diagnosis with hard signs
  - Follow-up of nonoperatively managed arterial injuries
  - Penetratring trauma with hard signs plus:
    - Multiple potential sites of injury (i.e., shotguns)
    - Missile parallels vessel over long distance
    - Chronic vascular disease
    - Extensive bone or soft-tissue injury

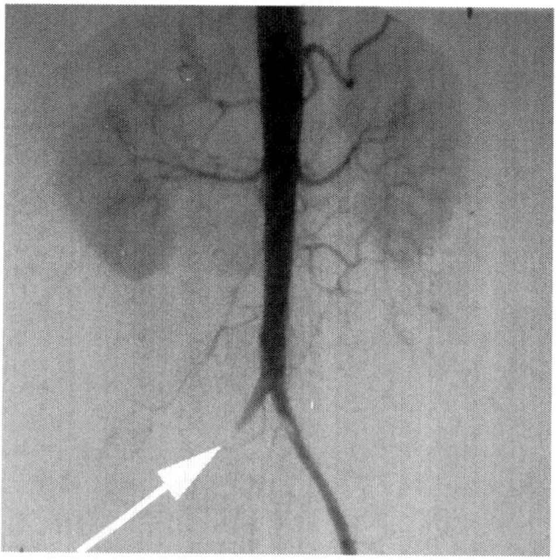

Figure 18.1. Blunt trauma to right common iliac with small defect to left internal iliac.

- Although most patients with PET who present with hard signs of vascular injury will undergo immediate operative management, a few situations will prompt initial arteriography.
  - Patients with multiple potential sites of injury, a missile that parallels the vessel over a long distance, extensive bony or soft-tissue injury, and thoracic outlet injuries will usually require arteriography even with hard signs because the surgical approach will vary depending on the location of arterial injury (Fig. 18.2).

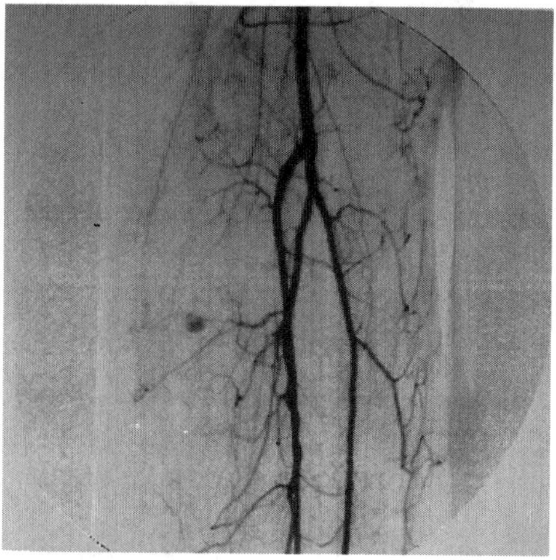

Figure 18.2. Right popliteal artery dividing into the posterior tibial artery and peroneal artery.

Aortic Dissection

- Retrograde aortography was long considered the study of choice for evaluating suspected thoracic aortic dissection simply because it was the only way to accurately diagnose dissection antemortem.
  - The introduction of CT, MRI and transesophageal echocardiography (TEE) have all proven to be safer studies for aortic dissection while demonstrating that aortography was probably not as sensitive as previously thought.
  - The sensitivity of aortography is commonly listed as 80-90% with a specificity of 90-95%.
    - False-negative angiograms may occur in cases of thrombosis of the false lumen, faint opacification of the false lumen, or opacification of both true and false lumens so that an intimal flap is not visualized.
- Plain radiography will miss 10-20% of aortic dissections so if there is a suspicion of aortic dissection another study is necessary.
- Based on their prospective blinded comparison of MR, TEE and CT, Nienaber et al recommend MR for the evaluation of aortic dissection in stable patients and TEE for unstable patients.

Penetrating Trauma to the Neck

- Angiography is the gold standard to evaluate penetrating vascular injuries to the neck.
- Preiviously angiography was routinely performed for all penetrating injuries to zone 1 (clavicles to cricoid cartilage) and zone 3 (above the angle of the mandible) of the neck regardless of exam or symptoms, with surgical exploration or angiography for zone 2 injuries.
- Although mandatory angiography or surgical exploration is considered the safest way to avoid missing vascular injuries, there is a growing body of evidence to support the conservative approach of ultrasonography and/or 24 h clinical observation (Fig. 18.3).
- Like penetrating extremity trauma, the majority of penetrating neck injuries without hard signs will not have vascular injuries, and those that do will rarely require surgery.
- Ultimately the majority of these patients will be admitted to a surgeon, and the decision for angiography will be made in consultation with that surgeon (Fig. 18.4).

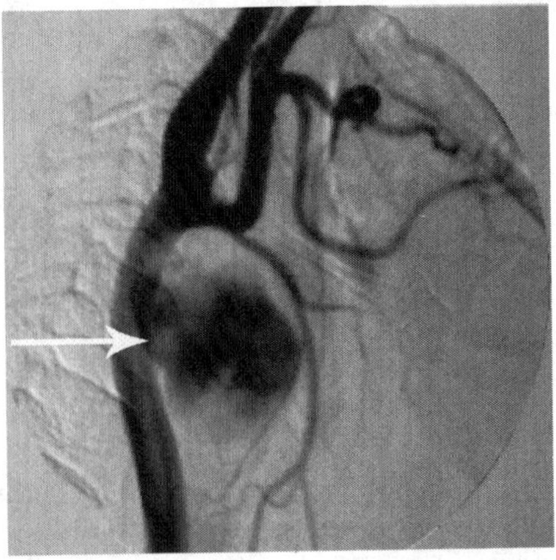

Figure 18.3. Penetrating injury to common carotid artery with extravasation of contrast.

18

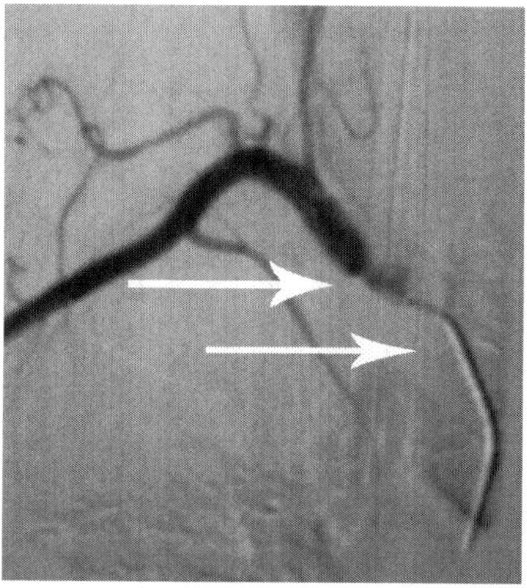

Figure 18.4. Innominate artery—right arrow shows an angio catheter in the innominate artery with distal stenosis and contrast extravasation.

Pelvic Fracture Hemorrhage

- The mortality rates in patients with pelvic fractures range from 9%-20% but are reportedly as high as 50% in patients who present with pelvic fractures and hemodynamic instability.
- Angiography's primary role in pelvic fracture hemorrhage is a therapeutic one because of the potential for embolectomy of the injured small arteries.
- Large vessel injuries (aorta, iliacs, femorals) in pelvic fractures are fairly rare (1%) and are not amenable to embolization so these injuries, if discovered on angiogram, are an indication for surgical repair.
- 90% of bleeding in pelvic fractures comes from low pressure venous plexus and fractured cancellous bone surfaces which is not amenable to venous embolectomy because of the extensive anastamoses and valveless collateral flow.
- Only 10% of pelvic fracture hemorrhage is due to arterial bleeding (branches of the internal iliac) and this is seen most commonly with anterior-posterior injuries—APC II (symphisis widening) and APC III (symphisis and SI disruption).
- Arterial embolectomy will help control venous hemorrhage by slowing the arterial supply.
  - Embolectomy is achieved using various embolic agents including hemostatic absorbable gelfoam cakes, autologous clots, muscle, detachable balloons, polystyrene spheres, and wire coils which are injected from the angiography catheter.
- Because the majority of bleeding in pelvic fractures can be controlled by tamponade, angiography is deferred at many trauma centers until the external fixator is applied.

Blunt Aortic Injury

- Blunt aortic injury is the second most common cause of death in blunt trauma patients with only 13-15% of patients arriving at the hospital with signs of life.
- The most commonly noted signs are pseudocoarctation (kinking of aorta at the ligamentum arteriosum with decreased blood pressure in the lower extremities) and an intrascapular murmur.

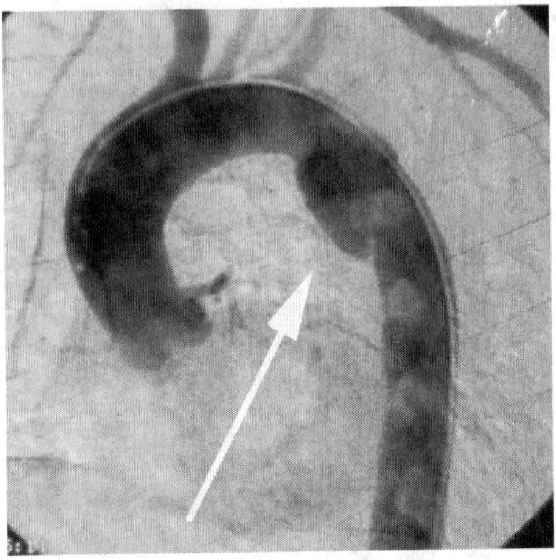

Figure 18.5. Blunt deceleration injury to descending aorta with aneurysmal dilatation.

- Chest radiograph is the initial screening test and will detect at least one of the following abnormalities in most aortic injuries: widened mediastinum, indistinct aortic knob, depression of the left mainstem bronchus, deviation of the NG tube, opacification of the AP window, widened paratracheal and paraspinous stripes, and apical capping.
- Angiography is the gold standard but contrast-enhanced helical or spiral CT has recently been shown to have excellent sensitivity with 100% negative predictive value (Fig. 18.5).

Pulmonary Angiography

- Pulmonary angiography (PA) is rarely indicated as an emergent diagnostic test from the emergency department.
- Use of pulmonary angiography to diagnose pulmonary embolism is generally reserved for patients with conflicting pretest and scan probabilities (i.e., high pretest probability and low probability VQ scans or vice versa).
- Angiography is now rarely ordered before spiral CT because a positive CT (either for PE or an alternative diagnosis) negates the need for angiogram.
  - Although spiral CT has acceptably high specificity to confidently rule-in PE, three metaanalyses published in 2000 demonstrated that CT's sensitivity is too low to safely exclude PE so patients with negative CT need further tests (i.e., duplex, V/Q, or angiogram) to exclude it (Fig. 18.6).

Upper Gastrointestinal Bleeding

- Although endoscopy is the primary diagnostic modality to evaluate, localize and treat upper GI bleeding, approximately 10-20% of patients with massive hemorrhage (hemodynamic instability, requiring 4-6 units of blood in 24 h) will require angiography to control bleeding.
- Endoscopy is always performed before angiography which will often negate the need for angiography or at the very least help guide which artery to cannulate first at angiography (celiac, superior mesenteric, left gastric, gastroduodenal, pancreaticoduodenal, and splenic).
- Therapeutic angiography is most strongly indicated in poor surgical candidates (elderly, severely ill patients) but is increasingly offered to all acute GI bleeders who continue to bleed after endoscopy.

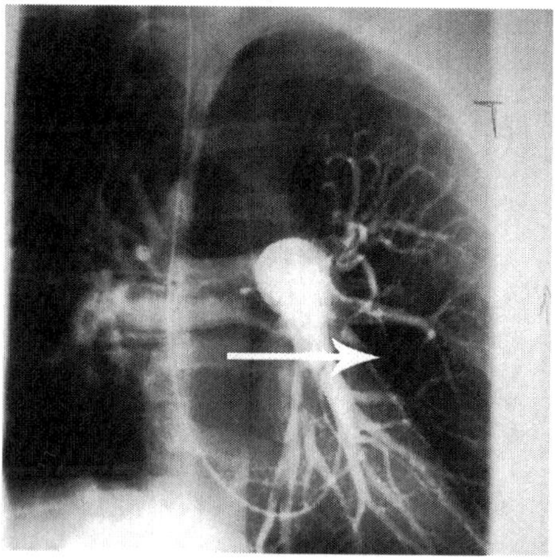

Figure 18.6. Pulmonary embolus demonstrated by abrupt cutoff in mid lower left lung zone with paucity of vessels distal to cutoff.

- Selective intra-arterial vasopressin (usually 12-24 h infusion) is the standard angiographic therapy although embolotherapy is increasingly popular because of new microtherapy which reduces the chance of large vessel occlusion and consequent bowel infarction.

Lower Gastrointestinal Bleeding

- Because colonoscopy of an unprepared colon is more difficult than upper endoscopy and lesions may be missed in a dirty colon because of poor preparation or active bleeding, the evaluation of lower GI bleeding is less straightforward.
- The three diagnostic modalities employed emergently for evaluation of acute lower GI bleeding are colonoscopy/sigmoidoscopy, technetium bleeding scans and angiography.
- As with upper GI bleeding, angiography is typically reserved for heavy bleeding (>1 ml/min), and has two advantages as a precedent to surgery—it may stabilize a patient through vasopressin and embolization to avoid surgery or allow for elective surgery, and angiography can localize the bleeding site to reduce the extent of bowel resection.

Subarachnoid Hemorrhage

- The gold standard for the diagnosis of intracranial aneurysms remains angiography.
- Emergent angiography is recommended in patients who have a positive CT scan, a negative CT scan but positive lumbar puncture, or in those with a very suggestive history despite negative studies.
- The study should be done urgently, certainly within 24 h of presentation (ideally within 6-8 h), unless the patient is of poor Hunt-Hess grade (see Table 18.1)

Abdominal Aortic Aneurysm

- Although angiography is helpful in depicting the anatomy of the aorta, CT, MRI, and U/S are generally preferred for AAA because angiography is invasive, costly, fraught with complications, and not entirely reliable.
- In some hospitals, an arteriogram is necessary for surgery.

Nontraumatic Limb Ischemia

- Although angiography is the gold standard for evaluating peripheral vascular disease, the diagnosis of acute limb ischemia is usually made reliably by clinical exam.

Table 18.1. Grading system for SAH

0	=	Asymptomatic; unruptured aneurysm
1	=	Mild symptoms (headache)
2	=	Severe headache with nuchal rigidity =/- Cranial nerve deficit
3	=	Confusion/AMS +/- decerebrate rigidity
4	=	AMS with moderate to severe hemiparesis +/- decerebrate rigidity
5	=	Comatose with posturing or flaccidity

- The emergency physician will consult a vascular surgeon as soon as acute peripheral ischemia is suspected.
- The vascular surgeon may order angiography in three scenarios: (1) the diagnosis of acute arterial occlusion is uncertain, (2) consideration of emergency vascular bypass grafting, (3) characterization of the vascular abnormality before emergency surgical correction.

V/Q Scan (Fig. 18.7)

Summary

- V/Q is still the recommended initial imaging test for the workup of PE although many institutions have replaced V/Q with CT because of uncertainty with how to manage nondiagnostic V/Q scans.
- CT is an ideal initial test in those patients with high clinical suspicion in whom you may choose to perform CT first or in those with a specific abnormality on CXR in whom CT may diagnose the abnormality, explain the patient's symptoms and exclude PE.

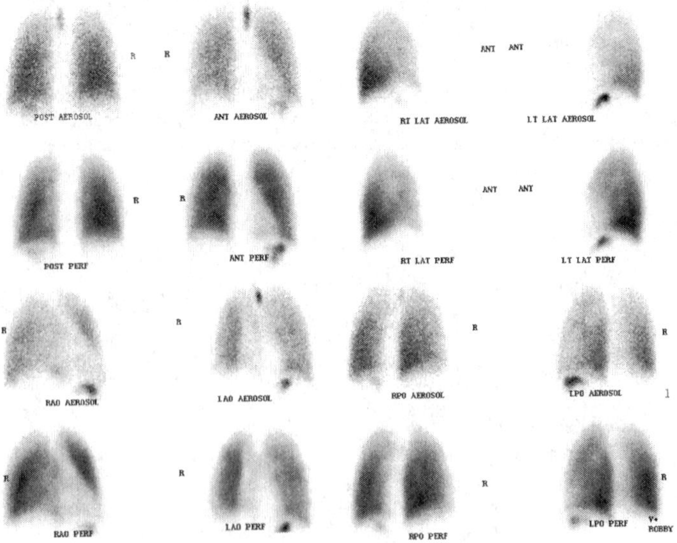

18

Figure 18.7. Normal ventilation-perfusion scan with homogenous distribution of ventilation and perfusion. Top row of images is always ventilation, following row always perfusion.

    - The caveat of this approach is that a negative CT in these patients will require follow-up V/Q or duplex since CT cannot rule-out PE.
- Nondiagnostic (low and intermediate probability) scans are not an acceptable end-point for workup of PE except in patients with low probability V/Q scan and low clinical suspicion who have only a 4% risk of PE and thus could be safely discharged home.
- The management of nondiagnostic scans depends entirely on the pretest clinical suspicion which must be determined before V/Q scan results.
- Patients with nonhigh prestest probability and nonhigh V/Q scans may be candidates for serial duplex scans as outpatients or can be admitted for angiography depending on the patient's stability and patient/physician preferences.
- Patients with high pretest probability and a low probability V/Q scan result or a low pretest probaility and a high probability V/Q scan should receive CT followed by angiography if CT is negative.
- Patients with high pretest probability and intermediate V/Q scans have a 66% probability of PE and are typically given the presumptive diagnosis of PE.

Modified PIOPED Scan Criteria

- High Probability (Fig. 18.8)

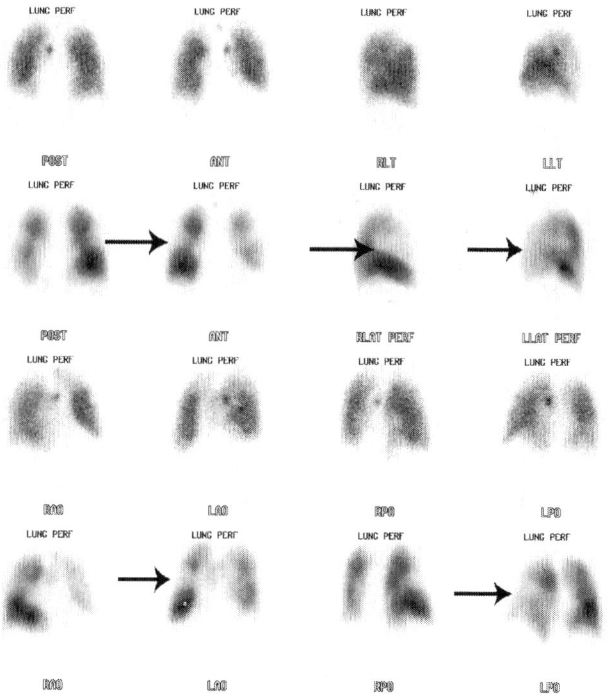

Figure 18.8. High probability lung scan. Large perfusion defects noted without matching ventilation defects.

18

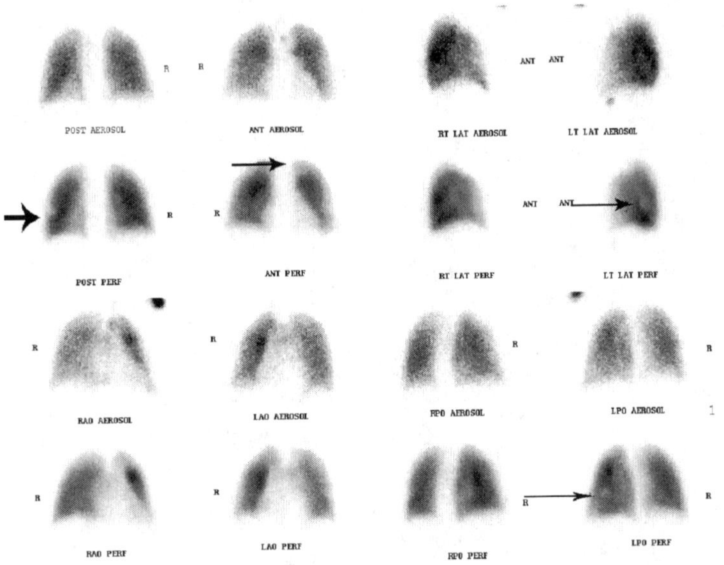

Figure 18.9. Intermediate probability scan with small mismatched perfusion defects not large enough to qualify as a high probability scan.

- Two or more large (>75% of a segment) mismatched segmental perfusion defects or the arithmetic equivalent in moderate (25-75% of a segment) or large and moderate defects.
- Intermediate Probability (Fig. 18.9)
    - <Two large mismatched segmental perfusion defects or the arithmetic equivalent in moderate or large and moderate defects.
    - One matched V/Q defect plus a normal CXR
    - Difficult to categorize as low or high or not described as low or high.
- Low probability (Fig. 18.10)
    - Any perfusion defect with a substantially larger CXR abnormality
    - >One matched V/Q defects plus some normal perfusion plus normal CXR
    - Any number of small perfusion defects plus normal CXR
    - Nonsegmental perfusion defects (e.g., pleural effusion, cardiomegaly, enlarged mediastinal structures, raised hemidiaphragm)
- Normal
    - No perfusion defects seen

Indeterminate Results

- The main criticism of the V/Q scan is the high percentage (60-70%) of indeterminate results (low probability or intermediate probability).
    - Unless the V/Q scan is low probability and the clinical suspicion is low, the diagnosis of PE cannot be excluded or made based on a nondiagnostic study.
    - A reasonable approach to nondiagnostic V/Q scans in patients with nonhigh pretest probability is serial duplex scans of the lower extremities (repeat duplex over a 1-2 wk period) which has a lower false negative rate than pulmonary angiogram (Fig. 18.10).

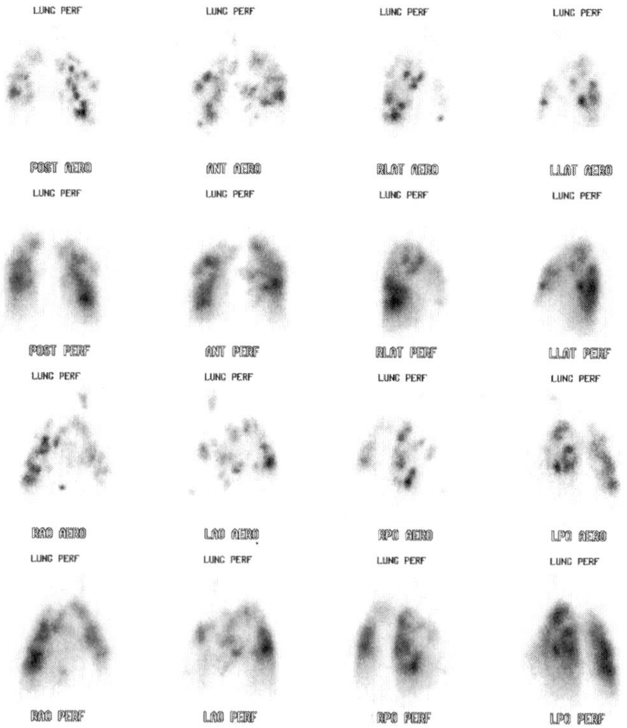

Figure 18.10. Low probability scan with marked nonhomogenous ventilation consistent with chronic pulmonary disease.

MRI

Summary

- As the speed of image acquisition improves and nonmagnetic critical care monitoring equipment becomes available, MRI will become an increasingly important diagnostic tool of the emergency physician.
- Currently there are several emergent indications for MRI where there is no appropriate substitute, and other indications to consider MRI on a semiemergent basis, and finally indications to substitute MRI for angiography and CT because of contrast contraindications.
- Emergent indications:
  - Spinal cord compression—gold standard for nontraumatic cord compression; time-delays in diagnosis may result in irreversible damage.
  - Traumatic spinal cord injury—MRI is the recommended modality for trauma patients with neurologic symptoms and normal plain films.
  - Aortic dissection—One of four possible studies which can be used emergently to diagnose dissection; considered the best all-around study as long as the patient is not unstable.

- Semiemergent indications:
  - Dural venous sinus thrombosis—a rare cause of stroke but one which affects primarily young people and has disastrous consequences; although CT is the emergent test in stroke patients to exclude bleed, MRI should be done soon after if suspicious for DVST because anticoagulation and antibiotics may improve outcome.
  - Cervical artery dissections—another rare cause of stroke seen in younger patients; MRI/MRA is now the recommended first line study although surgery is rarely performed and anticoagulation is of unclear benefit.
  - Spinal cord infections—although antibiotics are typically started based on clinical findings, information gathered from MRI can elucidate whether emergent surgery is necessary (for expanding abscess) to avoid imminent cord compression.
  - Encephalitis/vasculitis—a diagnosis of Herpes encephalitis or cerebral vasculitis can be missed and appropriate treatment not started if MRI is not performed so consider in patients with clinical clues for these processes.
- Substitute for angiography:
  - Subarachnoid hemorrhage—although angiography is the gold standard for aneurysm location and potential embolization, patients who have contraindications to angiography may undergo MRA as a highly accurate alternative.

Complications

- 5% of the population experience claustrophobia in the MR scanner which can be reduced by mild sedation.
- Unlike CT, movement of the patient during MRI will distort all images so patient cooperation is very important.
- The only serious complication of MRI is from the interaction of the high magnetic fields with internal ferromagnetic objects (aneurysm clips, pacemakers, etc) or external metal objects which can be attracted to the magnet and act as missiles if brought into the room.
- Gadolinium, the contrast agent used in many MRI studies of the CNS, does not cause renal failure and allergic reactions are extremely rare.
- Common contraindications for MRI
  - Cardiac pacemaker or permanent pacemaker leads
  - Internal defibrillatory devices
  - Cochlear prostheses
  - Bone growth stimulators
  - Implanted spinal cord stimulators
  - Electronic infusion devices
  - Intracranial aneurysm clips (some)
  - Ocular implants (some) or ocular metallic foreign body
  - Omniphase penile implant
  - Swan-Ganz catheter
  - Magnetic stoma plugs
  - Magnetic dental implants
  - Magnetic sphincters
  - Ferromagnetic IVC filters, coils, stents—considered safe at 6 wk after implant.
  - Tattooed eyeliner

Indications for Emergent MRI

Spinal Cord Compression

- The most compelling reason to order an MRI from the ED is for suspected spinal cord compression.
- MRI has replaced CT myelography as the imaging study of choice for evaluation of the spinal canal because of superior contrast resolution, lack of ionizing radiation, lack

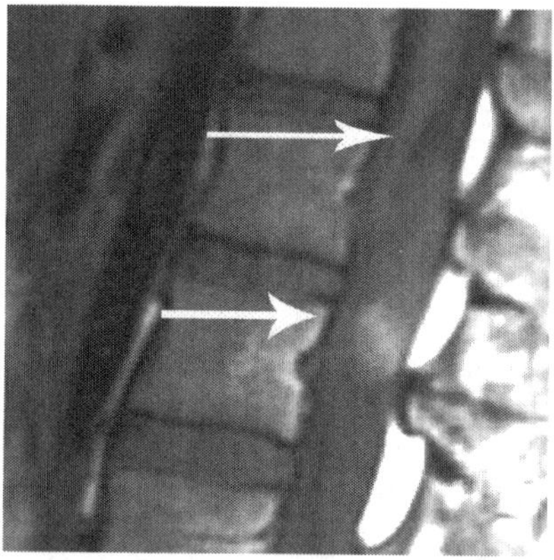

Figure 18.11. Conus ependymoma. Ependymoma in lower lumbar spinal canal with anterior displacement of the conus medullaris.

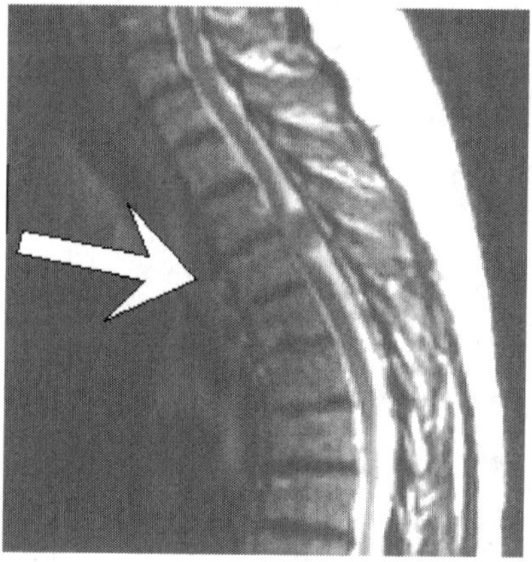

Figure 18.12. Thoracic meningioma compressing the cord with metastasis noted posterior to the vertebrae (arrow).

of contrast material risk, absence of need for spinal needle placement and ability to detect multiple lesions with minimal patient manipulation (Figs. 18.11 and 18.12).

Cervical Spine Trauma

- The American College of Radiology (ACR) published appropriateness criteria for ordering various imaging tests in cervical spine trauma.
- Their only indication for MRI is a patient with neurological signs and symptoms whose plain films are normal (Figs. 18.13 and 18.14).

Dural Venous Sinus Thrombosis

- DVST is a thrombus formation within any of the major dural venous sinuses that drain blood from the brain.
- DVST most commonly affects young and middle-aged individuals (accounts for 1-2% of strokes in patients under 45) and may occur spontaneously or in association with risk factors.
- MRI and MR venography (a noninvasive technique using contrast) has replaced angiography as the study of choice for the evaluation of suspected DVST (Fig. 18.15).

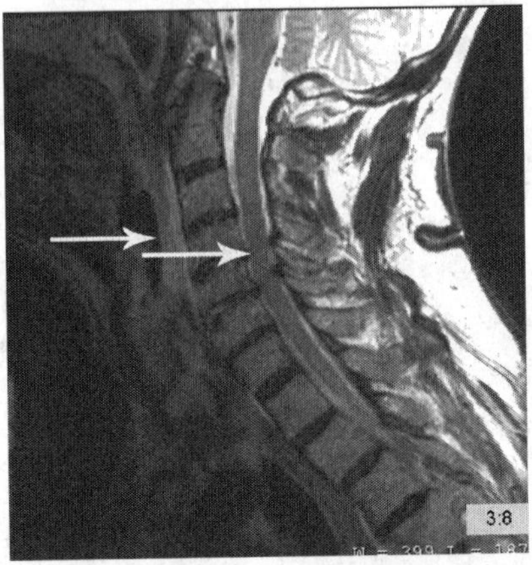

Figure 18.13. Cervical trauma. Left arrow shows increased signal representing blood and edema of anterior vertebral ligamentous injury; right arrow shows bony intrusion into the spinal canal.

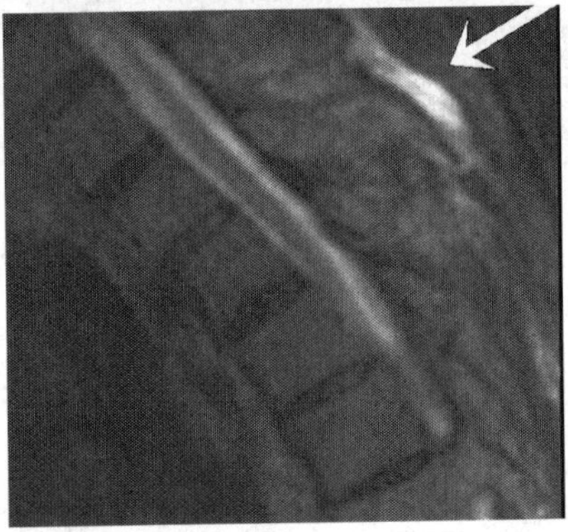

Figure 18.14. Cervical trauma. Arrow shows increased signal of blood and edema along posterior vertebral ligament.

Carotid and Vertebral Artery Dissection

- Carotid and vertebral artery dissections are uncommon but potentially devastating complications of blunt trauma, hypertension, vasculopathy and infections.
- MRI is now an attractive alternative to angiography for the diagnosis of cervical arterial dissections due to its nonivasive nature as well as the ability to simultaneously evaluate the brain, soft tissues and cervical and cerebral vasculature.
  - Characteristic findings on MRI are a periarterial rim of abnormal signal (due to intramural hemorrhage) that expands the outer diameter of the artery and narrows the arterial lumen (Fig. 18.16).

Stroke

- Earlier detection of areas of infarction and areas of reversible ischemia is possible using conventional and advanced MR imaging techniques (Fig. 18.17).
  - In the first 24 hours, more than 90% of MR images show evidence of stroke compared with 60% of CT scans.
  - MR is superior in detection of lacunar stroke and stroke in the posterior cranial fossa, whereas CT is limited by beam hardening artifacts from the adjacent skull base (Figs. 18.18 and 18.19).

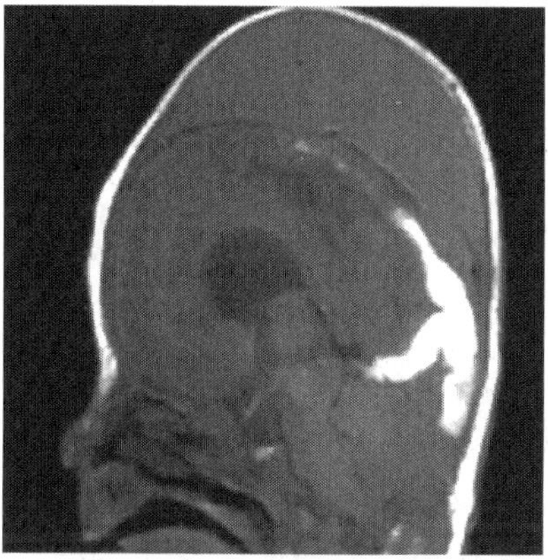

Figure 18.15. Sinus thrombosis. Distended sagittal sinus filled with thrombus, concurrent cephalohematoma.

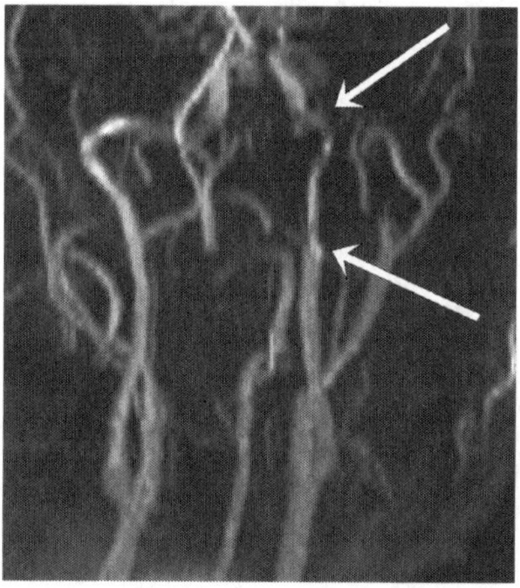

Figure 18.16. Carotid artery dissection. 38 yo male with nontraumatic left internal carotid artery dissection.

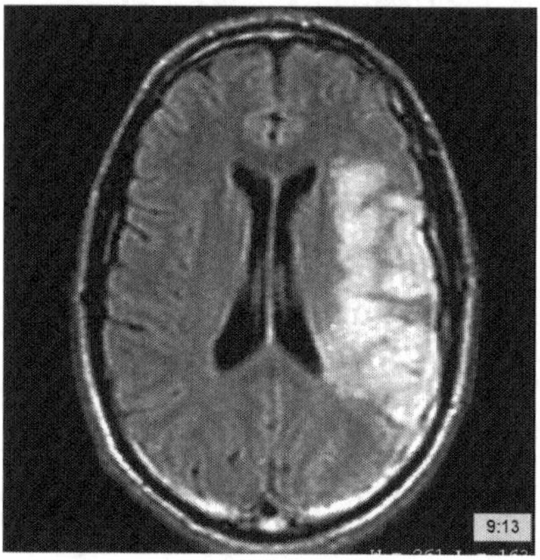

Figure 18.17. Cortical infarct. Same 38 yo male with subsequent large left cortical infarct.

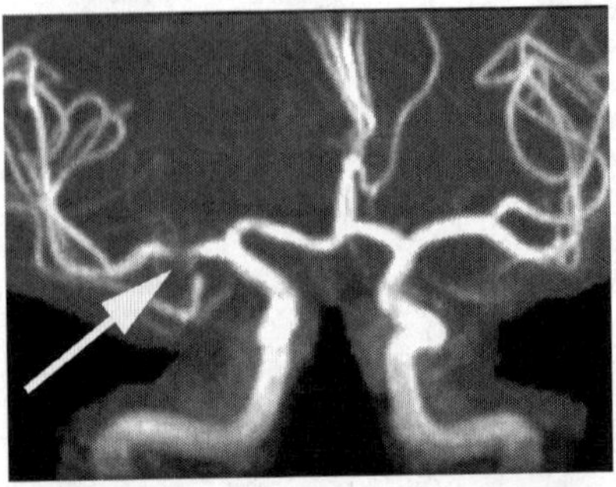

Figure 18.18. MCA occlusion. 25 yo male with right middle cerebral artery occlusion secondary to dissection.

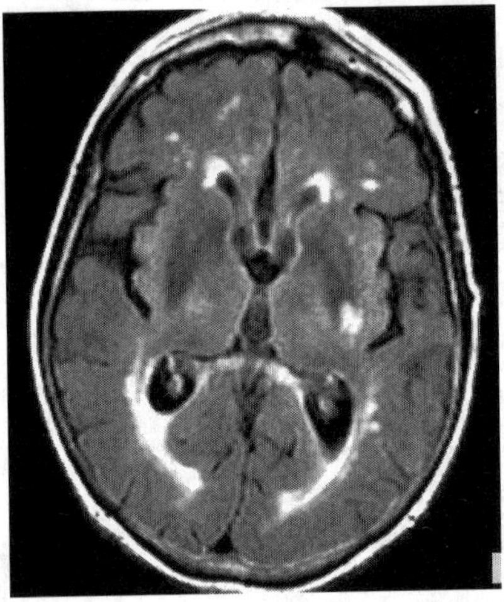

Figure 18.19. Watershed infarct.

Aortic Dissection

- Magnetic resonance imaging is emerging as the premier imaging method for the diagnosis of diseases of the thoracic aorta.
    - In certain applications, a single study can provide information similar to that obtained from a combination of echocardiography, computed tomography, and angiography.
    - The MRI scan shows the site of intimal tear, type and extent of dissection, presence of aortic insufficiency, and differential flow velocities in the true and false channels and in the aortic side branches.

Encephalitis and CNS Vasculitis

- MRI is considered the most sensitive imaging test for detecting CNS lesions caused by infection or vasculitis.
- Although the initial workup for patients with fever and headache or confusion may include CT and LP, the results of these can be nonspecific.
- Since the treatment of herpes encephalitis or a CNS vasculitis differs from bacterial meningitis and since MRI in many cases is the most expedient way to make this diagnosis and initiate the appropriate treatment, emergent MRI may be considered in certain scenarios, although not necessarily from the ED (Figs. 18.20-18.22).

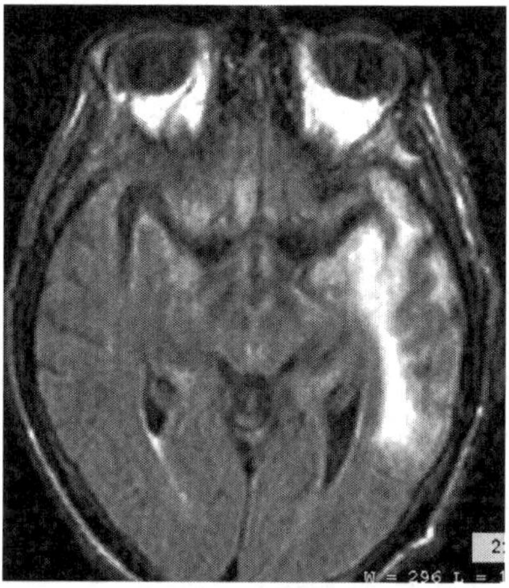

Figure 18.20. PML. Progressive multifocal leukoencephaly seen as high signal asymmetric coalescing white-matter lesions.

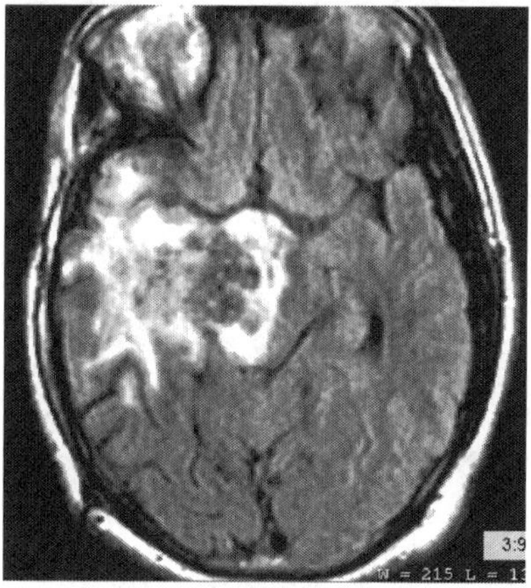

Figure 18.21. Tuberculoma. Contrast-enhanced ring lesions.

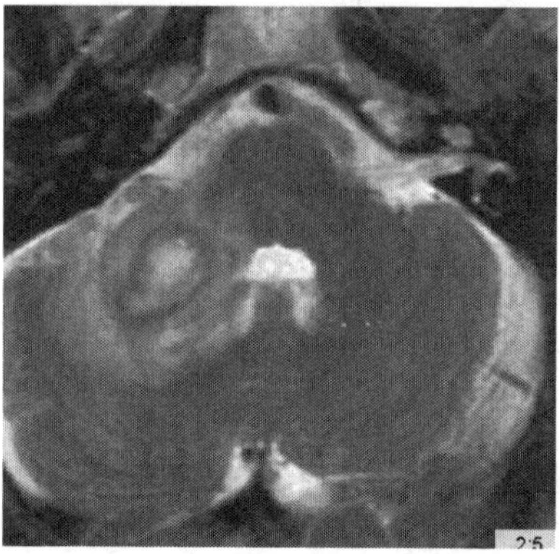

Figure 18.22. Cerebellar abscess.

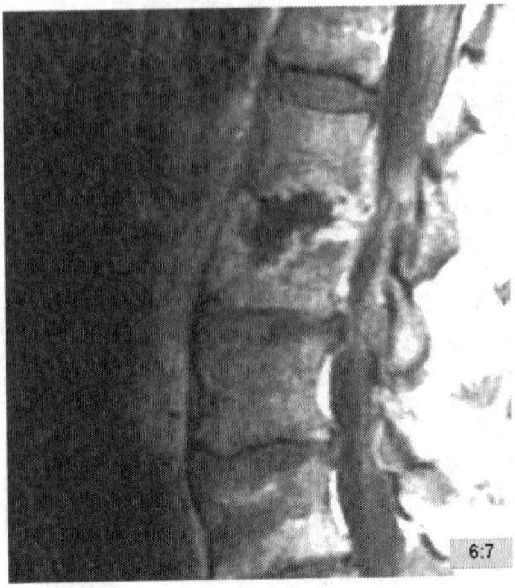

Figure 18.23. Discitis and osteomyelitis.

Spinal Cord Infections

- MRI is the study of choice in the evaluation of infectious and inflammatory disease of the spinal cord and spinal column including epidural abscess, myelitis, vertebral osteomyelitis, and spondylodiscitis.
- The sensitivity of MR for infectious processes of the spine, especially when used with gadolinium, is excellent (Fig. 18.23).

Sestamibi Myocardial Perfusion Scan

Summary

- In summary, sestamibi myocardial perfusion imaging (MPI) using single photon emission computed tomography (SPECT) has rapidly emerged as an effective tool to help limit the number of atypical chest pain admissions.
- A large randomized controlled trial published in 2002 demonstrated that SPECT reduced admissions for noncardiac causes of chest pain by 20% while not effecting the rate of admission for true cardiac chest pain.
- A review of 14 trials with over 12,000 patients with low risk chest pain found that the incidence of hard cardiac events after a normal resting SPECT scan was 0.6%.
- Recent reviews have come to similar conclusions: the *Annals of Emergency Medicine* published a report by the National Heart Attack Alert Program Working Group which concluded "it is unclear whether the technique (sestamibi) will be of value as a screening test in lower risk patients, until more evidence is available, it cannot be recommended at this stage for general use".
- The availability of SPECT and the specific protocol used is institution dependent.
- The sensitivity of SPECT has been further increased by the addition of provocative testing using exercise or in patients who cannot exercise, dipyridamole or adenosine.
- To lower the risk of stress-induced ischemia, patients are typically studied at rest, ideally with the injection of technetium. Once the patient has had two negative troponins and a normal resting perfusion scan, they undergo stress imaging.
- One caveat regarding the use of sestamibi in the disposition decision is that some patients with atypical chest pain may ultimately prove to have a life-threatening noncardiac cause of their pain for which monitored observation is the safest strategy.

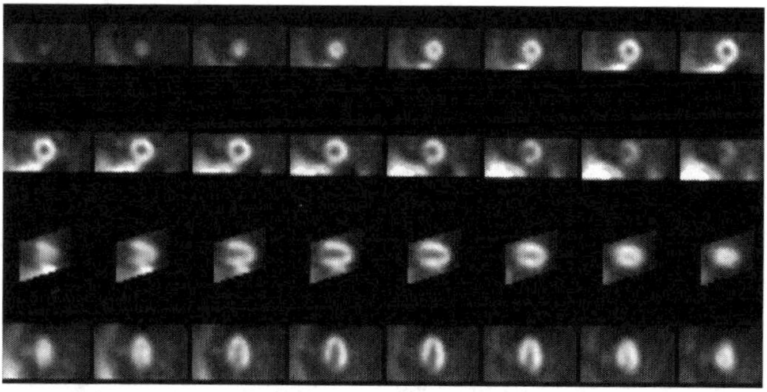

Figure 18.24. Normal sestamibi.

Technique

- Like thallium, technetium is a radioisotope perfusion agent which distributes into the myocardium as a function of viability and blood flow.
- Unlike thallium, technetium has a short half-life so higher doses can be injected to improve image quality without risk of excessive radiation.
- Also unlike thallium, technetium does not immediately redistribute which allows for delayed imaging (1-6 h) so that the patient's pain can be treated with antianginals after technetium injection but prior to imaging.
- This was the critical hurdle which had prevented thallium from being used in the acute evaluation and relegated the thallium stress test to the elective evaluation setting (Fig. 18.24).

Sensitivity and Specificity

- Over the last 10 yr, at least 12 studies of technetium 99m-sestamibi have been performed on patients presenting to the ED with acute chest pain and nondiagnostic ECGs.
- These studies had two objectives: to prove that sestamibi has a high sensitivity for acute coronary syndrome to allow safe discharge of patients with normal scans and to show that sestamibi's specificity for acute coronary syndrome is high enough to prompt intervention: catheterization, thrombolysis, or simply a higher level of care.
- The second of these objectives has clearly been unsuccessful.
- Despite study designs which were often biased to enhance the results, the specificity of sestamibi scans for acute coronary syndrome has been disappointing, with positive predictive values as low as 12%, partly due to the scan's inability to distinguish between new and old perfusion defects.
- Due to the inability of a sestamibi scan to distinguish between new and old injury, an abnormal perfusion scan's specificity for acute injury is far too low to initiate thrombolysis or catheterization in the absence of accepted EKG changes.
- In contrast to the poor specificity, the sestamibi perfusion scan had demonstrated impressive sensitivity for acute coronary syndrome in initial studies.
- Four studies demonstrated 100% sensitivity for AMI with low rates of revascularization (a marker for unstable angina) among patients with normal perfusion scans.
- Because of these impressive results, some institutions are now using the sestamibi scans liberally in patients with acute chest pain to determine disposition.
- There are several reasons to be cautious using the sestamibi scan as a send-home test.

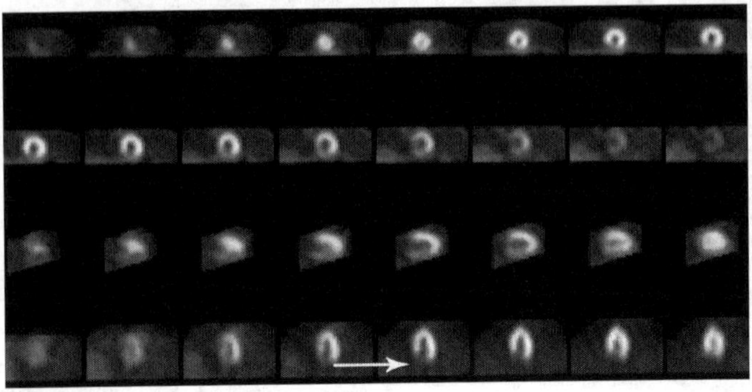

Figure 18.25. Positive sestamibi.

- The primary reason is the absence of a large study with narrow confidence intervals.
- Because the studies with high sensitivities had relatively small sample sizes and rare cardiac events, the confidence intervals, if published, were wide (60%-100%), (a sensitivity for AMI of 60% is obviously too low to send home a patient with possible ischemia).
- The worrisome sensitivity results (<60%) of some small studies and the lower sensitivity (91%-93%) for AMI of three large studies adds more doubt to the sensitivity of the sestamibi scan. If in fact the sestamibi scan's sensitivity is closer to 90% than 100%, then it markedly underperforms the ED missed MI rate (2%) using simply history, EKG and troponin (Fig. 18.25).

Suggested Reading

1. Frykberg ER et al. A reassessment of the role of arteriography in penetrating proximity extremity trauma: A prospective study. J Trauma 1989; 29:1041-1052.
2. Frykberg ER et al. Nonoperative observation of clinicall occult arterial injuries: A prospective evaluation. Surgery 1991; 109:85-96.
3. Reid JD et al. Assessment of proximity of a wound to major vascular structures as an indication for arteriography. Arch Surg 1988; 123(8):942-6.
4. Rose SC et al. Emergency trauma angiography: Accuracy, safety, and pitfalls. AJR Am J Roentgenol 1987; 148(6):1243-6
5. Demaerel P et al. MRI of herpes simplex encephalitis. Neuroradiology 1992; 34:490-493.
6. Hurst RW et al. Neuroradiology of central nervous system vasculitis. Semin Neurol 1994; 14:320-340.
7. Smith AS et al. MR of infectious and inflammatory diseases of the spine. Crit Rev Diagn Imaging 1991; 32(3):165-89.
8. Pretre R et al. Aortic dissection. Lancet 1997; 349(9063):1461-1464.
9. Quint, D. Indications for emergent MRI of the central nervous system. JAMA 2000; 283(7):853-855.
10. American college of radiology: Appropriateness criteria for imaging and treatment decisions. Reston: American College of Radiology, 1995.
11. Bogousslavsky J et al. Ischemic stroke in patients under age 45. Neurol Clin 1992; 10:113-124.
12. Provenzale JM et al. Brain infarction in young adults: Etiology and imaging findings. AJR 1997; 169:1161-1168.
13. Gottschalk A et al. Ventilation-perfusion scintigraphy in the PIOPED study. Part II: Evaluation of the scintigraphic criteria and interpretations. J Nucl Med 1993; 34:1119-1126.

14. Bergstein JM et al. Pitfalls in the use of color-flow duplex U/S for screening of suspected arterial injuries to penetrated extremities. J Trauma 1992; 33:395-402.
15. Erbel R et al. Echocardiography in diagnosis of aortic dissection. Lancet 1989; 1:457-61.
16. Pretre R et al. Aortic dissection. Lancet 1997; 349(9063):1461-1464.
17. Nienaber CA et al. The diagnosis of thoracic aortic dissection by noninvasive imaging procedures. NEJM 1993; 328(1):1-9.
18. Cirarroa, JE et al. Medical progress: Diagnostic imaging in the evaluation of suspected aortic dissection- old standards and new directions. NEJM 1993; 328(1):35-43.
19. Moreno C et al. Hemorrhage associated with pelvic fracture: A multispecialty challenge. J Trauma 1986; 26:987.
20. Gilliland MD et al. Factors affecting mortality in pelvic fractures. J Trauma 1982; 22:691.
21. Rothenberger DA et al. Major vascular injuries secondary to pelvic fractures: An unsolved clinical problem. Am J Surg 1978; 136:660.
22. Ben-Menachem Y. Exploratory angiography and trancatheter embolization for control of arterial hemorrhage in patients with pelvic ring disruption. Tech Orthop 1995; 9:271-274.
23. Wolinsky P. Assessment and management of pelvic fracture in the hemodynamically unstable patient. Ortho Clin of North America 1997:28(3).
24. Fabian TC et al. Prospective study of blunt aortic injury: Multicenter trial of the american association for the surgery of trauma. J Trauma 1997; 42:374.
25. Bynoe RP et al. Noninvasive diagnosis of vascular trauma by duplex U/S. J Vasc Surg 1991; 14:346-352.
26. Knudson MM et al. The role of duplex U/S arterial imaging in patients with penetrating extremity trauma. Arch Surg 1993; 128:1033-1038.
27. Iskander S, Iskandrian AE. Risk assessment using single-photon emission computed tomographic technetium-99m sestamibi imaging. J Am Coll Cardiaol 1998; 32:57-62.
28. Udelson JE. Myocardial perfusion imaging for evaluation and triage of patients with suspected acute cardiac ischemia: a randomized controlled trial. JAMA 2003; 289(2):178.

CHAPTER 19

Computed Tomography: Emergency Department Applications

Sujal Mandavia and Diku Mandavia

Computed Tomography (CT) of the Head

CT scanning has replaced most other modalities of imaging in the acute evaluation of neurologic injury or disease. No longer are we taught to look for pineal gland calcification on skull film as a way of detecting midline shift, nor do we order plain skull films to identify skull fractures, when what we are really concerned about is underlying brain injury. (Although there may still be a role for skull films in selected cases.)

As CT technology has evolved, so has the acute treatment of many neurologic illnesses. Neurologic injuries or illnesses are typically time-sensitive diagnoses, much as myocardial infarction, with survival dependent on time to treatment. A good example of this is treatment of subdural hematoma, whose mortality goes from 90% to 30% if operative treatment is within 4 h. Rapid, available and accurate CT scanning is no longer the property of tertiary academic centers but now ubiquitous throughout the healthcare landscape. For this reason, it is important for everyone to understand the utility and limitations of the technology.

This chapter will review the utility of CT scan in selected diagnoses and touch on some of the pitfalls associated with each. The value of magnetic resonance imaging (MRI) will be discussed in a separate section.

CT in Nontraumatic Conditions

Subarachnoid Hemorrhage (SAH)

- CT preferred over MRI for acute SAH
- Detects over 90% within 24 h. (corollary is that 10% are missed)
- May become negative after 3 days with change in blood density.
- Etiology most commonly aneurysmal.
- Lumbar puncture required to rule out acute SAH after negative CT scan.
- Findings include blood in the sulci, interpeduncular cisterns, sylvian fissures, or posterior falx cerebri.
  - Pitfalls:
    - Failure to perform lumber puncture after negative head CT to rule out occult SAH (false negative rate up to 5% with CT alone.)
    - CT scan will be negative after 3 days, but lumbar puncture shows xanthochromia.
    - Sensitivity may be decreased in setting of anemia.

Cerebral Vascular Accident (CVA)

- May not be evident on initial scan within 24 h.
- Early signs include hyper-dense artery (usually MCA distribution) due to thrombus in the artery, loss of basal ganglia, loss of grey-white border distinction.
- Subacute stroke will often show slight mass effect.

Emergency Medicine, edited by Sean Henderson. ©2006 Landes Bioscience.

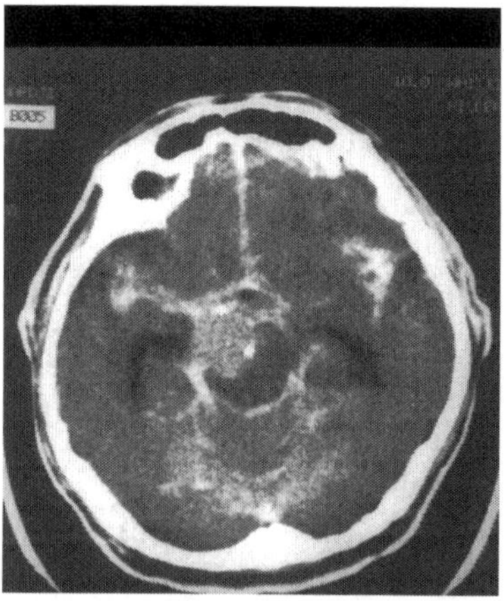

Figure 19.1. Subarachnoid Hemorrhage. Note the bright signal of the subarachnoid hemorrhage collecting in the sulci and cisterns.

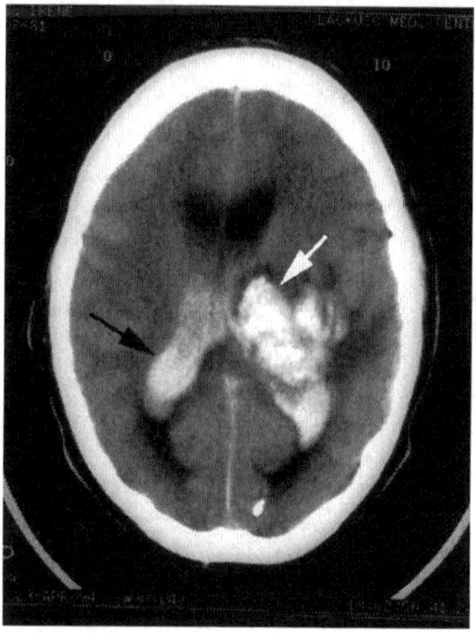

Figure 19.2. Hemorrhagic basal ganglia cerebral vascular accident (white arrow) with rupture into ventricular space (black arrow).

- Subacute to chronic CVA will gradually become hypo-dense to normal brain tissue.
- Acutely, CT scan is primarily useful to rule out associated hemorrhage or other stroke mimics, such as subdural hematoma, tumor, etc.
- Nontraumatic hemorrhagic CVA most commonly hypertensive in etiology with the following locations most common:
  - Putamen
  - Thalamus
  - Pons
  - Cerebellum
- Pitfalls in CVA:
  - Acute CVA may not be evident on CT within 24 h.
  - Cerebellar anatomy not clearly visualized on CT and may require thin cuts through the posterior fossa or MRI.

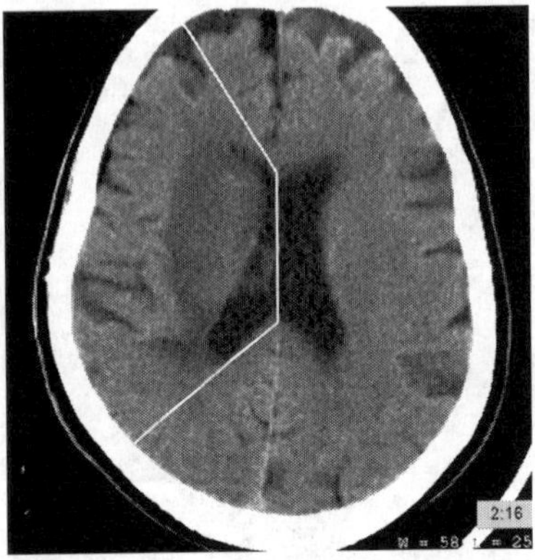

Figure 19.3. Rigł.t middle cerebral artery territory cerebral vascular accident. Note the edema and loss of gray-white junction seen in the highlighted area.

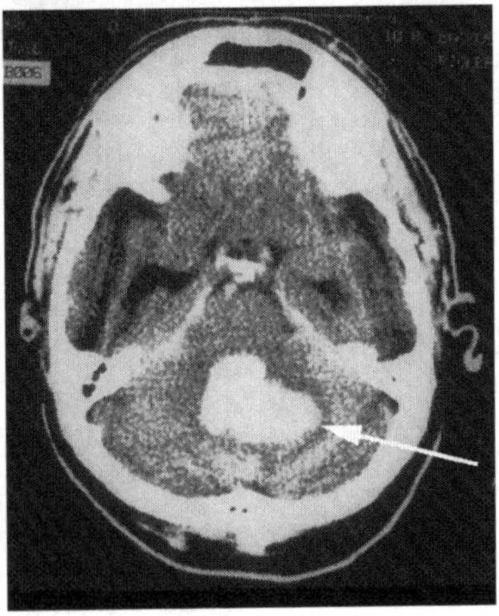

Figure 19.4. Cerebellar hemorrhage.

Infectious Processes

Toxoplasmosis

- Seen in AIDS or other immunocompromising illnesses.
- Most commonly focal, but may be diffuse.
- Focal disease ring enhanced with intravenous contrast.
- Commonly found at basal ganglia and gray-white border of the cerebral hemispheres.
- Differential includes lymphoma and may be impossible to distinguish without biopsy or empiric treatment for Toxoplasmosis.

Cysticercosis

- Parasitic infection arising from larvae of pork tapeworm, most often occurring in Latin America.
- Headache or seizures are common presenting symptoms.
- Lesions are most often found at the gray-white border and in the ventricular system.
- Has four distinct phases, which can co-exist: vesicular, colloidal, granular and nodular.

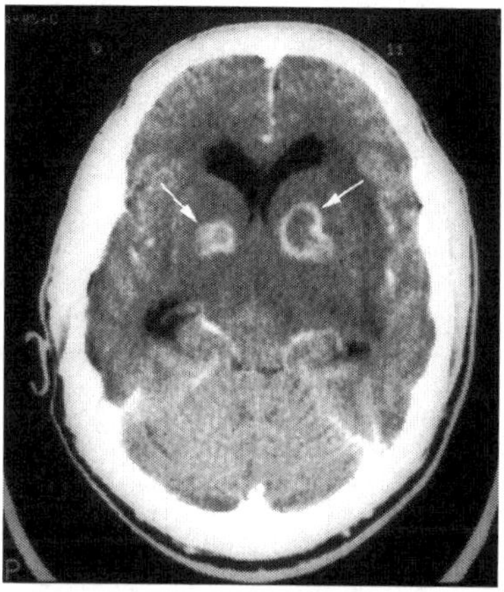

Figure 19.5. Toxoplasmosis. Bilateral ring enhancing basal ganglia lesions are seen on this contrast CT scan.

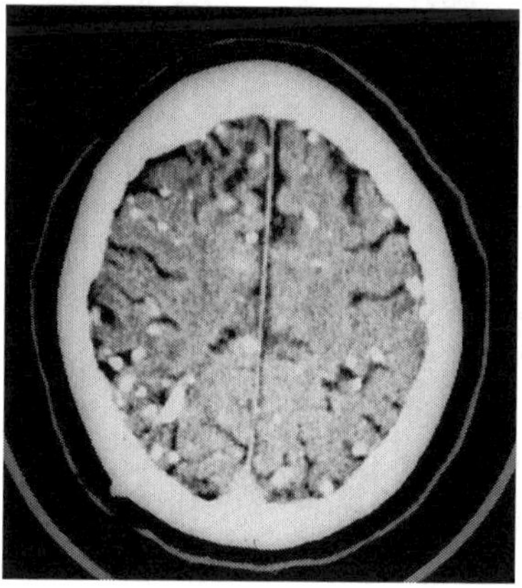

Figure 19.6. Neurocysticercosis. Multiple calcified lesions are seen on this non-contrast CT scan.

- Vesicular lesions are cyst-like in form and minimally enhance with contrast. The parasite dies during the colloidal phase and produces an inflammatory process with surrounding edema and slight contrast enhancement. The granular phase produces a ring-enhancing lesion, and the nodular phase involves a calcified lesion with no contrast enhancement.

Hydrocephalus

- May be classified as communicating (problem with CSF re-absorption) or non-communicating (obstruction within the ventricular system.)
- CT scan shows dilation of the ventricular system. In communicating hydrocephalus, the whole ventricular system will be dilated, whereas in non-communicating, the system will be dilated proximal to the level of obstruction.
- Acute hydrocephalus will show dilation of the ventricular system with the lateral temporal horns being an early indicator.
- Periventricular edema is also an indicator of acute hydrocephalus.

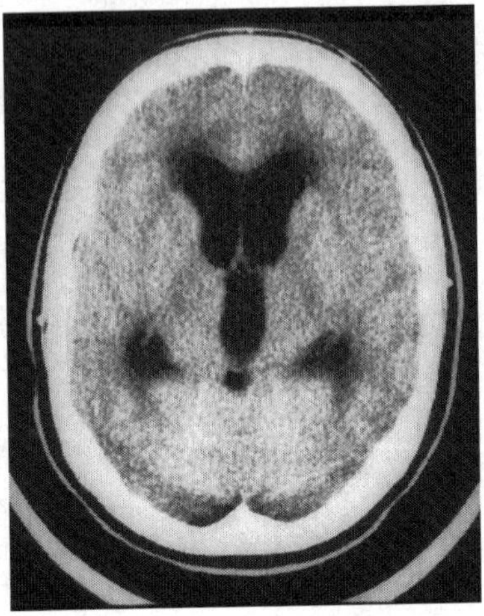

Figure 19.7. Hydrocephalus. Ventricular enlargement is seen at the level of the lateral ventricles and temporal horns, and the third ventricle.

- Non-communicating hydrocephalus may be caused by hemorrhage, mass or congenital stenosis of the ventricular system.
  - CT scan may be used to evaluate the patency and function of ventricular shunt, and should always be compared with prior films to determine if ventricular size has increased, indicating shunt failure.

Trauma

Epidural Hematoma

- Collection of blood in the potential space between the dura and skull. It is most commonly found in the temporal region and most often has an overlying skull fracture. This type of injury has a better prognosis (90% survival) than subdural because the skull has often absorbed the majority of the force involved. The epidural hematoma has distinct characteristics based on its pathophysiology:
  - It is lenticular (lens shaped) in form as it separates the dura from the skull.
  - It does not cross suture lines as the dura is attached to the skull at these points.
  - It has an associated fracture in 95% of cases seen on the bone windows of CT.
  - It may exhibit "swirling", an inhomogeneous appearance of the bleed, due to active arterial bleeding within the epidural hematoma.
  - Less common in infants and the elderly as the dura is firmly attached in both these populations.
- Pitfalls in epidural hematoma:
  - Failure to diagnose small epidural which can either expand or rebleed, causing neurologic deterioration.
  - Obtaining the CT scan too early. Really. Although an uncommon problem, especially in inner-city hospitals, a CT scan obtained too quickly may not detect small epidural hematomas, which can expand and cause secondary brain injury.
- Subdural hematoma:
  - Collection of blood in the potential space between the dura and the pia mater. The source of bleeding is usually due to the bridging veins as they empty in the dural sinuses. Subdural hematomas may occur without direct head injury in the case of severe acceleration/deceleration injury. Subdural hematoma has a 30% mortality

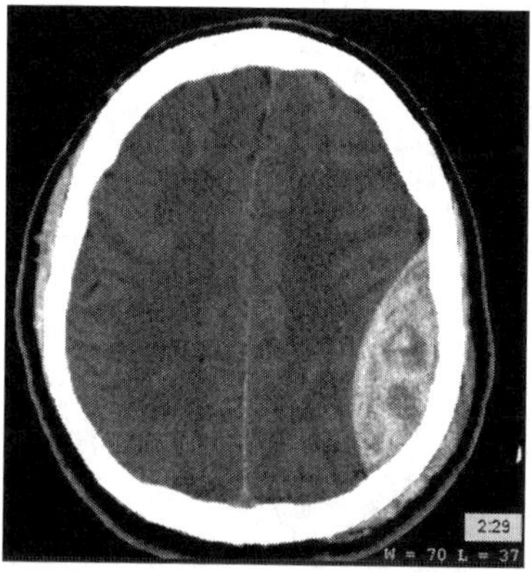

Figure 19.8. Epidural hematoma. The lenticular-shaped hemorrhage is seen in the left temporal-occipital region.

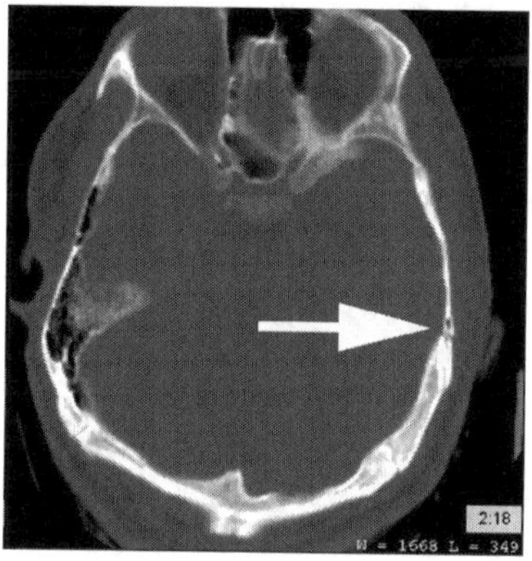

Figure 19.9. Underlying temporal skull fracture associated with the above epidural hematoma.

rate (with appropriate treatment) since underlying brain injury is both common and severe.

- Characteristics of subdural hematoma on CT scan include:
  - Crescent shaped.
  - Confined by dural attachments at the midline.
  - May be acute, subacute or acute and chronic.
  - May show a hematocrit effect, where the red cells settle from the serum and produce a fluid level on the CT scan. This occurs in a patient who remains supine for a prolonged period of time.
- Pitfalls in subdural hematoma:
  - A small subdural may only appear as increased density of the tentorium as the subdural blood tracks along it.
  - Failure to recognize the iso-dense subdural. Often the only clues may be subtle mass effect or loss of sulci on the side of the subdural.
  - Failure to recognize bilateral injuries. Subdurals will be bilateral in 15% of cases, and with iso-dense or chronic blood the diagnosis can be very difficult.
- As mentioned above, the appearance of blood will change on CT scan depending on age of the bleed. This differentiation is clinically helpful in determining age of injury.

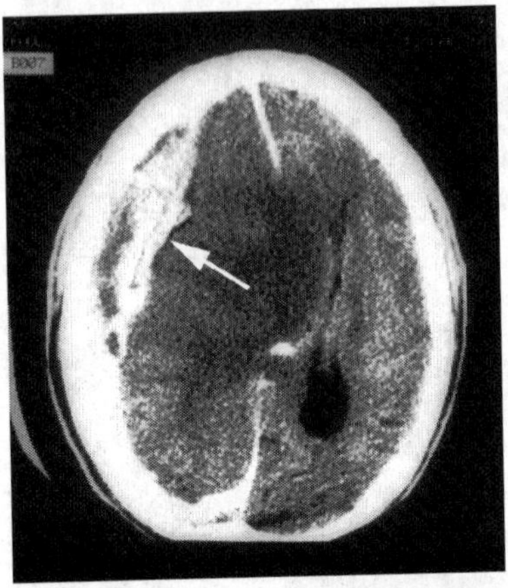

Figure 19.10. Acute subdural hematoma. A right frontal-temporal subdural hematoma with midline shift is demonstrated in this CT scan of a fall victim.

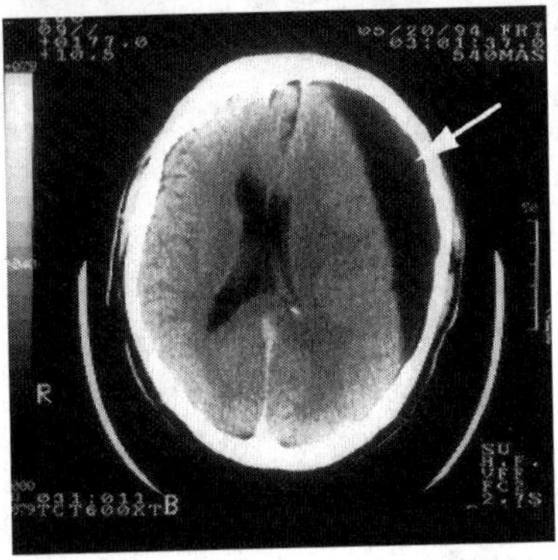

Figure 19.11. Chronic subdural hematoma. A left chronic (hypo-dense) subdural with midline shift is seen on this CT.

Table 19.1. Comparison of CT findings of EDH and SDH

Finding	EDH	SDH
Shape:	Lenticular	Crescent
Bleed confined to:	Will not cross suture lines	Will not cross midline falx
Source of bleed:	Arterial (may see swirling)	Venous
Skull fracture associated:	Yes	No
Rebleed:	Yes	Less common

- Acute blood (<7 days) is hyper-dense to the brain tissue and often easily recognizable.
- Subacute blood (between 7 days and 3 wk) is often iso-dense with the brain tissue and has subtle findings of mass effect on CT.
- Chronic blood (over 3 wk of age) will be hypo-dense in comparison with brain tissue.

Other Injuries

- Contusions are classified as either hemorrhagic or non-hemorrhagic and are relatively common in head trauma.

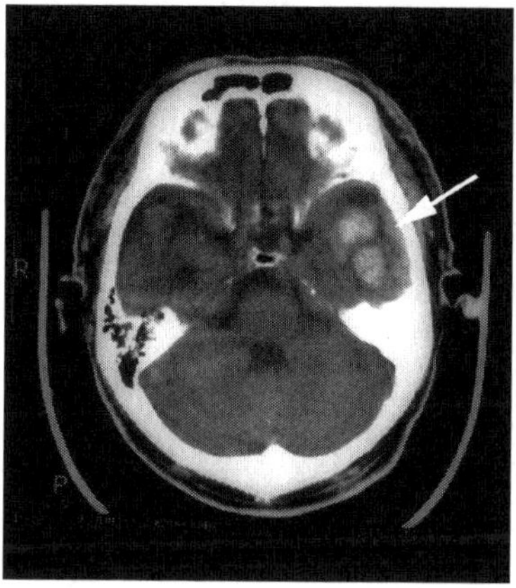

Figure 19.12. Left temporal hemorrhagic contusion.

- Contusions are most common in the frontal, occipital and temporal lobes and may be the result of depressed skull fractures, direct head trauma ("coup" injury), or opposite the site of injury, ("contre-coup" injury).
- Non-hemorrhagic contusions are often not visualized on initial CT scan but will tend to show mild edema or hemorrhagic conversion over 24-48 h. Hemorrhagic contusions are seen on initial CT scan and will also demonstrate edema over the following couple of days.

Diffuse Axonal Injury (DAI)

- Though patients usually present clinically in deep coma, the CT scan is often normal in appearance (70-80%).
- This type of shearing injury leads to disruption of the grey-white border, corpus callosum or brainstem. This accounts for the profound neurologic deficits. Unless there are associated small hemorrhages, the initial CT scan may reveal no abnormality.
- MRI is sensitive for the detection of diffuse axonal injury, though not often available or practical in the acute poly-trauma patient.

Skull Fractures

- Skull fractures are readily identified on the bone windows of CT scan, but small non-displaced linear skull fractures may be missed on CT. Most nondisplaced fractures are not clinically significant but for underlying brain injury.
- Skull fractures depressed more than one bony table width may require elevation and are more often associated with underlying injuries.
- Basilar skull fractures are the exception to CT sensitivity. The clinical exam is still more reliable than CT scan in the detection of basilar skull fractures. Clinical signs include:
  - Raccoon eyes: bruising of the eyelids sparing the tarsal plate, signifying bleeding tracking from behind the orbits rather than direct trauma.
  - Battle's sign: bruising of the mastoid due to associated fracture. This sign may be delayed up to 48 h and is not reliable in excluding fracture.
  - Hemotympanum: Blood behind the tympanic membrane is seen immediately after basilar skull fracture.
  - CSF leak: either otorrhea or rhinorrhea may be associated with basilar skull fracture and may lead to meningitis. Prophylactic antibiotics are not usually recommended but remain controversial.
  - Cranial nerve deficits: Look for cranial nerve deficits of the facial, auditory or trigeminal or ophthalmic nerves.

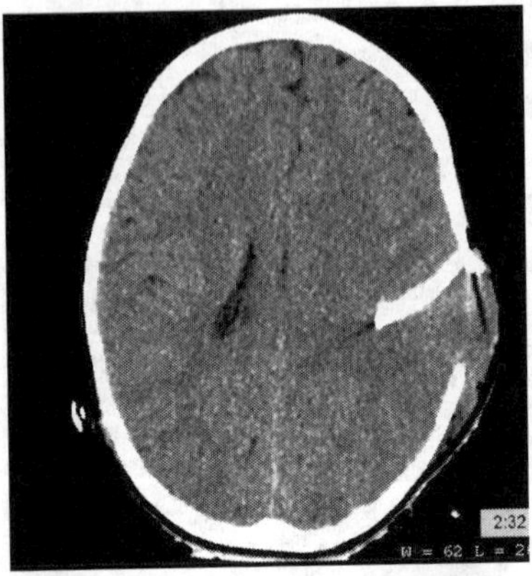

Figure 19.13. Left temporal depressed skull fracture on CT.

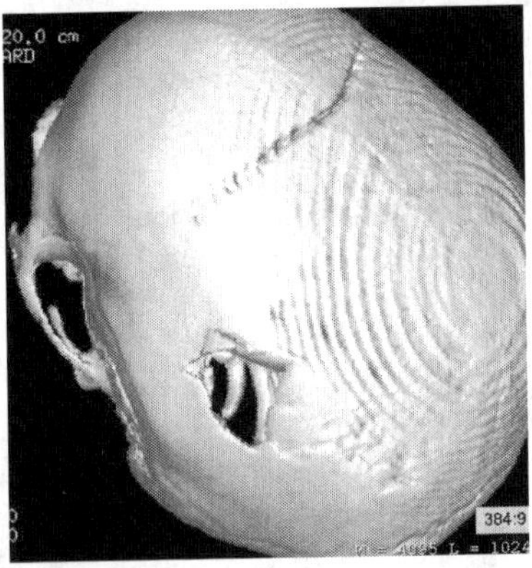

Figure 19.14. Three dimensional reconstruction of the same temporal skull fracture.

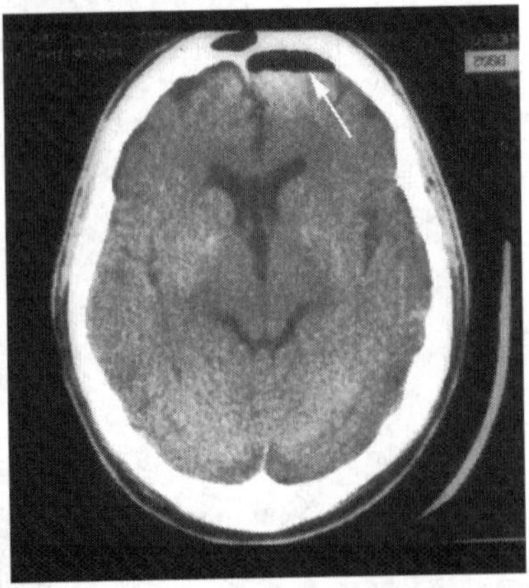

Figure 19.15. Pneumocephalus appears as black space on CT scan. This should prompt a search for an underlying basilar skull or sinus fracture.

Chest CT

Trauma

Although CT scan is much more sensitive for thoracic injuries than is chest X-ray, the CT does not usually add clinical information that changes management. There are often very small pneumothoraces or contusions picked up incidentally, which require only observation. The exception to this is the mediastinal hematoma, which may be a marker for underlying aortic injury.

Preparation of Patient

- Patients should have BUN/creatinine checked prior to intravenous contrast if there is risk of renal insufficiency.
- The oral hypoglycemic metformin may cause lactic acidosis when renal failure is present. Therefore, it should be held for 48 h after a contrast CT scan in case renal failure develops.
- No oral contrast is necessary for chest CT.
- Generally any patient undergoing chest CT for either traumatic or non-traumatic conditions will require ACLS-capable escort to CT.

Specific Conditions

Acute Aortic Disruption

- This condition is fatal in the majority of patients, who will die on scene of trauma. The patients that survive are usually harboring a silent but fatal condition requiring a high index of suspicion. As will be discussed, plain radiography and even CT scan are used as screening tests instead of more invasive diagnostics such as aortography.
- Diagnosis of aortic disruption starts with suspicion of the injury, usually because of mechanism of injury, associated injuries or chest X-ray findings. Any high-speed acceleration or deceleration injury or fall leads to the aortic tear, which is at the root of this injury. The tear is most often at the descending aorta just past the left subclavian artery. The aorta is fixed at this level by the ligamentum arteriosum.
- Injury patterns seen in association with aortic disruption include flail chest, first rib fractures, sternal fracture or pelvic ring disruption. Chest X-ray findings are numerous and are covered elsewhere in this handbook.
- CT scan serves as a screening tool for mediastinal hematoma in the diagnosis of traumatic aortic disruption. The mediastinal hematoma is rarely from an aortic tear itself, but more often from small mediastinal vessels. If no hematoma is visualized, then no aortography is necessary. Mediastinal hematoma requires either CT aortography or conventional aortography to rule out aortic injury.

Pitfalls in Aortic Disruption

- Failure to consider the diagnosis. Any patient with high-risk mechanism or symptoms/signs should undergo evaluation for aortic disruption.
- Failure to order CT scans in the presence of a normal mediastinum on chest X-ray. Aortic disruption or its precursor, aortic intimal flap is well described with normal appearance on chest X-ray.

Non-Trauma

Aortic Dissection

- Aortic dissection is another potential disastrous condition that is often delayed in diagnosis due to variable presentation. The prognosis is good with early diagnosis and treatment.

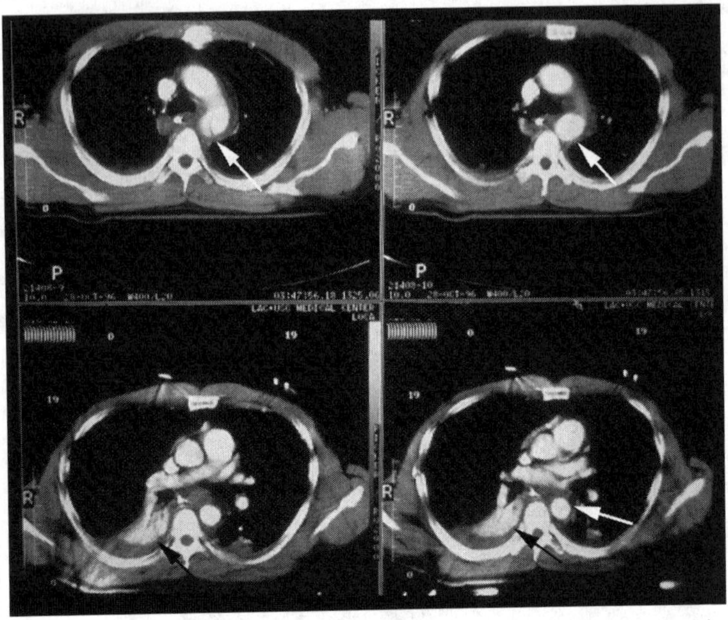

Figure 19.16. These four images reveal an aortic disruption. The white arrows reveal the intimal tear and surrounding mediastinal hematoma. The black arrows reveal an associated hemothorax.

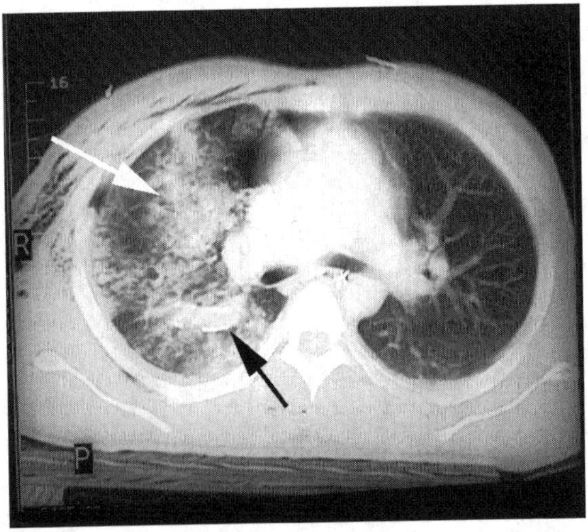

Figure 19.17. Pulmonary contusion. White arrow shows the pulmonary contusion. The black arrow outlines a chest tube previously inserted for a pneumothorax.

- The pathogenesis of aortic dissection begins with an intimal tear which will eventually dissect and in its worst form will either rupture proximally into the pericardium or rupture the surrounding adventia into the thorax, resulting in death.
- Aortic dissection is generally classified as either Stanford type A arising in the proximal aorta or Stanford type B arising in the aorta distal to the ligamentum arteriosum.
- Risk factors for dissection include hypertension, smoking, atheroscelerotic disease, collagen disorders and pregnancy.
- Clinical presentation is quite variable depending on site of dissection or occlusion of vessels. Pain, either located in the chest or back is present in almost all cases. The clinical presentation is covered in more detail elsewhere in the manual.
- There are many imaging options for aortic dissection, each with particular strengths or weaknesses:
  - CT scan: CT aortography is a sensitive and specific test (80-100% and 90-100% respectively) for aortic dissection and is readily available in most facilities on a 24 h basis. It is noninvasive and may be performed quickly. It has replaced aortography for the initial evaluation of aortic dissection in most facilities.
  - Disadvantages of CT scan include:
    - Need for intravenous contrast dye.
    - Inability to identify small intimal flap.
    - Inability to identify involvement of the coronary arteries.
  - Aortography: CT scan has almost replaced invasive conventional aortography as the primary method of investigating aortic dissection. Aortography may still have a role in questionable CT results or if detail is necessary regarding location of intimal flap or coronary artery involvement. It has a sensitivity close to that reported for CT scan.
  - Disadvantages of aortography include:
    - Invasive test.
    - May not visualize dissection in the setting of thrombus in the false lumen.
    - Not readily available, and angiography teams often take 1-2 h to assemble even under best circumstances.
    - Large intravenous contrast dye load.
  - MRI: Magnetic resonance imaging is covered elsewhere in this handbook; however it deserves mention as a very sensitive test for aortic dissection with sensitivities reported as 100%. It has a number of key disadvantages that limit clinical use.

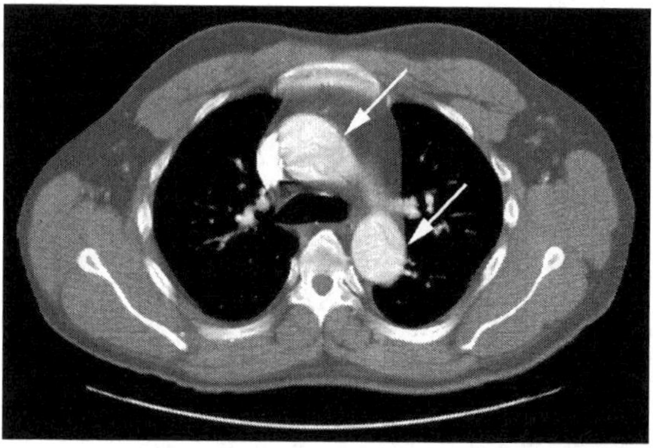

Figure 19.18. Aortic dissection is seen in the proximal ascending and descending aorta. The white arrows identify the intimal tear.

- Disadvantages of MRI include:
  - Lack of availability.
  - The MRI scan is time-consuming, and proper monitoring of these critical patients is difficult.
- Transesophageal ultrasound (TEE): TEE is an excellent diagnostic tool for evaluating aortic dissection. It may be performed at the bedside, making it an excellent choice for unstable patients. It has excellent sensitivity (100%) and specificity and requires no intravenous contrast. It provides detail of the pericardium and the aorta as well as the coronary arteries. As availability becomes more widespread, we may see its role increased.
- Disadvantages of TEE include:
  - Minimally invasive.
  - Availability, especially on a 24 h basis.

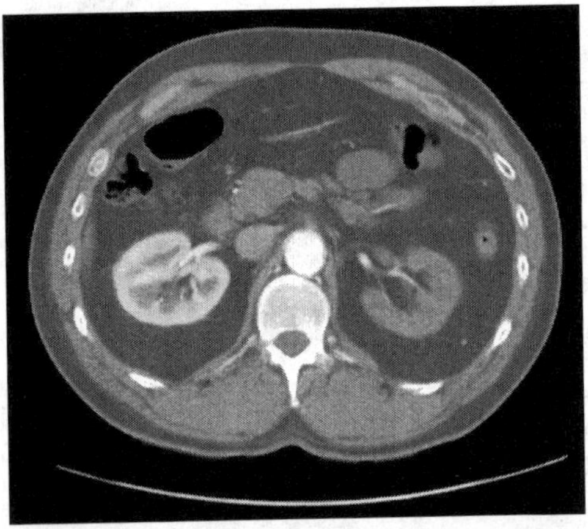

Figure 19.19. This image from the same patient shows a non-enhancing left kidney due to renal artery occlusion from the dissection.

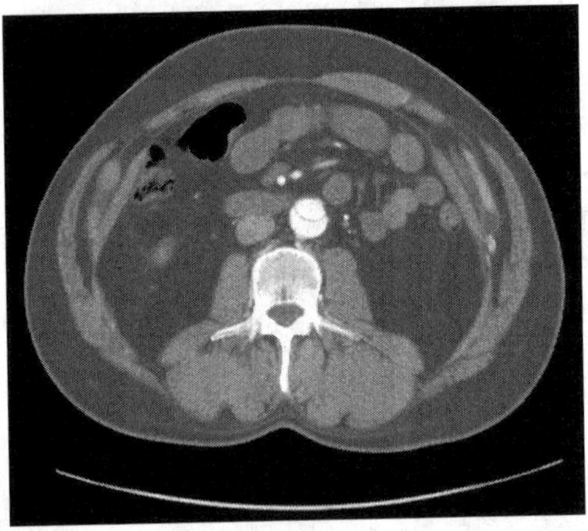

Figure 19.20. This CT image of the descending aorta shows the intimal flap of the dissection in the mid-lumen.

Pulmonary Embolus

Advances in helical CT scanning have allowed CT to become quickly incorporated into the algorithm of evaluation of pulmonary embolus. While details of its role are being resolved in clinical trials, there is no doubt it will be a major tool in the diagnosis of pulmonary embolus.

Current evaluation of pulmonary embolus depends heavily on Bayesian algorithms involving indirect testing for PE. The history is paramount in the primary investigation of pulmonary embolus as it will determine the pre-test probability of all other diagnostics to follow. Traditionally, suspected pulmonary embolus is investigated using ventilation-perfusion nuclear scanning utilizing labeled radioisotopes to assess pulmonary arterial perfusion defects, indirect evidence of pulmonary embolism. Duplex ultrasound of the deep veins of the legs may also be used to search for venous thrombosis, which would also require anticoagulation and hopefully obviate the need for further investigation for pulmonary embolus. V-Q scanning is very sensitive and therefore a normal scan will give a negative predictive value of 100%, ruling out the disease. A high probability scan as determined by criteria involving visualized perfusion defects not matched with ventilation defects, indicates presence of pulmonary embolus in over 90% of patients with high clinical suspicion of pulmonary embolus. A low probability V-Q scan in combination with low clinical suspicion yields <5% incidence of pulmonary embolus.

The problem in evaluating suspected pulmonary embolus patients comes with any combination of clinical suspicion and scans results other than low/ low or high/ high. This is the indeterminate gray zone, comprising the majority of patients. Traditionally pulmonary angiography is used to rule out pulmonary embolus; however most patients do not undergo this test and may either be anticoagulated inappropriately or not treated for their pulmonary embolus. Spiral CT scanning of the chest may help decrease the need for angiography.

Helical, or spiral CT scanning as it is termed, is a recent advance in CT technology whereby the patient is moved through the CT beam as it circles the patient thereby cutting scan time to 20 seconds with newer generation scanners. The patient will ideally hold her breath during the scan and scan cuts are usually 2 mm in width for pulmonary embolus evaluation. Patients receive intravenous contrast bolus at the time of scanning to provide adequate visualization of the pulmonary vasculature. Sensitivities in trials have been reported in a range of 50-100% but usually with high specificities. The origin of such variability is probably with difference in scan techniques or whether subsegmental pulmonary emboli were considered.

Inherent Advantages of Helical CT Scanning

- Direct visualization of the embolus rather than indirect evidence of pulmonary embolus as with perfusion scanning.
- It may suggest an alternative diagnosis that may change treatment plan, i.e., you may not want to anticoagulate the patient with aortic dissection.
- Fast and readily available, usually on a 24 h basis.
- Highest inter-observer reliability when compared with V-Q scan or angiography.

Disadvantages of Helical CT Scanning

- May miss small subsegmental pulmonary emboli.
- Large intravenous contrast load.
- Breath-holding and contrast variation may affect results of scan.
- Negative predictive value has not yet been adequately established but will probably not match a normal perfusion scan.

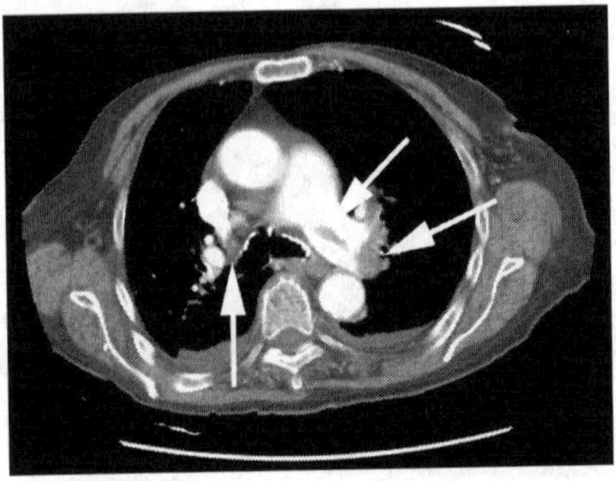

Figure 19.21. This large saddle embolus causes filling defects bilaterally as shown by the white arrows.

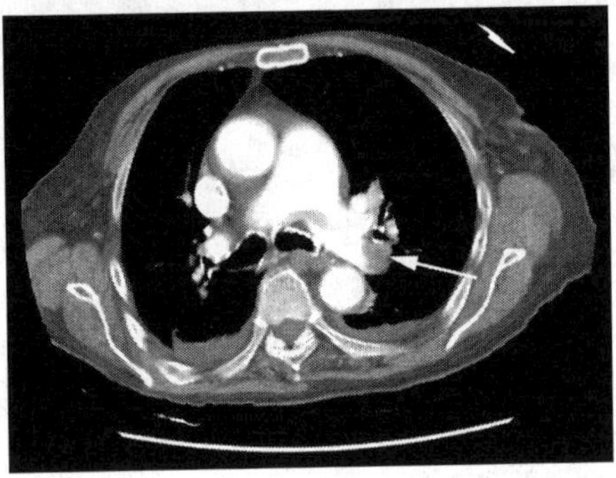

Figure 19.22. This view from the same patient also reveals the large filling defect occluding the left main pulmonary artery (white arrow).

Today's role of helical CT in evaluation of suspected pulmonary embolus is primarily in those patients with significant cardiopulmonary disease or abnormal chest X-ray, as these groups are most likely to have an indeterminate perfusion scan.

Abdominal/Pelvic CT Scans

CT scanning has drastically changed the evaluation of abdominal disorders and has decreased dramatically the number of unnecessary laparotomies, both traumatic and nontraumatic. It has allowed us to open the "black box" which hides many catastrophic processes, guiding or changing management. While a complete discussion of use of abdominal CT is beyond the scope of this chapter, selected emergency conditions will be presented.

Overview

Patient Preparation

- Patients undergoing abdominal CT for either traumatic or nontraumatic conditions should ideally have oral contrast prep with adequate bowel transit time to allow for visualization of the bowel. The oral contrast is not necessary for investigation of renal colic or abdominal aneurysm, and is optional for traumatic conditions if time does not permit.
- Intravenous contrast is usually given for abdominal CT, but is not necessary for the majority of renal stone CTs, unless more detail of the ureteral anatomy is needed. Intravenous contrast should be avoided in the setting of renal insufficiency for fear of renal contrast damage or in the presence of the oral hypoglycemic metformin for fear of lactic acidosis.

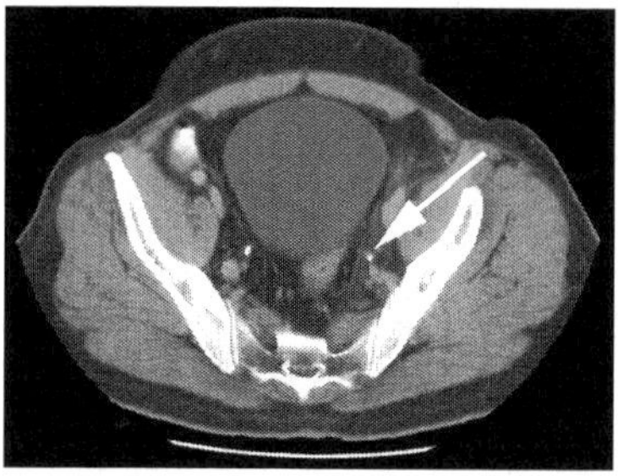

Figure 19.23. Distal left ureter stone (white arrow).

Renal Colic

Since the advent of helical CT, it has quickly replaced intravenous pyelography (IVP) as the initial investigation of suspected renal colic. It can be performed quickly, without the need for patient preparation, or laboratory investigations such as BUN/ creatinine and costs less than IVP.

- Findings of renal calculus on unenhanced abdominal CT:
  - Renal calculus visualized.
  - Hydronephrosis will be seen with an obstructing ureteral stone.
  - Perinephric stranding or fluid may be seen with increasing grades of obstruction.
- Advantages of helical CT:
  - Fast scanning time.
  - No intravenous contrast exposure.
  - No need for patient preparation such as precontrast labs or bowel prep.
  - May identify alternative diagnoses, changing management such as abdominal aneurysm, appendicitis, and diverticulitis.
  - Cost is lower in most institutions than IVP.
- Disadvantages of helical CT include:
  - Radiation exposure is greater than with IVP.
  - May not identify small nonobstructing stones.
  - May not identify some noncalcified stones such as uric acid stones, medication-related stones such as indinivir.
  - Does not show renal function as with IVP.
- Alternative imaging:
  - Ultrasound
  - Intravenous pyelography

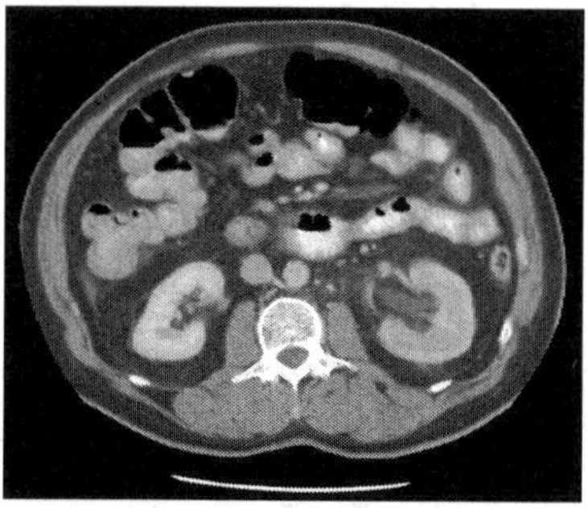

Figure 19.24. Left hydronephrosis seen on CT scan due to obstructing ureter stone. Note the enlarged renal pelvis compared with the right kidney.

Acute Appendicitis

Appendicitis is an excellent example of how imaging has changed the practice of medicine. While it has been acceptable to have negative laparotomy rates of up to 30%, this number has come down substantially in those institutions utilizing CT scanning in the investigation of right lower abdominal pain. While initial studies were performed on selected populations, adding selection bias, high sensitivities and specificities have seemed to hold up in the general population. Like its use in renal colic, it may also reveal alternate diagnoses, thereby obviating exploratory laparotomy/laparoscopy.

Appendiceal CT scanning is performed using helical technique with thin slices through the region of the appendix after the patient has received adequate oral contrast. Rectal contrast is often used to ensure filling of the cecum. Intravenous contrast is commonly used but may be omitted if contraindicated.

- Signs of appendicitis on CT scan include:
  - Thickened appendiceal wall greater than 2 mm.
  - Distended appendix measuring greater than 6 mm.
  - Appendix not filling with contrast despite adequate terminal ileal and cecal visualization.
  - Inflammatory changes surrounding the appendix are suggestive, but not diagnostic, of appendicitis as this may occur with nonsurgical conditions.
  - Fluid collection or abscess in the right lower quadrant is also suggestive of appendicitis.
- As with any other test, it is important to understand its limitations and not rule out appendicitis solely on a negative CT, in the face of high clinical suspicion. Possible reasons for a false negative CT scan include:
  - Inadequate patient preparation not allowing for visualization or filling of the appendix.
  - Thin patient not providing enough fat stranding to appreciate inflammatory changes.
  - CT scan performed too early!! Not enough time may have lapsed to allow the inflammatory changes to be seen on CT.
  - As with any imaging study, its power lies in those interpreting the scan. Reading appendiceal CT requires skill and experience.
- Advantages of CT for appendicitis:
  - May provide alternate, nonsurgical diagnosis thereby obviating surgery.
  - Noninvasive diagnosis, unlike laparoscopy.
  - May be more cost-effective than in-hospital observation or surgical exploration.
- Disadvantages of CT scan in appendicitis:
  - Radiation exposure is significant, averaging around 3- 5 rems.
  - May delay treatment in those with obvious clinical diagnosis if ordered inappropriately.

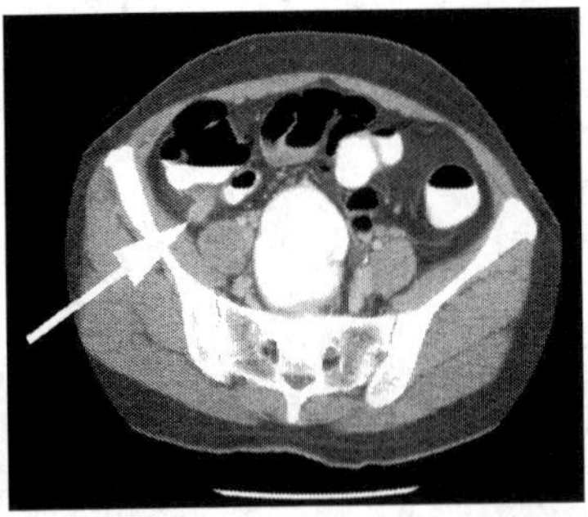

Figure 19.25. Acute appendicitis. The nonfilling appendix is seen entering the cecum.

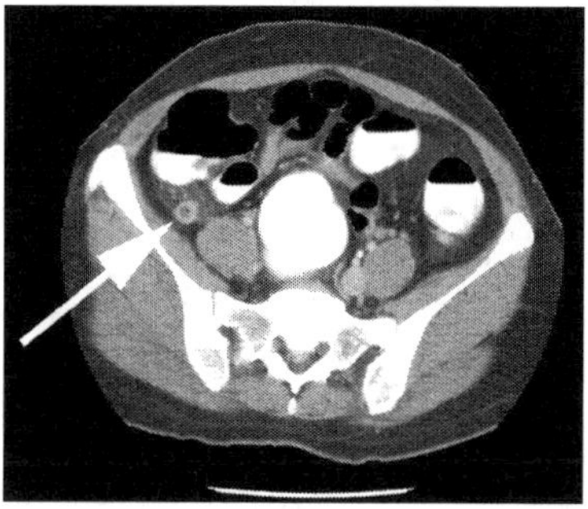

Figure 19.26. Acute appendicitis. This cut shows the appendix with associated inflammatory fat stranding.

- Alternate imaging studies in appendicitis:
  - Ultrasound
  - Laparoscopy

Abdominal Aneurysm

- Although treatment decisions regarding abdominal aneurysm repair are often made using clinical symptoms and signs, CT scanning can be an appropriate imaging study.
- CT scanning is very useful for evaluating suspected or known abdominal aneurysm patients without indications for immediate operative management. It provides detail of the aneurysm dimensions and location, and can often reliably comment on leaking or rupture.
- In the setting of operative indications, especially with a known aneurysm, obtaining a CT scan will only delay definitive surgical treatment.
- Although an abdominal aneurysm will be identified without the aid of intravenous contrast, details such as leaking, rupture, thrombosis or dissection may not be identified.
- Findings of abdominal aneurysm on CT scan:
  - Abdominal aorta measuring greater than 2.5 cm in diameter or greater than 1.5 times the diameter of adjacent aorta.
  - Calcifications are common in the lumen due to associated atherosclerotic disease.
  - Associated thrombus may be seen lining the lumen of the aneurysm.
  - An intimal flap may be seen with associated dissection.
  - Retroperitoneal or free peritoneal leak may be seen.

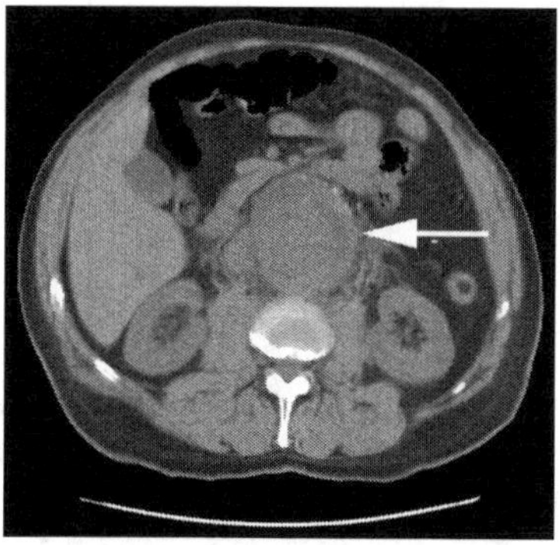

Figure 19.27. Large abdominal aneurysm is identified on this noncontrast CT scan of the abdomen.

- The location and extent of aneurysm (i.e., involvement of renal arteries) as well as associated leak are important in surgical planning and should be conveyed to the consultant.
- Advantages of CT scan in abdominal aneurysm evaluation:
  - High sensitivity.
  - Able to provide dimensions and extent of aneurysm.
  - Unlike ultrasound, can often identify complications such as leak, rupture or dissection.
- Disadvantages of CT scan in abdominal aneurysm evaluation:
  - Usually requires patient to be transported out of the department.
  - Since it provides helpful details, consultant surgeons may try to "push" for a CT before definitive treatment.
- Pitfalls in abdominal CT scan for abdominal aneurysm:
  - Ordering CT scan inappropriately for patient with clear operative indications.
  - Sending a potentially unstable patient to CT scan when a bedside ultrasound would identify the aneurysm.
- Alternative imaging studies:
  - Bedside ultrasound
  - MRI
  - Angiogram

Small Bowel Obstruction

- The majority of small bowel obstructions can be diagnosed on plain radiographs and occur in the setting of clear risk factors such as previous surgery or incarcerated hernia.
- When small bowel obstruction is suspected and is either not clear on abdominal radiographs or occurs without a discernable etiology, CT scanning can provide the necessary details.
- CT scanning with bowel contrast preparation is about 90% sensitive and specific for bowel obstruction and may reveal uncommon etiologies such as internal incarcerated hernias, intussusception, or volvulus.

Undifferentiated Elderly Abdominal Pain

- As previously discussed, abdominal CT scanning has allowed clinicians to open the "black box" of the abdomen without need for surgical exploration. One clinical presentation in which this is most evident is undifferentiated abdominal pain, especially in the elderly.
- The elderly are notoriously difficult to evaluate by conventional means as they often lack both symptoms and clinical and laboratory signs found in serious disease.
- Despite this, the elderly are much more likely to have a surgical condition causing abdominal pain.

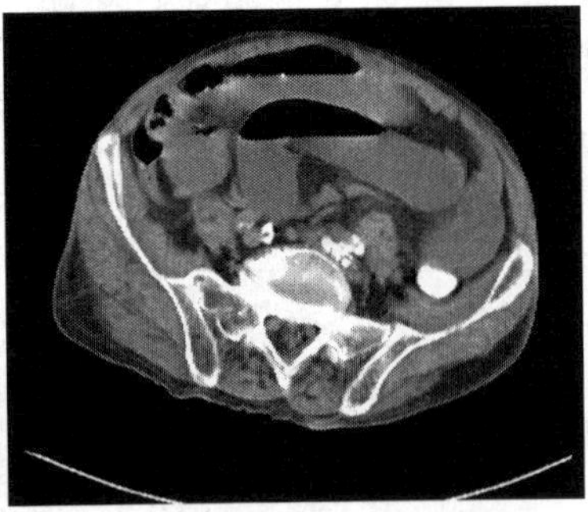

Figure 19.28. Loops of distended small bowel are seen on this CT scan of a small bowel obstruction.

19

- Plain abdominal radiographs are often not helpful in young healthy patients with abdominal pain. The elderly, however, often reveal silent abnormalities that may change management, and ordering of plain X-rays should be liberal in this population.
- CT scan is able to identify many of the surgical conditions causing elderly abdominal pain and is often helpful when conventional investigation and consultation has not produced a diagnosis.
- A caveat is needed, however, in order that the CT not be used to "rule out" disease in the setting of a "normal" plain abdominal film. It has limitations, as any study, and the reader is cautioned to use the CT scan as one more piece of information and clinical judgement above all.

Abdominal Trauma

The combination of bedside ultrasound and CT scanning has replaced diagnostic peritoneal lavage in the evaluation of blunt abdominal trauma. While the camps of ultrasound and CT advocates are often mutually exclusive, they are very complementary resources. Ultrasound has advantages of being a bedside exam, easily repeated, and sensitive to intra-peritoneal free fluid (a sensitive indicator of need for laparotomy). CT scan is able to detect solid organ injuries, retroperitoneal injuries, and may detect hollow viscus injuries.

In the emergency department, the clinician is often only given a report of CT findings so that you are often at the mercy of the radiologist's ability. While learning to formally read an abdominal CT is beyond the objective of this chapter, pertinent injuries will be presented in image form. In accepting a report from a radiologist, it is important to understand the ability and limitations of CT.

- CT scanning is very sensitive to free peritoneal air or pneumothorax, which may be detected on higher abdominal cuts.
- CT is also very sensitive to intraperitoneal fluid or hemorrhage, with a "sentinel clot" often pointing out the source of injury. CT cystogram will aid in detection of bladder injuries.
- CT is very good at detecting solid organ injuries such as splenic fractures or liver fractures, even when not associated with free fluid.
- CT is also the best imaging study for retroperitoneal organs such as pancreas, kidneys or retroperitoneal hemorrhage.
- CT is an excellent study for detecting bony fractures, such as rib, spine or pelvic ring fractures.

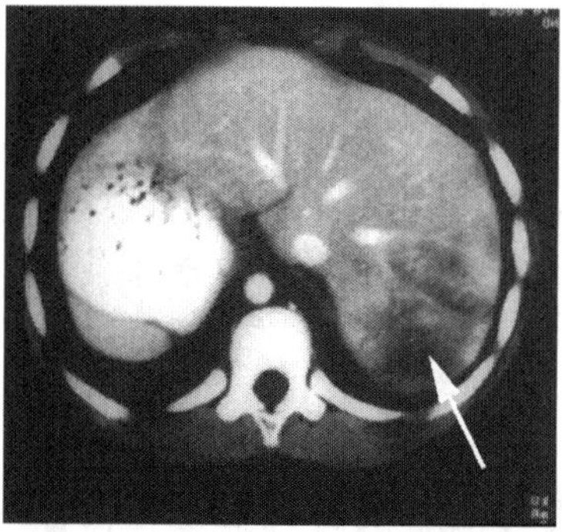

Figure 19.29. Liver laceration on contrast enhanced CT scan (white arrow).

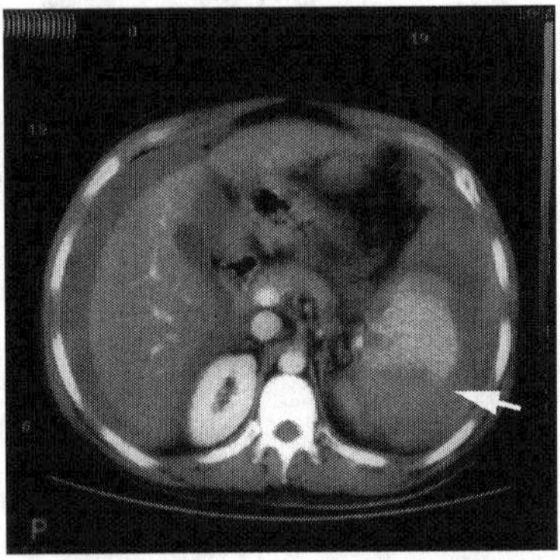

Figure 19.30. Contrast-enhanced CT showing splenic rupture with free fluid also surrounding the liver.

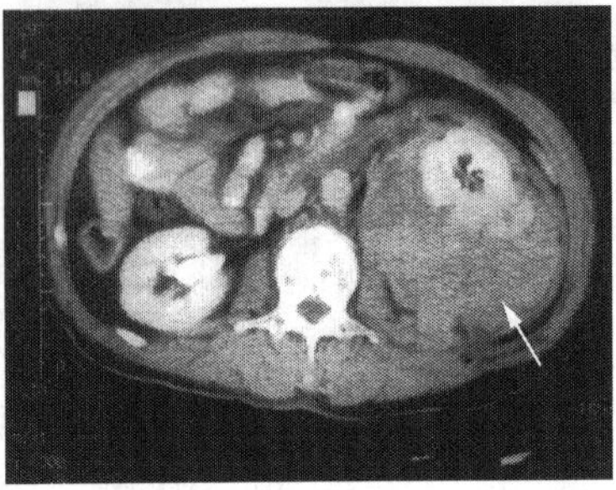

Figure 19.31. Left renal fracture with associated hematoma.

- CT cannot adequately rule out hollow viscus injuries, such as small bowel or colon injuries with small perforation. These are often only detected after observation or exploratory laparotomy.

Cervical Spine

- Although not common, cervical spine fractures raise the hairs of emergency physicians due to the potential for catastrophic medical and legal complications.
- Not only are cervical spine fractures potentially unstable and could lead to spinal cord injury, they are often multiple and noncontiguous in up to 16% of patients. This makes the identification of a spinal fracture the beginning of a search for the next one.
- The evaluation of the cervical spine after trauma begins with a clinical evaluation. Since 800,000 plain radiographs are done each year in the US to evaluate cervical spine injuries, clinical decision guidelines have been derived and prospectively evaluated by the NEXUS study group. This multi-center randomized prospective trial was able to detect 99.6% (sensitivity) of all significant cervical spine fractures. They evaluated patients using the following five criteria:
  1. No midline tenderness
  2. No focal neurological deficit
  3. Normal alertness
  4. No intoxication
  5. No painful, distracting injury

  If none of these were present they were able to exclude a significant fracture with a negative predictive value of 99.9%.

19

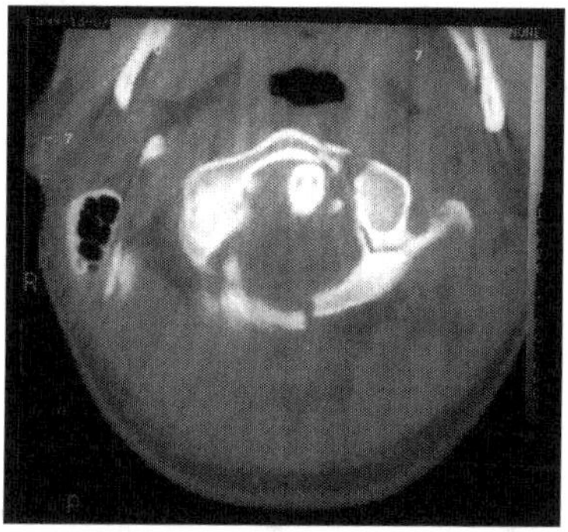

Figure 19.32. CT scan of burst fracture of the ring of C1 (Jefferson fracture).

- For those patients not meeting criteria to be "cleared clinically", plain radiography should be performed using at least three views (AP, lateral, and odontoid). A single lateral X-ray has only a sensitivity of about 80%, while this increases to 93% with three views.
- CT scanning is increasingly common in the evaluation of the cervical trauma patient, especially since the advent of helical CT, which removes much of the movement artifact, and the scanning time, previously limiting its use.
- Technique for CT scanning of cervical injuries usually involves helical CT for the reasons mentioned, although any CT scan may be used. Thin cuts between 2-3 mm are generally used to detect small fractures.
- CT scanning is used in those patients in whom adequate plain films are not obtainable or if there is a question of a fracture on plain film. Berne et al studied a protocol of using helical CT to primarily screen for CSI and found CT sensitivity of fracture detection to be about 90% versus 60% for plain films. Of note was that this study population were ICU patients and had a mean GCS of about 8; thus were probably at higher risk for CSI and difficult to evaluate clinically.
- Although CT scan provides great detail and is inarguably more sensitive than plain films in fracture detection, there are limitations:

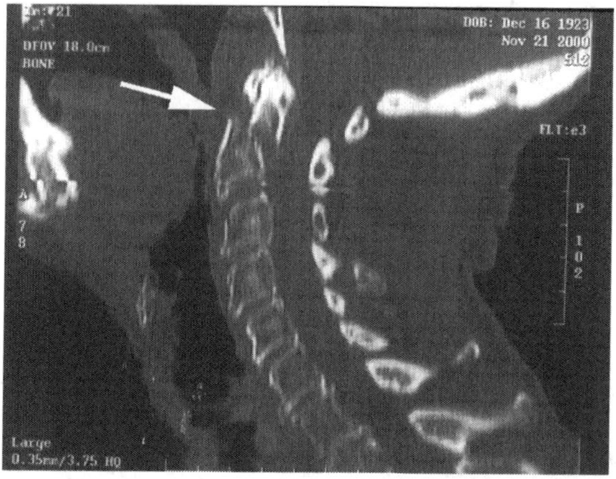

Figure 19.33. CT sagittal reconstruction of type 2 odontoid cervical fracture.

- CT scanning is axial and therefore insensitive to subluxation or dislocation, without three-dimensional reconstruction.
- CT scanning will not detect ligamentous injuries, and MRI should be performed when there is suspicion.

Suggested Reading

1. Berne, JD, Velmahos GC, El-Tawil Q et al. Value of complete cervical helical computed tomographic scanning in identifying cervical spine injury in the unevaluable blunt trauma patient with multiple injuries: A prospective study. J Trauma 1999; 47(5):896.
2. Curry JD, Recine CA, Snavely E et al. Periaortic hematoma on abdominal computed tomographic scanning as an indicator of thoracic aortic rupture in blunt trauma. J Trauma 2002; 52(4):699-702.
3. Daneshmand S, Hedley CG, Stain SC. The utility and reliability of computed tomography scan in the diagnosis of small bowel obstruction. Am Surg 1999; 65(10):922-6.
4. Demetriades D, Gomez H, Velmahos GC. Routine helical computed tomographic evaluation of the mediastinum in high-risk blunt trauma patients. Arch Surg 1998; 133(10):1084-8.
5. Edlow JA, Caplan LR. Avoiding pitfalls in the diagnosis of subarachnoid hemorrhage. NEJM 2000; 342:29-36.
6. Hoffman J, Mower W, Wolfson A et al. Validity of a set of clinical criteria to rule out injury to the cervical spine in patients with blunt trauma. NEJM 2000; 343:94-99.
7. Remy-Jardin M, Remy J, Deschildre F et al. Diagnosis of pulmonary embolism with spiral CT: Comparison with pulmonary angiography and scintigraphy. Radiology 1996; 200:699-706.
8. Sommer T, Fehske W, Holzknecht N. Aortic dissection: A comparative study of diagnosis with spiral CT, multiplanar transesophageal echocardiography, and MR imaging. Radiology 1996; 199(2):347-52.
9. Streitwieser DR, Knopp R, Wales LR. Accuracy of standard radiographic views in detecting cervical spine fractures. Ann Emerg Med 1983; 12(9):538-42.
10. Thompson NW, Thompson TJ, Love MH et al. Drugs and intravenous contrast media. BJU International. 2000; 85(3):219-21.
11. Vieweg J, Teh C, Freed K. Unenhanced helical computerized tomography for the evaluation of patients with acute flank pain. J Urology 1998; 160(3-I):679-684.

CHAPTER 20

Bedside Ultrasound in Emergency Medicine

Diku Mandavia and Sujal Mandavia

Introduction

- Bedside ultrasound is rapidly being integrated into emergency medicine practice with more graduates of emergency medicine training programs and practicing EPs learning this important diagnostic technique.
- Bedside ultrasound performed by emergency physicians offers significant benefits including the following:
  - Faster diagnoses with better patient outcomes
  - Increased emergency department throughput
  - Decreased medical-legal liability because of quicker diagnostic testing and expedited treatment of life-threatening conditions
- *Emergency physicians perform focused bedside ultrasound.* With focused exams, EP performed ultrasounds are simpler, easier to learn and more accurate. A focused exam is designed to answer a simple clinical question at the bedside e.g.
  - Does this patient have life-threatening hemoperitoneum?
  - Does this patient have an abdominal aortic aneurysm?
- A focused emergency ultrasound is not a replacement for a formal comprehensive ultrasound performed by radiology. Rather, these different exams should be considered complimentary with the former answering key management questions at the bedside. The information in this chapter concentrates on emergency physician performed bedside ultrasound.

Equipment Considerations

- Bedside ultrasound in emergency medicine does not require high-end or expensive machinery, and thus EM ultrasonography can be performed with equipment costing as little as $25,000.
- Several important features are important when selecting equipment:
  - Portability—Equipment should be small and portable enough to be moved around the ED and to the bedside.
  - Durability—An emergency department can be a hostile location for delicate equipment, and thus an ultrasound machine should be durable. Look for enclosed probe covers, protective side bumpers and spill-proof keyboards.
  - Probes—Ultrasound probes are the most important part of your equipment *and the most delicate.* Most transducers today are electronic and are multifrequency. A 3.5 MHz small footprint probe which allows for intercostal and abdominal scanning is ideal for most emergency applications. Small parts scanning and vaginal ultrasound will require separate special transducers.
  - Output—All ultrasound images should be captured and the easiest and most convenient way is by printing using a portable thermal printer. Other important output methods include video and digital storage to disks and via hospital networks.

20

- Service agreements—Because of the delicate nature of ultrasound equipment especially probes, it is highly recommended that a service agreement be included as part of your initial purchase.

Clinical Applications

- The primary applications are well recognized to be within the scope of emergency medicine. These emergency indications include the following:
  - Detection of hemoperitoneum following acute abdominal trauma
  - Detection of pericardial effusions/tamponade or cardiac activity in pulseless electrical activity
  - Detection of abdominal aortic aneurysms
  - Detection of live early intrauterine pregnancy in first trimester abdominal pain or bleeding in rule out ectopic pregnancy algorithms
  - Detection of acute hydronephrosis in acute renal colic
  - Detection of cholelithiasis
- Secondary applications are increasingly becoming important and include:
  - Procedural indications including:
    - Vascular access—central and peripheral
    - Paracentesis
    - Pericardiocentesis
    - Thoracentesis
    - Pacemaker placement and capture
    - Bladder aspiration
    - Arthrocentesis
  - Foreign body detection
  - Gestational dating and fetal viability
  - Deep vein thrombosis
  - Endotracheal tube placement
  - Abscess detection

Basic Terminology

- Acoustic window–organ or tissue that facilitates viewing of structures beyond it. Acoustic windows are necessary when imaging certain organs that are less amenable to ultrasound scanning, e.g., heart.
- Echogenic–describes structures that appear as echoes on an ultrasound monitor. Differences in the level of echogenicity are described by the following:
  - hypoechoic—less echoes (darker in appearance)
  - hyperechoic—more echoes (brighter in appearance) e.g., gallstones
  - anechoic–no echoes (completely black in appearance) e.g., bladder
- Power—modifies amount of energy leaving the transducer
- Gain—modulates the amount of received echo signal (analogous to a volume control)
- Time Gain Compensation (TGC)—controls echo amplification at various depths and usually present as a slide control. TGC allows increasing the gain at different depths without changing the overall gain
- Near field and Far field—refers to imaged areas that are nearer and farther from the probe surface

Primary Applications

Abdominal Trauma

Background

- Ultrasound has been used to evaluate emergency patients since the 1970s, but only in the last 10 yr has there been significant interest in the United States. Many prospective studies done by emergency physicians and surgeons here in the United States confirm that this modality can be used by nonradiologists with the reported sensitivity for free intraperitoneal fluid varying from 80-90% and the specificity 95-100%.

- The noninvasive and time-sensitive nature of bedside ultrasound has now led to the replacement of diagnostic peritoneal lavage at most major US trauma centers.

Indications

- Bedside ultrasound is very useful for the rapid detection of:
  - hemoperitoneum,
  - pericardial effusions, and
  - pleural effusions.
- Ultrasound's greatest utility is in the evaluation of blunt abdominal trauma for hemoperitoneum and in penetrating chest injuries for the detection of pericardial effusions.
- Below are suggested algorithms for incorporating bedside ultrasound in trauma (Figs. 20.1 and 20.2)

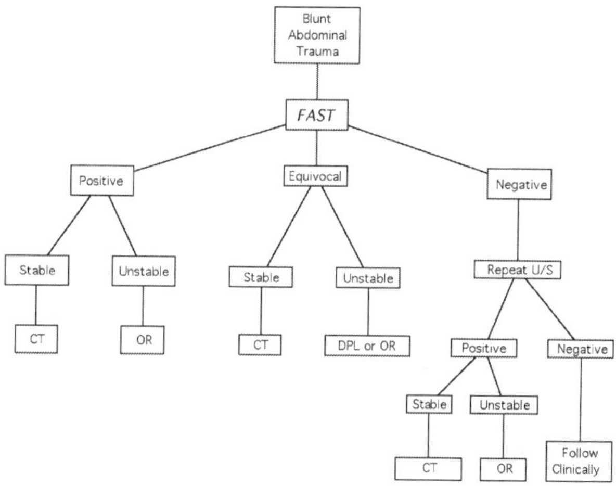

Figure 20.1. Blunt abdominal trauma algorithm.

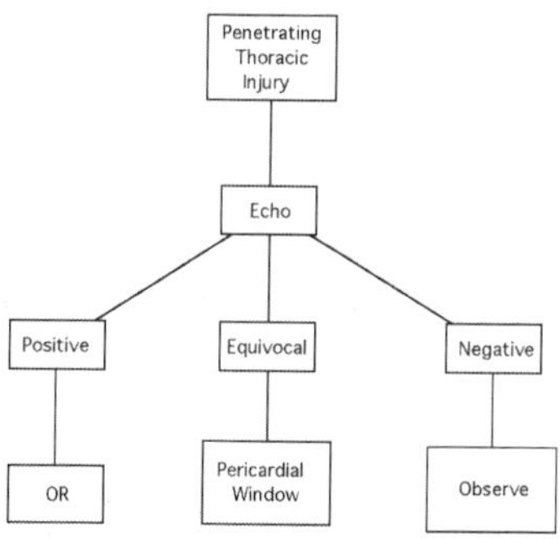

Figure 20.2. Penetrating chest injury algorithm.

20

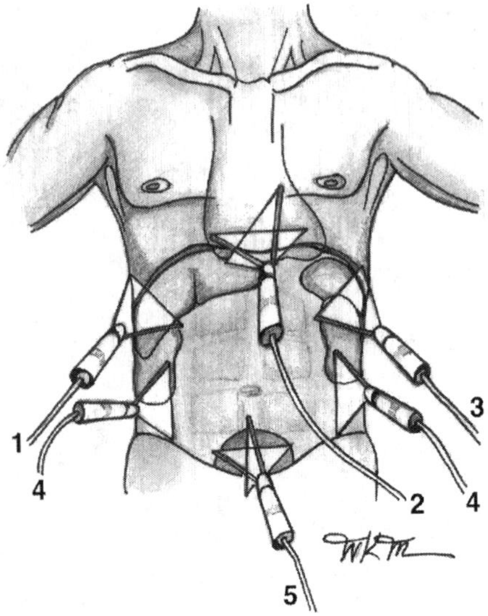

Figure 20.3. Transducer positions for the FAST exam.

The Focused Examination

- Focused abdominal sonography for trauma (FAST) consists of focused views of the abdomen including the pericardium and is performed after the primary survey. Multiple views greatly increase the sensitivity of ultrasonography and standard areas examined include the following (Fig. 20.3):
    1. Morison's pouch
    2. Pericardium
    3. Perispenic space
    4. Paracolics
    5. Suprapubic view

Morison's Pouch

- Morison's pouch is a very useful initial view in the ultrasound evaluation of the trauma victim. The exact amount of free fluid detected in Morison's pouch varies but is as little as 250 ml. This view is easily obtained within ~30 seconds as the landmarks are easy to find.
- The probe is placed in the mid to posterior axillary line at the just below the nipple level. The liver is identified and the kidney will be adjacent (Fig. 20.4). The space between these two organs is Morison's pouch and is a potential space that can fill with fluid.
- *Free fluid appears as a anechoic or as a black stripe in this area* (Fig. 20.5). With time, hemoperitoneum loses its anechoic consistency and becomes more hyperechoic, thus the fluid will have a grayer color and an inconsistent appearence.
- Hyperechoic (white or gray areas) that surround the kidney represent normal perinephric fat and Gerota's fascia and are not to be confused with free fluid. Patient positioning in Trendelenburg can improve sensitivity by making this area more dependent.
- *Once Morison's pouch is adequately examined, angle the probe cephalad and examine the diaphragm for fluid above or below. This will be evident by black areas and small hemothoraces can easily be detected with a little practice* (Fig. 20.6).

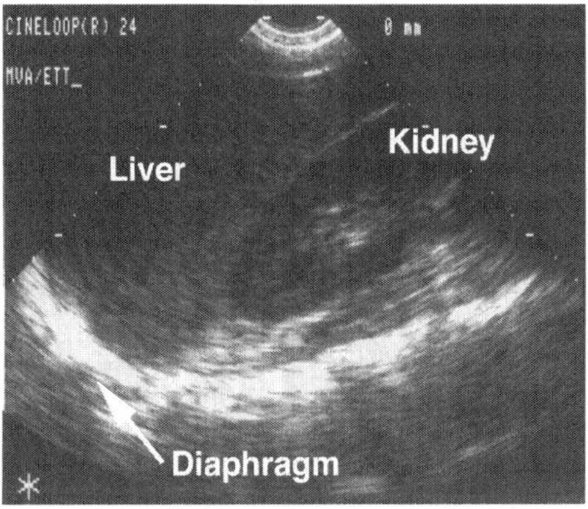

Figure 20.4. Normal Morison's pouch.

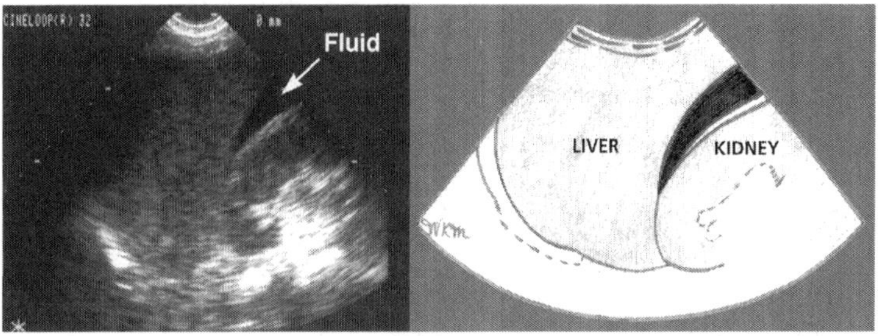

Figure 20.5. Free intraperitoneal fluid demonstrated by an anechoic stripe at Morison's pouch.

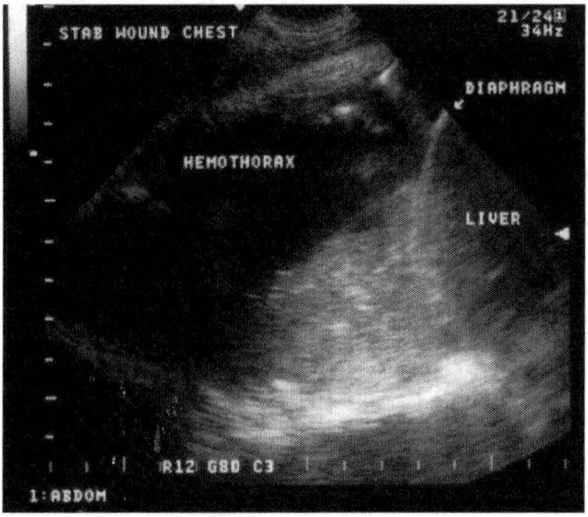

Figure 20.6. Hemothorax demonstrated by an anechoic area above the diaphragm.

20

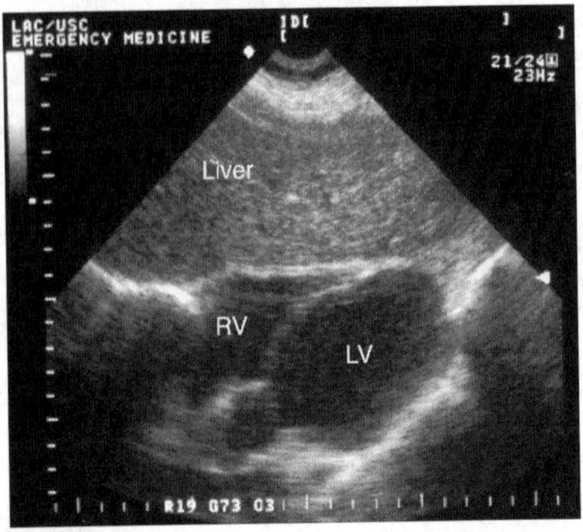

Figure 20.7. Normal subcostal view of the heart. Pericardium is seen as a hyperechoic line surrounding the heart.

Pericardium

- The pericardium is especially important to evaluate in penetrating thoracic injuries to rule out a pericardial effusion and tamponade.
- For this view, the probe is placed in the subcostal area just to the right of the xiphisternum and the probe angled toward the patient's left shoulder. To view the heart adequately, you will need to increase the depth of penetration at this point. A coronal section of the heart should give you a good four chamber view of the heart.
- The normal pericardium is seen as a hyperechoic (white) line intimately surrounding the heart (Fig. 20.7).
- A pericardial effusion is seen as an anechoic (or black area) surrounding the heart within the pericardium (Fig. 20.8).

Perisplenic Area

- The perisplenic view is obtained by placing the probe at the posterior axillary line at the 9-10th interspace. A common mistake when doing this view is not placing the probe posterior enough to adequately see the kidney. Once the kidney is found, angle the probe slightly cephalad to find the spleen and carefully look for free fluid surrounding it usually from above (Fig. 20.9).

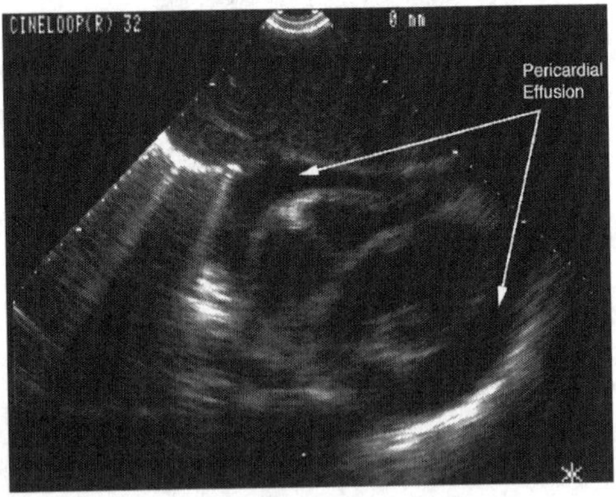

Figure 20.8. Pericardial effusion demonstrated by an anechoic area surrounding the heart.

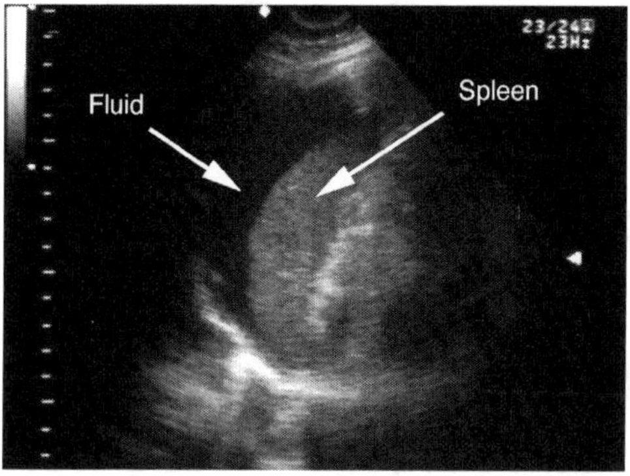

Figure 20.9. Hemoperitoneum noted above the spleen.

- Once the spleen and kidney are fully scanned, angle the probe more cephalad to examine the diaphragm. As with Morison's pouch, the diaphragm should be visualized to see a pulmonary effusion or subdiaphragmatic fluid.

Paracolic Views

- The paracolic views can be done in conjunction with Morison's pouch and the perisplenic view. Simply place the probe in the paracolic area and examine for free fluid and/or free floating loops of bowel. Fluid is often detected first on other views limiting the usefulness of the paracolic view, and thus this view is eliminated in some protocols (Fig. 20.10).

Suprapubic View

- Ideally this exam is done prior to the placement of a Foley catheter. The full bladder is easily found by placing the probe just cephalad to the pubis. Once the bladder is found, look for free fluid anterior, posterior and lateral to the bladder. In females, the uterus will be seen posterior to the bladder. The cul-de-sac is a very dependent area of the peritoneal cavity and should be carefully examined for free fluid (Fig. 20.11).

Pitfalls

- **Failure to do a multiple view examination**. Sensitivity is highly dependent on the number of views obtained; thus a full exam is necessary.

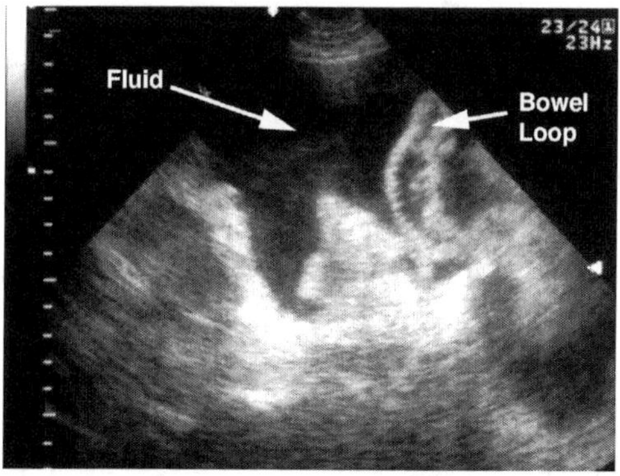

Figure 20.10. Free intraperitoneal fluid in the paracolic gutters.

20

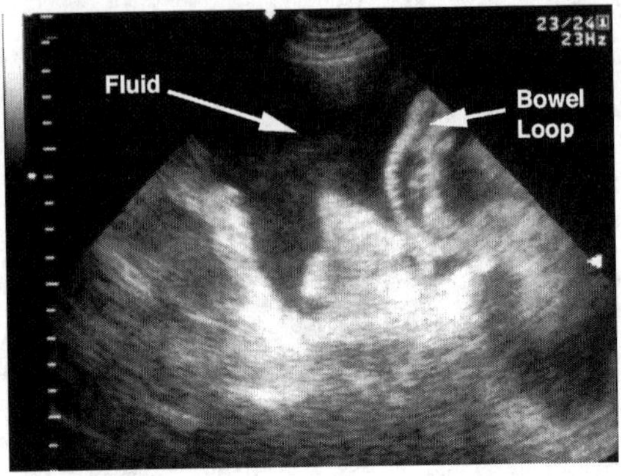

Figure 20.11. Hemoperitoneum noted on the suprapubic view in a female trauma victim.

- **Failure to consider other etiologies of free intraperitoneal fluid.** Intestinal fluid and intraperitoneal bladder rupture mimic hemoperitoneum, but both require laparotomy. Ascites will mimic hemoperitoneum but is easily differentiated with needle aspiration.
- **Failure to do serial exams when the initial examination is negative.** Trauma patients are extremely dynamic and contained injuries may later release causing a positive ultrasound exam. Consider serial exams in those with high clinical suspicion and those with changing vital signs and/or hematocrits.
- **Overreliance on ultrasonography.** Use ultrasonography as a single data point in the entire clinical picture. Use it in conjunction with other data such as mechanism of injury, vital signs, hematocrits, radiographs and clinical suspicion.

Echocardiography

Background

1. Echocardiography provides a bedside noninvasive manner to examine the anatomy and physiology of the heart. Cardiac anatomy including chambers, valves and pericardium and surrounding structures are well visualized using today's equipment. Furthermore valvular pathology, pericardial effusions and tamponade, wall motion abnormalities and hemodynamic information can also be provided using this technique.
2. *Echocardiography for the emergency physician focuses on the presence of:*
    - Pericardial effusions and tamponade
    - Gross cardiac activity in pulseless electrical activity (PEA). The absence of cardiac activity in PEA implies true electromechanical dissociation and thus an extremely poor prognosis.
3. Used in this focused manner, bedside echocardiography meets the criteria for emergency physician ultrasound use.
    - The condition in question is life-threatening.
    - Ultrasound is the primary diagnostic modality.
    - Ultrasound decreases the time or cost of the evaluation.
    - Ultrasound serves as an adjunct for a procedure.

Indications

1. Suspicion of a pericardial effusion with or without tamponade
2. Evaluation of cardiac activity in PEA

The Focused Exam

- To learn this technique you must first understand the concept of cardiac windows and planes. The heart, being surrounded by air-filled lungs, poses a natural challenge for ultrasound technology. Special acoustic windows give us a view into the chest to examine the heart by ultrasound.
- The main acoustic windows include the following:
  - Subxiphoid window
  - Parasternal window
  - Apical window
- Through these three windows the heart is examined through three different planes. Since the heart is not aligned along standard anatomical planes, the following convention is used:
  - Four chamber plane (coronal)
  - Long axis plane (sagittal)
  - Short axis plane (transverse)
- A combination of the above 3 windows and 3 planes is used in emergency echocardiographic exams using a 2.5-3.5 MHz phased array or microconvex probe. At times patient body habitus and cooperation will allow only one of the below views to be completed, and thus knowledge of multiple views is necessary.
  - Subcostal window—place the probe just right of the subxiphoid area and scan in a coronal plane with the tip of the probe pointing toward the left shoulder. You will need to maximize the depth for this view, and you will examine the heart in a four chamber plane (Fig. 20.7).
  - Parasternal window—place the probe at the 2nd-4th intercostal space just left lateral of the sternum. The marker will be toward the patient's right shoulder and thus the beam which you scan the heart is in a long axis plane. In this view you will see the right and left ventricles, left atrium, aortic outflow tract and the pericardium. *This is the single best view to examine for pericardial effusions* (Figs. 20.12 and 20.13).
  - The short axis view is obtained by rotating the probe 90 degrees such that the marker is now toward the left shoulder. On this view you will see the right and left ventricles and pericardium.
  - Apical window—have the patient turn to their left side and place the probe at the point of maximal impulse (PMI). Keep the marker toward the patients left and you will obtain a four chamber plane of the heart (Fig. 20.14).

4. Pericardial effusions—*effusions will be identified as anechoic areas surrounding the heart within the pericardium* (Fig. 20.13). Pericardial fat can have the same echogenicity of fluid. On the parasternal long axis view, anechoic areas seen only anteriorly usually represent pericardial fat as fluid should be seen posteriorly as well because of gravity. Pericardial effusions are graded as follows:

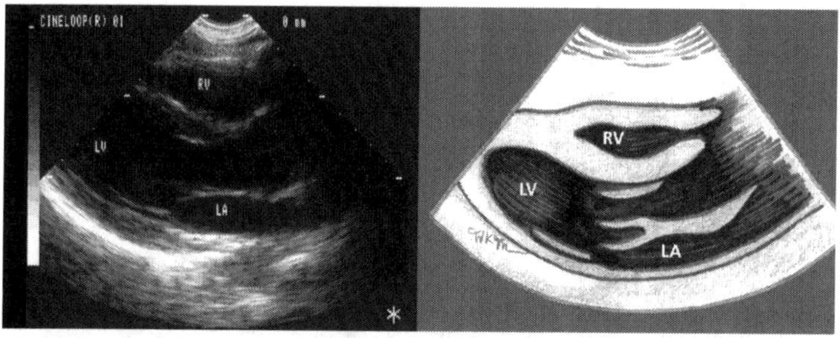

Figure 20.12. Normal long axis parasternal view of the heart.

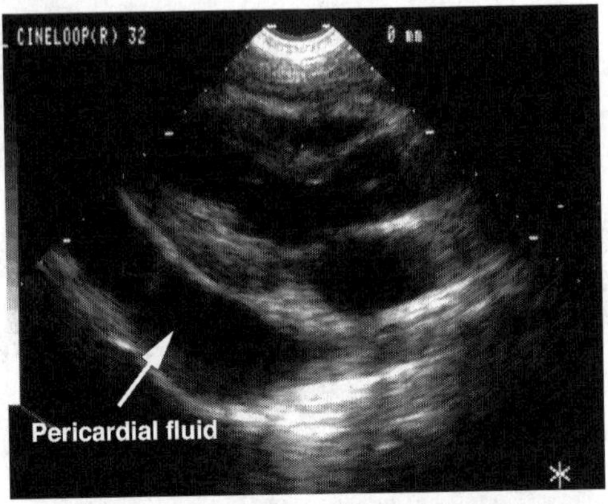

Figure 20.13. Pericardial effusion seen on the long axis parasternal view of the heart.

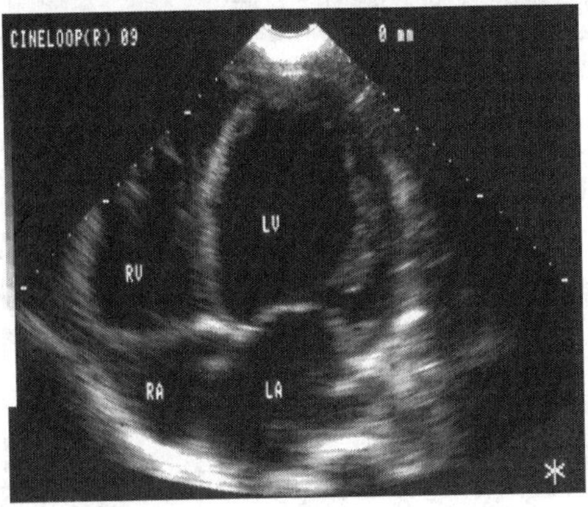

Figure 20.14. Normal apical four chamber view of the heart.

1. Small—seen posteriorly only
2. Moderate—circumferential but <1 cm in width
3. Large—circumferential and >1 cm in width

3. Pericardial effusions that have diastolic collapse of the right ventricle represent patients with tamponade physiology.

Pitfalls

- Obese patients and those with COPD are more difficult to ultrasound.
- Epicardial fat and pleural effusions can be confused with pericardial fluid.
- Loculated effusions will be more difficult to detect.

Biliary Tract Disease

Background

- Ultrasound is well suited to ultrasonography and is the primary imaging modality of the biliary tract system. First, the solid and homogenous liver provides an excellent acoustic window into the right upper quadrant. Second, the cystic bile filled gallbladder conducts ultrasound sound waves well and provides a contrasting medium. Finally gallstone pathology is echogenic and is easily identified by ultrasound.

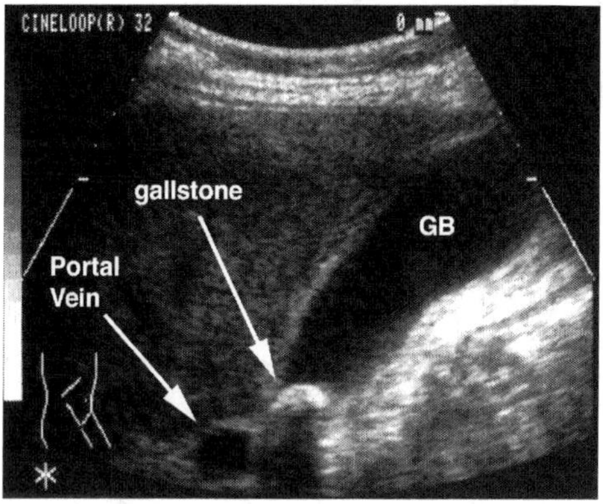

Figure 20.15. Single large gallstone at the neck of the gallbladder.

Emergency Indications

- Ultrasound provides valuable information on emergency patients with following suspected conditions:
  - acute biliary colic
  - acute cholecystitis
  - acute pancreatitis

The Focused Exam

- *The focused exam is to detect the presence or absence of cholelithiasis*
- Other important sonographic signs of cholecystitis include:
  - Thickened gallbladder wall—>5 mm
  - Decreased echogenicity of gallbladder wall
  - Increased transverse diameter of the gallbladder—>4-5 cm
  - Ultrasonographic Murphy's sign
  - Pericholecystic fluid
  - Dilated common bile duct—normal = 1 mm/decade of life
- Place the patient in a slight left lateral decubitus or in a supine position. The upright or sitting position may be necessary in some patients. Body habitus, presence of abdominal gas and patient cooperation will may play a significant role in good image acquisition.
- Use a 3.5 MHz probe with a microconvex head which allows intercostal scanning and obtain the following:
  - Long axis view of the gallbladder—Place the axis of your probe perpendicular to the costal margin and scan either intercostal or subcostal. On this view the portal vein should easily be seen inferior to the gallbladder (Fig. 20.15).
  - Short axis view of the gallbladder—After obtaining and reviewing the long axis view, rotate your probe so that the marker is toward the patient's right and scan the gallbladder in a transverse fashion from superior to inferior.
- Carefully scan the entire gallbladder for the presence of cholelithiasis. Gallstones lie within the gallbladder, are echogenic, cast shadows and are often mobile (Figs. 20.15 and 20.16). Bowel shadows will lie outside the gallbladder but can mimic gallstones to the novice.

Pitfalls

- Not scanning the entire gallbladder in all planes—Stones at the gallbladder neck may be easily overlooked unless the gallbladder is fully examined.
- Abdominal gas may make gallbladder visualization difficult or may mimic cholelithiasis.

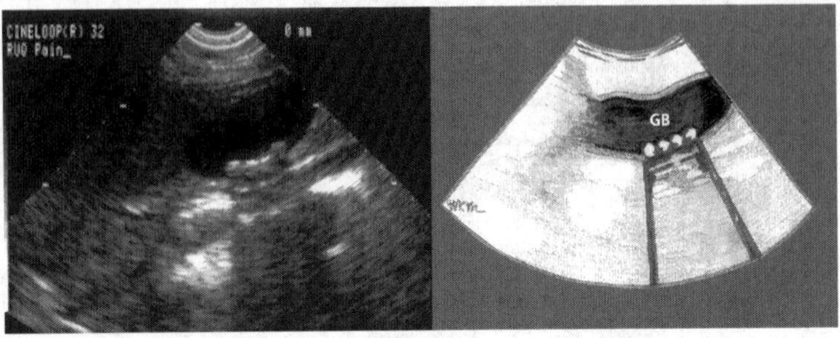

Figure 20.16. Multiple echogenic areas within the gallbladder which cast a shadow and represent gallstones.

- Chronic cholecystitis with shrunken gallbladder
- A gangenous gallbladder will be less defined and harder to image.

Renal Ultrasound

Background

- Renal ultrasound is a useful adjunct when evaluating patients for acute obstructive uropathy. Typically acute urinary tract obstruction causes hydronephrosis of the calyceal system, and this dilatation is easily detected by renal ultrasound.
- Though computed tomography is increasingly being used to evaluate obstructive uropathy, ultrasound has the advantage of using no radiation or contrasts agents. Furthermore it is fast, inexpensive and can be done at the bedside.

Indications

- Acute flank pain with suspicion of renal colic

The Focused Exam

- The focused exam is the detection of acute hydronephrosis.
- Occasionally renal, ureteral or bladder calculi may be seen as echogenic structures with shadowing (similar to gallstones).
- Place the patient in the left or right decubitus position for evaluating the right and left kidneys respectively. The large acoustic window of the liver makes the right kidney easier to visualize than the left, and thus this kidney is often well examined in the supine position. More probe manipulation may be required to obtain good left kidney images.
- Use a 3.5 Mhz microconvex probe though a 5 MHz probe may be used in thin patients.
- Initially place the probe at the base of the costal margin laterally until the kidney can be seen in coronal view. Obtain a good long axis view of the kidney, and then rock the transducer gently side to side to examine the entire central calyceal system (Fig. 20.17).
- Rotate the probe such that a good transverse view can be seen. Scan the kidney superior to inferior and look for hydronephrosis which will be evident by an anechoic area within the pelvis of the kidney (Fig. 20.18).
- Always scan both kidneys and grade the presence of hydronephrosis as none, mild, moderate or severe.

Pitfalls

- Parapelvic cysts and an extrarenal pelvis can be confused with hydronephrosis.
- Overdistended bladder—can cause a normal amount of mild hydronephrosis.

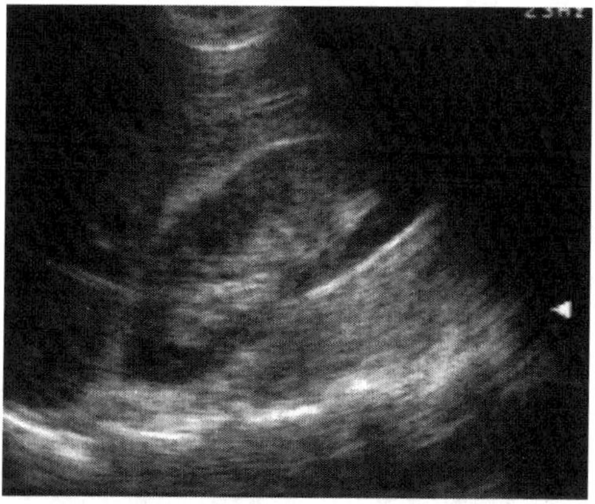

Figure 20.17. Normal long axis view of the right kidney.

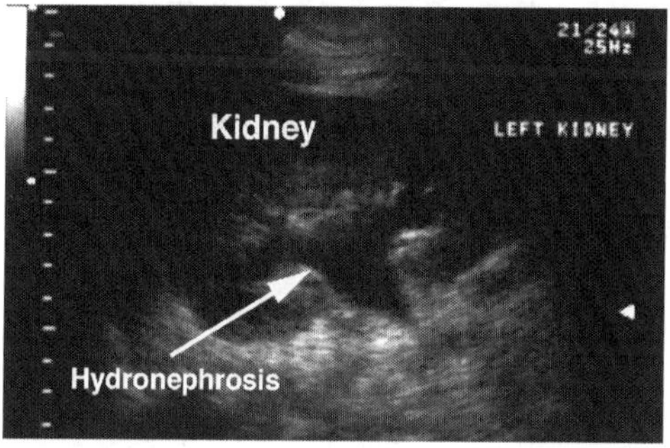

Figure 20.18. Anechoic area within the renal pelvis representing hydronephrosis.

- Renal pyramids in children and young adults can be hypoechoic and should not be mistaken for calyceal dilation.
- Advanced pregnancy will cause mild bilateral hydronephrosis.
- Dehydrated patients may not demonstrate hydronephrosis and thus hydration will be required in these patients.
- Medical renal disease–Patients with underlying chronic renal disease will often have scarred shrunken kidneys that will be difficult to evaluate.

Ectopic Pregnancy

Background

- Ectopic pregnancy is the leading cause of maternal death in the first trimester and is increasing in incidence. Thus all females of child-bearing age should be assumed to be pregnant and one of the essential components of the emergency evaluation is to ensure an intrauterine pregnancy.
- Ultrasound is a safe, integral and essential component of the evaluation of these patients. This bedside technique can quickly and reliably detect an intrauterine pregnancy and hemoperitoneum from a ruptured ectopic pregnancy.

Indications

- First trimester abdominal pain or bleeding in which an intrauterine pregnancy has yet not been established.

20

The Focused Exam

- *In the clinical approach to first trimester abdominal pain or vaginal bleeding our main focus is the detection of an intrauterine pregnancy.* In the algorithm to this condition, identification of an early intrauterine pregnancy essentially rules out an ectopic pregnancy. The risk of a heterotopic pregnancy is low at approximately 1:25,000 but increases with the use of fertility medications to as high as 1:5000.
- Patients that have an intrauterine pregnancy (IUP) identified may be treated clinically as the situation dictates.
- For those with no IUP identified, a quantitative B-HCG can help further stratify these patients.
  - Those with quantitative measurements <2000 and who are *clinically asymptomatic,* may be discharged to 48 h follow-up with repeat sonography and quantitative B-HCG. They need appropriate aftercare instructions to return if pain or bleeding recur.
  - Patients that are symptomatic or those with B-HCG levels of >2000 are at higher risk and need emergency gynecology consultation.
- The two approaches to pelvic sonography are transabdominal and transvaginal. Outlined below are key differences in these techniques.

	Transabdominal	**Transvaginal**
Probe type	3.5-5.0 MHz	5.0-7.5 MHz
Sonographic planes	Longitudinal and transverse	Sagittal and coronal
Bladder preparation	Full bladder	Empty bladder
Field of view	Wide	Narrow
Resolution	Less resolution	Better resolution
Gestation	Later gestation	Early gestation

- Transvaginal sonography detects early embryonic structures indicative of pregnancy approximately 1 wk before that seen on transabdominal sonography. Thus transvaginal sonography is a superior technique and should be used whenever possible. Transabdominal sonography used alone is only appropriate if the technique clearly identifies an IUP.
- Early embryonic structures sonographically appear in a sequential pattern as outlined below. *The first definitive sign of an IUP is the presence of a yolk sac within the gestational sac* (Fig. 20.20).

Embryonic Structure	**Transabdominal (weeks)**	**Transvaginal (weeks)**
Gestational sac	5.5-6	4.5-5
Yolk sac	6-6.5	5-5.5
Fetal pole	7	5.5-6
Cardiac activity	7	6
Fetal parts	>8	8

- Transabdominal technique—Using a 3.5-5.0 MHz probe, identify the uterus posterior to the bladder and scan in a sagittal and transverse view. Carefully examine the entire uterus looking for a gestational sac and evidence of an IUP including yolk sac and fetal pole (Figs. 20.19 and 20.20). Examine the cul-de-sac and morison's pouch for evidence of free intraperitoneal fluid.
- Transvaginal technique—Place ultrasound gel on the probe tip followed by a protective condom over the entire transvaginal probe. Gently insert the probe into the vagina. Scan the uterus in a sagittal and coronal fashion examining for evidence of an IUP. Also examine the adnexa for unusual masses and the cul-de-sac for free intraperitoneal fluid. After the exam, carefully clean and disinfect the vaginal probe. (Fig. 20.21)

Pitfalls

- Failure to perform both transvaginal and transabdominal sonography when evaluating for ectopic pregnancy.

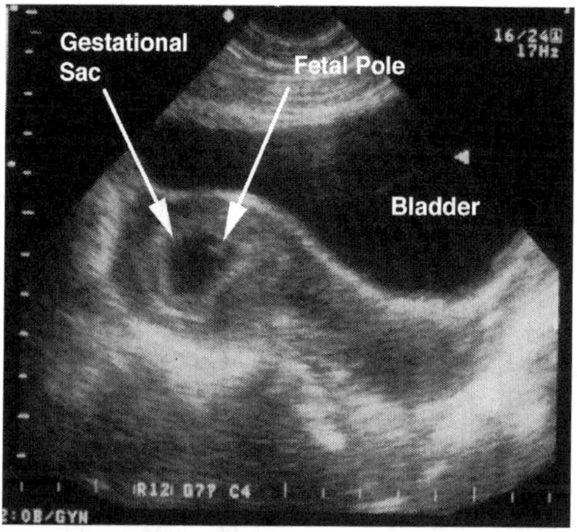

Figure 20.19. Transabdominal view of the uterus in long axis showing a gestational sac and fetal pole.

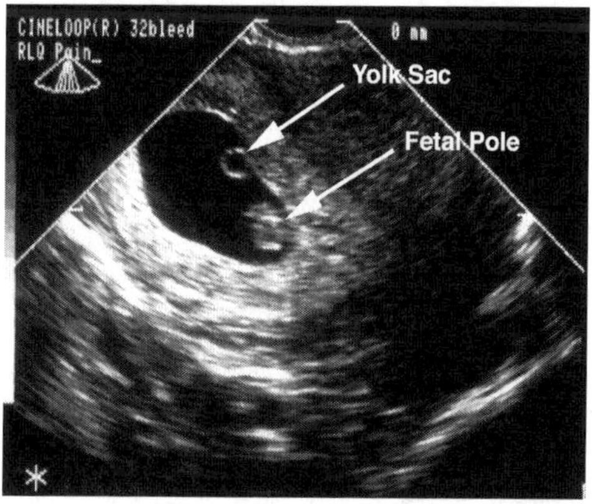

Figure 20.20. Transvaginal view of the uterus in long axis showing a yolk sac and gestational sac within the uterus.

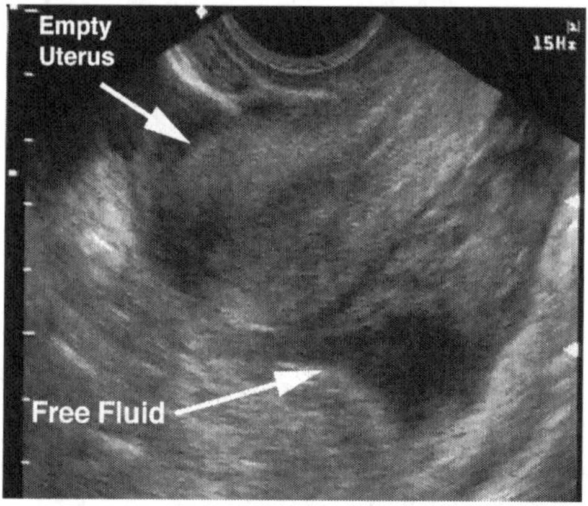

Figure 20.21. Empty uterus in long axis with free fluid on the transvaginal view. This in conjunction with a positive pregnancy test and free intraperitoneal fluid is indicative of an ectopic pregnancy.

20

- Falsely interpreting a gestional sac as a definite indicator of an IUP.
- Failure to recognize the position of the gestational sac and pole.
- Failure to look for free intraperitoneal fluid in the upper quadrants.
- Over-reliance on B-HCG measurements in symptomatic patients.

Abdominal Aortic Aneurysms

Background

- There are few medical emergencies that have consequences equal to that of a ruptured aortic aneurysm. Early detection and operative therapy are paramount in survival of this acute life threatening condition.
- Though computed tomography is an excellent imaging modality in stable patients, it has no role in acutely symptomatic patients. Unstable patients are best evaluated at the bedside with EP performed ultrasonography. Ultrasound provides an excellent screen of aortic diameter and can help direct management at the bedside.

Indications

- Abdominal pain suspicious for abdominal aortic aneurysm

The Focused Exam

- *Our primary goal is the identification of an abdominal aortic aneurysm.*
- The patient should be in the supine position.
- A 2.5-3.5 MHz abdominal probe will be required.
- Start by examining the aorta in the transverse plane by placing the ultrasound probe at the subxiphoid region. Identify the aorta on the left, the inferior vena cava on the right and the vertebral body inferiorly (Fig. 20.22). The aorta will also be thick-walled and more echogenic while inferior vena cava is tear drop-shaped and thin-walled.
- Slowly scan the entire abdominal aorta from superior to inferior until the iliac bifurcation is reached (usually at the level of the navel). If bowel gas obscures the aorta, place gentle steady pressure in that area as often this will move bowel to allow visualization of the aorta.
- Do sequential measurements of the entire aorta in the transverse plane. Measure from outer to outer wall. Any measurement >3.5 cm should be considered abnormal (Fig. 20.23).
- Examine the entire abdominal aorta in a sagittal plane and look for aneurysmal areas.
- If an aneurysm is identified, examine all quadrants for free intraperitoneal fluid. The presence of fluid suggests the free intraperitoneal rupture.

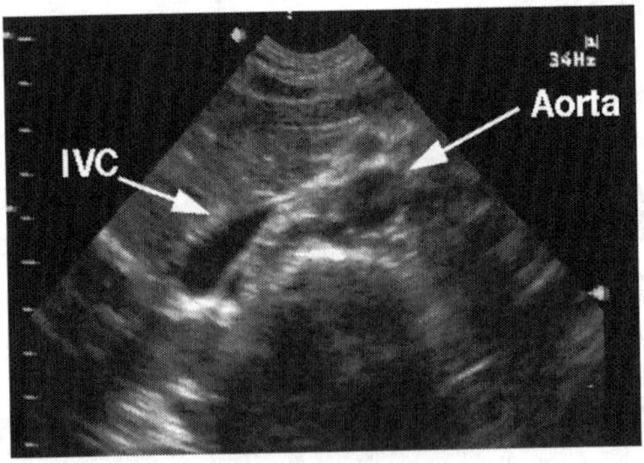

Figure 20.22. Normal transverse (short axis) view of the aorta.

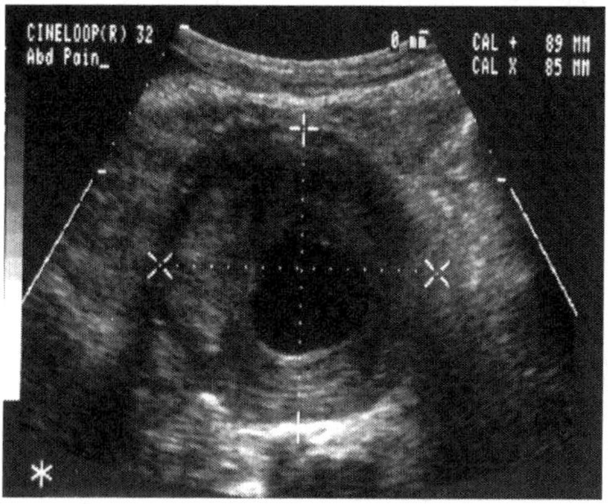

Figure 20.23. Large abdominal aortic aneurysm in transverse view (8.9 cm x 8.5 cm).

Pitfalls

- Bowel gas may impair your ability to fully examine the aorta. If the entire aorta cannot be visualized, then an alternate imaging modality such as computed tomography will be necessary.
- Retroperitoneal nodes or masses may displace the aorta making it more difficult to examine.
- Patients with previous aortic repair will be more technically difficult and thus formal imaging will be required.

Suggested Reading

1. Blaivas M, Harwood RA, Lambert MJ. Decreasing length of stay with emergency ultrasound examination of the gallbladder. Acad Emerg Med 1999; 6(10):1020-3.
2. Boulanger BR, McLellan BA, Brenneman FD et al. Prospective evidence of the superiority of a sonography-based algorithm in the assessment of blunt abdominal injury. J Trauma 1999; 47(4):632-7.
3. Chan D. Echocardiography in thoracic trauma. Emerg Med Clin North Am 1998; 16(1):191-207.
4. Durham B, Lane B, Burbridge L et al. Pelvic ultrasound performed by emergency physicians for the detection of ectopic pregnancy in complicated first-trimester pregnancies. Ann Emerg Med 1997; 29(3):338-47.
5. Heller M, Melanson SW. Applications for ultrasonography in the emergency department. Emerg Med Clin North Am 1997; 15(4):735-44.
6. Henderson SO, Hoffner RJ, Aragona JL et al. Bedside emergency department ultrasonography plus radiography of the kidneys, ureters, and bladder vs intravenous pyelography in the evaluation of suspected ureteral colic [see comments]. Acad Emerg Med 1998; 5(7):666-71.
7. Jehle D, Davis E, Evans T et al. Emergency department sonography by emergency physicians [see comments]. Am J Emerg Med 1989; 7(6):605-11.
8. Ma OJ, Mateer JR, Ogata M et al. Prospective analysis of a rapid trauma ultrasound examination performed by emergency physicians. J Trauma 1995; 38(6):879-85.
9. Mandavia D, Aragona J, Chan L et al. Ultrasound training for emergency physicians-A prospective study. Acad Emerg Med 2000; 7:1008-14.
10. Mandavia DP, Mallon WK. Using bedside ultrasonography to evaluate trauma patients. J Crit Illness 2000; 15(7):387-94.
11. Mateer J, Plummer D, Heller M et al. Model curriculum for physician training in emergency ultrasonography. Ann Emerg Med 1994; 23(1):95-102.

20

12. Mateer JR, Valley VT, Aiman EJ et al. Outcome analysis of a protocol including bedside endovaginal sonography in patients at risk for ectopic pregnancy. Ann Emerg Med 1996; 27(3):283-9.
13. Mayron R, Gaudio FE, Plummer D et al. Echocardiography performed by emergency physicians: Impact on diagnosis and therapy [see comments]. Ann Emerg Med 1988; 17(2):150-4.
14. Plummer D, Brunette D, Asinger R et al. Emergency department echocardiography improves outcome in penetrating cardiac injury. Ann Emerg Med 1992; 21(6):709-12.
15. Schlager D, Lazzareschi G, Whitten D et al. AB. A prospective study of ultrasonography in the ED by emergency physicians. Am J Emerg Med 1994; 12(2):185-9.
16. Shuman WP, Hastrup WJ, Kohler TR et al. Suspected leaking abdominal aortic aneurysm: Use of sonography in the emergency room. Radiology 1988; 168(1):117-9.
17. Mandavia D, Hoffner J, Mahaney K et al. Bedside echocardiography by emergency physicians. Ann Emerg Med 38:377-382,2001.

CHAPTER 21

Disaster Management

Mark Hollinger

Definition

- A disaster is any event (man-made or natural) that causes enough devastation or destruction that it cannot be managed by the usual resources.
- FEMA definition: An occurrence of a severity and magnitude that normally results in deaths, injuries, and property damage that cannot be managed through routine procedures and resources.
- A disaster is not defined by the number of injuries or deaths or by its destructive damage. If the resources are overwhelmed, it is a disaster.
- "Disaster Management" is the activities before, during, and after a disaster, which attempt to maintain control, lessen the impact, and aid in recovery.

Types of Disasters

Natural

- Weather (hurricane, drought, typhoon, cyclone)
- Topographic (landslide, avalanche, flood, mud slide)
- Underground (earthquake, tsunami, volcanic eruption)
- Biological (communicable disease outbreak)

Man-Made

- Warfare (nuclear/biologic/chemical terrorism, blockade, siege)
- Civil disturbance (riot, demonstration)
- Accidental:
  - Transportation accident (airplane crash, train wreck, sinking ship, traffic accident)
  - Structural collapse (building, mine, dam)
  - Explosion, fire, hazardous materials release, nuclear accident
  - Biological (inadequate sanitation)

Disaster Levels

- Level I: local resources adequate
- Level II: requires regional aid
- Level III: requires Federal aid

Objectives of Disaster Management

- To reduce or avoid human suffering and loss
- To reduce or avoid physical and economic losses
- To speed recovery back to normal (or as close to normalcy as possible)

These objectives are best accomplished through preparation before the disaster.

Three Stages of Disaster Management

Preparedness

- This includes having equipment and supplies in place and having a well thought-out, rehearsed plan. Teaching disaster response and basic first aid to the public is invaluable.

- Hazard mitigation: building stability and safety, code enforcement, safety measures (e.g., bolting down bookcases in earthquake areas) and possession of insurance policies to cover disaster-related damage

Response

21

- Provision of emergency medical care (basic first aid or more advanced care depending on the number of injuries and the availability of staff and supplies)
- Provision of psychological support
- Provision of basic needs including food, water, and shelter

Recovery

- Getting things back to normal (continued medical care, cleanup, and rebuilding).

Field Treatment Following a Disaster

The Goals of Field Medical Care

- Maintenance of an open airway
- Control of bleeding
  - Most external hemorrhage will stop with direct pressure. If not, elevate the affected area and/or compress the nearest pressure point.
  - Cover all open wounds. The sight of blood after a disaster increases anxiety among the public. If dressing supplies are inadequate, bed sheets (or other material) can be cut into strips for use.
  - Irrigate all wounds prior to closing. Use forceps to remove all obvious foreign material in the wound.
- Immobilization of suspected fractures
  - Use cardboard, magazines, or padded pieces of wood.
  - Another option is to splint two body parts together ("buddy splint"). Make sure that there is adequate padding in between the extremities
  - Once the extremity is splinted, elevate it and apply a cold pack, if available.
  - Always check distal nerve and circulatory function before and after splinting. If distal circulation is compromised after splinting, remove the splint and start over. If distal circulation is compromised prior to splinting, a closed reduction is indicated.
- Recognizing and treating shock
- Pain relief
- Recognition of crush injury
  - Crush syndrome occurs secondary to prolonged (>4 h) continuous compression of the extremities and can lead to rhabdomyolysis, life-threatening myoglobinuria, renal failure, hyperkalemia, and disseminated intravascular coagulation.
  - Prehospital care begins before the victim is removed. After extrication, the hemodynamic status of the injured victim may rapidly deteriorate and the victim may develop severe hypovolemia (extremity edema develops and upon release of the extremity, a redistribution of body fluid occurs). Treatment prior to removal of the trapped extremity includes bolus intravenous hydration using normal saline. Albuterol via hand-held nebulizer will help mitigate hyperkalemia. Also consider calcium chloride (1 g IV push) and sodium bicarbonate (1 mEq/kg added to a liter of normal saline).

Disaster Triage

- The primary goal of disaster triage is to do the most good for the most number of victims. This triage is designed to give maximal medical results with available resources.
- The main function of the triage team is to SORT the victims, not treat them. Therefore, the triage team should not spend an inordinate amount of time on any single victim.

- Disaster triage consists of two phases:
  - The first phase is to initially categorize the victims ("immediate", "delayed", "dead", or "walking wounded" as defined below).
  - If treatment is to begin immediately on scene, a second triage will be needed to determine which of the "immediate" victims will be seen first, then which "delayed" patients will be seen first, etc.

Triage Categories

- Immediate
  - Airway compromise
  - Suspected internal bleeding
  - Severe uncontrolled external bleeding
  - Serious fractures (pelvis, femur, neurovascular compromise)
- Delayed
  - Stable fractures
  - Spinal cord injuries
  - Minor burns
- Dead/Expectant
  - Obvious dead
  - Severe burns (80-100% full thickness)
  - Cardiac arrest
  - Severe head injury (brain matter showing or evidence of increased ICP)

The START System

- START stands for Simple Triage and Rapid Treatment. This is a widely accepted system developed by a hospital in California. It is a simple step-by-step method employed by the first qualified person who arrives on scene at a disaster. Although designed for the prehospital setting, it can be used in the hospital as well.
- START allows first responders to triage victims based on only three assessments:
  - Is ventilation adequate?
  - Is perfusion adequate?
  - Is the brain injured?

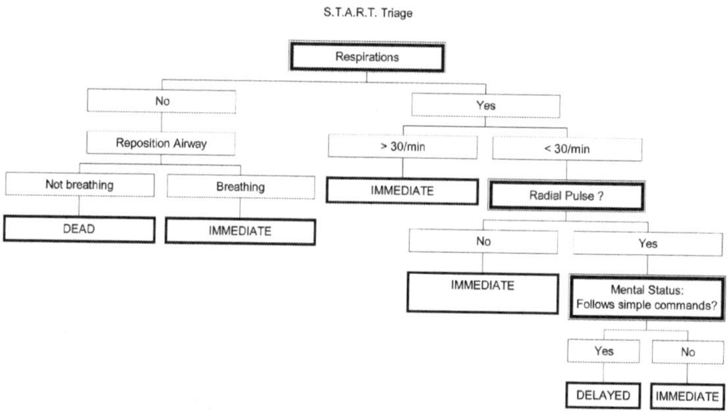

Figure 21.1. START triage chart.

- How START works:
  - Assess the scene to make sure it is safe to enter and begin triage.
  - Identify the mechanism of disaster (e.g., building collapse, fire, electrocution, etc)
  - Separate the "walking wounded" by announcing "If you can walk, move over there" (designated area). These victims are considered MINOR and are designated with *green* tags.
  - Triage and tag the remaining victims
    - Immediate: *red* tag
    - Delayed: *yellow* tag
    - Dead: *black* tag
  - Once all the victims have been tagged, preparation for transport and/or field treatment can begin.

Disaster Management within the Hospital

- After a major disaster, a hospital must assess the damage, care for the injured, assess the ability of the hospital to function, and prepare for the potential influx of victims.
- Some hospitals have an organized system to determine their operational status. The following is an example of the system used for hospitals in Los Angeles County:
  - "Green": the hospital is able to carry out both emergency and inpatient services in a normal manner.
  - "Amber": some reduction in patient care services, but overall, the hospital is able to continue providing emergency and inpatient services.
  - "Red": significant reductions in patient care services. Only emergency services being provided.
  - "Black": hospital is severely impacted and unable to provide emergency or inpatient services.

Damage Assessment

- This is mandatory after any disaster that is capable of causing structural damage (e.g., earthquake). This can be done locally by various hospital employees or globally by the hospital's safety or engineering departments. The following are types of damage that need to be assessed and reported:
  - Structural: this includes structural supports (posts, pillars), beams, floors, roofs, slabs and decks, load-bearing walls, and foundations.
  - Gas: if a gas leak is suspected, evacuations must occur and the fire department notified immediately. All fire and heat sources must be promptly discontinued.
  - Water: insufficient water supply, loss of water pressure, contaminated water supply, or flooding will all impact hospital functions.
  - Telephones: loss of phone lines will require a backup system. Options include walkie-talkies, HAM radios, or a "runner" messenger system.
  - Power: hospitals have a backup generator system for critical areas. Those areas without backup power may require evacuation and relocation. Preparation includes having additional portable generators and lights as well as flashlights in all offices and patient care areas.

Patient Census

- There must be a system in place to account for all patients including those being relocated or transferred.
- An inventory of open beds within the hospital must be determined. This includes critical care beds and specialized beds (e.g., neonatal, burn, "step down" units, etc.).
- It is important to identify all patients who can be discharged early and patients that can be "downgraded" from a critical care bed to a ward bed.

Personnel Inventory

- There must be a way to rapidly identify all available personnel. This includes medical staff, nursing personnel, ancillary staff, clerical staff, and volunteers.
- Determine if additional personnel will be needed. If this is necessary, a predetermined callback system should be utilized.
- Volunteers from outside the hospital may arrive and ask to help. The hospital must be able to keep track of these additional "personnel".
- Other staffing issues:
  - What should be done if there is too much staff ("staff triage")?
  - How are overtime issues handled?
  - How long can you keep staff on duty?
  - If staff desires to leave after their assigned shift, can they?
  - How will the increased staff be fed?
  - Can on-site childcare be provided for staff who volunteers to work?
  - If calling in staff from home, is there enough parking available? Can they get through police line on the way to work? Can they bring their children?

Command Post Organization

- Hospitals need to have an organized system so that those managing the incident know their roles and responsibilities and so that chain-of-command issues are addressed.
- One system is called the Incident Command System (ICS). The ICS has been in place for many years and is used by the military, police, fire personnel, and disaster managers. The ICS has been modified for hospital use in some areas and is referred to as the Hospital Emergency Incident Command System (see Fig. 21.2). This system allows everyone to communicate through a common language. It also allows for a more dependable chain-of-command and facilitates mutual aid with other hospitals and agencies.
  - Positions
    - Incident Commander
      - Has overall responsibility for the event
      - Sets objectives and priorities and determines strategy.
      - This is usually a hospital administrator with disaster management experience
      - Operations: the "hands on" people who carry out the objectives (caregivers, custodial staff, engineering, etc)

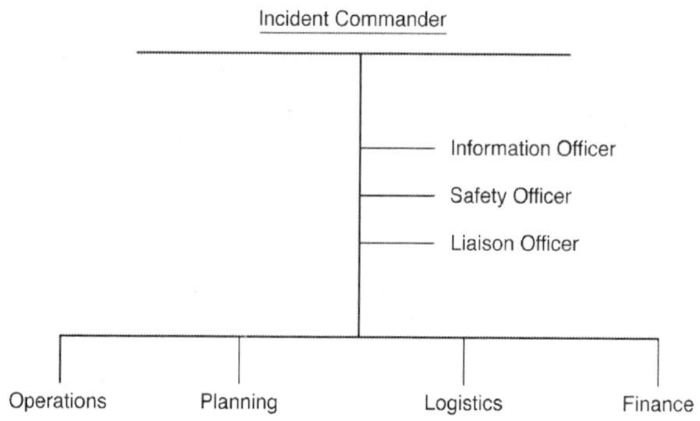

Figure 21.2. Hospital Emergency Incident Command System.

21

- Planning
    - The "brains of the operation"
    - Develops specific plans of action
- Logistics
    - Provides the resources to meet the needs of the incident
    - Requests all personnel and equipment
    - Supplies transportation, housing, and meals
- Finance: monitors all cost related to the incident (money spent, items loaned, overtime accrued, etc.)
- Information Officer
    - Provides information to the public including how to access medical care, what precautions to take, etc
    - Provides news media with appropriate information
- Safety Officer
    - Makes sure the operation runs smoothly in relation to the safety and welfare of the workers, patients, and visitors
    - Identifies potential or real hazards and takes steps to correct them
- Liaison
    - Communicates with others outside the hospital including formal requests to or from outside agencies
    - Coordinates interagency activities
- Emergency Operations Center (EOC)
    - Houses the command staff (incident commander, operations chief, finance officer, logistics chief, planning chief, safety officer)
    - Purpose: allows the command staff to monitor the situation/event and establish a clear chain-of-command
    - Location
        - Centrally located in an uncongested area
        - Easily relocated if necessary
        - Relatively safe from further damage
        - Familiar to all personnel/easy to find
        - Communications capability
        - Large enough to accommodate the command functions

Three Ingredients for a Successful Emergency Operation

- Management: leadership and organization
- Resources: personnel and supplies/equipment
- Communication: radios, telephones, and person-to-person

Terrorism and Weapons of Mass Destruction

- The FBI defines terrorism as the unlawful use of force or violence against persons or property to intimidate or coerce a government or civilian population in furtherance of political or social objectives.
- The goal of the terrorist is to create fear. Even if a threat is not carried out but makes people scared enough to prepare, the terrorist has won.
- Fortunately, the attack most likely to occur, the use of conventional explosives, is the one with the lowest destructive power. Conversely, the weapon with the highest destructive power, nuclear attack, has the lowest likelihood of occurrence. In the middle are chemical and biological weapons, both of which have moderate destructive power and moderate likelihood of use.
- Unfortunately, the majority of hospitals are unprepared for large terrorist activities since there is little financial incentive to prepare for an event that may never occur.

Additionally, in the civilian population, little is known about how to handle most of these events. The current medical literature is not sufficient and provides little or no practical guidance. Most information is gained from military sources and is of limited value in the civilian setting.

- Decontamination
  - A lot of time, money and energy are spent on preparing a hospital to decontaminate victims of a chemical or biological attack. In reality, decontamination by hospital personnel is not useful and can be dangerous. Furthermore, training hospital personnel is expensive, risky, a potential legal liability. Hospitals must arrange for trained "hazardous materials" teams to respond as needed.
  - Some decontamination is unnecessary. People can be either exposed or contaminated by a chemical or radioactive material. Exposed victims (i.e., inhalation) do not require decontamination procedures. Secondary exposure to the health care worker is minimal. Contaminated victims (liquid or powder on clothing or skin) require decontamination, which includes removing clothes and rinsing with copious amounts of water.
- Categories: the following list represents only a fraction of potential agents that could be used by terrorists. A detailed discussion of all agents is beyond the scope of this chapter although nerve agents and radiation will be reviewed.
  - Chemicals
    - Nerve agents: Tabun, Sarin, Soman, VX
    - Irritant gases: chlorine, phosgene, and others
    - Vesicant gases: mustard and lewisite
    - Cyanide
    - Riot control agents: mace and pepper spray
  - Biological
    - Bacteria: Anthrax, plague, Brucellosis, Q-fever, Tularemia, Salmonella
    - Viruses: smallpox, Venezuelan equine encephalitis, viral hemorrhagic fevers
    - Fungi: mycotoxins
    - Bacterial toxins: Staphylococcus enterotoxin B, ricin, botulinum toxin
  - Nuclear
  - Incendiaries and explosives

Nerve Agents

- Nerve agents are organophosphate esters that can be inhaled, ingested, or absorbed through the skin. Toxicity is related to their ability to bind with the enzyme acetylcholinesterase, which prevents the breakdown of acetylcholine at the neuromuscular junction. All nerve agents are liquids but can be aerosolized for inhalation.
- In general, victims exposed to a lethal amount will die on scene if untreated. Victims who arrive at a hospital without decontamination procedures do not have a lethal exposure. These victims have been exposed to the chemical and do not require a shower. Removal of clothing and the changing into a hospital gown is all that is required since the clothing may contain trapped vapor.
- Clinical presentation: can be divided into muscarinic, nicotinic and CNS effects:
  - Muscarinic: the "sludge syndrome" consisting of salivation, lacrimation, urination, defecation, GI distress, and emesis. Other effects include bronchoconstriction, laryngospasm, chest tightness, bradycardia and heart block.
  - Nicotinic: includes muscle fasciculations, twitching, jerking, weakness, flaccid paralysis, hypertension, tachycardia, mydriasis and diaphoresis.
  - Central nervous system (CNS): includes seizures, altered mental status, agitation and coma.

21

- Death is usually due to respiratory failure from diaphragmatic paralysis and or increased bronchial secretions.

Treatment

Atropine

- Antagonizes the muscarinic effects but has no effect on nicotinic symptoms.
- Administer at first sign of symptoms.
- Can be given IV, IM, or by endotracheal tube (IV preferred).
  - Dose
    - Adults: begin with 2 mg with repeat doses given as needed.
    - Children: begin with 0.02 mg/kg IV (0.5 mg minimum and 2 mg maximum)
  - *The dose should be repeated as clinically indicated with the endpoint being the drying of secretions.*

Pralidoxime (Protopam, 2-PAM)

- Most effective when given early. Start therapy as soon as you make the diagnosis.
- Removes nerve agent from acetylcholinesterase.
- Acts at both muscarinic and nicotinic sites. It is synergistic with atropine, thus decreasing the atropine requirements.
  - Dose
    - Adults: 1-2 g slow IV (dilute with 100 ml of normal saline and infuse over 15-20 min). May require repeat doses or an infusion.
    - Children: 20-40 mg/kg IV (IV preferred but may be given IM) to a maximum of 1 g.
  - The dose is repeated in 1 h if muscle weakness persists. Additional doses or an infusion may be necessary.
  - *Endpoint of therapy is the absence of signs / symptoms (weakness) of poisoning after the 2-PAM is stopped.*

Benzodiazepines

- Used to control associated seizures.
- Decreased neurologic morbidity in patients treated with benzodiazepines. Administer in all severe cases to prevent seizures.

Radiation Injuries

- Radiation is typically expressed in terms of *rem* when referring to biological exposure. The dose that is lethal for ~50% of subjects is 400 rem. Doses above 600 rem have nearly 100% mortality. Radiation doses received over a long period of time are less toxic than doses received over a short period of time.

Ionizing vs. Nonionizing

- Ionizing radiation includes gamma rays, X-rays, α particles, and β particles.
- Nonionizing radiation includes UV-rays, visible light, infrared, radio waves, and microwaves. Nonionizing radiation usually only causes mild problems related to local heat production.

Types of Ionizing Radiation

- Alpha particles: relatively harmless if intact skin; penetration limited to the epidermis.
- Beta particles: can cause deeper skin damage but are completely blocked by clothing.
- Gamma rays: deep tissue penetration and are the primary cause of acute radiation syndrome.
- Note that otherwise harmless α and β particles are dangerous if inhaled, ingested, or located in open wounds.

Irradiation vs. Contamination

- Irradiation: a radioactive substance has passed through a person, but that person is not made radioactive.
- Contamination: radioactive particles remain on the person or within any wounds that are present. This is generally seen with α/β particles and can pose a risk to health care personnel.

Acute Radiation Syndrome (ARS)

- Results from internal or external exposure over a short period of time.
- Tissues with higher rate of cell turnover are more sensitive (gastrointestinal, hematologic).
- Severity of ARS can be estimated based upon clinical symptoms and certain blood markers.
- Note that the radiation dose is inversely related to time of symptom onset (rather than symptom severity).
  - Nausea and vomiting (N/V) within 2 h of exposure indicates dose of >400 rem. N/V after 2 h suggests dose <200 rem.
  - Diarrhea indicates exposure >400 rem with bloody diarrhea being a bad prognostic sign.
  - CNS symptoms and seizures indicate very high dose (>2000 rem) as this is the organ system that is least sensitive to radiation injury.
- Management
  - Determine dose/type of radiation if possible.
  - Determine if patient is contaminated as this will pose a risk for all health care personnel.
    - Decontamination includes removal of clothing followed by rinsing with water. This should be started in the field but may need to take place in the hospital for severely injured patients. Hospital decontamination should follow the specifications of the existing disaster plan. Ideally this involves use of a separate ED entrance and decontamination room with closed drainage and ventilation systems.
    - Contaminated wounds may require surgical debridement or amputation.
  - Reverse isolation for exposure >200 rem.
  - Supportive care as indicated (there is little that one can do after decontamination). There are chelators/blocking agents which can be used for certain types of radiation: potassium iodide for radioactive iodine; EDTA for radioactive lead; DTPA for heavy metals.
- Prognosis: depends upon the dose and duration of exposure and can be estimated by absolute lymphocyte count at 48 h after exposure.
  - >1200: good prognosis and nonlethal exposure.
  - Between 300 and 1200: fair prognosis with possibility of lethal exposure.
  - <300: poor prognosis and likely lethal exposure.

Biological Terrorism

- There are a number of agents and toxins to consider (see Table 21.1). A detailed discussion is beyond the scope of this chapter.
- Use of universal precautions can prevent transmission of most agents to healthcare providers. For example, an N-95 filter mask (orange duck bill as used for tuberculosis) is effective for many organisms. Some infections (e.g., inhalation anthrax) are not transmitted person-to-person.
- The best defense is an alert health care system. Identifying suspicious outbreaks of illness or unusual presentations of disease points to the possibility of a biological terrorism event.

Table 21.1. Biological warfare agents

Agent / Disease	Transmit: Man-to-Man	Incubation Period	Illness Duration	Lethality	Clinical	Prophylaxis	Treatment
Anthrax (inhalational) *Bacillus anthracis*	No	1-6 days (up to 30-60 days)	3-5 days	High	Initial: fever, malaise, fatigue, cough Later: respiratory distress, stridor, cyanosis; shock and death 24-36 h after onset severe symptoms CXR*: widened mediastinum, pleural effusion	Ciprofloxacin or doxycycline x 60 days	Ciprofloxacin, doxycycline, or penicillin
Brucellosis *Brucellae sp.*	No	5-60 days	Wk to mo	<5% (un-treated)	Fever, headache, fatigue, myalgias, anorexia, cough, GI**/joint/skeletal symptoms (CXR usually normal)	Doxycycline + rifampin	Doxycycline + rifampin
Plague, pneumonic *Yersinia pestis*	High	1-6 days	1-6 days	≈ 100% unless Rx w/in 24 h	Fulminant course: fever, chills, headache, hemoptysis, dyspnea, cyanosis, stridor, DIC***, respiratory failure, shock	Ciprofloxacin or doxycycline	Streptomycin, gentamicin, or doxycycline
Q Fever *Coxiella burnetii*	Rare	2-14 days	Days to wk	Very low	Fever, cough, pleuritic chest pain, and myalgias Self-limited even without treatment	Tetracycline or doxycycline	Tetracycline or doxycycline
Tularemia *Francisella tularensis*	No	1-21 days (avg 3-5 d)	Days to wk	Rare if treated	Ulcer formation, lymphadenopathy, fever/chills, pharyngitis, non-productive cough, pneumonia	Ciprofloxacin, doxycycline, or tetracycline	Streptomycin, ciprofloxacin or gentamicin
Smallpox Variola virus	High	7-19 days (avg 12 d)	2-4 wk	High	Prodrome: malaise, fever/chills, vomiting, backache Rash: macules → papules → pustules (2-3 days after prodrome), central spread	Vaccine w/in 7 days (best w/in 24 h)	None
Venezuelan equine encephalitis (VEE)	No	1-6 days	Days to wk	Low	Fever, chills, severe headache, photophobia, myalgias, GI symptoms	None	Supportive

continued on next page

Table 21.1. Continued

Agent / Disease	Transmit: Man-to-Man	Incubation Period	Illness Duration	Lethality	Clinical	Prophylaxis	Treatment
Viral hemorrhagic fever	High	4-21 days	7-16 days	Moderate to high	Fever, coagulopathy, petechiae, hypotension, multi-organ involvement	None	Supportive (Ribavirin—compassionate use)
Botulism (*Clostridium botulinum* toxin)	No	Symptom onset: 18 h to days	Days to mo	High	Ptosis, weakness, dizziness, dry mouth, blurred vision, diplopia, dysphagia, dysphonia, urine retention, ileus, descending flaccid paralysis, respiratory failure; patient remains afebrile with normal mental status	None	Antitoxin (give early), supportive care
Ricin (toxin from castor beans)	No	4-8 h	36-72 h	High	Weakness, fever, cough, hypotension, cardiovascular collapse, airway necrosis/edema (ingestion: GI bleed)	None	Supportive, lavage if ingested
Staph enterotoxin B (SEB)	No	3-12 h (inhalation)	Days to wk	Low	Fever, chills, headache, myalgias, nonproductive cough, pulmonary edema, GI symptoms (if swallowed)	None	Supportive

* CXR: Chest radiograph. **GI: gastrointestinal. ***Disseminated intravascular coagulation.

21

- The Public Health Department should be notified immediately upon initial suspicion or confirmation of a bioterrorism event. Hospital planning should include a designated ward/area to be used to house and treat victims. This area must be separated from other wards and have its own access.
- Prognosis for bioterrorism is largely unknown due to lack of recent experience with some of the infectious agents. Although many resources are available, the statistics were formulated during a time when there was less sophisticated intensive care and a more limited choice of antibiotics.

Suggested Reading

1. Textbook of Military Medicine. Washington, DC: Chemical and Biological Warfare Office of the Surgeon General, US Army, 1997.
2. Auf Der Heide E. Disaster Response: Principles of Preparation and Coordination. Mosby, CV: St Louis Mo. 1989.
3. Sullivan J. Hazardous materials toxicology: Principles of environmental health. Lippincott: Wilkey and Williams., 1992.
4. Morres C. A Disaster Medicine ≅, Emergency Medicine Clinics of North America, Saunders Publ., 1996.
5. Chemical and Biological Terrorism. Natl Academy of Sciences, Natl Academy Press Publ., 1999.

Chapter 22

Special Pediatric Considerations

David Leader, Marilyn Hicks, Russell Karten, Lance Brown and Olly Duckett

Part A: Pediatric Cardiopulmonary Arrest

David Leader

The outcome of cardiopulmonary arrest in children is generally poor, reaching a 90% mortality rate in some studies.

- Most of these patients (87%) have an underlying disease.
- Respiratory arrest has been associated with a mortality rate of 33%, with most affected children being <1 yr of age.
- Mortality is the same regardless of whether the arrest occurs in the prehospital or hospital setting.
- In adults, cardiac arrest usually is the result of underlying cardiac disease. In contrast, children usually develop cardiac arrest secondary to respiratory arrest and shock syndromes.

Clinical Presentation

- Signs of respiratory failure can be recognized but may be quite subtle initially.
  - These include noting abnormal respirations that are characterized by tachypnea, bradypnea, apnea, or increased work of breathing.
- Cardiac function decompensation then leads to a shock state that is associated with impaired cardiac output. This is usually manifested by tachypnea, hypotension, and poor peripheral circulation.
- As cardiac functions worsens, impaired perfusion causes delayed capillary refill time (>2 seconds), mottling, cyanosis, cool skin, altered level of consciousness, poor muscle tone and finally decreased urine output.
- The most common rhythm seen in pediatric arrest situations is bradycardia leading to asystole.
- By recognizing this typical pattern, the emergency physician can intervene and potentially prevent full cardiopulmonary arrest. The precipitating causes of cardiopulmonary arrest are many, and often these can be identified and prevented (See Table 22A.1).

ED Management

- Pediatric cardiopulmonary resuscitation is best when performed by a coordinated team of physicians, nurses, respiratory therapists, and other support personnel.
  - For an overview of resuscitation, please see Chapter 1.
  - Specific concerns re: pediatric resuscitation are discussed here.
- Children usually respond to initial airway, ventilation and fluid therapy interventions. In other words, the first treatment for bradycardic arresting young child is bagging and then intubating them, not atropine as with adults.
  - All drugs, medications, invasive catheters and tubes must be age- and weight-specific (discussed later in this chapter).

Emergency Medicine, edited by Sean Henderson. ©2006 Landes Bioscience.

22

Table 22A.1. Causes of cardiopulmonary arrest in children

Respiratory disease	Upper airway obstruction (croup, epiglottitis, foreign body, suffocation, trauma) lower airway disease (pneumonia, asthma, bronchiolitis, foreign body, near drowning, smoke inhalation, pulmonary edema)
Infection	Sepsis, meningitis
Cardiac disorders	Congenital heart disease, myocarditis, pericarditis, dysrhythmia
Shock	Cardiogenic, hypovolemic, distributive
Traumatic/evironmental	Multiple organ failure, child abuse, hypothermia, hyperthermia, near-drowning
Metabolic	Hypoglycemia, hypocalcemia, hyperkalemia
CNS	Meningitis, encephalitis, head trauma, anoxia, hypoventilation
Sudden infant death syndrome	
Intoxication	Narcotic depressants, sedatives, antidysrhythmics

- Also, children must be carefully monitored with frequent vital sign assessment, cardiac monitoring, and pulse oximetry.
- Fluid calculations need to be made based on an estimated weight based on age, a measured weight, or measured length using a resuscitation tape (e.g., Broselow tape).

- The key to any pediatric resuscitative effort is adequate and timely airway management.
  - Careful attention must be given to positioning the infant or young child as the prominent occiput can cause neck flexion and airway occlusion. Mild extension of the head to the "sniffing" position will most often provide a patent airway.
  - Avoid over-extension as this may cause airway obstruction by compressing the flexible trachea. Avoidance of neck movement with in-line stabilization must be done if there is any suspicion of neck trauma.
  - In an unconscious patient, hypotonia may result with the tongue falling against the posterior pharyngeal wall. The chin lift and jaw thrust maneuver will open the airway in this case.
  - Once the airway has been opened manually, nasopharyngeal tubes for conscious children or oropharyngeal airways for unconscious children may be beneficial.
- If the airway cannot be maintained or other indications exist (respiratory failure, decreased mental status with inability to protect the airway, or cardiopulmonary arrest), endotracheal intubation is indicated.
  - Preparations for intubation can be started while the child is being properly positioned and oxygenated.
    - A bag valve-mask device can provide ventilation until ready to intubate.
    - As a general rule, a straight (Miller) blade is preferred in children younger than 5 yr of age and a curved (Macintosh) blade should be used for older children.
    - Endotracheal tubes without cuffs should be used in children younger than 10 yr of age since the cricoid ring is small enough to produce an air seal in this age group.
    - Selection of the appropriate size endotracheal tube can be determined by several means.
    - The correct tube size can be approximated by using a simple formula based on the patients age:
      - Inside diameter (ID) in mm = (16 + age in years) divided by 4
    - Two additional tubes should be immediately available, 1/2 size smaller and 1/2 size larger since this is an estimate.
    - Using a table of approximate sizes, based on an accurate age or weight, an appropriate sized endotracheal tube can be determined.

- Estimation of the tube size based on the diameter of the child's fifth finger has been shown to be inaccurate. Length has been shown to be the most accurate predictor of the correct endotracheal tube size.
- The Broselow tape (a length-based tape) provides accurate equipment selection as well as resuscitation drugs with a single length measurement. The proper depth of the endotracheal tube insertion (in centemeters) is approximately three times the internal diameter of the endotracheal tube.

- Following intubation, correct tube placement should be confirmed using end-tidal CO_2 monitoring, auscultation of breath sounds bilaterally, pulse oximetry monitoring and a stat portable chest X-ray to check endotracheal tube placement.
- The most common problem identified on chest X-ray after endotracheal intubation is a right mainstem bronchial intubation. These cases of placing the tube "too deep" may lead to a marked degree of left lung collapse.
- Since endotracheal tubes can easily be dislodged in children, the tube should be secured in place with tape.
- In general, oxygen should initially be supplied at 100% concentration.
- Alternative management of the airway including esophageal/tracheal tubes, laryngeal mask airways, and transtracheal ventilation have not been well tested and researched in children.
- These modalities remain as techniques worth considering in settings where standard methods have failed.
- The assessment of circulation is made on clinical grounds. The primary indication for assisting circulation with chest compressions is pulselessness.
- In the newborn chest compressions are also indicated when the heart rate remains <60 bpm despite effective airway intervention.
- In the newborn the compression rate should be 100-120/min, and in older children 80-100/min.
- Compression depth should be sufficient to produce a brachial or femoral pulse.
- First line drugs are given to virtually all patients who suffer a cardiopulmonary arrest and who do not respond to initial airway and ventilatory support.
- Drug dosing can be based on body weight estimation or the length-based resuscitation (Broselow) tape (see Tables 22A.2 and 22A.3).

Length Based Resuscitation Tape (Broselow Tape)

- The Broselow tape provides accurate equipment selection as well as resuscitation drug doses with single length measurement.
- The drug side of the Broselow resuscitation tape includes 25 precalculated weight zones for resuscitation drugs based on length measurements.
- On the equipment side of the Broselow resuscitation tape are included seven color equipment zones including laryngoscope size and type, endotracheal tube sizes, stylet, suction catheter, nasogastric tube, Foley catheter size, chest tube size as well as CPR guidelines.

Table 22A.2. Body weight estimation guidelines

Age	Weight (kg)
Term infants	3.5 kg (birth wt-BW)
6 mo	7 kg (2 x BW)
1 yr	10 kg (3 x BW)
4 yr	16 kg (1/4 adult wt of 70 kg)
10 yr	35 KG (1/2 adult wt)

22

Table 22A.3. Weight-based drug dosing for drugs commonly used during treatment for pediatric cardiopulmonary arrest

Drug	Dose
Epinephrine	0.1 ml/kg initial dose 1:10,000 preparation (some suggest 1:1,000 preparation for subsequent doses)
Atropine	0.01-0.02 mg/kg (0.1 mg minimum)
Lidocaine	1 mg/kg
Adenosine	0.1 mg/kg
Dextrose	2-4 ml/kg of D25W (5-10 ml/kg of D10W in young infants)
Succinyl choline	1.5 mg/kg
Dopamine	5-20 mcg/kg/min infusion

Vascular and Intraosseous Access

- Vascular access for medication or intravenous fluid administration is of critical importance in the resuscitation situation and may be the most difficult to obtain quickly.
- Peripheral vascular access in the infant or small child can be obtained in the dorsal hand veins, dorsal foot veins, antecubital veins, superficial scalp veins, or the external jugular veins.
- If peripheral IV access cannot be obtained, intraosseous (IO) needle placement should proceed without delay.
- Central line placement is a more time consuming option.
- Resuscitation drugs (especially epinephrine, atropine, lidocaine) can be given via the endotracheal tube until vascular access is obtained.
- The femoral vein, subclavian vein, or internal jugular vein may provide central venous access, but these should not be first line attempts. When the patient is stable, these lines may be replaced under more controlled conditions.
- Intraosseous (IO) needle placement is a quick method for obtaining vascular access during CPR, impending arrest, or severe shock.
- The IO route is effective for rapid drug delivery, and some studies suggest that the intraosseous route is superior to the intravenous route, with higher central venous and arterial drug concentrations obtained.
  - The sites of choice are the proximal tibia (1 fingerbreath distally from the tibial tuberosity), and the distal femur.
  - The technique for placement includes anesthetizing the skin with 1% lidocaine in the nonarrested patient (not necessary in the unconscious patient) and cleansing the skin insertion site with betadine solution to prevent introducing an infection.
  - The needle is then placed perpendicular to the bone and using a controlled forceful screwing motion until a pop is felt as the marrow is entered.
  - The inner stylet is removed. Bone marrow can be obtained on aspiration.
  - If the IO needle flushes with 10 ml of NS without extravasation and the needle remains in an upright position without support, it is probably in place.
  - Contraindications to IO placement include osteogenesis imperfecta, osteoporosis, as the brittle bones in these cases are susceptible to bone fractures.
  - Other contraindications for IO placement include fractured long bones or placement into the same extremity where a needle hole was previously made.
  - Complications of IO placement include osteomyelitis (seen in about 1% of cases), anterior compartment syndrome, tibial fractures, and fat or marrow emboli to the lungs.
- If there is any question of hypovolemia, a rapid infusion of normal saline or lactated Ringers Solution at 20 ml/kg over a 10-20 min should be given. Repeat doses will almost certainly be required in children in shock.

Trauma as a Cause of Pediatric Cardiopulmonary Arrest

- Injuries are the dominant problems encountered in prehospital pediatric emergency care and are the most common cause of death in children over 1 yr of age.
- Events involving motor vehicle crashes (occupant, bicyclist, or pedestrian), fires and falls comprise the primary mechanics of injury.
- Pediatric trauma is mainly blunt, although penetrating trauma is increasing in the urban areas.
- Fatal injuries often include the head, chest and abdomen.
- Airway management is of vital importance as are resuscitation with crystalloid fluids and blood when needed.
- Rapid transfer to a tertiary care facility or trauma team evaluation is also important.

Electrical Countershock/Defibrillation

- It is unusual for a child's heart to fibrillate, therefore defibrillation is an uncommon intervention in pediatric resuscitation.
- The typical pattern is to progress from tachycardia to bradycardia and finally asystole.
- Since fibrillation is uncommon in children, it should be confirmed prior to defibrillation.
- A precordial thump for witnessed arrest is not recommended in children.
- The initial defibrillation dose is 2 J/kg, with a second dose at the same dose of energy, immediately if the first was unsuccessful.
- The first dose lessens resistance, thereby making the second dose more effective.
- If this first defibrillation sequence is unsuccessful, CPR is continued 3-5 min (with epinephrine, lidocaine given) and then the next defibrillation dose is doubled to 4 J/kg and repeated twice if needed.
- The dose can be doubled to 4 J/kg for two successive shocks if necessary.

Disposition

- Preliminary evidence has shown that if cardiac muscle is not responsive to the first three doses of epinephrine when adequate oxygenation and ventilation have been supplied, then there is no hope for survival.
- Additional research is needed to confirm this, particularly in children.
- Efforts should be terminated if there is obvious brain death.
- Resuscitation should be continued if there is evidence of drug depression or hypothermia.
- Those children surviving resuscitation should be admitted to a pediatric intensive care unit.
- Resuscitation of an infant or young child is an emotionally stressful event that can leave an indelible impression on the mind and emotions of health care providers.
- Consequently, organization and an understanding and preparation for pediatric resuscitation should be a high priority for Emergency Departments.

Suggested Reading

1. Hazinski MF, Zaritsky AL, Nadkarni VM et al. PALS provider manual dallas. American Heart Association 2002.
2. Carpenter TC, Stenmark KR. High dose epinephrine is not superior to standard-dose epinephrine in pediatric In-hospital cardiopulmonary arrests. Pediatrics 1997; 99:403-408.
3. Seidel J. Pediatric cardiopulmonary resuscitation: An update based on the New American Heart Association guidelines. Pediatric Emerg Care 1993; 9:98.
4. Zaritsky A. Pediatric Resuscitation pharmacology. Ann Emerg Med 1993; 22:445.
5. King BB, Baker MD, Braitman LE et al. Endotracheal tube Selection In children: A comparison of four methods. Ann Emerg Med 1993; 22:530.
6. Luten RC, Wears RL, Broselow J et al. Length-based endotracheal tube and emergency equipment in pediatrics. Ann Emerg Med 1992; 21:900.

7. Fisher D. Intraosseous infusion. N Engl J Med 1990; 322:1579.
8. Arkins DL, Kerber. Pediatric defibrillation. Pediatrics 1994; 94:90.
9. Johnston C. Endotracheal drug delivery. Pediatr Emerg Care 1992; 8:94.
10. Ludwig S. Resuscitation- Pediatric basis and advanced life support. Textbook of Pediatric Emergency Medicine. 4th ed. Lippincott, Philadelphia: Williams & Wilkins, 2000:3-31.
11. Zaritsky A, Luten R. Pediatric resuscitation. In: Rosen P, ed. Emergency Medicine: Concepts and Clinical Practice. 4th ed. St. Louis: 1998:66-85.

Part B: Cardiac Problems in Children

Marilyn Hicks

Supraventricular Tachycardia

Incidence

Supraventricular tachycardia (SVT) is the most common pediatric tachyarrhythmia.

- Infants <4 mo of age account for about 40% of initial episodes of SVT. Reentrant tachycardias are the most common with only 10% of patients having atrial ectopic tachycardias.
- Reentrant tachycardias are caused by electrical excitation around a fixed pathway obstruction. During normal conditions, electrical excitation occurs simultaneously down both ventricular pathways, each of which has different conduction velocities.
- Tachycardia occurs when antegrade excitation is blocked in one of the pathways. If conduction velocity in the intact pathway is slow enough, excitation can occur in the retrograde direction in the blocked pathway.
- There are two forms of reentrant supraventricular tachycardia of which atrioventricular reciprocating tachycardia (AVRT) is the most common.
- The AV node is the antegrade limb, and the accessory pathway is the retrograde limb. The accessory pathway allows conduction to occur between the atria and ventricles via a pathway other than the AV node. Wolff-Parkinson White syndrome (WPW) exists when conduction occurs via the accessory pathway during sinus rhythm.
- The other form of reentrant SVT, atrioventricular nodal tachycardia (AVNRT), occurs when both limbs of the circuit are within the AV node.
- Atrial ectopic tachycardia occurs when the abnormal impulse is generated within the atria.
- WPW patients are also prone to develop atrial flutter with rapid conduction down the accessory pathway. There is a significant risk of sudden death associated with WPW that involves antegrade conduction down the accessory pathway. Any child with syncope associated with WPW supraventricular tachycardia should be considered for a catheter ablation procedure.

Clinical Presentation

The clinical presentation depends upon the age of the patient and the duration of the arrhythmia.

- Neonates and infants often present with vague complaints such as poor feeding or irritability. If the arrhythmia has gone unnoticed for hours or days, the infant may develop heart failure and present with respiratory distress and shock.
- Older children may complain of tachycardia or palpitations as well as dizziness, fatigue, or nausea.
- Occasionally children complain of chest pain, which may be the result of insufficient coronary perfusion.

- Unless there is significant hemodynamic compromise, the physical examination is often unremarkable except for the tachycardia.
  - Neonates often have a heart rate of 200-300 beats/min. Those infants presenting with congestive heart failure and/or shock will respond rapidly to the restoration of normal sinus rhythm.
- The hallmark of evaluation is a 12-lead EKG and should be obtained prior to any therapy unless there is a significant hemodynamic instability.
  - Reentrant SVT is a regular tachycardia with rates of 220-300 beats/min.
  - The width of the QRS complex is almost always normal and inverted P waves are generally present in lead II, III, and AVF.
  - EKG findings associated with WPW syndrome are a very short PR interval and slurring of the upstroke of QRS or delta waves.
  - Atrial ectopic tachycardias are often difficult to distinguish from sinus rhythm and the rate is typically only moderately increased for age. Subtle abnormalities of the P-wave configuration may be the only clue to the diagnosis.
  - Echocardiography should be performed for all infants with the first episode of SVT to rule out associated CHD.
- Chest X-ray and laboratory tests are of little benefit except in the presence of congestive heart failure, CHD, or other comorbid diseases.

ED Management

- Initial treatment of SVT should be directed toward the termination of the tachycardia.
- In unstable patients synchronized DC cardioversion at 0.25-1.0 joules/kg should be applied.
- In more stable patients, the initial drug of choice is adenosine at a dose of 50-100 μg/kg as a rapid IV push. The dose can be doubled as needed to a maximum of 12 mg/dose.
- Other drugs for acute treatment of SVT include propranolol, flecainide, and procainamide (see Table 22B.1).
- Verapamil can cause significant hypotension and bradychardia and should be avoided in infants less than 1 yr of age and used with caution in older children.
- Digoxin is contraindicated in WPW syndrome and has no place in the acute management of SVT.
- Infants and children with SVT should be referred to a pediatric cardiologist for chronic management.

Table 22B.1. Medical therapy for supraventricular tachycardia

Drug	Route	Dose
Adenosine	IV, rapid push	0.1 Mg/kg May double dose to max 0.5 mg/kg Max dose = 12 mg
Propranolol	PO	0.5-1 mg/kg QID
	IV (over 10 min)	0.01-0.15 mg/kg Max = 1 mg/dose
Flecainide	PO	1-2 mg/kg Max = 100 mg
Procainamide	PO	4-12 mg/kg Max = 1 g
	IV	Load: 5-15 mg/kg Max 100 mg Maintenance: 20-80 μg/kg/min Max = 2 g/24 h

22

Table 22B.2. Causes of congestive heart by age of presentation

First days of life	Congenital heart disease Supraventricular tachycardia, complete heart block Asphyxia / ischemia Sepsis Hypocalcemia, hypoglycemia Myocarditis Anemia, hyperviscosity syndrome (hematocrit >65)
First wk of life	Congenital heart disease Supraventricular tachycardia, complete heart block Renal failure, hypertension Hyperthyroidism, adrenal insufficiency
First 2 mo of life	Congenital heart disease Congenital cardiomyopathy, endocardial fibroelastosis, myocarditis, glycogen storage disease Pulmonary disease—bronchopulmonary dyspasia, upper airway, obstruction, central hypoventilation Renal failure, hypertension Hyperthyroidism, adrenal insufficiency
Childhood and adolescence	Acquired heart disease Unoperated congenital heart disease Postoperative congenital heart disease

Congestive Heart Failure in Infants

The approach to congestive heart failure (CHF) in infants is largely dependent upon the etiology.

- There are multiple causes of CHF in infants and age at presentation can be a useful guide (see Table 22B.2).
- Congenital heart disease is the most common cause of CHF in infants; however sepsis, metabolic and endocrine abnormalities, rhythm disturbances, and myocarditis are frequent etiologies.
- The pathophysiology of CHF is basically that of ventricular dysfunction and the clinical manifestations the result of inadequate cardiac output or increased systemic or pulmonary venous pressure.

Clinical Presentation

- The clinical presentation of CHF may be insidious and manifest as poor feeding, poor weight gain, sweating during feeding, or irritability with the subsequent development of cardiorespiratory symptoms over several days.
- Heart failure can also present acutely with respiratory failure and cardiogenic shock; when this occurs the clinician must have a high index of suspicion for left-sided obstructive cardiac disease.
- The four major categories of left-sided obstructive heart disease are hypoplastic left heart, critical coarctation, critical aortic stenosis, and interrupted aortic arch. All of these lesions are dependent upon the patency of the ductus arteriosus and severity of symptoms are directly related to the closure of the ductus.
- Clinical symptoms vary and are dependent upon the age of the infant, the cause of ventricular dysfunction, and the severity and time course of deterioration.
- Systemic manifestations include growth failure due to poor feeding and increased caloric requirements, tachypnea and sweating during feeding due to increased autonomic activity.

- Cardiac manifestations are tachycardia (resting heart rate >150 beats/min in infants), gallop rhythm, and altered perfusion. Symptoms of altered perfusion include cool, mottled extremities, diminished pulses, and delayed capillary refill.
- The infant with sustained tachycardia, loud cardiac murmur and precordial thrill or heave should be suspected of having a structural cardiac lesion.
- The most reliable respiratory symptom associated with CHF is tachypnea, the principle manifestation of interstitial pulmonary edema.
- Other respiratory symptoms include cough, wheezing, and cyanosis; rales are infrequently heard in infants and indicate the presence of severe edema.
- Signs of systemic venous congestion are not frequently seen in infants with CHF. Hepatomegaly, pedal edema, and jugular venous distention are difficult to assess but when present are an indication of severe right-sided heart failure.

Diagnostic Evaluation

- The evaluation of the infant suspected to have CHF should begin with the chest X-ray, electrocardiogram (EKG), and echocardiogram.
- Typical chest X-ray findings are cardiomegaly (cardiothoracic ratio >0.65 in infants), increased pulmonary vascular workings, and hyperexpansion with flattened diaphragms.
- The absence of cardiomegaly should prompt questioning of the diagnosis of CHF, except with restrictive cardiomyopathy and obstructed total anomalous pulmonary venous return.
- Pleural effusions resulting from heart failure are uncommon in infants except with severe right-sided failure.
- The EKG should be evaluated for the presence of dysrhythmias, as well as for atrial or ventricular enlargement; however it is rarely useful in making the diagnosis of CHF.
- Echocardiography is extremely helpful in delineating anatomic abnormalities and quantifing ventricular dysfunction.
- Treatment of heart failure should never be delayed while awaiting an echocardiogram.
- Additional laboratory testing should include arterial blood gases, electrolytes, BUN, creatinine, glucose, and complete blood count. These allow a more complete evaluation of the heart failure as well as elucidate causes of CHF other than structural heart disease.
- A septic work-up should be performed in very young infants with nonspecific findings and those infants with symptoms suggestive of infection.
- The evaluation of CHF often requires investigation of several potential etiologies simultaneously if the history and physical examinations fail to provide an obvious cause.

ED Management

- The management of the infant with CHF who presents with acute respiratory failure and shock requires rapid intervention.
- This should begin with the ABCs in an effort to minimize myocardial oxygen demand and maximize myocardial performance.
- Rapid sequence intubation with subsequent sedation and mechanical ventilation should occur quickly.
- IV fluid therapy should be initiated in an effort to improve perfusion. These infants are often volume depleted due to poor po intake, vomiting, or dehydration; however fluid administration should be cautious in the presence of CHF. Smaller fluid boluses of 5-10 ml/kg with frequent monitoring of vital signs is a more judicious approach.
- The presence of hypothermia, hypoglycemia, and acidosis should be determined and rapidly corrected.
- Sustained tachycardia with heart rates >200 beats/min suggest the possibility of supraventricular tachycardia.
- Severe anemia (Hct <25) should be corrected by slow administration of packed RBCs (10 ml/kg) over 1-2 h.

Table 22B.3. Medical therapy of congestive heart failure

Drug	Route	Dose
Dopamine	Infusion	Intropic dose: 5-15 µg/kg/min
Dobutamine	Infusion	2-20 µg/kg/min
		max = 40 µg/kg/min
Amrinone	Infusion	Load: 0.5 mg/kg; may be repeated q 20 min
		Maintenance: 5-20 µg/kg/min
Milrinone	Infusion	Load: 50 µg/kg over 10-15 min
		Maintenance: 0.1-1 µg/kg/min
Digoxin	PO	Load: 10 µg/kg
		max = 375 µg
		Maintenance: 5 µg/kg
		max 125 µg
		(Premature infants 1/2 leading and maintenance doses)
Lasix	PO	2-4 mg/kg
	IV	1-2 mg/kg max = 80 mg
	Infusion	0.1-1 mg/kg/h
Nitroglycerin	Infusion	Intial: 0.25-0.5 µg/kg/min
		Maintenance: 0.5-10 µg/kg/min
		(usual dose 1-3 µg/kg/min)
Nitroprusside	Infusion	0.5-10 µg/kg/min
		(usual dose 3 µg/kg/min)
		(neonates max = 6 µg/kg/min)
Captopril	PO	Neonates: 0.05-0.1 mg/kg tid
		Infants and children:
		Initially 0.15-0.3 mg/kg tid,
		Then titrate to max of 2 mg/kg
Enalapril	PO	0.1 mg/kg qd, bid
		max = 40 mg/day
Atenolol	PO	0.8-2 mg/kg
Esmolol	Infusion	Load: 0.5 mg/kg over 1 min
		Maintenance: 50 µg/kg/min, increase by
		Increments to max of 300 µg/kg/min
		Reload with each increase—titrate to effect

- The medical therapy of CHF (see Table 22B.3) should begin shortly after the institution of life-saving measures or can be more of a methodical approach in the infant who is less ill.
- The aim of medical therapy is to improve the relationship of cardiac contractility and left ventricular end diastolic filling pressure (LVEDP), thereby improving cardiac output.
- Inotropic agents increase the force of myocardial contraction.
  - Dopamine (3-20 mg/kg/min) and dobutamine (5-20 mg/kg/min) are sympathomimetic agents that may be used alone or in combination to augment renal blood flow and for inotrophy.
  - Other drugs also have specific roles in the treatment of CHF in children.
  - Amrinone and milrinone are selective inhibitors of phosphodiesterase that increase cardiac output and reduce cardiac filling pressure. Rapid infusion may cause hypotension and should be used cautiously in the presence of significant hypotension.
  - Digitalis has long been a therapy for congestive heart failure but in infants has a very narrow therapeutic window as well as a long half-life. It should be administered only in consultation with a pediatric cardiologist.
  - Diuretics reduce LVEDP by decreasing the circulating blood volume.
  - Lasix is the diuretic of choice in the initial management of CHF in infants.

- Vasodilators that cause both arteriolar and venodilation enhance myocardial performance by changing both the resistance and the capacity of the peripheral vasculature.
- Nitroglycerin and nitroprusside are administered intravenously and have short half-lives that allow for rapid titration of effect.
- Angiotensin-converting enzyme (ACE) inhibitors have become the cornerstone of the management of chronic heart failure. Not only are ACE inhibitors potent vasodilators but reverse left ventricular hypertrophy associated with heart failure.
- Beta-blockers have been shown to improve outcome in adults with heart failure, but there is little data to support their use in children.

Chest Pain in Children

Chest pain is a frequent complaint of children and adolescents and is usually a symptom of benign disease.

- The causes of chest pain in children are multiple (see Table 22B.4).
- The most common diagnostic category is idiopathic chest pain. A large number of children with idiopathic chest pain have psychogenic pain related to recent stressful life events.
- More than 50% of the children have other complaints associated with stress (headache, abdominal pain, limb pain).
- The second largest group is that of musculoskeletal pain, most often related to trauma, muscle strain, or costochondritis.
- Pulmonary causes include pneumonia and exercise induced reactive airway disease. A frequent gastrointestinal cause of chest pain is esophageal reflux or spasm.
- Biliary and pancreatic disorders may cause chest pain, but are usually accompanied by more serious, systemic symptoms.
- Cardiovascular chest pain occurs at a rate of 4-15% in children. Although a relatively uncommon cause of childhood chest pain, cardiovascular disorders are usually serious and may be potentially life threatening.
- Common cardiovascular causes of chest pain in children include pericardial disease, left heart obstructive disease, coronary artery disease, aortic dissection, hypertrophic cardiomyopathy, and tachyarrhythmias.

Table 22B.4. Causes of chest pain on children

Idiopathic Disorder	**Psychogenic, Hyperventilation**
Musculoskeletal disorder	Costochondritis
	Chest wall syndrome
	Tietze syndrome
	Xiphoid cartilage syndrome
	Stitch
	Precardial catch syndrome
	Slipping rib syndrome
Breast disorder	Gynecomastia
	Mastitis
	Mass
Pulmonary disorder	Pneumonia
	Reactive airway disease
	Pneumothorax
Cardiovascular disorder	Myocarditis
	Structural heart disease
	Dysrhythmia
Gastrointestinal disorder	Esophogitis
Vertebral/radicular disorder	

22

- Other miscellaneous causes of chest pain in children include breast disorders, vertebral disease, drug use, and collagen vascular disease.

Clinical Presentation

- The clinical presentation of children with chest pain is largely dependent upon the etiology.
  - Rarely are the presenting symptoms life-threatening, except when associated with serious trauma. Often the only complaint is that of pain.
  - A thorough history should be obtained with emphasis on time and activity of onset, location of pain, and associated symptoms.
  - Inquiry should be made about aggravating or precipitating factors. The character of the pain is of less importance, as children often cannot adequately describe pain.
  - The child and family should be questioned about psychological, situational, and environmental stressors.
  - Inquiry about family history should include coronary heart disease and sudden death.
  - Specific questions on review of systems should address pulmonary and gastrointestinal symptoms.
  - The occurrence of syncope in association with chest pain is often a sign of cardiovascular disease and should be aggressively evaluated.
- Abnormal vital signs and constitutional signs (i.e., fever, cough, vomiting) indicate an organic cause for pain.
- Palpable chest wall tenderness and reproducible pain in association with a negative history and otherwise normal physical exam suggest a musculoskeletal etiology.
- Physical findings suggestive of heart disease are significant murmur (> grade II/VI), precordial heave or thrill, pericardial rub, hypertension, and sustained tachycardia.

Diagnostic Evaluation

- Laboratory testing is seldom helpful unless the history and physical exam suggest serious organic disease. Testing may be occasionally necessary to reassure the patient and family.
- Laboratory tests which may be helpful include chest X-ray, EKG, and echocardiogram.
  - The chest X-ray is indicated for the evaluation of trauma or thoracic spine problems, or when cardiac or pulmonary disease is suspected.
  - The EKG is helpful for the evaluation of arrhythmia and for suspected myocarditis, pericarditis, or myocardial ischemia.
  - Echocardiography should be reserved for those children with positive cardiac signs and symptoms and ordered in conjunction with a pediatric cardiology consult.
- Other laboratory testing has limited usefulness except in the presence of comorbid or systemic disease.
- Cardiac enzymes have little or no value in the evaluation of children with chest pain.
- Drug screening may be indicated in selected patients who exhibit signs of drug abuse.

ED Management

The management of children with chest pain is dependent upon the cause.

- Those who have idiopathic, psychogenic, or musculoskeletal chest pain rarely need more than reassurance and mild analgesia.
- Specific etiologies such as pneumonia, reactive airway disease, or gastroesophogeal reflux disease should be treated accordingly.
- Children with signs of structural or undiagnosed congenital heart disease should be managed according to the severity of symptoms.
- Those who are stable and unlikely to decompensate may be discharged to follow-up with cardiology.
- Children with signs and symptoms of more serious disease such as pericarditis, myocarditis, or cardiomyopathy require prompt evaluation as an inpatient.

- Rarely a child will present with myocardial ischemia or acute myocardial infarction as a result of anomalous coronary arteries, Kawasaki disease, or accelerated artherosclerosis.
- Case reports of such children exist who were treated with thromoblytic agents and emergency angioplasty.

Hypertension in Children

Definitions

- Normal blood pressure (BP) in children is defined as systolic and diastolic BP less than the 90th percentile for age and sex.
- Hypertension is defined as average systolic and diastolic BP greater than or equal to the 95th percentile for age and sex measured on at least three separate occasions.
- Recent evidence also suggests that body size is an important determinant of BP in children (see Tables 22B.5 and 22B.6).
- Almost all hypertension in children is due to a secondary cause although there has been an increase of essential hypertension in children over recent years.
- Children with essential hypertension tend to be overweight and have family histories positive for high blood pressure.
- All children age 3 yr or greater presenting to the ED should have routine BP measurement and all acutely ill children should have continuous noninvasive BP monitoring regardless of age.
- Common secondary causes of hypertension in children include renovascular and cardiovascular disease.
- Initial BP measurements should be taken in both upper and lower extremities as coarctation of the aorta is the most common cardiovascular cause of hypertension which may occasionally present in later childhood or adolescence.
- Children rarely present with severe hypertension and those that do often have chronic renal disease.

Table 22B.5. Blood pressure: 95th percentile: Girls

Age	Systolic BP	Diastolic BP
Neonate	76	68
6 mo	106	66
1 yr	109	60
2 yr	108	64
5 yr	113	72
10 yr	121	80
15 yr	130	85

Table 22B.6. Blood pressure: 95th percentile: Boys

Age	Systolic BP	Diastolic BP
Neonate	87	68
6 mo	105	66
1 yr	106	59
2 yr	109	63
5 yr	115	73
10 yr	122	81
15 yr	134	85

- Hypertensive urgency is an elevated BP that is potentially harmful but is without evidence of end-organ damage or dysfunction.
- Treatment of hypertensive urgency can occur gradually over 24-48 h.
- Hypertensive emergency is a clinical syndrome in which elevated BP is associated with evidence of end-organ damage.
- Organs most often affected are the central nervous system, the cardiovascular system, and the kidneys.
- Common presentations of hypertensive emergency are congestive heart failure, acute renal failure, and hypertensive encephalopathy.

Clinical Presentation

- Children who present with hypertension as an incidental finding should have a thorough history and physical examination.
- Diagnostic testing should include BUN, electrolytes, serum creatinine, urinalysis, urine culture, and EKG.
- In the absence of end-organ dysfunction these children may be discharged from the ED to follow-up with a physician familiar with the evaluation and management of childhood hypertension.
- Children with hypertensive urgency should be carefully monitored in the ED and undergo thorough evaluation for end-organ dysfunction.
- If hypertension has been previously undiagnosed, blood pressure should be gradually lowered over 24-48 h while the child undergoes inpatient evaluation.
- Children with chronic hypertension who present with hypertensive urgency may often be safely treated in the ED and released to follow-up in 8-12 h.
- Hypertensive emergency requires immediate intervention and evaluation to prevent further end-organ damage.
- Hypertensive encephalopathy is characterized by headache, seizures, altered mental status, and focal neurologic signs.
- The differential diagnosis includes intracranial hemorrhage, stroke, and encephalitis.
- Acute left ventricular failure and acute renal failure manifest as volume overload, and these children may present with acute pulmonary edema and respiratory failure.
- Evaluation of children with hypertensive emergency should include BUN, electrolytes, creatinine, urinalysis, urine culture, EKG, chest X-ray, and CT of the brain, as well as other tests appropriate to the clinical presentation.

ED Management

- Children with the incidental finding of hypertension require no immediate pharmacological intervention.
- Appropriate follow-up should be arranged to provide for further evaluation and treatment. The importance of follow-up should be emphasized to the parent/caretaker.
- Treatment for hypertensive urgency should be initiated in the ED but does not require immediate reduction in BP.
- Multiple drugs are available for the treatment of chronic hypertension in children (see Table 22B.7).
- Most children with chronic hypertension are followed by a pediatric nephrologist or cardiologist. If appropriate subspecialties are available, they should be consulted regarding the management of these children.
- Children who present with hypertensive urgency as the initial manifestation of hypertension should be admitted for evaluation and pharmacological therapy.
- The goal of management of hypertensive emergency is the expedient reduction of blood pressure while maintaining vital organ function. The too rapid reduction of BP may result in permanent neurologic damage. Gradually controlled reduction by incremental infusion is the safest approach.

Table 22B.7. Medical therapy for chronic hypertension in children

Drug	Dose	Interval
Labetalol	1 mg/kg/d max: 3mg/kg/d	Every 6-12 h
Prazosin	0.05-0.1 mg/kg/d max: 0.5 mg/kg/d	Every 6-8 h
Atenolol	1 mg/kg/d max: 8mg/kg/d	Every 12-24 h
Propranolol	1 mg/kg/d max: 8 mg/kg/d	Every 6-12 h
Clonidine	0.05-0.1 mg max 0.5-0.6 mg	Every 6 h
Nifedipine	0.25 mg/kg/d max: 3 mg/kg/d	Every 4-6 h
Nifedipine XL	0.25 mg/kg/d max: 3 mg/kg/d	Every 12-24 h
Captopril	1.5 mg/kg/d max: 6 mg/kg/d	Every 8 h
	neonates: 0.03-0.15 mg/kg/d max: 2 mg/kg/d	Every 8-24 h
Enalapril	0.15 mg/kg/d	Every 12-24 h
Furosemide	1 mg/kg/d max: 12 mg/kg/d	Every 4-12 h
Bumetanide	0.02-0.05 mg/kg/d max: 0.3 mg/kg/d	Every 4-12 h
Hydrochlorothiazide	1 mg/kg/d max: 3 mg/kg/d	Every 12 h
Spironolactone	1mg/kg/d max: 3 mg/kg/d	Every 6-12 h
Triamterene	2 mg/kg/d max: 3 mg/kg/d	Every 6-12 h
Hydralazine	0.75 mg/kg/d max: 7.5 mg/kg/d	Every 6 h
Minoxidil	0.1-0.2 mg/kg/d max: 1 mg/kg/d	Every 12 h

- Initial management should be directed toward the ABCs.
- Continuous cardiac and pulse oximetry monitoring should be instituted and IV access established.
- In the presence of severely altered mental status or signs of increased intracranial pressure, treatment should begin with neuroprotective rapid sequence intubation.
- Intravenous agents are preferable for the reduction of BP.
- Common drugs include sodium nitroprusside, nicardipine, labetalol, and fenoldopam (see Table 22B.8).
  - Nicardipine, a calcium channel antagonist, offers easy titration of BP. Common side effects are due to excessive vasodilation and include dizziness, hypotension, headache, flushing, tachycardia, nausea, and vomiting.
  - Sodium nitroprusside is a potent vasodilator with an extremely short half-life of 2-4 min. Infusions given over a prolonged period of time may cause thiocyanate and cyanide toxicity.
  - Nitroprusside should be used with caution in patients with renal disease or hepatic dysfunction.

22

22

Table 22B.8. Medical therapy for hypertensive emergencies in children

Drug	Route	Dose
Nitroprusside	Infusion	0.5-1.0 µg/kg/min max: 8 µg/kg/min
Nicardipine	Infusion	1-10 µg/kg/min (usual dose 1-5 µg/kg/min)
Labetolol	IV	Bolus: 0.3-1.0 mg/kg max: 20 mg (may repeat q 10 min)
	Infusion	0.4-1.0 mg/kg/h max: 3 mg/kg/h
Fenoldopam	Infusion	0.1 µg/kg/min
Esmolol	IV	Bolus: 100-150 µg/kg over 1 min
	Infusion	25-100 µg/kg/min (usual dose 50-500 µg/kg/min)

  - Labetalol is a competitive α- and β-adrenergic blocking agent. It has a rapid onset of action within 2-5 min and duration of action of 2-4 h. It is contraindicated in asthmatics or patients with chronic obstructive pulmonary disease.
  - Fenoldopam mesylate (Corlopam) is a peripherally acting, selective dopamine-I receptor agonist. It acts by causing vasodilation primarily involving the renal, mesenteric, coronary, and cerebral vascular beds. It is particularly useful in patients with renal disease.
  - Esmolol is a short-acting β-1 selective adrenergic blocking agent with a half-life of 10 min. It may cause bronchospasm and should be avoided in patients with pulmonary disease.
- There is no place for the use of oral agents in the treatment of hypertensive emergency. Onset of action is prolonged and unpredictable.
  - Oral nifedipine has long been recommended as treatment for acute hypertension in children, especially those with renal disease. This drug has recently been reported to cause serious adverse effects in adults with acute hypertension. There have been no serious adverse effects reported in children, but it is recommended it be used with extreme caution in children with severe hypertension.

Disposition

All children with hypertensive emergency should be admitted to the intensive care unit under the care of a pediatric intensivist and nephrologist or cardiologist.

Suggested Reading

1. Case CL. Diagnosis and treatment of pediatric arrhythmias. Ped Clin N Amer 1999; 46:347-354.
2. Losek JD, Endom E, Dietrich A et al. Adenosine and pediatric supraventricular tachycardia in the emergency department: Multicenter study and review. Ann Emerg Med 1999; 33:185-191.
3. Checchia PA, Dietrich A, Perkin RM. Current concepts in the recognition and management of pediatric cardiogenic shock and congestive heart failure. Ped Emergency Med Reports 2000; 5:42-60.
4. Abdulla R, Young S, Barnes SD. The pediatric cardiology pharmacopoeia. Ped Cardiology 1997; 18:162-183.
5. Westaby S, Franklin O, Burch M. New developments in the treatment of cardiac failure. Arch Dis Child 1999; 81:276-277.

6. Zavaras-Angelidou KA, Weinhouse E et al. Review of 180 Episodes of Chest Pain in 134 Children. Ped Emerg Care 1992; 8:189-193.
7. Hirsch R, Landt Y et al. Cardiac tropinin 1 in pediatrics: Normal values and potential use in the assessment of cardiac injury. J Ped 1997; 130:872-877.
8. National high blood pressure education program working group on hypertension control in children and adolescents. Update on the 1987 task force report on high blood pressure in children and adolescents: A working group report from the national high blood pressure education program. Pediatrics 1996; 98:649-658.
9. Swift JD, Perkin RM et al. Evaluation and management of children with hypertensive emergencies. Ped Emerg Med Reports 1998; 3:39-50.

Part C: Genitourinary Problems in Children

David Leader

Urinary Tract Infections

Urinary tract infections (UTIs) pose particular problems in diagnosis and management.

- Infants present with varied symptomatology not always related to the urinary tract. Also, the significance of the infection and necessary work up will vary depending on age and sex.
- Urinary tract infections can range from being asymptomatic to causing systemic disease associated with pyelonephritis.
- The morbidity associated with UTI can be significant.
- Febrile UTIs reveal pyelonephritis, which can be detected by nuclear scanning.
- The main clinical consequences of recurring renal damage caused by pyelonephritis are arterial hypertension and renal insufficiency. Reflux nephropathy, renal injury attributed to the combination of vesicouretural reflux, and recurring infection, are responsible for up to 15% of the cases of end-stage renal failure in children in the United States.
- Because of these potentially serious consequences, it is imperative for the emergency physician to consider UTI as a cause of fever in the young child, to perform the appropriate diagnostic studies, to render adequate therapy, and arrange for follow-up.

Incidence

- The presence of UTI should be strongly considered in infants and young children 2 mo to 2 yr of age with unexplained fever. According to the American Academy of Pediatrics (AAP) and several recent evidence-based studies on pediatric UTIs, the prevalence of UTIs in this setting is about 5%.
- Stratifying by age, sex and presence or absence of circumcision significantly impacts these numbers. The relative risk in a febrile girl is more than twice that of a boy.
  - Under 1 yr of age, the prevalence is 6.5% in girls and 3.3% in boys.
  - In the age group between 1 and 2 yr of age, the prevalence is 8.1% for girls vs. 1.9% for boys.
  - In circumcised boys, the rate is low (0.2-0.4%), while the literature suggests that the rate is 5-20 times higher in uncircumcised boys.
  - In the age group of 2-5 yr of age, the incidence of UTIs is approximately 2% of boys, and up to 5% of the school-aged girls.
  - In two independent studies the prevalence of UTI in febrile white female infants under 1 yr of age that presented to the pediatric emergency department with a fever was found to be 16%. This racial difference may be due to differences in blood group antigens on the surface of uroepithelial cells that are thought to affect *E. coli* adherence.

- In the post HIB vaccine era, the rate of occult UTI is much greater than the prevalence of occult bacteremia.
- For this reason, physicians should consider "occult UTI" more common than occult bacteremia in the work-up of the "febrile child".
- The bacteria causing most infections of the urinary tract in infants and young children are from the normal rectal or perineal flora.
- By far the most common organism is *Escherichia coli*, Klebsiella, enteric streptococci, and Proteus are also common pathogens.

22

- Rarely, *Staphylococcus aureus*, *S. epidermidis*, viral, fungal, and mycobacterial agents are the cause.
- Enterococcus species, *S. aureus*, and group B streptococci are the most common Gram-positive organisms and these are more commonly seen in neonates than in other age groups.
- The urinary tract infection usually is a result of retrograde contamination of the lower urinary tract from these normal rectal and perineal organisms.
- In contrast, neonates usually develop UTI after seeding of the renal parenchyma from hematogenous spread.
- Predisposing factors include poor perineal hygiene, the short urethra of females, infrequent voiding with voluntary urinary retention, congenital urinary tract abnormalities, vesicoureteral reflux, and urolithiasis.
- Also, virulence of the pathogen and host immunity are important factors. As mentioned previously, noncircumcised males also increases the males risk for the development of UTI.

Clinical Presentation

- The clinical features of pediatric UTI vary considerably with age.
- The symptoms in younger children are often nonspecific.
- Neonates may have fever or hypothermia, poor feeding, failure to thrive, septic appearance, jaundice, irritability or lethargy and cyanosis.
- Infants and young children often present with gastrointestinal complaints including abdominal pain, fever, vomiting, diarrhea, and a poor appetite.
- Older school-aged children have the most classic signs and symptoms of fever, abdominal pain, vomiting, urinary frequency, dysuria, and urgency, and possible enuresis. Systemic toxicity usually accompanies upper urinary tract disease (i.e., pyelonephritis) with symptoms of high temperature, chills, back pain (costovertebral angle tenderness), vomiting, and dehydration.

Differential Diagnosis

- Although the infant will present with nonspecific symptoms, careful questioning may uncover additional symptoms that would direct attention to the urinary tract.
- Symptoms such as dribbling, poor stream, frequency, and malodorous urine may be present, although not the presenting complaint.
- When the preschool-aged and older child present with urgency, dysuria or frequency, the differential should include chemical urethritis, pinworms, vaginitis, and possible child abuse.

Diagnostic Evaluation

- Urine culture is the gold standard in the diagnosis of UTI. No element (or combination of elements) of the urinalysis is considered as sensitive or as specific as a urine culture in making the diagnosis of a UTI.
- Urinalysis performed on urine obtained by any method may yield a negative finding in the face of a positive culture.

- Cultures from bag-obtained urine specimens should not be obtained as they can be expected to yield unacceptably high false positive rates (as high as 85% false positives). Because of these two facts, only cultures of specimens obtained by catheterization or suprapubic aspiration are considered the gold standard upon which the diagnosis of UTI in a febrile child is to be made.
- Since urine culture results are not immediately available on the initial ED visit, microscopic urinalysis and rapid urine (dip) chemical strips are used for screening and diagnostic purposes.
- The dipstick components most relevant for UTI screening are leukocyte esterase, which indicates the presence of WBC breakdown proteins, and nitrites, which are converted from nitrates by Gram-negative urinary pathogens.
- In the standard microscopic analysis, a centrifuged urine sample is examined for pyuria (expressed as WBC per high-powered field) and bacteriuria (typically described using a scale of 1+ to 4+). The sensitivities of a positive leukocyte esterase test (rage 64-89%) and a positive nitrite test (range 16-72%) for a positive urine culture is <50%.
- If you combine pyuria (>5WBC/high powered field) and bacteriuria on urine microscopy, a sensitivity of approximately 65% can be obtained.
- It is important to remember that neonates and young infants may not produce sufficient pyuria to have a positive test strip on dipstick screening.
- Other findings of dipstick screening in UTIs can include pyuria, hematuria, and proteinuria, but these can be present in the absence of infection. None of these tests has sufficiently high sensitivity to confidently rule out UTI.
- It is therefore prudent to send a confirming urine culture in all cases of suspected UTI in febrile infants and young children. Diagnostic criteria for a positive urine culture is defined as $>5 \times 10^4$ colony forming units (CFU)/m; of a single urinary pathogen. With suprapubic aspiration, growth of any number of a urinary pathogen is considered a positive culture.
- As mentioned previously, in infants and children who are not toiled trained, bladder catheterization is the preferred method of collection.
- In toilet-trained children, a midstream clean catch specimen can be considered.
- Previous studies have indicated that a cup-caught, spontaneously voided specimen is, for all practical purposes, as clean as a catheter-obtained specimen in little boys.
- Transurethral catheterization should also be considered in toilet-trained children under 5 yr of age as the contamination rates from midstream, clean catch urine specimens sent for culture may be as high as 50% without proper positioning and supervision.
- In one recent study, girls aged 2-5 yr of age old who were cleansed with wipes, followed by betadine, and seemingly correct backwards positioning on a toilet to separate the labia, the contamination rate was found to approach 50%.

ED Management

- Early treatment of UTIs decreases the risk of kidney damage.
- Intravenous antibiotics should be instituted for any child <2 mo of age with a febrile UTI, for those children who have significant dehydration or appear toxic, for those children vomiting and unable to take oral medication, or when outpatient compliance and follow-up is questionable.
- Likewise, hospital admission is necessary for immunocompromised patients, those with renal insufficiency, and those with urinary stents or other urinary foreign bodies.
- In a recent multicenter, randomized clinical trial comparing the efficacy of oral cefixime to IV cefotaxime for 3 days and followed by oral cefixime, results were very similar. Both groups required 24 h for defervescence, with similar reinfection rates, and subsequent renal scarring rates that were within 3 percentage points. (7%,10%).

22

Antibiotic Selection

- In the neonate parenteral antibiotics that are useful for treating UTIs include a combination of ampicillin and gentamicin or a combination of ampicillin and cefotaxime.
- For children over 3 mo of age, cefotaxime, ceftriaxone, or gentamicin may be used as a single agent.

Oral Therapy

- A list of standard antibiotics is presented in no particular ranking order.
- These include amoxicillin, amoxicillin/clavulante, trimethoprim-sulfamethoxazole, cephalexin, and cefixime.
- It is noted that ampicillin, amoxicillin, and trimethoprim-sulfamethoxazole have become less effective due to *E. coli* resistance.
- Agents that are concentrated in the urine but which do not reach therapeutic blood levels (e.g., nalidixic acid, and nitrofurantion) are specifically advised against since renal involvement is likely in febrile children.
- Infants and young children 2 mo to 2 yr of age, including those whose treatment initially was administered parenterally, should complete a 7-14 day antimicrobial course orally.
- Short-course 3 day or single-day therapy has been advocated in adult patients; however, the infection recurrence rate in pediatric patients precludes the use of these short regimens.
- Adolescent females with cystitis alone may be treated as adults with a 3 day oral antibiotic regimen.

Disposition

- Although imaging studies for UTI are rarely indicated as part of the diagnostic work-up of UTI in the ED (except in the case of a palpable mass), it is important to arrange for follow-up for all children with UTI.
- Radiographic evaluation for functional and/or structural causes of UTI begin with voiding cystourethrogram (VCUG) to demonstrate reflux and a renal ultrasound to show anatomic abnormalities.
- Renal cortical scans can also be very useful in detecting evidence of pyelonephritis and focal renal scarring. The primary care provider should be notified to ensure that those patients of highest risk can have this testing arranged as an outpatient or during hospitalization.
- The diagnosis of a urinary tract infection should be considered in all febrile children, especially when the diagnosis is unclear.
- Especially with *Haemophilus influenzae* type b essentially eradicated by vaccine and now with pneumococcal vaccine use becoming more widespread, occult UTI should be considered much more prevalent than occult bacteremia.
- The decision to admit or treat as an outpatient requires careful consideration of the child's overall appearance, oral intake and adequacy of hydration.
- Urinalysis and culture results must be evaluated carefully as potential errors in collection and sample management can result in spurious results.
- Finally, because pediatric urinary tract infections can have potentially serious sequelae, appropriate and timely patient follow-up by a primary care provider is a necessity.

Male Genital Problems

Testicular Torsion

- The incidence of torsion of the testis is 1 per 4000 males.
- While it may occur at any age, it is most common during puberty and the second decade of life.

- Testes that are predisposed to torsion have an anatomic abnormality, where the tunica vaginalis attaches high (above the epididymis) rather than attaching to the posterior aspect of the testis. This abnormality is referred to as the "bell clapper" deformity, and is usually bilateral and allows the testicles to rotate within the tunica vaginalis. This turning causes subsequent occlusion of the venous system. Arterial flow is compromised due to the compression of twisting and edema. Damage to the testis relates to the duration and extent of vascular obstruction, as partial arterial flow may persist with <540-720 degrees of rotation.
- Salvage of the testicle has been correlated with the time of onset to detorsion. Continuous pain for 24 h is usually associated with an infarcted testis. With surgical detorsion a salvage rate of 100% can be obtained if performed within 6 h of the onset of symptoms.
- The diagnostic findings include intermittent or consistent testicular pain, which is sudden and intense on onset. Rarely, it may be gradual in onset with progression of severity. Often there is a history of strenuous physical activity just prior to the onset of scrotal pain; however, a fair number of cases occur during sleep when spontaneous unilateral cremaster muscle contraction results in ischemic torsion.
- The testicle is tender, high riding and often horizontal (this horizontal axis can be quite subtle and is best detected with the patient in a standing position).
- The patient reports severe testicular pain that may radiate to the abdomen.
- Nausea and vomiting are common.
- Elevation of the scrotum to the symphysis does not relieve the pain (absent Prehn's sign) and the cremasteric reflex is often absent as well.
- The differential diagnosis includes acute epididymitis, orchitis, testicular or epididymal appendix torsion, appendicitis, hernia, tumor or hydrocele.
- Laboratory studies are of limited use.
- A CBC may reveal leukocytosis. Pyuria is absent on urinalysis. Radionuclide scanning is the test of choice with 89-100% accuracy in determining testicular blood flow but may not be readily available.
- Doppler ultrasound estimation is rapid, although it has a high false-negative rate.
- Color doppler flow is 86-100% sensitive and 100% specific for torsion.
- False negative exams occur in small testicles (infants), spontaneous detorsion, or early torsion.
- When the diagnosis of testicular torsion is suspected, a urologist should be consulted immediately.
- Manual or surgical detorsion must occur within 4-6 h, and treatment should not be delayed to await ancillary test results.
- Sedation and analgesia are useful after the diagnosis is certain and the decision made to operate.
- Spermatic cord blocks with 1% lidocaine high in the scrotum may be useful, but it eliminates the end-point of pain relief with manual detorsion. Initial attempts at reducing may be done cautiously.
- Attempts at reduction should be toward the ipsilateral thigh because the torsion usually occurs toward the midline. (i.e., left testicle rotated clockwise and right testicle rotated counterclockwise as the physician faces the patient).

Torsion of the Appendix Testis

- This condition usually occurs in prepubescent boys.
- The clinical features include a sudden onset of pain near the location of the appendage of the testicle.
- This location is at the head of the epididymis or the superior pole of the testicle. A "blue dot" sign is seen as the engorged, cyanotic appendage approaches the overlying scrotal skin.

- The differential diagnosis includes testicular torsion, epididymitis, orchitis, and testicular trauma.
- The diagnosis is often clear following physical examination.
- Doppler ultrasound may be helpful, and if the diagnosis is still uncertain, then urologic consultation with possible surgical exploration may be required.
- The management of this condition is usually conservative. Relief of pain may be provided with oral analgesics.
- Some urologists may consider excision.

22
- Following discharge from the ED, these patients should see a urologist or primary care physician within 1 wk.

Acute Epididymitis

- Inflammation or infection limited to the epididymis that causes acute pain and swelling of rapid onset is termed acute epididymitis.
- Epididymo-orchitis implies extension of the inflammation or infection to the testis. This seldom occurs before puberty and most often results from infection caused by retrograde spread of organisms up the vas deferens.
- This infection in the prepubertal patient is usually caused by uncommon organisms including coliform bacteria and Pseudomonas.
- In older children and adolescents, the etiology of infection is commonly caused by *Neisseria gonorrhoeae* and *Chlamydia trachomatis.*
- When these sexually transmitted organisms are identified in the younger child, sexual abuse should be considered.
- Viruses can be another cause of epididymitis, as can trauma. A chemical epididymitis can occur, caused by a reflux of sterile urine into the ejaculatory duct.
- Other systemic diseases such as Kawasaki syndrome, Henoch-Schonlein purpura, and sarcoid can also be causative factors.
- The clinical findings usually are consistent with a gradual onset of scrotal and groin pain with edema.
- Initially, the epididymis is tender and swollen, with an associated soft, normal testicle. This can then progress to a diffuse tenderness and edema of the epididymis and testicle, and sometimes involve the scrotal wall as well.
- The patient may present with a history of trauma, lifting, or heavy exercise.
- Other symptoms may include dysuria, frequent urination, and urethral discharge may be present.
- Laboratory testing should include a urinalysis and urine culture and Gram stain with culture of any urethral discharge. Pyuria may be present it 25-60% of cases. The WBC may be elevated.
- Any child below 2 yr of age and older patients with reflux should have urological consult and investigation.
- The differential diagnosis for epididymitis includes testicular torsion, trauma, hydrocele or tumor.
- In the management of epididymitis, the most important aspect is to be certain of the diagnosis.
- In patients who present with testicular pain, testicular torsion must be excluded.
- A testicular ultrasound may be helpful by demonstrating epididymal swelling/inflammation, epididymal abscess, and normal testicular blood flow.
- If there is any question of torsion of the testicle, urologic consultation should be obtained.
- For treatment of epididymitis, scrotal support with bedrest and elevation, and pain relief with analgesics are helpful.
- Antibiotic therapy should be initiated. For boys more than 45 kg, give amoxicillin 3 g PO once or ceftriaxone 250 mg IM. This should be followed by doxycycline 100 mg

(for boys older than 8 yr of age) every 12 h for 10 days. For boys younger than 8 yr of age, oral erythromycin 50 mg/kg/24 h every 6 h may be used.
- If a specific organism is identified, antibiotic selection may be altered accordingly.

Phimosis
- Phimosis is a tightness or fibrotic contraction of the foreskin orifice that does not permit the retraction of the foreskin.
- The problem with severe phimosis is that urinary flow can be impeded with bulging of the foreskin during micturition.
- In most newborns, the foreskin is not retractile and many boys still do not have retractile foreskins by 6 yr of age.
- This is usually not a problem because by age 17 yr, 97-99% of males will have retractile foreskins. This condition rarely brings patients to the ED.

Posthitis and Balanitis
- Posthitis is foreskin inflammation and cellulitis, and if the infection progresses to the glans it is called balanitis.
- Treatment of mild balanitis is typically achieved with topical ointments, local hygiene and oral antibiotics.
- If the infection involves the penile shaft parenteral agents are preferred.
- If balano-posthitis is recurrent, diabetes mellitis should be excluded as a cause.

Paraphimosis
- Paraphimosis results when the foreskin is retracted proximal to the coronal sulcus of the glans and cannot be reduced distally. The retracted foreskin becomes edematous due to the lymphatic and venous congestion which can further progress to arterial compromise. Necrosis of the glans can occur if this arterial compromise is prolonged.
- By applying pressure to the glans for 5 min, edema can be reduced to the tissue, sometimes allowing reduction of the foreskin.
- If this pressure reduction technique is not successful, making a dorsal slit following local infiltration with 1% lidocaine will allow for reduction of the foreskin over the glans.
- Urology referral is typically indicated for this procedure.

Constriction with Hair Tourniquets
- Hair or clothing threads may accidentally encircle the penis, usually of the coronal sulcus in circumcised young males.
- This constriction may lead to local tissue damage, urethral injury, and venous/arterial blood supply interruption with eventual tissue necrosis.
- If urethral inury is suspected, a retrograde urethrogram should be obtained.
- For removal, 1% lidocaine without epinephrine may be required with gentle removal/incision of the offending object.
- If there is any evidence of tissue necrosis or urethral injury, urologic consultation should be obtained.

Zipper Entrapment of the Foreskin or Penis
- Local anesthesia with infilration of 1% lidocaine without epinephrine will allow the zipper to be unzipped.
- An alternative technique after local anesthetic infiltration with 1% lidocaine, is cutting of the zipper cross bar using wire cutters. This allows the zipper mechanism to be pulled apart releasing the entrapped foreskin.
- No further treatment is usually necessary, as any swelling/ecchymosis resolves without complications.

22

- Instruct the caregiver to observe the penile skin/shaft for any signs of infection (redness, swelling, increased pain).

Urethral Foreign Bodies

- Urethral foreign bodies may be placed in boys of all ages and include a variety of objects.
- The placement of these foreign bodies may be related to curiousity or sexual experimentation, but abuse should always be considered.
- The patient may present with complaint of penile pain, dysuria, or hematuria. A physical exam, including a rectal exam, should be carried out.
- Laboratory studies including urinalysis and urine culture should be performed.
- Radiographic views of the penis and a KUB should be obtained.
- Immediate urologic referral for foreign body removal should be obtained, followed by retrograde urethrography or endoscopy.

Female Genital Problems

Labial Adhesions

- Labial adhesions are common in young children, usually secondary to irritation or inflammation.
- They may spontaneously separate by 12 mo, but this process may be facilitated by topical application of Premarin cream nightly for 2 wk.
- These spontaneously resolve with puberty, and surgical intervention is rarely required.

Vulvovaginitis and Vaginal Discharge

- Vulvovaginitis and vaginal discharges are common problems of both prepubertal and pubescent females.
- The patient should be evaluated for the presence of an accompanying pruritis and odor and establish the quality, duration and volume of the discharge.
- Other illnesses (diabetes, infections) need to be excluded, and investigate medications and hygiene products, (e.g., contraceptives, deodorant sprays, douches, antibiotics), and menstrual history and any sexual activity should be ascertained.
- In prepubescent females, the causes are usually nonspecific. Poor hygiene or an allergic reaction are usually causative.
- Cultures show a variety of organisms, most of which are normal flora. Bubble bath and soaps may be responsible for allergic reactions in younger children.
- In adolescents and adults, sanitary napkins, chemical douches, deodorant and contraceptive sprays may be responsible.
- In postpubescent females the causes for vulvovagniitis can be the same, but other infections (e.g., candida) and sexually transmitted infections (gonorrhea, chlamydia, and trichomonas) are more common.
- In younger children, sexual abuse should be considered if these sexually transmitted infections are present (particularly gonorrhea).
- The treatment of nonspecific vaginitis includes good perineal hygiene, clean cotton underwear, loose fitting clothes and nightgowns.
- Also sitz baths 3 times daily with patting of perineum to dry the region and leaving the legs open for 10-20 min to dry the perineal region.
- For severe vulvovaginitis, sitz baths 4 times daily or more, Tucks for wiping, A&D ointment, and 1% hydrocortisone may be helpful.
- If vulvovaginitis persists, continue hygiene measures; amoxicillin or amoxicillin-clavulanate for 10 days or topical metronidazole or clinadamycin antibacterial cream may also be helpful.
- In some cases topical estrogen cream nightly for 2-3 wk may be helpful. For specific treatment of vaginal infections (see Table 22C.1).

Table 22C.1. Treatment for specific causes of vaginal infections (obtained by culture/wet prep)

S. pygenes	PCN V 250 mg TID for 10 days
H. influenzae	amoxicillin/augmentin 40 mg/kg/day for 7 days
S. aureus	cephalexin 50 mg/kg/day for 7 days
S. pneumoniae	PCN V 250 mg TID for 10 days
Shigella	trimethroprim-sulfasoxazole 40 mg/kg for 5 days
C. trachomatis	<8 yr, erythromycin 50 mg/kg/day for 14 days >8 yr, doxycycline 100 mg BID for 7-10 days (other option: azithromycin)
N. gonorrhea	ceftriaxone 125 mg IM, Cefixime 400 mg PO
Candida	topical nystatin, clotrimazole
Trichomonas	metronidazole 15 mg/kg/day TID, max 250 mg TID for 7 days
Pinworms	mebendazole 100 mg chewable once, repeat in 2 wk.

22

Physiologic Leukorrhea

- At the onset of puberty, a clear, sticky, nonirritating discharge may develop secondary to estrogen stimulation and irritation; it ceases once menarche occurs.
- It is common for infant females in the first few days of life to have a thin vaginal discharge. This discharge may become bloody at 5-10 days because of withdrawal of maternal estrogens.

Vaginal Foreign Bodies

- Commonly found articles in the vagina include paper products, vegetable matter, and in older patients, forgotten tampons.
- The discharge is often blood-tinged and foul smelling.
- Physical exam should reveal the foreign body and can usually be removed without difficulty.
- It can often be palpated on rectal examination.

Suggested Reading

1. America Academy of Pediatrics, Committee on Quality Improvements, Subcommittee on Urinary Tract Infections. Practice parameter: The diagnosis, treatment, and evaluation of the initial urinary tract infection in febrile infants and young children. Pediatrics 1999; 103:843-859.
2. Baraff LJ, Bass JW, Fleisher GR et al. Practice guidelines for the management of infants and children 0-36 mo of age with fever without a source. Ann Emerg Med 1993; 22(7):1198-1210.
3. Baucher H, Phillip B, Dashefsky B et al. Prevalence of bacteriuria in febrile children. Pediatr Infect Dis J 1987; 6:239-242.
4. Gorelick MH, Shaw KN. Screening tests for UTI in children: A meta analysis. Pediatrics 1999; 140(5).
5. Gorelick MH, Shaw KN. Clinical decision rule to identify young febrile children at risk for UTI. Arch Pediatr Adol Med 2000; 154:386-390.
6. Hoberman A, Han-Pu C, Keller CM et al. Prevalence of urinary tract infection in febrile infants. J Pediatr 1993; 123:17-23.
7. Hoberman A, Wald ER, Hickey RW et al. Oral versus initial intravenous therapy for urinary tract infections in young febrile children. Pediatrics 1999; 104:79-86.
8. Jantausch BA, Criss VR, O'Donnel R et al. Association of Lewis blood group phenotypes with urinary tract infections in children. J Pediatr 1994; 124:863-868.
9. Hoberman A, Wald ER, Reynolds EA et al. Is urine culture necessary to rule out urinary tract infection in young febrile children? Pediatr Infect Dis J 1996; 15:304-309.

10. Dick PT, Feldman W. Routine diagnostic imaging for childhood urinary tract infections: A systemic overview. J Pediatr 1996; 128:15.
11. Hellerstein S. Urinary tract infections: Old and new concepts. Pediatric Clin North Am 1995; 42:1433.
12. Kass ET, Lundak B. The acute scrotum. Pediatr Clin North Am Oct 1997; 44(5):1251-6.
13. Lewis AG, Bukowski TP, Jarvis PD et al. Evaluation of the acute scrotum in the emergency department. J Pediatr Surg Feb 1995; 30(2):277-81.
14. Haynes BE, Bassam HA. Diagnosis of testicular torsion. JAMA 1983; 249:2522.
15. Middleton WE, Siegal BA, Melson GL et al. Acute scrotal disorders, prospective comparison of color doppler ultrasound and testicular scintigraphy. Radiology 1990; 177:177-181.
16. Yazbeck S, Patriquin HB. Accuracy of doppler sonography in the evaluation of acute condition of the scrotum in children. J Pediatr Surg 1994; 29:1270-1272.
17. Brown MR, Cartwright PC et al. Common office problems in pediatric urology and gynecology. Pediatr Clin North Am 1997; 44:1091.
18. Harrison BP. Pediatric penile swelling. Acad Emerg Med 1996; 3:384.
19. Schneider RE, Hedges JR. Emergency urologic proceedures. In: Roberts JR, Hedges JR, eds. Clinical Prodeedures in Emergency Medicine, 3rd Ed. Philadelphia: Saunders, 1998:947-986.
20. Fleisher G, Lundwig S, Fuddy R et al, eds. Textbook of Pediatric Emergency Medicine. 4th ed. Baltimore: Williams & Wilkins, 2000.
21. Al-Orifi F, McGillivray D, Tange S et al. Urine culture from bag specimens in young children: Are the risks too high? J Pediatr 2000; 137(2):221-226.

22

Part D: Orthopedic Problems in Children

Russell Karten

Salter-Harris Classification of Fractures

- In children fractures often occur at the physis (growth plate).
- These injuries are more common in adolescents than in other children, with a peak incidence at 11 to 12 yr of age.
- Most growth plate injuries occur in the upper limb, especially the radius and ulna.
- The Salter-Harris classification system of physeal injuries describes five types of injuries. It provides a practical classification based on the mechanism of injury, the relationship of the fracture line to the physis, and the prognosis for subsequent bone growth.
    - Salter-Harris type I physeal fracture occurs when the epiphysis separates from the metaphysis through the hypertrophic cells of the physis.
        - There are no associated bony fragments, and the periosteum surrounding the physis remains intact.
        - There is little chance of growth disturbance if a near anatomic reduction is achieved.
        - Salter-Harris type I injuries of the proximal radius, proximal and distal femur, and the proximal tibia are subject to premature physeal closure and growth arrest.
        - If radiographic studies are negative but examination is suggestive of Salter-Harris type I injury with point tenderness over a growth plate, splint immobilization and a follow-up examination are essential.
    - Salter-Harris type II fractures are the most common type of pediatric physeal fractures, representing 75% of physeal injuries.
    - Salter-Harris type II fractures are those injuries where a portion of metaphyseal bone is displaced with the epiphyseal fragment.
    - Radiographically the triangular fragment of the metaphysis is noted without associated injury to the epiphysis.

- Reduction is usually easily achieved, and these fractures generally carry a good prognosis. Subsequent treatment includes immobilization, elevation, and ice.
- Referral to orthopedics for follow-up evaluation is essential.

- Salter-Harris type III fractures are an intra-articular fractures of the epiphysis extending to the physis with the cleavage plane continuing along the physis.
  - This injury usually involves the proximal or distal tibia and is caused by severe intra-articular shearing.
  - Salter-Harris type III fractures account for <10% of growth plate injuries, and the prognosis for subsequent bone growth is related to the preservation of circulation.
  - Because of the increased incidence of growth disturbance and functional disability following a Salter-Harris type III fracture, an orthopedic consultation is usually obtained in the emergency department.
- Salter-Harris type IV fractures are also intra-articular and extend through both the epiphysis and the metaphysis.
  - These fractures represent approximately 10% of physeal injuries and are most commonly seen involving the distal humerus.
  - Excellent reduction alignment is required to prevent growth arrest and to preserve normal joint mechanics.
  - An urgent orthopedic consultation in the emergency department is indicated.
- Salter-Harris type V fractures usually involve the knee or ankle and are the result of severe axial compression on the physis.
  - Typically, there is minimal or no displacement at the physis, but the cells responsible for further bony growth are crushed.
  - Radiographs may appear deceptively normal, and initial diagnosis is often difficult.
  - Treatment consists of cast support of the involved joint, nonweight-bearing for at least 3 wk, and close orthopedic follow-up.
  - Focal bone growth arrest is nearly inevitable, and these injuries are often diagnosed after a growth arrest becomes evident.

Slipped Capital Femoral Epiphysis

Clinical Presentation

- Slipped capital femoral epiphysis (SCFE) describes a separation from the epiphysis and metaphysis of the proximal femur.
- Affected children are typically between 8 and 15 yr of age.
- There is a male predominance, and this condition is more common in African Americans than in whites.
- Obesity appears to be a significant risk factor.
- Slipped capital femoral epiphysis can be acute or chronic, but approximately 85% are chronic.
- The onset of symptoms of a chronic slipped capital femoral epiphysis is generally insidious.
- Many patients have been symptomatic longer than 2 wk prior to diagnosis, with symptoms generally developing over wk to mo as the separation of the epiphysis from the metaphysis progresses.
- An acute SCFE may be the result of an acute traumatic event or may represent an acute-on-chronic slip.
- The patient presents with abrupt onset of severe pain, decreased range of motion, and inability to bear weight.
- The patient may complain of a discomfort in the groin, a limp, or knee pain. At the hip, pain with abduction and external rotation is common.
- 3/4 of these children with acute events will have had mild symptoms in the involved hip before the acute episode.

22

- These symptoms include hip weakness, a limp with exertion, pain in the knee or groin, and a decrease in internal rotation.

Diagnostic Evaluation

- Diagnosis is made by physical and radiographic examination.
- With an acute presentation, the patient almost always holds the hip in abduction and flexion and has limited internal rotation.
- Range of motion in all directions is painful, and care should be taken not to force the hip into maximum range of motion that may aggravate the bony displacement.
- 10-25% of children with an SCFE will have bilateral disease.
- Anteroposterior views as well as frog-leg lateral films of both hips are recommended for radiographic diagnosis and comparison. Both views are necessary to determine the presence of medial and posterior slips, and usually the lateral view is more helpful.
- In general, SCFE injuries are Salter-Harris type I injuries.
- Radiographically, there may be slippage of the epiphysis compared to the contralateral side, irregular widening of the epiphyseal line, and widening of the growth plate.
- If a fracture is clinically suggested but not seen on plain radiographs, further evaluation is recommended with the use of a CT scan, an MRI scan, or a bone scan.

ED Management

- After confirmation of the diagnosis with examination and radiographs, immediate orthopedic consultation is warranted.
- All patients should be admitted and should be made strictly nonweight-bearing.
- Orthopedic management of a slipped capital femoral epiphysis is operative reduction and fixation.
- The prognosis depends on the initial degree of slippage.
- Those cases where displacement of the epiphysis is <1/3 the diameter of the femoral neck will generally have a good prognosis.
- An important complication of this injury is necrosis of the femoral head.

Radial Head Subluxation (Nursemaid's Elbow)

Clinical Presentation

- Radial head subluxation is an extremely common injury, representing 20% of upper extremity injuries in children.
- Peak incidence occurs between 1 and 4 yr of age; however it has been seen in patients from 6 mo to 15 yr of age.
- The classic mechanism of injury is longitudinal traction on the arm with the wrist in pronation, as occurs when the child is lifted up by the wrist.
- This classic history, however, is by no means universally illicited.
- Often, the parents simply report that the child is unable or unwilling to use the arm.
- Stretching of the annular ligament allows fibers to become entrapped between the capitellum and the head of the radius.
- This results in an inability to supinate the arm, and the arm is often held in the "Nursemaid's position" with the arm partially flexed at the elbow, forearm held in pronation, and the arm adducted close to the trunk.
- With the exception of their arm, these patients are typically playful and well appearing.

Diagnostic Evaluation

- The diagnosis is made clinically based on history and examination.
- In clear cases radiographs are not required.
- The physician should examine the clavicle, humerus, radius, and wrist prior to reduction attempts.

- Point tenderness, soft tissue swelling, deformity, or ecchymosis of the elbow suggests a different diagnosis and radiographs should be considered.
- Mild tenderness on direct palpation over the radial head may be present.
- Without clinical suggestion of fracture, other injury, or history of a fall, it is appropriate to attempt reduction without prior radiographs.

ED Management

- Radial head subluxation can usually be easily reduced.
  - After positioning the child on an assistant's lap, the physician grasps the child's hand and places it in extension and forearm pronation.
  - Subsequently, three simultaneous maneuvers are performed: pressure on the child's radial head by the physician's thumb, full supination of the child's forearm, and then passive full flexion of the child's elbow.
  - An audible or palpable click signifies successful reduction, but it may not always be noted.
  - If this method is unsuccessful, supination with concomitant extension at the elbow is recommended.
  - Two or more attempts are required to produce the click in up to 30% of patients.
  - Many patients are asymptomatic within 5-10 min and 90% of patients will regain use of the arm within 30 min.
- Radiographs are suggested if the patient continues to avoid movement >30 min post-reduction.
- X-ray technicians sometimes reduce radial head subluxations when they flex the child's arm for the lateral elbow X-ray.
- In these instances, the child returns from X-ray using the involved arm normally.
- Immobilization is almost never indicated.
- Complications are rare, although some children do experience multiple episodes of radial head subluxation in early childhood.
- It is important to document full unrestricted and painless use of the arm prior to discharge.
- Follow-up is usually with the primary care provider as needed.

22

Approach to the Child with Limp

In is not uncommon for children with a limp to present to the emergency department for evaluation and treatment.

- The differential diagnosis for a child with a limp or lower extremity pain is broad, and fortunately only a few conditions require urgent treatment.
- Although a definitive diagnosis in the emergency department can be difficult, a systematic approach to these patients is critical to identify conditions requiring immediate treatment.
- Most children walk independently by 14-18 mo and develop a mature walking pattern by age 3.
- Any limp is considered abnormal.
- Limping is due to three general causes: pain, weakness, or structural abnormalities.
  - Pain may be secondary to infection, inflammation, neoplasm, or trauma.
  - Weakness may be secondary to a neuromuscular disease, CNS disorder, or atrophy.
  - Structural abnormalities may include limb-length discrepancies or articular deformities, and these may present with the pelvis and trunk depressed on the affected side during the stance phase.

Diagnostic Evaluation

- Diagnosis not only includes a directed history and physical examination but also knowledge of the most common diagnoses in different age groups.

- Common diagnoses for limp in children include: septic arthritis, osteomyelitis, congential hip dislocation, child abuse, Legg-Calve-Perthes disease, leukemia, fractures, slipped capital femoral, epiphyses, synovitis of hip and tumors.
- The two most helpful laboratory studies are complete blood cell count (CBC) and erythrocyte sedimentation rate (ESR).
- Infectious or inflammatory disorders are 8 times as likely if the temperature is ≥38° C or the erythrocyte sedimentation rate ≥30 mm/h.
- Synovial fluid analysis is required in all patients with suspected cases of septic arthritis and should be sent for Gram stain, aerobic and anaerobic cultures, cell count, glucose, protein, and crystal analysis.
- Blood cultures are necessary in suspected infectious conditions.
- Plain radiographs are recommended in children with suspected orthopedic disorders, trauma, bony tenderness, abnormal joint examination findings, deformity, or any patient suspected of suffering child abuse.
- In older children X-rays of the affected joint are usually adequate.
- An important caveat to this is in slipped capital femoral epiphysis (SCFE) where the pathology is in the hip, but the patient complains of knee pain.
- Children with chronic SCFEs may have had multiple X-rays of the knee by unsuspecting clinicians prior to the true diagnosis being made.
- In young children, particularly toddlers with poorly developed language skills, X-rays of the entire lower extremity are usually required as the patient is developmentally unable to localize their pain.
- A bone scan is reserved for patients with normal radiographs but suspected fractures or other pathology such as osteomyelitis, avascular necrosis, and metastatic diseases.

ED Management

- The following conditions require immediate consultation and potential admission: septic arthritis, osteomyelitis, slipped capital femoral epiphysis, congenital dislocation of hip, malignant neoplasms, sickle cell crisis, and child abuse injuries.
- Often a diagnosis is uncertain, but emergent conditions have been excluded by history, examination and diagnostic work-up.
- In these cases not requiring immediate consultation or admission, over-the-counter analgesia and discharge with close appropriate follow-up is appropriate.

Legg-Calve-Perthes Disease

Clinical Presentation

Legg-Calve-Perthes disease is an idiopathic form of avascular necrosis of the femoral head seen in growing children.

- It is caused by a vascular occlusion of the blood supply to the capital femoral epiphysis (CFE).
- Is most commonly seen in boys between the ages of 2 and 13 with 80% of patients between 4 and 9 yr old. It is bilateral in approximately 10 to 20% of children.
- Vascular occlusion precedes resorption of bone within the femoral head, which is then followed by the production of new bone.
- With the early infarction of the epiphysis, the articular cartilage continues to grow because it is nourished from the synovial fluid.
- Over time, widening of the articular space, subchondral fracture in the femoral head, and increased ossification with distortions of the femoral head develop.
- The child presents with pain in the hip or groin with radiation to the thigh or knee.
- Often the history includes a "painless limp" with symptoms developing over wk to mo.
- The pain is usually mild, dull, and chronic with exacerbation on exertion and relief with rest.

- The patient may also demonstrate a limp.
- Common findings include restrictive range of motion especially with abduction and internal rotation, muscle spasm, proximal thigh atrophy, and mild shortness of stature.

Diagnostic Evaluation

- Anteroposterior and lateral frog radiographs of the pelvis should be obtained to establish the diagnosis.
- The radiographic findings depend on the stage of the disease process.
  - During the incipient stage in the first 3 mo there is a widening of the cartilage space of the affected hip and a smaller ossific nucleus of the femoral head.
  - The second stage involves the subchondral stress fracture line in the femoral head (Caffey's sign).
  - The third finding is increased opacification of the femoral head that signifies deposition of new bone upon avascular trabeculae.
  - Ultimately the femoral head and neck become deformed and distorted, which increases the likelihood of subluxation and extrusion from the acetabulum.
- MRI scan and technetium-99m bone scan are commonly employed to aid in the diagnosis.
- These studies can be helpful to distinguish Legg-Calve-Perthes disease from other disease processes.
- The differential diagnosis includes toxic tenosynovitis of the hip, acute rheumatic fever, tuberculous arthritis of the hip, and unilateral tumors of the hip.

ED Management

- Consideration or establishment of Legg-Calve-Perthes disease in the emergency department warrants orthopedic consultation.
- Treatment efforts are aimed at restoring hip range of motion, stabilization of the femoral head, and preventing CFE collapse or subluxation.
- Treatment may either be surgical or nonsurgical, and almost all affected children are hospitalized initially and treated with traction to decrease hip irritability and prevent long term complications.

Fracture Patterns in Child Abuse

Pediatric skeletal injuries can be caused by both accidental and nonaccidental trauma.

- Correlation of the child's age, motor abilities, and the mechanism of injury with the injury pattern are important in evaluation.
- Fractures due to abuse should be suspected in a young child if the caretaker brings in the child for evaluation and reports no accident but does report a change in the child such as swelling or decreased limb movement.
- Although not diagnostic of child abuse, certain injury patterns are encountered consistently as a result of child abuse and are strongly suggestive of abuse.
  - Fractures inconsistent with history
  - Fractures inconsistent with developmental stage of child
  - Fractures in various stages of healing
  - Multiple, complex, or depressed skull fractures
  - Epiphyseal-metaphyseal rib fractures
  - Spiral fractures of the femur or tibia in preambulating children
  - Spiral fractures of the humerus
  - Metaphyseal chip fractures

22

Bucket-Handle Fractures and Corner Fractures

The corner and bucket-handle metaphyseal fracture patterns are pathognomonic fractures of child abuse, and they are referred to as the classis metaphyseal lesion (CML).

- Radiographically there is no gross displacement of the epiphysis, but a corner or a chip fracture is seen at the edge of the involved metaphysis.
- The radiographic appearance varies, depending on whether this fragment is viewed tangentially as a corner fracture or obliquely as a bucket-handle fracture.
- Significant distraction to the long bones creates hemorrhagic separation of the distal metaphyses with lucency parallel and proximal to the physis.
- The result is a disc-like fragment of bone and calcified cartilage that is wider peripherally than centrally.
- The CML can occur throughout the infant skeleton, with a predilection for the femur, tibia, and humerus.
- These fractures are most likely due to violent shaking or traction injuries to the extremity.
- They are specifically suggestive of child abuse and are frequently the only radiographically detectable evidence of abuse; however, they are the minority of fractures seen in abuse.

Long Bone Fractures

Multiple studies have shown that there is no predominant pattern of diaphyseal fracture in child abuse.

- In ambulatory children, long bone fractures may be transverse, oblique or spiral. Metaphyseal and epiphyseal fractures of the long bones are classically associated with child abuse.
- In nonambulatory children, twisting injuries can create spiral fracture in the long bones.
- Spiral femur fractures in newborns and preambulatory infants are highly suggestive of nonaccidental trauma.
- Toddlers commonly suffer spiral fractures of the lower portion of the tibia accidentally, but spiral fractures of the tibia may also occur with child abuse.

Rib Fractures

Rib fractures in child abuse may be due to squeezing of the chest, hitting the child from behind, or stepping on the chest.

- Posterior rib fractures are the most common in child abuse, and this is believed to be from severe shaking of an infant that causes front-to-back chest compression.
- Posterior rib fractures are difficult to diagnose acutely because of a lack of displacement. Rib fractures in child abuse may occur anywhere along the arc of the rib.
- Lateral rib fractures are also likely due to anteroposterior chest compression. Lateral compression of the chest likely causes or rib fractures along its posterior arc and disruption at the costochondral junction.
- Particular attention should be given to multiple rib fractures and rib fractures in various stages of healing.
- Evaluation of the associated soft tissues as well as the radiographic search for a fracture line or callus formation may provide important information to aid in detecting discrepancies.

Suggested Reading

1. Bachman D, Santora S. Orthopedic trauma. In: Fleisher GR, Ludwig S, eds. Textbook of Pediatric Emergency Medicine. 4th Ed. Philadelphia: Lippincott Williams & Wilkins, 2000:1435-1478.

2. Campbell RM. Problem injuries in unique conditions of the musculoskeletal system. In: Rockwood Jr CA, Wilkins KE, Beaty JH, eds. Fractures in Children. 4th ed. Philadelphia: Lippincott-Raven, 1996:167-320.
3. Christoph RA. Musculoskeletal disorders in children. In: Tintinalli JE, Kelen GD, Stapczynski JS, eds. Emergency Medicine: A Comprehensive Study Guide. 5th ed. New York: McGraw-Hill, 2000:905-918.
4. Thompson GH, Scoles PV. Common fractures. In: Behrman RE, Kliegman RM, Jenson HB, eds. Nelson Textbook of Pediatrics, 16th ed. Philadelphia: W.B. Saunders Company, 2000:2095-2098.
5. Cwinn AA. Pelvis and Hip. In: Rosen P, Barkin R, eds. Emergency Medicine: Concepts and Clinical Practice. 4th ed. St. Louis: Mosby-Year Book Inc., 1998:757-761.
6. Flynn FC, Zink WP. Fractures and dislocations of the elbow. In: MacEwen GD, Kasser JR, Heinrich SD, eds. Pediatric Fractures: A Practical Approach to Assessment and Treatment. Baltimore: Williams & Wilkins, 1993:133-164.
7. Geiderman JM, Magnusson AR. Humerus and elbow. In: Rosen P, Barkin R, eds. Emergency Medicine: Concepts and Clinical Practice. 4th ed. St. Louis: Mosby-Year Book Inc., 1998:690-708.
8. Gruber JE. Injuries of the proximal femur. In: Rosen P, Barkin R, eds. Emergency Medicine: Concepts and Clinical Practice. 4th ed. St. Louis: Mosby-Year Book Inc., 1998:763-785.
9. Heinrich SD. Proximal femur fractures: Hip dislocations. In: MacEwen GD, Kasser JR, Heinrich SD, eds. Pediatric Fractures: A Practical Approach to Assessment and Treatment. Baltimore: Williams & Wilkins, 1993:263-277.
10. Lawrence LL. The limping child. Emerg Med Clin North Am 1998; 16:911-929.
11. McNamara R. Management of common dislocations. In: Roberts JR, Hedges JR, eds. Clinical Procedures in Emergency Medicine. 3rd ed. Philadelphia: W.B. Saunders Company, 1998:818-852.
12. Nimkin K, Kleinman PK. Imaging of child abuse. Pediatr Clin North Am 1997; 44(3):615-35.
13. Peterson HA. Physeal and Apophyseal Injuries. In: Rockwood Jr CA, Wilkins KE, Beaty JH, eds. Fractures in Children. 4th ed. Philadelphia: Lippincott-Raven, 1996:103-163.
14. Thompson GH, Scoles PV. The hip. In: Behrman RE, Kliegman RM, Jenson HB, eds. Nelson Textbook of Pediatrics. 16th ed. Philadelphia: W.B. Saunders Company, 2000:2077-2082.

22

Part E: Infectious Disease Problems in Children

Lance Brown

Fever in the Well Appearing Child

One of the most controversial areas in the care of infants and young children in the emergency department is the management of fever in the well-appearing child.

- Because of large studies looking at blood culture results in large numbers of febrile children, it is known that some percentage of febrile young children will have bacteremia or a urinary tract infection.
- A subset of bacteremic children will progress to develop a serious bacterial infection.
- The approach to identifying these children and developing a rational management plan is an area of great controversy.
- A helpful organizational approach is to look at various age groups as the clinical utility of the physical examination, epidemiology, and prognosis depend most heavily on the age of the child.
- It is important to point out that the following discussion deals with well-appearing, febrile infants and children.
- Ill-appearing children or those with focal infections require a different approach.

22

Neonates (birth to 28 days of age)

- Neonates are the most difficult age group to assess as the physical examination and clinical signs and symptoms are the most difficult to evaluate even for the experienced physician.
- Fortunately, the management of this age group is the least controversial.
- Neonates with a rectal temperature of 38°C (100.4°F) or higher have an overall risk of a serious bacterial infection of about 13%.
- The disease processes may be difficult to diagnose initially and include meningitis, urinary tract infections, sepsis, osteomyelitis, cellulitis, and Salmonella gastroenteritis.
- The bacteriology of these infections in neonates involves a relatively unusual group of organisms including *Listeria monocytogenes*, Gram-negative enteric organisms (mostly *Escherichia coli*, but also Klebsiella) and group B streptococci.
- These young patients have the least physiologic reserve and progress fairly rapidly to sepsis and potentially death if not treated appropriately. Therefore, physicians are in general conservative and relatively aggressive with this age group.
- In general, all febrile neonates with a rectal temperature of 38° C (100.4° F) or higher should undergo a septic work-up (including lumbar puncture), receive intravenous antibiotics, and be admitted to the hospital.
- The septic work-up typically includes a complete blood count (CBC), a blood culture, a catheterized urinalysis, a urine culture, cerebrospinal fluid analysis, and some physicians include a chest X-ray.
- Appropriate antibiotics include the combination of ampicillin and cefotaxime or the combination of ampicillin and gentamicin.

Young Infants (29-89 days of age)

- The approach to young infants remains an area of evolving research and understanding.
- In general, this age group can be divided into the 29 through 56 days of age group and the 56 through 89 days of age group.
- Infants between 29 and 56 days of age can be divided into low risk and high risk groups.
- Low risk infants typically have a low grade fever and a negative septic work up including lumbar puncture.
- One set of criteria used to determine if an infant is low risk is the Philadelphia criteria. The Philadelphia criteria includes a white blood cell count $<15 \times 10^9$ cells/L, a band-to-neutrophil ratio lower than 0.2, a urinalysis with fewer than 10 white cells per high-power field, cerebrospinal fluid with fewer than 8 white cell/microliter, no evidence of discrete infiltrate on chest X-ray, and stool that is negative for occult blood by bedside testing.
- Infants that meet this formal definition of low risk have approximately a 1-5% risk of harboring an occult serious bacterial infection.
- Etiologic agents for this age group include Gram-negative rods (predominantly *E. coli*), group B streptococci, *Streptococcus pneumoniae*, *Neisseria meningitidis*, and Salmonella.
- If the infant appears well and is truly low risk by a formal negative septic work up, there are three currently acceptable management plans.
- The infant can be discharged home without antibiotics as long as close follow-up in 12 to 24 h is arranged with the primary care physician. Or the infant could be given parenteral ceftriaxone and discharged home with close follow-up.
- Alternatively, the infant can be admitted to the hospital and antibiotics can be withheld pending culture results or a change in clinical condition.
- High risk infants are those that fail any of the low risk criteria. All high risk infants should undergo a septic work up including lumbar puncture, be given parenteral ceftriaxone, and be admitted to the hospital.
- Infants 56-89 days of age are not well represented in the literature.

- In general, clinicians either treat these infants as they would the 29-56 day old infant or as they would an infant >89 days of age.
- There is clearly no consensus for the treatment of this age group. It is recommended that those physicians with the least familiarity in treating young infants should be more conservative and treat these infants in a fashion similar to the 29-56 day old infant.
- However, physicians with experience in treating young infants may feel more comfortable in approaching the 56-89 day old infant as they would a 3 or 4 mo old (as described below).

Older Infants and Young Children (3-36 mo of age)

- There is a great deal of controversy in interpreting the literature about fever in well appearing children in this age group. The same data can be interpreted to support a fairly aggressive approach where testing and treating with parenteral antibiotics is commonly undertaken or to support a relatively minimalist approach using no tests (except perhaps urinalysis and culture), no treatment other than antipyretics, and close follow-up.
- One key question to ask is, "What degree of risk am I willing to take given the epidemiology of this patient population?" Overwhelming cases of *Neisseria meningitidis* are a great example of an occult illness that progresses so quickly that the initial presentation may be that of a well-appearing child with perhaps "the flu" only to develop into full blown life-threatening sepsis with petechiae within hours.
- If there were a large number of these cases, most physicians would give parenteral ceftriaxone relatively frequently. If these cases were rare, giving parenteral ceftriaxone would cause unnecessary and unpleasant side effects including rash, anaphylaxis, and diarrhea in many children, and perhaps most importantly, would promote microbial resistance (especially in *Streptococcus pneumoniae*). What are the odds that a given patient has an occult serious bacterial illness?
- Classically, the organisms responsible for bacterial infections in young children have included *E. coli* in the urine and *Haemophilus influenzae*, Salmonella, *N. meningitidis*, and *S. pneumoniae* in the blood and other areas (e.g., osteomyelitis, epiglotitis, septic arthritis, infectious diarrhea, meningitis).
- The effectiveness of the *H. influenzae* or Hib vaccine has been remarkable, and the organism has essentially been eradicated as a cause of serious bacterial infections in children in the continental United States. Previous studies that placed the risk of occult bacteremia in the 4% range now place the risk as low as 1.6% since the widespread use of the Hib vaccine began.
- With the elimination of *H. influenzae*, *S. pneumoniae* is now the dominant organism of concern and antibiotic resistance in *S. pneumoniae* is rising.
- However, a relatively new conjugate vaccine against *S. pneumoniae* is now available, and as this vaccine moves into widespread use the incidence of serious infections developing from *S. pneumonia* would be expected to decrease.
- There is also a vaccine against *N. meningitidis*; however, widespread use of this vaccine is not expected as it is currently indicated only for high risk groups.
- The major concern in missing occult disease is in missing meningitis that is accompanied by a bad neurologic outcome. Fortunately, this is quite uncommon.
- From the available data, it appears that although many children with bacteremia (especially with *S. pneumoniae*) will clear their bacteremia spontaneously, about 3% will develop meningitis. About one-third of these children will have a bad neurologic outcome.
- Therefore, if about 2% of febrile young children have occult bacteremia and 3% of these children go on to develop meningitis and 33% of these children have a bad neurologic outcome, this represents one bad neurologic outcome in every 5,000 well appearing febrile young children.

22

- If all of these children were treated with parenteral antibiotics, between about 400 and 1,500 children would suffer from side effects of the antibiotics. If oral antibiotics are given, the risk in one preHib study showed a decrease in risk of meningitis from about 3% to about 1%.
- The administration of oral antibiotics does not prevent meningitis from occurring. The data reflecting the impact of the *S. pneumoniae* vaccine is expected in a few years and will undoubtedly impact such an analysis.
- One area where our understanding of the risk of serious bacterial infections has increased is in the area of urinary tract infections.
- It has been estimated that in febrile young children under the age of 2 yr the risk of occult urinary tract infection is about 4%.
- Urinary tract infections in children in diapers are difficult to detect as the urine is seldom described as foul smelling by parents, and there is really no way for a preverbal child to express urgency, hesitancy, frequency, or dysuria without simply crying.
- Generally, the clinical picture of these children is misleading.
- Vomiting or apparent abdominal pain are fairly common presentations and may lead the unsuspecting physician away from the diagnosis of urinary tract infection.
- Uncircumcised boys under the age of 6 mo and girls under the age of 2 yr are at relatively high risk.
- The highest risk, which may be as high as 15% is in white, female infants younger than 1 yr of age with fever of 2 or more days duration, no other source of fever, and a temperature >39° C (102.2° F).
- Those children with urinary tract infections are at higher risk for anatomic abnormalities and subsequent renal injury.
- Based on a careful analysis of the risk, a reasonable plan can be developed for well-appearing febrile children. All neonates (from birth to 28 days of age) should undergo a septic work up, receive parenteral antibiotics, and be admitted. All infants from 29-56 days of age should undergo a septic work up including lumbar puncture.
- High risk infants should receive parenteral antibiotics and be admitted.
- Low risk infants may have a management plan that includes admission and observation without antibiotics or discharge home with or without antibiotics but with a 12 to 24 h follow-up.
- Infants 56-89 days of age should either be treated like those from 29-56 days of age or in a manner similar to older infants based on the clinician's experience.
- Infants between 3 and 6 mo should all have urinalyses and urine cultures obtained. Pyuria or bactiuria should be treated with antibiotics.
- Girls between 6 mo of age and 2 yr should probably undergo urinalysis and urine culture to identify and treat those cases of urinary tract infections, facilitating the identification of anatomic abnormalities and avoiding further renal injury.
- Uncircumsized boys up to 12 mo of age should probably undergo urinalysis and urine culture. Circumsized boys in this age group are very low risk.
- The use of various tests has been proposed to help further assess the risk of bacteremia.
- The most commonly used is the WBC count in the blood. Unfortunately, this test is not very useful. Attempts to use various cut off values to determine when other actions (such as drawing blood cultures or giving antibiotics) should take place is fraught with difficulty. If a cut-off of 15 x 10^9 cells/L is used, this test has only an 86% sensitivity and 77% specificity for bacteremia. If a cut off of 20 x 10^9 cells/L is used, then the WBC has a sensitivity of only 48% and 92% specificity. If too low a cut-off is used, the WBC fails to stratify children and if a high WBC cut-off is chosen, more than half of bacteremic children will not be identified.
- With the exception of urinalysis and urine culture, no other testing is usually required of well-appearing, febrile infants and children between 3 and 36 mo of age.

- Reliable and close follow-up should be sufficient to identify those children who deteriorate. And there is little controversy as the process is no longer occult.

Characteristic and Common Rashes with Fever

There are distinct and classic patterns of rashes of which the emergency physician should be aware. Familiarity with the classic presentations of rashes in children will allow for the prompt recognition and treatment of many commonly seen conditions in the emergency department. Regrettably, many rashes look alike. Fortunately, many rashes require the same treatment (i.e., antipyretics). This section will review some commonly seen rashes in children and those with distinct clinical patterns or appearance.

Impetigo

- Skin infections with staphylococci and group A β-hemolytic streptococci may arise, especially in young children under warm and humid conditions.
- There are two major forms of the disease: Bullous impetigo and impetigo contagiosum (the latter being referred to most commonly as just "impetigo").
  - Bullous impetigo is caused by staphylococci that produce an epidermolytic toxin.
    - This toxin causes separation of the skin layers.
    - This results in the development of thin-walled bullae with pale yellow fluid that easily rupture leaving a shiny denuded base.
    - Involvement may range from a single bullous lesion to extensive areas of skin involvement.
    - The diagnosis is made on clinical appearance.
    - Treatment is an oral first generation cephalosporin and a topic antimicrobial (e.g., neosporin).
  - Impetigo (contagiosum) results from an infection of staphylococci or group A β-hemolytic streptococci.
    - This form is very common and usually strikes young children.
    - The infection can be a site of minor trauma, a bug bite, or have no apparent etiology.
    - This infection is commonly seen on the face around the nose and mouth.
    - The lesions are small papules and vessicles that burst to release a golden, crusting fluid.
    - Scratching at the crusts may spread the infection.
    - The diagnosis is made on clinical grounds.
    - Treatments include an oral first generation cephalosporin, erythromycin, dicloxacillin, and/or topical mupirocin (Bactroban).

Erysipelas

- Erysipelas usually presents as an intense local facial cellulitis.
- It is caused by group A β-hemolytic streptococci.
- The patient is usually systemically ill with fever, chills, malaise, localized pain, headache, and vomiting.
- Diagnosis is made on clinical grounds.
- The concern with erysipelas is that the infection will spread, lead to bacteremia, or develop into a facial abscess.
- Intravenous antibiotics (e.g., a macrolide, first generation cephalosporin, or oxacillin) and admission are usually warranted in the hope of avoiding local and systemic complications.
- Cold panniculitis:
  - One fairly common condition that should not be confused with erysipelas is cold panniculitis also known as popsicle panniculitis.
  - With this condition a child's face is exposed to cold either environmentally or by holding a popsicle against their face while they eat it.

- The subcutaneous fat of the face solidifies and the skin turns somewhat red.
- The child has no other symptoms and is well-appearing.
- Fever is, of course, not a part of this localized condition.
- There is no treatment needed and the area is expected to soften and return to normal in a week or so.

Scarlet Fever

- A toxin produced by a phage-infected group A streptococcus is responsible for the rash of scarlet fever.
- The most common association is with strep pharyngitis (i.e., strep throat), but may occur with other streptococcal infections such as an infected wound.
- The rash is fine, raised, and generalized.
- A coarse, sandpaper feel to a maculopapular rash is most characteristic.
- The area around the mouth is typcially spared.
- The tongue may become erythematous with hypertrophied papillae (also called a "strawberry tongue").
- A characteristic but certainly not universal finding is the presence of Pastia's lines, bright linear discolorations (usually red or purplish) in the axillae or anticubital fossae.
- The rash typically lasts 3-5 days.
- Interestingly, after about 2 wk, the skin may peel in sheets of various sizes.
- The classic description of this peeling is "brawny desquamation."
- The treatment is penicillin or a macrolide in the penicillin-allergic patient.

Hand-Foot-and-Mouth Disease

- Caused by Coxsackie viruses, hand-foot-and-mouth disease has a distinct appearance.
- These infections are most commonly seen in the late summer and early fall.
- Older children usually complain of a sore throat and infants may appear fussy while feeding.
- This is due to small lesions on the uvula, palate, and tonsils that are painful and on exam appear as shallow white to gray ulcerations.
- In addition, the palms and soles are typically involved.
- Initially small, tender, erythematous macules develop. These progress to become oval, gray, somewhat flat vessicles. These lesions may itch somewhat and be tender enough to interfere with walking in young children.
- The frequent presence of similar lesions throughout the diaper area has suggested to some that the proper name for this disease should be "hand-foot-mouth-and-butt" disease.
- An unsuspecting physician may even diagnose candida diaper rash if the hands and feet are not examined.
- Less commonly, and in more extensive cases, the lesions may be found on the dorsa of the hands and feet and progress to include the distal arms and legs.

Chickenpox

- Chickenpox is very common.
- Even with the advent of the varicella vaccine, cases are still quite common.
- There are three stages to the lesions with chickenpox.
  - Initially the lesions appear as small, red macules.
  - Later, these lesions become small vesicles that have a "dew drop on a rose petal" appearance.
  - Finally, these vesicles crust over.
- Characteristic of chickenpox is the presence of multiple stages of the lesions on the same patient at the same time.
- Although there are many variations on the classic pattern, the classic description of the rash is that it starts centrally on the chest and progresses outwardly.

- The diagnosis is made clinically by the appearance of the rash.
- If the patient presents early in the disease course with just fever and red macules, the diagnosis will be obscure.
- Mucous membranes may be involved and may interfere with feeding.
- Previously healthy children typically do not have difficulties with this illness; however, picking at the crusts should be strongly discouraged as permanent scarring will occur.
- Two major complications of chickenpox include varicella pneumonia and bacterial superinfection of the lesions. These are uncommon.

Erythema Infectiosum (Fifth disease)

- Erythema infectiousum is caused by parvovirus and typically strikes school-aged children in the spring.
- The rash essentially has two phases.
  - The first stage involves the abrupt appearance of a fiery red rash on the cheeks with closely grouped fine papules.
  - The rash is classically described as a "slapped-cheek appearance" or "butterfly wings" and is most intense on the cheeks sparing the perioral area, the eyelids, and the chin. This rash lasts 4-5 days.
  - Within a couple of days of the appearance of the facial rash, an erythematous, generally macular rash appears on the trunk and limbs.
  - This rash begins clearing centrally which results in a lacy appearance to the evolving rash.
- Associated symptoms are variable but may include sore throat, cough, vomiting and diarrhea, headache and muscle aches.
- Previously healthy children require only supportive care such as antipyretics and the encouraged intake of fluids.

Roseola (Roseola infantum)

- Roseola is currently thought to be due to herpes simplex virus-6.
- The typical pattern for this illness is a very high fever that resolves and is then followed by the rash.
- There is typically a very rapid development of a high fever that may be in the 104° F (40° C) range. This rapid rise in temperature may be accompanied by a febrile seizure.
- This febrile stage may last 3 or 4 days.
- Upon defervescence or shortly thereafter, a rash develops.
- The characteristic rash consists of fine pink macules that start on the trunk and spread peripherally.
- The diagnosis is made clinically primarily based on the time-course of the illness and not on the fairly nondescript rash appearance.
- There is no specific treatment other than treatment of the fever with antipyretics and management of the febrile seizure should one occur.

Measles (Rubeola)

- Since vaccination began in the 1960s, the number of cases of this once common pediatric illness has decreased dramatically.
- Most recently trained physicians have never seen a case of measles. However, sporadic and clusters of cases occur each year, most notably in college-aged individuals.
- This viral illness typically starts as fever (which may rise as high as 104° F (40° C)) and malaise.
- Within the first day an upper respiratory tract infection develops which typically includes rhinorrhea, conjunctivitis, and cough.
- On or around the third day of illness, Koplik's spots, small whitish lesions on the buccal mucosa of the mouth the size of grains of sand, develop.

- The typical skin rash develops around the 4th day of illness.
  - The rash progresses from head to toe.
  - The face and neck are initially and most heavily involved and the erythematous macules and papules may become confluent.
  - As the rash progresses toward the lower parts of the body, the facial and neck rash begins to fade along with the Koplik spots.
- The diagnosis is made clinically and typical treatment is supportive (antipyretics and oral fluids).

22

- Complications include otitis media, pneumonia, and encephalitis. Measles encephalitis has a fatality rate of 15%.

Rocky Mountain Spotted Fever

- Rocky Mountain spotted fever (RMSF) is caused by *Rickettsia rickettsii* and is transmitted by ticks. Despite the name, most cases occur in the East Coast states of the United States. There are about 1000 cases reported in the United States each year and about two-thirds of these patients are children or adolescents.
- The clinical findings in cases of RMSF are related to the diffuse vasculitis that develops.
- The time-course of the illness classically begins with headache and malaise and then fever.
- On the 3rd or 4th day of the illness, the rash usually develops.
- Classically, the rash begins as erythematous, blanching maculopapular lesions on the wrists and ankles that becomes petechial and spreads centrally over the subsequent 1 to 2 days.
- Complications include DIC, hyponatremia, a prolonged PR interval on the EKG, seizures, myocarditis, and septic shock.
- Treatment is chloramphenicol for children under 8 yr of age and tetracycline in older children.

Lyme Disease (Erythema migrans)

- Lyme disease is caused by the spirochete *Borrelia burgdorferi* and is transmitted by the *Ixodes* species of ticks commonly found on deer.
- The disease caused by this spirochete is classically divided into three stages.
  - Stage 1 Lyme disease primarily presents as the erythema migrans rash.
    - Starting as a small red papule at the site of the tick bite, a spreading erythematous plaque with central clearing occurs.
    - This rash occurs in about 75% of cases.
    - The rash slowly expands in diameter to as large as 15 cm and spontaneously resolves in about 1 mo.
  - If untreated, Lyme disease progresses to the second stage, disseminated disease.
    - The symptoms of this stage include fever, adenopathy, and flulike symptoms.
    - Multiple, ring-like skin lesions develop in about half of these patients. Migratory arthralgias, myalgias, and debilitating fatigue may develop.
    - Severe headache, meningismus, and Bell's palsy may be seen.
    - A small number of patients (about 5%) develop cardiac involvement that may include conduction abnormalities and pancarditis.
  - Persistent infection leads to stage three Lyme disease.
  - A chronic scleroderma type rash, erosive oligoarthritis, and subtle neurologic findings may develop long after other symptoms have resolved.
- The diagnosis of Lyme disease is primarily clinical.
- Serologic testing is available, but difficult to perform, confusing (cross-reacting with viral and autoimmune conditions), and may not be positive until the second month after the tick bite.
- Management primarily consists of supportive measures and antibiotics.

- In stage 1, treatment consists of amoxicillin for 2-3 wk for children under 9 yr of age and doxycycline for 2-3 wk for older children and adolescents.
- Stage 2 is treated with amoxicillin or doxycycline as above, but for 3-4 wk.
- Stage 3 is treated with parenteral ceftriaxone or penicillin for 2-4 wk.

Kawasaki Disease

- Kawasaki, otherwise known as mucocutaneous lymph node syndrome, is an idiopathic vasculitis of small- and medium-sized vessels.
- Kawasaki disease is the most common cause of acquired heart disease in children in the United States.
- The formal diagnosis of Kawasaki disease includes fever duration of 5 or more days unresponsive to antibiotics.
- This diagnostic feature makes the formal diagnosis in the emergency department problematic.
- At best, particularly with a fairly early presentation, the suggestion of Kawasaki disease is the most that can be done in the emergency department.
- This is important because the early initiation of treatment (intravenous γ-globulin and aspirin) can decrease the incidence of coronary artery aneurysms that develop during the second or third week of the illness and are the major cause of morbidity with Kawasaki disease.
- Other features of Kawasaki disease include fever, conjunctivitis, rash, mucosal inflamation, lymphadenopathy, and extremity changes.
- The rash is fairly nondescript, polymorphous, nonvesicular, and primarily involves the trunk. There may be areas of desquamation and target lesions.
- The mucosal inflamation presents as cracked, red lips, a strawberry tongue, and an injected pharynx.
- Lymphadenopathy is classically cervical and some nodes are >1.5 cm in diameter.
- The extremity changes include erythema, edema, and periungual desquamation.
- Suspected cases of Kawasaki disease require admission and consultation.
- If any cardiac manifestations are present, admission to a pediatric intensive care unit should be arranged.

Suggested Reading

1. Brown LA. The well-appearing febrile child. Critical Decisions in Emergency Medicine 2000; 15(3):1-6.
2. Baker MD, Bell LM. Unpredictability of serious bacterial illness in febrile infants from birth to 1 mo of age. Arch Pediatr Adolesc Med 1999; 153:508-511.
3. Baker MD, Avner JR. Management of fever in young infants. Clin Pediatr Emerg Med 2000; 1:102-108.
4. Green SM, Rothrock SG. Evaluation styles for well-appearing febrile children: Are you a "risk-minimizer" or a "test-minimizer"? Ann Emerg Med 1999; 33:211-214.
5. Sadow KB, Derr R, Teach SJ. Bacterial infections in infants 60 days and younger: Epidemiology, resistance, and implications for treatment. Arch Pediatr Adolesc Med 1999; 153:611-614.
6. McCollough M. Progress toward eliminating Haemophilus influenzae type b disease among infants and children- United States, 1987-1997. Ann Emerg Med 1999; 34:110-111.
7. Lee GM, Harper MB. Risk of bacteremia for febrile young children in the post-Haemophilus influenzae type b era. Arch Pediatr Adolesc Med 1998; 152:624-628.
8. Segal GS, Chamberlain JM. Resource utilization and contaminated blood cultures in children at risk for occult bacteremia. Arch Pediatr Adolesc Med 2000; 154:469-473.
9. Gruskin K. Rash-maculopapular. In: Fleisher GR, Ludwig S, eds. Textbook of Pediatric Emergency Medicine. 4th ed. Philadelphia: Lippincott Williams & Wilkins, 2000:507-524.

10. Weinstock MS, Rosenan AM. Pediatric exanthems. In: Tintinalli JE, Kelen GD, Stapczynski JS, eds. Emergency Medicine: A Comprehensive Study Guide. 6th ed. New York: McGraw-Hill, 2004:869-876.
11. Fleisher GR. Infectious disease emergencies. In: Fleisher GR, Ludwig S, eds. Textbook of Pediatric Emergency Medicine. 4th ed. Philadelphia: Lippincott Williams and Wilkins, 2000:776-781.
12. Meredith JT. Zoonotic infections. In: Tintinalli JE, Kelen GD, Stapczynski JS, eds. Emergency Medicine: A Comprehensive Study Guide. 6th ed. New York: McGraw-Hill, 2004:969-999.
13. Sundel RP. Rheumatologic Emergencies. In: Fleisher GR, Ludwig S, eds. Textbook of Pediatric Emergency Medicine. 4th ed. Philadelphia: Lippincott Williams & Wilkins, 2000:1191-1228.

Part F: ENT and Respiratory Problems in Children

Lance Brown

Foreign Bodies

Ear

Foreign body removal from the external auditory canal can be one of the most satisfying or most frustrating experiences in the emergency department. Many physicians enjoy the heroic sense of accomplishment when a bead or cockroach is removed from a child's ear canal to the relief of the parent and child. However, no emergency physician wants to have a bloodied ear canal or dizziness from cold water irrigation to be their "reward" for trying to remove a foreign body from the ear canal. Proper preparation and planning can minimize the risk to the child while increasing the likelihood of success for the physician.

- There are multiple types of objects that young children find or place in their ear canals.
- Some of the most common include beads, plastic toys, pebbles, insects, popcorn kernels, and paper. However, the list of things found is limited only by the imagination.
- Although not always possible, objects should be classified to generate a plan for removal.
- Many older children will do well with gentle attempts to remove objects with nothing more than continuing verbal reassurance.
- Younger children usually require physical restraint.
- Other children will be so intolerant of the procedure as to require procedural sedation (e.g., with ketamine) to achieve cooperation and minimize risk to the ear canal and the tympanic membrane.
- One of the most emotionally disturbing is the live insect or spider. These bugs will cause a sense of movement and a loud noise for the child. These children may run around screaming or hold their ear and cry.
- It is usually best to kill the bug with a small quantity of 1% lidocaine and then remove the bug with a small alligator forceps or irrigation. Irrigation should be attempted only if the tympanic membrane is intact (usually the case with bugs unless the parent has gone digging around in the ear with a cotton swab).
- Irrigation should only be performed with warm solution (such a tap water) as cold water will cause profound discomfort, dizziness, and nystagmus.
- It is important to reexamine the ear canal after the bug has been removed as barbed legs may remain behind and cause a delayed inflamation within the ear canal.
- Other objects, such as beans, are soft and absorb water. It is best to try to take these objects out with forceps or a cerumen loop as irrigation will make these objects swell and become more difficult to remove.

- Hard, round objects such as beads can be particularly difficult. If the tympanic membrane can be reasonably assured to be intact, irrigation is usually the best method.
- On occasion, the hole of the bead will be facing the examiner and a the tip of small alligator forceps can be placed within the hole. When the forceps are opened, the force is exerted against the inner aspect of the hole and the object can then be removed.
- In general, tightly wedged objects, objects deep within the canal or sitting against the tympanic membrane, or objects with sharp edges such as pieces of glass require referral to an ENT specialist.
- One other type of object that may require referral is putty. Sometimes used as an ear plug, putty can be very difficult to remove and may require an operating microscope and general anesthesia.
- One special object to mention is the disc battery that is most commonly used in watches and hearing aids. Disc batteries can cause a liquifaction necrosis of the external ear canal and tympanic membrane. Significant damage can result. Irrigation should never be attempted when a disc battery is present. Unless the disc battery is in the most lateral aspect of the ear canal and atraumatic removal is quite likely in the hands of an experienced emergency physician, these objects are best removed by the ENT specialist on an urgent basis.
- Besides unsuccessful removal, external auditory canal abrasion or laceration is the most common complication of attempted removal of a foreign body. If this occurs, the patient should have dry ear precautions (using a vaseline coated cotton ball at the entrance of the ear canal during bathing and avoiding swimming until healed) and use an antibiotic otic suspension.
- Follow-up with an ENT specialist is recommended for significant lacerations and any other complications including tympanic membrane perforation.

Nose

Just as with the ear, the list of objects found in the noses of young children is long. Some of the most common include foods such as beans, seeds, and kernels of corn, toys such as beads, and other small objects including wads of paper and pebbles. In general, these objects can be removed in the emergency department. Many parents present stating that the child reported putting something up his nose or can see the object themselves. However, the presentation may simply be that of persistent unilateral nasal discharge without a known history of foreign body insertion. This situation is so common that any young child with persistent unilateral nasal discharge should be examined for a foreign body as this is the diagnosis until proven otherwise.

- Two techniques of the most commonly applied techniques include forceps removal and Foley catheter removal. The forceps removal is straight forward. The child is placed supine and the object is grabbed with the forceps. Some preparation with suction may make visualization easier and avoid damage to the turbinates.
- The Foley catheter technique can be quite atraumatic and elegant. A 5 or 6 French Foley catheter is lubricated and slid past the object with the balloon deflated. The balloon is then inflated to a variable degree depending on the size of the child's nose and the object. The catheter and the object are then withdrawn with the balloon inflated. This technique can remove many objects including fairly friable ones such as wads of paper.
- Complications of these techniques include pushing the object into the nasopharynx with the potential for subsequent aspiration and epistaxis. Epistaxis is controlled with direct pressure and seldom requires further intervention.

22

- Other techniques include the nasal wash technique and the positive-pressure technique. With the nasal wash technique a bulb syringe is filled with 7 ml of sterile normal saline and the fluid is forcefully administered into the contralateral nostril (the side without the foreign body) with the child seated.
- The positive-pressure technique is also known as "the big kiss." In this technique the child is placed supine, the contralateral nostril is occluded and the parent is instructed to blow forcefully into the child's mouth to expel the object.
- It is often the case that the parent is rewarded with a cheek covered in snot.
- Alternatively, this technique can be performed with an ambu bag. Each of these techniques has their supporters and each physician should develop a cadre of useful techniques for various situations.

Aspirated

Aspirated foreign bodies can be life-threatening in children. The most common age group is in the second year of life when physical skills are developing quickly, exploration is vigorous, language is minimal, and judgment is poor. It sometimes seems that these young toddling explorers will put almost anything in their mouths except key foods like vegetables at meal time. On occasion, these young children will choke on something and breath it in to their respiratory tract (i.e., aspirate the object).

- The most common of these objects are peanuts. This has been demonstrated in quite a few studies. Other nuts, food and small toys are also commonly aspirated objects.
- In older children, pieces of broken balloons and other objects that conform to the airway are particularly deadly. The clinical presentation depends on the location of the object.
- Objects high in the pharynx may occlude the entire airway and lead to complete or near respiratory arrest. These children will present apneic or with stridor.
- If the child is still breathing and able to oxygenate, prompt consultation with both an anesthesiologist and an ENT specialist are required to arrange removal of the object in the operating room under controlled conditions.
- If the child is unable to ventilate, is unconscious or otherwise grossly unstable, the airway must be rapidly managed in the emergency department. Direct examination with a laryngoscope and attempted retrieval with pediatric Magill forceps should be attempted.
- If the object is lower in the respiratory tract, the most common presentation is of the sudden onset of choking and intractable coughing with or without vomiting. There may be unilateral wheezing or the appearance of respiratory distress. Air trapping is seen in about half of the cases of aspiration in children (compared with fewer than 20% of adults) and an inspiratory and expiratory chest X-rays may reveal persistent hyperexpansion on the affected side on the expiratory film.
- In the stable child with a classic presentation, a specialist with skill in removing such foreign bodies (e.g., a pediatric surgeon or ENT surgeon) should be consulted or the patient should be transferred to a tertiary care facility with these specialists available.
- In cases of a suspected foreign body aspiration, but in the stable child lacking a classic presentation, adjunct studies are usually indicated. Inspiratory and expiratory chest X-rays may reveal hyperexpansion on the affected side due to air trapping behind the object.
- Similarly, as coordinating breathing with taking X-rays in young children can be quite challenging, the child can have right and left decubitus chest X-rays performed.
- Again, air trapping will cause the affected dependent side to remain expanded despite the pressure of the weight of the heart and chest contents. More recently, magnetic resonance imaging has come to be used to identify retained foreign bodies.
- The gold standard for suspected foreign bodies is for the child to undergo bronchoscopy to identify and remove the object.

Swallowed

Swallowed foreign bodies are commonly encountered in emergency departments. Coins are the most commonly ingested objects. Surprisingly, in one home-based survey it was estimated that only 15% of children with a history of coin ingestion were evaluated by a physician. From this data, it would appear that many families are managing swallowed foreign bodies at home without seeking medical attention.

- The most common age group for ingesting foreign bodies are preschool-aged children. Most of the children who ingest foreign bodies like coins will pass the coins spontaneously.
- Special circumstances such as known esophageal disease, mental retardation, or anatomic alterations (e.g., ileostomies) pose unique situations that deserve individualized management plans.
- Objects can be classified as metallic or nonmetallic, sharp or not sharp and localized as in the esophagus, stomach, or anywhere in the intestine.
- Esophageal foreign bodies require the greatest understanding and a certain degree of urgency in their management. A special circumstance is the esophageal button battery that must be promptly removed to avoid liquefaction necrosis, mediastinitis, and mortality.
- Metallic objects that are not sharp (e.g., coins) are the best studied. All such metallic objects are expected to be visible on X-ray. The coin may be located in the upper esophagus, middle esophagus, or lower esophagus. Upper esophageal coins are most commonly symptomatic with respiratory distress, stridor, and drooling. Vomiting may occur.
- Coins in the upper esophagus are the least likely to pass spontaneously and prompt consultation (usually with an ENT specialist) should be made in the emergency department to have the coin removed endoscopically.
- Coins in the middle or lower third of the esophagus may or may not be symptomatic. Between 1/3 and 2/3 of these patients are expected to have coins that will pass into the stomach spontaneously within 24 h of ingestion.
- A period of observation with repeat X-rays to assess coin migration is recommended in coordination with the consulting specialist.
- Other objects in the esophagus may require an individualized plan. Fish and chicken bones can be diagnostically challenging as a minority of these bones are visible on X-ray. If a patient has persistent symptoms, endoscopy can be diagnostic and therapeutic.
- Sharp objects lodged in the esophagus require prompt removal to avoid perforation and the resultant morbidity.
- Objects that traverse the esophagus and make it to the stomach and intestine rarely cause problems.
- Metallic objects that are not sharp such as coins can be managed expectantly as symptoms are rare, and the coins are expected to pass spontaneously.
- Sharp objects, even needles, can successfully navigate the intestine and may cause no problems, passing spontaneously. Unusually large and sharp objects as are occasionally ingested by developmentally delayed children or older children with psychiatric disorders may require surgical removal.
- In general, expectant management looking for symptoms such as abdominal pain, bowel obstruction, and the passage of bloody stool is all that is required.

Croup and Bacterial Tracheitis

Croup

Croup, otherwise known as laryngotracheobronchitis, is a viral infection of the upper respiratory tract. The most common etiologic agent is parainfluenza virus.

22

Croup is most commonly seen in the winter months and is by far the most common cause of fever and stridor in young children.

- Older infants, toddlers, and preschoolers are the most commonly affected.
- The classic presentation is that of a young child who has had a couple of days of rhinorrhea and mild cough. There may be some low-grade fever. Then, in the late evening or nighttime hours the child begins to cough somewhat vigorously with a seal-like barking sound accompanied by stridor.
- Upon packing up, going outside into the cool nighttime air, and making it to the emergency department, the child will have shown remarkable improvement. This improvement may be so significant that the parents express embarrassment at even coming to the emergency department. Other cases are more serious.
- As croup is a self-limited, viral illness, management depends on the clinical appearance and presentation of the child.
- The two main tools in treating croup are steroids and nebulized epinephrine. In most cases of very mild croup with only mild stridor with agitation (e.g., being examined), a single dose of 0.6 mg/kg of dexamethasone either intramuscularly (IM) or orally (PO) can be given.
- If stridor persists at rest or with mild activity, nebulized epinephrine should be given. Either 1:1000 L-epinephrine or racemic epinephrine can be used. Some children will only transiently improve and then return to having stridor.
- Therefore, children who receive nebulized epinephrine should be held in the emergency department for at least 2 h to assess the lasting effectiveness of the treatment.
- Children who have hypoxia require a second epinephrine nebulized treatment; those in whom stridor persists despite nebulized epinephrine will need to be admitted.
- The relatively rare child with respiratory failure will require intubation and admission to a pediatric intensive care unit.

Bacterial Tracheitis

Bacterial tracheitis is a relatively rare diagnosis usually made at intubation. Bacterial tracheitis, also known as membranous tracheitis, initially presents as an aggressive type of "croup."

- The children are usually older infants or toddlers who typically have a toxic appearance, a barking cough that may be productive, stridor, and high fever. The trachea can become occluded with the copious exudate produced.
- This is a true airway emergency and these children frequently require rapid intubation for airway control.
- On intubation a subglottic membrane and airway narrowing is typically identified.
- The most common etiologic agent is *Staphylococcus aureus*. Antistaphylloccal, broad-spectrum antibiotics should be administered intravenously and the child should be admitted or transferred to the pediatric intensive care unit.

Bronchiolitis

Brochilitis is a wheezing syndrome of infants and young children. The etiology is commonly from the respiratory syncytial virus (RSV), but other viruses including influenza virus, parainfluenza virus, and adenovirus have been implicated.

- Clinical presentation usually begins with an upper respiratory tract infection with rhinorrhea, cough, and fever.
- In older children and adults, the clinical manifestations remain that of a "bad head cold."
- In infants and young children, bronchiolitis can progress to involve tachypnea, tachycardia, wheezing, increased work of breathing, retractions, and coughing. Younger infants may have difficulty with tolerating this degree of respiratory distress and

fatigue. Respiratory distress, respiratory failure, and respiratory arrest may ensue fairly rapidly in young infants.

- Children with underlying cardiac or pulmonary problems are at particularly high risk.
- Young infants may also have difficulty feeding with the increased work in breathing and may become dehydrated.
- The mainstay of treatment is the use of bronchodilators, particularly β-agonists such as nebulized albuterol.
- Clinically there appear to be two groups of children, those who respond promptly to nebulized albuterol and those who don't. A trial of nebulized albuteral is suggested in cases of clinically suspected bronchiolitis.
- There is some new work being done on the role of nebulized epinephrine in the treatment of bronchiolitis. The role of nebulized epinephrine has yet to be determined.
- Cases may range from very mild to severe requiring prompt intubation.
- There is currently no role for steroids or antiviral therapy in the emergency department.
- Children who are adequately hydrated, do not have an excessive work of breathing, are older that 3 mo of age, are oxygenating well on their own, are not markedly tachypneic, have no history of significant cardiac or pulmonary disease, are able to feed, and have responsible care providers may be discharged home with 12-24 h follow-up.

Asthma

Asthma is a chronic condition that involves bronchospasm, airway inflamation, increased mucous production, and airway edema. Acute exacerbations may occur and these events may lead to an emergency department visit.

- The primary agents in the treatment of acute exacerbations of asthma in children involve bronchodilators (e.g., albuterol), systemic steroids (e.g., prednisolone, prednisone, or methylprednisolone), and anticholinergic agents (e.g., ipratropium bromide).
- General treatment is similar to that of young adults. Recent work indicates that inhalers with spacers are as effective as nebulizers to deliver bronchodilators and anticholinergics.

Pneumonia

This section will deal with community-acquired acute pneumonia in previously healthy children with intact immune systems. For example, the technology-dependent child, the child with a congenital immunocompromised state, a transplant recipient, and children with sickle cell disease and acute chest syndrome will not be discussed here.

- The age of the child greatly influences the clinical presentation, the likelihood of various etiologic agents, and the management. One feature that is common to all children (especially those under 10 yr of age) is that sputum is rarely if ever produced as a useful specimen and is therefore not generally used to make management decisions.

Neonates

- Like most serious conditions in neonates, the presentation is usually nonspecific.
- The infant may appear to have lethargy, irritability, poor feeding, fever without an initially apparent source, hypothermia, vomiting, or poor muscle tone. In other words, pneumonia is one cause of the septic-appearing infant.
- Cough and abnormal breath sounds are not usually the dominant clinical picture.
- Neonates have immature immunologic systems and are prone to develop sepsis from a relatively localized infection. Because of this, neonates with pneumonia should undergo a full septic work up including blood, urine, and cerebrospinal fluid analysis and culture.
- Antibiotics should be initiated and the neonate should be admitted to the hospital.

- The most common organisms causing pneumonia in neonates are bacteria and include the same organisms responsible for most serious neonatal infections. These include group B streptococcus, Gram-negative enteric bacteria (*Esherichia coli* and Klebsiella), *Listeria monocytogenes*, and other Gram-positive cocci.
- Appropriate antibiotic choices include the combination of ampicillin and cefotaxime or the combination of ampicillin and gentamicin.

Infants

22

- There are two general forms of pneumonia in young infants, afebrile pneumonitis and bacterial pneumonia.
- Afebrile pneumonitis is classically caused by *Chlamydia trachomatis* acquired during delivery. The infant usually becomes symptomatic with nasal congestion followed by a prominent dry staccato cough between 4 and 12 wk of age. Typically these infants are well-appearing, active, feeding well, and are afebrile.
- There are more severe presentations, but this is relatively uncommon.
- Treatment is usually given on an outpatient basis and consists of oral erythromycin. Notably, these cases usually have a history of conjunctivitis that resolved prior to the onset of the respiratory symptoms.
- If the conjunctivitis is noted in early infancy, oral erythromycin should be administered instead of topical eye ointment or drops.
- It is thought that by giving oral erythromycin, the later complications of the *Chlamydia* infection, including afebrile pneumonitis, can be avoided.
- Bacterial pneumonia in young infants (1-3 mo of age) can result in serious clinical conditions including sepsis.
- The clinical picture does not necessarily include classic symptoms of lower respiratory tract infection. Cough and rales as seen in older children and adolescents are rarely seen in this age group.
- The organisms responsible for pneumonia in this age group include those seen in the neonatal period and *Streptococcus pneumoniae, Haemophilus influenzae*, and *Staphylococcus aureus*.
- In general, these young infants are admitted to the hospital and started on a second generation cephalosporin such as cefuroxime or a third generation cephalosporin such as cefotaxime.
- Additional coverage for *S. aureus* is gained by adding nafcillin or oxacillin.

Older Infants Toddlers and Young Children

- Pneumonia in older babies and young children typically presents with fever and cough. However, a gastrointestinal "picture" may predominate the initial clinical presentation with prominent vomiting and abdominal pain.
- A lower lobe pneumonia may even mimic an acute abdomen (leading to a young child getting an appendectomy instead of a chest X-ray and antibiotics).
- The dominant etiologic agent for pneumonia in this age group is viral. Common viruses include respiratory syncytial virus (RSV), parainfluenze, adenovirus, influenza, enterovirus, and rhinovirus.
- These children appear well, usually have rhinorrhea, may have some abnormal breath sounds but are not tachypneic, and have a "viral pattern" on chest X-ray (nonlobar increase in interstitial markings).
- In general these children require no antimicrobial treatment. Reasonable (24-48 h) follow-up is important as these viral pneumonias may act as a focus for the development of bacterial pneumonia.
- Bacterial pneumonia in young children has a range of severity from well-appearing and playful to ill and septic-appearing.

- The predominant etiologic bacterial organisms responsible for pneumonia in this age group include *S. pneumoniae* and *H. influenzae.*
- Appropriate outpatient antibiotics include amoxicillin, amoxicillin/clavulanate, or a macrolide (e.g., azithromycin).
- The ill appearing or hypoxic young child requires intravenous antibiotics and admission. The responsible etiologic agents include *S. pneumoniae, H. influenzae, S. aureus,* and rarely *Neisseria meningitidis.*
- Appropriate antibiotic selection includes a broad spectrum cephalosporin such as cefuroxime.

School Aged Children and Adolescents

- These older children typically present much more like adults with chest discomfort, cough, and fever.
- The most common etiologic agents are the atypical organisms such as mycoplasma.
- Appropriate antibiotic treatment can be achieved using a macrolide such as azithromycin.

Suggested Reading

1. Ansley JF, Cunningham MJ. Treatment of aural foreign bodies in children. Pediatrics 1998; 101(4):638-641.
2. Kadish HA, Corneli HM. Removal of nasal foreign bodies in the pediatric population. Am J Emerg Med 1997; 15:54-56.
3. Backlin SA. Positive-pressure technique for nasal foreign body removal in children. Ann Emerg Med 1995; 25:554-555.
4. Lichenstein R, Giudice EL. Nasal wash technique for nasal foreign body removal. Pediatr Emerg Care 2000; 16(1):59-60.
5. Finkelstein JA. Oral ambu-bag insufflation to remove unilateral nasal foreign bodies. Am J Emerg Med 1996; 14(1):57-58.
6. Cataneo AJM, Reibscheid SM, Ruiz RL et al. Foreign body in the tracheobronchial tree. Clin Pediatr 1997; 36:701-706.
7. Baharloo F, Veyckemans F, Francis C et al. Tracheobronchial foreign bodies: Presentation and management in children and adults. Chest 1999; 115:1357-1362.
8. Quinones FR, Saez MM, Serrano EMP et al. Magil forceps: A vital forceps. Pediatr Emerg Care 1995; 11(5):302-303.
9. Rimell FL, Thome A, Stool S et al. Characteristics of objects that cause choking in children. JAMA 1995; 274:1763-1766.
10. Imaizumi H, Kaneko M, Nara S et al. Definitive diagnosis and location of peanuts in the airways using magnetic resonance imaging techniques. Ann Emerg Med 1994; 23:1379-1382.
11. Conners GP, Chamberlain JM, Weiner PR. Pediatric coin ingestion: A home-based survey. Am J Emerg Med 1995; 13:638-640.
12. Cheng W, Tam PKH. Foreign-body ingestion in children: Experience with 1,265 cases. J Pediatr Surg 1999; 34:1472-1476.
13. Conners GP, Chamberlain JM, Ochsenschlager DW. Symptoms and spontaneous passage of esophageal coins. Arch Pediatr Adolesc Med 1995; 149:36-39.
14. Smith SA, Conners GP. Unexpected second foreign bodies in pediatric esophageal coin ingestions. Pediatr Emerg Care 1998; 14(4):261-262.
15. Soprano JV, Fleisher GR, Mandl KD. The spontaneous passage of esophageal coins in children. Arch Pediatr Adolesc Med 1999; 153:1073-1076.
16. Seikel K, Primm PA, Elizondo BJ et al. Handheld metal detector localization of ingested metallic foreign bodies. Arch Pediatr Adolesc Med 1999; 153:853-857.
17. Emslander HC, Bonadio W, Klatzo M. Efficacy of esophageal bougienage by emergency physicians in pediatric coin ingestion. Ann Emerg Med 1996; 27:726-729.

22

18. Potsic WP, handler SD. Otolaryngologic emergencies. In: Fleisher GR, Ludwig S, eds. Textbook of Pediatric Emergency Medicine. 4th ed. Philadelphia: Lippincott Williams & Wilkins, 2000:1569-1583.
19. Dowell SF, Butler JC, Giebink GS et al. Acute otitis media: Management and surveillance in an era of pneumococcal resistance-A report from the Drug-resistant *Streptococcus pneumoniae* Therapeutic Working Group. Pediatr Infect Dis J 1999; 18:1-9.
20. Bosker G. Acute otitis media (AOM) Year 2000 Update: A rational and evidence-based analysis of current controversies in antibiotic therapy and drug selection for AOM. Part II: Evidence-based trials, CDC recommendations, divergent opinion, and the PPPD approach. Emergency Medicine Reports 2000; 21(8):85-94.
21. Bitar CN, Kluka EA, Steele RW. Mastoiditis in children. Clinical Pediatrics 1996; 35(8):391-395.
22. Wang NE, Burg JM. Mastoiditis: A case-based review. Pediatr Emerg Care 1998; 14(4):290-292.
23. Harley EH, Sdralis T, Berkowitz RG. Acute mastoiditis in children: A 12-yr retrospective study. Otolaryngol Head Neck Surg 1997; 116:26-30.
24. Khafif A, Halperin D, Hochman I et al. Acute mastoiditis: A 10-yr Review. Amer J Otolaryngology 1998; 19(3):170-173.
25. Ueda D, Yoto Y. The 10-day mark as a practical diagnostic approach for acute paranasal sinusitis in children. Pediatr Infect Dis J 1996; 15:576-579.
26. Nelson LP, Needleman HL, Padwa BL. Dental trauma. In: Fleisher GR, Ludwig S, eds. Textbook of Pediatric Emergency Medicine. 4th ed. Philadelphia: Lippincott Williams and Wilkins, 2000:1417-1426.
27. Woods WA, Carter CT, Schlager TA. Detection of group A streptococci in children under 3 yr of age with pharyngitis. Pediatr Emerg Care 1999; 15(5):338-40.
28. Tsevat J, Kotagal UR. Management of sore throats in children: a cost-effectiveness analysis. Arch Pediatr Adolesc Med 1999; 153(7):681-688.
29. Nussinovitch M, Finkelstein Y, Amir J et al. Group A β-hemolytic streptococcal pharyngitis in preschool children aged 3 mo to 5 yr. Clinc Pediatr (Phila) 1999; 38(6):357-360.
30. Feder Jr HM, Gerber MA, Randolph MF et al. Once-daily therapy for streptococcal pharyngitis with amoxicillin. Pediatrics 1999; 103(1):47-51.
31. Schwartz RH, Gerber MA, McKay K. Pharyngeal findings of group A streptococcal pharyngitis. Arch Pediatr Adolesc Med 1998; 152(9):927-928.
32. Tarlow MJ. Macrolides in the management of streptococcal pharyngitis/tonsillitis. Pediatr Infect Dis J 1997; 16(4):444-448.
33. Super DM, Cartelli NA, Brooks LJ et al. A prospective randomized double-blind study to evaluate the effect of dexamethasone in acute laryngotracheitis. J Pediatr 1989; 115(2):323-329.
34. Cruz MN, Stewart G, Rosenberg N. Use of dexamethasone in the outpatient management of acute laryngotracheitis. Pediatrics 1995; 96(2 Pt 1):220-223.
35. Klassen TP. Croup. A current perspective. Pediatr Clin North Am 1999; 46(6):1167-1178.
36. Rizos JD, DiGravio BE, Sehl MJ et al. The disposition of children with croup treated with racemic epinephrine and dexamethasone in the emergency department. J Emerg Med 1998; 16(4):535-539.
37. Kunkel NC, Baker MD. Use of racemic epinephrine, dexamethasone, and mist in the outpatient management of croup. Pediatr Emerg Care 1996; 12(3):156-159.
38. Donnelly BW, McMillan JA, Weiner LB. Bacterial tracheitis: report of eight new cases and review. Rev Infect Dis 1990; 12(5):729-735.
39. Leversha AM, Campanella SG, Aickin RP et al. Costs and effectiveness of spacer versus nebulizer in young children with moderate and severe acute asthma. J Pediatr 2000; 136(4):497-502.
40. Mace SE. Diagnosis and Management of Pneumonia in Infants and Children. Pediatric Emergency Medicine Reports 1998; 3(7):51-64.
41. Strange GR, ed. APLS: The Pediatric Emergency Medicine Course (Textbook). 3rd ed. Dallas, TX: American College of Emergency Physicians, 1998.

Part G: Sickle Cell Disease in Children

Olly Duckett

Definitions

Sickle cell anemia is one of the most common genetic conditions in childhood.

- Complications from sickle cell disease commonly prompt visits to the emergency department.
- Any physician practicing in an urban emergency department should be prepared to diagnose and treat the numerous complications of this disease.
- With aggressive management and newer therapies, the life expectancy of sickle cell patients has increased greatly, and the early mortality has been diminished.

Homozygous disease with hemoglobin SS is an autosomal recessive inherited disorder of hemoglobin structure.

- The exact point mutation is a β-globulin chain amino acid substitution (valine for glutamic acid at the 6th position on the β-globulin gene).
- This substitution leads to the formation of hemoglobin S (HgS) that polymerizes when deoxygenated causing sickling of red blood cells (RBCs).
- The sickling of RBCs leads to intermittent episodes of vascular occlusion causing tissue ischemia and acute and chronic organ dysfunction.
- Vaso-occlusion can result in injury to virtually every organ in the body; furthermore, sickle cell patients are at increased risk for aplastic crises and gallstones because of the hemolysis and shortened RBC lifespan.
- Injury to spleen, lungs and brain are responsible for much of the morbidity and mortality of childhood sickle cell disease.

Homozygous Hg SS is the most common sickling disorder in the US (60-70% of sickle cell disease).

- Other disorders are due to coinheritance of another abnormal hemoglobin gene along with sickle hemoglobin.
- The most common coinherited disorders are sickle β-thalassemia, SC disease and sickle cell trait.

Clinical Presentations and Their ED Management

Sickle cell disease most commonly presents in people of African, Mediterranean, Middle-Eastern and Indian ancestry.

- Today the majority of cases are picked up in infancy by newborn screening test.
- However, sickle cell can present in infancy, most commonly with dactylitis (hand-foot syndrome).
- A child may present with recurrent severe infections and possibly simply an unexplained microcytic anemia with sickle cells seen on blood smear.
- Hemoglobin electrophoresis will make the definitive diagnosis.
- More commonly, the emergency physician will have to deal with the complications of sickle cell disease, for the complications are the reason sickle cell patients seek emergency care.

Infection and Fever

Because of the tortuous microcirculation of the spleen, it is an organ particularly susceptible to damage in sickle cell anemia. Hence, the spleen is susceptible to autoinfarction, making the child with sickle cell functionally asplenic and susceptible to severe infections.

- Splenic injury may begin as early as 3-6 mo of life.

22

22

- The child with sickle cell anemia is at increased risk for life-threatening infection, especially encapsulated organisms such as *Streptococcus pneumoniae, Haemophilus influenzae* type b, Salmonella, and Klebsiella.
- Historically, 10% of children with sickle cell disease died from bacterial infections before 3 yr of age.
- Children are most susceptible to fulminant septicemia by encapsulated organisms during the early years of life (<5 yr of age).
- All children with Hg SS and sickle β-thalessemia presenting to the emergency department with a febrile illness (Temp >101) should be managed as follows:
  1. Rapid triage
  2. Brief history, physical examination
  3. Laboratory: Stat CBC, reticulocyte count, blood culture
  4. Consider type and crossmatch (extreme pallor, respiratory or neurologic symptoms, or acute splenic enlargement)
  5. Consider urine, CSF, other cultures
  6. Consider CXR (high fever, respiratory symptoms, toxic appearance)
  7. Exclude coexisting complications (splenic sequestration, aplastic crisis, acute chest)
  8. Prompt administration of IV or IM ceftriaxone or cefotaxime, consider vancomycin for severe illness, especially in areas of resistant pneumococci.
  9. Observation
- The presence of a "focus" of infection does not change the importance of full work-up and parenteral antibiotics.
- With administration of ceftriaxone, a well-appearing child with a normal work-up may be safely managed as an outpatient, provided reliable follow-up is available within 24 h.
- Admission is recommended in children <1 yr of age.
- Children with sickle cell disease should be on prophylactic penicillin started by their primary care physicians.
- The risk of bacterial infection my be reduced by as much as 85% with this intervention.
- Children with sickle cell disease should also receive pneumovax (the immunization against strains of *S. pneumoniae*).
- Children not followed at a center that specializes in the care of children with sickle cell disease should be referred by the emergency physician to ensure proper preventative and maintenance care.

Splenic Sequestration Crisis

Another cause of morbidity and mortality in children with sickle cell disease is splenic sequestration.

- Splenic sequestration is characterized by rapid enlargement of the spleen causing a fall in the Hg level >2 g/dl.
- Platelets may also be sequestered causing a thrombocytopenia.
- The child usually presents with pallor, lethargy, increased thirst, tachycardia, left upper quadrant pain and a palpable spleen.
- An episode may progress to severe anemia and circulatory collapse within a few hours.
- Splenic sequestration occurs in 10-30% of children with Hg SS, usually between the ages of 6 mos to 3 yr, frequently in patients with previously palpable spleens.
- Splenic sequestration may also occur in older children and adolescents.
- Precipitating factors include viral illnesses.
- Up to 50% of children who have one episode of splenic sequestration will have another episode.

The treatment of splenic sequestration crisis is transfusion of packed red blood cells.

- In rapidly progressive cases exchange transfusion may be required.

- Exchange transfusion should be done in consultation with a pediatric hematologist.
- On presentation to the emergency department the patient should have two relatively large bore IV's placed, type and crossmatch of packed RBCs, a crystalloid fluid bolus and oxygen.
- Transfusion should be initiated promptly. Surgical splenectomy following the first episode of splenic sequestration remains a controversy.

Acute Chest Syndrome

Approximately 50% of pediatric patients with Hg SS will experience acute chest syndrome, and it is the second leading cause for hospitalization.

- Acute chest syndrome refers to any acute illness with a new pulmonary infiltrate on chest X-ray; the hallmarks of the disease are fever, cough, tachypnea, chest pain, and hypoxemia.
- The etiology is unclear, but believed to be infection secondary to pulmonary infarction.
- Acute chest is a common postoperative complication of general anesthesia, possibly due to excessive hydration.
- Acute chest syndrome may be life-threatening.

The child with acute chest syndrome should be managed in a monitored setting.

- Therapy includes oxygen to avoid hypoxemia, judicious use of fluids, and incentive spirometry to avoid atelectasis.
- The patient should be started on a broad spectrum antibiotic.
- Packed red blood cell transfusions should be used in moderately severe crisis.
- Exchange transfusions are reserved for those patients refractory to other treatment and should be conducted in consultation with a pediatric hematologist.
- Proper management of acute chest syndrome is important as many patients with repeated episodes of acute chest syndrome go on to develop chronic lung complications such as restrictive lung disease, pulmonary hypertension, and cor pulmonale.

Pain Crisis

Acute, painful vaso-occlusive crisis is the most common reason for the presentation of a child with Hg SS to the emergency department.

- The pain crisis can be severe and unpredictable.
- Pain crisis results from sickling of cells, occlusion of blood flow in the microcirculation, and tissue ischemia.
- Microinfarction also occurs.
- Pain may be widespread and migratory, often presenting as bone pain.
- Commonly affected areas include the chest, low back, femoral shaft, hips, ribs and knees.
- Adolescents frequently complain of chest, back or abdominal pain.
- Priapism occurs and is one of the most difficult complications to treat.

Opioid analgesia is the primary treatment of sickle cell pain crisis.

- Often large doses of opioids are required.
- Acute pain crisis is a medical emergency that requires potent opiate analgesia and anti-inflammatory medication.
- The physician should avoid undertreatment by adjusting the dose of analgesic based on the patient's reports of pain.
- Intravenous doses of morphine at intervals of 1-2 h should be given along with non-steroidal anti-inflammatory medication (ibuprofen or ketorolac).
- This author recommends consideration of admission after management in the emergency department with q 1 h morphine if 0.1 mg/kg does not control pain.

- Patient controlled analgesia (PCA) devices optimally manage pain and, if available, should be started in the emergency department once the decision is made to admit.
- The physician should start morphine, 20 μg/kg/bolus, at rate of 5 boluses/h with 6 min intervals.
- Hydromorphone may be substituted for morphine; the bolus dose is 3-4 μg/kg.
- Continuous infusion is also recommended (morphine, 20-30 μg/kg/h, or hydromorphone, 3-4 μg/kg/h).

The emergency physician should obtain a complete blood count (CBC) and reticulocyte count as well as beginning intravenous hydration and oxygen therapy. Intravenous hydration should begin with D5 two normal saline solution at 1-2 times the maintenance requirement. The clinician must be careful to recognize signs of infection as septic arthritis and osteomyelitis may be misdiagnosed as a pain crisis. Furthermore, acute chest may mimic pain crisis in adolescents who often have pain crisis involving the chest. Management of pain crisis requires a high degree of vigilance to rule out other complications of sickle cell disease.

The patient with priapism should be emergently managed similar to the patient in pain crises.

- If the erection persists, the patient should be hospitalized.
- Intravenous hydration should be initiated.
- Transfusion of red cells is recommended if the swelling does not decrease.
- Aspiration of the corpora should be considered in refractory cases of priapism.
- Institute exchange transfusion to reduce HgS to <30% of total hemoglobin if priapism is refractory to simple transfusion.

Stroke

Approximately 5-15% of patients with sickle cell disease will develop a stroke.

- The etiology of stroke in sickle cell patients remains somewhat unclear.
- Cerebral angiography show that some children with sickle cell disease have lesions in intracerebral arteries similar to lesions seen in adults.
- Pathologly studies demonstrate that these lesions are due to fibrous hyperplasia of the intima with superimposed thrombosis.
- The distal internal carotid, middle and anterior cerebral arteries are most commonly involved.
- Presenting symptoms include hemiparesis, aphasia or dysphasia, and seizures. Children may present with coma or change of mental status. Others may experience TIAs.
- Stroke has been described following severe episodes of priapism.
- Management begins with computed tomography in patients suspected of having a stroke.
- The child should be hospitalized and neurologic status monitored.
- Many authors recommend immediate exchange transfusion to lower sickle hemoglobin to <30%.
- Blood transfusion and aggressive management lead to complete or nearly complete recovery in approximately half of these patients.
- Patients with one episode of stroke have a 60-90% rate of recurrence within 3 yr.
- Chronic transfusion therapy reduces that risk to 10%.

Suggested Reading

1. Lane PA. Sickle Cell disease. Pediatr Clin North Am 1996; 43(3):639-664.
2. Duncan GW. Ischemic stroke in sickle cell disease: A review. Tenn Med 1997; 90(12):498-499.
3. Embury SH, Hebbel RP et al. Sickle cell disease: Basic Principles and Clinical Practice. New York, Raven Press, 1994.

4. Castro O. Management of sickle cell disease: Recent advances and controversies. Br J Haematol 1999; 107(1):2-11.
5. Vichinsky EP, Styles LA, Colangelo LH et al. Acute chest syndrome in sickle cell disease: clinical presenation and course. Cooperative Study of Sickle Cell Disease. Blood 1997; 89(5):1787-92.
6. Castro O, Brambilla DJ, Thorington B et al. The acute chest syndrome in sickle cell disease: Incidence and risk factors. The cooperative study of sickle cell disease. Blood 1994; 84(2):643-649.
7. McHugh D. Enhancement of pain control with ketorolac tromethamine in patients with sickle cell vaso-occlusive crisis. Am J Hematol 1994; 46(1):43-47.
8. Poncz M, Kane E, Gill FM. Acute chest syndrome in sickle cell disease: etiology and clinical correlates. J Pediatr 1985; 107(6):861-866.
9. Poncz M, Greenberg J, Gill FM et al. Hematologic changes during acute chest syndrome in sickle cell disease. Am J Pediatr Hematol Oncol 1985; 7(1):96-99.
10. Emond AM, Collis R, Darvill D et al. Acute splenic sequestration in homozygous sickle cell disease: natural history and management. J Pediatr 1985; 107(2):201-206.
11. Yaster M, Kost-Byerly S, Maxwell LG. The management of pain in sickle cell disease. Pediatr Clin North Am 2000; 47(3):699-710.

Part H: Neurologic Problems in Children

Marilyn Hicks

Status Epilepticus

Definitions

- The traditional definition of SE is a single seizure lasting at least 30 min or recurrent seizures lasting more than 30 min during which the patient does not regain consciousness.
- A more useful definition is that of continuous or intermittent seizures lasting 10 min without recovery.
- Status epilepticus may be either convulsive or nonconvulsive; in children more than 90% are of the convulsive type.
- Common causes of acute symptomatic SE in children include acute withdrawal of medications, infection, anoxia, and vascular events.
- Causes of remote symptomatic SE in children include brain malfunctions, cerebral palsy, and prior head trauma, or meningitis.
- The adverse systemic consequences of SE may be metabolic (hypoxemia, acidemia, hypoglycemia, and lactic acidemia), cardiorespiratory (tachycardia, hypertension, dysrhythmias, airway obstruction, and hypercarbia), or neurologic (increased intracranial pressure, uncontrolled increases in cerebral blood flow, and neuronal death).

Incidence

- Status epilepticus (SE) affects 25,000-50,000 children yearly in the United States. Between 10-20% of patients with seizure disorders have at least one episode of SE.
- The mortality rate for children as a result of SE in currently about 5%.

Clinical Presentation

- The most common presentation of SE is that of generalized tonic clonic convulsions.
- Clinical features may vary with different age groups. Neonates and young infants may present with subtle manifestations such as chewing, bicycling movements, and eye deviation or more dramatically with episodes of apnea. The child may also have signs

and symptoms of underlying disease such as petechiae or neurocutaneous stigmata (i.e., café au lait spots, ash leaf spots).

- The emergent nature of status epilepticus is such that the evaluation and treatment must occur simultaneously.
- An initial rapid assessment should be done to determine airway adequacy, hemodynamic stability, and to look for signs of trauma.
- A rapid determination of vital signs including temperature should occur (25% of SE is caused by febrile seizures).
- A bedside glucose should be done; hypoglycemia may be the inciting problem or a secondary manifestation of SE.
- Blood for laboratory testing should be obtained with the establishment of IV access and sent for measurement of anticonvulsant drug levels. Other laboratory tests should be ordered based on patient history and physical findings.
- Cranial computed tomography may be indicated after cardiopulmonary stabilization and termination of seizure, especially if there are signs of significant trauma, focal neurologic deficits or increased intracranial pressure.

ED Management

- The goals of treatment of SE are termination of seizure activity, hemodynamic stabilization and prevention of systemic complications.
  1. O_2 should be administered via nasal canula or face mask after quick assessment of airway adequacy. In the presence of apnea or significant hypoxia nonresponsive to bag mask ventilation, rapid sequence intubation should occur.
  2. A short-acting paralytic (e.g., succinylcholine 1.5-2.0 mg/kg) should be used. Longer-acting paralytics should be avoided.
  3. Treatment is directed toward stopping seizure activity that is gauged by motor manifestations. Prolonged paralysis will mask underlying brain seizure activity by artificially stopping the motor manifestations of the seizure.
  4. Early intubation should also be considered for SE secondary to head trauma or if there is evidence of increased intracranial pressure.
  5. Medical therapy of SE (see Table 22H.1) should begin with a benzodiazepine. The three most commonly used drugs are diazepam, lorazepam, and midazolam. Lorazepam is the drug of choice due to longer anticonvulsant half-life and less respiratory depressant effect than diazepam. Lorazepam may be repeated in 3-5 min for a total of 3 doses for continuing seizure activity.
  6. The risk of respiratory depression increases with each dose necessitating the need to assist ventilations.
  7. If unable to establish IV access within 5-10 min, rectal medications (see Table 22H.1) or intramuscular (IM) midazolam may be used. Establishment of an intraosseous line may be necessary if ongoing efforts to obtain IV access fail.
  8. When first line drugs fail or when a very short acting agent (midazolam) is used, a second line drug should be administered. Phenytoin and phenobarbital are the most commonly used second line agents. Phenytoin is highly efficacious and has been widely used in children. However, phenytoin contains 40% propylene glycol and has a pH of 12. The high alkalinity prevents rectal or intramuscular use and due to the propylene glycol rapid IV administration (>1 mg/kg/min) can cause hypotension, dysrhythmias, and asystole. Also, propylene glycol is highly toxic to tissues. Extravasation of phenytoin into soft tissue causes severe tissue necrosis. Fosphenytoin is a prodrug that is converted to phenytoin after administration. It is water soluble which gives it many advantages over other second line drugs. It is not toxic to tissues and does not cause hypotension or arrhythmias and can be given IM if venous access is difficult.

Table 22H.1. Drug treatment of status epilepticus

Drug	Dose	
Diazepam	0.1-0.5 mg/kg IV	
	0.22.0 mg/kg PR	max dose 10 mg
Lorazepam	0.05-0.2 mg/kg IV	
	0.1-0.4 mg/kg PR	
Midazolam	0.1-0.3 mg/kg IV	
	0.15-0.2 mg/kg IM	
Phenytoin	15-20 mg/kg IV	at a rate of 25-50 mg/min
Fosphenytoin	15-20 mg/kg IV, IM	at a rate of 100-150 mg/min
Phenobarbital	15-20 mg/kg IV	at a rate of 30-100 mg/min
	Additional bolus of 10 mg/kg as needed to total dose of 40 mg/kg	

9. If seizure activity continues after the administration of therapeutic doses of phenytoin or fosphenytoin or if the patient has hypersensitivity to phenytoin, phenobarbital should be given. However due to the high level of sedation and long half-life of phenobarbital, intubation is often required during or shortly after its administration for respiratory depression or apnea. Phenobarbital may also cause hypotension and therefore blood pressure should be continuously monitored during and after administration.
10. Refractory SE occurs when seizures persist beyond aggressive management with first and second line drugs. Continuous infusion of anticonvulsant agents should begin with midazolam, pentobarbital, or propofol (see Table 22H.2). All of these agents produce respiratory depression and hypotension. Children undergoing continuous infusion anticonvulsant therapy with these drugs should be intubated and mechanically ventilated with continuous EEG monitoring in an intensive care unit setting.

- The goal of medical therapy should be termination of seizure activity within 30 min of onset. Experimental data suggests occurrence of detrimental cerebral changes at 30 min of continuous seizures and definitely at 1 h.
- A helpful timeline for management of SE is as follows:
  0-10 min
    ABCs, O_2
    IV/IO
    Bedside glucose check
    CR monitoring
    Labs
  10-15 min
    Lorazepam/Diazepam
    Rectal drugs
    Phenytoin/Fosphenytoin

Table 22H.2. Drug treatment of refractory status epilepticus

Drug	Dose
Midazolam	0.15 mg/kg bolus, then 0.05-0.4 mg/kg/h
Pentobarbital	5-15 mg/kg bolus, then 0.5-5.0 mg/kg/h
Propofol	3-6 mg/kg/h

15-30 min
  Repeat phenytoin/fosphenytoin mini doses
  Phenobarbital
  Evaluate for intubation
30-45 min
  Intubate and ventilate
  Midazolam/pentobarbital/propofol infusion
  Admit PICU
22 EEG monitoring

CSF Shunt Complications

Definitions

- The placement of cerebrospinal fluid (CSF) shunts is the most commonly performed neurosurgical procedure in children.
- CSF shunts have a high complication rate including shunt malfunction, obstruction, and infection. Failure to recognize shunt complications can result in irreversible neurologic damage or death.
- Hydrocephalus is the primary indication for placement of a CSF shunt and is usually caused by an obstruction of CSF flow between the point of production within the ventricles and its point of absorption in the arachnoid villi.
- The CSF shunt system usually consists of three parts: a radiopaque ventricular catheter, a one-way valve, and a distal portion that is also radiopaque. The distal portion most often drains into the peritoneal cavity, but occasionally drains into the right atrium, pleural space, lumbar peritoneal space, or the ureter.
- The most common complication of CSF shunts is mechanical malfunction, which can cause obstruction of flow or excessive outflow of CSF. Obstruction may occur at either the proximal or distal portion of the shunt. Proximal obstruction results from tissue debris, choroid plexus in the shunt, fibrosis, infection, or catheter migration. Distal obstruction occurs as a consequence of kinking, clogging, envelopment with omentum, infection, migration of the catheter, or thrombosis.
- Chronic overdrainage of CSF can result in the slit ventricle syndrome. Symptoms of intermittent increased intracranial pressure (ICP) occur as a result of intermittent catheter occlusion by the collapsing ventricles. Minor consequences of overdrainage include small hygromas and recurrent low-pressure headaches.
- Shunt infections may involve the shunt equipment, the scalp, the CSF, or the distal site into which the shunt drains. Most shunt infections present within the first 2 mo following placement. Organisms most frequently involved are staphylococci, with approximately 40% caused by *Staphylococcus epidermidis* and 20% by *Staphylococcus aureus*. Other organisms causing shunt infections are streptococci, enterococci, Gram-negative organisms, and fungi. Less common complications of ventricular peritoneal shunts are spontaneous perforation of viscera, abdominal pseudocysts, and peritonitis.
- Complications of ventriculoatrial shunts are more severe because of involvement of the cardiovascular system. They may include arrhythmias, cardiac perforation and tamponade, pulmonary embolism, and sepsis or endocarditis from transient bacteremia.

Clinical Presentation

- Clinical presentation of shunt obstruction is largely related to increased intracranial pressure (see Table 22H.3). Symptoms usually develop over several days, but may develop rapidly resulting in death within 24 h.

Table 22H.3. Clinical symptoms of shunt malfunction

Fever	Gait disturbance
Nausea, vomiting	Autonomic instability
Headache	Seizures
Irritability	Altered mental status
Lethargy	Coma

Table 22H.4. Pathogens causing CSF shunt infections

Common	Uncommon
Methicillin-susceptible *Staphylococcus aureus*	*Haemophilus influenzae* type B
Methicillin-resistant *Staphylococcus aureus*	*Enterococcus faecalis*
Coagulase-negative Staphylococcus	Klebsiella
Haemophilus influenzae non-typable	*Streptococcus pneumoniae*

- Infections account for a major cause of morbidity and mortality in children with CSF shunts. Clinical signs are dependent upon the site of the infection (i.e., cellulitis, meningitis, peritonitis) and the infecting organism (see Table 22H.4).
- Often the symptoms of shunt infection are subtle and nonspecific such as nausea, vomiting, lethargy or irritability, and feeding problems.
- Not all children will have fever and those infected with *S. epidermidis* may be rather well appearing. Only about one-third of patients with shunt infection will have meningial signs.

ED Management

- The evaluation of the child with a CSF shunt should always begin with a thorough history and physical examination. Other causes of illness should be sought or excluded; however a high index of suspicion regarding shunt malfunction should be maintained.
- An important part of the exam is to visualize and palpate the shunt valve and catheter. Any evidence of cellulitis overlying the course of the palpable catheter is an indication of shunt infection. The valve patency should be tested; if easily depressed there is usually distal patency, and a rapid refill of the chamber implies proximal patency. However, an apparently functioning valve does not exclude shunt malfunction.
- Imaging studies should include a shunt series of X-rays. This includes a lateral X-ray of the skull and a X-ray of the neck, thorax, and abdomen; the purpose is to visualize the location and connections of the shunt apparatus.
- A head CT should be done, but a negative CT does not rule out shunt obstruction. As many as one-third of patients with shunt malfunction will not have the diagnosis supported by CT scan. It is difficult to interpret CT findings without previous studies with which to compare.
- The evaluation of a patient with a suspected shunt infection should include a complete blood count, blood culture, and examination of the cerebrospinal fluid (CSF). CSF obtained by lumbar puncture is rarely positive.
- The best samples are obtained by percutaneous aspiration of the shunt reservoir. Potential complications of shunt tapping include bleeding or CSF leak from the puncture site, introduction of skin organisms, and mechanical disruption of the system. Shunt taps should only be done in consultation with a neurosurgeon.
- Consultation with a neurosurgeon should occur early in the evaluation and management of the child with suspected shunt problems. Many children will require transfer to a tertiary center for definitive management.

22

- The child who presents in extremis or with a rapidly deteriorating course should undergo neuroprotective rapid sequence intubation with subsequent sedation, paralyzation, and mechanical ventilation with hyperventilation to maintain pCO_2 between 30 and 35 mm Hg.
- Medical therapy for increased ICP (isosorbide, glycerol, or mannitol) should be used with caution. Children with increased ICP and vomiting tend to be dehydrated and use of osmotic agents will worsen the dehydration.
- Diuretics (furosemide or acetazolamide) decrease CSF production but also may exacerbate dehydration. Also acetazolamide may cause hypochloremic metabolic acidosis, diarrhea, and tachypnea. Decisions regarding medical therapy of increased ICP should be made in consultation with a neurosurgeon.
- Empiric antibiotic therapy for presumed shunt infections include vancomycin for *Staphylococcus* coverage and a third generation cephalosporin or aminoglycoside for Gram-negative coverage. Intraventricular antibiotics are often necessary and nearly all patients require shunt removal.

Disposition

- A high index of suspicion should exist for complications involving CSF shunts, even in children with mild symptoms.
- Neurosurgical consultation should be sought early in the evaluation of patients with CSF shunts, even if initial studies are negative or inconclusive.
- All children with shunt infections or shunt malfunction should be admitted to the hospital with the decision to admit the patient to the intensive care unit based on the severity of the patient's symptoms.

Apparently Minor Head Injury

- Minor head injury in children accounts for approximately 600,000 ED visits per year.
- The management of head injury remains controversial and clear guidelines do not exist. However, recent literature suggests parameters that may be used in management.
- Mechanisms of injury vary among age groups.
  1. Older children most often incur head injury in pedestrian or bicycle related accidents, falls, or assaults.
  2. Motor vehicle crashes (MVC) are the most common cause of head injury in adolescents, followed by sports and recreational accidents.
  3. Infants and young children most often incur head injury as a result of falls or nonaccidental trauma.
  4. The skull of an infant is thinner than the skull of an older child and more prone to fracture as a result of minor trauma. The skull is also more pliable and provides less protection for the brain and intracranial vessels, increasing the likelihood of intracranial injury (ICI) due to minor trauma.
  5. Nonaccidental trauma accounts for as much as one-half of intracranial injury in infants <1 yr of age and as much as one-third in infants <3 yr of age.
  6. In older children abuse is a much less likely mechanism for head injury.

Clinical Presentation

- The history should begin with an attempt to ascertain the severity of the mechanism of injury, i.e., the height of the fall or the velocity of the MVC.
- Important historical questions should include: Whether there was a loss of consciousness (LOC) and if so, for how long; is the child amnestic to the event; were there mental status changes at the time of the injury or subsequently; was there a post-impact seizure; has there been persistent headache or vomiting?

Table 22H.5. Glascow coma scale

Eye Opening	Verbal Activity	Motor Activity
4. Spontaneous	5. Oriented	6. Obeys commands
3. To command	4. Confused	5. Localize pain
2. To pain	3. Inappropriate	4. Withdraws to pain
1. None	2. Incomprehensible	3. Flexion to pain
	1. None	2. Extension to pain
		1. None

- Additional questions relative to young infants should include changes in feeding and sleep-wake cycles or subtle behavioral changes.
- Important physical findings include focal neurologic deficits, signs of basilar skull fracture such as hemotympanum, "raccoon" eyes, Battle's sign, CSF otorrhea or CSF rhinorrhea. The head and neck should be carefully examined for evidence of trauma such as hematoma or depressed skull fracture.
- Even subtle signs of trauma such as minor contusions are important in the evaluation of the infant. Recent studies have shown a significant correlation between minor scalp trauma and significant intracranial injury in infants.
- The level of alertness of the child should be assessed initially and monitored throughout the evaluation. The Glasgow coma scale (GCS) is often used for this purpose (see Table 22H.5).
- The child coma scale (CCS) is a modified version of the GCS which uses a different assessment of the best verbal response for preverbal children (see Table 22H.6).
- Also when assessing alertness in infants, developmental capabilities must be taken into account as well as environmental stressors such as fear and anxiety.

Diagnostic Evaluation

- Radiological evaluation of the older child with head injury should always be computed tomography. Skull X-rays have no place in the evaluation of children >2 yr of age. Indications for head CT in children with minor head injury are as follows:
  1. Suspicion of child abuse
  2. Mechanism of injury consistent with significant trauma
  3. Evidence of depressed skull fracture or basilar skull fracture
  4. Altered mental status (GSC <15), post-traumatic seizure, focal neuorologic deficit, or loss of consciousness >5 min.
- A head CT should be performed when the child has severe or persistent headache, repetitive vomiting, amnesia of the event, or even a brief period of loss of consciousness.
- Virtually all infants require sedation for head CT. Conscious sedation is a procedure not without risk. In those infants (<2 yr of age) who do not exhibit characteristics or

Table 22H.6. Children's coma scale

Eye Opening	Best Verbal Response		Best Motor Response
4. Spontaneous	5. Smiles, oriented to sound Follows objects, interacts		6. Spontaneous
3. Reacts to speech	CRYING	INTERACTS	5. Localizes pain
2. Reacts to pain	4. Consolable	Inappropriate	4. Withdraws to pain
1. None	3. Inconsistently consolable	Moaning	3. Abnormal flexion to pain
	2. Inconsolable	Irritable, restless	2. Abnormal extension to pain
	1. None	None	1. None

22

findings indicative for CT, skull X-rays may be a reasonable screening procedure. There is a significant correlation between skull fractures and intracranial injury (ICI). As many as 40% of infants with skull fractures have ICI. Recent studies have shown the sensitivity of the presence of skull fracture for identifying ICI to be high.

- An infant with an intracranial injury would be unlikely to have normal skull films. Also, recent studies suggest the sensitivity of minor scalp injury to the presence of skull fracture to be significant. A reasonable approach to the infant with minor mechanism of injury who appears well and has only minor scalp injury (contusion or edema) may initially be skull X-rays followed by head CT for those who have fractures.

ED Management and Disposition

- Neurosurgical consultation should be sought early in the evaluation of the child with altered mental status or focal neurologic deficit.
- Children with positive findings of ICI on head CT should be admitted with neurosurgical consultation. Children with GCS <15, persistent vomiting, or severe headache need admission for continued monitoring, rehydration, and pain control. If symptoms persist, head CT should be repeated in 24-36 h.
- Infants with isolated linear skull fracture and negative head CT with a CCS of 15 may be discharged home after consultation with neurosurgery. These infants require close follow-up for complications such as leptomeningeal cysts. All infants with skull fractures suspected to be the result of child abuse should be admitted for observation, further evaluation, and investigation.

Febrile Seizures

Definitions

- Febrile seizures are typically brief (<15 min), generalized, and occur only once during the febrile illness.
- Complicated febrile seizures are characterized by atypical features such as focal neurologic deficits, more than one seizure per illness, and prolonged seizure activity (>15 min). Approximately 30-40% of febrile seizures will be complex.
- Febrile seizures often occur at the onset or during the first 24 h of illness and are usually associated with high fever; however some children will experience seizures at lower temperatures (101°-102°).
- Certain predisposing factors may increase the risk of febrile seizures by 6-10%:
  1. Prolonged and complicated neonatal course
  2. Parental or sibling history of febrile seizures
  3. Developmental delay
  4. Attendance at daycare

Incidence

- Approximately 2-4% of children between the ages of 3 mo and 6 yr will have a febrile seizure. The peak incidence occurs at 18 mo of age.

Clinical Presentation

- Most children who suffer a simple febrile seizure will be asymptomatic upon arrival in the emergency department or at most experience a brief postictal period after arrival.
- Once the child has returned to baseline neurologic status, the evaluation should proceed as indicated for any child with a febrile illness.
- A thorough history and physical examination should be directed toward the determination of the etiology of the fever.
- Laboratory evaluations should be based upon specific findings as indicated by history and physical examination.
- Routine laboratory tests add little to the evaluation of the seizure.

- Lumbar puncture is rarely indicated in children with simple febrile seizures. Indications for LP are:
  1. Focal, prolonged, or multiple seizures
  2. Neurological signs or prolonged alteration of mental status
  3. Findings on physical exam suggestive of serious disease (i.e., petechiae, dehydration, poor perfusion).
- Meningitis does not present as a simple febrile seizure only.
- Children with meningitis who experience seizures have obvious signs and symptoms of serious disease and will not return to a normal neurologic state post seizure.
- Lumbar puncture should be considered in infants 3-6 mo of age since meningitis has a high incidence in this age group and is more clinically difficult to diagnose.
- There is no indication for cranial computed tomography or magnetic resonance imaging unless the child has persistent neurologic deficit or signs of increased intracranial pressure.
- Electroencephalography (EEG) provides no useful information in the evaluation of febrile seizures and may be misleading. Approximately 15-55% of children who experience febrile seizures will have focal epileptiform abnormalities on EEG, but these abnormalities have no correlation with recurrent febrile or nonfebrile seizures.

ED Management

- Most children with a febrile seizure will fully recover within 1 h with no signs of serious illness and therefore require no intervention.
- Occasionally children present in status epilelpticus during a febrile seizure.
- ABCs of resuscitation should be initiated and IV access achieved. The seizure may be treated with intravenous diazepam or lorazepam (see Table 22H.7).
- If IV access cannot be readily obtained, diazepam may be given rectally as diastat (see Table 22H.7). Rarely will further medical therapy be required; however if seizure activity persists after adequate doses of benzodiazepine, treatment should proceed as in status epilepticus caused by nonfebrile seizures.
- Antipyretics are the only effective treatment for fever; however antipyretics are not always effective in preventing febrile seizures. Adequate doses of acetaminophen or ibuprofen should be used, but physicians must be careful not to promote "fever phobia" in families.
- Phenobarbital and valproic acid may be effective in preventing febrile seizures. But the daily use of these medications with their inherent side effects and complications far outweigh their use in such a benign condition.
- The use of intermittent rectal diazepam is controversial. There may be some indications for its use in those children who have recurrent complex febrile seizures.
- The risk of recurrence after the first febrile seizure is 30-40%. There are four independent factors predictive of recurrent febrile seizures:
  1. Age <18 mo at first event
  2. Family history
  3. Onset of febrile illness <1 h prior to seizure
  4. Low temperature (101° or <) at time of the event.
- The vast majority of children who have febrile seizures require no physician interventions or subsequent evaluations.

Table 22H.7. Drugs used in the management of febrile seizures

	IV Dose	Rectal Dose
Lorazepam	0.05-0.2 mg/kg	0.1-0.4 mg/kg
Diazepam	0.1-0.5 mg/kg	0.2-2.0 mg/kg (max dose 10 mg)

- There are no long-term sequelae associated with uncomplicated febrile seizures and only 2-4% of children subsequently experience nonfebrile seizures.
- Parents should be reassured and educated about the natural history and the benign nature of the disease.

Disposition

- In general, children with uncomplicated febrile seizures are discharged home.

Nonfebrile Seizures

Clinical Presentations and Definitions

- Seizures are classified as either partial or generalized based on the site of onset of seizure activity in the brain.
  1. Partial seizures have onset of abnormal electrical activity in a specific area of one cerebral hemisphere. Seizures that remain limited to one hemisphere do not cause loss of consciousness and are referred to as simple partial seizures. When the abnormal electrical activity involves both cerebral hemispheres, there is a loss of consciousness associated with generalized convulsive movements.
  2. Generalized seizures manifest as an immediate loss of consciousness with impairment of sensorium alone, as in absence seizures, or with tonic and clonic, tonic, clonic, or myoclonic activity in a generalized fashion.
- In addition to partial or generalized seizures, there are several epileptic syndromes common to children and include benign childhood epilepsy, absence epilepsy (petit mal epilepsy), juvenile myoclonic epilepsy, complex partial epilepsy, infantile spasm, and Lennox-Gastaut syndrome.
- Benign childhood epilepsy occurs in children between 3 and 15 yr of age and is characterized by generalized motor seizures occurring predominately during sleep. Partial seizures often begin with sensory symptoms involving one side of the face or tongue, progress to anarthria, drooling, and clonic motor activity of the facial and lingual muscles. The partial seizure can occasionally spread to become a generalized tonic-clonic seizure. Children with benign childhood epilepsy are otherwise healthy and have a normal neurologic examination.
- Absence seizures occur typically in healthy school-aged children and are characterized by frequent "absences" of short duration. The seizures occur without warning and are periods of loss of consciousness of a few seconds. Occasionally the episodes may be prolonged and accompanied by automatism, such as eyelid blinking or clonic jerking of the limbs.
- Juvenile myoclonic epilepsy is characterized by the occurrence of bilaterally synchronous single jerks or clusters of myoclonic jerks. These typically occur upon falling asleep or as the child is waking up. The jerks predominately involve the arms, often causing the child to drop or throw objects. The seizures begin during adolescence and teen years and most patients experience at least one generalized tonic-clonic episode.
- Complex partial epilepsy affects all ages and is the most common form of idiopathic epilepsy. It is characterized by impairment of sensorium associated with psychic, autonomic, olfactory, and occasionally gustatory symptoms. Automatisms occur in patterns of semi-purposeful activity such as lip smacking, sucking movements, fumbling of the hands, moaning, or verbal perseveration. Following the seizure, there is often a period of confusion or nondirected aggressive behavior.
- Infantile spasms are seen only in infancy and are the response of the developing brain to a variety of insults. The syndrome may be idiopathic or secondary to underlying disease. The spasms always begin before 1 yr of age and are characterized by either sudden flexion or extension of the trunk. They typically occur in clusters at the time of waking or falling asleep; often accompanied by a sharp cry.

- The symptom complex of the Lennox-Gastaut syndrome consists of intractable seizures of multiple types and psychomotor retardation. The seizure types include drop attacks, absence seizures, myoclonic jerks, and generalized tonic-clonic seizures.
- Over three-quaarters of childhood epilepsy is idiopathic; the remaining quarter is due to developmental abnormalities or secondary to CNS trauma or infection. Common conditions that mimic seizures in children include syncope, psychiatric disorder, migraine headache, breath holding spells, and night terrors. These disorders can often be elucidated by a careful history and physical examination.

Inicidence

- The prevalence of epilepsy within the pediatric population is 4.3-9.3 per 1000 children.

Diagnostic Evaluation

- The evaluation of the infant or child presenting with new onset seizures should begin with the identification of life-threatening disorders such as central nervous system infection, metabolic disorders, head injury, or intoxication.
- The next step is to establish if the activity in question is a seizure or a manifestation of other paroxysmal disorders. A detailed description of the activity is crucial as well as any precipitating or associated events.
- Older children may describe an aura which indicates a partial seizure. The convulsive activity should be noted as focal, generalized, or beginning focally and spreading (Jacksonian seizure). Any focal component also suggests a partial seizure.
- History for loss of muscle tone, automatism, and the frequency and duration of abnormal motor activity are helpful in the diagnosis. The presence of a postictal period occurs following generalized motor and partial seizures, but not with absence seizures. Focal weakness (Todd's paralysis) is typical of partial seizures.
- Past medical history of illness or injury as possible causes of seizures should be sought as well as developmental history and school performance.
- Physical examination should document growth parameters, especially head circumference in infants.
- Abnormalities of the eyes or skin may suggest a neurocutaneous disorder.
- Hepatosplenomegaly associated with infantile seizures should prompt an evaluation for glycogen-storage disease. The neurologic examination should include evaluation of cranial nerves, deep tendon refluxes, muscle strength and tone, coordination, mental status and development.
- Laboratory evaluation is seldom useful in children with recurrent, unprovoked seizures who are otherwise healthy.
- The most important studies are the EEG and neuroimaging scans. The EEG should be obtained as soon as possible after the diagnosis of seizures is suspected. However, a normal EEG does not exclude the diagnosis of epilepsy.
- Neuroimaging studies are indicated in all children who have partial seizures, focal neurologic signs, and all neonates and infants because of the frequent association of congenital and/or progressive disease. Computed tomography (CT) may be utilized emergently to rule out life threatening conditions, but magnetic resonance imaging is preferred, especially for patients who have complex partial or simple partial seizures.

ED Management

- Hospital admission should be considered for all infants presenting with a first seizure because of the high association with congenital or chronically progressive disease. Also any child with a prolonged postictal period or focal neurologic deficits should be admitted.

22

Table 22H.8. Treatment of nonfebrile seizures

Seizure Type	First Line Therapy	Initial Dose (po)	Maintence Dose (po)
Partial (simple and complex)	Carbamazepine	5 mg/kg/d	20-40 mg/kg/d
Generalized tonic-clonic	Valproic acid	10-15 mg/kg/d	15-30 mg/kg/d
	Phenytoin	3-10 mg/kg/d	same
Absence	Ethosuxinide	10 mg/kg/d	10-20 mg/kg/d
	Valproic acid	10-15 mg/kg/d	15-30 mg/kg/d
Atypical absence myoclonic	Valproic acid	10-15 mg/kg/d	15-30 mg/kg/d
Lennox-Gastaut	Carbamazepine	5 mg/kg/d	20-40 mg/kg/d
	Valproic acid	10-15 mg/kg/d	15-30 mg/kg/d
	Phenytoin	3-10 mg/kg/d	same
	Phenobarbital	3-5 mg/kg/d	same
	Primidone	2-4 mg/kg/d (adults)	10-15 mg/kg/d (adults)

- Older children with a first seizure who are healthy and have a normal examination do not require admission and may be safely followed by their primary care physician.
- Children with known seizure disorders who return to their baseline neurologic status and have no concurrent illness requiring admission may also be discharged after laboratory measurement of anticonvulsant drug levels and an appropriate period of observation.
- Plans for follow-up should be made in consultation with the primary care physician or pediatric neurologist.
- Anticonvulsant medications are not routinely prescribed following a single first seizure as many children experience a nonfebrile seizure as a single lifetime event.
- Children who present after a second or third seizure should begin anticonvulsant therapy. The choice of the appropriate drug therapy (see Table 22H.8) should be based upon the seizure type and made in consultation with a pediatric neurologist or the primary care physician. Drug side effects, as well as risks and benefits should be discussed with the patient and parents.
- In addition to standard first time therapy drugs, there are several new anticonvulsants for treatment of complicated seizure disorders (see Table 22H.9). These drugs are not

Table 22H.9. Second line therapy for nonfebrile seizures

Seizure Type	Alternative Therapy	Initial Dose (po)	Maintinence Dose (po)
Partial (simple and complex)	Felbamate	15 mg/kg/d	45 mg/kg/d
	Gabapentin	1200 mg/d (adults)	3600 mg/d(adults)
	Lamotrigine	50-100 mg/d (adults)	100-400 mg/d (adults)
	Tiagabine	4 mg/d	16-32 mg/d
	Topiramate	1 mg/kg/d	3-9 mg/kg/d
Generalized tonic-clonic	Felbamate	15 mg/kg/d	45 mg/kg/d
	Gabapentin	1200 mg/d(adults)	3600 mg/d(adults)
	Lamotrigine	50-100 mg/d (adults)	100-400 mg/d (adults)
Absence	Clonazepam	0.05-0.10 mg/kg/d	
Atypical absence myoclonic	Clonazepam	0.05-0.10 mg/kg/d	
Lennox-Gastaut	Felbamate	15 mg/kg/d	45 mg/kg/d
	Lamotrigine	50-100 mg/kg/d	100-400 mg/d (adults)
	Topiramate	1 mg/kg/d	3-9 mg/kg/d

recommended for initial therapy, and their use should always be discussed with a pediatric neurologist.
- Neurology consultation should occur for any neonate or infant with confirmed or possible seizures, and older children with complicated nonfebrile focal seizure activity.
- Children with uncomplicated seizure disorders may be safely managed by their primary care physician.

Meningitis and Encephalitis

Definitions

- Meningitis is an inflammation of the membrane surrounding the brain and spinal cord.
- Encephalitis is inflammation involving the cerebral cortex, and meningoencephalitis involves both the meninges and the cortex. Inflammation of the meninges occurs when the capillary endothelium of the central nervous system (the blood-brain barrier) is disrupted by an offending agent. Offending agents include bacteria, viruses, fungi, and tuberculosis. They enter the central nervous system (CNS) by either hematogenous spread or direct invasion.
- Etiologic agents are common to specific age groups (see Table 22H.10).
- Preterm infants are considered immunocompromised, and any agent isolated from their CSF is significant.
  - Common bacterial pathogens during the first 3 mo of life include Group B Streptococcus and Gram-negative enteric bacilli, where as *Streptococcus pneumoniae* and *Neisseria meningitidis* account for nearly all bacterial meningitis beyond 3 mo of age.
  - The peak incidence of pneumococcal meningitis occurs between 3 and 24 mo of age.
- Viral pathogens are common to all age groups and can be particularly virulent in the preterm infant and neonate.
  - These viruses include herpes simplex virus (HSV), enteroviruses, and cytomegalovirus (CMV).
  - HSV infection usually occurs within the first 7-10 days of life and is commonly associated with a history of maternal herpes infection.
  - Enteroviral infections are common and may be identified by culture or polymerase chain reaction (PCR).

Table 22H.10. Common etiologic agents of meningitis

0-3 mo	Group B Streptococcus *Escherichia coli* *Listeria monocytogenes* Virus
3-36 mo	*Streptococcus pneumoniae* *Neisseria meningitidis* Tuberculosis Virus
3-21 yr	*Streptococcus pneumoniae* *Neisseria meningitidis* Virus
Immunoincompetent	*Streptococcus pneumoniae* *Neisseria meningitidis* Virus Cryptococcus Fungus Tuberculosis

- Neonates with meningoencephalitis caused by these agents can present with hemodynamic instability, seizures, and disseminated intravascular coagulopathy (DIC).
- Viral CNS infections in older infants and children include enteroviruses, HSV, human herpes virus 6, and arboviruses and tend to have a more benign presentation and course than in neonates.

Clinical Presentation

- Clinical presentation of meningitis and encephalitis differs with the age of child.
  - Neonates and infants typically present with nonspecific symptoms such as irritability, lethargy, poor feeding, vomiting, or respiratory distress. Temperature instability can manifest as either hypo- or hyperthermia.
  - Older children and adolescents frequently may have fever, headache, vomiting, altered sensorium, and meningismus.
  - Children with encephalitis present with severe headache, vomiting, ataxia, and often profound alteration of mental status.
  - It is not uncommon for children with encephalitis to have seizures and occasionally present in status epilepticus.
- The most important diagnostic procedure for the diagnosis of meningitis and encephalitis is the lumbar puncture (LP).
  - Cerebrospinal fluid (CSF) should be tested for cell count with differential, glucose, protein, Gram stain, and culture.
  - CSF results generally differ with bacterial versus viral etiologies (see Table 22H.11); however when findings are equivocal, bacterial meningitis should be presumed. In HIV positive children, India ink, cryptococcal antigen, acid fast bacillus smear and culture, and fungal cultures should be obtained.
  - Traumatic or bloody LPs can present a diagnostic challenge. A general estimate of white blood cell (WBC) to red blood cell (RBC) ratio is 1-2 WBCs per 1000 RBCs. Bacterial antigen tests (latex agglutination and counter immunoelectrophoresis) need not routinely be ordered but may be helpful for those patients pretreated with antibiotics who have negative Gram stains or cultures.
  - Other laboratory testing should include complete blood count, electrolytes, BUN, creatine, glucose, coagulation studies, and cultures of blood and urine.
  - Contraindications for LP include hemodynamic instability, increased intracranial pressure, and coagulopathy. In these cases, empiric antibiotics should be initiated and the LP deferred until the patient's condition has been stabilized.
- Cranial computed tomography (CT) is indicated for the child or infant with trauma, focal neurologic signs, or increased intracranial pressure.
- Antibiotics for suspected meningitis should never be delayed while awaiting laboratory or CT results.
- A window of 2-6 h exists before antibiotics significantly effect the CSF.

ED Management

- Children with meningitis/encephalitis can present with altered mental status, coma, shock, respiratory distress, or status epilepticus.

Table 22H.11. Normal values for CSF

	WBC	Glucose	Protein
Newborn	<29	30-40	<170
>1 wk of age	<10	>40	<40

Table 22H.12. Empiric drug therapy for meningitis (initial dose)

0-60 days	Ampicillin	50-100 mg/kg	**plus**
	Gentamycin	2.5 mg/kg	**or**
	Cefotaxime	50-75 mg/kg	
60 days to adulthood	Ceftriazone	100 mg/kg	**or**
	Cefotaxime	50-75 mg/kg	
Vancomycin	15 mg/kg		
Dexamethasone	0.15 mg/kg		

- Management should always begin with the ABCs and stabilization of life-threatening problems. Any child with altered mental status should have a bedside glucose and hypoglycemia corrected with $D_{25}W$ at 2 ml/kg.
- Children with coma, severe respiratory distress, or signs of increased intracranial pressure should always undergo neuroprotective rapid sequence intubation to insure adequate airway and ventilation.
- Seizures should be treated as in any other etiology depending upon presentation (i.e., febrile, nonfebrile recurrent, or status epilepticus).
- Intravenous fluid administration for the treatment of dehydration or shock associated with suspected meningitis should be 10 ml/kg boluses of normal saline or lactated Ringers solution, titrated to hemodynamic stabilization.
- After resuscitation, fluids should be given at two-thirds of calculated maintenance.
- Empiric antibiotics (see Table 22H.12) should be given immediately upon suspecting meningitis, without awaiting CSF or CT results.
- Vancomycin is indicated for any child suspected of having infection with pneumococcus due to the rising incidence of cephalosporin resistance.
- Also preterm, low birth weight neonates should receive vancomycin if methicillin-resistant *Staphylococcus aureus* is suspected.
- Dexamethasone (see Table 22H.12) may provide some protection against neurologic sequelae (neurosensory hearing loss) in pneumococcal meningitis; however it must be given prior to antibiotic administration to be helpful. Dexamethasone is not indicated in infants <6 wk of age.

Disposition

- Infants and children with bacterial meningitis or encephalitis need admission to a pediatric intensive care unit.
- Older children with a benign clinical presentation and CSF findings unequivocally consistent with viral meningitis can be considered for discharge and close outpatient follow-up.

Suggested Reading

1. Haafiz A, Kissoon N. Status epilepticus: Current concepts. Pediatric Emergency Care 1999; 15:119-129.
2. Wolf SM, Ochoa JG et al. Seizure management in pediatric patients for the nineties. Pediatric Annals 1998; 27:653-664.
3. Morton KD, Rizkallah E et al. New drug therapy for acute seizure management. Seminars in Pediatric Neurology 1997; 4:51-63.
4. Key CB, Rothrock SG, Falk JL. Cerebrospinal fluid shunt complications: An emergency medicine perspective. Pediatric Emergency Care 1995; 11:265-272.
5. Madikaus A, Conway EE. Cerebrospinal fluid shunt problems in pediatric patients. Pediatric Annals 1997; 26:613-620.

22

6. Roman A, Hogg GG, Klug GL. Cerebrospinal fluid shunt infections in children. Pediatric Infectious Disease 1995; 14:782-786.
7. Gruskin KD, Schutzman SA. Head trauma in children younger than 2 Yr. Archives of Pediatric and Adolescent Medicine 1999; 153:15-20.
8. Greenes DS, Schutzman SA. Occult intracranial injury in infants. Annals of Emergency Medicine 1998; 32:680-686.
9. Greenes DS, Schutzman SA. Infants with Isolated skull fracture: What are their clinical characteristics, and do they require hospitalization? Annals of Emergency Medicine 1997; 30:253-259.
10. Davis RL, Hughes M, Gubler KD et al. The use of cranial CT scans in the triage of pediatric patients with mild head injury. Pediatrics 95:345-349.
11. Camfield PR, Camfield CS. Management and treatment of febrile seizure. Current Problems in Pediatrcs January 6-13, 1997.
12. Green SM, Rothrock SG et al. Can seizures be the sole manifestation of meningitis in febrile children? Pediatrics 1993; 92:527-534.
13. Verlanarayanan VV. Diagnosis of epilepsy in children. Pediatric Annals 1999; 28:218-224.
14. Russell RJ, Parks B. Anticonvulsant medications. Pediatric Annals 1999; 28:238-245.
15. Wubbel L, McCracken GH. Management of bacterial meningitis. Pediatrics in Review 1998; 19:78-84.
16. Nozicka CA, Schafermeyer R. Pediatric meningitis: Clinical guidelines, issues, and update. Pediatric Emergency Medicine Reports 1997; 2:5.
17. Negrini B, Kelleher KJ, Wald ER. Cerebrospinal fluid findings in aseptic versus bacterial meningitis. Pediatrics 2000; 105:316-319.
18. Ahmed A. A Critical evaluation of vancomycin for treatment of bacterial meningitis. Pediatric Infectious Disease Journal 1997; 16:895-903.
19. McIntyre PB, Berkey CS et al. Dexamethasone as adjunctive therapy in bacterial meningitis: A Meta-Analysis of randomized clinical trials since 1989. JAMA 1997; 278:925-931.

A

B

C

D

E

F

G

H

Index

I

J

K

L

M

N

Index

R

Index

S

T

U

V

W

Z